Polymers for Agri-Food Applications

Tomy J. Gutiérrez
Editor

Polymers for Agri-Food Applications

Springer

Editor
Tomy J. Gutiérrez
Thermoplastic Composite Materials (CoMP) Group
Faculty of Engineering
Institute of Research in Materials Science and Technology (INTEMA)
National University of Mar del Plata (UNMdP) and National Scientific
and Technical Research Council (CONICET)
Mar del Plata, Buenos Aires, Argentina

ISBN 978-3-030-19415-4 ISBN 978-3-030-19416-1 (eBook)
https://doi.org/10.1007/978-3-030-19416-1

This Springer imprint is published by the registered company Springer Nature Switzerland AG
The registered company address is: Gewerbestrasse 11, 6330 Cham, Switzerland

I would like to dedicate this book:

To my God (Father, Son, and Holy Spirit), to the Virgin Mary, and to my Guardian Angel, Their energy stimulates me to enjoy the landscape we call life, and their peace encourages me to always continue toward the future, a place where we will all go and each one will be under the law of the creative father.

To my Mother, Dr. Mirian Arminda Carmona Rodríguez,
For forming my character and attitude toward life.

To my Grandmother, Mrs. Arminda Teresa Rodríguez Romero,
A person who unfortunately left before this world's time, but I am sure that she is up watching me and supporting me in all facets of my life. You are in my most beautiful memories.

To all the anonymous persons,
Those who give me their love, friendship, patience, and support in various situations.

To Venezuela and Argentina,
The first for giving me my academic and professional training, and the second for welcoming me with love and friendship before the dictatorship that my country (Venezuela) is experiencing today.

Tomy J. Gutiérrez, Ph.D.
Editor

Preface

The science and technology of polymers for agri-food applications is often known as "plasticulture." This agricultural science and technology (Plasticulture Science and Technology) is growing daily, and is particularly aimed at developing alternative materials that avoid the use of non-biodegradable products obtained from the petrochemical industry. Strong concerns related to climate change, as well as new sustainable and ecological concepts in emerging economies have driven the development of natural polymers for agri-food applications. Synthetic polymers also have a prominent role to play in agro-industry, and as such, they will also be discussed by recognized researchers. This book addresses the different applications of natural and synthetic polymers in agri-industrial processes with the goal of improving food crop yields while reducing the carbon footprint.

Tomy J. Gutiérrez
National Scientific and Technical Research Council (CONICET)
Institute of Research in Materials Science and Technology (INTEMA)
Thermoplastic Composite Materials (CoMP) Group
Mar del Plata, Argentina

Contents

About the Editor

Tomy J. Gutiérrez has a degree in chemistry (Geochemical option) and a degree in education (Chemical mention) from the Central University of Venezuela (UCV) (December 2007 and July 2008, respectively) and has a specialization in the International Negotiation of Hydrocarbons from the National Polytechnic Experimental University of the National Armed Forces (UNEFA), Venezuela (July 2011). He also has a Master's and PhD degree in Food Science and Technology obtained in October 2013 and April 2015, respectively, and PhD studies in Metallurgy and Materials Science from the UCV, and postdoctoral studies at the Research Institute in Materials Science and Technology (INTEMA). He has been a professor and researcher at the UCV both at the Institute of Food Science and Technology (ICTA) and the School of Pharmacy. He is currently an adjunct researcher in the INTEMA, National Scientific and Technical Research Council (CONICET), Argentina. He has at least 20 book chapters and 40 publications in international journals of high-impact factor. He has been a guest editor of several international journals, such as *Journal of Food Quality*, *Polymers for Advanced Technology*, *Current Pharmaceutical Design*, and *Frontiers in Pharmacology*. He, today, is developing a line of research in nanostructured materials based on polymers (composite materials), which are obtained on a pilot scale to be transferred to the food, pharmaceutical, and polymer industry. He is also a collaborator of international projects between Argentina and Brazil, Colombia, France, India, Iran, Spain, Sweden, and Venezuela.

Chapter 1
Trends in Polymers for Agri-Food Applications: A Note from the Editor

Tomy J. Gutiérrez

Abstract An expanding world population and a planet limited in natural resources are two of the main reasons that have propelled the development of scientific and technological solutions that enable the efficient, sustainable, and massive production of food and its subsequent preservation. In this regard, natural and synthetic polymers have found a place in agri-industry. In this first chapter, we introduce the main polymeric systems and their potential applications in agri-industrial processes.

Keywords Agriculture · Agrochemicals · Controlled release systems · Crops · Edible coatings and films · Foams and emulsions · Horticultural produces · Hydrogels · Micro- and nanoencapsulation · Mulch films · Postharvest

1.1 Present and Future Perspectives

The future of food presents us with several real challenges, some of which require an immediate response. The solutions to others, however, must be built up more slowly, especially when contemplating the large amount of food that will be required to feed the growing world population, coupled with the limited availability of productive lands and water (FAO 2009). Every day science and technology make giant steps forward. Even when the answers or solutions are not quite found, progress continues to be made by informing us of the limitations of some of the possible alternatives or by signposting paths that we should not follow. There are many ways that researchers in agri-food can contribute to the pool of knowledge, but increasingly a more complex and holistic approach is sought. This, in turn, requires highly trained inter- and intradisciplinary professionals who are also extremely creative, as

T. J. Gutiérrez (✉)
Thermoplastic Composite Materials (CoMP) Group, Faculty of Engineering, Institute of Research in Materials Science and Technology (INTEMA), National University of Mar del Plata (UNMdP) and National Scientific and Technical Research Council (CONICET), Colón 10850, Mar del Plata 7600, Buenos Aires, Argentina
e-mail: tomy.gutierrez@fi.mdp.edu.ar

T. J. Gutiérrez (ed.), *Polymers for Agri-Food Applications*,
https://doi.org/10.1007/978-3-030-19416-1_1

traditional thought processes can limit progress towards the re-creation and redefinition of some ideas, and even science itself. With this in mind, the high importance given to natural and synthetic polymer systems in agricultural production today would have been unthinkable a few decades ago. Now, however, we can see how polymers have carved a space for themselves in the agri-food sector. For example, as micro- and nano-systems encapsulating phytohormones and agrochemicals which allow the controlled release of these substances giving food crops what they require for growth, as well as protecting them against weeds and insects, whilst avoiding damage to the environment (El-Ramady et al. 2018). Some natural polymers have even demonstrated elicitor properties, i.e. they can stimulate plantations to create their own means of defense (Merino et al. 2018b). Hydrogel systems have also been created as moisture regulators for the long periods of drought that are being currently experienced due to global climate changes (Cheng et al. 2018). Polymers have also been used in regulatory agrochemical systems (Xiang et al. 2018). Agricultural mulch films protect crops from UV radiation, regulate moisture levels and control pests (Merino et al. 2019a,b; Merino et al. 2018a; Merino et al. 2018c; Touchaleaume et al. 2018). Foams, emulsions, membranes, films and coatings have also made significant contributions to the preservation and extension of the shelf life of fruits and vegetables post-harvest (Saberi and Golding 2018). All these polymeric systems have been proposed and studied with the aim of increasing the production of foods, or avoiding losses during their cultivation, harvest, transport and sale, until they reach the final consumer. In this book, the aforementioned polymeric systems will be discussed by several research groups worldwide.

Acknowledgements The author would like to thank the Consejo Nacional de Investigaciones Científicas y Técnicas (CONICET) (Postdoctoral fellowship internal PDTS-Resolution 2417), Agencia Nacional de Promoción Científica y Tecnológica (ANPCyT) (grant PICT-2017-1362), Universidad Nacional de Mar del Plata (UNMdP) for financial support, and Dr. Mirian Carmona-Rodríguez.

Conflicts of Interest The author declares no conflict of interest.

References

Cheng, D., Liu, Y., Yang, G., & Zhang, A. (2018). Water- and fertilizer-integrated hydrogel derived from the polymerization of acrylic acid and urea as a slow-release N fertilizer and water retention in agriculture. *Journal of Agricultural and Food Chemistry, 66*(23), 5762–5769. https://doi.org/10.1021/acs.jafc.8b00872.

El-Ramady, H., Abdalla, N., Alshaal, T., El-Henawy, A., Elmahrouk, M., Bayoumi, Y., & Schnug, E. (2018). Plant nano-nutrition: Perspectives and challenges. In K. M. Gothandam, S. Ranjan, N. Dasgupta, C. Ramalingam, & E. Lichtfouse (Eds.), *Nanotechnology, food security and water treatment* (pp. 129–161). Cham: Springer International Publishing. https://doi.org/10.1007/978-3-319-70166-0_4.

Food and Agriculture Organization of the United Nations (FAO). (2009). 2050: A third more mouths to feed. Retrieved August 11, 2018, from http://www.fao.org/news/story/en/item/35571/icode/.

Merino, D., Gutiérrez, T. J., Mansilla, A. Y., Casalongué, C., & Alvarez, V. A. (2018a). Critical evaluation of starch-based antibacterial nanocomposites as agricultural mulch films: Study on their interactions with water and light. *ACS Sustainable Chemistry & Engineering, 6*(11), 15662–15672. https://doi.org/10.1021/acssuschemeng.8b04162.

Merino, D., Mansilla, A. Y., Casalongué, C. A., & Alvarez, V. A. (2018b). Preparation, characterization, and in vitro testing of nanoclay antimicrobial activities and elicitor capacity. *Journal of Agricultural and Food Chemistry, 66*(12), 3101–3109. https://doi.org/10.1021/acs.jafc.8b00049.

Merino, D., Mansilla, A. Y., Gutiérrez, T. J., Casalongué, C. A., & Alvarez, V. A. (2018c). Chitosan coated-phosphorylated starch films: Water interaction, transparency and antibacterial properties. *Reactive and Functional Polymers, 131*, 445–453. https://doi.org/10.1016/j.reactfunctpolym.2018.08.012.

Merino, D., Gutiérrez, T. J., & Alvarez, V. A. (2019a). Potential agricultural mulch films based on native and phosphorylated corn starch with and without surface functionalization with chitosan. *Journal of Polymers and the Environment, 27*(1), 97–105. https://doi.org/10.1007/s10924-018-1325-1.

Merino, D., Gutiérrez, T. J., & Alvarez, V. A. (2019b). Structural and thermal properties of agricultural mulch films based on native and oxidized corn starch nanocomposites. *Starch-Stärke, 71*(7–8), 1800341. https://doi.org/10.1002/star.201800341.

Saberi, B., & Golding, J. B. (2018). Postharvest application of biopolymer-based edible coatings to improve the quality of fresh horticultural produce. In T. J. Gutiérrez (Ed.), *Polymers for food applications* (pp. 211–250). Cham: Springer International Publishing. https://doi.org/10.1007/978-3-319-94625-2_9.

Touchaleaume, F., Angellier-Coussy, H., César, G., Raffard, G., Gontard, N., & Gastaldi, E. (2018). How performance and fate of biodegradable mulch films are impacted by field ageing. *Journal of Polymers and the Environment, 26*(6), 2588–2600. https://doi.org/10.1007/s10924-017-1154-7.

Xiang, Y., Zhang, G., Chen, C., Liu, B., Cai, D., & Wu, Z. (2018). Fabrication of a pH-responsively controlled-release pesticide using an attapulgite-based hydrogel. *ACS Sustainable Chemistry & Engineering, 6*(1), 1192–1201. https://doi.org/10.1021/acssuschemeng.7b03469.

Chapter 2
Polymer Based Micro- and Nanoencapsulation of Agrochemicals

Tanushree Sinha, Prachi Bhagwatwar, Chandan Krishnamoorthy, and Ramalingam Chidambaram

Abstract Agrochemicals such as pesticides and fertilizers have been used to increase agricultural production. Most agrochemicals although highly beneficial, lead to major environmental and ecological damage. At present, controlled release systems such as "encapsulation" are developed to control agrochemicals availability, reduce wastage for sustenance and increased production for agricultural plants. Polymers, due to their tailorable properties can be used in encapsulation. In recent years, micro and nano-encapsulation have shown great promise for a better design release of agrochemicals. New sustainable technologies are being designed *via* different encapsulation techniques, in order to avoid the loss of crops and increase agricultural productivity.

Keywords Agrochemicals · Micro- and nano-encapsulation · Polymer

2.1 Introduction to Agrochemicals

Agrochemicals have a pivotal role, in maintaining the rising demand for food. However, the harmful effects of agrochemicals on the environment and health is a major problem with their use. Hence, careful handling of these chemicals is also necessary. Agrochemical is a general term that describes the various chemicals used in agriculture including pesticides, fertilizers, and other growth agents. Among these, pesticides play a vital role in agricultural development by removing harmful

---

T. Sinha · P. Bhagwatwar
Department of Biotechnology, School of Bioscience & Technology, Vellore Institute of Technology (VIT), Vellore, Tamilnadu, India

C. Krishnamoorthy · R. Chidambaram (✉)
Department of Biotechnology, School of Bioscience & Technology, Vellore Institute of Technology (VIT), Vellore, Tamilnadu, India

Instrumental & Food Analysis Laboratory, Industrial Biotechnology Division, School of Bioscience & Technology, Vellore Institute of Technology (VIT), Vellore, Tamilnadu, India
e-mail: cramalingam@vit.ac.in

T. J. Gutiérrez (ed.), *Polymers for Agri-Food Applications*,
https://doi.org/10.1007/978-3-030-19416-1_2

pests because greater than 30–40% of the food production goes to waste due to pests like insects and weeds (Abhilash and Singh 2009; Campos et al. 2014b). Fertilizers also play a crucial role in agriculture, affecting between 35 and 40% food production (Ram et al. 2014).

In conventional agrochemicals, the active component is combined with inert materials for their safe and easy handling as well as for their effective use in the field. However, the immediate release of these active components is the biggest issue faced in the field. This is caused due to the quick loss of the agrochemical that happens through degradation processes like photolytic, hydrolytic and microbial degradation as well as phenomena like volatilization, evaporation, and leaching. Thus, accumulation of the active component in the field drops fast, below the concentration required to have an effective impact. Since the desired concentration should be maintained, agrochemicals must be added repeatedly, eventually resulting in various health and environmental issues. Excess pesticides may be toxic to plants as well. Newer formulations of agrochemicals have been synthesized that address these shortcomings. These formulations release agrochemicals in a controlled, gradual and sustained manner (Prasad et al. 2017; Campos et al. 2014b).

2.2 Agrochemicals

2.2.1 Pesticides

Pesticides are agrochemicals that can effectively target the elimination of plant pests and diseases for increased crop production. Although the traditional pesticides formulations are very effective, they are used in the form of water-miscible formulations or granules or dust which are sprayed on the soil or the plants directly which is not a specific site. There is an excess loss of pesticides due to poor dispersion, heavy or small dust drifts and high concentration of organic solvent contents with only 1% of the used pesticide finally reaching the target organism, contributes to serious environmental pollution, excess wastage and very high expenses for farmers. To avoid these excess losses and unwanted exposure to other organisms and the environment, a controlled release system which releases the pesticides in a sustained manner at a suitable location and an at a suitable time is desired. Intense research has opened a gateway to many diverse controlled release systems amongst which encapsulation of pesticides with polymers is majorly assessed as the most advantageous.

To date, studies on pesticides can be differentiated based on the target organism to be eliminated. They are:

i. Insecticides-As the name suggests, they are used to kill insects by targeting their habits by specific formulations in the form of aerosols, solids, liquids or granules. For example, flying pests are a target by using aerosol sprays or fogs, while surface powders, sprays or granules for dermal and/or oral entry is best suited for crawling insects. Through different paths the insecticides reach inside the insect's body, thus, poisoning them.

ii. Herbicides- targets plants that are unwanted and competitively inhibits the growth of the agricultural plants, thus, affecting the agricultural produce.
iii. Bactericides- targets bacteria.
iv. Fungicides- targets fungi.
v. Larvicides- targets larvae.

Thus, pesticides target different organisms based on the major infestation or the risk of an infestation of a specific organism.

2.2.2 *Fertilizers*

Fertilizers including macro- and micro-nutrients are some of the most important agrochemicals for crop production, in terms of both quality and quantity (Wen et al. 2016). The characteristics an ideal fertilizer should have are:

i. It should allow optimal growth for plant and enough nutrients with a one application per season.
ii. It should give the maximum percentage recovery and absorption of nutrients to achieve highest returns from the expenses on raw materials and input.
iii. It should have minimum detrimental side effects on the soil, water and atmospheric environment (González et al. 2015; Neamtu et al. 2015).

Nitrogen is the most vital nutrient for soil fertility and plant growth and plays a major role in agricultural productivity, although only about 30–50% is efficiently utilized. Phosphorus (P) is the second most important macronutrient for plants after nitrogen. When phosphorus is applied as fertilizer, plant uptake is limited to 10–25% and the rest is lost to leaching in soil. Urea is an economically favorable fertilizer due to a high nitrogen content and low production cost. The consumption of urea has increased by about 29% after the green revolution in India. However, the major disadvantages of urea are ammonia volatilization and nitrate leaching which lead to a high loss rate and low efficiency, eventually resulting in economic loss and environmental hazards. The use of nitrogen fertilizers has been useful but has had harmful effects on the beneficial microbes that live in the soil. Simultaneously, large quantities of the fertilizers are not taken up by plants due to run-off and cause pollution (Wen et al. 2016; Navarro-Guajardo et al. 2017; Duhan et al. 2017).

Biofertilizers are beneficial microorganisms like fungal mycorrhizae, Azotobacter, Rhizobium, phosphate solubilizing bacteria like Pseudomonas and Bacillus, Azospirillum and cyanobacteria. Microorganisms break down complex organic matter into simple compounds containing micro and macronutrients, improve soil fertility, and increase crop yield. The use of biofertilizers has drawbacks like temperature sensitivity, short shelf life, and storage issues (Duhan et al. 2017).

Micro and nano-fertilizers are encapsulated or coated fertilizer particles that can deliver nutrients to crops in three possible ways; the nutrient can be encapsulated inside nanomaterials (nanotubes or nanoporous materials), coated with a thin polymer film and delivered as nanoparticles or nanoemulsions. The use of nano-fertilizers in soil maximizes the efficiency of the absorption of nutrients while reducing their loss, avoids unwanted interaction of fertilizers with microorganisms, soil, and water, reduces the toxicity to the soil, provides more residence time of the fertilizer in the soil, and reduces the frequency of application of fertilizers. Thus, nano-fertilizers help in improving crop yield. Nanoparticles contain fertilizers better due to their higher surface tension than conventional surfaces and provide surface protection for larger particle sizes (González et al. 2015; Neamtu et al. 2015; Duhan et al. 2017; Navarro-Guajardo et al. 2017; Nagula and Usha 2016).

Synthetic polymers have been used for encapsulation due to their availability and variety. These are composed of urea-aldehyde resins, hydrogel nanocomposites, kaolin-chitosan (Cs) systems, polyamides, polyesters, polyvinylidene chloride, polyvinyl alcohol (PVA), polyurethanes, and other hybrid systems. However, the use of non-biodegradable synthetic matrices, due to their persistence and slow decomposition, leads to serious environmental issues associated with the accumulation of plastic residues. These concerns have motivated research towards the development of encapsulated fertilizers with biodegradable natural or synthetic polymers/blends derived from renewable natural resources or biopolymers, like matrices of Cs, cellulose, starch, and natural rubber (Tolescu et al. 2014; González et al. 2015; Neamtu et al. 2015; Gutiérrez 2017, 2018a; Herniou--Julien et al. 2019).

Cost is another major factor to be considered when analyzing the viability of the synthesis of nanoparticles. The production hydrogels as carrier matrices for nanoparticles is still more expensive than the use of conventional mineral fertilizers. Some polymers have lower costs due to lower processing requirements, are abundant in nature or are byproducts/ waste materials from other industries. The utilization of hydrophilic biopolymers for the synthesis of agrochemicals has several benefits. However, due to their higher cost, their use is still widely hindered in agriculture (Perez and Francois 2016).

The rising demand for fresh water, increasing manufacturing costs of agrochemicals as well as an awareness of their harmful ecological effects have thus helped in creating a need for better formulations of agrochemicals.

2.2.3 Phytohormones

In the recent year's encapsulation of gibberellic acid and growth promoting bacterias such as rhizobacteria have been achieved, but they have not been explored much. A lot of research and development is required in the field of encapsulation of phytohormones.

2.3 Agrochemical Formulations

An active ingredient or compound targeting specific pests needs the addition of different materials called inert ingredients, such as solvents, surface active ingredients, stabilizers, etc. to stabilize, improve its solubility and the mode of action. Active and inert ingredients together are formulated into these formulations. These are made in the form of different concentrates like emulsifiable concentrates, suspension concentrates, ready-to-use low concentrate solutions, dry flowable, soluble concentrates, and ultra-low volume concentrates etc. They are also made in the form of powders like soluble powders and, wettable powders. Other than different concentrates and powders, they are found in the form of dust, pellets, granules tablets, microencapsulates, water dispersible granules, baits, gels, aerosol, etc. (Liu et al. 2012).

Either direct application of these formulations or prior processing before the application in the form of a solution, emulsion or a suspension is done, based on their nature and properties. Suspension concentrates have become quite popular whereas emulsifiable concentrates and wettable powders are no longer commonly used by farmers.

Ultimately, effective delivery systems which improve pesticide handling and application extends shelf life, decrease pesticide toxicity, control the release rate, improves bioactivity, is targeted and expands the control spectrum and delays development of resistance is required (Liu 2012). Although the above formulations are quite effective their release rate and targeted delivery are insufficient, thus, they are used in excess leading to wastage, pollution, and extremely high expenses.

As discussed, to increase the release rate and control spatiotemporal release of agrochemicals, encapsulation is done.

2.4 Encapsulation

Nanotechnology can be defined as the engineering of functional systems through the manipulation of molecules and atoms in the nanoscale to create nanoparticles or nanomaterials, which exhibit useful and distinct chemical and physical properties. The term 'nano' describes a billionth of a meter and implies that structures manufactured with a circumference ranging between 1 and 100 nm form a part of nanotechnology (Ammar 2018).

Encapsulation can be defined as the process in which a coating or matrix is used to surround or embed the active chemical or ingredient (Gutiérrez and Álvarez 2017; Gutiérrez 2018b). The advantages of encapsulation are protection from unfavorable environments, for controlled and sustained release, and for precise targeting. There are several reasons for its use and is applied in different industries such as pharmaceuticals, chemical production, cosmetics, biotechnology, etc. (Hack et al. 2012; Campos et al. 2014b).

Agrochemicals are incorporated into a polymer carrier for controlled release. These carriers can deliver the agrochemicals gradually and for longer periods to a specified target at a desired rate, continuously (Nair et al. 2010; Gogos et al. 2012; Campos et al. 2014b).

The main parameters for the release of agrochemicals from capsules are material properties, capsule properties, and some external parameters. Material properties govern the controlled release from the capsules. These include density, crystallinity, solubility, orientation, cross-linkage, polarity, chemical degradation, and biodegradability. The capsule properties control the controlled release and include size, shell-to-core ratio, configuration, shell thickness, concentration, etc. External parameters, which decide the controlled release, consisting of temperature, pH, humidity, partial pressure difference, etc. among others (Hack et al. 2012).

The following classes of agrochemicals and related materials may be prepared by encapsulation of suspensions:

i. Synthetic origin-pesticides (insecticides, herbicides, etc.) and fertilizers.
ii. Biological origin-microbial biopesticides, pheromones, plant hormones, seeds and cells (Hack et al. 2012).

2.5 Need for Encapsulation for Agrochemicals

Food security is a global concern that needs to be addressed with sustainable methods that allow future generations to satisfy their demands. Most developing countries rely on agriculture to generate income. Diminishing arable land and reduction in organic matter in the soil, as well as a drop-in crop yields, are amongst these challenges. To maximize crop yields from their fields and to cope with nutrient deficiencies, farmers resort to excessive fertilizers utilization and agro-chemical products (Tolescu et al. 2014; Ammar 2018).

The green revolution led to excessive use of pesticides and chemical fertilizers, leading to a loss of soil biodiversity. Pathogens and pests have become resistant to them, along with the occurrence of biomagnification of the agrochemicals. The use of excessive amounts of mineral fertilizers led thus, to the growth of crops containing excess quantities of chemicals, ultimately ending up in the food chain and affecting the consumers' health along with having a detrimental impact on the environment (Tilman et al. 2002; Auffan et al. 2009; Duhan et al. 2017).

Efforts are being made to improve agricultural yield through research in nanotechnology. There is a growing demand for the development of new technologies that could seamlessly work with various agrochemicals, potentially increase yield efficiency and quality of farmlands, preserve agro-ecosystems, and diminish the negative health risks imposed by conventional practices. Smart delivery systems that utilize encapsulated agrochemicals in micro or nanoscale carriers, capsules, or particles, could enable the accurate and targeted delivery and dosage for fertilizers, pesticides and growth promoters to crops. They reduce the impact of the harmful chemicals by reducing loss due to leaching, hydrolysis, photolysis, volatilization, and bacterial degradation (Nair et al. 2010; Gogos et al. 2012; Campos et al. 2014b; Ammar, 2018).

The main goal of using encapsulation is the slow or gradual release of an active material from a polymeric matrix with the purpose of getting higher yields during a longer time period. Their physical characteristics (reservoir or matrix system) and the technique controlling the release (diffusion, swelling or degradation of the polymeric matrix), can control and change the agrochemical release pattern. Encapsulation would also make it possible to create new product features, like improved storage and transport properties, ability to adjust particle size, synthesizing formulations that are combinations of different substances and changing the material properties. It also allows protection for the agrochemicals against external forces like oxidation, light, water, and evaporation. The environment itself can also be protected against the agrochemicals that are encapsulated (Hack et al. 2012; González et al. 2015; Perez et al. 2016).

2.6 Types of Encapsulation Based on Their Size

2.6.1 Microencapsulation

Microencapsulation is the process in which solid, liquid or gaseous substances are combined into another material, usually a polymer, to give useful properties to the product. Due to their varying structures, microcapsules are normally classified into two basic forms:

i. In matrix structures, the agrochemical is embedded over the entire matrix of the capsule
ii. The core/ shell structures consist of a capsule shell and a core where the agrochemical is contained

Some types of microcapsules among these include double- and multi-walled particles, surface eroding particles and bulk eroding particles, and self-aggregated structures, like micelles, emulsions, and liposomes (Hack et al. 2012).

Techniques for microcapsule synthesis can be divided into chemical and physical (or mechanical) methods. Chemical methods involve strong intermolecular forces (electrostatic interaction, hydrogen bonding, and hydrophobic interactions) or chemical reactions. Physical methods are based on shape or structure. Techniques like wet/dry grinding, condensation coating, spray drying, pan coating, extrusion and membrane emulsification are normally used. In general, encapsulation depends on the technology, composition of the shell, physicochemical properties of the active compound, release mechanism, and particle size (Hack et al. 2012).

Microencapsulation involves alteration of colloidal and surface properties of a coating material that encapsulates an active compound such as agrochemicals, in the diameters ranging between 1 to 1000 μm. Microencapsulation leads to the control of the spatiotemporal release of active compounds such as agrochemicals, protection from environmental factors such as pH, temperature, light, humidity etc.

Table 2.1 Differences between microencapsulation and nanoencapsulation

Microencapsulation	Nanoencapsulation	References
Dependent on both technology and material properties	More dependent on properties of materials used	(Hack et al. 2012; Prasad et al. 2017)
Capsule dispersions and powders are usually obtained	Nanoparticles are easily obtained	(Hack et al. 2012; Prasad et al. 2017)
Smaller surface area to volume ratio	Larger surface area to volume ratio	(Hack et al. 2012; Prasad et al. 2017)
Greater shell thickness	Smaller shell thickness	(Hack et al. 2012; Prasad et al. 2017)

which affects the encapsulated agent. Environment-friendly and biodegradable materials which allow the sustained release of the agrochemicals is required as an encapsulating agent.

2.6.2 Nanoencapsulation

Nanoencapsulation is an innovative and promising nanotechnology in which the active compounds are efficiently released in a controlled, gradual manner from the capsules or particles. It is like microencapsulation, except that it involves particles of a size on the nanoscale. Different release mechanisms such as diffusion, dissolution, or biodegradation could be used for the delivery of agrochemicals in nano-encapsulated materials (Hack et al. 2012; Ammar 2018). A few major differences between microencapsulation and nanoencapsulation are shown in Table 2.1.

2.7 Types of Encapsulations Based on Their Application

Plant growth promoting bacteria and nitrogen-fixing bacteria (NFB) used as biofertilizers have gained importance due to their potential sustainability, especially over chemical fertilizers. However, their main drawback at the moment is the low inoculum survival rates in the field. Encapsulation techniques to control the biofertilizer and pesticidal activity are now gaining popularity, due to their ability to increase cell survival or viability, provide protection against the environment and improve release kinetics of the microorganisms (Campos et al. 2014b).

2.7.1 Water-In-Oil Emulsion

Water-in-oil emulsions are a method to embed microorganisms in liquid formulations. The microorganism is trapped in water by a layer of oil, which is favorable to prevent dehydration, along with improving cell viability and release kinetics.

However, sedimentation during storage can create problems that affect its release properties. In a study carried out by Vandergheynst et al. (2007), hydrophobic silica nanoparticles were used to reduce cell sedimentation and improve cell viability and survival by thickening the oil phase during storage.

2.7.2 *Spray Drying*

Spray drying represents a fast, economic process widely used in the large-scale production of biofertilizers. It involves the dissolution or suspension of the active material in a polymer solution, followed by its drying, while trapped in the matrix. A study was conducted by Campos et al. (2014a) in order to demonstrate the use of spray drying for the microencapsulation of plant nodule forming and putative NFB associated with Lupin nodules in polymer matrices to obtain a fertilizer with controlled release, reduced cost and improved efficiency on and protection against the environment.

2.7.3 *Ionotropic Gelation*

Ionotropic gelation occurs when uric acid in alginate crosslinks with multivalent cations in metals or substrates. He et al. (2015) microencapsulated *Raoultellaplanticola* Rs-2 with sodium alginate (NaAlg)-bentonite (Bent) composites for formulating a biofertilizer for controlled release and to minimize production costs. All release curves of the bacteria from the NaAlg-Bent microcapsules showed an initial burst followed by a gradual increase. The presence of Bent in microcapsules regulated the continuous release of bacteria and minimized the corresponding burst release effect. NaAlg-Bent composites could thus be ideal materials for encapsulation for bacterial fertilizers in farmlands. These were found to possess a long shelf life and could be stored at room temperatures for longer periods of time. Overall, the results indicated that Bent could be used as a filler to enhance the compatibility and structural properties of the polymer beads.

2.8 Polymers Used

Cs nanoparticles are being used in agriculture in seed treatment and also as biopesticides that help the plants fight off fungal infections and their effect on the growth of plants and agrochemical uptake depends on the plants being cultivated (Puoci et al. 2008; Duhan et al. 2017).

Polymers (especially as hydrogels and beads) are pivotal in controlled-release systems. These formulations provide slow-release properties and have excellent water-retention properties. They can also be used as compost after their degradation.

Natural polymers are gaining more popularity and approval over synthetic polymers as materials for encapsulation due to their eco-friendly nature, cost-effectiveness, easy availability, and biodegradability (Campos et al. 2014b).

2.9 Encapsulation Methods

There are many different types of encapsulation methods formulated to encapsulate agrochemicals in recent years (Fig. 2.1). A few common methods are:

2.9.1 Interfacial Polymerization

Interfacial polymerization is a process of microencapsulation in which polycondensation reaction takes place at the phase interface of two different solvents having one reactive monomer each (Liu et al. 2012). For example, natural pyrethinnanocapsules (Guo et al. 2014).

2.9.2 Solvent Evaporation

Microcapsule formation takes place by precipitation of the solution containing the dispersion of encapsulating material and core material in the organic phase along with a solution immiscible with the encapsulating material (Zhou et al. 2007). For example, Spinosad nanocapsules (Cao et al. 2009).

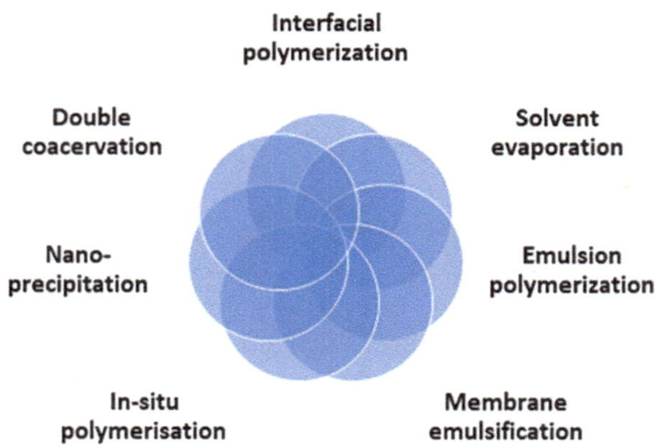

Fig. 2.1 Different encapsulation methods

2.9.3 *Emulsion Polymerization*

In emulsion polymerization, a uniform emulsion, made by dispersing an emulsifier in a solvent-insoluble monomer, is used for initiating a polymer reaction which ultimately forms the nanocapsules (Xu 2013). For example, abamectin nanocapsule suspension, natural pyrethrin nanocapsules, etc. (Shang et al. 2007; Song et al. 2009).

2.9.4 *Membrane Emulsification*

A shirasu porous glass (SPG) membrane under inert gas pressure, is used for the entry of the dispersion phase into the continuous phase, leading to breakage of continuous phase into droplets due to shear force between the membrane and the droplets themselves. This leads to the formation of small nanocapsules such as chlorantraniliprole and avermectin nanocapsules (Wu et al. 2008; Zhu-Zhu Li et al. 2006).

2.9.5 In situ *Polymerization*

Water-insoluble polymer formed by polymerization of two or more soluble monomers takes place to encapsulate the active compounds within the polymer (Feng et al. 2011). For example, acetochlor, methotrexate etc.

2.9.6 *Nano-Precipitation*

The dispersion of the formulation along with the polymer occurs from the oil phase into the aqueous phase by interaction at the interface of a solvent and a non-solvent. As a result, the formulation gets wrapped into a nanocapsule *via* precipitation (Feng et al. 2011). For example, pyrazole azoxystrobin and azoxystrobin nanocapsules, among other (Zhou et al. 2014; Wang et al. 2018).

2.9.7 *Double Coacervation*

Pesticide which is specifically water insoluble is encapsulated with a wall of two polymers of opposite charges (Feng et al. 2011). For example, azoxystrobin microcapsules (Ma et al. 2008).

2.10 Major Techniques for Encapsulation of Agrochemicals Inside the Polymer/Embedded in a Matrix

2.10.1 Interfacial Polycondensation

In interfacial polycondensation, the two reacting compounds or substances meet at an interface and react rapidly. One study used to synthesize encapsulated fertilizers involved a microencapsulation technique in which polycondensation was done between urea and formaldehyde following a stepwise process, along with maintaining a strict control of room temperature. The main goal of this study was to show the effectiveness of a controlled release fertilizer registered under the name of FERLENT®, which was obtained from a polymeric material. The resulting material acted as an alternative source of nutrients, allowing plants to consume nutrients when required (González et al. 2015).

Another study involved the synthesis of a urea-formaldehyde polycondensate using interfacial polycondensation in a microemulsion. Urea-formaldehyde resins were formed when formaldehyde was reacted with excess concentration urea at controlled temperature and pH. Polycondensation was done after the initial prepolymerization to synthesize the fertilizer. This yielded medium molecular weight products from which water and residual solvents were removed (Neamtu et al. 2015).

Another microencapsulation study involving urea-formaldehyde systems was used to synthesize encapsulated fertilizers containing N, P, Ca, and Mg as active nutrients. In order to perform tests to study the biological activity and leaching from the microcapsules, it was necessary to agglomerate the capsules using specific ingredients. The aim was to synthesize and test a new type of N-P-Ca-Mg encapsulated fertilizer, which had a higher availability in *Arabidopsis thaliana*, even though lower doses than conventional granular fertilizers were applied. The *in vitro* results showed good slow-release properties for both macronutrients and micronutrients (Neamtu et al. 2015).

2.10.2 Ionotropic Gelation

The technique of ionotropic gelation or crosslinking involves the interaction of an ionic polymer with an oppositely charged ion or polymer to create crosslinking.

Perez and Francois (2016) conducted a study to examine the application of macrospheres prepared with Cs and Cs-starch blends, using a sodium tripolyphosphate aqueous solution as the crosslinking agent. These biopolymers formed hydrogels that were capable of controlling release kinetics of the encapsulated fertilizers. Potassium nitrate encapsulated inside beads was also found to be useful as a controlled-release fertilizer. On using starch-Cs blends, lesser concentrations of Cs have been found to be needed to encapsulate the agrochemicals.

2.10.3 Thermally Induced Polymerization

In thermal polymerization, polymerization is carried out directly on the agrochemical particles' surface, using an external heat source. Microwave irradiation (MW) has been considered and studied as a method to prepare SAPs. This has various advantages like uniform heating, increased reaction rate, simple experimental apparatus, and convenient process control, along with saving energy and time (Wen et al. 2016).

In a study, an encapsulated urea fertilizer was synthesized in a single step, by incorporating it in a polymer matrix of NaAlg, acrylic acid (AA), acrylamide (AM) and Bent. The proposed microwave-assisted method yielded high reaction rates with a very low reaction time. This also showed that Bent helped increase water absorption, improved water-holding ratio and water-retention capacity of the soil, along with causing the system to release the nutrient in a more sustained manner. Microwave irradiation could thus be used to produce encapsulated fertilizers (Wen et al. 2016).

2.10.3.1 Spray Chilling

The spray chilling process is one of the most used encapsulation technologies in the pharmaceutical, cosmetics and food additives fields. It consists of the atomization of a matrix, containing a suspension or emulsion of an active ingredient, into an environment maintained at temperatures below the melting point of the matrix. The atomization leads to the formation of small droplets, which then solidify on cooling and result in microencapsulated structures with spherical shapes and smooth surfaces. This technology is easy to use and scale-up, due to no need for the use of organic solvents and the use of high temperatures. However, this technology has not been as widely studied for agrochemicals (Ammar 2018).

Navarro-Guajardo et al. (2017) made a study to explore a bioinspired structure based on the plant cuticle model and its membrane properties for water and nutrient transport control. Candelilla wax, extracted from Candelilla wild plants (*Euphorbia antisyphilitica*), was studied as a renewable, biodegradable matrix for a controlled release phosphate fertilizer, microencapsulated via a modified spray chilling process. The choice of a cuticular wax as encapsulating matrix conferred slow-release characteristics that displayed the regulated water and nutrient transport features on use in plants.

2.10.4 In situ *Free Radical Polymerization*

In *in situ* free radical polymerization, direct polymerization of the basic units or monomers is carried out on the surface of the active components to be encapsulated, using free radicals. In a study, slow-release NPK fertilizers were encapsulated by SAPs, synthesized *via in situ* free radical polymerization of NaAlg, AA, AM, and montmorillonite to coat fertilizer particles. It was found that these had good slow-release properties, a water adsorption capacity, which could reduce fertilizer loss and increased utilization of water in agriculture (Campos et al. 2014a).

2.11 Role of Polymer in Different Types of Nanoencapsulation

Different structures of nanoencapsulation can be obtained (Fig. 2.2):

2.11.1 *Nanocapsules*

Polymers form a membrane around the active compounds, i.e. pesticide formulations, sometimes, along with another polymeric matrix which is present mainly to absorb the active compounds for spontaneous formation and encapsulation of the nanocapsules (Mora-Huertas et al. 2010; Ezhilarasi et al. 2013). Nanocapsules are prepared by nanoprecipitation, double-emulsification, emulsion-diffusion, layer-by-layer deposition, emulsification-coacervation and solvent evaporation (Pinto Reis et al. 2006).

Natural polysaccharides are biodegradable, non-toxic and biocompatible which makes them more preferable over polymers such as PEG, PCL etc. For example, imidacloprid, an insecticide which acts on the central nervous system of insects, is encapsulated in NaAlg using emulsion cross-linking technology (Kumar et al. 2014). Also, the self-linking property of copolymers found amongst natural polysaccharides has given rise to a vast number of possible synthesis techniques for nanoencapsulation such as encapsulation of imidacloprid crystals with Cs and NaAlg. Oppositely charged polymers like Cs and NaAlg are cross-linked due to electrostatic interactions and absorption of oppositely charged polyelectrolytes. Amphiphilic polymers are cross-linked more easily and thus, also reduces the need of a surfactant which aids in self-assembly of polymers. For example, methomyl is loaded in photo-crosslinked nanocapsules of amphiphilic carboxymethyl Cs (Az-CMCs) (Yin et al. 2010).

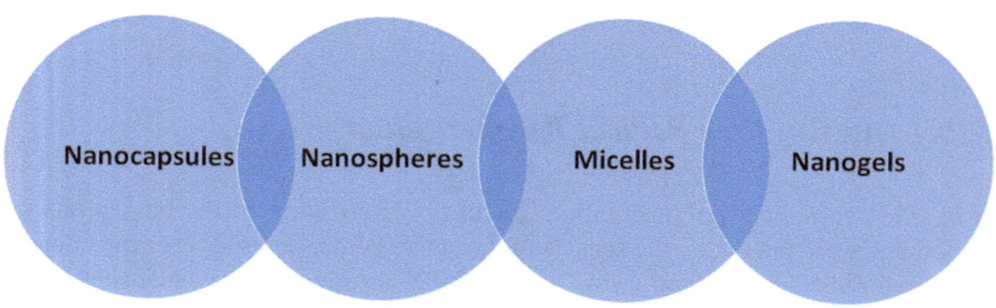

Fig. 2.2 Role of polymer in different types of nanoencapsulation

Encapsulation efficiency and the size of the nanocapsules formed is decided mainly by the ratio of masses of pesticide to be loaded and the encapsulating copolymers. Lesser the mass ratio of pesticide is to the polymer, the greater is the nanocapsule's size and lesser is the efficiency while the size of the nanocapsule decreases and the efficiency of encapsulation increases as the mass ratio increases (Perlatti et al. 2013).

2.11.2 Nanospheres

Nanospheres are synthesized in a similar way as nanocapsules but instead of having a cross-shell arrangement, the active ingredients, i.e. pesticides are uniformly embedded in the whole polymeric nanocapsule. Polymerization techniques such as emulsion or interfacial polymerization play a major in nanosphere synthesis (Forim et al. 2012; Zhang et al. 2013).

Many polymers such as PCL used for azadirachtin encapsulation to produce nanospheres and nanocapsules in the range of 150–200 nm diameter have shown great potential in the encapsulation of active ingredients such as pesticides like azadirachtin (Trivedi et al. 2010).

2.11.3 Micelles

Self-assembly property of amphiphilic polymers is used to form micelles in an aqueous solution. Amphiphilic block copolymers, surfactants, etc. together form spherical micelles in an aqueous solution due to self-assembly of amphiphilic polymers which formation of an outer hydrophilic micellar shell and an inner hydrophobic core. Water-soluble pesticides are easier to encapsulate to form micelles. While water-soluble copolymers form micelles either through direct dissolution or through film casting methods. The direct dissolution method involves the addition of the copolymers above their critical micelle concentration to the aqueous solution which in turn, leads to the micelle formation along with the entrapment of pesticides forming the hydrophobic core. Whereas, the film casting method involves using different approaches to solubilize previously formed films of copolymers with pesticides (Trivedi et al. 2010; Zhang et al. 2013).

2.11.4 Nanogels

Polymers are cross-linked physically or chemically into a group of networks of aqueous dispersions of hydrogel particles, called nanogels which are in the nanoscale size. Prospects of targeted delivery perhaps could not be established with these preparations, thus are not of much use in the controlled release of pesticides.

2.12 Agrochemical Delivery from Polymer Encapsulated Materials

Polymers help in the delivery of various agrochemicals by altering their physical and chemical structure, in response to chemical and physical changes in the environment. It is divided into following kinds of stimulus-based deliveries (Fig. 2.3).

2.12.1 Light Sensitivity

Some formulations are light sensitive and undergo decomposition on light exposure leading to rapid release and reduced effectiveness of the pesticides. Due to this property, pesticide formulations need to be frequently added, increasing the overall costs and wastage. To protect the pesticides from decomposition, they are encapsulated with or within photo-responsive polymers. Photo-responsive polymers absorb light energy from light stimulation which induces changes either in the intramolecular or intermolecular arrangement leading to changes in solubility, color and conductivity of the polymers, or it can induce various chemical reactions like polymerization, dimerization, photolysis or isomerization reactions (Wang 2014). These changes due to light stimulation are utilized to have a controlled and targeted transmission of pesticides.

The photoresponsive polymer encapsulated pesticides can be made to respond in the following ways:

- Change in the physical and chemical properties of luminophore groups added to carrier materials leads to a controllable change in the properties of the polymers on light exposure.
- Decomposition of light-sensitive polymers into ions on light stimulation can be used to controlled pesticide release by tailoring the polymers while considering the role of environmental factors in its decomposition.
- Some polymers become light sensitive on exposure to heat which can be generated by conversion of light energy to heat energy by some additional light-sensitive molecules. Thus, indirectly controlling the release of the pesticide.

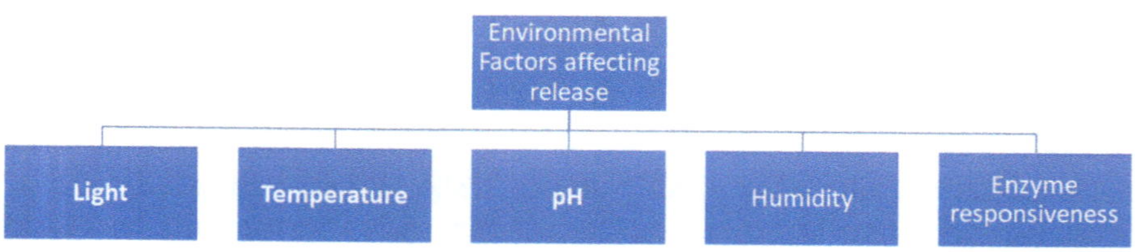

Fig. 2.3 Agrochemical release because of environmental factors on the encapsulating polymer material

Recently, a photoresponsive herbicide formulation of 2,4-D using a coumarin copolymer was formulated which effectively transmitted the herbicide into the plant body while showing strong fluorescence along with a high light response (Singh et al. 2015).

2.12.2 Temperature Sensitivity

Alteration in physical and chemical properties of temperature-responsive polymers due to temperature change have been proposed to be one of the properties of temperature sensitive polymers that can be utilized in the agriculture field. Temperature sensitive polymers are used along with the agrochemicals in the form of gels, liposomes, polymers, and nanoparticles etc.

Phase separation property of certain polymers on temperature change is used as one of the release mechanisms in the active ingredient loaded polymers. The property that plays a major role in the determination of response mechanism of a polymer encapsulated pesticide is lower critical solution temperature (LCST). LCST is the minimum temperature found on the phase separation curve which forms in the plot of polymer concentration and the surrounding temperature of a homogeneous polymer. The release response is inhibited at a temperature lower than the LCST of a polymer as strong hydrogen bonding interactions between the water molecules and the polymers take place at that temperature. Whereas pesticide release occurs at higher temperatures than LCST when all the polymer becomes water soluble which leads to hydrophobicity, since the polymer side chains are simultaneously intensified. This makes the polymer amphiphilic which changes the irregularly coiled structure of polymer encapsulated pesticide into a tight colloidal structure, since the hydrophobic chains of the side chains also influence the backbone to form cross-linked network structures. In the case of rice farming, a temperature sensitive pesticide formulation of NIPAM and butyl acrylate has shown promising applications as pyraclostrobin microcapsules whose LCST is 28.2 °C above which the release of pyraclostrobin takes place. Toxicity of pyraclostrobin in pyraclostrobin microcapsule was observed to be much lower than the original pyraclostrobin thus, protecting the environment with encapsulation (Otsuka et al. 2012).

Currently, temperature-responsive pesticide encapsulations need additional research in the areas related to improving the response and controlling the spatial-temporal release of pesticides. The difference in the indoor and the outdoor temperatures affects the release mechanism, thus, controlling the release mechanism under indoor and the outdoor conditions must also be optimized.

2.12.3 pH Sensitivity

Amongst the environmentally responsive polymers, pH-responsive polymers have the most potential as they are easily soluble, have volume and allows good permeability (Xu 2014). They are comprised of weakly acidic or weakly alkaline groups, and many acidic groups such as carboxyl groups or alkaline groups such as amine groups are found within the molecular segments of these polymers.

As these polymers in response to a change in the external pH value, release and capture protons, accordingly, i.e. protonate and de-protonate on change in the pH of the environment (Sun et al. 2016), thus, the change in the degree of ionization of these pH-sensitive polymers have been mainly exploited to encapsulate pesticides and direct its release mechanisms.

The change in the solubility of these polymers depends on their ionization capacity. Above the polymer's isoelectric point, the ionization capacity changes which is due to the pH change. Polyacrylic acid (PAA), polyvinyl pyridine, polymethacrylic acid-2-(diethylamino) ethyl methacrylate, polymethacrylic acid (PMAA), polypropylamine hydrochloride, etc. are some of the known pH sensitive polymers.

pH-responsive polymers are divided into two types based on pH range and ionization and deionization at different pH ranges (Yang et al. 2012; Lee et al. 1999). They are polyacids and polybases.

Polyacids such as PAA and PMAA are ionized/deionized at low pH and accept protons at neutral or higher pH range of 4–8. Whereas polybases, pH-responsive polymers under acidic conditions, turns into cations by accepting protons by amino groups which are typically present on their side chains. Protons are released under alkaline conditions which makes them less soluble.

Although a lot of systematic research is required for the field application of these pesticide carriers, pH-responsive polymers tend to have high purity, properties that can be easily controlled and chemically modified with good selectivity, thus, they can be used to encapsulate pesticides (Lin 2016).

Recent developments in the pH-responsive polymer encapsulated agrochemicals includes PAA or aminated or chlorpyrifos mesoporous silicone incorporated encapsulations where release rates were the highest at acidic conditions, while their slow release in the neutral environmental conditions could be used for sustained release of pesticides (Li et al. 2003).

2.12.4 *Humidity*

Material resistance, dielectric constant, and volume are some properties of certain polymers, which are altered with the change of humidity of the local environment (Liu 1997).

Polymer humidity-sensitive material interacts to form hydrogen bonds with water molecules and van der Waals force comes into play to make the polymers more hygroscopic (Klier et al. 2000). High sensitivity, desired surface areas, fast response, small size, much lower weight along with easier incorporation allows humid sensitive polymers such as polyimide to be applied in the agricultural field for sustained release of agrochemicals.

In the recent years, formulations such as BASF Seltima® microcapsule which are released in contact with leaves to control rice blast, whereas is water resistant to control the release of pesticides with respect to the surrounding humidity (Huang et al. 2018).

2.12.5 *Enzyme-Responsive*

Enzyme-responsive polymers undergo physical and chemical changes as a response when a pest interacts with a plant. The stimulation is generated in the form of a series of changes which includes changes in the plant's enzymes (Bai et al. 2016).

These enzyme responsive pesticides are biodegradable and biocompatible. The enzyme responsive polymer encapsulated pesticides degrade on contact with the enzymes to release the agrochemicals, e.g. crop endangering Lepidoptera insects breaks the walls of pesticide microcapsules with the cutinase enzyme present in them, thus coming in contact with the pesticides. Further development in this field is necessary to control the morphology and chemical reversibility of these encapsulations.

2.13 Some Specific Polymers Used for Encapsulation of Fertilizers

2.13.1 *Silicon-Based Nano-Fertilizers*

Mesoporous silica and aluminosilicates have greatly defined channels with high porosity, active sites for adsorption, large surface areas, and ion exchange and catalysis. Silicon-based fertilizers can improve the growth and development of seedlings and roots. To increase food production, titanium oxide or titanium that are non-toxic have been tested as additives in fertilizers, in order to increase water retention properties of soil (Abhilash and Singh 2009; Duhan et al. 2017).

2.13.2 *Superabsorbent Polymers*

The growth and quality of the plants depend on fertilizers and water. Superabsorbent polymers (SAP) have been studied as materials for water management. They are cross-linked 3D hydrophilic materials that can absorb fluids much higher than their original weight without dissolution, and can, even at high pressure, contain liquids. It also helps improve the water retention capacity and soil fertility. However, synthetic SAPs derived from non-degrading petroleum-based polymers like polyacrylate or polyacrylamide play a major role in the market and lead to environmental problems. Newer studies are focused on the incorporation of biodegradable and eco-friendly polysaccharides into SAPs, instead of using petroleum-related products (Wen et al. 2016).

One such polymer is NaAlg, which is inert in aqueous solutions within a matrix, undergoes gelatinization, and has a higher gel porosity besides being water-soluble, non-toxic, biodegradable, and compatible with biological systems. However, NaAlg is not practical due to high cost, low modulus, thermal stability and gas permeability.

To overcome the drawbacks, clay/polymer nanocomposites containing a higher proportion of clay were synthesized by Bashidzadeh and Olad (2014). The use of clay reduces production costs, along with the improvement of basic NaAlg properties, while ensuring mechanical and thermal stability. The need for sustainable NaAlg-based clay/polymer materials has consistently increased. A range of nanofertilizers with controlled-release kinetics have been developed consisting of synthetic or biological polymer composites made from NaAlg and clays. Adsorbent clays like zeolite, montmorillonite and Bent have been used to design them (Wen et al. 2016; Duhan et al. 2017).

Bent is derived from sedimentary rocks and is mainly composed of montmorillonite, which has good water absorption and swelling properties. Addition of low-cost Bent to NaAlg increases the concentration of solids present and improves the mechanical strength due to increase viscosity and enhanced stability. Bent can also act as a modifying agent in alginates, which makes the encapsulation more efficient and controls the agrochemical release profiles (He et al. 2015).

2.13.3 Biodegradable Polymers

Two major hydrophilic, biodegradable polymers that are widely used are Cs and starch, polysaccharides that have the desired characteristics like lower cost, environment-friendly and ability to regulate fertilizer release (Perez and Francois 2016).

Cs is a biodegradable, linear, cationic, and non-toxic polysaccharide. It is a derivative of chitin, which is usually present in the exoskeletons of crustaceans. It is synthesized by partial deacetylation of chitin under alkaline conditions or enzymatic hydrolysis in the presence of a chitin deacetylase. It forms ionic complexes with various water-soluble polymers and crosslinking agents, thus forming insoluble gels. It also acts as a growth promoter in plants and guards' agrochemicals against harmful microbial activities, making it a good polymer to be used with agrochemicals (Perez and Francois 2016).

Cs polymer synthesis has been done with PMAA for encapsulation of NPK fertilizers. Wu and Liu (2008) developed and characterized Cs particles for the controlled release of NPK fertilizers, and water retention. The particles were formed by three layers: a water-soluble core containing the fertilizer, Cs as an internal coating, and an absorbent polymer acting as an external coating. Addition of the particles significantly enhanced the water retention capacity of the soil (Duhan et al. 2017; Campos et al. 2014b).

Starch is a complex, non-ionic polysaccharide that exists in nature, with properties such as hydrophilicity, biodegradability and low cost (Suárez and Gutiérrez 2017; Toro-Márquez et al. 2018; Gutiérrez 2018c). Han et al. (2009) developed a starch and PVA film for the encapsulation of soluble fertilizer granules, due to their permeability and biodegradability.

Acknowledgments The authors wish to acknowledge Vellore Institute of Technology, Vellore, India for providing the opportunity to write this book chapter.

Conflicts of Interest The authors declare no conflict of interest.

References

Abhilash, P. C., & Singh, N. (2009). Pesticide use and application: An Indian scenario. *Journal of Hazardous Materials, 165*(1–3), 1–12. https://doi.org/10.1016/j.jhazmat.2008.10.061.

Ammar, A. S. (2018). Nanotechnologies associated to floral resources in agri-food sector. *Acta Agronómica, 67*(1), 146–159. https://doi.org/10.15446/acag.v67n1.62011.

Auffan, M., Rose, J., Bottero, J. Y., Lowry, G. V., Jolivet, J. P., & Wiesner, M. R. (2009). Towards a definition of inorganic nanoparticles from an environmental, health and safety perspective. *Nature Nanotechnology, 4*, 634–641. https://doi.org/10.1038/nnano.2009.242.

Bai, J. K., Zhang, Y., & Wang, J. X. (2016). Progress intelligent materials based on enzyme response. *International Materials Reviews, 30*, 134–139.

Bashidzadeh, A., & Olad, A. (2014). Slow-released NPK fertilizer encapsulated by NaAlg-*g*-poly(AA-*co*-AAm)/MMT superabsorbent nanocomposite. *Carbohydrate Polymers, 114*, 269–278. https://doi.org/10.1016/j.carbpol.2014.08.010.

Campos, D. C., Acevedo, F., Morales, E., Aravena, J., Amiard, V., Jorquera, M. A., Inostroza, N. G., & Rubilar, M. (2014a). Microencapsulation by spray drying of nitrogen-fixing bacteria associated with lupin nodules. *World Journal of Microbiology and Biotechnology, 30*(9), 2371–2378. https://doi.org/10.1007/s11274-014-1662-8.

Campos, E. V. R., de Oliveira, J. L., Fraceto, L. F., & Singh, B. (2014b). Polysaccharides as safer release systems for agrochemicals. *Agronomy for Sustainable Development, 35*(1), 47–66. https://doi.org/10.1007/s13593-014-0263-0.

Cao, M., Guan, P., Hu, L., & Chen, X. P. (2009). The research progress and influential factors on preparation of microcapsules by w/o/w multiple emulsion-solvent evaporation method. *Chemical Bioengineering, 9*, 002.

Duhan, J. S., Kumar, R., Kumar, N., Kaur, P., Nehra, K., & Duhan, S. (2017). Nanotechnology: The new perspective in precision agriculture. *Biotechnology Reports, 15*, 11–23. https://doi.org/10.1016/j.btre.2017.03.002.

Ezhilarasi, P. N., Karthik, P., Chhanwal, N., & Anandharamakrishnan, C. (2013). Nanoencapsulation techniques for food bioactive components: A review. *Food and Bioprocess Technology, 6*(3), 628–647. https://doi.org/10.1007/s11947-012-0944-0.

Feng, J. G., Xu, Y., Luo, X. R., Yan, H., & Wu, X. M. (2011). Discussion on the solvent evaporation method for preparation of microcapsules and the development of the pesticides microcapsules. *Chinese Journal of Pesticide Science, 13*, 568–575.

Forim, M. R., Silva, M. F.d. G. F. d., & Fernandes, J. B. (2012). Secondary metabolism as a measurement of efficacy of botanical extracts: The use of *Azadirachtaindica* (Neem) as a model, insecticides. In F. Perveen (Ed.),. ISBN: 978–953–307-780-2 *Advances in integrated pest management. InTech.* https://doi.org/10.5772/27961.

Gogos, A., Knauer, K., & Bucheli, T. D. (2012). Nanomaterials in plant protection and fertilization: Current state, foreseen applications, and research priorities. *Journal of Agricultural and Food Chemistry., 60*(39), 9781–9792. https://doi.org/10.1021/jf302154y.

González Hurtado, M., Rieumont Briones, J., Quintana Owen, P., Bartolo-Perez, P., Soares, B. G., & Cossa, M. M. (2015). Ferlent® a controlled release fertilizer produced from a polymeric material with agronomic benefits. *Environmental Engineering and Management Journal, 14*(12), 2913–2917.

Guo, Y. Z. (2014). Study on the preparation and the release mechanism of acetochlor microcapsules produced by interfacial polymerization method. Master's Thesis, Southwest University, Chongqing, China, April 2014.

Gutiérrez, T. J. (2017). Chapter 8. Chitosan applications for the food industry. In S. Ahmed & S. Ikram (Eds.), *Chitosan: Derivatives, composites and applications* (pp. 183–232). WILEY-Scrivener Publisher. EE.UU. ISBN: 978–1–119-36350-7. https://doi.org/10.1002/9781119364849.ch8.
Gutiérrez, T. J. (2018a). Biological macromolecule composite films made from sagu starch and flour/poly(ε-caprolactone) blends processed by blending/thermo molding. *Journal Polymers and the Environment, 26*(9), 3902–3912. https://doi.org/10.1007/s10924-018-1268-6.
Gutiérrez, T. J. (2018b). Chapter 55. Processing nano- and microcapsules for industrial applications. In C. M. Hussain (Ed.),. Editorial Elsevier. EE.UU. ISBN: 978–0–12-813351-4 *Handbook of nanomaterials for industrial applications* (pp. 989–1011). https://doi.org/10.1016/B978-0-12-813351-4.00057-2.
Gutiérrez, T. J. (2018c). Are modified pumpkin flour/plum flour nanocomposite films biodegradable and compostable? *Food Hydrocolloids, 83*, 397–410. https://doi.org/10.1016/j.foodhyd.2018.05.035.
Gutiérrez, T. J., & Álvarez, K. (2017). Chapter 6. Biopolymers as microencapsulation materials in the food industry. In M. Masuelli & D. Renard (Eds.), *Advances in physicochemical properties of biopolymers: Part 2* (pp. 296–322). Bentham Science Publishers. EE.UU. ISBN: 978–1–68108-545-6. eISBN: 978–1–68108-544-9. https://doi.org/10.2174/978168108544911 7010009.
Hack, B., Egger, H., Uhlemann, J., Henriet, M., Wirth, W., Vermeer, A. W. P., & Duff, D. (2012). Advanced agrochemical formulations through encapsulation strategies? *Chemie-Ingenieur-Technik, 84*(3), 223–234. https://doi.org/10.1002/cite.201100212.
Han, X., Chen, S., & Hu, X. (2009). Controlled-release fertilizer encapsulated by starch/polyvinyl alcohol coating. *Desalination, 240*(1–3), 21–26. https://doi.org/10.1016/j.desal.2008.01.047.
He, Y., Wu, Z., Tu, L., Han, Y., Zhang, G., & Li, C. (2015). Encapsulation and characterization of slow-release microbial fertilizer from the composites of bentonite and alginate. *Applied Clay Science, 109-110*, 68–75. https://doi.org/10.1016/j.clay.2015.02.001.
Herniou--Julien, C., Mendieta, J. R., & Gutiérrez, T. J. (2019). Characterization of biodegradable/non-compostable films made from cellulose acetate/corn starch blends processed under reactive extrusion conditions. *Food Hydrocolloids, 89*, 67–79. https://doi.org/10.1016/j.foodhyd.2018.10.024.
Huang, B., Chen, F., Shen, Y., Qian, K., Wang, Y., Sun, C., Zhao, X., Cui, B., Gao, F., Zeng, Z., & Cui, H. (2018). Advances in targeted pesticides with environmentally responsive controlled release by nanotechnology. *Nanomaterials, 8*(2), 102. https://doi.org/10.3390/nano8020102.
Klier, J., Weichhard, C., & Leiderer, P. (2000). Wetting behaviour of solid and liquid hydrogen films. *Physics B: Condensed Matter, 284*, 391–392. https://doi.org/10.1016/S0921-4526(99)02784-2.
Kumar, S., Bhanjana, G., Sharma, A., Sidhu, M. C., & Dilbaghi, N. (2014). Synthesis, characterization and on field evaluation of pesticide loaded sodium alginate nanoparticles. *Carbohydrate Polymers, 101*, 1061–1067. https://doi.org/10.1016/j.carbpol.2013.10.025.
Lee, A. S., Gast, A. P., Butun, V., & Steven, P. A. (1999). Characterizing the structure of pH dependent polyelectrolyte block copolymer micelles. *Macromolecules, 32*(13), 4302–4310. https://doi.org/10.1021/ma981865o.
Li, Y. H., Huang, Y. D., & Liu, Y. Y. (2003). Humidity sensing polymer materials. *Materials Science and Technology, 3*, 332–336.
Lin, Y. S. (2016). Synthesis, modification and application in slow-release pesticide of mesoporous MCM-41. Master's Thesis, Zhongkai University of Agriculture and Engineering, Guangzhou, China, June 2016.
Liu, C. J. (1997). General situations of high polymer humidity sensors and its developing direction. *Instrumentation Technology Sensing, 12*, 1–4.
Liu, G. W. (2012). *Modern pesticide formulations processing technology* (pp. 1–7). Beijing: Chemical Industry Press.
Ma, C. J., & Tao, L. H. (2008). Study on the technology of preparing azoxystrobin microcapsules by complex conservation. *Journal of Anhui Agricultural Sciences, 36*, 3496–3508.

Mora-Huertas, C. E., Fessi, H., & Elaissari, A. (2010). Polymer-based nanocapsules for drug delivery. *International Journal of Pharmaceutics, 385*, 113–142.
Nagula, S., & Usha, P. B. (2016). Application of nanotechnology in soil and plant system with special reference to nanofertilizers. *Advances in Life Sciences, 5*(14), 5544–5548.
Nair, R. M., Whittall, A., Hughes, S. J., Craig, A. D., Revell, D. K., Miller, S. M., Powell, T., & Auricht, G. C. (2010). Variation in coumarin content of *Melilotus* species grown in South Australia. *New Zealand Journal of Agricultural Research., 53*(3), 201–213. https://doi.org/10.1080/00288233.2010.495743.
Navarro-Guajardo, N., García-Carrillo, E. M., Espinoza-González, C., Téllez-Zablah, R., Dávila-Hernández, F., Romero-García, J., Ledezma-Pérez, A., Mercado-Silva, J. A., Pérez Torres, C. A., & Pariona, N. (2017). Candelilla wax as natural slow-release matrix for fertilizers encapsulated by spray chilling. *Journal of Renewable Materials, 6*(3), 226–236. https://doi.org/10.7569/JRM.2017.634164.
Neamaţu, C., Popescu, M., Oancea, F., & Dima, Ş. O. (2015). Synthesis optimization and characterization of microencapsulated N-P-K slow-release fertilizers. *Open Chemistry, 13*(1), 813–823. https://doi.org/10.1515/chem-2015-0098.
Neamtu, C., Popescu, M., & Dima, Ş.-O. (2015). Leaching and *in vitro* agrochemical screening for new slow release fertilizers containing N, P, Ca, and Mg. *Academic Research Journal of Agricultural Science and Research., 3*(3), 45–53. https://doi.org/10.14662/arjasr2015.003.
Otsuka, I., Travelet, C., Halila, S., Fort, S., Pignot-Paintrand, I., Narumi, A., & Borsali, R. (2012). Themosponsive self-assemblies of cyclic and branched oligosaccharide-*block*-poly(*N*-isopropylamide) diblock copolymers into nanoparticles. *Biomacromolecule, 13*(5), 1458–1465. https://doi.org/10.1021/bm300167e.
Perez, J. J., & Francois, N. J. (2016). Chitosan-starch beads prepared by ionotropic gelation as potential matrices for controlled release of fertilizers. *Carbohydrate Polymers, 148*, 134–142. https://doi.org/10.1016/j.carbpol.2016.04.054.
Perlatti, B., Bergo, P. L. d. S., Silva, M. F. d. G. F. d., Fernandes, J. B., & Forim, M. R. (2013). Polymeric nanoparticle-based insecticides: a controlled release purpose for agrochemicals. In S. Trdan (Ed.), *Insecticides – Development of safer and more effective technologies, Prof* (pp. 523–550). INTECH Open Access Publisher. Available from: http://www.intechopen.com/books/insecticides/development-of-safer-and-more-effective-technologies/polymeric-nanoparticle-based-insecticides-a-controlled-release-purpose-for-agrochemicals.
Pinto Reis, C., Neufeld, R. J., Ribeiro, A. J., Veiga, F., & Nanoencapsulation, I. (2006). Methods for preparation of drug-loaded polymeric nanoparticles. *Nanomedicine, 2*, 8–21. https://doi.org/10.1016/j.nano.2005.12.003.
Prasad, R., Bhattacharyya, A., & Nguyen, Q. D. (2017). Nanotechnology in sustainable agriculture: Recent developments, challenges, and perspectives. *Frontiers in Microbiology, 8*, 1–13. https://doi.org/10.3389/fmicb.2017.01014.
Puoci, F., Iemma, F., Spizzirri, U. G., Cirillo, G., Farmaceutiche, S., Calabria, U., & Cs, R. (2008). Polymer in agriculture. *A review, 3*(1), 299–314. https://doi.org/10.3844/ajabssp.2008.299.314.
Ram, P., Vivek, K., & Kumar, S. P. (2014). Nanotechnology in sustainable agriculture: Present concerns and future aspects. *African Journal of Biotechnology, 13*(6), 705–713. https://doi.org/10.5897/ajbx2013.13554.
Shang, Q., Shang, Z. H., & Ci, S. Y. (2007). HPLC analysis of abamectin nanocapsules suspension concentrate. *Agrochemicals, 46*, 185–186.
Singh, P. N. D., Atta, S., Bera, M., Chattopadhyay, T., Paul, A., Ikbal, M., & Maiti, M. K. (2015). Nano-pesticide formulation based on fluorescent organic photoresponsive nanoparticles: for controlled release of 2,4-D and real time monitoring of morphological changes induced by 2,4-D in plant Systems. *RSC Advances, 5*, 86990–86996. https://doi.org/10.1039/c5ra17121k.
Song, Q., Mei, X. D., Huang, Q. L., Wang, Z. Y., & Ning, J. (2009). Preparation of abamectin microcapsules by means of emulsion polymerization and it bioactivity. *Chinese Journal of Pesticide Science, 11*, 392–394.
Suárez, G., & Gutiérrez, T. J. (2017). Chapter 15. Recent advances in the development of biodegadable films and foams from cassava starch. In C. Klein (Ed.), *Handbook on Cassava: Production, potential uses and recent advances* (pp. 297–312). New York. EE.UU. ISBN: 978-1-53610-307-6: Editorial Nova Science Publishers, Inc..

Sun, C. J., Cui, H. X., Wang, Y., Zeng, Z. H., Zhao, X., & Cui, B. (2016). Studies on applications of nanomaterial and nanotechnology in agriculture. *Journal of Agricultural Science and Technology, 18*, 18–25. https://doi.org/10.13304/j.nykjdb.2015.240.

Tilman, D., Cassman, K. G., Matson, P. A., Naylor, R., & Polasky, S. (2002). Agricultural sustainability and intensive production practices. *Nature., 418*, 671–677. https://doi.org/10.1038/nature01014.

Tolescu, C., Fierascu, I., Neamtu, C., Anton, I., & Fierascu, R. C. (2014). Microencapsulated fertilizers for improvement of plant nutrition. *Journal of the Serbian Chemical Society, 79*(6), 659–668. https://doi.org/10.2298/jsc131004147t.

Toro-Márquez, L. A., Merino, D., & Gutiérrez, T. J. (2018). Bionanocomposite films prepared from corn starch with and without nanopackaged Jamaica (*Hibiscus sabdariffa*) flower extract. *Food and Bioprocess Technology, 11*(11), 1955–1973. https://doi.org/10.1007/s11947-018-2160-z.

Trivedi, R., & Kompella, U. B. (2010). (2010). Nanomicellar formulations for sustained drug delivery: Strategies and underlying principles. *Nanomed-UK, 5*(3), 485–505. https://doi.org/10.2217/nnm.10.10.

Vandergheynst, J., Scher, H., Guo, H. Y., & Schultz, D. (2007). Water-in-oil emulsions that improve the storage and delivery of the biolarvacide *Lagenidium giganteum*. *BioControl, 52*(2), 207–229. https://doi.org/10.1007/s10526-006-9021-9.

Wang, S.P. (2014). Construction of photosensitive nanocarriers and their photo-responsive characteristic. Master's Thesis, Zhejiang University, Zhejiang, China, March 2014.

Wang, A. Q., Wang, Y., Wang, C. X., Cui, B., Sun, C. J., Zhao, X., Zeng, Z. H., Yao, J. W., Liu, G. Q., & Cui, H. X. (2018). Research progress on nanometer microencapsulation of pesticide. *Journal of Agricultural Science and Technology, 20*, 10–18.

Wen, P., Wu, Z., He, Y., Ye, B. C., Han, Y., Guan, X., & Wang, J. (2016). Microwave-assisted one-step synthesis and characterization of a slow release nitrogen fertilizer with inorganic and organic composites. *RSC Advances, 6*(44), 37337–37346. https://doi.org/10.1039/c5ra27828g.

Wu, L., & Liu, M. (2008). Preparation and properties of chitosan coated NPK compound fertilizer with controlled release and water retention. *Carbohydrate Polymers, 72*(2), 240–247. https://doi.org/10.1016/j.carbpol.2007.08.020.

Wu, J., Zhou, Y. F., Chen, J., Nie, W. Y., & Shi, R. (2008). Preparation of natural pyrethrum nanocapsule by means of microemulsion polymerization. *Polymer Materials Science and Engineering, 24*, 35–38.

Xu, L. (2013). Preparation and characteristic analysis of microcapsule controlled releasing agent of pesticide chitosan copolymer. Master's Thesis, Northeast Agricultural University, Harbin, China, June 2013.

Xu, B. B. (2014). Preparation and characterization of magnetic/temperature/ph responsive nanocapsules. Master's Thesis, Anhui University, Anhui, China, May 2014.

Yin, Y., Xu, S., Chang, D., Zheng, H., Li, J., Liu, X., Xu, P., & Xiong, F. (2010). One-pot synthesis of biopolymeric hollow nanospheres by photocrosslinking. *Chemical Communications, 46*, 8222–8224. https://doi.org/10.1039/c0cc03129a.

Zhang, J., Li, M., Fan, T., Xu, Q., Wu, Y., Chen, C., & Huang, Q. (2013). Construction of novel amphiphilic chitosan copolymer nanoparticles for chlorpyrifos delivery. *Journal of Polymer Research, 20*(3), 107. https://doi.org/10.1007/s10965-013-0107-7.

Zhou, Y. F., Wu, J., Chen, J., & Nie, W. Y. (2007). Applied research of APEP-type polymerizable emulsifier in the preparation of pesticide nanocapsule. *Journal of Anhui University, 31*, 65–68.

Zhou, X. Q., Cao, L. D., Liu, Y. J., & Li, F. M. (2014). Preparation and performance characteristics of azoxystrobin microcapsules. *Chinese Journal of Pesticide Science, 16*, 213–219.

Zhu-Zhu Li, Shi-Ai Xu, Li-Xiong Wen, Fan Liu, An-Qi Liu, Qing Wang, Hai-Yan Sun, Wen Yu, Jian-Feng Chen. (2006). Controlled release of avermectin from porous hollow silica nanoparticles: Influence of shell thickness on loading efficiency, UV-shielding property and release. *Journal of Controlled Release, 111* (1–2),81–88.

Chapter 3
Nano and Microencapsulated Nutrients for Enhanced Efficiency Fertilizer

Débora França, Lucas Luiz Messa, Claudinei Fonseca Souza, and Roselena Faez

Abstract Enhanced Efficiency Fertilizer (EEF) is a current and very important subject. These systems offer an effective way to improve nutrient efficiency, minimize fertilizer losses by physical, chemical and biological processes, and reduce environmental impact. Examples of such materials are: Slow Release Fertilizer (SRF), which releases the nutrient slower than a common fertilizer; and Controlled Release Fertilizer (CRF), when factors as release rate, pattern and release period can be controlled during its use. Ideally, it should be faster at the beginning and slower during cultivation, to meet the nutritional need of the plant, which changes over the growing period. Those materials have been developed in diverse platforms such as: (i) the matrix itself acts as a fertilizer, or (ii) the fertilizer is dispersed in a matrix, and/or (iii) the fertilizer is covered or encapsulate by a matrix, typically, polymeric or composites materials. The first EEF developed were made with synthetic polymers, mainly acrylamide-based ones. However, recent studies have proposed the use of environmentally friendly materials such as biodegradable polymers, clay minerals and calcium carbonate. The processes used to obtain EEFs are numerous, e.g. chemical and physical methods as polymerization, coacervation, coating, and (micro and nano) encapsulation. As the name suggests, encapsulation technique involves capsules preparation composed of a shell material completely around a chemical core material (nutrient). This shell provides a prolongation of nutrient release from the core. Once this shell is dissolved or ruptured the nutrient encapsulated is released.

D. França · L. L. Messa
Laboratory of Polymeric Materials and Biosorbents, Federal University of Sao Carlos, UFSCar, Araras, SP, Brazil

Graduate Program in Materials Science and Engineering, University of Sao Paulo, USP-FZEA, Pirassununga, SP, Brazil

C. F. Souza
Research Group on Water, Soil and Environment Engineering, Federal University of São Carlos, UFSCar, Araras, São Paulo, Brazil

R. Faez (✉)
Laboratory of Polymeric Materials and Biossorbents, Federal University of Sao Carlos, UFSCar, Araras, SP, Brazil
e-mail: faez@ufscar.br

T. J. Gutiérrez (ed.), *Polymers for Agri-Food Applications*,
https://doi.org/10.1007/978-3-030-19416-1_3

Even though encapsulation is not a novelty, its industry has rapidly innovated with new technologies applied to pharmaceutical, agricultural, cosmetic and food industries. In this chapter, we discuss the techniques to improve nutrient efficiency according to the materials and encapsulation methods used.

Keywords Agricultural · Biodegradable · Encapsulation · Natural polymers

3.1 Introduction

The great population growth and the scarcity of natural resources are determining factors for the food production to be more efficient and with fewer losses of nutrients. In this scenario, development of promising technologies in plant nutritional effectiveness linked to sustainable materials is of great importance for the agricultural and economic sector. The requirement of intensive use of fertilizers, water and pesticides, together with new technologies, has led to enormous developments in modern agriculture over the years. Agricultural production (unit of productive land) has developed greatly, contributing to population growth and economic development. However, while these developments were significant, environmental impacts were not fully considered and measured. Such impacts associated with agricultural expansion have resulted from the excessive application of agrochemicals, which have led to processes such as eutrophication, water toxicity, groundwater pollution, air pollution, soil degradation and even ecosystem change, bring up questions on the sustainability of modern agriculture. This scenario is no longer acceptable and alternatives for increasing agricultural production without compromising the environment should be considered.

The efficiency of the agrochemical and water is then a fundamental requirement for the sustainable management of the agricultural systems. An alternative to a better plant nutrient utilization has been attained with enhanced efficiency fertilizers (EEF) in which refine nutrient management in agricultural systems (Dubey et al. 2011; Davidson and Gu 2012; Azeem et al. 2014; Mukerabigwi et al. 2015). The enhanced efficiency fertilizer is dependent on polymer matrices and the methodology to cover, disperse or encapsulate the nutrient. Those factors drive the type of EEF, i.e. slow/controlled release and/or environmentally friendly fertilizer as shown in Fig. 3.1.

3.2 Types of EEF: Slow/Controlled Released and Environmentally Friendly Fertilizer

An EEF is defined as any system where it is possible to reduce the use of agrochemicals in order to not impact the environment and the economy. Figure 3.1 shows a scheme of three different ways to propose an EFF, i.e. controlled or slow release and environmentally friendly fertilizers. In a slow release fertilizer system,

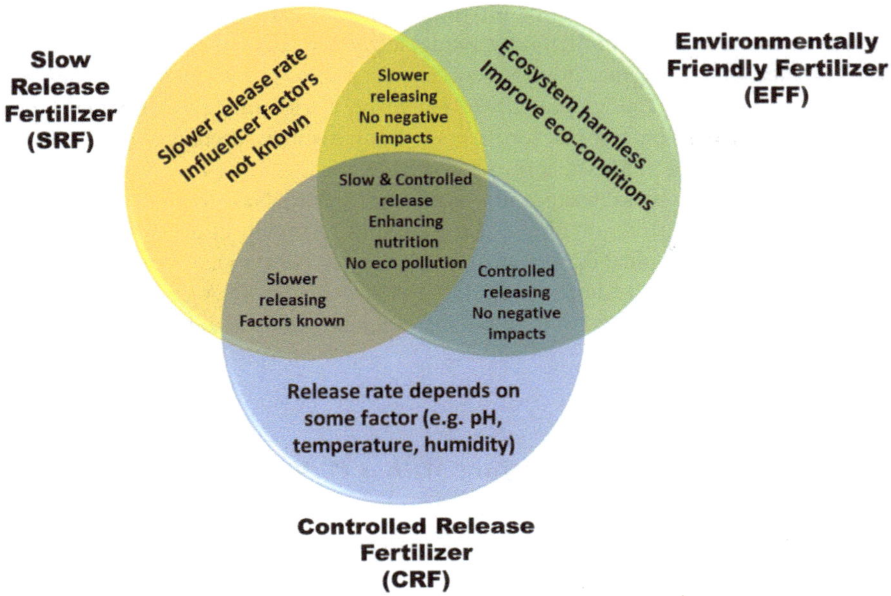

Fig. 3.1 Slow and controlled release and environmentally friendly fertilizers and their intersections

the nutrients are released at a slower rate than a common fertilizer. However, release rate, pattern and release duration are not well controlled and can be influenced by the conditions of use and the soil type. On the other hand, controlled release fertilizer the settings that influence the rate and release pattern are known and can be controlled during its use (Trenkel 1997, 2010; Shaviv 2001). To be considered as an environmentally friendly fertilizer the materials used to cover, encapsulate or dispersed the fertilizer must be degraded after the fertilizer release and cause no harm to the environment. After that, to be slowed or controlled release not necessarily biodegradable polymers have been used, but the best relationship among EFF systems are based on aspects of slow and/or controlled together with its sustainability.

3.3 How to Prepare EEF

The development of new fertilizers has increased considerably in recent years due to the need to achieve sustainable systems, i.e. to improve economic efficiency and minimize socio-environmental problems due to excessive use of fertilizers. EEFs offer an effective way to improve nutrient efficiency, minimize leach losses and volatilization of fertilizers, and reduce negative environmental impact (Fu et al. 2018). Many efforts have been made to develop environmentally friendly coating materials, which can be degraded in the soil and converted to carbon dioxide, water, methane, inorganic compounds or microbial biomass (Naz and Sulaiman 2016). The most common and commercially available methodology is to cover the fertilizer granules with a polymer or sulfur coating in order to produce a low permeability and compact film (Liang et al. 2007). However, a wide range of strategies have been used such as dispersions of fertilizers in a microsphere prepared by coacervation (Santos et al. 2015; Messa et al. 2016; Feng et al. 2017), or microencapsulated by spray drying (Tomaszewska and Jarosiewicz 2006; Navarro-Guajardo et al. 2018; França et al. 2018), or film and membranes (Ariyanti et al. 2012; Bortolin et al. 2013) and mechanical or melt-processed materials of polymers and agrochemicals (Pereira et al. 2015; Xiang et al. 2018; Souza et al. 2018). A description of the different methodologies used to improve the efficiency of fertilizers can be detailed in Table 3.1.

3.3.1 Cover Up Particle Surface With a Polymeric Coating

To cover up the surface of a particle a layer of polymer, biopolymer, sulfur and others is applied on the surface of the granules in order to regulate the availability of nutrients (Gutiérrez, 2018a). In this way, the fertilizer must have a regular effect and, mainly, be adequate to the nutritional needs of the crop. The types of the coating materials as well their thickness determine the gradual release of the coated nutrient. Additionally, knowing the chemical composition of these coatings is fundamental to understand the characteristics that determine the mechanism of release of each type of coating (Shaviv 2001). According to Majeed et al. (2015), fertilizers can be coated with sulfur and polymers of various natures, including synthetic, natural and biodegradable (Majeed et al. 2015). More recent studies are devoted to the development of polymers from both renewable and biodegradable, and those most cited in the literature include lignin, starch, chitin, cellulose, and chitosan (Cs) (Azeem et al. 2014; Gutiérrez & Álvarez, 2017).

In a recent study, Treinyte et al. (2018) developed a biodegradable polymer coating to assess the applicability of different combinations of by-products and residues. Composites of poly(vinyl alcohol), horn meal (waste materials of slaughter houses), phosphogypsum (by-product of the phosphate fertilizer industry) and crude glycerin (by-products of biodiesel production) were proposed as matrices for fertilizer encapsulation. Poly(vinyl alcohol) was used as a binder and the other components were added in order to reduce the costs. The material contains phosphorus, nitrogen, calcium, potassium and sulfur, useful nutrients for plants and also good mechanical properties. The encapsulation resulted in a prolonged release time and after tested in a tomato culture a positive effect on the plant roots development was observed (Treinyte et al. 2018).

Qiao et al. (2016) demonstrated the development of a double-coated fertilizer using ethyl cellulose as an inner coating and superabsorbent starch-based polymer from three distinct botanical sources as the outer coating. The granules of the fertilizer coated with starch presented a slow release and the differences on nitrogen

Table 3.1 Techniques to prepare EEF

Techniques	Type of material shaped	Images	References
Cover up the fertilizer	Particle surface covered with polymeric film	 Single layer	Treinyte et al. (2018) Qiao et al. (2016) Lubkowski (2014)
		 Double layer	Bortolin et al. (2013) Ariyanti et al. (2012) Ni et al. (2011)
Coacervation	Macro and microspheres		Feng et al. (2017)
Fertilizer is dispersed in a matrix phase	Fertilizer dispersions during coacervation process		
	Fertilizer sorption after coacervation process	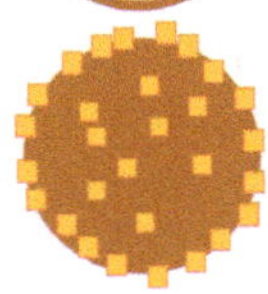	Messa et al. (2016) Santos et al. (2015)
Mechanical and melt-processing	Films, pellets, particles, etc.	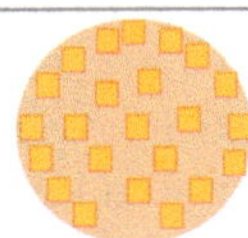	Xiang et al. (2018) Souza et al. (2018) Xiao et al. (2017) Harmaen et al. (2016) Pereira et al. (2015) Pereira et al. (2012)
Spray-dryer (2-fluid nozzle)	Nano and microspheres		França et al. (2018) Tomaszewska and Jarosiewicz (2006)
Spray-dryer (3-fluid nozzle) or Core-Shell	Microcapsules		França et al. (2018) Wu and Liu (2008) Liang et al. (2007) Xie et al. (2011) Pillai (2010) Zohuriaan-Mehr et al. (2010)

release behavior can be attributed to the botanical origins of the starch, which give the starch different water-absorbing properties. In particular, those with potato starch exhibited a constant release behavior for more than 96 h (Qiao et al. 2016).

The polymer coating is one of the most used technology to cover up the fertilizer solid particle (> 0.1 mm). However, the film layer should not rupture during transportation and application, since the cracks disable the release properties of the material.

3.3.2 Coacervation

Coacervation also is very common for producing EEF based mainly on polysaccharides. The production of microspheres is applicable in the case of the of retention water-soluble compounds such as inorganic fertilizers. For this, the fertilizers can be dispersed in a polymeric matrix during the coacervation process (Messa et al. 2016). Another way to retain the nutrients in the composite matrix could be fertilizer sorption after the coacervation process, as mentioned by Santos et al. (2015). Coacervation method consists on adding a solution of hydrocolloid (a polymer formed by more than one monosaccharide type) to an electrolyte with an opposite charge of the colloidal species. Thus, an insoluble derivative is formed as a result of the electrostatic interactions, producing solid-walled microspheres around the dispersed fertilizer in the same reaction medium (Schrooyen et al. 2001). In addition, the coacervation process depends on the net charge of the system and is consequently influenced by parameters such as the conformation and chain length of the polymers and media conditions that include: pH, ionic strength, temperature and nature of the reactants. These variables are directly related to the efficiency of microspheres production and their physical characteristics such as structure, size, and porosity (Yeo et al. 2005).

Messa et al. (2016) prepared double layer microspheres based on Cs and clay-potassium nitrate (KNO_3). The addition of lamellar silicates such as montmorillonite clay worked to increase the water absorption capacity and retention of chemical compounds. The microspheres were obtained in a two-step, first, the Cs-clay-fertilizer solution was casting and the film was ground to obtain solid particles. The particles were then incorporated into a fresh Cs solution which was dripped onto a sodium hydroxide solution yielding the double layer microspheres using the coacervation method. These two steps were necessary to decrease the dissolution of KNO_3 during the washing stage until neutral pH. The fertilizer release was carried out in water and soil, whereas in soil the release values were higher in the first 10 days but still keep the delivery for 60 days (Messa et al. 2016).

Santos et al. (2015) described the preparation of potassium-containing microspheres based on Cs and montmorillonite clay using coacervation method. The microspheres were prepared by dripping a solution of Cs/clay into a solution of sodium hydroxide. After the microspheres were added to a fertilizer solution so that the KNO_3 fertilizer could be trapped in the microspheres by sorption process. As described the montmorillonite clay provides better sorption properties than the Cs microspheres because of the rough and porous surface. Due to these properties, high levels of KNO_3 fertilizer were sorbed in the material. The potassium content of the microspheres presented two specific release periods: the first hold approximately 3 days and was attributed to the external fertilizer in the microspheres, while the second was attributed to the internal fertilizer (Santos et al. 2015).

Coacervation is an easy way to prepare microspheres, however, the solubility of the nutrient influences the formulation. Then both methodologies used for Messa et al. (2016) and Santos et al. (2015) were efficient to retain the fertilizer in the composite matrix, in spite of double coacervation and coacervation-sorption step process used, respectively.

3.3.3 Mechanical and Melt-Processing

The mechanical process is one of the most interesting methods to produce EEF in a large-scale production. It is important to emphasize that, as an economical, efficient and environment-friendly technology, the thermomechanical process has a great potential to be applied in horticulture and agriculture. In this case, the fertilizer is homogeneously dispersed in a matrix phase during the processing. The dispersity, homogeneity and phase interaction are important parameters to be considered (Xiang et al. 2018). Considering the source of the polymer and also its biodegradability the polymeric materials must fit into a thermodegradation window that enables it to be processed before its degradation. Examples of biodegradable thermoplastic that has been used to process EEF are polyhydroxyalkanoates such as polyhydroxybutyrate (Souza et al. 2018), polyhydroxybutyrate-*co*-valerate (Harmaen et al. 2016), poly(vinyl alcohol) (Julinová et al. 2018), thermoplastic starch (Xiao et al. 2017; Merino et al. 2018a, b; 2019a, b), poly(caprolactone) (Pereira et al. 2015; Gutiérrez 2018b).

Mechanical (melt-) processing can be performed continuously or by batch, however, these are differentiated by characteristics inherent to the type of process. Rotation speed, mixing type, feed rate, residence time and temperature profile are processing variables that have a significant influence on the microstructures and rheological properties of the final product (Kokini 1993).

In the work of Xiao et al. (2017), a one-step method was established to prepare the slow release fertilizer incorporated in starch-based superabsorbent polymer (SBSAP). A modified double rotor blender was used to study chemical and physical reactions during thermal processing by monitoring temperature and torque. Three corn starches with different amylose/amylopectin ratios were used to understand the impact on the superabsorbent polymer structure and release performance. Using the SBSAPs prepared by processing, the slow release of urea in water was achieved. A lower value of 15% of the fertilizer was released in 1 day, the release rate after 30 days exceeded 80% and the release lasted more than 45 days. Furthermore, the processing model demonstrated in this work provides an economical solution to prepare a slow urea release technology, which was promising for agricultural applications (Xiao et al. 2017).

Xiang et al. (2018) developed a granular slow release fertilizer based on urea-formaldehyde (UF), ammonium polyphosphate (APP) and amorphous silica gel (ASG) using cold-extrusion. A strong interaction *via* hydrogen bonds between the components generates compatibility and improved the mechanical properties, as well as the slow release properties of the fertilizer, mainly after the addition of APP and ASG in the UF matrix. The extrusion technique improved the performance of the material and the nitrogen was efficiently used favoring the plant growth (Xiang et al. 2018).

Rychter et al. (2016) obtained films by extrusion of potato starch with urea as a plasticizer for plants fertilization. Extrudable films of 1 mm thickness and 100 mm width were obtained from pellets prepared with starch/glycerol/urea. They had tested the releasing efficiency of urea from the starch films in water conditions. They also made an effort to utilize extruded samples as a fertilizer for agriculture purposes. Despite, results indicated the fertilizer release time from obtained films, in proposed conditions was not satisfactory, urea-plasticized starch was successfully used as a fertilizer. Plant growth assessment, including parameters determination of fresh and dry matter of the plants and their visual evaluation, was proved the stimulating effect of using extruded films on the growth and development of plants (Rychter et al. 2016).

The rheometer mixer is also extremely useful melt-processing method for a wide range of functional polymers and composites (Qiao et al. 2012). Some studies reported the use of rheometer mixers for chemical modification of polymers, blends, and composites which have shown their utility to produce materials with desirable performance. In the work of Souza et al. (2018), composites based on polyhydroxybutyrate (PHB), starch, glycerol and montmorillonite (PHBSGMMt) had been proposed as a carrier for controlled fertilizer release, such as KNO_3 and NPK. PHBSGMMt with KNO_3 or NPK was processed by melting in a mixer chamber of torque rheometer. Clay, KNO_3, and NPK increased the PHB/starch compatibility and homogeneity, resulting in a well-interfaced material obtained for PHBSGMMTNPK. $MMtKNO_3$ was mainly on PHB phase and MMtNPK on the compatible polymeric phase. These behaviors explain the faster release for incompatible blends, $PHBSGMMtKNO_3$, in which KNO_3 was completely released after 60 h, than compatible PHBSGMMtNPK that released 60% of the total amount of fertilizer during the test time (Souza et al. 2018).

3.3.4 Core-Shell Material

A core-shell type material can be modulated according to the both phases composition and hydrophobic-core and hydrophilic-shell would guarantee the nutrients will not dissociate rapidly in the medium. A hydrophilic shell enables the material to swell, slowly releasing the active agent inside the core by diffusion. The preparation of functional materials with core-shell structures aims to modulate the absorption and release of water and nutrients avoiding mainly the burst effect. Few reports in the literature deal with such a structure to be used in EEF. Wu and Liu (2008) impregnated N, P and K in a core-shell type structure using poly(acrylic acid-*co*-acrylamide) and kaolin. The release of the nutrient and water absorption was satisfactory; however, the copolymer synthesis methodology is cumbersome and is accomplished together with the impregnation of fertilizer. In addition, by using organic solvents it cannot be evaluated as a sustainable technique and this should be taken into consideration. Liang et al. (2007) also prepared N, P and K doubly coated with poly (acrylic acid) as an internal coating and Cs as external one. N, P and K have been recovered by a film composed of polymers using both the spray process (Liang et al. 2007; Wu and Liu 2008). Thus, as discussed above, the methodology employed is also laborious which increases the value cost of the EEF.

Xie et al. (2011) prepared a core-shell structure material in which the urea fertilizer is coated by attapulgite-alginate composite and the shell consists of a superabsorbent composite based on wheat straw-*g*-poly(acrylic acid)/attapulgite containing urea and boron (Xie et al. 2011). Despite several reports about different formulations of controlled release materials with superabsorbent properties, only few can achieve the important sustainable material features, i.e. non-toxicity, biodegradability, simply obtaining method, efficiency in the fertilizer delivery, control of the real agrochemical amount release in the soil, among others. In addition, due to economic reasons only a few formulations have reached the market (Pillai 2010; Zohuriaan-Mehr et al. 2010).

3.3.5 Spray-Drying

Spray drying is a technique used in many industrial areas, the most well-known are food and pharmaceutical, with few studies on its use in agricultural approaches as to EEFs. The first spray dryer equipment was built in 1878 (Hayashi 1989), but the first report of its use to encapsulate a material (an aroma at this time) was in 1930 (Chávarri et al. 2012). Some of the advantages of using the spray drying technique are: (i) versatility of the applications for several raw materials and sensitive or temperature resistant materials; (ii) its reproducible production of particles; and (iii) process flexibility (Oliveira 2005). The particle properties of the final product will depend on the process parameters which involves four main steps: the feed through an atomizer nozzle, followed by the atomization of the solution and contact of the droplets produced with the pressurized air, the drying process, and the collection of the obtained particles (Fig. 3.2). He et al. (1999) evaluated the influence of atomization process parameters, such as the type of atomizer nozzle, the inlet and outlet temperature, the flow rate of the nozzle peristaltic pump and drying air flow, on the properties of the particles (size, morphology, zeta potential, among others). Particle size is mainly affected but not the zeta potential. They considered the type of spray nozzle and the inlet temperature used as the major influencers. A variety of atomizing nozzles, e.g. the centrifugal, pressure or 2-fluid nozzle (2FN) have been applied. Recently, the 3-fluid nozzle (3FN) (Fig. 3.2) has been used especially for the encapsulation of immiscible liquids such as oils or aromas (Legako and Dunford 2010; Kašpar et al. 2013a). Other advantages of the spray drying technique for microencapsulation application are its productivity, speed, reproducibility and ease of scale-up (Tokárová et al. 2013).

Spray drying technique applied to agricultural purposes has few articles in the literature. França et al. (2018) obtained microspheres and microcapsules of controlled release fertilizer by a swelling mechanism. They used the 2-fluid nozzle to obtain microsphere where the nutrient was dispersed on the polymeric matrix of Cs. To obtain microcapsules, they used the 3-fluid nozzle and the nutrient was concentrated in the core and covered by a polymeric shell of Cs (Fig. 3.3).

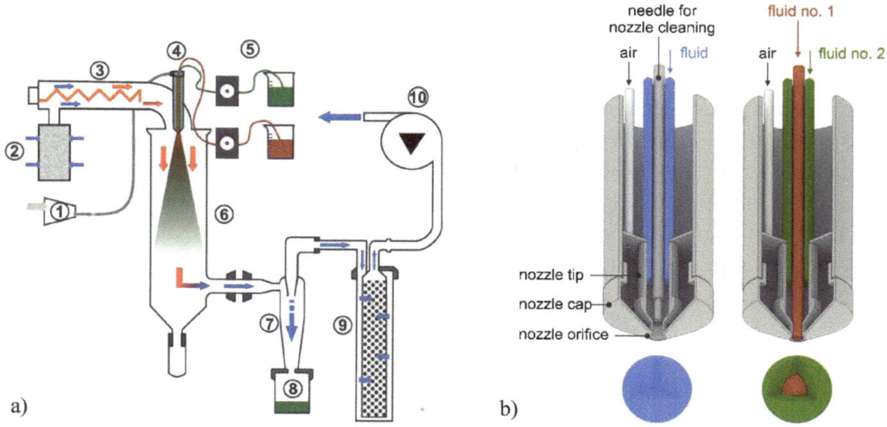

Source: (Kašpar et al. 2013b)

Fig. 3.2 (**a**) Scheme of spray drier Buchi B-290 with 3-fluid nozzle (1) compressor; (2) inlet filter; (3) heating coil; (4) nozzle; (5) stock solutions; (6) drying chamber; (7) cyclone; (8) dry product; (9) outlet filter; (10) aspirator. (**b**) Scheme of droplets

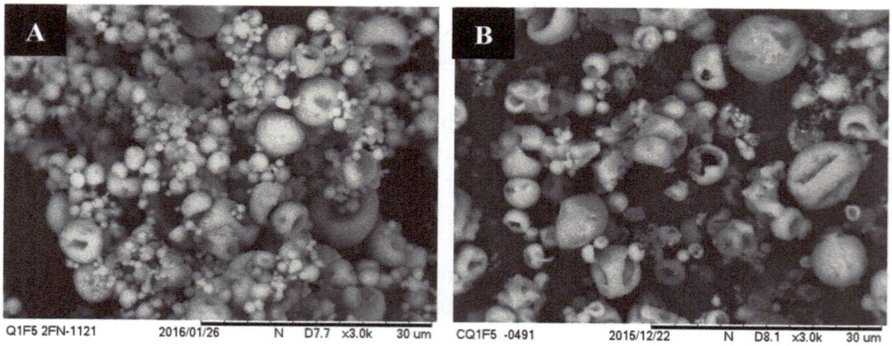

Fig. 3.3 MEV of (**A**) microspheres and (**B**) microcapsules obtained by spray dryer using 2FN and 3FN, respectively

3.3.6 Microfluidics

In a sense of encapsulation, microfluidics can also offer an opportunity to encapsulate active species such as agrochemicals. Microfluidics systems are characterized for manipulate and process small amounts of fluids using channels with dimensions of a hair human (10–100 μm). It has the advantage of useful capabilities as the: using very small quantities of samples and reagents, and carrying out separations and detections with high resolution and sensitivity; low cost; low energy consumption; short times for analysis; control of molecule concentrations in space and time; and small footprints for the analytical devices (Whitesides 2006). Microfluidics is a science that involves unique physical phenomena that use of effects in the micro-domain as laminar flow, capillary forces, and diffusion. It is mainly conducted by the law-Reynolds number flow rule which dictates that all fluid flow is essentially laminar. Continuous-flow based systems have exploited this phenomenon to create many novel micro-environments (Teh et al. 2008).

Microfluidics devices are divided in closed and open chips. Closed chips are conducted by (i) continuous flow that uses external pressure, capillary effect and electro kinetic mechanisms; (ii) droplet based (or digital microfluidics) which uses electrowetting-on-dielectric, surface acoustic waves, optoelectrowetting. On the other hand, open chips takes advantages of microfluidic probes (MFP) or are used for micro-arrays as it is applied to ink-jet printing, pin-printing technology, micro-contact printing (Ramli et al. 2013). Kim et al. (2007) used the capillary microfluidic technique to obtain monodisperse core-shell microgels based on poly(*N*-isopropylacrylamide). This technique allows incorporation of additional materials into the core-shell gels and precisely controls the particle size in the range of 10–1000 mm. It can be applied in drug delivery, artificial muscles, and cancer therapy (Kim et al. 2007). Chen et al. (2012) associated the parameters fluid flow rates and device geometry with size and shell thickness of the microcapsules made from double emulsion templates in microfluidic devices. They have showed the permeability and microstructure of the capsule shell can also be controlled through the addition of crosslinkers and silica nanoparticles in the middle phase of the double emulsion templates (Chen et al. 2012). Microfluidics used to encapsulate applied to agriculture are not reported on the literature until now. Nguyen et al. (2009) showed examples as a sensor for insecticides detection in agriculture.

3.4 Polymers and Agrochemicals Used in EEF

Besides the methods, the polymeric systems are also an important issue that affect the behavior of fertilizer release and the biodegradability of the material. Table 3.2 shows examples of polymers-agrochemicals association that have been studied for controlled-release in agriculture in the past 10 years.

3.5 Final Consideration

Different methods to prepare an EEF (coating, dispersion or encapsulation) using diverse strategies (coacervation, mechanical or melt-processing, spray-dryer, chemical synthesis of core-shell, microfluidics) were widely exposed in the text and each one has some advantages and disadvantages. Despite the methods for fertilizer nano- and micro-encapsulation has not been yet well explored in the literature it opens up new opportunities for both academia and industry since the results achieved suggest the best pattern for the efficiency of fertilizer perform.

Table 3.2 Polymers and agrochemical used into slow/controlled-release materials (Milani et al. 2017)

Agrochemical used	Polymer used
Urea	Cs
	Polyhydroxybutyrate (PHB), ethyl cellulose
	Polyethylene, polyvinyl acetate, polyurethane, polyacrylic, polylatic acid
KH_2PO_4	Cs, gellan gum
NPK	Cs
	Cellulose, natural gum, rosin, waxes
	Paraffins, ester copolymers, urethane composites, epoxy, alkide resins, polyolefines,
$CaH_4P_2O_8$	Cs
KNO_3	Cs
	Cs-clay (montmorillonite)
	Xanthan
Paraquat ($[(C_6H_7N)_2]Cl_2$)	Alginate, Cs
Hexazinone ($C_{12}H_{20}N_4O_2$)	Cs – clay
Clopyralid ($C_6H_3Cl_2NO_2$)	(montmorillonite)
2, 4-d ($C_8H_6Cl_2O_3$)	Polysaccharides
2-chloro-;4-chloro-	Cellulose, agarose, dextran, alginates, carrageenans, starch, Cs, gelatin,
2,4,5-Trichloro-phenoxyacetates	Albumin
Validamycin ($C_{20}H_{35}NO_{13}$)	Polystyrene, polyacrylamide, polymethylacrylate, polyamides, polyesters,
Bifenthrin ($C_{23}H_{22}ClF_3O_2$)	Polyanhydrides, polyurethanes, amino resins, polycyanoacrylates
Chlorpyriphos ($C_9H_{11}Cl_3NO_3PS$)	
Bifenthrin ($C_{23}H_{22}ClF_3O_2$)	
Azadirachtin ($C_{35}H_{44}O_{16}$)	

Acknowledgements The authors would like to thank the Fundação de Amparo à Pesquisa do Estado de São Paulo (FAPESP, Proc.2014/06566-9, 2017/24595-4 and 2017/03980-7).

Conflicts of Interest The authors declare no conflict of interest.

References

Ariyanti, S., Man, Z., & Bustam, M. A. (2012). Improvement of hydrophobicity of urea modified tapioca starch film with lignin for slow release fertilizer. *Advances in Materials Research, 626*, 350–354. https://doi.org/10.4028/www.scientific.net/AMR.626.350.

Azeem, B., KuShaari, K., Man, Z. B., Basit, A., & Thanh, T. H. (2014). Review on materials & methods to produce controlled release coated urea fertilizer. *Journal of Controlled Release, 181*, 11–21. https://doi.org/10.1016/j.jconrel.2014.02.020.

Bortolin, A., Aouada, F. A., Mattoso, L. H. C., & Ribeiro, C. (2013). Nanocomposite PAAm/Methyl cellulose/montmorillonite hydrogel: Evidence of synergistic effects for the slow release of fertilizers. *Journal of Agricultural and Food Chemistry, 61*(31), 7431–7439. https://doi.org/10.1021/jf401273n.

Chávarri, M., Marañón, I., & Villarán, M. C. (2012). Encapsulation technology to protect probiotic bacteria. In *Probiotics* (pp. 501–540). Rijeka: InTech. https://doi.org/10.5772/50046.

Chen, P. W., Erb, R. M., & Studart, A. R. (2012). Designer polymer-based microcapsules made using microfluidics. *Langmuir, 28*(1), 144–152. https://doi.org/10.1021/la203088u.

Davidson, D., & Gu, F. X. (2012). Materials for sustained and controlled release of nutrients and molecules to support plant growth. *Journal of Agricultural and Food Chemistry, 60*(4), 870–876. https://doi.org/10.1021/jf204092h.

dos Santos, B. R., Bacalhau, F. B., Pereira T dos, S., Souza, C. F., & Faez, R. (2015). Chitosan-montmorillonite microspheres: A sustainable fertilizer delivery system. *Carbohydrate Polymers, 127*, 340–346. https://doi.org/10.1016/j.carbpol.2015.03.064.

Dubey, S., Jhelum, V., & Patanjali, P. K. (2011). Controlled release agrochemicals formulations: A review. *Journal of Scientific and Industrial Research (India), 70*, 105–112. http://nopr.niscair.res.in/handle/123456789/10970.

Feng, D., Bai, B., Wang, H., & Suo, Y. (2017). Novel fabrication of biodegradable superabsorbent microspheres with diffusion barrier through thermo-chemical modification and their potential agriculture applications for water holding and sustained release of fertilizer. *Journal of Agricultural and Food Chemistry, 65*(29), 5896–5907. https://doi.org/10.1021/acs.jafc.7b01849.

França, D., Medina, A. F., Messa, L. L., Souza, C. F., & Faez, R. (2018). Chitosan spray-dried microcapsule and microsphere as fertilizer host for swellable – controlled release materials. *Carbohydrate Polymers, 196*, 47–55. https://doi.org/10.1016/j.carbpol.2018.05.014.

Fu, J., Wang, C., Chen, X., Huang, Z., & Chen, D. (2018). Classification research and types of slow controlled release fertilizers (SRFs) used – a review. *Communications in Soil Science and Plant Analysis, 49*(17), 2219–2230. https://doi.org/10.1080/00103624.2018.1499757.

Gutiérrez, T. J. (2018a). Processing nano- and microcapsules for industrial applications. In C. M. Hussain (Ed.), *Handbook of nanomaterials for industrial applications* (pp. 989–1011). https://doi.org/10.1016/b978-0-12-813351-4.00057-2.

Gutiérrez, T. J. (2018b). Biological macromolecule composite films made from sagu starch and flour/poly(ε-caprolactone) blends processed by blending/thermo molding. *Journal Polymers and the Environment, 26*(9), 3902–3912. https://doi.org/10.1007/s10924-018-1268-6.

Gutiérrez, T. J., & Álvarez, K. (2017). Biopolymers as microencapsulation materials in the food industry. In M. Masuelli & D. Renard (Eds.), *Advances in physicochemical properties of biopolymers* (pp. 296–322). https://doi.org/10.2174/9781681085449117010009.

Harmaen, A. S., Khalina, A., Ali, H. M., & Azowa, I. N. (2016). Thermal, morphological, and biodegradability properties of bioplastic fertilizer composites made of oil palm biomass, fertilizer, and poly(hydroxybutyrate-co-valerate). *International Journal of Polymer Science, 2016*, 1–8. https://doi.org/10.1155/2016/3230109.

Hayashi, H. (1989). Drying technologies of foods – their history and future. *Drying Technology, 7*(2), 315–369. https://doi.org/10.1017/cbo9781107415324.004.

He, P., Davis, S. S., & Illum, L. (1999). Chitosan microspheres prepared by spray drying. *International Journal of Pharmaceutics, 187*(1), 53–65. https://doi.org/10.1016/S0378-5173(99)00125-8.

Julinová, M., Slavík, R., Vyoralová, M., Kalendová, A., & Alexy, P. (2018). Utilization of waste lignin and hydrolysate from chromium tanned waste in blends of hot-melt extruded pva-starch. *Journal of Polymers and the Environment, 26*(4), 1459–1472. https://doi.org/10.1007/s10924-017-1050-1.

Kašpar, O., Jakubec, M., & Štěpánek, F. (2013a). Characterization of spray dried chitosan-TPP microparticles formed by two- and three-fluid nozzles. *Powder Technology, 240*, 31–40. https://doi.org/10.1016/j.powtec.2012.07.010.

Kašpar, O., Tokárová, V., Nyanhongo, G. S., Gübitz, G., & Štěpánek, F. (2013b). Effect of cross-linking method on the activity of spray-dried chitosan microparticles with immobilized laccase. *Food and Bioproducts Processing, 91*(4), 525–533. https://doi.org/10.1016/j.fbp.2013.06.001.

Kim, J. W., Utada, A. S., Fernández-Nieves, A., Hu, Z., & Weitz, D. A. (2007). Fabrication of monodisperse gel shells and functional microgels in microfluidic devices. *Angewandte Chemie (International Ed. in English), 46*(11), 1819–1822. https://doi.org/10.1002/anie.200604206.

Kokini, J. L. (1993). The effect of processing history on chemical changes in single- and twin-screw extruders. *Trends in Food Science and Technology, 4*(10), 324–329. https://doi.org/10.1016/0924-2244(93)90102-G.

Legako, J., & Dunford, N. T. (2010). Effect of spray nozzle design on fish oil-whey protein microcapsule properties. *Journal of Food Science, 75*(6), E394–E400. https://doi.org/10.1111/j.1750-3841.2010.01708.x.

Liang, R., Liu, M., & Wu, L. (2007). Controlled release NPK compound fertilizer with the function of water retention. *Reactive and Functional Polymers, 67*(9), 769–779. https://doi.org/10.1016/j.reactfunctpolym.2006.12.007.

Lubkowski, K. (2014). Coating fertilizer granules with biodegradable materials for controlled fertilizer release. *Environmental Engineering and Management Journal, 13*(10), 2573–2581.

Majeed, Z., Ramli, N. K., Mansor, N., & Man, Z. (2015). A comprehensive review on biodegradable polymers and their blends used in controlled-release fertilizer processes. *Reviews in Chemical Engineering, 31*(1), 69–95. https://doi.org/10.1515/revce-2014-0021.

Merino, D., Gutiérrez, T. J., Mansilla, A. Y., Casalongué, C. A., & Alvarez, V. A. (2018a). Critical evaluation of starch-based antibacterial nanocomposites as agricultural mulch films: Study on their interactions with water and light. *ACS Sustainable Chemistry & Engineering, 6*(11), 15662–15672. https://doi.org/10.1021/acssuschemeng.8b04162.

Merino, D., Mansilla, A. Y., Gutiérrez, T. J., Casalongué, C. A., & Alvarez, V. A. (2018b). Chitosan coated-phosphorylated starch films: Water interaction, transparency and antibacterial properties. *Reactive and Functional Polymers, 131*, 445–453. https://doi.org/10.1016/j.reactfunctpolym.2018.08.012.

Merino, D., Gutiérrez, T. J., & Alvarez, V. A. (2019a). Potential agricultural mulch films based on native and phosphorylated corn starch with and without surface functionalization with chitosan. *Journal Polymers and the Environment, 27*(1), 97–105. https://doi.org/10.1007/s10924-018-1325-1.

Merino, D., Gutiérrez, T. J., & Alvarez, V. A. (2019b). Structural and thermal properties of agricultural mulch films based on native and oxidized corn starch nanocomposites. *Starch-Stärke, 71*(7–8), 1800341. https://doi.org/10.1002/star.201800341.

Messa, L., Froes, J. D., Souza, C., & Faez, R. (2016). Híbridos de quitosana-argila para encapsulamento e liberação sustentada do fertilizante nitrato de potássio. *Quim Nova, 39*(10), 1215–1220. https://doi.org/10.21577/0100-4042.20160133.

Milani, P., França, D., Balieiro, A. G., & Faez, R. (2017). Polymers and its applications in agriculture. *Polímeros, 27*(3), 256–266. https://doi.org/10.1590/0104-1428.09316.

Mukerabigwi, J. F., Wang, Q., Ma, X., Liu, M., Lei, S., Wei, H., Xueing, H., & Cao, Y. (2015). Urea fertilizer coated with biodegradable polymers and diatomite for slow release and water retention. *Journal of Coating Technology and Research, 12*(6), 1085–1094. https://doi.org/10.1007/s11998-015-9703-2.

Navarro-Guajardo, N., García-Carrillo, E. M., Espinoza-González, C., Téllez-Zablah, R., Dávila-Hernández, F., Romero-García, J., Ledezma-Pérez, A., Mercado-Silva, J. A., Torres, C. A. P., & Pariona, N. (2018). Candelilla wax as natural slow-release matrix for fertilizers encapsulated by spray chilling. *Journal of Renewable Materials, 6*(3), 226–236. https://doi.org/10.7569/jrm.2017.634164.

Naz, M. Y., & Sulaiman, S. A. (2016). Slow release coating remedy for nitrogen loss from conventional urea: A review. *Journal of Controlled Release, 225*, 109–120. https://doi.org/10.1016/j.jconrel.2016.01.037.

Nguyen, N-T., Kwok, Y-C., Tan, H. Y., Loke, W. K. (2009). Polymeric labs on a chip for sustainable development, Paper No. MNHMT2009-18156, 53–57. https://doi.org/10.1115/MNHMT2009-18156.

Ni, B., Liu, M., Lü, S., Xie, L., & Wang, Y. (2011). Environmentally friendly slow-release nitrogen fertilizer. *Journal of Agricultural and Food Chemistry, 59*(18), 10169–10175. https://doi.org/10.1021/jf202131z.

Oliveira BF (2005) Preparação de microesferas de quitosana por spray drying com diferentes tipos de reticulação para uso na vacinação gênica. http://repositorio.unicamp.br/jspui/handle/reposip/267440

Pereira, E. I., Minussi, F. B., Cruz, C. C. T., Bernardi, A. C., & Ribeiro, C. (2012). Urea – Montmorillonite-extruded nanocomposites: A novel slow- release material. *Journal of Agricultural and Food Chemistry, 60*(21), 5267–5272. https://doi.org/10.1021/jf3001229.

Pereira, E. I., Da Cruz, C. C. T., Solomon, A., Le, A., Cavigelli, M. A., & Ribeiro, C. (2015). Novel slow-release nanocomposite nitrogen fertilizers: The impact of polymers on nanocomposite properties and function. *Industrial and Engineering Chemistry Research, 54*(14), 3717–3725. https://doi.org/10.1021/acs.iecr.5b00176.

Pillai, C. K. S. (2010). Challenges for natural monomers and polymers: Novel design strategies and engineering to develop advanced polymers. *Designed Monomers and Polymers, 13*(2), 87–121. https://doi.org/10.1163/138577210X12634696333190.

Schrooyen, P. M. M., van der Meer, R., De Kruif, C. G. (2001) Microencapsulation: its application in nutrition. Proceedings of the Nutrition Society, 60 (4):475–479.

Qiao, D., Zou, W., Liu, X., Yu, L., Chen, L., Liu, H., & Zhang, N. (2012). Starch modification using a twin-roll mixer as a reactor. *Starch/Stärke, 64*(10), 821–825. https://doi.org/10.1002/star.201200044.

Qiao, D., Liu, H., Yu, L., Bao, X., Simon, G. P., Petinakis, E., & Chen, L. (2016). Preparation and characterization of slow-release fertilizer encapsulated by starch-based superabsorbent polymer. *Carbohydrate Polymers, 147*, 146–154. https://doi.org/10.1016/j.carbpol.2016.04.010.

Ramli, R. A., Laftah, W. A., & Hashim, S. (2013). Core–shell polymers: A review. *RSC Advances, 3*, 15543–15565. https://doi.org/10.1039/c3ra41296b.

Rychter, P., Kot, M., Bajer, K., Rogacz, D., Šišková, A., & Kapuśniak, J. (2016). Utilization of starch films plasticized with urea as fertilizer for improvement of plant growth. *Carbohydrate Polymers, 137*, 127–138. https://doi.org/10.1016/j.carbpol.2015.10.051.

Shaviv, A. (2001). Advances in controlled-release fertilizers. *Advances in Agronomy, 71*, 1–49. https://doi.org/10.1016/S0065-2113(01)71011-5.

Souza J de, L., Chiaregato, C. G., & Faez, R. (2018). Green composite based on phb and montmorillonite for KNO_3 and NPK delivery system. *Journal of Polymers and the Environment, 26*(2), 670–679. https://doi.org/10.1007/s10924-017-0979-4.

Teh, S.-Y., Lin, R., Hung, L.-H., & Lee, A. P. (2008). Droplet microfluidics. *Lab on a Chip, 8*, 198–220. https://doi.org/10.1039/b715524g.

Tokárová, V., Kašpar, O., Knejzlík, Z., Ulbrich, P., & Štěpáneka, F. (2013). Development of spray-dried chitosan microcarriers for nanoparticle delivery. *Powder Technology, 235*, 7970–7805. https://doi.org/10.1016/j.powtec.2012.12.005.

Tomaszewska, M., & Jarosiewicz, A. (2006). Encapsulation of mineral fertilizer by polysulfone using a spraying method. *Desalination, 198*(1–3), 346–352. https://doi.org/10.1016/j.desal.2006.01.032.

Treinyte, J., Grazuleviciene, V., Paleckiene, R., Ostrauskaite, J., & Cesoniene, L. (2018). Biodegradable polymer composites as coating materials for granular fertilizers. *Journal of Polymers and the Environment, 26*(2), 543–554. https://doi.org/10.1007/s10924-017-0973-x.

Trenkel, M. E. (1997). *Controlled-Release and stabilized fertilizers in agriculture*. Paris: IFA.

Trenkel, M. E. (2010). *Slow- and controlled-release and stabilized fertilizers: An option for enhancing nutrient use efficiency in agriculture* (Second ed.). Paris: IFA.

Whitesides, G. M. (2006). The origins and the future of microfluidics. *Nature, 442*, 368–373. https://doi.org/10.1038/nature05058.

Wu, L., & Liu, M. (2008). Preparation and properties of chitosan-coated NPK compound fertilizer with controlled-release and water-retention. *Carbohydrate Polymers, 72*(2), 240–247. https://doi.org/10.1016/j.carbpol.2007.08.020.

Xiang, Y., Ru, X., Shi, J., Song, J., Zhao, H., Liu, Y., & Zhao, G. (2018). Granular, slow-release fertilizer from urea-formaldehyde, ammonium polyphosphate, and amorphous silica gel: A new strategy using cold extrusion. *Journal of Agricultural and Food Chemistry, 66*(29), 7606–7615. https://doi.org/10.1021/acs.jafc.8b02349.

Xiao, X., Yu, L., Xie, F., Bao, X., Liu, H., Ji, H., & Chen, L. (2017). One-step method to prepare starch-based superabsorbent polymer for slow release of fertilizer. *Chemical Engineering Journal, 309*, 607–616. https://doi.org/10.1016/j.cej.2016.10.101.

Xie, L., Liu, M., Ni, B., Zhang, X., & Wang, Y. (2011). Slow-release nitrogen and boron fertilizer from a functional superabsorbent formulation based on wheat straw and attapulgite. *Chemical Engineering Journal, 167*, 342–348. https://doi.org/10.1016/j.cej.2010.12.082.

Yeo, Y., Bellas, E., Firestone, W., Langer, R., & Kohane, D. S. (2005). Complex coacervates for thermally sensitive controlled release of flavor compounds. *Journal of Agricultural and Food Chemistry, 53*(19), 7518–7525. https://doi.org/10.1021/jf0507947.

Zohuriaan-Mehr, M. J., Omidian, H., Doroudiani, S., & Kabiri, K. (2010). Advances in non-hygienic applications of superabsorbent hydrogel materials. *Journal of Materials Science, 45*(21), 5711–5735. https://doi.org/10.1007/s10853-010-4780-1.

Chapter 4
Potential Use of Polymeric Particles for the Regulation of Plant Growth

Anderson E. S. Pereira, Bruno T. Sousa, María J. Iglesias, Vera A. Alvarez, Claudia A. Casalongué, Halley C. Oliveira, and Leonardo F. Fraceto

Abstract Plant growth regulators (PGRs) are molecules widely applied in the agriculture, leading to increased crop yield and improved quality of agricultural products. These compounds act as plant hormones, affecting the plant hormonal homeostasis, and thus control plant growth and development. Recently, the development of polymer-based modified release systems for PGRs has emerged as a promising alternative for increasing the efficacy of these compounds. This review will focus on polymeric particles that are used as carrier systems for PGRs, allowing their controlled release and protecting them from degradation. Successful examples include the phytohormone gibberellic acid(GA_3)-loaded nanoparticles, which showed higher efficacy than the non-nano active ingredient in promoting seed germination and seedling growth, and salicylic acid (SA) and nitric oxide (NO)-releasing nanoparticles as effective plant protection agents against stresses. Polymeric nanomaterials *per se* such as chitosan (Cs) can also alter plant signaling pathways and promote plant growth and development. Despite their great potential in improving the plant production with less damage to the environment, relatively few studies have focused on the use of these nanomaterials for the development of modified release systems for PGRs. In this scenario, this review discusses on the major advances and obstacles in the area.

Keywords Carrier system · Chitosan · Nanoparticles · Plant growth regulator

---

A. E. S. Pereira · L. F. Fraceto (✉)
Department of Environmental Engineering, Institute of Science and Technology of Sorocaba (ICTS), São Paulo State University (UNESP), Sorocaba, SP, Brazil
e-mail: leonardo@sorocaba.unesp.br

B. T. Sousa · H. C. Oliveira
Department of Animal and Plant Biology, State University of Londrina, Londrina, PR, Brazil

M. J. Iglesias · C. A. Casalongué
Instituto de Investigaciones Biológicas UE CONICET Universidad Nacional de Mar del Plata, Facultad de Ciencias Exactas y Naturales, Mar del Plata, Argentina

V. A. Alvarez
Facultad de Ingeniería, CoMP Instituto de Investigaciones en Ciencia y Tecnología de Materiales UE CONICET Universidad Nacional de Mar del Plata, Mar del Plata, Argentina

T. J. Gutiérrez (ed.), *Polymers for Agri-Food Applications*,
https://doi.org/10.1007/978-3-030-19416-1_4

4.1 Introduction

Currently, agrochemicals are of extreme importance in the agriculture, especially for their large production and uses (Prasad et al. 2017b). However, the excessive use of these compounds has caused environmental damage, resulting in soil degradation and contamination of natural resources (Mishra et al. 2017). These negative factors present major challenges for today's agriculture and also open some questions, such as how agricultural practices could increase global production without causing damage to the environment (Ciura and Kruk 2018).

In this context, new technologies for the controlled release of agrochemicals can revolutionize the agricultural sector (Mishra et al. 2017; Duhan et al. 2017). They include the development of microparticles and nanoparticles as active substance carrier systems in order to improve their biological action and reduce environmental impact (Chen and Yada 2011; Ghormade et al. 2011; Khot et al. 2012; de Oliveira et al. 2014; Campos et al. 2015; Grillo et al. 2016; Fraceto et al. 2016; Athanassiou et al. 2017).

Several advantages are associated with the use of micro and nanoparticles, among them, greater protection against premature degradation, slower release of the active ingredient, extension of its duration of action, and improved uptake of the active ingredient by target species (Kah et al. 2013; Kah and Hofmann 2014; Valletta et al. 2014; Nguyen et al. 2016; Tripathi et al. 2017; Prasad et al. 2017a). These micro and nanoparticle characteristics allow reductions not only in the dosage of the active ingredient but also on the application frequency, decreasing also the environmental contamination and the risk of harming non-target organisms (Kah et al. 2013; Kah and Hofmann 2014).

Among the active ingredients used in the agriculture, we have the plant growth regulators (PGRs), which are natural or synthetic substances that are applied exogenously to alter plant hormonal homeostasis and/or signaling (Rademacher 2015). Phythormones (also called plant hormones) may be used as PGRs, as well as their precursors and synthetic analogues. PGRs also include compounds that inhibit the biosynthesis, the translocation or the signaling pathway of phytohormones (Basra 2000; Rademacher 2015). Phytohormones are substances of plant metabolism that act at low concentrations to regulate physiological processes of plant growth, development and responses to the environment (Ordaz-Ortiz et al. 2015; Rademacher 2015). According to their chemical structure and functions in plant physiology, nine major groups of phytohormones are found in plants: auxins, cytokinins, gibberellins, ethylene, abscisic acid, brassinosteroids, jasmonic acid, SA and strigolactones (Fig. 4.1). Recently, other signaling substances with functions similar to those of plant hormones have been described. An important example is NO, a gaseous signaling molecule that acts in many developmental processes and in plant responses to biotic and abiotic stresses (Lindermayr and Durner 2018).

PGRs have wide applications in agriculture and horticulture, being applied from seed germination and seedling production to grain filling and fruit ripening (Table 4.1). Therefore, PGRs provide important benefits that include enhanced crop

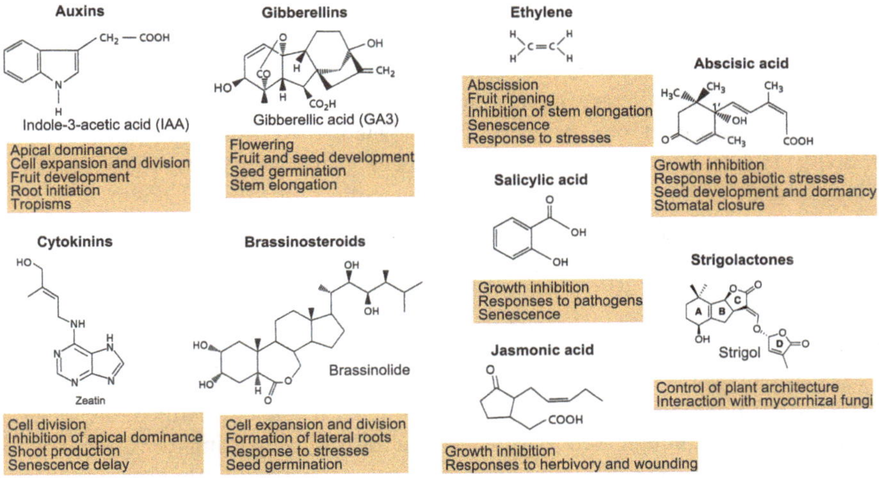

Fig. 4.1 Major groups of phytohormones. Chemical structure of a representative compound from each group and some of their respective physiological functions are indicated

Table 4.1 Major PGR types for agriculture applications

PGR types	Examples	Applications	References
Auxins	2,4-D, indole acetic acid (IAA), naphthalene-1-acetic acid (NAA)	Induction of rooting of cuttings, cell culture, herbicide	Cardoso et al. (2011); Dibax et al. (2013); Schulz and Segobye (2016)
Cytokinins	Kinetin,6BA	Cell culture	Dibax et al. (2013)
Ethylene releaser	Ethephon	Induction of fruit ripening and flowering	Hussain et al. (2015); Espinosa et al. (2017)
Ethylene inhibitor	AVG, 1-MCP	Delay of senescence and fruit ripening	Petri et al. (2007); Steffens et al. (2009); Grozeff et al. (2010)
Gibberellin	GA_3	Induction of flowering, seed germination and fruit growth	Peixoto et al. (2011); Cardoso et al. (2012); Camara et al. (2018)
Gibberellin inhibitor	Trinexapac-ethyl, Calcium proexadione, Clormequat chloride, Mepiquat chloride	Reduction of shoot height	Rodrigues and Fioreze (2015)
NO donor	Sodium nitroprusside, *S*-nitrosothiols	Tolerance to abiotic stress	Seabra and Oliveira (2016)
JA	n-Propyl dihydrojasmonate	Improvement of fruit quality	Kondo (2010)

yield and quality, facilitated crop management and extended storage of perishable products (Basra 2000; Rademacher 2015). However, despite the high number of studies regarding PGRs and their systematic agricultural use since the 1930s, PGRs currently represent a relatively small portion of the agrochemical market, especially if compared with pesticides (Rademacher 2015). One factor that hinders the application of PGRs is their degradation when exposed to field conditions of light and temperature, which compromises their biological activities (Silva et al. 2013; Dong et al. 2016; Yang et al. 2018). In addition, when applied at supra-optimal concentrations, PGRs may exert phytotoxic effects, thereby acting as an herbicide rather than a hormone (Skůpa et al. 2014). Another aspect is the poor water solubility of some PGRs, which may hamper their applications (Ambrogi et al. 2006; Ge et al. 2011; Yang et al. 2018).

This review was focused on the major advances and obstacles regarding the use of polymeric micro and nanomaterials for the development of modified release systems for PGRs. Although there are multiple publications reporting the synthesis and physico-chemical characterization of micro and nanocarriers of PGRs particularly in the chitosan (Cs) polymeric matrix with a potential use in agriculture, few studies have demonstrated their mode of action and biological effects in plants (Gutiérrez 2017). This important issue constitutes a challenge for the next years. The actions of some polymers *per se*e specially Cs in altering plant signaling pathways and promoting growth will also be discussed.

4.2 Production of Polymeric Nanoparticles

The development of nanoparticles becomes a valuable strategy in the field of active ingredient vectorization. Nanoparticles allow a wide variety of molecules to be targeted to different parts by releasing them in a controlled manner over time, protecting them from degradation, increasing their half-life and decreasing its toxicity. Figure 4.2 shows the possible structure for polymeric nanoparticles.

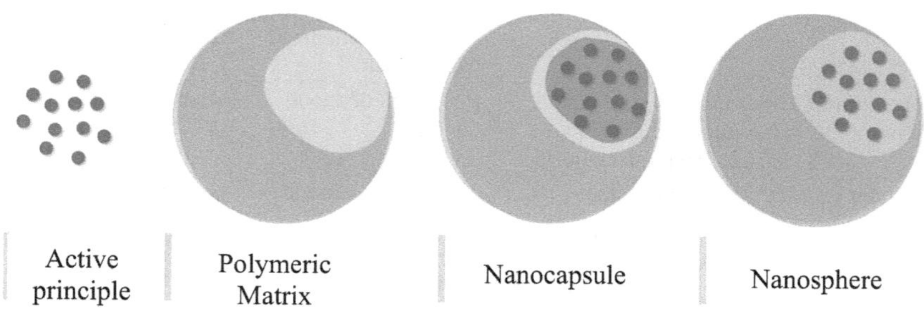

Fig. 4.2 Possible structures of a polymeric nanoparticle. The nanocapsules show a core-shell structure and nanospheres present a polymeric matrix

4.2.1 *Biodegradable Polymers Used as Active Principle Carriers*

During the past decades, significant advances have been made in the development of biodegradable polymeric materials as active principle vehicles. Degradable polymeric biomaterials are preferred candidates for developing carriers. A wide range of natural or synthetic polymers are being investigated for agricultural applications. Biodegradable polymers can be derived from different sources. The number of such materials that are used in or as adjuncts in delivery has increased dramatically over the past decade. The different kinds of biodegradable polymers used as vehicles are summarized in Fig. 4.3.

Many biopolymers such as alginate, Cs, cellulose, pectin and cellulose have been used to the development of carriers systems for agrochemicals (Campos et al. 2015) as well as to the coating of metallic nanoparticles (Navarro et al. 2015; Barrios et al. 2016; López-Moreno et al. 2018). The most widely used polymer for the development of nanocarrier systems has been Cs (Kashyap et al. 2015). Cs is a polysaccharide derived from chitin. It has great characteristics including the biodegradability and biocompatibility as well as fungicidal properties, Cs has been also use in the pharmaceutical, cosmetics and food fields (Zargar et al. 2015; Malerba and Cerana 2016). In the case of biopolymers used as a stimulator for plant development, Cs has been one of the most cited in order to promote plant growth resulting in increased production.

In addition, many types of polymers and agents can be used for the coating nanoparticles such as lactate, polyvinylpyrrolidone, polyethylene glycol, gelatin, sodium dodecylbenzenesulfanate, citrate, dexpanthenol and carbonate (Barrios et al. 2016;

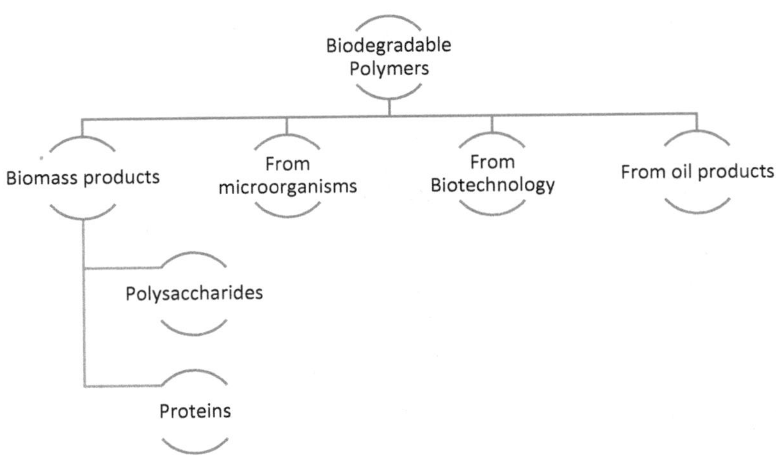

Fig. 4.3 Different kinds of biodegradable polymers used as vehicles of active ingredients. The main sources of biodegradables polymeric materials come from biomass products (polysaccharides and proteins), from microorganisms or obtained from biotechnology routes and oil byproducts

Navarro et al. 2015). When these systems are coated, their properties such as size, zeta potential are altered, as well as the biological effects on plants. Thus, this issue opens up a wide field in the design of new nanoparticles or nanocarrier systems in order to increase the biological activity of a nanoparticle system or even reduce toxic effects on plants.

Depending on the mode of degradation, polymeric biomaterials can be classified as: hydrolytically and enzymatically degradable polymers. It is important to remark that most of the naturally occurring polymers undergo enzymatic degradation.

4.2.2 Common Techniques to Prepare Polymeric Nanoparticles

There are numerous methods for producing polymeric nanoparticles carrying hydrophobic or hydrophilic molecules, simple or complex (Rao and Geckeler 2011). These methods can be classified into two categories: those that involve polymer synthesis or those that involve preformed polymers. Among the methods involving polymer synthesis, mention may be made of polymerization/emulsion (Thickett et al. 2007) and interfacial polymerization (Crespy et al. 2007). Considering the methods that use preformed polymers, simple (Solans et al. 2005; Fryd and Mason 2012) or double (Hanson et al. 2008; Iqbal et al. 2015) emulsion with subsequent evaporation of solvent, and nanoprecipitation (Hornig et al. 2009; Martín-Saldaña et al. 2016, 2017). The simple emulsion method with subsequent solvent evaporation is the most widely technique used to accommodate hydrophobic active principles in nanoparticulate polymer matrix systems (Gutiérrez 2018). It consists in forming a stable emulsion from two immiscible phases: an aqueous or continuous phase provided with an appropriate stabilizing agent and an organic dispersed phase containing the drug and the matrix polymer. In most cases, an ultrasound probe is responsible for generating an emulsion that guarantees nanometric droplets composed of the dispersed phase (Gutiérrez and Álvarez 2017). Then, the polymer contained in the droplets precipitates in the form of nanoparticles trapping the drug as a result of the evaporation of the organic solvent, which must naturally be volatile. The main advantage of this method is the high efficiency of encapsulation of hydrophobic active ingredients (Gómez-Gaete 2014).

Differently, double emulsion is a strategy to house hydrophilic molecules in hydrophobic polymeric nanoparticles. Emulsions are a type of dispersed phase systems, depending on the type of dispersion the emulsions are classified as those of the water in oil (w/o) or oil in water (o/w) type. It is possible to obtain more complex dispersions for more specific purposes, such as multiple emulsions of the w/o/w or o/w/o type, which require, first, the formation of a stable primary emulsion and then its dispersion in the phase external by dispersing two immiscible liquid phases, which have a high attraction force between their own molecules, a large interface area is generated producing a thermodynamically unstable system, which entails the breakdown of the emulsion in a certain time. To stabilize the dispersed systems or emulsions, an agent having interfacial activity must be added, which allows decreasing the interfacial tension and the attractive interactions between the droplets that are dispersed. These agents that are called surfactants, are amphipathic chemical species, which by various mechanisms prevent the collapse of the droplets, preventing their coalescence or flocculation. Double emulsions are complex systems in which the droplets of the dispersed phase contain one or more types of smaller scattered droplets (Iqbal et al. 2015).

Double emulsions have the potential to transport both hydrophobic and hydrophilic active principles. However, this technique is more commonly used to encapsulate hydrophilic molecules, which suffer from a low loading efficiency due to the rapid partition of the drug in the external aqueous phase when using simple emulsions (Iqbal et al. 2015).

4.3 Polymeric Nanoparticles as Carrier Systems for PGRs

Polymeric nanoparticles have been developed as carrier systems for different types of PGRs (Table 4.2).

Liu et al. (2013) developed the first reported material for the controlled delivery of GA_3, which is the most representative gibberellin. GA_3 is not dissolved in water and is easily degraded under neutral and alkaline conditions as well as by light and temperature, properties that affect its efficiency in formulations for the use in the field (Kah and Hoffman 2014). Hence, GA_3-Cs conjugate efficiently protects the phytohormone from photo and thermal degradation. Its release properties can be achieved by controlling pH, temperature and UV irradiation. Pereira et al. (2017a,b) described the properties of Cs nanocarrier systems for GA_3. These particles showed a sustained release of 58% of GA_3 in 2 days and enhanced properties compared with free GA_3 hormone in the promotion of seed germination, root and leaf development and also, increased the photosynthetic pigments in *Phaseolus vulgaris*.

Quiñones et al. (2010) described the encapsulation of two synthetic brassinosteroid analogues (DI31 and S-7) in tripolyphosphate (TPP)-Cs microparticles. Higher loading capacity and release from microparticles were obtained when the steroids were dissolved in ethanol. Both steroids show a sustained and constant release rate for the first 10 h. Until now the biological activity of these microparticles has not been assayed in plants (Quiñones et al. 2010).

IAA constitutes the widespread natural auxin in plants, however, there are different synthetic auxins such as 2,4-D or NAA, which are used as phytohormones to promote auxin-mediated processes but also as herbicides at higher doses in which auxins inhibit growth and trigger plant death (Enders and Strader 2015). The development of Cs-based particles using glutaraldehyde as a crosslinker for the controlled IAA and NAA delivery has been performed by Fan et al. (2012). These particles efficiently encapsulated around 60% of auxins and released the bioactive by a super Case-II transport diffusion mechanism. The NAA release from NAA-Cs derivative synthetized by protecting amino groups of Cs with phthalic anhydride and then mixed with 1-naphthylacetyl chloride has been studied (Tao et al. 2012). NAA release depended on pH and temperature. At pH 12.0 and 60 °C a sustained release of the hormone for

Table 4.2 Nanoparticle and nanocarrier systems for PGRs for crop applications. The table contains information about the nanoparticle or nanocarrier systems, nanoparticles characterization, target organism and biological effects

	Nanoparticle/ Microspheres/Polymer	PGRs	Characterization	Target plant	Biological effects	Author
Carrier systems for plant growth regulators	Cs/alginate and Cs/ tripolyphosphate nanoparticles	**GA_3 (Gibberellin)**	Spherical nanoparticles, nanoparticles of alginate/Cs with average size of 450 nm, zeta potential of −29 mV, and Cs/tripolyphosphate with 195 nm, zeta potential of −27 mV. Sustained release of the PGR	***Phaseolus vulgaris***	The effects depend on the concentration. Increase of plant growth and of the content of photosynthetic pigments	Pereira et al. (2017b)
	Cs/polyglutamic acid nanoparticles	**GA_3**	Spherical nanoparticles with average size of 117 nm, zeta potential of −29 mV. Sustained release of the PGR	***Phaseolus vulgaris***	Increase of seed germination and root development	Pereira et al. (2017a)
	GA3-Cs conjugate	**GA_3**	Conjugate with 60% w/w modification degree for Cs and good solubility in water at pH 6. Sustained release of the PGR.	***No evaluation***		Liu et al. (2013)
	Cs microparticles	**IAA and NAA (Auxins)**	Spherical microparticles with average size of 20 and 100 μm for IAA and NAA, respectively. Sustained release of the PGRs.	***No evaluation***		Fan et al. (2012)
	Cs nanoparticles	**NAA (Auxin)**	Authors demonstrated the chemical interaction between O-naphthylacetyl hormone with Cs. Sustained release of the PGR.	***No evaluation***		Tao et al. (2012)

	Cs microspheres	**D-31 analogue (Brassinosteroids)**	Microspheres with average size of 790–1490 μm. Sustained release of the PGR.	***No evaluation***		Quiñones et al. (2010)
	Mesoporous silica nanoparticles	**ABA**	Smart system with average size of 20 nm. The nanoparticles containing gatekeepers though gluthathione. Sustained release of the PGR.	***Arabidopsis thaliana***	Reduction of drought stress. Reduction of leaf stomatal aperture and reduction of water loss	Sun et al. (2014)
	Cs nanoparticles	***S*-nitroso-MSA(NO donor)**	Average size of 39 nm, zeta potential of −18 mV. Sustained NO release.	***Zea mays***	Protection against salt stress	Oliveira et al. (2016)
	Cs microparticles	**SA**	Spherical particles with average size of 2 μm. Sustained release of the PGR.	***Lactuca sativa***	Enhancement of root growth and expression of defense proteins	Martin-Saldaña et al. (2018)
	Mesoporous silica/gold core nanoparticles	**2,4-D (Auxin)**	Mesoporous nanostructures ranging from 40 to 60 nm, au core between 10–15 nm.	***Linumusitatissimum***	Biotechnological application in plant cell culture. Increase of ploidy numbers, embryogenesis.	Kokina et al. (2017)
Nanoparticles systems with plant growth effects	Cs-cooper	**No hormone**	Nanoparticles with average size 326 nm and zeta potential of 22.1 mV. Release profile of cooper from Cs.	***Zea mays***	Growth effects as increase of height, steam diameter, root length, root number, chlorophyll content and increase of production. Additional effect, defense responses against *Curvularia leaf spot*	Choudhary et al. (2017)

(continued)

Table 4.2 (continued)

	Nanoparticle/ Microspheres/Polymer	PGRs	Characterization	Target plant	Biological effects	Author
	Cs-cooper nanoparticles	**No hormone**	Nanoparticles with average size of 374 nm and zeta potential of +22.6 mV	***Zea mays***	In seeds, treatments increase the α-amylase and protease enzymes and total pro protein content in seeds with the decrease of starch and protein.	Saharan et al. (2016)
	Cs-cooper nanoparticles	**No hormone**	Nanoparticles with average size of 88.21 nm and zeta potential of −29 mV.	***Eleusinecoracana***	Increase plant development and production. Increase of defense enzymes. Suppression of blast disease after seed and foliar treatment.	Sathiyabama and Manikandan (2018)
	Zinc nanoparticles coated with phycomolecules	**No hormone**	Spherical nanoparticles with average size 0f 2–54 nm.	***Gossypiumhirsutum***	Growth promotion effects with increase of biomass, levels of chlorophyll, carotenoids and soluble proteins.	Venkatachalam et al. (2017)
	Silver-Cs nanoparticles	**No hormone**	Nanoparticles with average size of 59 nm and zeta potential of +24 mV.	***Cicerarietinum***	Seeds treatments increase the seed germination, seedlings length fresh and dry weight. Increase of α and β-amylase, ascorbate peroxidase, peroxidase, catalase activity and chlorophyll content	Anusuya and Banu (2016)

	Cs	**No hormone**	Cs nanoparticles with different average sizes (420, 750 and 970 nm).	***Coffeacanephora***	Increase of the chlorophyll content, nutrient uptake and plant growth, regardless the particle size.	Nguyen Van et al. (2013)
	Cs	**No hormone**	Nanoparticles with average size of 100 nm.	***Hordeumvulgare***	Increase of leaf area and grain production. Protection against drought stress	Behboudi et al. (2018)
	Cs	**No hormone**	Spherical nanoparticles with average size of 80–180 nm.	***Cammeliasinensis***	Improvement of plant innate immune response (induction of defense-related genes, antioxidant enzymes phenolic production).	Chandra et al. (2015)
	Cs	**No hormone**	Nanoparticles with average size of 143 nm and zeta potential of 55.7 mV	***Triticumaestivum***	Increase of leaf gas exchange parameters and grain protein, iron and zinc contents.	Xue et al. (2018)
	Cs	**Nitrogen, phosphorus and potassium**	Nanoparticles with average size o 330–580 nm)	***Triticumaestivum***	Improve plant development and increase of harvest, crop and mobilization index.	Abdel-Aziz et al. (2016)

55 days *in vitro* could be achieved. However, their biological actions in plants have not yet been assayed. Alternatively, non-polymeric silica nanoparticles for the control delivery of NAA with proved biology action in the modulation of root development in wheat plants where described (Ao et al. 2013).

The phytohormone SA triggers local and systemic defense responses against pathogen attack. The synthesis of SA-Cs particles with different doses of immobilized SA has been recently described (Martin-Saldaña et al. 2018). SA1%-Cs particles showed very low cytotoxicity and enhanced root growth in *Lactuca sativa* seedlings. In accordance with the activation of SA signaling *in planta,* SA-Cs particles promoted the induction of NPR1 and PR2 protein levels required for plant defense responses. However, the action of SA-Cs nanosystem has not yet been assayed in the protection of plants against environmental stress.

The NO donor *S*-nitroso-mercaptosuccinic acid (*S*-nitroso-MSA) was also encapsulated in Cs for the generation of nanoparticles (Oliveira et al. 2016). The sustained release of *S*-nitroso-MSA from Cs nanoparticles enhanced the efficiency of NO donor compared with non-encapsulated compound. *S*-nitroso-MSA-Cs nanoparticles developed a better performance in the protection of *Zea mays* plants against salt stress, evidenced by higher levels of chlorophyll and reduced inhibition of root and shoot growth. Hence, the nanoparticles of Cs for the controlled delivery of GA, SA and S-nitroso-MSAas bioestimulants/growth promoter and stress protection agents with proved action in plants constitute promising biomaterials for agricultural applications (Pereira et al. 2017a, b; Oliveira et al. 2016; Martin-Saldaña et al. 2018).

4.4 Potential of Polymeric Nanoparticles to be Used as PGRs

Unlike nanocarrier systems whose activity is related to the active ingredient, some nanoparticles have direct effects on plants, being able to alter their metabolism. Little is known about the phytotoxic or stimulate effects of polymeric nanoparticles systems without an ingredient active on plants, however, some studies have shown that these nanomaterials are capable of being uptake by vegetables and transported, as well uptake by vegetal cells (Valletta et al. 2014; Nguyen et al. 2016; Prasad et al. 2017a).

The evaluations of these nanoparticles are of extreme importance mainly for agricultural application, in which these systems cannot cause phytotoxic effects. Studies conducted by Nakasato et al. (2017) demonstrated the effects of solid lipid and Cs nanoparticles on the germination of *Zea mays*, *Brassica rapa* and *Pisumsativum* species. An inhibition of germination was observed depending on the concentration of the Cs nanoparticles while the lipid nanoparticles did not cause phytotoxic effects.

Novelty, Chandra et al. (2015) demonstrated that Cs nanoparticles function as an immune modulator. The foliar treatment with these nanoparticles increased the activity of the immune system inducing the production of defense enzymes and increasing the upregulation of genes linked to the vegetal immune system in *Cammeliasinensis*. In *Triticuma estivum*, the treatment with Cs nanoparticles favored the leaf gas exchange and the grains showed increase of protein and micronutrient levels (Xue et al. 2018). Cs nanoparticles have shown fungicidal properties, mainly when bound to copper as metal ion (Saharan et al. 2016). For example, *Zea mays* seeds treated with copper-containing Cs nanoparticles resulted in physiological and biochemical changes including, high germination rates and increase of dry mass and activation of amylases and proteases enzymes (Saharan et al. 2016).

Choudhary et al. (2017) demonstrated that seeds treated with copper-containing Cs nanoparticles increased antioxidant enzyme activities such as superoxide dismutase, peroxidase and polyphenol oxidase and phenylalanine ammonia-lyase showed protection against the fungus *Curvularia* leaf spot. In addition to these effects, a promoter stimulus was observed in the development of the *Zea mays* plants treated with the Cs nanoparticles. This is an important point of view for products that aim at a more sustainable agriculture, since systems that promote plant development resulting in seed vigor, plant development and increased production can also improve the immunological activity of the plant and resistance against pathogens (Anusuya and Banu 2016; Venkatachalam et al. 2017; Choudhary et al. 2017; Sathiyabama and Manikandan 2018).

Also, as example of hybrid systems, many metal nanoparticle systems have potential as a plant growth promoter. Silver nanoparticles may have bactericidal or fungicidal action but they have phytotoxic effects. However, in order to maintain their biological activity and reduce phytotoxic effects one of the alternatives is the coating of these nanomaterials with polymers. These systems have great applications, many of which are capable of increasing the uptake of plant nutrients, the immune system, alleviating adverse effects under stress conditions, as well as increasing production in the field (there is a lack in the development of nanocarrier systems for PGRs, and these systems may have many applications such as flowering, fruiting and fruit ripening). However, for the use of these systems to be safe, it is important to have a broad spectrum of evaluation in different plants, as well as in other living organisms. In summary, the use of polymeric nanocarrier systems for PGRs such as the coating of metal nanoparticles by polymers has a great potential for use in field applications and the stimulation of plant growth. These systems can be used for different stages, such as in the treatment of seeds or during plant development resulting in greater plant development, increase of production and quality of agricultural products.

4.5 Uptake, Transport and Distribution of Nanoparticles

PGR-loaded system can be extracellularly sensed or incorporated from the extracellular matrix to be metabolized or secreted. Nanoparticles are highly effective as carrier systems for phytohormone; they can confer high stability and enhanced and prolonged delivery to the target cell (Revell 2006). Thus, currently, controlled release systems are improving stability, efficiency and minimizing the applied doses of traditional PGR in plants. Some PGR are unstable and have quick metabolism limiting their application *in planta*. This is the case of two unstable synthetic analogues of brassinosteroids D121 and S7 that have been recently reported to be loaded in polyethylene glycol micelles to extend their stability (Pérez Quiñones et al. 2018). However, the underlying mechanisms of adhesion and transportation onto plant tissues have not yet been explored.

On the other hand, bioengineered polymeric nanoparticles exert positive or negative effects on growth and development by regulating endogenous PGR homeostasis and metabolism in the plant (Vankova et al. 2017). The exposition of rice shoots to mesoporous carbon nanoparticles showed negative effects on growth and increased the concentrations of the phytohormones brassinosteroids, indole propionic acid, and dihydrozeatinriboside (Hao et al. 2018).

In general, physico-chemical properties of PGR-loaded nanoparticles can affect its behavior including adherence, penetration and circulation along the plant. Once applied on plant, PGRs-loaded polymeric nanoparticles could also have different adherence depending on target plant cells. Apparently, size and shape are key parameters for penetration into plant tissues (Pérez-de-Luque 2017). Adhesion to the plant cell mainly takes place in the epidermic tissues of different organs such as leaves, shoots or roots (Khutoryanskiy 2011). Nanoparticles might enter into the plant by apoplastic or symplastic routes. Nanocapsules containing herbicides penetrate through cuticles and tissues, allowing the slow and constant release of the active substances (Pérez-de-Luque and Rubiales 2009).

Once internalized onto the plant, the PGR-loaded nanoparticles can modulate growth, development and morphogenesis depending on the level, distribution and sensing. Thus, in addition to get know ledge on the action of PGR-loaded nanoparticles on physiology processes it is important to investigate how they can be sensed, taken up by cell and then, transported to other plant tissues or organs (Wang et al. 2016). Recently, a hormone-like activity has been assigned to Cs/TPP nanoparticles (Asgari-Targhi et al. 2018; Fu et al. 2018). This work highlights the positive effect of Cs nanoparticles for growth and *in vitro* micropropagation of *Capsicum annuum* plants. However, the authors do not describe by which mechanism nanoparticles are absorbed in these *in vitro* plants. In Arabidopsis, the lysin motif (LysM)-containing chitin elicitor receptor kinase 1 (CERK1) has been shown to sense chitin and Cs (Petutschnig et al. 2010). CERK1 has an extracellular LysM motif–containing a transmembrane and intracellular kinase domains that is critical in chitin perception (Wan et al. 2008). Recently, a new model involving LysM-containing receptor complexes has been proposed (Gubaeva et al. 2018). Chitooligomers could also be generated from Cs nanoparticles in the apoplastic space by extracellular chitinases (Grover 2012). In this sense, the wall-associated protein W5G2U8 has found to be a chitooligomer receptor in wheat plants (Liu et al. 2018).

The plant cell wall is composed primarily of polysaccharides of which cellulose is the major component (Stavolone and Lionetti 2017). Being the pore diameter of the cell wall from 5 to 50 nm it can exclude the entry of any larger polymeric nanoparticle into cells. Cellulose microfibrils separation may be affected by a number of factors including cross-linking polysaccharides, spacing by interpenetrating polysaccharides and even by the water content in the cell wall. The water has a substantial effect on separations of cellulose microfibrils (Thompson 2007). Thus, application of nanoparticles might modify microfibril depositions by modifying water availability in the cell wall, consequently modifying its entry onto plant cells. Hence, Cs-loaded nanoparticles might physically affect cell wall architecture facilitating their incorporation into the cell. The size of Cs nanoparticle has been found to affect its viscosity, adding other level of regulation to the incorporation of nanoparticles into plant cells (Chattopadhyay and Inamdar 2010).

In the case of leaves, stomata are other possible point for nanoparticle penetration (Corredor et al. 2009). It could be investigated whether the passage of the particle through it does not imply its functional deregulation. Cytological observation and complementary approaches measuring stomatal conductance and infrared thermographic can be used to investigate stomatal opening/closure in response to nanoparticles application (Allègre et al. 2007). Since interactions between the plant cell wall and membrane trafficking have been reported (Kim and Brandizzi 2014; Ebine and Ueda 2015), endocytic pathways involved in the cellular uptake of nanoparticles might represent other level of specific regulation in plant cell. Although most of current studies have revealed that PGR-loaded nanoparticles are highly promising for agriculture use, there is still much to learn about how nanoparticles are incorporated, translocated and distributed in vascular plants. Studies on the mechanism of interactions between plant cells and polymeric nanoparticles must be performed.

4.6 Gap, Obstacles and Challenges

There are many challenges for the use of PGR in nanocarrier systems. First is to increase the interest of the scientific community in the development of controlled release systems for PGRs or even nanoparticles with potential plant growth promotion effects. Second is the evaluation and the comprehension of these systems, to understand how different polymers, methodologies of preparation and the characteristics of the nanomaterials could affect the biological effect of the PGR, and how these different systems interact with specific targets of the plant. Finally, evaluation using different biological models and their responses, whether in relation to their activity or even possible toxic in order to understand the applicability of these systems. The understanding of how these systems act in different plant tissues and at the cellular level related to the physico-chemical characteristics of these nanomaterials will help in the development of more efficient and intelligent systems for the immobilization/control delivery of the different types of PGRs.

4.7 Conclusions and Remarks

PGRs have great potential for agricultural application. The use of nanocarrier systems associated with PGRs and nanoparticles that have a promoter effect or even relieve vegetable stress are of great interest for field application. These systems can be applied in different stages of plant development in order to increase field performance. However, it is still necessary to develop and exploit the applications of these PGRs associated with polymer nanoparticles which can not only potentiate the production, but also increase the plant immune system and alleviate adverse environmental conditions. In short, the association of PGRs with polymer systems constitutes a promising strategy in order to increase productivity without causing major damage to the environment.

Acknowledgments Authors would like to thank São Paulo Research Foundation (FAPESP grant #2017/21004-5) and Agencia Nacional de Promoción Científica y Tecnológica (PICT 0008, PICT 0959, PICT 2421), Universidad Nacional de Mar del Plata and CONICET from Argentina.

Conflicts of Interest The authors declare no conflict of interest.

References

Abdel-Aziz, H. M. M., Hasaneen, M. N. A., & Omer, A. M. (2016). Nano chitosan-NPK fertilizer enhances the growth and productivity of wheat plants grown in sandy soil. *Spanish Journal of Agricultural Research, 14*(1), 0902. https://doi.org/10.5424/sjar/2016141-8205.

Allègre, M., Daire, X., Héloir, M.-C., Trouvelot, S., Mercier, L., Adrian, M., & Pugin, A. (2007). Stomatal deregulation in *Plasmopara viticola*-infected grapevine leaves. *The New Phytologist, 173*(4), 832–840. https://doi.org/10.1111/j.1469-8137.2006.01959.x.

Ambrogi, V., Famiani, F., Perioli, L., Marmottini, F., Di Cunzolo, I., & Rossi, C. (2006). Effect of MCM-41 on the dissolution rate of the poorly soluble plant growth regulator, the indole-3-butyric acid. *Microporous and Mesoporous Materials, 96*(1–3), 177–183. https://doi.org/10.1016/j.micromeso.2006.06.033.

Anusuya, S., & Banu, K. N. (2016). Silver-chitosan nanoparticles induced biochemical variations of chickpea (*Cicer arietinum* L.). *Biocatalysis and Agricultural Biotechnology, 8*, 39–44. https://doi.org/10.1016/j.bcab.2016.08.005.

Ao, M., Zhu, Y., He, S., Li, D., Li, P., Li, J., & Cao, Y. (2013). Preparation and characterization of 1-naphthylacetic acid–silica conjugated nanospheres for enhancement of controlled-release performance. *Nanotechnology, 24*(3). https://doi.org/10.1088/0957-4484/24/3/035601.

Asgari-Targhi, G., Iranbakhsh, A., & Ardebili, Z. O. (2018). Potential benefits and phytotoxicity of bulk and nano-chitosan on the growth, morphogenesis, physiology, and micropropagation of *Capsicum annuum*. *Plant Physiol Biochem PPB, 127*, 393–402. https://doi.org/10.1016/j.plaphy.2018.04.013.

Athanassiou, C. G., Kavallieratos, N. G., Benelli, G., Losic, D., Usha Rani, P., & Desneux, N. (2017). Nanoparticles for pest control: Current status and future perspectives. *Journal of Pest Science, 91*(1), 1–15. https://doi.org/10.1007/s10340-017-0898-0.

Baldwin, E. A. (1994). Edible coatings for fresh fruits and vegetables: Past, present, and future. In J. M. Krochta, E. A. Baldwin, & M. O. Nisperos-Carriedo (Eds.), *Edible coatings and films to improve food quality* (pp. 25–64). Lancaster: Technomic Publishing.

Barrios, A. C., Rico, C. M., Trujillo-Reyes, J., Medina-Velo, I. A., Peralta-Videa, J. R., & Gardea-Torresdey, J. L. (2016). Effects of uncoated and citric acid coated cerium oxide nanoparticles, bulk cerium oxide, cerium acetate, and citric acid on tomato plants. *Science of the Total Environment, 563–564*, 956–964. https://doi.org/10.1016/j.scitotenv.2015.11.143.

Basra AS (2000) Plant growth regulators in agriculture and horticulture: Their role and commercial uses. Plant growth Regul Agric Hortic their role Commer uses.

Behboudi, F., Tahmasebi Sarvestani, Z., Kassaee, M. Z., Modares Sanavi, S. A. M., Sorooshzadeh, A., & Ahmadi, S. B. (2018). Evaluation of chitosan nanoparticles effects on yield and yield components of barley (*Hordeum vulgare* L.) under late season drought stress. *Journal of Water and Environmental Nanotechnology, 3*(1), 22–39. https://doi.org/10.22090/jwent.2018.01.003.

Camara, M. C., et al. (2018). Current advances in gibberellic acid (GA3) production, patented technologies and potential applications. *Planta, 248*, 1049–1062.

Campos, E. V. R., de Oliveira, J. L., Fraceto, L. F., & Singh, B. (2015). Polysaccharides as safer release systems for agrochemicals. *Agronomy for Sustainable Development, 35*(1), 47–66. https://doi.org/10.1007/s13593-014-0263-0.

Cardoso, C., et al. (2011). AIB e substratos no enraizamento de estacas de pessegueiro 'Okinawa' coletadas no outono. *Semina: Ciências Agrárias, 32*, 1307–1314.

Cardoso, J. C., Ono, E. O., & Rodrigues, J. D. (2012). Ácido giberélico na indução e qualidade do florescimento de orquídea Phalaenopsis 'White Dream'. *Revista Brasileira de Horticultura Ornamental*, 135–140. https://doi.org/10.14295/rbho.v18i2.381.

Chandra, S., Chakraborty, N., Dasgupta, A., Sarkar, J., Panda, K., & Acharya, K. (2015). Chitosan nanoparticles: A positive modulator of innate immune responses in plants. *Scientific Reports, 5*(15195). https://doi.org/10.1038/srep15195.

Chattopadhyay, D. P., & Inamdar, M. S. (2010). Aqueous behaviour of chitosan. *International Journal of Polymer Science, 2010*, 1–7. https://doi.org/10.1155/2010/939536.

Chen, H., & Yada, R. (2011). Nanotechnologies in agriculture: New tools for sustainable development. *Trends in Food Science and Technology, 22*(11), 585–594. https://doi.org/10.1016/j.tifs.2011.09.004.

Choudhary, R. C., Kumaraswamy, R. V., Kumari, S., Sharma, S. S., Pal, A., Raliya, R., Biswas, P., & Saharan, V. (2017). Cu-chitosan nanoparticle boost defense responses and plant growth in maize (*Zea mays* L.). *Scientific Reports, 7*, 9754. https://doi.org/10.1038/s41598-017-08571-0.

Ciura, J., & Kruk, J. (2018). Phytohormones as targets for improving plant productivity and stress tolerance. *Journal of Plant Physiology, 229*, 32–40. https://doi.org/10.1016/j.jplph.2018.06.013.

Corredor, E., Testillano, P. S., Coronado, M.-J., González-Melendi, P., Fernández-Pacheco, R., Marquina, C., Ibarra, M. R., de la Fuente, J. M., Rubiales, D., Pérez-de-Luque, A., & Risueño, M.-C. (2009). Nanoparticle penetration and transport in living pumpkin plants: *in situ* subcellular identification. *BMC Plant Biology, 9*, 45. https://doi.org/10.1186/1471-2229-9-45.

Crespy, D., Stark, M., Hoffmann-Richter, C., Ziener, U., & Landfester, K. (2007). Polymeric nanoreactors for hydrophilic reagents synthesized by interfacial polycondensation on miniemulsion droplets. *Macromolecules?, 40*(9), 3122–3135. https://doi.org/10.1021/ma0621932.

de Oliveira, J. L., Campos, E. V. R., Bakshi, M., Abhilash, P. C., & Fernandes Fraceto, L. (2014). Application of nanotechnology for the encapsulation of botanical insecticides for sustainable agriculture: Prospects and promises. *Biotechnology Advances, 32*(8), 1550–1561. https://doi.org/10.1016/j.biotechadv.2014.10.010.

Dibax, R., de Alcantara, G. B., Machado, M. P., Filho, J. C. B., & Oliveira, R. A. de. (2013). Protocol optimization and histological analysis of in vitro plant regeneration of RB92579 and RB93509 sugarcane cultivars. *Ciência Rural, 43*, 49–54.

Dong, C., Xu, N., Ding, C., Gu, H., Zhang, W., & Li, X. (2016). Suitable gibberellic acid treatment for double-purpose rice (*Oryza sativa* L.) varieties at different harvest times. *Field Crops Research, 193*, 178–185. https://doi.org/10.1016/j.fcr.2016.04.038.

Duhan, J. S., Kumar, R., Kumar, N., Kaur, P., Nehra, K., & Duhan, S. (2017). Nanotechnology: The new perspective in precision agriculture. *Biotechnology Reports, 15*, 11–23. https://doi.org/10.1016/j.btre.2017.03.002.

Ebine, K., & Ueda, T. (2015). Roles of membrane trafficking in plant cell wall dynamics. *Frontiers in Plant Science, 6*. https://doi.org/10.3389/fpls.2015.00878.

Enders, T. A., & Strader, L. C. (2015). Auxin activity: Past, present, and future. *American Journal of Botany, 102*, 180–196.

Espinosa, M. E. Á., et al. (2017). Early histological, hormonal, and molecular changes during pineapple (Ananas comosus (L.) Merrill) artificial flowering induction. *Journal of Plant Physiology, 209*, 11–19.

Fan, W., Yan, W., Xu, Z., & Ni, H. (2012). Formation mechanism of monodisperse, low molecular weight chitosan nanoparticles by ionic gelation technique. *Colloids and Surfaces. B, Biointerfaces, 90*, 21–27.

Fraceto, L. F., Grillo, R., de Medeiros, G. A., Scognamiglio, V., Rea, G., & Bartolucci, C. (2016). Nanotechnology in agriculture: Which innovation potential does it have? *Frontiers in Environmental Science, 4*. https://doi.org/10.3389/fenvs.2016.00020.

Fryd, M. M., & Mason, T. G. (2012). Advanced Nanoemulsions. *Annual Review of Physical Chemistry, 63*, 493–518. https://doi.org/10.1146/annurev-physchem-032210-103436.

Fu, L., Shen, Q., Kuang, L., Yu, J., Wu, D., & Zhang, G. (2018). Metabolite profiling and gene expression of Na/K transporter analyses reveal mechanisms of the difference in salt tolerance between barley and rice. *Plant Physiological Biochemistry PPB, 130*, 248–257. https://doi.org/10.1016/j.plaphy.2018.07.013.

Ge, X., He, J., Yang, Y., Qi, F., Huang, Z., Lu, R., Huang, L., & Yao, X. (2011). Study on inclusion complexation between plant growth regulator 6-benzylaminopurine and β-cyclodextrin: Preparation, characterization and molecular modeling. *Journal of Molecular Structure, 994*(1–3), 163–169. https://doi.org/10.1016/j.molstruc.2011.03.013.

Grozeff, G., et al. (2010). 1-Methyl cyclopropene extends postharvest life of spinach leaves. *Postharvest Biology and Technology, 55*, 182–185.

Ghormade, V., Deshpande, M. V., & Paknikar, K. M. (2011). Perspectives for nano-biotechnology enabled protection and nutrition of plants. *Biotechnology Advances, 29*(6), 792–803. https://doi.org/10.1016/j.biotechadv.2011.06.007.

Grillo, R., Abhilash, P. C., & Fraceto, L. F. (2016). Nanotechnology applied to bio-encapsulation of pesticides. *Journal of Nanoscience and Nanotechnology, 16*(1), 1231–1234. https://doi.org/10.1166/jnn.2016.12332.

Grover, A. (2012). Plant chitinases: Genetic diversity and physiological roles. *Critical Reviews in Plant Sciences, 31*(1), 57–73. https://doi.org/10.1080/07352689.2011.616043.

Gubaeva, E., Gubaev, A., Melcher, R. L. J., Cord-Landwehr, S., Singh, R., El Gueddari, N. E., & Moerschbacher, B. M. (2018). 'Slipped sandwich' model for chitin and chitosan perception in *Arabidopsis. Molecular Plant-Microbe Interactions, 31*(11), 1145–1153. https://doi.org/10.1094/mpmi-04-18-0098-r.

Gutiérrez, T. J. (2017). Chapter 8. Chitosan applications for the food industry. In S. Ahmed & S. Ikram (Eds.), *Chitosan: Derivatives, composites and applications* (pp. 183–232). Hoboken. EE.UU. ISBN: 978-1-119-36350-7: WILEY-Scrivener Publisher. https://doi.org/10.1002/9781119364849.ch8.

Gutiérrez, T. J. (2018). Chapter 55. Processing nano- and microcapsules for industrial applications. In C. M. Hussain (Ed.), *Handbook of nanomaterials for industrial applications* (pp. 989–1011). Amsterdam: Editorial Elsevier. EE.UU. ISBN: 978–0–12-813351-4. https://doi.org/10.1016/B978-0-12-813351-4.00057-2.

Gutiérrez, T. J., & Álvarez, K. (2017). Chapter 6. Biopolymers as microencapsulation materials in the food industry. In M. Masuelli & D. Renard (Eds.), *Advances in physicochemical properties of biopolymers: Part 2* (Vol. 2017, pp. 296–322). Bentham Science Publishers. EE.UU. ISBN: 978–1–68108-545-6. eISBN: 978–1–68108-544-9. https://doi.org/10.2174/978168108544911 7010009.

Hanson, J. A., Chang, C. B., Graves, S. M., Li, Z., Mason, T. G., & Deming, T. J. (2008). Nanoscale double emulsions stabilized by single-component block copolypeptides. *Nature, 455*(7209), 85–88. https://doi.org/10.1038/nature07197.

Hao, Y., Xu, B., Ma, C., Shang, J., Gu, W., Li, W., Hou, T., Xiang, Y., Cao, W., Xing, B., & Rui, Y. (2018). Synthesis of novel mesoporous carbon nanoparticles and their phytotoxicity to rice (*Oryza sativa* L.). *Journal of Saudi Chemical Society, 23*(1), 75–82. https://doi.org/10.1016/j.jscs.2018.05.003.

Hornig, S., Heinze, T., Becer, C. R., & Schubert, U. S. (2009). Synthetic polymeric nanoparticles by nanoprecipitation. *Journal of Materials Chemistry, 19*, 3838–3840. https://doi.org/10.1039/b906556n.

Hussain, I., Ahmad, S., Amjad, M., & Ahmed, R. (2015). Thephon application at kimri stage accelerates the fruit maturation period and improves phytonutrients status (Hillawi and Khadrawi (c.v.)) of date palm fruit. *52*(1), 10.

Iqbal, M., Zafar, N., Fessi, H., & Elaissari, A. (2015). Double emulsion solvent evaporation techniques used for drug encapsulation. *International Journal of Pharmaceutics, 496*(2), 173–190. https://doi.org/10.1016/j.ijpharm.2015.10.057.

Kah, M., Beulke, S., Tiede, K., & Hofmann, T. (2013). Nanopesticides: State of knowledge, environmental fate, and exposure modeling. *Critical Reviews in Environmental Science and Technology, 43*(16), 1823–1867. https://doi.org/10.1080/10643389.2012.671750.

Kah, M., & Hofmann, T. (2014). Nanopesticide research: Current trends and future priorities. *Environment International, 63*, 224–235. https://doi.org/10.1016/j.envint.2013.11.015.

Kashyap, P. L., Xiang, X., & Heiden, P. (2015). Chitosan nanoparticle based delivery systems for sustainable agriculture. *International Journal of Biological Macromolecules, 77*, 36–51. https://doi.org/10.1016/j.ijbiomac.2015.02.039.

Khot, L. R., Sankaran, S., Maja, J. M., Ehsani, R., & Schuster, E. W. (2012). Applications of nanomaterials in agricultural production and crop protection: A review. *Crop Protection, 35*, 64–70. https://doi.org/10.1016/j.cropro.2012.01.007.

Khutoryanskiy, V. V. (2011). Advances in mucoadhesion and mucoadhesive polymers. *Macromolecular Bioscience, 11*(6), 748–764. https://doi.org/10.1002/mabi.201000388.

Kim, S.-J., & Brandizzi, F. (2014). The plant secretory pathway: An essential factory for building the plant cell wall. *Plant & Cell Physiology, 55*(4), 687–693. https://doi.org/10.1093/pcp/pct197.

Kong, D. D., & Cao, Y. S. (2018). Preparation and characterization of indole-3-butyric acid nanospheres for improving its stability and utilization. *Materials Science and Engineering C – Materials for Biological Applications, 89*, 175–181.

Kokina, I., Jahundoviča, I., Mickeviča, I., Jermaļonoka, M., Strautiņš, J., Popovs, S., Ogurcovs, A., Sledevskis, E., Polyakov, B., & Gerbreders, V. (2017). Target transportation of auxin on mesoporous Au/SiO_2 Nanoparticles as a method for somaclonal variation increasing in flax (*L. usitatissimum* L.). *Journal of Nanomaterials, 2017*, 1–9. https://doi.org/10.1155/2017/7143269.

Lindermayr, C., & Durner, J. (2018). Nitric oxide sensor proteins with revolutionary potential. *Journal of Experimental Botany, 69*(15), 3507–3510. https://doi.org/10.1093/jxb/ery193.

Liu, Y., Sun, Y., He, S., Zhu, Y., Ao, M., Li, J., & Cao, Y. (2013). Synthesis and characterization of gibberellin–chitosan conjugate forcontrolled-release applications. *International Journal of Biological Macromolecules, 57*, 213–217. https://doi.org/10.1016/j.ijbiomac.2013.03.024.

Liu, D., Jiao, S., Cheng, G., Li, X., Pei, Z., Pei, Y., Yin, H., & Du, Y. (2018). Identification of chitosan oligosaccharides binding proteins from the plasma membrane of wheat leaf cell. *International Journal of Biological Macromolecules, 111*, 1083–1090. https://doi.org/10.1016/j.ijbiomac.2018.01.113.

López-Moreno, M. L., Cedeño-Mattei, Y., Bailón-Ruiz, S. J., Vazquez-Nuñez, E., Hernandez-Viezcas, J. A., Perales-Pérez, O. J., Dela Rosa, G., Peralta-Videa, J. R., & Gardea-Torresdey, J. L. (2018). Environmental behavior of coated NMs: Physicochemical aspects and plant interactions. *Journal of Hazardous Materials, 347*, 196–217. https://doi.org/10.1016/j.jhazmat.2017.12.058.

Malerba, M., & Cerana, R. (2016). Chitosan effects on plant systems. *International Journal of Molecular Sciences, 17*(7). https://doi.org/10.3390/ijms17070996.

Martin-Saldaña, S., et al. (2018). Salicylic acid loaded chitosan microparticles applied to lettuce seedlings: Recycling shrimp fishing industry waste. *Carbohydrate Polymers, 200*, 321–331.

Martín-Saldaña, S., Palao-Suay, R., Aguilar, M. R., Ramírez-Camacho, R., & San Román, J. (2017). Polymeric nanoparticles loaded with dexamethasone or α-tocopheryl succinate to prevent cisplatin-induced ototoxicity. *Acta Biomaterialia, 53*, 199–210. https://doi.org/10.1016/j.actbio.2017.02.019.

Martín-Saldaña, S., Palao-Suay, R., Trinidad, A., Aguilar, M. R., Ramírez-Camacho, R., & San Román, J. (2016). Otoprotective properties of 6α-methylprednisolone-loaded nanoparticles against cisplatin: *In vitro* and *in vivo* correlation. *Nanomedicine: Nanotechnology, Biology and Medicine, 12*(4), 965–976. https://doi.org/10.1016/j.nano.2015.12.367.

Mishra, S., Keswani, C., Abhilash, P. C., Fraceto, L. F., & Singh, H. B. (2017). Integrated approach of Agri-nanotechnology: Challenges and future trends. *Frontiers in Plant Science, 8*(471). https://doi.org/10.3389/fpls.2017.00471.

Nakasato, D. Y., Pereira, A. E. S., Oliveira, J. L., Oliveira, H. C., & Fraceto, L. F. (2017). Evaluation of the effects of polymeric chitosan/tripolyphosphate and solid lipid nanoparticles on germination of *Zea mays*, *Brassica rapa* and *Pisum sativum*. *Ecotoxicology and Environmental Safety, 142*, 369–374. https://doi.org/10.1016/j.ecoenv.2017.04.033.

Navarro, E., Wagner, B., Odzak, N., Sigg, L., & Behra, R. (2015). Effects of differently coated silver nanoparticles on the photosynthesis of *Chlamydomonas reinhardtii*. *Environmental Science & Technology, 49*(13), 8041–8047. https://doi.org/10.1021/acs.est.5b01089.

Nguyen, M.-H., Nguyen, T.-H.-N., Hwang, I.-C., Bui, C.-B., & Park, H.-J. (2016). Effects of the physical state of nanocarriers on their penetration into the root and upward transportation to the stem of soybean plants using confocal laser scanning microscopy. *Crop Protection, 87*, 25–30. https://doi.org/10.1016/j.cropro.2016.04.012.

Nguyen Van, S., Dinh Minh, H., & Nguyen Anh, D. (2013). Study on chitosan nanoparticles on biophysical characteristics and growth of Robusta coffee in green house. *Biocatalysis and Agricultural Biotechnology, 2*(4), 289–294. https://doi.org/10.1016/j.bcab.2013.06.001.

Oliveira, H. C., Gomes, B. C. R., Pelegrino, M. T., & Seabra, A. B. (2016). Nitric oxide-releasing chitosan nanoparticles alleviate the effects of salt stress in maize plants. *Nitric Oxide, 61*, 10–19. https://doi.org/10.1016/j.niox.2016.09.010.

Ordaz-Ortiz, J. J., Foukaraki, S., & Terry, L. A. (2015). Assessing temporal flux of plant hormones in stored processing potatoes using high definition accurate mass spectrometry. *Horticulture Research, 2*(15002). https://doi.org/10.1038/hortres.2015.2.

Peixoto, C. P., & Passos, A. R. (2011). Ação da giberelina em sementes pré-embebidas de mamoneira. *Comunicata Scientiae, 6*.

Pereira, A. E. S., Sandoval-Herrera, I. E., Zavala-Betancourt, S. A., Oliveira, H. C., Ledezma-Pérez, A. S., Romero, J., & Fraceto, L. F. (2017a). γ-Polyglutamic acid/chitosan nanoparticles for the plant growth regulator gibberellic acid: Characterization and evaluation of biological activity. *Carbohydrate Polymers, 157*, 1862–1873. https://doi.org/10.1016/j.carbpol.2016.11.073.

Pereira, A. E. S., Silva, P. M., Oliveira, J. L., Oliveira, H. C., & Fraceto, L. F. (2017b). Chitosan nanoparticles as carrier systems for the plant growth hormone gibberellic acid. *Colloids and Surfaces. B, Biointerfaces, 150*, 141–152. https://doi.org/10.1016/j.colsurfb.2016.11.027.

Pérez Quiñones, J., Brüggemann, O., Kjems, J., Shahavi, M. H., & Peniche Cova, C. (2018). Novel brassinosteroid-modified polyethylene glycol micelles for controlled release of agrochemicals. *Journal of Agricultural and Food Chemistry, 66*(7), 1612–1619. https://doi.org/10.1021/acs.jafc.7b05019.

Pérez-de-Luque, A. (2017). Interaction of nanomaterials with plants: What do we need for real applications in agriculture? *Frontiers in Environmental Science, 5*. https://doi.org/10.3389/fenvs.2017.00012.

Pérez-de-Luque, A., & Rubiales, D. (2009). Nanotechnology for parasitic plant control. *Pest Management Science, 65*(5), 540–545. https://doi.org/10.1002/ps.1732.

Petri, J. L., Leite, G. B., & Argenta, L. C. (2007). Eficácia do tratamento de AVG no controle da queda e maturação dos frutos de maçã, cultivar imperial gala. *Revista Brasileira de Fruticultura, 29*, 239–244.

Petutschnig, E. K., Jones, A. M. E., Serazetdinova, L., Lipka, U., & Lipka, V. (2010). The Lysin motif receptor-like kinase (LysM-RLK) CERK1 is a major chitin-binding protein in *Arabidopsis thaliana* and subject to chitin-induced phosphorylation. *The Journal of Biological Chemistry, 285*, 28902–28911. https://doi.org/10.1074/jbc.m110.116657.

Prasad, A., Astete, C. E., Bodoki, A. E., Windham, M. K., Bodoki, E., & Sabliov, C. M. (2017a). Zein nanoparticles uptake and translocation in hydroponically grown sugar cane plants. *Journal of Agricultural and Food Chemistry, 66*(26), 6544–6551. https://doi.org/10.1021/acs.jafc.7b02487.

Prasad, R., Bhattacharyya, A., & Nguyen, Q. D. (2017b). Nanotechnology in sustainable agriculture: Recent developments, challenges, and perspectives. *Frontiers in Microbiology, 8*. https://doi.org/10.3389/fmicb.2017.01014.

Quiñones, J. P., García, Y. C., Curiel, H., & Covas, C. P. (2010). Microspheres of chitosan for controlled delivery of brassinosteroids with biological activity as agrochemicals. *Carbohydrate Polymers, 80*(3), 915–921. https://doi.org/10.1016/j.carbpol.2010.01.006.

Rademacher, W. (2015). Plant growth regulators: Backgrounds and uses in plant production. *Journal of Plant Growth Regulation, 34*(4), 845–872. https://doi.org/10.1007/s00344-015-9541-6.

Rao, J. P., & Geckeler, K. E. (2011). Polymer nanoparticles: Preparation techniques and size-control parameters. *Progress in Polymer Science, 36*(7), 887–913. https://doi.org/10.1016/j.progpolymsci.2011.01.001.

Revell P.A. (2006) The biological effect of nanoparticles, Nanotechnology perceptions? 2:283–229.

Rodrigues, & Fioreze. (2015). Reguladores são, para muitos cultivos, indispensáveis ao alcance de bons níveis. *Fisiologia compostos, 5*.

Saharan, V., Kumaraswamy, R. V., Choudhary, R. C., Kumari, S., Pal, A., Raliya, R., & Biswas, P. (2016). Cu-chitosan nanoparticle mediated sustainable approach to enhance seedling growth in maize by mobilizing reserved food. *Journal of Agricultural and Food Chemistry, 64*(31), 6148–6155. https://doi.org/10.1021/acs.jafc.6b02239.

Sathiyabama, M., & Manikandan, A. (2018). Application of copper-chitosan nanoparticles stimulate growth and induce resistance in finger millet (*Eleusine coracana* gaertn.) plants against blast disease. *Journal of Agricultural and Food Chemistry, 66*(8), 1784–1790. https://doi.org/10.1021/acs.jafc.7b05921.

Schulz, B., & Segobye, K. (2016). 2,4-D transport and herbicide resistance in weeds. *Journal of Experimental Botany, 67*, 3177–3179.

Seabra, & Oliveira. (2016). How nitric oxide donors can protect plants in a changing environment: What we know so far and perspectives. *AIMS Molecular Science, 3*, 692–718.

Silva, E. S. D., Wong-Wah-Chung, P., Burrows, H. D., & Sarakha, M. (2013). Photochemical degradation of the plant growth regulator 2-(1-naphthyl) acetamide in aqueous solution upon UV irradiation. *Photochemistry and Photobiology, 89*(3), 560–570. https://doi.org/10.1111/php.12050.

Skůpa, P., Opatrný, Z., & Petrášek, J. (2014). Auxin biology: Applications and the mechanisms behind. In P. Nick & Z. Opatrny (Eds.), *Applied plant cell biology: Cellular tools and approaches for plant biotechnology* (pp. 69–102). Heidelberg: Springer. https://doi.org/10.1007/978-3-642-41787-0_3.

Solans, C., Izquierdo, P., Nolla, J., Azemar, N., & Garcia-Celma, M. J. (2005). Nano-emulsions. *Current Opinion in Colloid & Interface Science, 10*(3–4), 102–110. https://doi.org/10.1016/j.cocis.2005.06.004.

Stavolone, L., & Lionetti, V. (2017). Extracellular matrix in plants and animals: Hooks and locks for viruses. *Frontiers in Microbiology, 8*. https://doi.org/10.3389/fmicb.2017.01760.

Steffens, C. A., do Amarante, C. V. T., Chechi, R., Silveira, J. P. G., & Brackmann, A. (2009). Aplicação pré-colheita de reguladores vegetais visando retardar a maturação de ameixas 'Laetitia. *Ciência Rural, 39*, 1369–1373.

Sun, C., Shu, K., Wang, W., Ye, Z., Liu, T., Gao, Y., Zheng, H., He, G., & Yin, Y. (2014). Encapsulation and controlled release of hydrophilic pesticide in shell cross-linked nanocapsules containing aqueous core. *International Journal of Pharmaceutics, 463*(1), 108–114. https://doi.org/10.1016/j.ijpharm.2013.12.050.

Tao, S., Pang, R., Chen, C., Ren, X., & Hu, S. (2012). Synthesis, characterization and slow release properties of O-naphthylacetyl chitosan. *Carbohydrate Polymers, 88*(4), 1189–1194. https://doi.org/10.1016/j.carbpol.2012.01.076.

Thickett, S. C., Gaborieau, M., & Gilbert, R. G. (2007). Extended mechanistic description of particle growth in electrosterically stabilized emulsion polymerization systems. *Macromolecules, 40*(13), 4710–4720. https://doi.org/10.1021/ma070837f.

Thompson, D. S. (2007). Space and time in the plant cell wall: Relationships between cell type, cell wall rheology and cell function. *Annals of Botany, 101*(2), 203–211. https://doi.org/10.1093/aob/mcm138.

Tripathi, D. K., Shweta, S. S., Singh, S., Pandey, R., Singh, V. P., Sharma, N. C., Prasad, S. M., Dubey, N. K., & Chauhan, D. K. (2017). An overview on manufactured nanoparticles in plants: Uptake, translocation, accumulation and phytotoxicity. *Plant Physiology and Biochemistry, 110*, 2–12. https://doi.org/10.1016/j.plaphy.2016.07.030.

Valletta, A., Chronopoulou, L., Palocci, C., Baldan, B., Donati, L., & Pasqua, G. (2014). Poly(lactic-co-glycolic) acid nanoparticles uptake by *Vitis vinifera* and grapevine-pathogenic fungi. *Journal of Nanoparticle Research, 16*(2744). https://doi.org/10.1007/s11051-014-2744-0.

Vankova, R., Landa, P., Podlipna, R., Dobrev, P. I., Prerostova, S., Langhansova, L., Gaudinova, A., Motkova, K., Knirsch, V., & Vanek, T. (2017). ZnO nanoparticle effects on hormonal pools in Arabidopsis thaliana. *Science Total Environmental, 593–594*, 535–542. https://doi.org/10.1016/j.scitotenv.2017.03.160.

Venkatachalam, P., Priyanka, N., Manikandan, K., Ganeshbabu, I., Indiraarulselvi, P., Geetha, N., Muralikrishna, K., Bhattacharya, R. C., Tiwarid, M., Sharmad, N., & Sahid, S. V. (2017). Enhanced plant growth promoting role of phycomolecules coated zinc oxide nanoparticles with P supplementation in cotton (*Gossypium hirsutum* L.). *Plant Physiology and Biochemistry, 110*, 118–127. https://doi.org/10.1016/j.plaphy.2016.09.004.

Wan, J., Zhang, X.-C., Neece, D., Ramonell, K. M., Clough, S., S-y, K., Stacey, M. G., & Staceya, G. (2008). A LysM receptor-like kinase plays a critical role in chitin signaling and fungal resistance in arabidopsis. *Plant Cell, 20*, 471–481. https://doi.org/10.1105/tpc.107.056754.

Wang, P., Lombi, E., Zhao, F.-J., & Kopittke, P. M. (2016). Nanotechnology: A new opportunity in plant sciences. *Trends in Plant Science, 21*(8), 699–712. https://doi.org/10.1016/j.tplants.2016.04.005.

Xue, W., Han, Y., Tan, J., Wang, Y., Wang, G., & Wang, H. (2018). Effects of nanochitin on the enhancement of the grain yield and quality of winter wheat. *Journal of Agricultural and Food Chemistry, 66*(26), 6637–6645. https://doi.org/10.1021/acs.jafc.7b00641.

Yang, R., Xiao, C.-F., Guo, Y.-F., Ye, M., & Lin, J. (2018). Inclusion complexes of GA3 and the plant growth regulation activities. *Materials Science and Engineering: C, 91*, 475–485. https://doi.org/10.1016/j.msec.2018.05.043.

Zargar, V., Asghari, M., & Dashti, A. (2015). A review on chitin and chitosan polymers: Structure, chemistry, solubility, derivatives, and applications. *ChemBioEng Reviews, 2*(3), 204–226. https://doi.org/10.1002/cben.201400025.

Chapter 5
Applications and Implications of Environmental-Responsive Polymers toward Agrochemicals

Yue Shen, Huaxin Zhu, Yan Wang, Haixin Cui, and Runcang Sun

Abstract Environmental-responsive polymers have many applications in biomedicine and nanotechnology, and have always been the hotspot of polymer science research. The stimuli from the external environment include both physical, chemical, and biological types, such as pH, temperature, ionic strength, light, enzyme or magnetic fields etc. For environmental-responsive polymers, different responsive groups are introduced into the polymers through various polymerization or other chemical means to meet different needs. This method cannot only accurately control the morphology of the assembly when external stimuli changes, but also can adjust the types and quantities of the groups according to actual needs. In this chapter, special attention has been paid to intelligent agrochemicals with precise controlled release modes that can respond to micro-ecological environment changes. Moreover, the establishing intelligent and controlled agrochemicals release technologies which could increase agrochemicals-loading, improve the dispersibility and stability of active ingredients, and promote target ability are reported.

Keywords Agrochemicals · Environmental response · Functional polymers

5.1 Introduction

Agrochemicals include fertilizers, herbicides, pesticides, fungicides and other chemicals imported in agricultural production (Scott et al. 2018). Among them, pesticides are an important material basis for safeguarding major biological disasters, ensuring national food production, and promoting sustained and steady growth of

Y. Shen · H. Zhu · Y. Wang · H. Cui (✉)
Institute of Environment and Sustainable Development in Agriculture,
Chinese Academic of Agriculture Sciences, Beijing, China
e-mail: cuihaixin@caas.cn

R. Sun (✉)
Center for Lignocellulose Science and Engineering, College of Light Industry
and Chemical Engineering, Dalian Polytechnic University, Dalian, China
e-mail: rcsun3@bjfu.edu.cn

T. J. Gutiérrez (ed.), *Polymers for Agri-Food Applications*,
https://doi.org/10.1007/978-3-030-19416-1_5

agricultural output (Zhao et al. 2018). According to the statistics of the Food and Agriculture Organization of the United Nations (FAO), the loss of agricultural products recovered by pesticides and pests worldwide is about 30% of total production. However, pesticides are a special kind of chemical. While defending against major biological disasters in agriculture, they also pose a serious threat to human health and the environment (Liang et al. 2018). With the improvement of global ecological environment and health awareness, the development of pesticide products is developing towards high efficiency, low toxicity and low residue. In the current situation that the creation of new pesticide compounds is becoming more and more difficult, improving the performance of pesticides through advanced formulation processing methods and giving full play to the biological activity and efficacy of active ingredients have become an important way to improve the control effect of pesticides, reduce residual pollution, and save the cost of application (Liu et al. 2018; Guan et al. 2018; Huang et al. 2018).

Conventional pesticide formulations such as emulsifiable concentrates, wettable powders, etc., do not have the controlled release ability of the active ingredients, so that the active ingredient is easily photolyzed, hydrolyzed or microbially metabolized in the environment near the target, and the effect of the ingredient is reduced due to the loss. Even if a highly active pesticide ingredient is used, its effective utilization rate is usually only 10%–30%, and the loss rate is as high as 70%–90% (Zhao et al. 2018). It is necessary sometimes to repeatedly spray to effectively control crop pests and diseases, which not only increases the cost of agricultural production, but also aggravates the impact of pesticides on agricultural products and the environment (Zhang 1998). There are many reasons for the low effective utilization rate of pesticides. The loss of target off-target and the backward use of existing pesticide formulations are two of the main reasons (Scott et al. 2018). Therefore, how to use advanced carrier materials and loading methods to overcome the functional defects of traditional formulations, improve the effectiveness and utilization of pesticides, and reduce their dosage and residual pollution in non-target areas and environments, has become a scientific problem that needs to be solved urgently in modern agriculture.

In recent years, the rapid development of nanotechnology has provided a new scientific methodology for modern agricultural science. It is promoting traditional agriculture to continually nurture new major breakthroughs in many cross-cutting fields and form the growth point of agricultural emerging strategic industries (Fu et al. 2018; Zhao et al. 2017a; Shen et al. 2017). The application of nanotechnology in agriculture has formed a very active and emerging interdisciplinary subject. Using the targeted transmission and control release function of nano-carrier materials, developing new formulations of green nano-pesticides, improving the effective utilization rate of agrochemicals such as pesticides and fertilizers, and reducing residues and pollution have become one of the research hotspots in the field of nanotechnology agricultural application in the world (Morris et al. 2011; Scott and Chen 2012; Ghormade et al. 2011; Khot et al. 2012), showing good application prospects in alleviating environmental pollution, pest resistance and biodiversity loss caused by pesticide abuse. The figure is shown in our previous paper (Fig. 5.1) (Huang et al. 2018). In 2003, the US Department of Agriculture launched the "Nanoscale Science and Engineering for Agriculture and

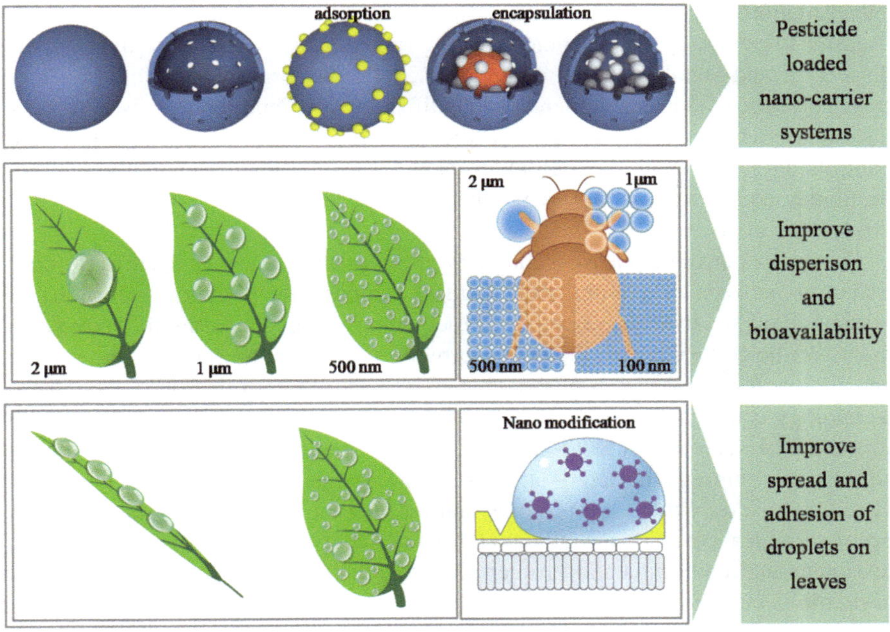

Fig. 5.1 Nano-based pesticide formulation increases bioavailability and efficiency

Food Systems" research program (Norman et al. 2002), which listed the intelligent transmission of agricultural chemicals as one of the six key research directions. Subsequently, the European Union, Brazil, India, Japan, Canada and other countries have also developed and started to implement relevant research plans (Observatory NANO 2010; Robinson and Morrison 2009). Some international organizations and institutions have also begun to pay close attention to the impact of nanotechnology on future agriculture. Since 2009, the FAO has organized high-level expert forums and technical round tables to discuss the application prospects of nanotechnology in food and agriculture, and edited and published strategic research reports (Food and Agriculture Organization 2010). The US Environmental Protection Agency (EPA) and the European Parliament have issued regulations on the production and use of nanopesticide (Morris et al. 2011). World-renowned agrochemical enterprises such as Bayer, DuPont and Syngenta have also attached great importance to the development and promotion of nano-pesticide technology and products (Norman et al. 2002; Observatory NANO 2010).

Currently, the key focus of nanopesticide research is to develop a smart nano-delivery system with target-oriented and controlled-release functions using nano-materials and technologies. In this way, the formulation function of the pesticide is improved, the utilization efficiency and biological activity of the active ingredients are maximized, and the loss and residual pollution are reduced. Compared with the conventional pesticide carrier, the nanocarrier has characteristics such as small particle size and large specific surface area. It loads pesticides by adsorption, coupling, encapsulation, inlaying, etc., and constructs a nano pesticide-loading system, which is easier to exert excellent pesticide-loading functions, is beneficial to improve the dispersion and stability of the components, promote target deposition and dose transfer, reduce the loss and increase the utilization rate, and reduce pesticide residues and environmental pollution. At the same time, the nano granules have the controlled release function of pesticides. They can protect environmentally sensitive pesticides, control release rate, and reduce loss and decomposition (Ghormade et al. 2011; Khot et al. 2012).

Pesticide controlled release technology provides a feasible way to solve the contradiction between agricultural yield increase and pesticide pollution control. It can reasonably adjust the release rate of the active ingredient according to the requirements of the control target, thereby improving the utilization rate of the pesticide and prolonging the duration of the effect. The core of controlled pesticide release technology is the carrier material. At present, common controlled pesticide release formulations such as microcapsules, microspheres and inclusion compounds can significantly improve the safety of users and non-target organisms, reduce the dosage of pesticides, reduce soil leaching and prolong the effective ingredients. It has become an ideal pesticide formulation that can effectively reduce pesticide residues, resistance and environmental problems. Research on new formulations of nano-pesticide is currently focused on these aspects, especially pesticide nanocapsules and nanospheres, which have good sustained-release properties (Frederiksen et al. 2003; Caboni et al. 2003; Boehm et al. 2003).

The environmental-responsive carrier material can respond to changes in environmental stimuli such as enzymes, redox, pH, light, temperature, electric field, magnetic field and ionic strength to achieve targeted controlled release of active ingredients, showing obvious advantages in controlled drug release. It has broad application prospects and has become a hot spot in the fields of medicine, food and environmental engineering (Manatunga et al. 2017; Zhang et al. 2017; Eswaramma and Rao 2017; Shao et al. 2017; Zhou et al. 2017). The application of environmental-responsive carrier materials in the field of pesticides, research and development of new environmental-responsive pesticide controlled release agents, has gradually become an important development direction of new pesticide formulations. This chapter intends to review the types of environmental-responsive carrier materials and their application and implication in the field of pesticides, aiming to highlight the importance and application potential of environmental-responsive carrier materials in the field of controlled pesticide release.

5.2 Environmental Stimuli-Responsive Carriers

Intelligent materials respond to changes in the environment and realize an artificial imitation of bio-intelligence. Now it has become an emerging multidisciplinary and cross-integrated science, and it is one of the international academic frontiers. Polymeric materials also play an important role in many fields. The intersection of the two, namely the environmental- responsive polymer, i.e. the intelligent polymer material, has been the frontier and hot spot in the research of polymer science in recent years. Such high-molecular polymers can receive small stimuli from the external environment, resulting in reversible or irreversible changes in the molecular structure, which in turn affects its chemical or physical properties. Stimulus signals from the external environment include both physical and chemical stimuli (Kwon et al. 1991; Gutowska et al. 1994; Pelton 2000; Zou et al. 2004; Tanaka et al. 1980; Sijbesma et al. 1997). The stimuli-responsive polymer may also be part of the polymer that responds to external stimuli (Abu-Lail et al. 2006; Xu et al. 2006). This change (or stimulus) of the external environment can be pH, temperature, ionic strength, light, electricity or magnetic field (Ayano et al. 2012; Palankar et al. 2009; Thomas et al. 2012; Nash et al. 2012; Dedinaite et al. 2010). Environmental-responsive polymers can be designed as a single response or as multiple responses (Cho et al. 2009; Debord and Lyon 2003; And and Lyon 2004). These materials have great industrial application prospects and have been used in the fields of chromatography, optoelectronics, drug release and biosensors (Liu et al. 2011; Sidorenko et al. 2007; Cai et al. 2012; Guo et al. 2012; Shim, and Kwon 2011; Tang et al. 2011).

5.2.1 Temperature-Responsive Polymer

Temperature-responsive polymers are a class of smart polymers that respond to changes in the external temperature. A common feature of this type of polymer is the presence of a lower critical solution temperature (LCST) in a particular solvent (Xu et al. 2007; Di et al. 2010; Qin et al. 2010). The LCST refers to the critical phase transition temperature of the polymer. When the ambient temperature is lower than the LCST, the polymer forms an extended coil structure, and the polymer molecular chain is soluble in the solvent. When the ambient temperature rises to the LCST, the polymer molecules are transformed into a compact colloidal structure from a loose coil structure, the polymer molecular chains are precipitated from the solvent, and the phase behavior changes. This phase change is the basis for the temperature response of the polymer (Xu et al. 2007; Liu and Fan 2005; Durand and Hourdet 1999; Yan et al. 2008). There are many kinds of monomers for synthesizing temperature-responsive polymers, but poly(*N*-isopropylacrylamide) (PNIPAM) is the most popular material. Although PNIPAM has been reported since the 1950s, its temperature-responsive nature was not discovered by Heshinks et al. until 1968, with a LCST of 32 °C (Schild 1992; Heskins and Guillet 1968). Polymer materials such as hydrogels that crosslink PNIPAM at the same time also have a LCST behavior. PNIPAM is capable of phase transitions due to temperature changes. When the temperature of the external environment changes, the molecular structure of PNIPAM changes with it, resulting in a reversible change in its hydrophily/hydrophobicity and volume. The PNIPAM molecular chain contains a large number of hydrophobic groups [-$CH(CH_3)_2$-] and hydrophilic groups [-CONH-]. When the temperature is lower than LCST, the hydrogen bond interactions between the hydrophilic amide groups and the water molecules cause the macromolecule to stretch, thereby exhibiting hydrophilicity, and PNIPAM is

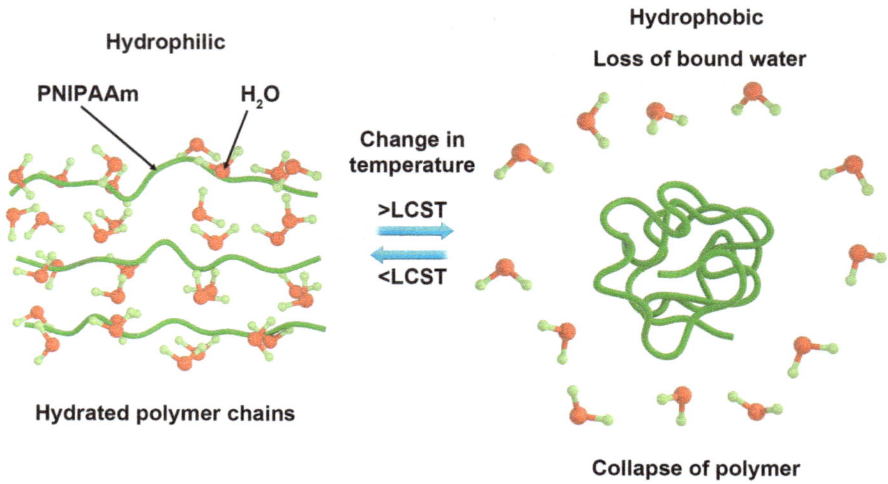

Fig. 5.2 Mechanism of PNIPAM hydrophilic and hydrophobic transition

dissolved in water (Long et al. 2009). When the temperature is higher than LCST, the hydrophobic group surrounds the hydrophilic group, causing the molecular chain to shrink, which makes PNIPAM exhibit hydrophobicity. Meanwhile, microphase separation occurs, and PNIPAM precipitates out of the water. The temperature-sensitive phase transition is shown in the Fig. 5.2.

5.2.2 pH-Responsive Polymer

A pH-responsive polymer is a type of polymer whose physical and chemical properties change as the pH of the environment changes. This change is generally based on stimulatory responses at the molecular and macromolecular levels. Such polymers generally contain a large number of weak electrolyte groups such as carboxylic groups or amino groups which allow them to accept or impart protons in response to changes in pH in the environment (Bae et al. 2003; Schmaljohann 2006). In addition, the pH-responsive polymer also responds to changes in ionic strength, and as the ionic strength increases, the electrostatic repulsion force also increases, causing the polymer chain to collapse. Bhattacharya et al. (2010) synthesized a composite microgel containing *N*-vinylimidazole (VIM) as a monomer, and explored the hydrodynamic diameter of the microgel under different pH conditions by light scattering. When pH was 4, the VIM unit combined a large number of protons and completely ionized. The charge density on the microgel structure became large, and strong electrostatic repulsion caused the microgel to swell. When the solution pH was 7, the degree of ionization of the VIM group became smaller. The charge density of the microgel structure became smaller, and the polymer chain changed from the extended state to the shrinkage collapse state, as shown in the Fig. 5.3.

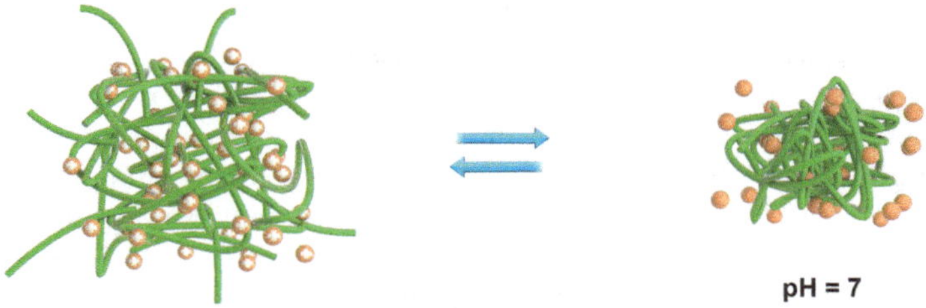

Fig. 5.3 Radius of microgels at different pH values

Manchanda and Nimesh (2010) prepared chitosan (Cs) nanoparticles through a reversed micelle system using glutaraldehyde as crosslinker for directional release of oligonucleotides. In the case of pH = 4, the electrostatic and ionic interaction between Cs and oligonucleotide were the strongest, and the release rate of oligonucleotide was slower than that at pH = 7. This is because when pH = 7, the deprotonation of Cs weakened the interaction between each other, and the release was faster. A similar situation occurs in the encapsulated ciprofloxacin hydrochloride Cs nanoparticles prepared by tripolyphosphate cross-linking (Hui and Gao 2009). The encapsulation and drug release in this system was affected by the environment pH value (Hui and Gao 2009; Gutiérrez and Álvarez 2017; Gutiérrez 2018). Compared with acidic conditions, the encapsulation efficiency was significantly improved at pH = 7.4. Though at higher pH conditions, the nanoparticles shrink. However, the carboxyl moiety of the ciprofloxacin hydrochloride molecule was deprotonated, a stronger interaction was formed with the positively charged Cs, and this action played a leading role. More ciprofloxacin hydrochloride was released in PBS buffer at pH = 7.4 compared to aqueous solution, because the presence of ions in the buffer can weaken the interaction between Cs and TPP or ciprofloxacin hydrochloride, thus speeding up the drug release rate. Kikuchi et al. (1997) used a six-arm block polymer poly(caprolactone)-*b*-polydimethylaminoethyl methacrylate to carry indomethacin after electrostatic interaction and then embedded in calcium alginate colloidal particles. This composite system had a significantly faster release rate at pH = 1.2, 6.8, or 7.4 compared to a pure drug-coated calcium alginate system. This phenomenon occured because the solubility of the drug in the particles was enhanced, so that more drug was loaded in the particles. The difference in concentration was more favorable for the drug to be released from the particles.

5.2.3 Photoresponsive Polymers

Relying on simple synthesis steps, fast response efficiency and excellent performance, photoresponsive polymers have been widely used in traditional chemistry and biochemistry research. The intensity and wavelength of the source can be easily controlled by using a filter or a light transmissive abrasive. Therefore, a range of wavelength ranges of light sources have been widely used in various fields from ultraviolet light to infrared light. As a basis for the synthesis of photoresponsive polymer materials, a large number of chemically synthesized photoresponsive monomer groups have been studied, and various photoreactive group monomers of various novel structures have emerged in an endless stream. Among them, the classical photoresponsive group of o-nitrotoluene type (o-NB) alcohol derivative has received great attention. In 1966, this photostructural o-NB alcohol derivative was first discovered by Woodward et al. under the irradiation of ultraviolet light, the photoisomerization reaction of the o-NB ester structure was cleaved into a nitrosobenzaldehyde structure and a carboxylic acid structure (Ding et al. 2016). The o-NB type monomer was embedded in the polymer chain by a simple chemical synthesis method, and under the stimulation of the external light source, the photoisomerization reaction of the o-NB monomer in the polymer chain gave the photoresponse properties of the polymer (Mok et al. 2016). Among the light-responsive polymer materials, photoresponsive hydrogels prepared around the o-NB type structure have been widely used. Such hydrogels are currently used more in cell preservation and in cell environment research, and their tunable sizes and novel geometries give them unique advantages. At the same time, a biocompatible base material such as polyethylene glycol, dextran, agarose or hyaluronic acid is crosslinked by the o-NB type monomer to prepare a biocompatible photoresponsive polymer material. Such materials have been successfully applied in the field of tissue engineering and controlled drug release (Il'ichev et al. 2004; Pellicciolі and Wirz 2002; Gohy and Zhao 2013). The photoresponsive self-assembled film prepared by modifying the o-NB type structure to the side chain of the copolymer has also been widely used. This side-chain modified o-NB type structure provide a new method for micro-scale photolithographic polymer film patterning. Photocracked self-assembled monolayers can be modified on nanoscale surfaces, such as nanoparticles in microfluidic channels (Kharkar et al. 2013; Tomatsu et al. 2011). The molecular micelles formed by the photo-induced decomposition of the block copolymer are then designed using an o-NB type structure to stimulate the release of the dye or drug at a specified time and place, such as inside the cell, by external light. This research direction has also become a recent hot spot.

The photoresponsive carrier material utilizes the photosensitivity phenomenon of the photosensitive polymer, and after absorbing the light energy, can cause physical changes (such as color, conductivity, and solubility, etc.) or chemical reactions (such as photolysis, dimerization, polymerization, and isomerization, etc.) between intramolecular and intermolecular interaction by the absorbed energy. The introduction of photosensitive groups in the main chain or side chain of a polymer, and the design and preparation of a photoresponsive carrier material are currently widely used methods. In terms of load and delivery systems, photoresponsive carrier materials have been shown to be easy to use, intelligent and efficient (Wang et al. 2016).

5.2.4 *Enzyme-Responsive Polymer*

Enzyme-responsive carrier materials include enzyme-responsive polymers (Wang et al. 2010), nanoparticles (Nguyen et al. 2015) and hydrogels (Hu et al. 2012), etc., and their research and development are receiving more and more attention (Gutiérrez 2019). At present, it has shown broad application prospects in the fields of drug controlled release (Li et al. 2016), optical sensing and imaging (Zha et al. 2013), biocatalysis tissue engineering (Teixeira et al. 2012), and clinical diagnosis (Roberto et al. 2012), especially in the field of medicine. The controlled drug release by utilizing enzymes related to the disease, and is a highly effective and highly selective drug-targeted controlled release strategy, which has good prospects in the field of cancer drug therapy (Andresen et al. 2010). In the process of harming crops, there are also a variety of related enzymes, such as cell wall degrading enzymes such as cellulase (Watanabe and Tokuda 2010), pectinase (Maisuria, Patel, and Nerurkar 2010) and protease (Agusti and Cohen 2000). Therefore, drug-targeted controlled release strategies can also be adopted to achieve precise and intelligent control of harmful organisms.

5.2.5 *Redox-Responsive Polymer*

The redox-responsive polymer responds to differences in the *in vivo* redox potential and thus has good biocompatibility. It usually contains disulfide bonds (S-S) or other substances with multiple oxidation states (such as iron and sulfur), the most common of which is the inclusion of disulfide bonds. It has redox responsiveness by a disulfide bond which is broken by the action of reduced glutathione *in vivo*. At present, redox-responsive carrier materials are also a hot topic (Zhao et al. 2017b).

5.2.6 *Pesticide Precision-Controlled Release Technology and Products*

The controlled pesticide release technology can reasonably adjust the release rate of the active ingredient according to the requirements of the target, thereby improving the utilization rate of the pesticide and prolonging the duration of the effect. Since the release of pesticides in the environment is controlled, the migration of pesticides in the environment can be reduced, and pesticide pollution can be controlled from the source. In addition, increased pesticide utilization will also help reduce the total amount of pesticides applied and agricultural production inputs. Therefore, controlled pesticide release technology provides a feasible way to solve the contradiction between agricultural yield increase and pesticide pollution control.

5.2.7 Temperature-Responsive Pesticide-Controlled Release Agent

Sheng et al. (2015) prepared avermectin controlled release agents with temperature and pH dual responsiveness by hydrazone bonding. The results show that the controlled release agent had a high loading capacity (97.4%) and good temperature and pH double stimuli response performance.

Xu et al. (2017) prepared a polydopamine (PDA)@PNIPAM nanocomposite with core-shell structure by using PDA microspheres as a photothermal agent and then coating a thermosensitive PNIPAM shell. It was used to load imidacloprid and characterized by transmission electron microscopy (TEM), scanning electron microscopy (SEM), infrared (IR) spectroscopy, dynamic light scattering and thermogravimetric analysis (TGA). It was found that the controlled agent release has high loading capacity and good temperature response performance.

Green et al. (1992) loaded active components in temperature-sensitive polymer microcapsules to prepare temperature-responsive diazinon, trifluralin and alachlor controlled release agents. It has also been found that these controlled release agents are effective in preventing the degradation of active ingredients until the soil temperature or temperature reaches the critical temperature at which the pests are most active. The pesticide were released at a predictable rate. The bioactivity evaluation test showed that the diazinon controlled release agent prolonged the duration of pest control compared to the conventional formulation. Trifluralin controlled release agent reduced the phytotoxicity of crops and eliminated the need for rapid soiling; the alachlor controlled release agent prolonged the duration of weed control and reduced the leaching of alachlor in the soil.

The United States 3 M Company (Minnesota Mining and Machine Manufacturing Company) used a phase separation method to encapsulate pesticides in temperature-sensitive lipid materials to produce a temperature-responsive controlled pesticide release agent. The controlled pesticide carrier released the compound when the ambient temperature was higher than the melting point of the lipid material, and when the temperature dropped, the lipid material hardened to stop the pesticide release (Thies, and Louis 2007).

Chi et al. (2017) prepared a core-shell structure temperature-responsive glyphosate controlled release agent with a mixture of attapulgite (ATP)-bicarbonate-glyphosate as the core and aminosilicone oil-polyvinyl alcohol as the shell. Since attapulgite had a porous micro-nano network structure, it could carry a large amount of glyphosate molecules, and ammonium bicarbonate acted as a foaming agent to generate carbon dioxide and ammonia gas bubbles. Thereby, a large number of micro-nanopores were formed on the amino silicone oil-polyvinyl alcohol shell, thus promoting the release of glyphosate. Moreover, the temperature change could effectively adjust the number of pores, and at the same time, the polyvinyl alcohol shell was easily soluble in an aqueous solution under high temperature conditions, so the release of glyphosate was easy to control. The controlled agent release could effectively reduce the loss of glyphosate under simulated rainfall conditions and improved the control effect on weeds. This study provides a promising method for controlling pesticide release and reducing the loss of active ingredients, and has potential application in improving pesticide utilization and reducing environmental pollution.

5.2.8 pH-Responsive Pesticide-Controlled Release Agent

Rudzmski et al. (2003) prepared acrylic acid-based copolymerized hydrogels by bulk and solution polymerization using benzoyl peroxide as initiator and ethylene glycol dimethacrylate as crosslinker. It was used to load cypermethrin to prepare a pH-responsive cypermethrin controlled release agent. It was characterized by IR spectroscopy and differential scanning calorimetry. The results of controlled release kinetics showed that the controlled release agent had good pH response. This study provides a new idea for the application of pH-responsive hydrogels in soil, which can be used to control the release of pesticides using alkaline conditions in the soil.

Poly(succinimide) (PSI) is easily hydrolyzed to hydrophilic polyaspartate (PSAP) under alkaline conditions. While plant phloem vascular bundle pH range from 8.0 to 8.5, and it is higher than that for surrounding plant tissues. In order to improve the transport conductivity and utilization of pesticide, an amphiphilic PSI-based star polymer nanocarrier was designed to support naphthylacetic acid (Chen et al. 2015). The products were characterized by TEM, dynamic light scattering and IR spectroscopy. The results show that the nano-controlled agent carrier can release the drug rapidly under the alkaline condition of the plant phloem. This study shows that such responsive biodegradable polymer nanocarriers can be further used as a controlled release delivery system for other pesticides and fertilizers.

Hill et al. (2015) prepared PSI-based pH-responsive nanoparticles by nanoprecipitation to load the hydrophobic model molecule nile red. The samples were characterized by nuclear magnetic resonance (NMR), TEM and dynamic light scattering. The results show that when the pH value was higher than 6, the hydrophobic succinimide unit in the PSI nanoparticles was hydrolyzed, releasing the mode molecule, and the release rate increased with the increase of pH value. Furthermore, the polymeric material had little toxic effect on plants. The prepared nanoparticles can be used for loading pesticides, fertilizers, etc., and can exhibit different release properties according to changes in soil acidity and alkalinity.

Lin et al. (2016) prepared aminated mesoporous silicon by copolycondensation method and was used to load chlorpyrifos. They then used polyacrylic acid (PAA) as a functional molecule to prepare a pH-responsive PAA/chlorpyrifos/aminated mesoporous silicon sustained-release system by electrostatic adsorption. The system was characterized by X-ray diffraction, nitrogen adsorption-desorption, TEM, SEM, TGA, zeta potential and IR spectroscopy. The sustained release kinetics study showed that when the $pH < 7$, the lower the pH value, the faster the release of chlorpyrifos. The release under slightly alkaline conditions was faster than that under neutral conditions, showing an obvious pH response.

Xiang et al. (2018) used nano-network structure of polydopamine-modified attapulgite to adsorb chlorpyrifos by hydrogen bonding and static electricity, and then cross-linked with calcium alginate to form porous hydrogel spheres and prepare a pH-responsive chlorpyrifos controlled release agent. Its polydopamine-modified attapulgite-calcium alginate hydrogel carrier material had high biosafety to *Escherichia coli* and showed good biocompatibility. The controlled release agent released chlorpyrifos under alkaline conditions and exhibited good pH response. It can also effectively protect the degradation of chlorpyrifos under ultraviolet light conditions. Its polydopamine-modified attapulgite-calcium alginate hydrogel carrier material had high biosafety to *E. coli* and showed good biocompatibility.

Xiang et al. (2017) constructed a pH-responsive controlled release pesticide (PRCRP) with magnetic collection performance in order to improve pesticide utilization. Glyphosate (Gly) and cypermethrin (Cyp) were separately loaded into the pores of the magnetic nanocarrier diatomaceous earth-Fe_3O_4. The obtained pesticide-diatomaceous earth-Fe_3O_4 was then coated with Cs. The Cs could be dissolved under acidic conditions to control the release of the pesticide. The magnetic properties of Fe_3O_4 imparted the magnetic collection performance of PRCRP. After the release of pesticides, PRCRP was separated from water and soil, and had a high recovery rate. PRCRP exhibited high adhesion on the surface of weeds and on the epidermis of pests, good controlled release properties, high utilization of pesticides, good control effects on weeds and pests, and potential collectability in practice. Therefore, this technology has broad application prospects, which can reduce pesticide residues and environmental risks (Fig. 5.4).

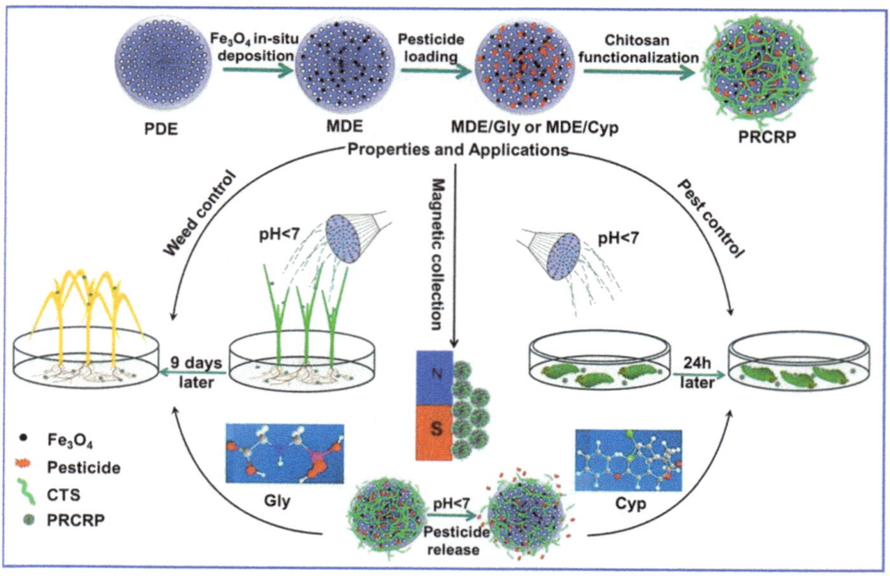

Fig. 5.4 Schematic diagram for the fabrication and application of PRCRP (Xiang et al. 2017). Abbreviation in figure: chitosan (CTS), pure diatomite (PDE) and magnetic PDE (MDE)

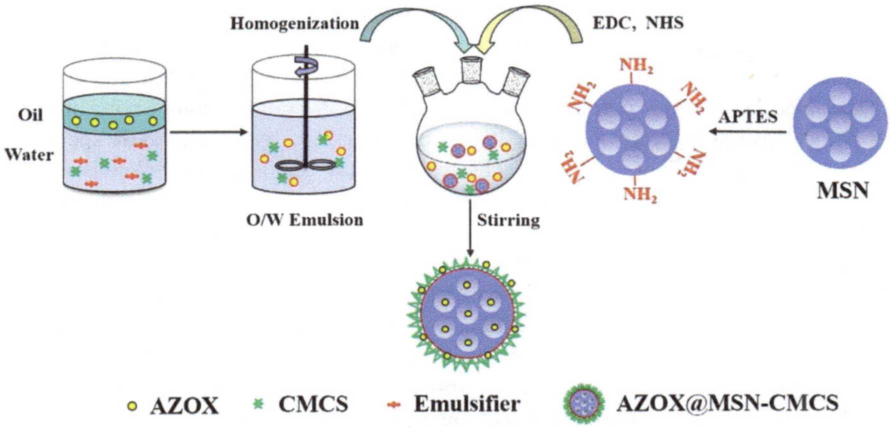

Fig. 5.5 Schematic illustration of emulsion-based synchronous AZOX encapsulation and MSN modification with CMCS (Xu et al. 2018a)

Amino-functionalized mesoporous silica nanoparticles (MSN) and carboxymethyl Cs (CMCS) were directly coupled and used as an oxytocin-loaded carrier (AZOX) with the loading capacity of 3.6% (Xu et al. 2018a). Simultaneous pesticide encapsulation based on CMCS emulsification and MSN surface modification can achieve 21% pesticide loading without affecting the pH intelligent response of CMCS control. At the dose of the same active ingredient, MSN-CMCS-loaded AZOX showed better fungicidal activity against the pathogenicity of tomato late blight compared to AZOX alone. In addition, fluorescein isothiocyanate-labeled MSN-CMCS was used to track the uptake of nanoparticles into target plants and fungi (Fig. 5.5).

5.2.9 Photoresponsive Pesticide-Controlled Release Agent

Atta et al. (2015) prepared a light-responsive 2,4-D nano-controlled release agent based on perylene-3-biphenylmethanol photoactive organic fluorescent nanoparticles. The results showed that the controlled release agent had good fluorescence, cell absorption and light stimulating response. Using the fluorescent properties of photoactive organic nanoparticles, the release and conduction of 2,4-D in plants can be monitored in real time. Bioassay experiments show that the nano-controlled release agent had significantly improved conductivity and herbicidal activity in plants. Subsequently, Atta et al. prepared a light-responsive 2,4-D controlled release agent based on the coumarin copolymer. The results showed that the conductivity, herbicidal activity and thermal stability of the controlled release agent were significantly improved. It also showed good fluorescence and photoresponse properties and reduced the leaching of 2,4-D.

CMCS to the labile 2-nitrobenzyl side group were bonded and self-assembled into polymer micelles under aqueous conditions (Ye et al. 2015). Dialdehyde was then added dropwise to form a crosslinked structure, and a light-responsive shell crosslinked micelle was prepared for loading diuron. A photoresponsive diuron controlled release agent was obtained. It was characterized by TEM, NMR spectroscopy and dynamic light scattering. The results show that the crosslinked micelles had a core-shell structure with an average diameter of 140 nm and were stable under the condition of no light and pH = 7 aqueous solution. The entrapment rate of the controlled release agent was 91.9%. Under the dark condition, diuron was not released, and under sunlight, the release rate of diuron was 96.8% after 8 h. This study provides a new idea for the application of photoresponsive polymer micelles in the controlled release of pesticides, i.e. the use of solar light-controlled photosynthetic inhibitor herbicides can greatly improve the utilization of pesticides.

Ding et al. (2016) bonded polyethylene glycol to a photolabile 0-nitrobenzyl group and then attached to 2,4-D to prepare a novel amphiphilic polymer-pesticide conjugate. It self-polymerized in an aqueous solution to form a photoresponsive micelle, thereby achieving controlled release of 2,4-D. The structure of the conjugate was confirmed by NMR spectroscopy and IR spectroscopy. TEM showed that the micelles were core-shell structures. Dynamic light scattering and TEM measurements showed that the average diameter of the micelles was reduced by 40 nm after 365 nm UV irradiation. It was not released by 2,4-D in the absence of light, and the cumulative release rate of 2,4-D was 99.6% after 8 h under sunlight.

Xu et al. (2018b) covalently cross-linked spirotetramat-enol with coumarin to prepare a photoresponsive spiropropanol-enol controlled release agent. They studied the photophysical and photochemical properties of the controlled release agent and its deprotection of photolysis and insecticidal activity. The controlled release agent was found to trigger the release of spiropropionol-enol at 420 nm blue or sunlight, and had good photo-stimulation response. The results of bioassay showed that the controlled release agent had no obvious insecticidal activity against bean aphid under dark conditions, and showed good insecticidal activity under light conditions.

The photo-responsive controlled-release herbicide particles (LCHP) with core-shell structure prepared were by Chen et al. (2018). It was a nanocomposite composed of biochar, ATP, glyphosate (Gly), azobenzene (AZO) and aminosilicone oil (ASO). Among them, ATP was uniformly distributed in the pores of the biochar to form a porous biochar-ATP compound, which was used as a carrier to efficiently carry a large amount of Gly and AZO and obtain porous biochar-ATP-Gly-AZO particles. Subsequently, the obtained biochar-ATP-Gly-AZO particles were coated with ASO to form LCHP rich in micropores in the ASO coating. Under UV-Vis radiation, the trans-cis and cis-trans isomers of the AZO molecule were converted, and AZO acted as a photoexcited "stirder" to promote the release of Gly from the nanopore of LCHP. Therefore, LCHP exhibited photo-responsive controlled release properties, as evidenced by pot experiments. It is important to note that, LCHP had good adhesion properties on the surface of weed leaves, which was beneficial to improve the control effect on weed control. In addition, coexisting ions

(CO_3^{2-}, SO_4^{2-}, and Cl^-) and pH had little effect on the release of LCHP in water, demonstrating the stability of the technique. Therefore, this work provides a method to control the release and improving the efficiency of herbicide utilization, which has potential application prospects.

5.2.10 Enzyme-Responsive Pesticide-Controlled Release Agent

The amino group of 3-aminopropyltriethoxysilane (APTES), which was a cross-linking agent, were bonded with the carboxyl group of kasugamycin to obtain APTES-modified compound (Ding et al. 2014). The nanospheres were then prepared by contacting the siloxy group of APTES with the silica formed by the hydrolysis of ethyl orthosilicate. The nanospheres can effectively prevent the photolysis of the kasugamycin, and triggered the release of the kasugamycin under the action of the amidase produced during the bacterial infection, thereby having good enzyme stimulating response performance. The results of biological activity assay showed that the prepared nano-controlled release agent had better and longer-lasting antibacterial activity against *E. coli* than the original pesticide.

By the catalysis of *N*-hydroxysuccinimide and 1-ethyl-(3-dimethylaminopropyl) carbodiimide, Liu et al. (2015) prepared a kasugamycin-pectin controlled release agent by directly bonding the amino group of the kasugamycin to the carboxyl group of the pectin. The controlled release agent was characterized by IR spectroscopy, ultraviolet spectroscopy and TGA. The results show that the prepared controlled release agent was stable under a wide range of pH, temperature and ultraviolet irradiation conditions. The controlled release agent could trigger the release of kasugamycin under the action of pectinase and amidase secreted by Pseudomonas, and thus had good enzyme stimulating response performance.

Guo et al. (2015) prepared emamectin benzoate silica microcapsules by emulsion polymerization using tetraethyl orthosilicate (TEOS) as the silicon source. Then the amino-functionalized silica microcapsules were modified by APTES, and finally the amino-functionalized silica microcapsules were cross-linked with the synthesized epichlorohydrin-modified carboxymethylcellulose (EMC). A novel enzyme-responsive emamectin benzoate microcapsule was prepared. The microcapsule had a high loading capacity and can effectively protect the degradation of emamectin benzoate under light and heat conditions. It triggered the release of the ingredient under the action of cellulase produced during the prickly sucking hazard, showing good enzyme stimulating response. Bioassay showed that the prepared microcapsules had better insecticidal activity against peach aphid at the same concentration than emulsifiable concentrate, and the genotoxicity to onion was lower than that of the technical material. The study proposed a novel, safe and sustainable crop protection strategy. The successful development of enzyme-responsive controlled release agents provides a method of application that can effectively control agricultural pests while reducing the risk to the ecological environment and human health.

Kaziem et al. (2017) loaded chlorantraniliprole by hollow mesoporous silica through emulsion method. Then the organic polymer cyclodextrin was cross-linked to encapsulate the surface mesopores and an enzyme-responsive chlorantraniliprole controlled release agent were prepared successfully. The controlled release agent was characterized by IR spectroscopy, SEM, TEM and TGA. The results show that it had a high loading capacity and can effectively protect the degradation of chlorantraniliprole under light and heat conditions. The controlled release agent triggered the action of the alpha-amylase in the saliva of the insects and the midgut to release the chlorantraniliprole, showing good enzyme stimulating response performance. The results of biological activity assay showed that the prepared controlled release agent had better insecticidal activity against *Plutella xylostella* than the suspension agent.

5.2.11 Redox-Responsive Pesticide-Controlled Release Agent

Amphiphilic self-assembled CMCS derivatives in aqueous solution formed a disulfide bond by sonication to prepare redox-responsive CMCS nanoparticles, which was used to load diuron (Yu et al. 2015). It was characterized by zeta potential and TEM. The results showed that the redox-responsive dexamethasone controlled release agent triggered the release of diuron by the action of reduced glutathione in plant tissues. It had good herbicidal activity against *Echinochloa crusgalli* (L.) Beauv. and was safe for crops.

Yi et al. (2015) grafted thiodecane to the outer surface of mesoporous silica nanoparticles to prepare functionalized redox-responsive mesoporous silica nanoparticles for loading salicylic acid, which were characterized by TEM, Raman spectroscopy and TGA. Controlled release kinetics studies showed that the release rate of salicylic acid was significantly higher in glutathione-containing environments than in glutathione-free environments. *In vivo* experiments showed that salicylic acid released from controlled release agents induced the sustained expression of plant resistance genes to 7 days after treatment under the action of glutathione in plant cells, while the control salicylic acid only caused the early peak of expression, and gradually decreased on the third day after treatment.

Guo (2016) cross-linked silica microspheres with pectin by disulfide bond to prepare a redox-responsive kasugamycin controlled release agent. The controlled release agent showed a high loading capacity (20%). It can effectively improve the stability of the kasugamycin to light and heat, and can trigger the release of the kasugamycin under the action of glutathione produced during the infection of soft rot of cabbage, showing good redox response performance. The controlled agent release had better and longer lasting bactericidal activity at the same concentration than the wettable powder. The onion chromosome aberration test showed that the controlled agent release had lower genotoxicity to onion than the technical material.

Sun et al. (2018) synthesized monodisperse mesoporous silica nanoparticles with a particle size of 20 nm and a pore diameter of 2.87 nm. The nanoparticles were then modified to coat the abscisic acid (ABA) in the pores of silica. A redox-responsive ABA-controlled release agent was prepared by encapsulating the surface mesopores by thiodecane. *In vitro* release kinetic tests have shown that the controlled agent release can effectively load and encapsulate ABA in the absence of glutathione triggering. *In vivo* experiments showed that under the triggering action of glutathione, the ABA released from the controlled release agent can significantly reduce the opening of the leaf pores, thereby inhibiting the water loss of the plants. In addition, the ABA released significantly prolonged the expression time of the marker gene *AtGALK2* compared with the control, and finally improved the resistance of *Arabidopsis* seedlings under drought stress.

5.3 Prospects

The stimuli-responsive carrier material preparation process and assembly technology are increasingly mature, and have good intelligent responsiveness and biocompatibility. They are very suitable for drug-loading and intelligent controlled-release fields. Because of its broad application prospects and potential application value in many fields such as medicine, food and environmental engineering, it has become a new hot spot of research. The application of stimuli-responsive carrier materials in the pharmaceutical field for drug delivery systems has been extensively studied, but its application in pesticide delivery systems is still limited.

In recent years, the research on new pesticide-controlled release agents based on stimulating responsive materials has become more and more popular, the development of such pesticide-controlled release agents has become an important development direction of new pesticide formulations. However, there is still a need for systematic research continuously on stimuli-responsive pesticide controlled release agents, especially in studies that can effectively respond to internal biostimulation systems. It is necessary to truly link the pesticide controlled release agent with a series of internal biostimulation changes generated during the harmful biological hazard process in order to design a precise and intelligent pesticide delivery system to intelligently respond to changes in the stimulus. Pesticides should only be released after pest activation to ensure optimal control and reduce the negative impact of pesticides on agricultural products and the environment. In addition, the current research on stimuli-responsive pesticide-controlled release agents is still in the basic stage, and there is still a certain distance from commercialization. It is still subject to many factors, such as the cost of carrier materials, pesticide stability, biological activity, and loading properties. Therefore, in order to realize the commercialization and large-scale application of stimuli-responsive pesticide controlled release agents, it is necessary to further clarify the relationship of the physical and chemical properties of controlled release agents, environmental factors and biological effects. Only by solving the above problems can we better play the role of stimulating carrier materials in the field of controlled release of pesticides.

Acknowledgements The authors would like to thank the Major Science and Technology Program for Water Pollution Control and Treatment (No.2017ZX07101-003) and the National Key Research and Development Program of China (2016YFD0200500, 2018YFD0200210).

Conflicts of Interest The authors declare no conflict of interest.

References

Abu-Lail, N. I., Kaholek, M., LaMattina, B., Clark, R. L., & Zauscher, S. (2006). Micro-cantilevers with end-grafted stimulus-responsive polymer brushes for actuation and sensing. *Sensors & Actuators B-Chemical, 114*(1), 371–378. https://doi.org/10.1016/j.snb.2005.06.003.

Agusti, N., & Cohen, A. C. (2000). *Lygus hesperus* and *L-lineolaris* (Hemiptera: Miridae), phytophages, zoophages, or omnivores: Evidence of feeding adaptations suggested by the salivary and midgut digestive enzymes. *Journal of Entomological Science, 35*(2), 176–186. https://doi.org/10.18474/0749-8004-35.2.176.

And, M. J. S., & Lyon, L. A. (2004). Optical and acoustic studies of pH-dependent swelling in microgel thin films. *Chemistry of Materials, 16*(22), 4373–4380. https://doi.org/10.1021/cm048844v.

Andresen, T. L., Thompson, D. H., & Kaasgaard, T. (2010). Enzyme-triggered nanomedicine: Drug release strategies in cancer therapy (invited review). *Molecular Membrane Biology, 27*(7), 353–363. https://doi.org/10.3109/09687688.2010.515950.

Atta, S., Bera, M., Chattopadhyay, T., Paul, A., Ikbal, M., Maiti, M. K., & Pradeep Singh, N. D. (2015). Nano-pesticide formulation based on fluorescent organic photoresponsive nanoparticles: For controlled release of 2,4-D and real time monitoring of morphological changes induced by 2,4-D in plant systems. *RSC Advances, 5*(106), 86990–86996. https://doi.org/10.1039/c5ra17121k.

Ayano, E., Karaki, M., Ishihara, T., Kanazawa, H., & Okano, T. (2012). Poly (*N*-isopropylacrylamide)-PLA and PLA blend nanoparticles for temperature-controllable drug release and intracellular uptake. *Colloids and Surfaces-Biointerfaces, 99*, 67–73. https://doi.org/10.1016/j.colsurfb.2011.10.003.

Bae, Y., Fukushima, S., Harada, A., & Kataoka, K. (2003). Design of environment-sensitive supramolecular assemblies for intracellular drug delivery: Polymeric micelles that are responsive to intracellular pH change. *Angewandte Chemie International Edition, 42*(38), 4640–4643. https://doi.org/10.1002/anie.200250653.

Bhattacharya, S., Eckert, F., Boyko, V., & Pich, A. (2010). Temperature-, pH-, and magnetic-field-sensitive hybrid microgels. *Small, 3*(4), 650–657. https://doi.org/10.1002/smll.200600590.

Boehm, A. L., Martinon, I., Zerrouk, R., Rump, E., & Fessi, H. (2003). Nanoprecipitation technique for the encapsulation of agrochemical active ingredients. *Journal of Microencapsulation, 20*(4), 433–441. https://doi.org/10.1080/0265204021000058410.

Caboni, P., Sammelson, R. E., & Casida, J. E. (2003). Phenylpyrazole insecticide photochemistry, metabolism, and gabaergic action: Ethiprole compared with fipronil. *Jounral of Agricultural and Food Chemistry, 51*(24), 7055–7061. https://doi.org/10.1021/jf030439l.

Cai, X. J., Dong, C. Y., Dong, H. Q., Wang, G. M., Pauletti, G. M., Pan, X. J., Wen, H., Mehl, I., Li, Y., & Shi, D. (2012). Effective gene delivery using stimulus-responsive catiomer designed with redox-sensitive disulfide and acid-labile imine linkers. *Biomacromolecules, 13*(4), 1024–1034. https://doi.org/10.1021/bm2017355.

Chen, M. S., Jensen, S. P., Hill, M. R., Moore, G., He, Z. L., & Sumerlin, B. S. (2015). Synthesis of amphiphilic polysuccinimide star copolymers for responsive delivery in plants. *Chemical Communication, 51*(47), 9694–9697. https://doi.org/10.1039/c5cc02726h.

Chen, C. W., Zhang, G. L., Dai, Z. Y., Xiang, Y. B., Liu, B., Bian, P., Zheng, K., Wu, Z., & Cai, D. (2018). Fabrication of light-responsively controlled-release herbicide using a nanocomposite. *Chemical Engineering Journal, 349*, 101–110. https://doi.org/10.1016/j.cej.2018.05.079.

Chi, Y., Zhang, G. L., Xiang, Y. B., Cai, D. Q., & Wu, Z. Y. (2017). Fabrication of a temperature-controlled-release herbicide using a nanocomposite. *ACS Sustainable Chemistry & Engineering, 5*(6), 4959–4975. https://doi.org/10.1021/acssuschemeng.7b00348.

Cho, J. K., Meng, Z. Y., Lyon, L. A., & Breedveld, V. (2009). Tunable attractive and repulsive interactions between pH-responsive microgels. *Soft Matter, 5*(19), 3599–3602. https://doi.org/10.1039/b912105f.

Debord, J. D., & Lyon, L. A. (2003). Synthesis and characterization of pH-responsive copolymer microgels with tunable volume phase transition temperatures. *Langmuir, 19*(18), 7662–7664. https://doi.org/10.1021/la0342924.

Dedinaite, A., Thormann, E., Olanya, G., Claesson, P., Nystrom, B., Kjøniksen, A. L., & Zhu, K. (2010). Friction in aqueous media tuned by temperature-responsive polymer layers. *Soft Matter, 6*(11), 2489–2498. https://doi.org/10.1039/c003320k.

Di, C. F., Jiang, X. S., & Yin, J. (2010). Synthesis of stimuli-responsive star-like copolymer H20-PNIPAm-r-PEGMA via the ATRP copolymerization technique and its micellization in aqueous solution. *Journal of Applied Polymer Science, 115*(3), 1831–1840. https://doi.org/10.1002/app.31309.

Ding, G., Li, D., Liu, Y., Guo, M., Duan, Y., Li, J., & Cao, Y. (2014). Preparation and characterization of Kasuga-silica-conjugated nanospheres for sustained antimicrobial activity. *Journal of Nano Research, 16*(11), 2671. https://doi.org/10.1007/s11051-014-2671-0.

Ding, K. K., Shi, L. Y., Ling, Z., Tian, Z., Yin, Y. H., & Ying, Y. (2016). Synthesis of photoresponsive polymeric propesticide micelles based on PEG for the controlled release of a herbicide. *Polymer Chemistry, 7*(4), 899–904. https://doi.org/10.1039/c5py01690h.

Durand, A., & Hourdet, D. (1999). Synthesis and thermoassociative properties in aqueous solution of graft copolymers containing poly(N-isopropylacrylamide) side chains. *Polymer, 40*(17), 4941–4951. https://doi.org/10.1016/S0032-3861(98)00698-3.

Eswaramma, S., & Rao, K. S. V. K. (2017). Synthesis of dual responsive carbohydrate polymer based IPN microbeads for controlled release of anti-HIV drug. *Carbohydrate Polymers, 156*, 125–134. https://doi.org/10.1016/j.carbpol.2016.09.023.

Food and Agriculture Organization. (2010). *International conference on food and agriculture applications of nanotechnologies: Report of technical round table sessions. June* (pp. 20–25). Brazil: São Pedro.

Frederiksen, H. K., Kristensen, H. G., & Pedersen, M. (2003). Solid lipid microparticle formulations of the pyrethroid gamma-cyhalothrin-incompatibility of the lipid and the pyrethroid and biological properties of the formulations. *Journal of Controlled Release, 86*(2–3), 243–252. https://doi.org/10.1016/S0168-3659(02)00406-6.

Fu, Z., Chen, K., Li, L., Zhao, F., Wang, Y., Wang, M., Shen, Y., Cui, H., Liu, D., & Guo, X. (2018). Spherical and spindle-like abamectin-loaded nanoparticles by flash nanoprecipitation for southern root-knot nematode control: Preparation and characterization. *Nanomaterials, 8*(6), 449. https://doi.org/10.3390/nano8060449.

Ghormade, V., Deshpande, M. V., & Paknikar, K. M. (2011). Perspectives for nano-biotechnology enabled protection and nutrition of plants. *Biotechnology Advances, 29*(6), 792–803. https://doi.org/10.1016/j.biotechadv.2011.06.007.

Gohy, J. F., & Zhao, Y. (2013). Photo-responsive block copolymer micelles: Design and behavior. *Chemical Society Reviews, 42*(17), 7117–7129. https://doi.org/10.1039/C3CS35469E.

Green, L. C., Meyers, A. A., Springer, J. T., & Banks, P. A. (1992). Biological evaluation of pesticides released from temperature-responsive microcapsules. *Journal of Agricultural and Food Chemistry, 40*(11), 2274–2278. https://doi.org/10.1021/jf00023a044.

Guan, W. X., Tang, L. M., Wang, Y., & Cui, H. X. (2018). Fabrication of an effective avermectin nanoemulsion using a cleavable succinic ester emulsifier. *Journal of Agricultural and Food Chemistry, 66*(29), 7568–7576. https://doi.org/10.1021/acs.jafc.8b01388.

Guo, M. C. (2016). *Preparation and biological efficacy evaluation of stimuli-responsive controlled release formulation of pesticide*. Beijing: China Agricultural University.

Guo, C., Chen, W., Lin, S., Li, H., Cheng, D., Wang, X., & Shuai, X. (2012). Synthesis and characterization of polycation block copolymer Poly(L-lysine)-*b*-poly[N-(N',N'-diisopropylaminoethyl)aspartamide] as potential pH responsive gene delivery system. *Polymer, 53*(2), 342–349. https://doi.org/10.1016/j.polymer.2011.12.008.

Guo, M., Zhang, W., Ding, G., Guo, D., Zhu, J., Wang, B., Punyapitak, D., & Cao, Y. (2015). Preparation and characterization of enzyme-responsive emamectin benzoate microcapsules based on a copolymer matrix of silica-epichlorohydrin-carboxymethylcellulose. *RSC Advances, 5*(113), 93170–93179. https://doi.org/10.1039/c5ra17901g.

Gutiérrez, T. J. (2018). Chapter 55. Processing nano- and microcapsules for industrial applications. In C. M. Hussain (Ed.), *Handbook of nanomaterials for industrial applications* (pp. 989–1011). Editorial Elsevier.. EE.UU. ISBN: 978–0–12-813351-4. https://doi.org/10.1016/B978-0-12-813351-4.00057-2.

Gutiérrez, T. J. (2019). Chapter 19. Antibiofilm enzymes as an emerging technology for food quality and safety. In M. Kuddus (Ed.), *Enzymes in food biotechnology: Production, applications, and future prospects* (pp. 321–342). Editorial Academic Press.. EE.UU. ISBN: 978-0-12813-280-7. https://doi.org/10.1016/B978-0-12-813280-7.00019-0.

Gutiérrez, T. J., & Álvarez, K. (2017). Chapter 6. Biopolymers as microencapsulation materials in the food industry. In M. Masuelli & D. Renard (Eds.), *Advances in physicochemical properties of biopolymers: Part 2* (pp. 296–322). Bentham Science Publishers.. EE.UU. ISBN: 978–1–68108-545-6. eISBN: 978–1–68108-544-9. https://doi.org/10.2174/9781681085449117010009.

Gutowska, A., Bae, Y. H., Jacobs, H., Feijen, J., & Kim, S. W. (1994). Thermosensitive interpenetrating polymer networks: Synthesis, characterization, and macromolecular release. *Macromolecules, 27*(15), 4167–4175. https://doi.org/10.1021/ma00093a018.

Heskins, M., & Guillet, J. E. (1968). Solution properties of poly(N-isopropylacrylamide). *Journal of Macromolecular Science: Part A-Chemistry, 2*(8), 1441–1455. https://doi.org/10.1080/10601326808051910.

Hill, M. R., Mackrell, E. J., Forsthoefel, C. P., Jensen, S. P., Chen, M., Moore, G. A., He, Z. L., & Sumerlin, B. S. (2015). Biodegradable and pH-responsive nanoparticles designed for site-specific delivery in agriculture. *Biomacromolecules, 16*(4), 1276–1282. https://doi.org/10.1021/acs.biomac.5b00069.

Hu, J. M., Zhang, G. Q., & Liu, S. E. (2012). Enzyme-responsive polymeric assemblies, nanoparticles and hydrogels. *Chemical Society Reviews, 41*(18), 5933–5949. https://doi.org/10.1039/c2cs35103j.

Huang, B., Chen, F., Shen, Y., Qian, K., Wang, Y., Sun, C., Zhao, X., Cui, B., Gao, F., Zeng, Z., & Cui, H. (2018). Advances in targeted pesticides with environmentally responsive controlled release by nanotechnology. *Nanomaterials, 8*(2), 102. https://doi.org/10.3390/nano8020102.

Hui, L., & Gao, C. Y. (2009). Preparation and properties of ionically cross-linked chitosan nanoparticles. *Polymers for Advanced Technologies, 20*(7), 613–619. https://doi.org/10.1002/pat.1306.

Il'ichev, Y. V., Schwörer, M. A., & Wirz, J. (2004). Photochemical reaction mechanisms of 2-nitrobenzyl compounds: Methyl ethers and caged ATP. *Journal of the American Chemical Society, 126*(14), 4581–4595. https://doi.org/10.1021/ja039071z.

Kaziem, A. E., Gao, Y. H., He, S., & Li, J. H. (2017). Synthesis and insecticidal activity of enzyme-triggered functionalized hollow mesoporous silica for controlled release. *Journal of Agricultural and Food Chemistry, 65*(36), 7854–7864. https://doi.org/10.1021/acs.jafc.7b02560.

Kharkar, P. M., Kiick, K. L., & Kloxin, A. M. (2013). Designing degradable hydrogels for orthogonal control of cell microenvironments. *Chemical Society Reviews, 42*(17), 7335–7372. https://doi.org/10.1039/C3CS60040h.

Khot, L. R., Sankaran, S., Maja, J. M., Ehsani, R., & Schuster, E. W. (2012). Applications of nanomaterials in agricultural production and crop protection: A review. *Crop Protection, 35*, 64–70. https://doi.org/10.1016/j.cropro.2012.01.007.

Kikuchi, A., Kawabuchi, M., Sugihara, M., Sakurai, Y., & Okano, T. (1997). Pulsed dextran release from calcium-alginate gel beads. *Journal of Controlled Release, 47*(1), 21–29. https://doi.org/10.1016/S0168-3659(96)01612-4.

Kwon, I. C., Bae, Y. H., & Kim, S. W. (1991). Electrically erodible polymer gel for controlled release of drugs. *Nature, 354*(6351), 291–293. https://doi.org/10.1038/354291a0.

Li, X., Burger, S., O'Connor, A. J., Ong, L., Karas, J. A., & Gras, S. L. (2016). An enzyme-responsive controlled release system based on a dual-functional peptide. *Chemical Communications, 52*(29), 5112–5115. https://doi.org/10.1039/c5cc10480g.

Liang, J., Yu, M., Guo, L., Cui, B., Zhao, X., Sun, C., Wang, Y., Liu, G., Cui, H., & Zeng, Z. (2018). Bioinspired development of P(St-MAA)-avermectin nanoparticles with high affinity for foliage to enhance folia retention. *Journal of Agricultural and Food Chemistry, 66*(26), 6578–6584. https://doi.org/10.1021/acs.jafc.7b01998.

Lin, Y. S., Zhou, H. J., Zhou, X. H., Gong, S., Xu, H., & Chen, H. Y. (2016). Preparation and properties of pH-responsive control release system of PAA/chlorpyrifos/amino functionalized mesoporous silica. *CIESC Journal, 67*(10), 4500–4507.

Liu, Y. Y., & Fan, X. D. (2005). Synthesis, properties and controlled release behaviors of hydrogel networks using cyclodextrin as pendant groups. *Biomaterials, 26*(32), 6367–6374. https://doi.org/10.1016/j.biomaterials.2005.04.011.

Liu, M., Liu, H., Liu, Y., Bai, L., Yang, G., Yang, C., & Cheng, J. (2011). Preparation and characterization of temperature-responsive poly(*N*-isopropylacrylamide-co-*N,N'*- methylenebisacrylamide) monolith for HPLC. *Journal of Chromatography A, 1218*(2), 286–292. https://doi.org/10.1016/j.chroma.2010.11.037.

Liu, Y., Sun, Y., Ding, G., Geng, Q., Zhu, J., Guo, M., Duan, Y., Wang, B., & Cao, Y. (2015). Synthesis, characterization, and application of microbe-triggered controlled-release kasugamycin- pectin conjugate. *Journal of Agriculture and Food Chemistry, 63*(17), 4263–4268. https://doi.org/10.1021/jf5055062.

Liu, B. X., Wang, Y., Yang, F., Cui, H. X., & Wu, D. C. (2018). Development of a chlorantraniliprole microcapsule formulation with a high loading content and controlled-release property. *Journal of Agricultural and Food Chemistry, 66*(26), 6561–6568. https://doi.org/10.1021/acs.jafc.7b01295.

Long, Q. D., Pan, C. Y., Meng, Y. H., Zhang, B. J., & Xu, C. X. (2009). Synthesis and property of temperature and pH sensitive xanthan-MA/PNIPAAm hydrogels. *Journal of Central South University of Technology, 16*(1), 66–72. https://doi.org/10.1007/s11771-009-0011-9.

Maisuria, V. B., Patel, V. A., & Nerurkar, A. S. (2010). Biochemical and thermal stabilization parameters of polygalacturonase from erwinia carotovora subsp. carotovora BR1. *Jounral of Microbiology and Biotechnology., 20*(7), 1077–1085. https://doi.org/10.4014/jmb.0908.08008.

Manatunga, D. C., de Silva, R. M., Nalin de Silva, K. M., de Silva, N., Bhandari, S., Yap, Y. K., & Pabakara Costha, N. (2017). pH responsive controlled release of anti-cancer hydrophobic drugs from sodium alginate and hydroxyapatite bi-coated iron oxide nanoparticles. *European Journal of Pharmaceutics and Biopharmaceutics, 117*, 29–38. https://doi.org/10.1016/j.ejpb.2017.03.014.

Manchanda, R., & Nimesh, S. (2010). Controlled size chitosan nanoparticles as an efficient, biocompatible oligonucleotides delivery system. *Journal of Applied Polymer Science, 118*(4), 2071–2077. https://doi.org/10.1002/app.32508.

Mok, Y., Noh, M., Kim, G. C., Song, Y., Kim, H., Kim, S., Yi, S., Seo, J.-H., & Lee, Y. (2016). Light-tunable thermoresponsive behavior of branched polyethylenimine derivatives in water. *Polymer, 107*, 37–43. https://doi.org/10.1016/j.polymer.2016.11.014.

Morris, J., Willis, J., De Martinis, D., Hansen, B., Laursen, H., Sintes, J. R., Kearns, P., & Gonzalez, M. (2011). Science policy considerations for responsible nanotechnology decisions. *Nature Nanotechnology, 6*(2), 73–77. https://doi.org/10.1038/nnano.2010.191.

Nash, M. A., Waitumbi, J. N., Hoffman, A. S., Yager, P., & Stayton, P. S. (2012). Multiplexed enrichment and detection of malarial biomarkers using a stimuli-responsive iron oxide and gold nanoparticle reagent system. *ACS Nano, 6*(8), 6776–6785. https://doi.org/10.1021/nn3015008.

Nguyen, M. M., Carlini, A. S., Chien, M. P., Sonnenberg, S., Luo, C. L., Braden, R. L., Osborn, K. G., Li, Y., Gianneschi, N. C., & Christman, K. L. (2015). Enzyme-responsive nanoparticles for targeted accumulation and prolonged retention in heart tissue after myocardial infarction. *Advanced Materials, 27*(37), 5547–5552. https://doi.org/10.1002/adma.201502003.

Norman, S., Chen, H. D., & Rutzke, C. J. (2002). Nanoscale science and engineering for agriculture and food systems: A report submitted to cooperative state research, education and extension service. *The united states department of agriculture: National planning workshop*, November 18–19, 2002, Washington, DC, USDA, 2003.

Observatory NANO. (2010). Nanotechnologies for nutrient & biocide application in agricultural production, working paper version.

Palankar, R., Skirtach, A. G., Kreft, O., Bédard, M., Garstka, M., Gould, K., Möhwald, H., Sukhorukov, G. B., Winterhalter, M., & Springer, S. (2009). Controlled intracellular release of peptides from microcapsules enhances antigen presentation on MHC class I molecules. *Small, 5*(19), 2168–2176. https://doi.org/10.1002/smll.200900809.

Pelliccioli, A. P., & Wirz, J. (2002). Photoremovable protecting groups: Reaction mechanisms and applications. *Photochemical & Photobiological Sciences, 1*(7), 441–458. https://doi.org/10.1039/b200777k.

Pelton, R. (2000). Temperature-sensitive aqueous microgels. *Advances in Colloid & Interface Science, 85*(1), 1–33. https://doi.org/10.1016/S0001-8686(99)00023-8.

Qin, S. H., Geng, Y., Discher, D. E., & Yang, S. (2010). Temperature-controlled assembly and release from polymer vesicles of poly(ethylene oxide)-*block*-poly(*N*-isopropylacrylamide). *Advanced Materials, 18*(21), 2905–2909. https://doi.org/10.1002/adma.200601019.

Roberto, D. L. R., Aili, D., & Stevens, M. M. (2012). Enzyme-responsive nanoparticles for drug release and diagnostics. *Advanced Drug Delivery Reviews, 64*(11), 967–978. https://doi.org/10.1016/j.addr.2012.01.002.

Robinson, D. K. R., & Morrison. M. (2009). Nanotechnology developments for the agri-food sector. Report of the Observatory NANO.

Rudzinski, W. E., Chipuk, T., Dave, A. M., Kumbar, S. G., & Aminabhavi, T. M. (2003). pH-sensitive acrylic-based copolymeric hydrogels for the controlled release of a pesticide and a micronutrient. *Journal of Applied Polymer Science, 87*(3), 394–403. https://doi.org/10.1002/app.11382.

Schild, H. G. (1992). Poly(*N*-isopropylacrylamide): Experiment, theory and application. *Progress in Polymer Science, 17*, 163–249. https://doi.org/10.1016/0079-6700(92)90023-R.

Schmaljohann, D. (2006). Thermo- and pH-responsive polymers in drug delivery. *Advanced Drug Delivery Reviews, 58*(15), 1655–1670. https://doi.org/10.1016/j.addr.2006.09.020.

Scott, M., & Chen, H. D. (2012). Nanoscale science and engineering for agriculture and food systems. *Industrial Biotechnology, 8*(6), 340–343. https://doi.org/10.1089/ind.2012.1549.

Scott, N. R., Chen, H. D., & Cui, H. X. (2018). Nanotechnology applications and implications of agrochemicals toward sustainable agriculture and food systems. *Journal of Agricultural and Food Chemistry, 66*(26), 6451–6456. https://doi.org/10.1021/acs.jafc.8b00964.

Shao, L., Hua, B., Sun, J., Li, Q., Yang, J., & Yu, G. (2017). A cucurbit[7]uril-based supra-amphiphile: Photo-responsive self-assembly and application in controlled release. *Tetrahedron Letters, 58*(19), 1863–1867. https://doi.org/10.1016/j.tetlet.2017.03.091.

Shen, Y., Wang, Y., Zhao, X., Sun, C., Cui, B., Gao, F., Zeng, Z., & Cui, H. (2017). Preparation and physicochemical characteristics of thermo-responsive emamectin benzoatemicrocapsules. *Polymers, 9*(9), 418. https://doi.org/10.3390/polym9090418.

Sheng, W. B., Ma, S. H., Li, W., Liu, Z. Q., Guo, X. H., & Jia, X. (2015). A facile route to fabricate a biodegradable hydrogel for controlled pesticide release. *RSC Advances, 5*(18), 13867–13870. https://doi.org/10.1039/c4ra15139a.

Shim, M. S., & Kwon, Y. J. (2011). Dual mode polyspermine with tunable degradability for plasmid DNA and sirna delivery. *Biomaterials, 32*(16), 4009–4020. https://doi.org/10.1016/j.biomaterials.2011.02.010.

Sidorenko, A., Krupenkin, T., Taylor, A., Fratzl, P., & Aizenberg, J. (2007). Reversible switching of hydrogel-actuated nanostructures into complex micropatterns. *Science, 315*(5811), 487–490. https://doi.org/10.1126/science.1135516.

Sijbesma, R. P., Beijer, F. H., Brunsveld, L., Folmer, B. J. B., Ky Hirschberg, J. H. K., Lange, R. F. M., Lowe, J. K. L., & Meijer, E. W. (1997). Reversible polymers formed from self-complementary monomers using quadruple hydrogen bonding. *Science, 278*(5343), 1601–1604. https://doi.org/10.1126/science.278.5343.1601.

Sun, D. Q., Hussain, H. I., Yi, Z. F., Rookes, J. E., Kong, L. X., & Cahill, D. M. (2018). Delivery of abscisic acid to plants using glutathione responsive mesoporous silica nanoparticles. *Journal of Nanoscience and Nanotechnology, 18*(3), 1615–1625. https://doi.org/10.1166/jnn.2018.14262.

Tanaka, T., Fillmore, D., Sun, S. T., Nishio, I., Swislow, G., & Shah, A. (1980). Phase transitions in ionic gels. *Physical Review Letters, 45*, 1636. https://doi.org/10.1103/physrevlett.45.1636.

Tang, R. P., Ji, W. H., & Wang, C. (2011). Synthesis and characterization of new poly(ortho ester amidine) copolymers for nonviral gene delivery. *Polymer, 52*(4), 921–932. https://doi.org/10.1016/j.polymer.2010.12.057.

Teixeira, L. S. M., Feijen, J., Van Blitterswijk, C. A., Dijkstra, P. J., & Karperien, M. (2012). Enzyme-catalyzed crosslinkable hydrogels: Emerging strategies for tissue engineering. *Biomaterials, 33*(5), 1281–1290. https://doi.org/10.1016/j.biomaterials.2011.10.067.

Thies, C., Louis, S. Pest controlling: US 7192603[P]. 2007–03–20.

Thomas, C. S., Xu, L. Z., & Olsen, B. D. (2012). Kinetically controlled nanostructure formation in self-assembled globular protein-polymer diblock copolymers. *Biomacromolecules, 13*(9), 2781–2792. https://doi.org/10.1021/bm300763x.

Tomatsu, I., Peng, K., & Kros, A. (2011). Photoresponsive hydrogels for biomedical applications. *Advanced Drug Delivery Reviews, 62*(14), 1257–1266. https://doi.org/10.1016/j.addr.2011.06.009.

Wang, C., Chen, Q. S., Wang, Z. Q., & Zhang, X. (2010). An enzyme-responsive polymeric superamphiphile. *Angewandte Chemie-internationgal Edition, 49*(46), 8612–8615. https://doi.org/10.1002/anie.201004253.

Wang, D. L., Jin, Y., Zhu, X. Y., & Yan, D. Y. (2016). Synthesis and applications of stimuli-responsive hyperbranched polymers. *Progress in Polymer Science, 64*, 114–153. https://doi.org/10.1016/j.progpolymsci.2016.09.005.

Watanabe, H., & Tokuda, G. (2010). Cellulolytic systems in insects. *Annual Review of Entomology, 55*, 609–632. https://doi.org/10.1146/annurev-ento-112408-085319.

Xiang, Y. B., Zhang, G. L., Chi, Y., Cai, D. Q., & Wu, Z. Y. (2017). Fabrication of a controllable nanopesticide system with magnetic collectability. *Chemical Engineering Journal, 328*, 320–330. https://doi.org/10.1016/j.cej.2017.07.046.

Xiang, Y. B., Zhang, G. L., Chen, C. W., Liu, B., Cai, D. Q., & Wu, Z. Y. (2018). Fabrication a pH-responsively controlled-release pesticide using an attapulgite-based hydrogel. *ACS Sustainable Chemistry & Engineering, Acssuschemeng, 6*(1), 1192–1201. https://doi.org/10.1021/acssuschemeng.7b03469.

Xu, C., Wu, T., Drain, C. M., Batteas, J. D., Fasolka, M. J., & Beers, K. L. (2006). Effect of block length on solvent response of block copolymer brushes: Combinatorial study with block copolymer brush gradients. *Macromolecules, 39*(9), 3359–3364. https://doi.org/10.1021/ma051405c.

Xu, J., Ye, J., & Liu, S. Y. (2007). Synthesis of well-defined cyclic poly(*N*-isopropylacrylamide) via click chemistry and its unique thermal phase transition behavior. *Macromolecules, 40*(25), 9103–9110. https://doi.org/10.1021/ma0717183.

Xu, X. H., Bai, B., Wang, H. L., & Suo, Y. R. (2017). A near-infrared and temperature-responsive pesticide release platform through core-shell polydopamine@PNIPAm nanocomposites. *ACS Applied Materials & Interfaces, 9*(7), 6424–6432. https://doi.org/10.1021/acsami.6b15393.

Xu, C., Cao, L., Zhao, P., Zhou, Z., Cao, C., Li, F., & Huang, Q. (2018a). Emulsion-based synchronous pesticide encapsulation and surface modification of mesoporous silica nanoparticles with carboxymethyl chitosan for controlled azoxystrobin release. *Chemical Engineering Journal, 348*, 244–254. https://doi.org/10.1016/j.cej.2018.05.008.

Xu, Z. P., Gao, Z. H., & Shao, X. S. (2018b). Light-triggered release of insecticidally active spirotetramat-enol. *Chinese Chemical Letters, 29*(11), 1648–1650. https://doi.org/10.1016/j.cclet.2018.01.025.

Yan, J. J., Ji, W. X., Chen, E. Q., Li, Z. C., & Liang, D. H. (2008). Association and aggregation behavior of poly(ethylene oxide)-*b*-poly(*N*-isopropylacrylamide) in aqueous solution. *Macromolecules, 41*(13), 4908–4913. https://doi.org/10.1021/ma7026726.

Ye, Z., Guo, J., Wu, D., Tan, M., Xiong, X., Yin, Y., & He, G. (2015). Photo-responsive shell cross-linked micelles based on carboxymethyl chitosan and their application in controlled release of pesticide. *Carbohydrate Polymers, 132*, 520–528. https://doi.org/10.1016/j.carbpol.2015.06.077.

Yi, Z., Hussain, H. I., Feng, C., Sun, D., She, F., Rookes, J. E., Cahill, D. M., & Kong, L. (2015). Functionalized mesoporous silica nanoparticles with redox-responsive short-chain gatekeepers for agrochemical delivery. *ACS Applied Materials & Interfaces, 7*(18), 9937–9946. https://doi.org/10.1021/acsami.5b02131.

Yu, Z. Y., Sun, X., Song, H. X., Wang, W. Q., Zhao, Y., Shi, L. Y., Cahill, D. M., & Kong, L. (2015). Glutathione-responsive carboxymethyl chitosan nanoparticles for controlled release of herbicides. *Materials Sciences and Applications, 6*(6), 591–604. https://doi.org/10.4236/msa.2015.66062.

Zha, Z. B., Zhang, S. H., Deng, Z. J., Li, Y. Y., Li, C. H., & Dai, Z. F. (2013). Enzyme-responsive copper sulphide nanoparticles for combined photoacoustic imaging, tumor- selective chemotherapy and photothermal therapy. *Chemical Communications, 49*(33), 3455–3457. https://doi.org/10.1039/c3cc40608c.

Zhang, Y. B. (1998). Development and recent trend of pesticide formulation technology. *Translated Collection of Pesticide, 20*(3), 49–55.

Zhang, C., Pan, D., Li, J., Hu, J., Bains, A., Guys, N., Zhu, H., Li, X., Luo, K., Gong, Q., & Gu, Z. (2017). Enzyme-responsive peptide dendrimer-gemcitabine conjugate as a controlled-release drug delivery vehicle with enhanced antitumor efficacy. *Acta Biomaterialia, 55*, 153–162. https://doi.org/10.1016/j.actbio.2017.02.047.

Zhao, X., Meng, Z., Wang, Y., Chen, W., Sun, C., Cui, B., Yu, M., Zeng, Z., Guo, S., Luo, D., Cheng, J. Q., Zhang, R., & Cui, H. (2017a). Pollen magnetofection for genetic modification with magnetic nanoparticles as gene carriers. *Nature Plants, 3*(12), 956–964. https://doi.org/10.1038/s41477-017-0063-z.

Zhao, S., Xu, M. M., Cao, C. W., Yu, Q. Q., Zhou, Y. H., & Liu, J. (2017b). A redox-responsive strategy using mesoporous silica nanoparticles for co-delivery of siRNA and doxorubicin. *Journal of Materials Chemistry B, 5*(33), 6908–6919. https://doi.org/10.1039/c7tb00613f.

Zhao, X., Cui, H. X., Wang, Y., Sun, C. J., Cui, B., & Zeng, Z. H. (2018). Development strategies and prospects of nano-based smart pesticide formulation. *Journal of Agricultural and Food Chemistry, 66*(26), 6504–6512. https://doi.org/10.1021/acs.jafc.7b02004.

Zhou, Y. J., Jie, K. C., & Huang, F. H. (2017). A redox-responsive selenium-containing pillar[5]arene-based macrocyclic amphiphile: Synthesis, controllable self-assembly in water, and application in controlled release. *Chemical Communications, 53*(59), 8364–8367. https://doi.org/10.1039/c7cc04779g.

Zou, S., Ma, Y. J., Hempenius, M. A., Schönherr, H., & Vancso, G. J. (2004). Grafting of single, stimuli-responsive poly(ferrocenylsilane) polymer chains to gold surfaces. *Langmuir, 20*(15), 6278–6287. https://doi.org/10.1021/la036341i.

Chapter 6
Starch Nanoparticles and Nanocrystals as Bioactive Molecule Carriers

Cristian C. Villa, Leidy T. Sanchez, and N. David Rodriguez-Marin

Abstract Starch nanoparticles and nanocrystals are small materials with high surface-to-volume ratio, which allows them to pass through biological barriers and encapsulate different types of bioactive molecules. Due to their biocompability, wide array of natural sources and its ease modification through physical, chemical and enzymatic methods, starch nanomaterials have generated a great interest in the food, agricultural, cosmetic and pharmaceutical industries. In recent year, research into nanoencapsulation of bioactive molecules in starch nanomaterials and their application in different agri-food fields has grown substantially. The objective of this chapter was to review and analyze the different advances in this research area.

Keyword Nanoencapsulation

6.1 Introduction

Starch is one of the most abundant polymers in nature, as it is used by many plants as a reserve of energy (Gutiérrez et al. 2017). This polymer is found in many roots of plants, seeds, tubers and cereals (Hoyos-Leyva et al. 2018; Kaur et al. 2018). The native starch granules are composed of two macromolecules based on glucose monomers: the linear amylose composed by glucose units and α-1-4 glycoside bonds while in the branched amylopectin are present α-1-6 glycoside bonds (Suárez and Gutiérrez 2017). Normally, amylose content is between 20–30%, while the amylopectin amount represents the remaining 70–80% (Kim et al. 2015; Wang et al. 2015; Wang et al. 2011). The reactivity of starch is one of the most interesting aspects, since it can be modified through several physical, chemical and enzymatic

C. C. Villa (✉)
Chemistry Department, Faculty of Basic Sciences and Technology, Quindío University, Armenia, Quindío, Colombia
e-mail: ccvilla@uniquindio.edu.co

L. T. Sanchez · N. D. Rodriguez-Marin
Food Engineering Department, Faculty of Agroindustrial Sciences, Quindío University, Armenia, Quindío, Colombia

T. J. Gutiérrez (ed.), *Polymers for Agri-Food Applications*,
https://doi.org/10.1007/978-3-030-19416-1_6

methods, which allows a wide range of applications in the food, agricultural, pharmaceutical, cosmetic and other industries (Ashogbon and Akintayo 2013). In recent years, the development of starch-based nanosystems has been applied from polymer reinforcements to encapsulation for controlled release of bioactive substances, which has led to the acquisition of many research articles and patents in this area (Hui et al. 2018).

6.2 Starch Based Nanosystems

In general, starch nanosystems can be classified according to their synthesis method as: starch nanocrystals (SNCs) and nanoparticles (SNPs). SNCs are commonly obtained through acid or enzymatic hydrolysis of the starch granules, removing layers (mostly amylose) from the starch granules until a nanosized particles is achieved. Since most amorphous regions have been hydrolysed from starch structure, the SNCs have a more organized and crystalline structure (> 40%) than common starch granules (~25%) (LeCorre et al. 2012b). Commonly, SNC are obtained by using either HCl or H_2SO_4 as hydrolyzing agents for long periods of time (usually beyond 5 days) and followed by a neutralization whit NaOH (Dai et al. 2018; LeCorre et al. 2011).

Despite acid hydrolysis is a simple way to obtain SNC, the high amount of hydrolyzing agent and the long periods of time necessary for a complete process have led to the development of new SNC synthesis methods. Several authors have reported that acid hydrolysis time can be considerably reduced when an enzymatic pretreatment is carried out. Hao et al. (2018) reported a two-step synthesis for the waxy potato SNC, using a 6-h glucoamylase pretreatment followed by conventional H_2SO_4 acid hydrolysis. The SNC subjected to enzymatic pretreatment reached its smallest particles size after 5 days of hydrolysis, while a similar particle size in SNCs was only achieved after 7 days without pretreatment. LeCorre et al. (2012b) observed a similar behavior in the synthesis of waxy maize SNC using the previous treatment of α-amylase combined with H_2SO_4 acid hydrolysis. More recently, Dai et al. (2018) reported that ball milling can be used as a pretreatment for SNC synthesis, which reduced hydrolysis in 3 days, thus obtaining a relative crystallinity and a size of particles similar to the conventional method by acid hydrolysis.

In contrast, SNPs are almost completely amorphous (crystallinity <10%), and are manufactured by controlled nanoprecipitation of gelatinized starch, using physical methods such as ultra-high-pressure treatments (Farrag et al. 2018; Le Corre et al. 2010). In general, SNP synthesis involves either hot or cold (addition of NaOH) gelatinization followed by a slow addition of an anti-solvent, mostly ethanol or acetone (Acevedo-Guevara et al. 2018; Kim et al. 2015; Ma et al. 2008; Qin et al. 2016). SNPs often have a larger particle size than SNCs, while their morphology can vary from spherical particles to more amorphous structures. However, the particle size, morphology and crystallinity of the SNCs can be controlled by simple modifications in the synthesis process such as temperature, reaction time, addition of anti-solvent and rates of mechanical agitation. It has also been reported that the botanical source does not play an important role in the physical properties (particle size, morphology and rheological properties) of SNC and SNP (LeCorre et al. 2011; LeCorre et al. 2012a).

6.3 Nanoencapsulation of Bioactive Molecules in SNPs and SNCs

As mentioned before, starch nanosystems have recently been used as nanocarriers of different bioactive molecules, especially those that can be used in food systems, such as vitamins and antioxidants (Qi and Tester 2019; Zhu 2017). Table 6.1 shows some of the most recent works involving native or modified nanosystems based on starch as nanocarriers of bioactive molecules. The most commonly used method for the synthesis of starch-based nanocarriers is the addition in a polar organic solvent (see Table 6.1), because the bioactive molecules can be dissolved in the organic solvent and then previously gelatinized starch can be added in a controlled manner. Due to the low water solubility of the bioactive molecules used, they can be easily entrapped in the hydrophobic interior of the SNPs. In addition, the load capacity can be increased by reducing the polarity of the starch through chemical reactions of cross-linking or esterification with other molecules (Acevedo-Guevara et al. 2018). In this sense, de Oliveira et al. (2017) reported that native and acetylated cassava SNPs can be loaded with two antioxidants that are commonly used in processed foods: gallic acid and butylated hidroxytoluene (BHT). These authors concluded

Table 6.1 Some recent examples of SNPs and SNCs used as bioactive molecule carriers

Synthesis method	Modification	Encapsulated bioactive molecule	References
Anti-solvent addition	STTP cross-linking	Diclofenac sodium	El-Naggar et al. (2015)
	$CaCO_3$ templates	Doxorubicin HCl	Yang et al. (2017)
	Ultrasound	Vitamin D_3	Hasanvand et al. (2018)
	Ultrasound	Vitamin D	Hasanvand et al. (2015)
	Ultrasound	Catechin	Ahmad et al. (2019)
	Acetylation	Gallic Acid BHT	de Oliveira et al. (2017)
	None	Curcumin	Chin et al. (2014)
	Acetylation	Curcumin	Acevedo-Guevara et al. (2018)
	Maleate	Curcumin	Pang et al. (2014)
	None	Quercetin	Farrag et al. (2018)
Acid hydrolysis	None	5-fluoroacyl	Ab'lah Norul et al. (2018)
	Ultrasound	Ascorbic Acid Oxalic Acid	Shabana et al. (2019)

that the acetylated starch allowed to load a greater amount of both antioxidant compounds compared to the native starch. This was given by the hydrophobic properties of the antioxidant molecule that allow a greater interaction between the molecule and the acetate chains incorporated in the starch molecules.

Numerous authors have reported on the production of curcumin (*Curcuma longa*)-loaded SNPs using native and modified SNPs derived from different botanical sources. Curcumin is a polyphenol present in the rhizomes of turmeric which is known for its anti-cancer, anti-oxidant, anti-inflammatory, anti-microbial and anti-viral activities (Mai et al. 2017; Menon and Sudheer 2007; Mirzaei et al. 2017; Nelson et al. 2017; Oliveira et al. 2015; Stanić 2017). However, use of curcumin is limited by its low water solubility, fast degradation and low bioavailability (Anand et al. 2007; Maghsoudi et al. 2017). One of the first research papers reporting on the loading of curcumin in SNP was carried out by Chin et al. (2014) using sago (*Metroxylon sagu*) SNP with a loading capacity of about 78% and particle sizes between 50 and 80 nm. Pang et al. (2014) also stablished that maleate ester modified SNP has a higher curcumin loading capacity (~85%) due to the increasing hydrogen bond interactions between the curcumin and ester molecules incorporated within the structure of the nanoparticles, nevertheless, chemical modification leads to larger particle sizes, around 120 nm. A similar behavior was also observed by Acevedo-Guevara et al. (2018) for native and acetylated green banana SNPs with curcumin loaded.

The encapsulation of vitamin D as a bioactive molecule within starch nanosystems, has caused great interest due to its important role in several metabolic processes. It is worth noting that vitamin D is susceptible to light and oxygen, for this reason its encapsulation is required (Ballard et al. 2007; Mahmoodani et al. 2017; Walia et al. 2017). High-amylose corn SNPs have been used as vitamin D carriers, showing values of encapsulation capacity of up to 78% and particle sizes of around 31 nm (Hasanvand et al. 2015). The results reported by Hasanvand et al. (2015) fit well with the trend observed by de Oliveira et al. (2017), i.e. the encapsulation of the bioactive compound within the SNPs was given by hydrogen bonding. This was followed by several techniques including differential scanning calorimetry and Fourier transform infrared spectroscopy. Hasanvand et al. (2018) also studied the effect of amylose content on the synthesis of starch nanocarries from high-amylose corn starch and low-amylose potato starch containing vitamin D (D-SNP). The results showed that the loading capacity was higher in low-amylose SNPs than in its counterpart higher-amylose SNPs. However, structural studies showed that this behavior was not directly related to the amylose content, but to the particle size, since the SNPs with low amylose content has a larger particle size and, therefore, can more vitamin D molecules be stored inside the SNPs.

One of the most interesting aspects of D-SNP based nanoencapsulation is its potential application in food fortification. Hasanvand et al. (2015) reported that these D-SNPs can be used to fortify milk without affecting its sensory properties.

Starch-based nanosystems have also proven to be an important vehicle for the encapsulation of antioxidant molecules, thus increasing their water solubility and chemical stability. Farrag et al. (2018) used three different starches (potato, pea and corn) with variations in their content of amylopectin to be used in the nanoencapsulation of quercetin (polyphenolic compound – flavonoid with antioxidant capacity). It was found that corn SNPs containing quercetin had the lowest percentage loading the bioactive substance, while potato SNPs showed a slightly higher encapsulation capacity than analogue SNPs derived from pea starch. Potato SNPs containing higher amylopectin content were capable of encapsulating a greater number of quercetin molecules compared to pea and corn SNPs. It was also reported that the encapsulation of quercetin using SNPs allowed to preserve the antioxidant activity of the bioactive compound, however, its antioxidant activity was related to the loading capacity of SNPs, i.e. a positive relationship between the carrying capacity of the antioxidant compound and the radical scavenging activity was observed.

Another interesting aspect of SNPs is that they can also achieve the controlled release of active substances under specific conditions, as well as their loading capacity can be modified (de Oliveira et al. 2017). Acevedo-Guevara et al. (2018) reported that native and acetylated banana SNPs can reduce the degradation of curcumin under acidic stomach conditions, thus increasing the release of the bioactive compound under intestinal conditions. Similarly, Ahmad et al. (2019) evaluated the release of encapsulated catechins within horse chestnut, water chestnut and lotus stem SNPs under simulated gastric conditions. They observed that starch-based nanoencapsulation is an effective tool in the conservation of the bioactive properties of catechin, such as the inhibition of lipase, cholesterol esterase and glucosidase under gastric conditions, thus decreasing its degradation due to low pH values of the stomach.

Finally, some examples of SNCs as nanocarriers of bioactive molecules can be found in the literature. Ab'lah Norul et al. (2018) reported that corn SNCs obtained by acid treatment can be used as a vehicle for 5-fluoracil, which is a synthetic molecule of great interest in the treatment of several cancers related to the gastric system (Jordan 2016). Once the encapsulation of 5-fluoracil has been given within the corn SNCs, the active compound has been shown to be especially resistant to the acid environment, thus increasing its stability under gastric conditions and allowing targeted drug release under intestinal conditions. More recently, Shabana et al. (2019) used potato SNCs obtained through ultrasound-assisted acid hydrolysis to encapsulate ascorbic and oxalic acid, noting that ultrasound treatment not only allows the development of smaller particles, but also can increase the antioxidant loading capacity. It is worth noting that although the encapsulation of agrochemicals from starch nanosystems has not yet been reported, it is an interesting area of research that should be investigated.

Acknowledgements The authors would like to thank the Vicerrectoria de Investigaciones, Facultad de Ciencias Basicas y Tecnologias, Programa de Química and Programa de Ingeniería de Alimentos from Universidad del Quindio, for their support.

Conflicts of Interest The authors declare no conflict of interest.

References

Ab'lah Norul, N., Venkata, N. K., & Wong, T. W. (2018). Development of resistant corn starch for use as an oral colon-specific nanoparticulate drug carrier. *Pure and Applied Chemistry, 90*(6), 1073–1084. https://doi.org/10.1515/pac-2017-0806.

Acevedo-Guevara, L., Nieto-Suaza, L., Sanchez, L. T., Pinzon, M. I., & Villa, C. C. (2018). Development of native and modified banana starch nanoparticles as vehicles for curcumin. *International Journal of Biological Macromolecules, 111*, 498–504. https://doi.org/10.1016/j.ijbiomac.2018.01.063.

Ahmad, M., Mudgil, P., Gani, A., Hamed, F., Masoodi, F. A., & Maqsood, S. (2019). Nano-encapsulation of catechin in starch nanoparticles: Characterization, release behavior and bioactivity retention during simulated *in-vitro* digestion. *Food Chemistry, 270*, 95–104. https://doi.org/10.1016/j.foodchem.2018.07.024.

Anand, P., Kunnumakkara, A. B., Newman, R. A., & Aggarwal, B. B. (2007). Bioavailability of curcumin: Problems and promises. *Molecular Pharmaceutics, 4*(6), 807–818. https://doi.org/10.1021/mp700113r.

Ashogbon, A. O., & Akintayo, E. T. (2013). Recent trend in the physical and chemical modification of starches from different botanical sources: A review. *Starch–Stärke, 66*(1–2), 41–57. https://doi.org/10.1002/star.201300106.

Ballard, J. M., Zhu, L., Nelson, E. D., & Seburg, R. A. (2007). Degradation of vitamin D3 in a stressed formulation: The identification of esters of vitamin D3 formed by a transesterification with triglycerides. *Journal of Pharmaceutical and Biomedical Analysis, 43*(1), 142–150. https://doi.org/10.1016/j.jpba.2006.06.036.

Chin, S. F., Mohd Yazid, S. N. A., & Pang, S. C. (2014). Preparation and characterization of starch nanoparticles for controlled release of curcumin. *International Journal of Polymer Science, 2014*(8), 1–8. https://doi.org/10.1155/2014/340121.

Dai, L., Li, C., Zhang, J., & Cheng, F. (2018). Preparation and characterization of starch nanocrystals combining ball milling with acid hydrolysis. *Carbohydrate Polymers, 180*, 122–127. https://doi.org/10.1016/j.carbpol.2017.10.015.

de Oliveira, N. R., Fornaciari, B., Mali, S., & Carvalho, G. M. (2017). Acetylated starch-based nanoparticles: Synthesis, characterization, and studies of interaction with antioxidants. *Starch–Stärke, 70*(3–4), 1700170. https://doi.org/10.1002/star.201700170.

El-Naggar, M. E., El-Rafie, M. H., El-sheikh, M. A., El-Feky, G. S., & Hebeish, A. (2015). Synthesis, characterization, release kinetics and toxicity profile of drug-loaded starch nanoparticles. *International Journal of Biological Macromolecules, 81*, 718–729. https://doi.org/10.1016/j.ijbiomac.2015.09.005.

Farrag, Y., Ide, W., Montero, B., Rico, M., Rodríguez-Llamazares, S., Barral, L., & Bouza, R. (2018). Preparation of starch nanoparticles loaded with quercetin using nanoprecipitation technique. *International Journal of Biological Macromolecules, 114*, 426–433. https://doi.org/10.1016/j.ijbiomac.2018.03.134.

Gutiérrez, T. J., Guarás, M. P., & Alvarez, V. A. (2017). Reactive extrusion for the production of starch-based biopackaging. In M. A. Masuelli (Ed.), *Biopackaging* (pp. 287–315). Miami., EE.UU. ISBN: 978-1-4987-4968-8: Editorial CRC Press Taylor & Francis Group.

Hao, Y., Chen, Y., Li, Q., & Gao, Q. (2018). Preparation of starch nanocrystals through enzymatic pretreatment from waxy potato starch. *Carbohydrate Polymers, 184*, 171–177. https://doi.org/10.1016/j.carbpol.2017.12.042.

Hasanvand, E., Fathi, M., & Bassiri, A. (2018). Production and characterization of vitamin D3 loaded starch nanoparticles: Effect of amylose to amylopectin ratio and sonication parameters. *Journal of Food Science and Technology, 55*(4), 1314–1324. https://doi.org/10.1007/s13197-018-3042-0.

Hasanvand, E., Fathi, M., Bassiri, A., Javanmard, M., & Abbaszadeh, R. (2015). Novel starch based nanocarrier for vitamin D fortification of milk: Production and characterization. *Food and Bioproducts Processing, 96*, 264–277. https://doi.org/10.1016/j.fbp.2015.09.007.

Hoyos-Leyva, J. D., Bello-Pérez, L. A., Alvarez-Ramirez, J., & Garcia, H. S. (2018). Microencapsulation using starch as wall material: A review. *Food Reviews International, 34*(2), 148–161. https://doi.org/10.1080/87559129.2016.1261298.
Hui, W., Tao, F., Haining, Z., Zhimin, X., Ran, Y., & Min, S. (2018). A review on patents of starch nanoparticles: Preparation, applications, and development. *Recent Patents on Food, Nutrition & Agriculture, 9*(1), 23–30. https://doi.org/10.2174/2212798410666180321101446.
Jordan, V. C. (2016). A retrospective: On clinical studies with 5-fluorouracil. *Cancer Research, 76*(4), 767–768. https://doi.org/10.1158/0008-5472.can-16-0150.
Kaur, J., Kaur, G., Sharma, S., & Jeet, K. (2018). Cereal starch nanoparticles—A prospective food additive: A review. *Critical Reviews in Food Science and Nutrition, 58*(7), 1097–1107. https://doi.org/10.1080/10408398.2016.1238339.
Kim, H.-Y., Park, S. S., & Lim, S.-T. (2015). Preparation, characterization and utilization of starch nanoparticles. *Colloids and Surfaces B: Biointerfaces, 126*, 607–620. https://doi.org/10.1016/j.colsurfb.2014.11.011.
Le Corre, D., Bras, J., & Dufresne, A. (2010). Starch nanoparticles: A Review. *Biomacromolecules, 11*(5), 1139–1153. https://doi.org/10.1021/bm901428y.
LeCorre, D., Bras, J., & Dufresne, A. (2011). Influence of botanic origin and amylose content on the morphology of starch nanocrystals. *Journal of Nanoparticle Research, 13*(12), 7193–7208. https://doi.org/10.1007/s11051-011-0634-2.
LeCorre, D., Vahanian, E., Dufresne, A., & Bras, J. (2012b). Enzymatic pretreatment for preparing starch nanocrystals. *Biomacromolecules, 13*(1), 132–137. https://doi.org/10.1021/bm201333k.
LeCorre, D. S., Bras, J., & Dufresne, A. (2012a). Influence of the botanic origin of starch nanocrystals on the morphological and mechanical properties of natural rubber nanocomposites. *Macromolecular Materials and Engineering, 297*(10), 969–978. https://doi.org/10.1002/mame.201100317.
Ma, X., Jian, R., Chang, P. R., & Yu, J. (2008). Fabrication and characterization of citric acid-modified starch nanoparticles/plasticized-starch composites. *Biomacromolecules, 9*(11), 3314–3320. https://doi.org/10.1021/bm800987c.
Maghsoudi, A., Yazdian, F., Shahmoradi, S., Ghaderi, L., Hemati, M., & Amoabediny, G. (2017). Curcumin-loaded polysaccharide nanoparticles: Optimization and anticariogenic activity against *Streptococcus mutans*. *Materials Science and Engineering: C, 75*, 1259–1267. https://doi.org/10.1016/j.msec.2017.03.032.
Mahmoodani, F., Perera, C. O., Fedrizzi, B., Abernethy, G., & Chen, H. (2017). Degradation studies of cholecalciferol (vitamin D3) using HPLC-DAD, UHPLC-MS/MS and chemical derivatization. *Food Chemistry, 219*, 373–381. https://doi.org/10.1016/j.foodchem.2016.09.146.
Mai, Z., Chen, J., He, T., Hu, Y., Dong, X., Zhang, H., Huang, W., Ko, F., & Zhou, W. (2017). Electrospray biodegradable microcapsules loaded with curcumin for drug delivery systems with high bioactivity. *RSC Advances, 7*(3), 1724–1734. https://doi.org/10.1039/c6ra25314h.
Menon, V. P., & Sudheer, A. R. (2007). Antioxidant and anti-inflammatory properties of curcumin. In B. B. Aggarwal, Y.-J. Surh, & S. Shishodia (Eds.), *The molecular targets and therapeutic uses of curcumin in health and disease* (pp. 105–125). Boston: Springer US. https://doi.org/10.1007/978-0-387-46401-5_3.
Mirzaei, H., Shakeri, A., Rashidi, B., Jalili, A., Banikazemi, Z., & Sahebkar, A. (2017). Phytosomal curcumin: A review of pharmacokinetic, experimental and clinical studies. *Biomedicine & Pharmacotherapy, 85*, 102–112. https://doi.org/10.1016/j.biopha.2016.11.098.
Nelson, K. M., Dahlin, J. L., Bisson, J., Graham, J., Pauli, G. F., & Walters, M. A. (2017). The essential medicinal chemistry of curcumin. *Journal of Medicinal Chemistry, 60*(5), 1620–1637. https://doi.org/10.1021/acs.jmedchem.6b00975.
Oliveira, A. S., Sousa, E., Vasconcelos, M. H., & Pinto, M. (2015). Curcumin: A natural lead for potential new drug candidates. *Current Medicinal Chemistry, 22*(36), 4196–4232.
Pang, S. C., Tay, S. H., & Chin, S. F. (2014). Facile synthesis of curcumin-loaded starch-maleate nanoparticles. *Journal of Nanomaterials, 2014*(7), 1–7. https://doi.org/10.1155/2014/824025.
Qi, X., & Tester, R. F. (2019). Starch granules as active guest molecules or microorganism delivery systems. *Food Chemistry, 271*, 182–186. https://doi.org/10.1016/j.foodchem.2018.07.177.

Qin, Y., Liu, C., Jiang, S., Xiong, L., & Sun, Q. (2016). Characterization of starch nanoparticles prepared by nanoprecipitation: Influence of amylose content and starch type. *Industrial Crops and Products, 87*, 182–190. https://doi.org/10.1016/j.indcrop.2016.04.038.
Shabana, S., Prasansha, R., Kalinina, I., Potoroko, I., Bagale, U., & Shirish, S. H. (2019). Ultrasound assisted acid hydrolyzed structure modification and loading of antioxidants on potato starch nanoparticles. *Ultrasonics Sonochemistry., 51*, 444–450. https://doi.org/10.1016/j.ultsonch.2018.07.023.
Stanić, Z. (2017). Curcumin, a compound from natural sources, a true scientific challenge – A review. *Plant Foods for Human Nutrition, 72*(1), 1–12. https://doi.org/10.1007/s11130-016-0590-1.
Suárez, G., & Gutiérrez, T. J. (2017). Recent advances in the development of biodegadable films and foams from cassava starch. In C. Klein (Ed.), *Handbook on cassava: Production, potential uses and recent advances* (pp. 297–312). New York. EE.UU. ISBN: 978-1-53610-307-6: Editorial Nova Science Publishers, Inc.
Walia, N., Dasgupta, N., Ranjan, S., Chen, L., & Ramalingam, C. (2017). Fish oil based vitamin D nanoencapsulation by ultrasonication and bioaccessibility analysis in simulated gastro-intestinal tract. *Ultrasonics Sonochemistry, 39*, 623–635. https://doi.org/10.1016/j.ultsonch.2017.05.021.
Wang, S., Li, C., Copeland, L., Niu, Q., & Wang, S. (2015). Starch retrogradation: A comprehensive review. *Comprehensive Reviews in Food Science and Food Safety, 14*(5), 568–585. https://doi.org/10.1111/1541-4337.12143.
Wang, S., Sharp, P., & Copeland, L. (2011). Structural and functional properties of starches from field peas. *Food Chemistry, 126*(4), 1546–1552. https://doi.org/10.1016/j.foodchem.2010.11.154.
Yang, J., Li, F., Li, M., Zhang, S., Liu, J., Liang, C., Sun, Q., & Xiong, L. (2017). Fabrication and characterization of hollow starch nanoparticles by gelation process for drug delivery application. *Carbohydrate Polymers, 173*, 223–232. https://doi.org/10.1016/j.carbpol.2017.06.006.
Zhu, F. (2017). Encapsulation and delivery of food ingredients using starch based systems. *Food Chemistry, 229*, 542–552. https://doi.org/10.1016/j.foodchem.2017.02.101.

Chapter 7
Biopolymer-Based Hydrogels for Agriculture Applications: Swelling Behavior and Slow Release of Agrochemicals

Bárbara Tomadoni, Claudia Casalongué, and Vera A. Alvarez

Abstract Hydrogels are cross-linked polymers that can absorb and hold large amounts of water and aqueous solutions due to the presence of hydrophilic functional groups in their 3D network. These materials also serve as vehicles of different active compounds because they can regulate their release rate. Because of such properties, hydrogels are currently used in many areas including agriculture, mainly as water retention agents for soil conditioning, and as carriers of agrochemicals for their slow or sustained-release. However, most of the hydrogels currently available on the market are based on polyacrylamide and acrylate derivatives, which means that they are not completely biodegradable, i.e. are considered as potential soil contaminants and present certain degree of toxicity. In this context, the development of biodegradable hydrogels for their use in agriculture is gaining interest worldwide. Biopolymers such as chitosan, cellulose, alginate and their derivatives are being explored due to their biocompatibility, biodegradability and low cost. Briefly, in this chapter, recent studies on biopolymer-based hydrogels for their use in agriculture are discussed in terms of their synthesis, swelling behavior, as well as their uses for slow and controlled release of agrochemicals.

Keywords Fertilizers · Herbicides

B. Tomadoni (✉) · V. A. Alvarez
Grupo de Materiales Compuestos Termoplásticos (CoMP), Instituto de Investigaciones en Ciencia y Tecnología de Materiales (INTEMA), Facultad de Ingeniería, Universidad Nacional de Mar del Plata (UNMdP) y Consejo Nacional de Investigaciones Científicas y Técnicas (CONICET), Mar del Plata, Argentina
e-mail: bmtomadoni@fi.mdp.edu.ar

C. Casalongué
Instituto de Investigaciones Biológicas. UE CONICET-UNMDP, Facultad de Ciencias Exactas y Naturales, Universidad Nacional de Mar del Plata, Mar del Plata, Argentina

T. J. Gutiérrez (ed.), *Polymers for Agri-Food Applications*,
https://doi.org/10.1007/978-3-030-19416-1_7

7.1 Introduction

Many regions worldwide are currently facing serious problems regarding lack of water and desertification (Vundavalli et al. 2015). A greater variability of rainfall, an increased frequency of droughts and floods, and soil temperature variations are some of the consequences of climate change. As a result of these, there are important crop losses every year, which constitute a threat to the world food security. Thus, in a world with ever-growing population it is essential to restore these degrading lands and to consciously use the natural resources available (Thombare et al. 2018). In this context, one of the main challenges for scientists of diverse fields is to achieve sustainable management of productive systems and their natural resources, and mostly to improve the water use efficiency in agriculture. A possible solution to this problem is to use materials with excellent water absorption and water retention properties.

Hydrogels are loosely cross-linked hydrophilic polymers with three-dimensional network, that have the capacity to absorb and retain large amounts of water or aqueous solutions (hundreds of times their weight) within their structure without dissolving (Thombare et al. 2018). Hydrogels have the ability to gradually release up to 95% of this stored water when the surroundings are dried out; and then rehydrate when exposed to water again (Kalhapure et al. 2016). Hydrogel research has grown vigorously over the last decades, because of their important and numerous applications in a variety of fields, such as medicine, pharmacy and agriculture. The applications of hydrogels in biomedicine are mainly in the fields of biosensors, controlled or sustained delivery of drugs, and tissue regeneration (Conte et al. 2019; Hamedi et al. 2018; Peppas and Van Blarcom 2016; Yegappan et al. 2018). With regards to the agriculture field, hydrogel applications are mainly divided into two groups (Fig. 7.1): (1) control of soil moisture and water supply of plants due to their water absorption capacity (soil conditioners); and (2) encapsulation and slow-release of agrochemicals (ACs) (both fertilizers and pesticides) (Elbarbary et al. 2017; Olad et al. 2018a; Teodorescu et al. 2009).

The conventional use of agrochemicals shows many drawbacks such as low efficiency and contamination of the soil and groundwater. The vast majority of the applied pesticides and fertilizers do not reach the target plant or crop, and end up being washed by irrigation water and rain, seriously polluting the environment. This low efficiency also implies a substantial economic loss (Teodorescu et al. 2009). The slow or sustained release of AC in the environment can solve this problem. To control the delivery of pesticides and fertilizers, they must be encapsulated into a polymer structure. Through this procedure, the active ingredient (a.i.) is allowed to diffuse towards its surroundings (Gutiérrez and Álvarez 2017; Gutiérrez 2018a). Thus, the AC is not released altogether, but in smaller amounts over a longer time period, increasing its efficiency. Among the different materials used for slow or sustained release of ACs, hydrogels have been widely studied over the last decades due to both their ability to slowly release a.i. from their matrix, and their water retention capacity in soils (Ramos Campos et al. 2014).

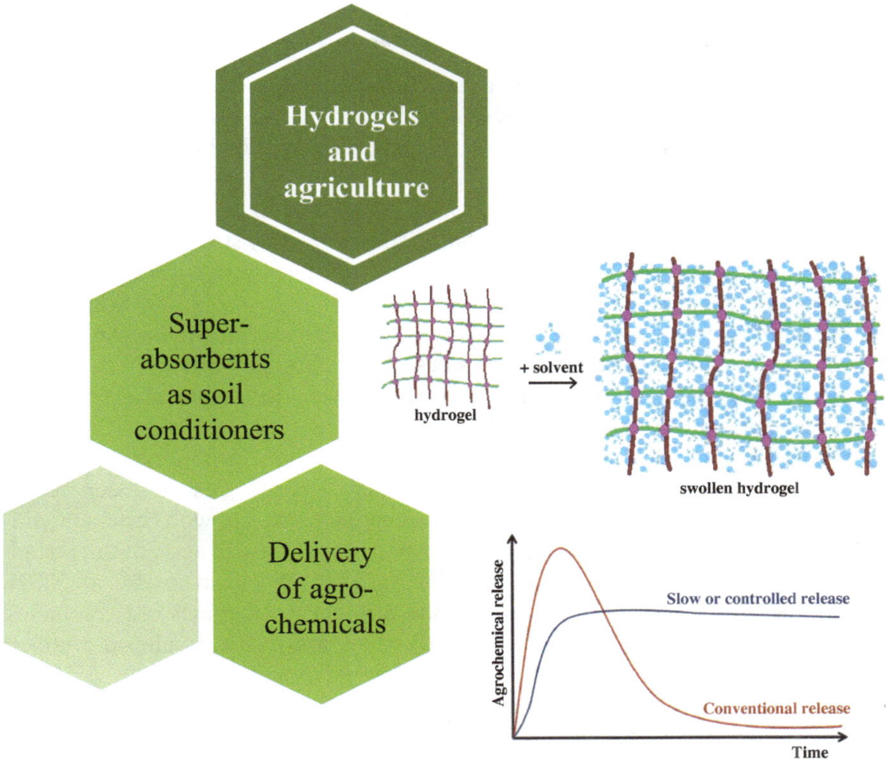

Fig. 7.1 Applications of hydrogels in agriculture. Adapted from Milani et al. (2017)

The use of hydrogels in agriculture has many advantages; however, most of the hydrogels currently available on the market are synthetized from polyacrylamide (PAAm) and acrylate derivatives. These hydrogels are not completely biodegradable and are considered as potential soil contaminants because they present a certain toxicity degree. Even though PAAm itself is relatively non-toxic, the commercially available PAAm contains residual amounts of acrylamide, which is a well-known neurotoxin and carcinogen. Therefore, the use of PAAm-based hydrogels in agriculture has raised many concerns because it may contaminate the soil and food products (Joshi and Abed 2017). In this scenario, the search for hydrogels based on biopolymers is gaining popularity since they are environmentally friendly, with good biodegradability and low production cost (Cannazza et al. 2014; Tanan et al. 2018).

In this chapter, briefly, relevant studies on biopolymer-based hydrogels for their use in agriculture are discussed in terms of their synthesis, swelling behavior, and their use as AC delivery systems.

7.2 Synthesis of Biopolymer-Based Hydrogels

Hydrogels are polymer networks that have hydrophilic properties. They are usually prepared from hydrophilic monomers but also hydrophobic monomers can be used in their synthesis in order to regulate specific properties required for a given application (Ahmed 2015).

Hydrogels can usually be produced from either natural or synthetic polymers. Most of synthetic polymers are hydrophobic and chemically stronger when they are compared to natural ones. They present good mechanical properties that result in low degradation rates, but high durability. Both properties could be balanced by optimizing the hydrogel design (Tabata 2009). Synthetic polymers can also be included in the preparation of hydrogels based on natural polymers in order to provide certain desire functional groups or to functionalize them with radical polymerizable groups (Shantha and Harding 2002).

Hydrogels can tune several properties and therefore, are commonly called "intelligent" or "smart" polymers. Their preparation technologies have been mainly developed for biomedical and pharmaceutical applications, but with certain modifications they could also be applied for active principle release in agriculture, including pesticides, herbicides, and nutrients (Kenawy et al. 1992; Rudzinski et al. 2002). In the case of agricultural applications, the formulation methods are commonly easier than those for drug delivery systems, which makes the developed products commercially viable.

Hydrogels are simply a hydrophilic polymeric network cross-linked in some way that can produce an elastic structure. In this sense, any technique able to produce a cross-linked polymer can produce hydrogels. Linear water-soluble polymers (natural or synthetic) can be cross-linked to produce hydrogels in several manners, which can be divided into two large groups: chemical cross-linking and physical cross-linking (Hennink and van Nostrum 2002). Due to their cross-linked structure, hydrogels can maintain stable networks even when they are in swollen state (Pourjavadi et al. 2004).

Gelation denotes the linking of macromolecules together, which firstly leads to branched (but still soluble) polymers, which are increasingly large, and depends on the conformation and structure of the initial material (Pethrick 2004). The mixture of these soluble branched polymers is named "sol". Extension of the linking process outcomes in branched polymer with higher size but lower solubility, called "gel" or "network", which is flooded with finite branched polymers. The transition from a system with finite branched polymer to infinite molecules is called "gelation" (or "sol-gel transition"), and the point at which the gel first appears is known as the "gel point". Gelation can take place either by physical or chemical cross-linking (Gulrez, Al-Assaf, and O Phillips 2011). The classification of gelation mechanisms is summarized in Table 7.1.

The most common chemical and physical cross-linking methods are summarized in the following paragraphs, and some examples of biopolymer-based hydrogels are highlighted:

Table 7.1 Classification of gelation mechanisms available for synthesis of hydrogels

Gelation mechanisms **Influences hydrogels properties** ✓ Water adsorption capacity ✓ Swelling kinetic ✓ Mechanical and rheological properties ✓ Degradation rate ✓ Porosity ✓ Toxicity	**Chemical cross-linking** *Formation of covalent bonds*	Complementary groups	*(cross-linking with aldehydes, addition reactions and condensation reactions)*
		High energy radiation	
		Free-radicals	
		Enzymes	
	Physical cross-linking *Not using cross-linking agents*	Ionic interactions, H-bonds, Van der Waals	
		Amphiphilic grafting	
		Crystallization	*(freeze-thawing, stereocomplex formation)*
		Protein interaction	

1. ***Chemical cross-linking:***

   1.1. *Complementary groups* (cross-linking with aldehydes, addition reactions and condensation reactions): Hydrophilic polymers have some hydrophilic groups that can be used to produce hydrogels. Those possessing –OH groups can be cross-linked with glutaraldehyde, e.g. poly(vinyl alcohol) (PVA) (Mansur et al. 2008). Besides, polymers having amine-groups (-NH_2) can be cross-linked using cross-linker under mild conditions in which Schiff's bases are formed. Hydrogels from chitosan (Cs) and PVA blends can also be obtained with this method (Mansur et al. 2004; Wang et al. 2004; Gutiérrez 2017). Polyamides and polyesters hydrogels can be prepared by condensation reactions (Ray et al. 2010), e.g. gelatin hydrogels have been successfully prepared by this method (Kuijpers et al. 2000).

   1.2. *By high energy radiation*: Unsaturated substance can be polymerized by electron beams and gamma rays (Terao et al. 2003). This technique has been used for different polymers, such as superabsorbent hydrogels based on polyvinylpyrrolidone (PVP)/carboxylmethyl cellulose (CMC) blends (Elbarbary et al. 2017) and poly(2-acrylamido-2-methylpropane-sulfonic acid)/Cs blends (Gad 2008), using gamma irradiation.

   1.3. *Free radicals*: Free radical polymerization can also be used to produce chemically cross-linked hydrogels (Bencherif et al. 2009). This technique has been used to synthesize hydrogels from synthetic, natural and mixed hydrophilic polymers. Wang and Wang (2010), e.g. developed a superabsorbent hydrogel based on sodium alginate-*g*-poly(sodium acrylate) (NaAlg-*g*-PNaA) network and linear PVP by free-radical solution polymerization.

   1.4. *Enzymes*: Enzyme-mediated cross-linking has been used to prepare hydrogels with different degrees of complexity (Moreira Teixeira et al. 2012), such as alginates with phenols moieties in the side chain (Sakai and Kawakami 2007).

2. ***Physical cross-linking:***

   2.1. *By ionic interactions*: Hydrogels can be formed as a result of ionic interaction between polyelectrolyte and oppositely charged multivalent ions. Alginate-based hydrogels can be ionically cross-linked by various cations (Gwon et al. 2015), such as Ca^{2+}. Polycations can also be cross-linked with anions. Cs-based hydrogels can be obtained, e.g. by cross-linking with glycerol-phosphate disodium salt (Chenite et al. 2000).
   2.2. *Amphiphilic grafting*: Block amphiphilic graft polymers can produce hydrogels in aqueous media by self- assembly (Förster and Antonietti 1998). Hydrogels from hydrophobized polysaccharides, such as dextran, Cs, carboxymethyl curdlan and pullulan, have been successfully prepared by this technique (Hennink and van Nostrum 2002). Cs-polylactide graft copolymer has also been obtained by this method (Wu et al. 2005).
   2.3. *By crystallization*: The synthesis of hydrogels by crystallization can be carried out through different methods. For example, freeze-thawing (F-T) procedure can be used repetitively to produce crystals that cause hydrogel formation. Dai et al. (2018) proposed novel carboxymethyl cellulose (CMC)/PVA eco-friendly hydrogels reinforced with bentonite synthesized by simple repeated F-T cycles.

The chosen cross-linking method has a great influence on several critical parameters of the hydrogel, mainly on its final properties and, therefore, on its potential applications (Ahmed 2015; Bordi et al. 2002), whichever available polymerization technique to prepare gels can be used (i.e. bulk, suspension or solution polymerizations). The selected method depends on the initial material, but also on the desired polymer (Rudzinski et al. 2002).

Although hydrogels can be based on natural or synthetic polymers (or both of them), due to growing environmental concerns, the use of free oil-based polymers has been stimulated. Alternatively, natural polymers are essentially used to replace or to be combined with synthetic polymers (Heinze et al. 2006; Mekonnen et al. 2013; Rinaudo 2006). Biodegradable polymers have progressively been more used, e.g. as plastic substitutes for different applications in agriculture (Gross and Kalra 2002; Nayak 1999; Wang et al. 2003). Degradable polymeric materials are ideal candidates to develop active principle carrier devices for their controlled or sustained release. These polymers display interesting characteristics such as abundance, renewability, low-cost and biodegradability. The classification of biodegradable polymers is indicated in Table 7.2.

The biodegradation process of these materials involves the cleavage of hydrolytically or enzymatically sensitive bonds in the polymer, leading to polymer erosion (Katti et al. 2002; Gutiérrez 2018b). Polymeric biomaterials can be classified as function of the mode of degradation into two categories: enzymatically and hydrolytically degradable materials. It is essential to note that most of the natural polymers suffer enzymatic degradation, mainly degrade by the enzymatic action of microorganisms, e.g. fungi, algae and bacteria. Enzymatically degradable polymers include protein-based biomaterials that suffer naturally-controlled degradation

Table 7.2 Classification of biodegradable polymers

Biodegradable Polymers	**Biomass products**	**Polysaccharides** *Starches (potato, cassava, maize, wheat…), Ligno-cellulosic products, and others*
		Proteins *Animal origin (gelatin, whey, cassein, collagen), Vegetal origin (Soya, Gluten, Zein)*
Advantages: ✓ abundance, ✓ renewability ✓ low-cost ✓ biodegradability	**From microorganisms**	**Obtained from extraction** *Polyhydroxy-Alkanoates (PHA), Polyhydroxybutirate (PHB) and derivatives*
	From biotechnology	**Conventional synthesis from bioderived monomers** *Polylactides: Polylactic acid (PLA)*
	From oil products	**Conventional synthesis from synthetic monomers** *Polycaprolactone (PCL), Polyesteramides, Aliphatic co-polyesters, Aromatic co-polyesters*

processes, such as collagen, Cs, alginic acid, among others. On the other hand, hydrolytically degradable polymers are polymers that have hydrolytically labile chemical bonds in their structure. Several of the functional groups susceptible to hydrolysis include esters, ortho esters, amides, carbonates, urethanes and anhydrides (Li 1999). Some of the most used hydrolytically degradable polymers are the poly(α-esters), polylactic acid (PLA), poly(lactic-*co*-glycolic acid) (PLGA), polyglycolide (PG) and polycaprolactone (PCL) (Gutiérrez and Alvarez 2017a, b; Gutiérrez 2018c).

7.3 Swelling Behavior of Biopolymer-Based Hydrogels

The lack of water and desertification are really serious problems in several regions of the world for different reasons, but the worst one is that they compromise the development of agriculture. An excellent tool to solve these problems could be the use of materials with high water absorption and retention capacities based on superabsorbent polymers (SAPs). The use of SAPs for agricultural applications has shown promising results: helping to reduce the irrigation water consumption and also, the death level of plants, improving the water retention in the soil and increasing the rate of plant growth (Puoci et al. 2008).

The swelling capacity of hydrogels after being immersed in water or an aqueous solution is the result of a diffusion process, which involves the solvent molecules into the 3D network due to the high hydrophilicity of the material. This phenomenon causes the expansion of the polymeric chains, which increases the average distance between cross-linking points. For most of the hydrogel's technical applications, its swelling could be considered as a disadvantage; however, the polymer swelling is extremely useful for agriculture applications as a device for controlled or sustained release of ACs as well as soil conditioners (Guilherme et al. 2015).

Usually, most of the SAPs used in agriculture are produced from synthetic polymers which are not biodegradable and pose a threat to both food safety and the environment. With the growing demand for environmentally friendly materials, biopolymer based-hydrogels (mainly, cellulose, Cs and alginate derivatives) are being studied for agriculture applications (Chang et al. 2010; França et al. 2018; Pang et al. 2011).

The water capacity of hydrogels is generally measured through the equilibrium swelling ratio (SR) or swelling degree (SD), which is easily calculated with Eq. 7.1.

$$SR = SD = \frac{M_{swollen} - M_{dry}}{M_{dry}} \tag{7.1}$$

where M represents hydrogel mass.

The swelling ratio is defined as the ratio of the mass of a fully swollen hydrogel (in equilibrium with an aqueous medium) ($M_{swollen}$) to the mass of a dehydrated hydrogel (M_{dry}) (Holback et al. 2011). Table 7.3 summarizes the latest results on water absorption of some biopolymer-based hydrogels.

The swelling behavior of the hydrogels is not only measured by their swelling ratio, but also by their swelling kinetics, which shows the capacity of the hydrogel to absorb water or aqueous solutions over time. Water retention capacity and water release over time are the key properties that swollen hydrogels must possess for application in agriculture for sustained irrigation purposes (Sharma et al. 2014). It is thus of high importance to study the behavior of the hydrogel through deswelling, and also, how the hydrogel performs after several swelling-deswelling cycles, and its effect directly on the soil or soil-less substrates (Abdel-Raouf et al. 2018; Olad et al. 2018a; Sharma et al. 2014) (Table 7.4).

Table 7.3 Swelling degree (SD) of different biopolymer-based hydrogels

Biopolymer	*Synthesis method*	*SD and conditions*	*Reference*
Alginate	Emulsion cross-linking method	1.65–3.85 g/g in distilled water	(Sharma et al. 2014)
Cellulose/CMC	Cross-linking with epichlorohydrin	200–1000 g/g at 25 °C in distilled water	(Chang et al. 2010)
Cs/cellulose	Grafting of PAA onto Cs/cellulose	150–400 g/g in distilled water	(Essawy et al. 2016)
Cs	Cross-linking by Sodium tripolyphosphate (TPP)	>1000% in distilled water at room temperature	(França et al. 2018)
CMC	Radiation-induced grafting of Acrylamide onto CMC (cross-linked with *N,N′*-methylenebisacrylamide).	50–200 g/g in distilled water	(Hemvichian et al. 2014)
Guar gum	Grafting copolymerization (cross-linked with different Acrylate monomers)	125–220 g/g at 25 °C in distilled water	(Abdel-Raouf et al. 2018)
Starch	Cross-linking reaction between starch and PVA molecules by maleic acid	500–1200% (pH 7, 25 °C)	(Pang et al. 2011)

Table 7.4 Evaluation of hydrogels swelling behavior for agriculture applications

Swelling behavior	Swelling degree or swelling ratio
	Swelling kinetics
Influenced by ✓ Cross-linking degree ✓ Porosity ✓ Chemical structure ✓ Ionic or neutral nature of side groups ✓ Environmental conditions (e.g. pH, temperature, analyte concentration)	Deswelling or retention capacity
	Re-swelling capacity or swelling-deswelling cycles
	Water holding capacity in soil/soil-less substrates

7.3.1 Effect of Hydrogel Composition on Swelling

Many factors influence the swelling behavior of the hydrogels, such as degree of cross-linking, porosity and chemical structure, and the ionic or neutral nature of the side groups attached to their backbone chains. Bajpai and Giri (2003) studied hydrogels based on CMC prepared by grafting PAAm chains cross-linked onto the CMC backbone by free radical polymerization. They evaluated the water absorption dynamics of the CMC hydrogels, and the impact of different variables on their swelling properties. For example, the authors analyzed the effects of the hydrogel composition on swelling. The results showed that when the amount of CMC increases in the hydrogel formulation, the SR also increases up to a certain point, and then decreases. This observation was explained by the fact that CMC is a polysaccharide with different hydrophilic functional groups. This means that by increasing the amount of CMC, the affinity for water increases, resulting in a greater swelling ratio. However, at much higher concentration of CMC the network chains density increases so much that both the diffusion of solvent molecules and relaxation of macromolecular chains are reduced. This explains the decrease in the swelling ratio of the hydrogel. Abdel-Raouf et al. (2018) studied Guar Gum hydrogels cross-linked with different acrylate monomers. These authors also found that the hydrogels swelling ratio decreased as the concentration of guar gum increases, and they explained that increasing the guar gum content affects the water absorption ability of the hydrogel due to the higher probability of physically crossed links.

The cross-linker content also significantly affects the swelling behavior of the hydrogels. Rohindra et al. (2004) studied the swelling properties of different Cs hydrogels (cross-linked with different amounts of glutaraldehyde). In this study, the authors evaluated the swelling kinetics of the hydrogels under different conditions (pH, temperature). With regards to the effect of the cross-linking agent, it could be seen that the more glutaraldehyde was added, the lower swelling ratio. The degree of crosslinking influences the available area for diffusion through the hydrogel network and, therefore, affects the hydrogels water absorption capacity (Holback et al. 2011). Abdel Raouf et al. (2018) also noted that the maximum swelling ratio decreased when the content of the crosslinking agent was increased in the synthesis of guar gum hydrogels. The high degree of crosslinking slows the mobility of polymeric chains and, thus, decreases the swelling capacity of hydrogels.

7.3.2 Hydrogels Swelling and External Stimulus

Hydrogels are also environmentally sensitive materials, which means they have the capability to imitate feedback mechanisms often observed in nature. The most common hydrogel systems swell in response to changes in pH, temperature and analyte concentration (Holback et al. 2011) (Fig. 7.2).

Hydrogel systems are usually pH-sensitive; thus, the pH of the immersion solution has a direct impact on the SD of the 3D network. This depends on the immersion medium pH and the dissociation constants of polymer side groups (pKa or pKb) (Holback et al. 2011). pH-responsive hydrogels contain acidic (e.g. carboxylic and sulfonic acids) or basic (e.g. ammonium salts) side groups that can accept or release protons in response to environmental pH changes. Therefore, pH variations in the immersion solution lead to conformational changes of the hydrophilic polymer chain, and when ionizable groups are linked to the polymer structure, the pH of the solution directly impacts on the hydrogel swelling behavior (Bawa et al. 2009).

The behavior of the hydrogel with pH depends on the nature of the polymer chain. When the anionic polymers are, e.g. immersed in acidic media (with $pH < pKa$), the "protonation" of anionic groups of the hydrogel network increases, thus decreasing the concentration of anionic groups. This reduces the electrostatic repulsion between the same or different polymer chains, and the hydrogel is thus contracted (decreases its SD). On the contrary, in basic media, the concentration of anionic groups increases relatively, and electrostatic repulsion causes the chain expansion, increasing the swelling ratio. The opposite occurs in the case of cationic polymer chains like Cs. Drozdov (2015) studied the swelling properties of Cs oligomers dissolved in acetic acid and cross-linked with glutaraldehyde by radiation. Contrary to the swelling behavior of anionic polymers, under acidic media, Cs hydrogels show a higher swelling ratio. This phenomenon was well explained by electrostatic repulsion.

With regards to the effect of temperature, generally the swelling ratio of hydrogels increases at higher temperatures due to the thermal expansion of the network and the destabilization of network connections. These cause the expansion of the hydrogel matrix after water diffuses into the porous structure (Guilherme et al. 2015). Usually, most polymers increase solubility with increasing temperature. However, in the case of polymers with a lower critical solution temperature (LCST), the solubility decreases as temperature increases and therefore, hydrogels based on this type of polymer show a decrease in their swelling ratio as the temperature

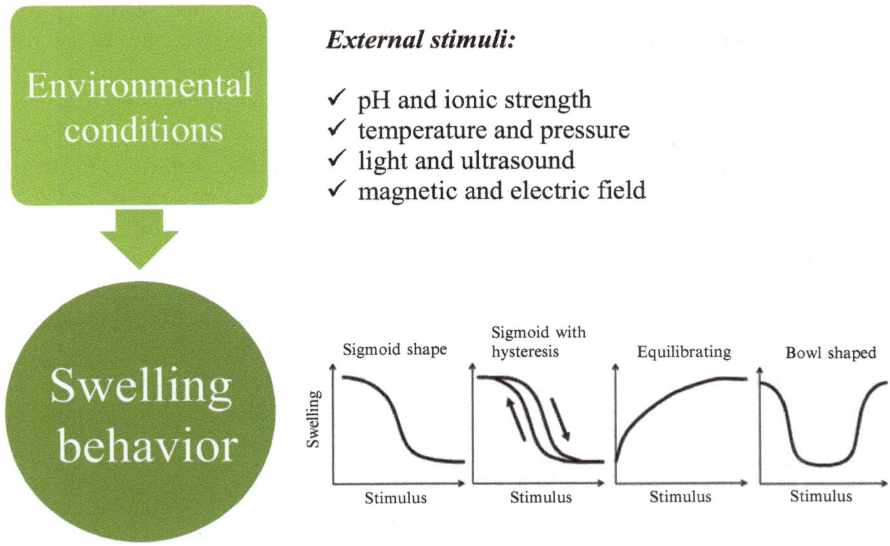

Fig. 7.2 Swelling behavior of hydrogels influenced by environmental conditions. Adapted from Ehrenhofer et al. (2018)

increases above the LCST. This type of swelling behavior is known as inverse dependence on temperature and occurs due to predominant hydrophobic interactions, and was observed by Abdel-Raouf et al. (2018) when studying hydrogels based on cross-linking of guar gum with different acrylate monomers. These authors monitored the swelling efficiency of the hydrogels prepared at different temperatures ranging from 15 to 60 °C. They observed that the highest swelling ratio was achieved at the lowest tested temperature (15 °C).

The ionic strength of the swelling medium has a significant effect on the swelling behavior of the hydrogel. The effect of the ion concentration on the swelling ability of the hydrogels is closely related to osmotic pressure. The osmotic pressure is the difference between the ion concentrations in the external immersion medium and in the hydrogel interior. Therefore, increasing the ionic concentration in the swelling medium reduces the osmotic swelling pressure, which produces the contraction of the hydrogel (Bajpai and Giri 2003). The effect of ionic strength in the SD of the hydrogels can be evaluated by the salt sensitivity factor (f), calculated with Eq. 7.2:

$$f = 1 - \frac{SD_{saline}}{SD_{water}} \tag{7.2}$$

where SD_{saline} and SD_{water} are the equilibrium SDs for saline solutions and pure water, respectively. The closer to 0 is the f value of the hydrogel in a certain saline solution, the lower is the ionic effect of said solution to its swelling ratio. The closer f is to 1, the stronger is the saline effect (Guilherme et al. 2015).

Different authors have studied the effect of ionic strength on the swelling ratio of different hydrogels based on biopolymers. Chang et al. (2010), e.g. developed cellulose/CMC hydrogels, and evaluated their swelling kinetics when immersed in different aqueous solutions of NaCl (concentration range between 0.0 and 0.1 mol/L). The authors observed that the swelling ratio of hydrogels decreased with an increase in the ionic strength of the solution. They also performed the same experiment with $CaCl_2$ solutions, and in this case the swelling ratio decreased even more quickly, because of the higher cationic charge of $CaCl_2$ compared with NaCl. Abdel-Raouf et al. (2018) also evaluated the effect of salinity of the swelling medium in the water absorption capacity of guar gum hydrogels. These authors reached the same conclusions, and stated that there was a swelling loss by increasing ionic strength due to a reduction in electrostatic repulsion.

7.3.3 Water Holding Capacity (WHC) in Soil

One of the main agriculture applications of hydrogels is to enhance the WHC of soil thanks to its outstanding swelling properties. Because hydrogel swelling behavior is influenced by the medium conditions in which it is immersed (mainly pH, temperature and ionic strength, as detailed above), it is very important to test its effects directly on the targeted substrate. For this purpose, different amounts of hydrogels have been mixed with the soil and placed in a plastic glass or tube, or soil-pot. A soil-pot without hydrogels is prepared as control. The pots are then exposed to different irrigation situations, and water content or moisture is measured in order to evaluate the WHC. Water content can be measured gravimetrically (Bai et al. 2015; Olad et al. 2018a) or with a digital moisture meter by inserting the instrument's sensor into the soil to a specific depth (Sharma et al. 2014).

Bai et al. (2015) prepared different carbendazim-loaded hydrogels based on starch (CLHs) by solution polymerization. These authors tested the WHC of the soil in tubes with CLHs (0, 2.6, 7.8, and 13.0 mg per 10 g of soil). The study indicated that starch-based hydrogels loaded with carbendazim were a good water absorbent for soil, and the WHC increased with greater amounts of hydrogels in the soil. They observed that 1.3 g of CLH/kg soil markedly improved WHC from 44.0% in the control sample to 52.2% in the soil mixed with CLHs. Sharma et al. (2014) also synthetized alginate-based hydrogels as nano-reservoirs of water for agriculture applications. They prepared the calcium alginate nanoparticles by the emulsion cross-linking method. These authors carried out a series of soil-pot experiments to test the WHC of the soil with and without the nanoparticles, but they also tested their effect on plant growth. The tests were performed with sandy soil collected from the cold desert areas of Leh-Laddakh (India). The authors measured the moisture content of the different samples at different times, and concluded that in the control sample (soil without nanoparticles) the water content decreased rapidly (about 100% of moisture loss after 7 days). On the other hand, in the soil sample mixed with alginate nanoparticles, the moisture content lasted longer (after 11 days, about 20% moisture was retained), thus providing a more favorable condition for plant growth. The results of plant growth experiments were also promising. The plants watered with mixed soil with alginate-based hydrogels exhibited better growth than control (soil without hydrogel). The control plants stopped growing after the first 14 days and showed signs of dehydration. Each sample of plant in the soil mixed with nanoparticles continued to grow for two weeks and were much healthier than control.

7.4 Slow Release of ACs through Biopolymer-Based Hydrogels

To resolve the water deficiency problems, significant progress has been made in improving the properties of hydrogels for their uses under different environments and crops. Hydrogel based formulations are being explored for fertilizers and pesticides to enhance the efficiency of these compounds and consequently, reduce the adverse effects on the environment compared with the standard and traditional application of ACs. In particular, hydrogels for controlled delivery of bioactive principles have expanded the range of targets and kinetics which can be used. The low application rate (2.5–5.0 kg/ha) of hydrogel is effective for almost all crops. Hydrogels have thus been accepted as a boon to increase agricultural productivity in water-stressed environments (Kalhapure et al. 2016). Within drug delivery systems, the smart stimulus-sensitive hydrogels such as soil temperature, salinity and pH, produce compounds with a prolonged residual activity between applications (Rudzinski et al. 2002; Soppimath et al. 2002).

In addition, hydrogel formulations based on biocompatible composite for the delivery of ACs are being currently investigated since they have high swelling ratio and they can also be degraded in the presence of soil and plant microbiome (Pereira et al. 2017). The biopolymer itself and its degradation products should not cause environmental pollution and should be stable during storage (Rudzinski et al. 2002). Biodegradable hydrogels contain labile bonds in the backbones of polymer itself or between cross-linker bonds. These bonds can be degraded enzymatically or chemically for a given period of time, resulting in CO_2, water and ammonia. The release rate of active compound from the polymer matrix is influenced by the chemical characteristics of the a.i. (structure, nature, distance of the active agent from the polymer backbone) and also biological or environmental agents such as UV, moisture and microorganisms (Ekebafe et al. 2011).

Smart hydrogels in addition to their ability to absorb and retain water can also be altered by external stimuli facilitating the release mechanism of the agrochemical. In this sense, an important aspect to consider is the analysis of the residual moisture and AC level in the hydrogel over time after the application. There are different approaches to combine the bioactive with the polymeric matrices mainly by physical encapsulation or heterogeneous dispersion and by chemical combination to act as a carrier (Ekebafe et al. 2011). Therefore, the type of hydrogel, the method of preparation, degree and dynamics of swelling, among other factors, affect the release of the active compound from the biopolymeric matrix (Mahdavinia et al. 2009). This statement is illustrated through the following representative examples of biomaterials that were developed thanks to the combined properties of some biopolymers and bioactive principles, including fertilizers and herbicides.

In this section, we attempt to review some general considerations about current devices to release fertilizers and some herbicides in particular from biopolymer-based hydrogels. We also consider the composition of different types of formulations for controlled release of fertilizers and herbicides mainly with alginate, cellulose and chitin derivatives (Table 7.5) which have shown efficient properties to be used as biopolymeric matrices in modern agriculture lands (Rudzinski et al. 2002).

7.4.1 Slow Release of Fertilizers

Fertilizers are natural or synthetic chemical compounds (inorganic or organic) that have the capacity to supply some of the essential nutrients required for plant growth and development (Chagas et al. 2018). Fertilizers have been applied extensively in agriculture to improve crop yields, which can have serious consequences for the environment. Several studies have shown that most of the fertilizers applied in the field are lost to the environment, which generates environmental pollution. Nutrients with nitrates and phosphates can be lost by leaching, which leads to eutrophication of surface and ground waters. If fertilizers are not applied correctly, then an imbalance of the ecosystem occurs, and the acute toxicity and carcinogenic sources increase (Chagas et al. 2018). The use of fertilizers is thus an activity that must be taken seriously into consideration for the economic, environmental and social sustainability of the agricultural sector.

In this context, there is an imperative necessity to develop devices for slow or controlled release of nutrients (also called, “smart fertilizers”) in agriculture systems (Calabi-Floody et al. 2018). These systems can deliver nutrients in a more controlled or slow manner for longer duration to a specified target and at a desired rate. These smart devices minimize the impact of these harmful chemicals on the environment by reducing losses due to leaching, volatilization and degradation of fertilizer, thereby maintaining biological efficacy of the a.i. (Ramos Campos et al. 2015).

Polymer-based hydrogels are widely studied for the development of slow-delivery systems. They can provide a controlled release of fertilizers and also increase the WHC of the soil (Ramos Campos et al. 2015). The development of hydrogels based on biopolymers as nutrient carriers is gaining popularity because of their eco-friendly nature, cost effectiveness, easy availability, and biodegradability. It should also be noteworthy that for agriculture applications, biocompatible matrices are preferred in order to avoid environmental contamination, after their degradation, biopolymer-based hydrogels also serve as compost in the field (Kumar et al. 2014; Gutiérrez 2018d; Gutiérrez et al. 2019; Herniou--Julien et al. 2019).

Table 7.5 Biopolymer-based hydrogels for slow release of ACs

Biopolymer	*Synthesis method*	*AC*	*References*
Alginate	Graft copolymerization of acrylic acid and acrylamide on NaAlg in the presence of clinoptilolite	(F) NPK (urea, potassium dihydrogen phosphate and ammonium dihydrogen phosphate)	(Rashidzadeh and Olad 2014; Rashidzadeh et al. 2014)
Alginate	Ionically cross-linked with $CaCl_2$	(H) Chloridazon and metribuzin	(Flores Céspedes et al. 2013)
Alginate	Ionically cross-linking with $CaCl_2$	(F) Urea	(González et al. 2015)
Alginate/Cs	Ionotrophic gelification process	(H) Paraquat	(dos Silva et al., 2011)
Carboxymethyl-Cs	Ionically cross-linked with calcium cations	(H) Atrazine and imidacloprid	(Li et al. 2012)
Cellulose acetate	Cross-linked by EDTA	(F) NPK (KCl and $NH_4H_2PO_4$)	(Senna and Botaro 2017; Senna et al. 2015)
Cellulose derivatives	Emulsification-solvent evaporation technique	(H) 2,4-D	(Belmokhtar et al. 2018)
Cellulose derivatives	Phase inversion technique	(F) Urea	(González et al. 2015)
Cs	Cross-linked by sodium tripolyphosphate (TPP)	(F) KNO_3	(França et al. 2018)
Cs	Ionic gelification technique using TPP	(H) Paraquat	(Grillo et al. 2014a, b)
Cs	Ionic gelification technique using TPP	(H) Imazapic and Imazapyr	(Maruyama et al. 2016)
Cs-starch	Ionotropic cross-linking of Cs using TPP	(F) KNO_3	(Perez and Francois 2016)
CMC	Radiation induced grafting of acrylamide onto CMC	(F) KNO_3	(Hemvichian et al. 2014)
CMC/PVP	In situ Polymerization method	(F) NPK (urea, ammonium dihydrogen phosphate, and potassium dihydrogen Phosphate)	(Olad et al. 2018a)
Ethylcellulose	Oil-in-water emulsion solvent evaporation technique	(H) Norflurazon	(Sopeña et al. 2011)

(F): Fertilizer; (H): Herbicide

In the following paragraphs, the use of biopolymer (mainly, alginate, cellulose and Cs derivatives) will be summarized for slow-release of nutrients.

Among the principal nutrients required for plant growth, nitrogen, phosphorus and potassium (generally referred to as NPK fertilizers) are needed. NPK are applied at higher concentrations (mmol/kg dry mass) than the other elements. Nitrogen is a crucial element in production of proteins, and it is necessary for the vigorous growth of plant. Phosphorous participates in different plant mechanisms, e.g. cell division, formation of cell structures in the plant, photosynthesis, stimulation of healthy root growth and seed germination. Finally, potassium plays an important role in the formation of fruits, resistance against diseases and cold weather, photosynthesis and maintaining water level in plants (Carson and Ozores-Hampton 2013). Many studies have been conducted in recent years, in which biopolymer-based hydrogels have been tested as carriers of NPK fertilizers for their slow or controlled release (Chagas et al. 2018; Essawy et al. 2016; Olad et al. 2018b; Rashidzadeh et al. 2014; Senna and Botaro 2017; Senna et al. 2015).

Urea is one of the most widely used fertilizer due to its high nitrogen content (46%), low cost and easy application (Azeem et al. 2014). Urea when applied to soil, a series of chemical transformations occur, both physical and biological to produce the nutrients available to the plant. However, the excess of these nutrients is lost because of leaching, since plants only need a small amount during their early growth. In addition, some urea reactions produce nitrogen in a form that is lost through gaseous emissions that are hazardous for the environment. Therefore, the development of materials for controlled or slow release of urea has been an issue of growing interest for decades. These materials would not only reduce nitrogen loss, but also alter the nitrogen release kinetic, which in turn will provide nutrients to the plant at a rate more compatible with their metabolic needs. Biocompatible polymer matrices are preferred in order to avoid environmental contamination (Azeem et al. 2014). Many have studied the controlled release of urea through biopolymer-based hydrogels (González et al. 2015; Olad et al. 2018a; Rashidzadeh and Olad 2014; Rashidzadeh et al. 2014). Bortolin et al. (2013), e.g. studied the synthesis of a hydrogel nanocomposite based on biopolymers. In this study, a polysaccharide of biodegradable MC and its compound containing PAAm, as well as with mineral clay calcic montmorillonite (Mt) as modifier was evaluated. The authors characterized the hydrogels in terms of their swelling behavior and performance as devices for slow release of fertilizers, in this case, by evaluating the desorption of the nutrient urea. The controlled desorption of urea was tested in different pHs (4, 7, and 9) on the hydrogel nanocomposites and was compared with the commercial urea, i.e. without being contained in the hydrogel structure. Bortolin et al. (2013) concluded that the presence of the mineral clay causes the hydrogel to slowly desorb the nutrient, but due to strong clay–urea interactions, the total amount of urea carried in the material was reduced. Therefore, in hydrogels with higher clay concentrations, the urea desorption rate is slower. This was observed in all the pHs tested, where the hydrogel without Mt. practically desorbed the whole urea content in 24–48 h. These values increased for nanostructured hydrogels at ~72 h.

With regards to K sources in crop production, the most common fertilizers are potassium nitrate (KNO_3), potassium chloride (KCl), mono potassium phosphate (KH_2PO_4) and potassium sulphate (K_2SO_4). The selection of the K fertilizer is strongly influenced by the nature of the accompanying anion (Chapagain et al. 2003). KNO_3 is the K fertilizer most used in the production of greenhouse vegetables, and many have studied its slow-release through biopolymer-based hydrogels (Hemvichian et al. 2014; Perez and Francois 2016). França et al. (2018) investigated

Cs as a biopolymeric matrix for the controlled release of fertilizers using a spray dryer. They tested KNO_3 as a nutrient, and studied the effect of TPP (as a cross-linker) and the incorporation of Mt. clay. The nutrient was trapped in the core and covered with an additional Cs layer (shell), which swells and then releases the nutrient. These authors evaluated the release of KNO_3 both in water and in soil. The release mechanism indicated a swelling-controlled nutrient delivery for all tested materials. Particularly, the microcapsules of Cs/KNO_3 and Cs/Mt-KNO_3 showed the highest swelling ratio (approximately 500%) and the lowest release of fertilizer in the water (2 h) and in the soil (40 days).

7.4.2 *Slow Release of Herbicides*

Weed management is also a fundamental step in the production of vegetable crops under commercial production systems. A sustainable crop weed alternative in vegetable production could thus be the use of hydrogels for herbicide application.

Herbicides are chemicals that inhibit plant growth and are used to control weeds. Most of them are highly toxic for human health and environment. Selective herbicides control specific weed species, while leaving the desired crop relatively unharmed. In contrast, non-selective herbicides kill all plant material with which they come into contact. In addition to its efficiency to killing plants, one of the most important traits of an herbicide is its persistence and residual action since these two traits could contribute to reduce the doses and the costs of the application. Herbicide formulations (a.i. plus additives) are mainly prepared to achieve a high biological effectiveness and to reduce negative environmental effects on organisms that are not the target for herbicide applications (Sopeña et al. 2009a).

The potential loss or degradation of soil-applied herbicide can occur by the action of different environmental factors such as light (photolysis), moisture and microbial degradation, as well as volatilization and leaching in the field. As a consequence, the concentration of active herbicidal in the target site can be disturbed by directly impacting the effectiveness in the control of weeds. The advantages of hydrogel-based herbicide applications include stabilization of a.i., its easy handling and application in the soil at the time (Kalhapure et al. 2016). Otherwise, the herbicidal persistence efficacy can depend on a variety of properties of the polymeric support but also, on the weed species assayed (Vasilakoglou and Eleftherohorinos 2003).

Dichlorophenoxy acetic acid or 2,4-D is one of the most widely applied herbicides in the world. More than 1500 herbicide products contain 2,4-D. This organic and synthetic compound is primarily used as a selective herbicide which kills many terrestrial and aquatic broadleaf weeds, but not grasses. Similarly to other herbicides, like paraquat (Aouada et al. 2010), the release of 2,4-D from a non-biodegradable polymer-based hydrogel was developed (Mahdavinia et al. 2009). Controlled release of 2,4-D has also been investigated from biopolymeric matrixes of corn starch, starch-*g*-poly(butyl acrylate) and alginate gels. In the case of the modified starch matrix, a low swelling capacity and slow release of herbicides was observed (Zhu and Zhuo 2000).

The non-toxic and biodegradable matrices of cellulose derivatives ethylcellulose/hydroxy propyl methyl cellulose (HPMC), cellulose acetate butyrate butyryle (CAB)/HPMC and pure CAB were used in microparticles to allow an initial burst of 2,4-D followed by a controlled bioactive release. This approach allowed to reach a high herbicide efficacy in early stages and also to be maintained for a long period of time (Belmokhtar et al. 2018). The delivery of 2,4-D provided a burst that varied from 7 to 50%, followed by a slow and controlled effect from all the formulations tested, ensuring an initial effectiveness of the herbicide and then,its functional preservation for a long period of time. These synthesized composites can acquire a great opportunity for their uses in the field; however, their properties as hydrogels have not yet been explored. The adjustment of the swelling behavior by these polymers alone or together with other polymers can be a good tactic to further improve the efficiency of the 2,4-D herbicide.

Alachlor (2-chloro-2′,6′-diethyl-*N*-methoxymethylacetanilide) is a chloroacetanilide herbicide used for annual grasses and control of broadleaf weeds in corn, soybeans and sorghum. Ethylcellulose is a hydrophobic polymer that is frequently used to prepare formulations of different herbicides by means of microencapsulation techniques (Dailey 2004; Pérez-Martínez et al. 2001). The encapsulation of alachlor in ethylcellulose microparticles managed to extend its concentration in the soil reducing the high rates of herbicide application rates and, consequently, decreasing the environmental damage. The highest levels of alachlor efficacy and persistence were observed in the soil with low microbial activity and organic matter content (Sopeña et al. 2009b). In the same work, two ethylcellulose 30–50 mPa s (Ethocel 40) and 10 mPa s (Ethocel Std 10) were assayed revealing that the polymer with a higher molecular weight produced a lower alachlor release rate, which suggests that the polymer size is also an important aspect to be considered for the controlled delivery of this herbicidal compound. The behavior of alachlor transported from these cellulose microspheres in soil was reported firstly by Sopeña et al. (2007). These authors demonstrated that mobility of alachlor from microspheres to water or soil was retarded compared to commercial formulations.

Norflurazon is a fluorinated pyridazinone herbicide that is used in soils cultivated with cotton, soybean, tree fruit and nut crops, citrus and cranberries, and its efficacy and persistence by controlled delivery through biopolymer-based hydrogels has been demonstrated (Sopeña-Vázquez et al. 2008). Ethylcellulose microencapsulated formulations of norflurazon and alachlor protected both of them against their dissipation in soil, and remained biologically active longer than the commercial formulations. Regarding particle size distribution, the authors reported higher release rates of norflurazon with a greater surface area of microspheres. However, particle porosity instead of size, could affect the alachlor release rate from the microspheres. All these findings allowed to conclude that, in addition to the weed species, the type and nature of components significantly affect the controlled release and biological efficacy of each formulation.

Photolysis on soil and leaves is an important mechanism for herbicide dissipation, resulting in a loss of its efficacy in controlling weeds. Thus, the photodegradation rate of the herbicide norflurazon from the ethyl cellulose microspheres was compared with a commercial formulation (Sopeña et al. 2011). The microspheres had slower delivery rate than the commercial formulation, and the gradual release allowed a protection of norflurazon against photolysis. The photodegradation rate was even lower when the microspheres contained colloidal soil components such as goethite and humic acids, than spheres alone.

Carboxymethyl Cs/bentonite (CM-chit/H-bent) composite gel was studied as a carrier of two pesticides: atrazine (1-chloro-3-ethylamino-5-isopropylamino-2,4,6-triazine) and imidacloprid (*N*-[1-[(6-chloro-3-pyridyl)methyl]-4,5- dihydroimidazol-2-yl]nitramide). Both are herbicides widely used in croplands and are often pollutants in the aquatic environment. The diffusion of these two compounds was restrained by the encapsulation in the polymer gel, and sorption with bentonite was further delayed the progress of the release thereof (Li et al. 2012). To obtain a slow release of atrazine, an agar/starch-based hydrogel (1% agar, 5% starch, 0.98 M acrylamide, and 12.97 mM *N*,*N*-methylenebisacrylamide) was also developed (Singh et al. 2015). Atrazine released from the hydrogels occurred after 144 hours through non-Fickian diffusion mechanism, which makes them suitable for agriculture. In the case of atrazine, its release profile and physic-chemical stability were modified when poly (ε-caprolactone) nanocapsules were coated with Cs (Grillo et al. 2014b).

The imazapic (IMC) and imazapyr (IMR) herbicides are representative members of the imidazolinone group. They are often used to control weeds in plantations of corn, soybean, and peanut but they have a high persistence in soil. In order to reduce the doses applied in soil, these herbicides were encapsulated in Cs/tripolyphosphate nanoparticles (400 nm) with efficiencies higher than 60% and with a polydispersity index of 0.3 and the zeta potentials of −30 and + 26 mV for CS/ALG and CS/TPP, respectively (Maruyama et al. 2016). After 300 min, the release obtained for IMC and IMR was 59% and 9%, respectively. The combined encapsulation of the compounds resulted in a decrease of ~35% compared to IMR encapsulated separately. Meanwhile, the combined encapsulation did not affect the release of IMC. It would also be interesting to analyze the water absorption of these polymer matrices.

Paraquat or *N*,*N*′-dimethyl-4,4′-bipyridinium dichloride is a non-selective herbicide, which kills the plant by contact. Due to its redox activity it is also toxic to human and animals. To develop an environmentally safe alternative for its use, paraquat was loaded onto Cs/TPP nanoparticles (Renato Grillo et al. 2014). This encapsulation showed changes in paraquat diffusion and release as well as its absorption by the soil compared with commercial formulations without affecting herbicidal efficiency. The loading and release of paraquat, and how they are strictly correlated with the swelling capacity of the PAAm and MC-based hydrogels were clearly described and schematized by Aouada et al. (2011). Slow delivery without losing paraquat the effectiveness has been also achieved from Cs nanoparticles (dos Silva et al. 2011).

Alginates-based formulations have also been investigated for controlled-release of a variety of herbicides, such as monolinuron, desmetryn, chloridazon, atrazine, simazine, and chloroxuron as a.i. (Calabi-Floody et al. 2018). For example, alginate hydrogels (with addition of bentonite and anthracite as sorbents) evidenced a slower release of the two herbicides: chloridazon and metribuzin, compared to their respective commercial formulations. This fact might allow less leaching of them in a calcareous soil. Therefore, its use is suggested in the most sustainable control of weeds (Flores Céspedes et al. 2013).

7.5 Future Perspectives and Conclusions

Biodegradable polymers have been used, mainly because of their abundance, renewability, relatively low-cost, and biodegradability to replace or reduce the use of synthetic polymers in the production of hydrogels for various technological applications and, in particular, in agriculture as nutrient carries or soil conditioners. There are several ways available to prepare said hydrogels, based on the crosslinking of polymer chains that can be classified into two main groups: chemical and physical crosslinked hydrogels. Main processing techniques have been developed for the crosslinking of polymer chains. Physically crosslinked are relevant in the formulation of labile bioactive principle vehicles, while chemically crosslinked ones have higher mechanical properties and stability.

Biopolymer-based hydrogels for the release of fertilizer/herbicidal must meet a series of conditions and properties that include technical and chemical aspects of the formulation, as well as its environmental and operational scope. Therefore, the challenge of developing disruptive technologies for the formulation of slow delivery of ACs requires fundamental knowledge of the structural components, as well as characterization of physicochemical properties that are established between the ingredients of the mixture, including the type of biopolymer, additives, sorbents, bioactives, among others. Nanotechnology can provide a great potential to enrich the developments of controlled release of ACs from hydrogels based on biopolymes.

There are still several challenges regarding these kinds of hydrogels, such as producing low-cost hydrogels with excellent absorption characteristics and also an excellent mechanical behavior; design of new eco-friendly techniques mainly for chemically cross-linked hydrogels; and obtaining bio-based nano-hydrogels that show a higher water absorption capability and, at the same time, low sensitivity to saline conditions present in the field. The potential of biopolymeric-based hydrogel formulation to successfully deliver the next generation of ACs at the desired rate and location in the target site is still in its infancy.

Acknowledgements The authors would like to thank Agencia Nacional de Promoción Científica y Tecnológica (ANPCyT), Consejo Nacional de Investigaciones Científicas y Técnicas (CONICET), and Universidad Nacional de Mar del Plata (UNMdP), Argentina. BT, CC and VAA are members of CONICET research staff.

Conflicts of Interest The authors declare no conflict of interest.

References

Abdel-Raouf, M. E., El-Saeed, S. M., Zaki, E. G., & Al-Sabagh, A. M. (2018). Green chemistry approach for preparation of hydrogels for agriculture applications through modification of natural polymers and investigating their swelling properties. *Egyptian Journal of Petroleum., 27*, 1345–1355. https://doi.org/10.1016/j.ejpe.2018.09.002.

Ahmed, E. M. (2015). Hydrogel: Preparation, characterization, and applications: A review. *Journal of Advanced Research, 6*(2), 105–121. https://doi.org/10.1016/j.jare.2013.07.006.

Aouada, F. A., de Moura, M. R., Orts, W. J., & Mattoso, L. H. C. (2010). Polyacrylamide and methylcellulose hydrogel as delivery vehicle for the controlled release of paraquat pesticide. *Journal of Materials Science, 45*(18), 4977–4985. https://doi.org/10.1007/s10853-009-4180-6.

Aouada, F. A., De Moura, M. R., & Mattoso, L. H. C. (2011). Biodegradable hydrogel as delivery vehicle for the controlled release of pesticide. In M. Stoytcheva (Ed.), *Pesticides – Formulations, effects, fate* (pp. 81–102). https://doi.org/10.5772/14139.

Azeem, B., Kushaari, K., Man, Z. B., Basit, A., & Thanh, T. H. (2014). Review on materials & methods to produce controlled release coated urea fertilizer. *Journal of Controlled Release, 181*(1), 11–21. https://doi.org/10.1016/j.jconrel.2014.02.020.

Bai, C., Zhang, S., Huang, L., Wang, H., Wang, W., & Ye, Q. (2015). Starch-based hydrogel loading with carbendazim for controlled-release and water absorption. *Carbohydrate Polymers, 125*, 376–383. https://doi.org/10.1016/j.carbpol.2015.03.004.

Bajpai, A. K., & Giri, A. (2003). Water sorption behaviour of highly swelling (carboxy methylcellulose-g-polyacrylamide) hydrogels and release of potassium nitrate as agrochemical. *Carbohydrate Polymers, 53*(3), 271–279. https://doi.org/10.1016/S0144-8617(03)00071-7.

Bawa, P., Pillay, V., Choonara, Y. E., & Du Toit, L. C. (2009). Stimuli-responsive polymers and their applications in drug delivery. *Biomedical Materials, 4*(2), 109–114. https://doi.org/10.1007/s10751-011-0451-0.

Belmokhtar, F. Z., Elbahri, Z., & Elbahri, M. (2018). Preparation and optimization of agrochemical 2,4-D controlled release microparticles using designs of experiments. *Journal of the Mexican Chemical Society, 62*(1), 1–21. https://doi.org/10.29356/jmcs.v62i1.579.

Bencherif, S. A., Siegwart, D. J., Srinivasan, A., Horkay, F., Hollinger, J. O., Washburn, N. R., & Matyjaszewski, K. (2009). Nanostructured hybrid hydrogels prepared by a combination of atom transfer radical polymerization and free radical polymerization. *Biomaterials, 30*(29), 5270–5278. https://doi.org/10.1016/j.biomaterials.2009.06.011.

Bordi, F., Paradossi, G., Rinaldi, C., & Ruzicka, B. (2002). Chemical and physical hydrogels: Two case systems studied by quasi elastic light scattering. *Physica A: Statistical Mechanics and its Applications, 304*(1–2), 119–128. https://doi.org/10.1016/s0378-4371(01)00537-4.

Bortolin, A., Aouada, F. A., Mattoso, L. H. C., & Ribeiro, C. (2013). Nanocomposite PAAm/methyl cellulose/montmorillonite hydrogel: Evidence of synergistic effects for the slow release of fertilizers. *Journal of Agricultural and Food Chemistry, 61*(31), 7431–7439. https://doi.org/10.1021/jf401273n.

Calabi-Floody, M., Medina, J., Rumpel, C., Condron, L. M., Hernandez, M., Dumont, M., & Mora, M. de la L. (2018). Smart fertilizers as a strategy for sustainable agriculture. In D. L. Sparks (Ed.), *Advances in agronomy* (Vol. 147, 1st ed.). Elsevier Inc.. https://doi.org/10.1016/bs.agron.2017.10.003.

Cannazza, G., Cataldo, A., de Benedetto, E., Demitri, C., Madaghiele, M., & Sannino, A. (2014). Experimental assessment of the use of a novel superabsorbent polymer (SAP) for the optimization of water consumption in agricultural irrigation process. *Water (Switzerland), 6*(7), 2056–2069. https://doi.org/10.3390/w6072056.

Carson, L. C., & Ozores-Hampton, M. (2013). Factors affecting nutrient availability, placement, rate, and application timing of controlled-release fertilizers for Florida vegetable production using seepage irrigation. *HortTechnology, 23*(5), 553–562. https://doi.org/10.21273/horttech.23.5.553.

Chagas, J. O., Gomes, J. M., de Cunha, I. C. M., de Melo, N. F. S., Fraceto, L. F., da Silva, G. A., & Lobo, F. A. (2018). Polymeric microparticles for modified release of NPK in agricultural applications. *Arabian Journal of Chemistry*. https://doi.org/10.1016/j.arabjc.2018.03.007.

Chang, C., Duan, B., Cai, J., & Zhang, L. (2010). Superabsorbent hydrogels based on cellulose for smart swelling and controllable delivery. *European Polymer Journal, 46*(1), 92–100. https://doi.org/10.1016/j.eurpolymj.2009.04.033.

Chapagain, B. P., Wiesman, Z., Zaccai, M., Imas, P., & Magen, H. (2003). Potassium chloride enhances fruit appearance and improves quality of fertigated greenhouse tomato as compared to potassium nitrate. *Journal of Plant Nutrition, 26*(3), 643–658. https://doi.org/10.1081/pln-120017671.

Chenite, A., Chaput, C., Wang, D., Combes, C., Buschmann, M. D., Hoemann, C. D., Leroux, J. C., Atkinson, B. L., Binette, F., & Selmani, A. (2000). Novel injectable neutral solutions of chitosan form biodegradable gels in situ. *Biomaterials, 21*(21), 2155–2161. https://doi.org/10.1016/S0142-9612(00)00116-2.

Conte, R., De Luise, A., Valentino, A., Di Cristo, F., Petillo, O., Riccitiello, F., Riccitiello, F., Di Salle, A., Calarco, A., & Peluso, G. (2019). Hydrogel nanocomposite systems: Characterization and application in drug-delivery systems. In S. S. Mohapatra, S. Ranjan, N. Dasgupta, R. K. Mishra, & S. Thomas (Eds.), *Nanocarriers for drug delivery* (pp. 319–349). https://doi.org/10.1016/B978-0-12-814033-8.00010-2.

Dai, H., Huang, Y., & Huang, H. (2018). Eco-friendly polyvinyl alcohol/carboxymethyl cellulose hydrogels reinforced with graphene oxide and bentonite for enhanced adsorption of methylene blue. *Carbohydrate Polymers, 185*, 1–11. https://doi.org/10.1016/j.carbpol.2017.12.073.

Dailey, O. D. J. (2004). Volatilization of alachlor from polymeric formulations. *Journal of Agricultural and Food Chemistry, 52*(22), 6742–6746. https://doi.org/10.1021/jf040034g.

Drozdov, A. D. (2015). Swelling of pH-responsive cationic gels: Constitutive modeling and structure–property relations. *International Journal of Solids and Structures, 64–65*, 176–190. https://doi.org/10.1016/j.ijsolstr.2015.03.023.

Ehrenhofer, A., Elstner, M., & Wallmersperger, T. (2018). Normalization of hydrogel swelling behavior for sensoric and actuatoric applications. *Sensors and Actuators, B: Chemical, 255*. (Part 2, 1343–1353. https://doi.org/10.1016/j.snb.2017.08.120.

Ekebafe, L. O., Ogbeifun, D. E., & Okieimen, F. E. (2011). Polymer applications in agriculture. *Nigerian Society for Experimental Biology, 23*(2), 81–89. https://doi.org/10.4314/biokem.v23i2.

Elbarbary, A. M., El-Rehim, H. A. A., El-Sawy, N. M., Hegazy, E. S. A., & Soliman, E. S. A. (2017). Radiation induced crosslinking of polyacrylamide incorporated low molecular weights natural polymers for possible use in the agricultural applications. *Carbohydrate Polymers, 176*, 19–28. https://doi.org/10.1016/j.carbpol.2017.08.050.

Essawy, H. A., Ghazy, M. B. M., El-Hai, F. A., & Mohamed, M. F. (2016). Superabsorbent hydrogels via graft polymerization of acrylic acid from chitosan-cellulose hybrid and their potential in controlled release of soil nutrients. *International Journal of Biological Macromolecules, 89*, 144–151. https://doi.org/10.1016/j.ijbiomac.2016.04.071.

Flores Céspedes, F., Pérez García, S., Villafranca Sánchez, M., & Fernández Pérez, M. (2013). Bentonite and anthracite in alginate-based controlled release formulations to reduce leaching of chloridazon and metribuzin in a calcareous soil. *Chemosphere, 92*(8), 918–924. https://doi.org/10.1016/j.chemosphere.2013.03.001.

Förster, S., & Antonietti, M. (1998). Amphiphilic block copolymers in structure-controlled nanomaterial hybrids. *Advanced Materials, 10*(3), 195–217. https://doi.org/10.1002/(sici)1521-4095(199802)10:3<195::aid-adma195>3.0.co;2-v.

França, D., Medina, Â. F., Messa, L. L., Souza, C. F., & Faez, R. (2018). Chitosan spray-dried microcapsule and microsphere as fertilizer host for swellable – controlled release materials. *Carbohydrate Polymers, 196*, 47–55. https://doi.org/10.1016/j.carbpol.2018.05.014.

Gad, Y. H. (2008). Preparation and characterization of poly(2-acrylamido-2-methylpropanesulfonic acid)/chitosan hydrogel using gamma irradiation and its application in wastewater treatment. *Radiation Physics and Chemistry, 77*(9), 1101–1107. https://doi.org/10.1016/j.radphyschem.2008.05.002.

González, M. E., Cea, M., Medina, J., González, A., Diez, M. C., Cartes, P., Monreal, C., & Navia, R. (2015). Evaluation of biodegradable polymers as encapsulating agents for the development of a urea controlled-release fertilizer using biochar as support material. *Science of the Total Environment, 505*, 446–453. https://doi.org/10.1016/j.scitotenv.2014.10.014.

Grillo, R., Pereira, A. E. S., Nishisaka, C. S., de Lima, R., Oehlke, K., Greiner, R., & Fraceto, L. F. (2014a). Chitosan/tripolyphosphate nanoparticles loaded with paraquat herbicide: An environmentally safer alternative for weed control. *Journal of Hazardous Materials, 278*, 163–171. https://doi.org/10.1016/j.jhazmat.2014.05.079.

Grillo, R., Rosa, A. H., & Fraceto, L. F. (2014b). Poly(ε-caprolactone) nanocapsules carrying the herbicide atrazine: Effect of chitosan-coating agent on physico-chemical stability and herbicide release profile. *International journal of Environmental Science and Technology, 11*(6), 1691–1700. https://doi.org/10.1007/s13762-013-0358-1.

Gross, R. A., & Kalra, B. (2002). Biodegradable polymers for the environment. *Science. (New York), 297*(5582), 803–807. https://doi.org/10.1126/science.297.5582.803.

Guilherme, M. R., Aouada, F. A., Fajardo, A. R., Martins, A. F., Paulino, A. T., Davi, M. F. T., Rubira, A. F., & Muniz, E. C. (2015). Superabsorbent hydrogels based on polysaccharides for application in agriculture as soil conditioner and nutrient carrier: A review. *European Polymer Journal, 72*, 365–385. https://doi.org/10.1016/j.eurpolymj.2015.04.017.

Gulrez, S. K. H., Al-Assaf, S., & Phillips, G. O. (2011). Hydrogels: Methods of preparation, characterisation and applications. In A. Carpi (Ed.), *Progress in molecular and environmental bioengineering IntechOpen progress in molecular and environmental bioengineering from analysis and modeling to technology applications Edited by Angelo Carpi Progress in molecular and environmental bioengineering*. https://doi.org/10.5772/24553.

Gutiérrez, T. J. (2017). Chitosan applications for the food industry. In S. Ahmed & S. Ikram (Eds.), *Chitosan: Derivatives, composites and applications* (pp. 183–232). WILEY-Scrivener Publisher. EE.UU. ISBN: 978-1-119-36350-7. https://doi.org/10.1002/9781119364849.ch8.

Gutiérrez, T. J. (2018a). Processing nano- and microcapsules for industrial applications. In C. M. Hussain (Ed.), *Handbook of nanomaterials for industrial applications* (pp. 989–1011). Editorial Elsevier. EE.UU. ISBN: 978-0-12-813351-4. https://doi.org/10.1016/b978-0-12-813351-4.00057-2.

Gutiérrez, T. J. (2018b). Biodegradability and compostability of food nanopackaging materials. In G. Cirillo, M. A. Kozlowski, & U. G. Spizzirri (Eds.), *Composite materials for food packaging* (pp. 269–296). WILEY-Scrivener Publisher. EE.UU. ISBN: 978-1-119-16020-5. https://doi.org/10.1002/9781119160243.ch9.

Gutiérrez, T. J. (2018c). Biological macromolecule composite films made from Sagu starch and flour/poly(ε-caprolactone) blends processed by blending/thermo molding. *Journal Polymers and the Environment, 26*(9), 3902–3912. https://doi.org/10.1007/s10924-018-1268-6.

Gutiérrez, T. J. (2018d). Are modified pumpkin flour/plum flour nanocomposite films biodegradable and compostable? *Food Hydrocolloids, 83*, 397–410. https://doi.org/10.1016/j.foodhyd.2018.05.035.

Gutiérrez, T. J., & Álvarez, K. (2017). Biopolymers as microencapsulation materials in the food industry. In M. Masuelli & D. Renard (Eds.), *Advances in physicochemical properties of biopolymers: Part 2* (pp. 296–322). Bentham Science Publishers.. EE.UU. ISBN: 978-1-68108-545-6. eISBN: 978-1-68108-544-9, 2017. https://doi.org/10.2174/9781681085449117010009.

Gutiérrez, T. J., & Alvarez, V. A. (2017a). Eco-friendly films prepared from plantain flour/PCL blends under reactive extrusion conditions using zirconium octanoate as a catalyst. *Carbohydrate Polymers, 178*, 260–269. https://doi.org/10.1016/j.carbpol.2017.09.026.

Gutiérrez, T. J., & Alvarez, V. A. (2017b). Films made by blending poly(ε-caprolactone) with starch and flour from Sagu rhizome grown at the Venezuelan amazons. *Journal Polymers and the Environment, 25*(3), 701–716. https://doi.org/10.1007/s10924-016-0861-9.

Gutiérrez, T. J., Toro-Márquez, L. A., Merino, D., & Mendieta, J. R. (2019). Hydrogen-bonding interactions and compostability of bionanocomposite films prepared from corn starch and nano-fillers with and without added Jamaica flower extract. *Food Hydrocolloids, 89*, 283–293. https://doi.org/10.1016/j.foodhyd.2018.10.058.

Gwon, S. H., Yoon, J., Seok, H. K., Oh, K. H., & Sun, J.-Y. (2015). Gelation dynamics of ionically crosslinked alginate gel with various cations. *Macromolecular Research, 23*(12), 1112–1116. https://doi.org/10.1007/s13233-015-3151-9.

Hamedi, H., Moradi, S., Hudson, S. M., & Tonelli, A. E. (2018). Chitosan based hydrogels and their applications for drug delivery in wound dressings: A review. *Carbohydrate Polymers, 199*, 445–460. https://doi.org/10.1016/j.carbpol.2018.06.114.

Heinze, T., Liebert, T., Heublein, B., & Hornig, S. (2006). Functional Polymers based on dextran. In D. Klemm (Ed.), *Polysaccharides II.* (pp. 199–291). Berlin/Heidelberg: Springer Berlin Heidelberg. https://doi.org/10.1007/12_100.

Hemvichian, K., Chanthawong, A., & Suwanmala, P. (2014). Synthesis and characterization of superabsorbent polymer prepared by radiation-induced graft copolymerization of acrylamide onto carboxymethyl cellulose for controlled release of agrochemicals. *Radiation Physics and Chemistry, 103*, 167–171. https://doi.org/10.1016/j.radphyschem.2014.05.064.

Hennink, W. E., & van Nostrum, C. F. (2002). Novel crosslinking methods to design hydrogels. *Advanced Drug Delivery Reviews, 54*(supplement), 13–36. https://doi.org/10.1016/j.addr.2012.09.009.

Herniou--Julien, C., Mendieta, J. R., & Gutiérrez, T. J. (2019). Characterization of biodegradable/non-compostable films made from cellulose acetate/corn starch blends processed under reactive extrusion conditions. *Food Hydrocolloids, 89*, 67–79. https://doi.org/10.1016/j.foodhyd.2018.10.024.

Holback, H., Yeo, Y., & Park, K. (2011). Hydrogel swelling behavior and its biomedical applications. In S. Rimmer (Ed.), *Biomedical hydrogels: Biochemistry, Manufacture and medical applications*. Woodhead Publishing Limited. https://doi.org/10.1016/b978-1-84569-590-3.50001-1.

Joshi, S. J., & Abed, R. M. M. (2017). Biodegradation of polyacrylamide and its derivatives. *Environmental Processes, 4*(2), 463–476. https://doi.org/10.1007/s40710-017-0224-0.

Kalhapure, A., Kumar, R., Singh, V. P., & Pandey, D. S. (2016). Hydrogels: A boon for increasing agricultural productivity in water-stressed environment. *Current Science, 111*(11), 1773–1779. https://doi.org/10.18520/cs/v111/i11/1773-1779.

Katti, D. S., Lakshmi, S., Langer, R., & Laurencin, C. T. (2002). Toxicity, biodegradation and elimination of polyanhydrides. *Advanced Drug Delivery Reviews, 54*(7), 933–961. https://doi.org/10.1016/S0169-409x(02)00052-2.

Kenawy, E. R., Chemistry, A., & Akelah, A. (1992). *Município de Odivelas, 28*(8), 2009. https://doi.org/10.1016/0014-3057(92)90310-x.

Kuijpers, A. J., van Wachem, P. B., van Luyn, M. J., Engbers, G. H., Krijgsveld, J., Zaat, S. A., Dankert, J., & Feijen, J. (2000). In vivo and in vitro release of lysozyme from cross-linked gelatin hydrogels: A model system for the delivery of antibacterial proteins from prosthetic heart valves. *Journal of Controlled Release : Official Journal of the Controlled Release Society, 67*(2–3), 323–336. https://doi.org/10.1016/s0168-3659(00)00221-2.

Kumar, S., Bhanjana, G., Sharma, A., Sidhu, M. C., & Dilbaghi, N. (2014). Synthesis, characterization and on field evaluation of pesticide loaded sodium alginate nanoparticles. *Carbohydrate Polymers, 101*, 1061–1067. https://doi.org/10.1016/j.carbpol.2013.10.025.

Li, J., Yao, J., Li, Y., & Shao, Y. (2012). Controlled release and retarded leaching of pesticides by encapsulating in carboxymethyl chitosan/bentonite composite gel. *Journal of Environmental Science and Health, Part B, 47*(8), 795–803. https://doi.org/10.1080/03601234.2012.676421.

Li, S. (1999). Hydrolytic degradation characteristics of aliphatic polyesters derived from lactic and glycolic acids. *Journal of Biomedical Materials Research, 48*(3), 342–353. https://doi.org/10.1002/(sici)1097-4636(1999)48:3<342::aid-jbm20>3.0.co;2-7.

Mahdavinia, G. R., Mousavi, S. B., Karimi, F., Marandi, G. B., Shahabivand, S., & Harati, M. (2009). Release behavior of 2,4-dichlorophenoxyacetic acid herbicide using novel porous polyacrylamide hydrogels. *E-Polymers, 9*(11), 1–13. https://doi.org/10.1515/epoly.2009.9.1.1718.

Mansur, H. S., Oréfice, R. L., & Mansur, A. A. P. (2004). Characterization of poly(vinyl alcohol)/poly(ethylene glycol) hydrogels and PVA-derived hybrids by small-angle X-ray scattering and FTIR spectroscopy. *Polymer, 45*(21), 7193–7202. https://doi.org/10.1016/j.polymer.2004.08.036.

Mansur, H. S., Sadahira, C. M., Souza, A. N., & Mansur, A. A. P. (2008). FTIR spectroscopy characterization of poly (vinyl alcohol) hydrogel with different hydrolysis degree and chemically crosslinked with glutaraldehyde. *Materials Science and Engineering: C, 28*(4), 539–548. https://doi.org/10.1016/j.msec.2007.10.088.

Maruyama, C. R., Guilger, M., Pascoli, M., Bileshy-José, N., Abhilash, P. C., Fraceto, L. F., & de Lima, R. (2016). Nanoparticles based on chitosan as carriers for the combined herbicides imazapic and imazapyr. *Scientific Reports, 6*, 19768. https://doi.org/10.1038/srep19768.

Mekonnen, T., Mussone, P., Khalil, H., & Bressler, D. (2013). Progress in bio-based plastics and plasticizing modifications. *Journal of Materials Chemistry A, 1*(43), 13379–13398. https://doi.org/10.1039/c3ta12555f.

Milani, P., Franca, D., Balieiro, A. G., & Faez, R. (2017). Polymers and its applications in agriculture. *Polímeros, 27*(3), 256–266. https://doi.org/10.1590/0104-1428.09316.

Moreira Teixeira, L. S., Feijen, J., van Blitterswijk, C. A., Dijkstra, P. J., & Karperien, M. (2012). Enzyme-catalyzed crosslinkable hydrogels: Emerging strategies for tissue engineering. *Biomaterials, 33*(5), 1281–1290. https://doi.org/10.1016/j.biomaterials.2011.10.067.

Nayak, P. L. (1999). Biodegradable polymers: Opportunities and challenges. *Journal of Macromolecular Science, Part C, 39*(3), 481–505. https://doi.org/10.1081/mc-100101425.

Olad, A., Zebhi, H., Salari, D., Mirmohseni, A., & Reyhani Tabar, A. (2018a). Slow-release NPK fertilizer encapsulated by carboxymethyl cellulose-based nanocomposite with the function of water retention in soil. *Materials Science and Engineering C, 90*, 333–340. https://doi.org/10.1016/j.msec.2018.04.083.

Olad, A., Zebhi, H., Salari, D., Mirmohseni, A., & Tabar, A. R. (2018b). Water retention and slow release studies of a salep-based hydrogel nanocomposite reinforced with montmorillonite clay. *New Journal of Chemistry, 42*(4), 2758–2766. https://doi.org/10.1039/c7nj03667a.

Pang, S. C., Chin, S. F., Tay, S. H., & Tchong, F. M. (2011). Starch-maleate-polyvinyl alcohol hydrogels with controllable swelling behaviors. *Carbohydrate Polymers, 84*(1), 424–429. https://doi.org/10.1016/j.carbpol.2010.12.002.

Peppas, N. A., & Van Blarcom, D. S. (2016). Hydrogel-based biosensors and sensing devices for drug delivery. *Journal of Controlled Release, 240*, 142–150. https://doi.org/10.1016/j.jconrel.2015.11.022.

Pereira, A. G. B., Martins, A. F., Paulino, A. T., Fajardo, A. R., Guilherme, M. R., Faria, M. G. I., Linde, G. A., Rubira, A. F., & Muniz E. C. (2017). Recent advances in designing hydrogels from chitin and chitin-derivatives and their impact on environment and agriculture: A review. *Revista Virtual de Quimica, 9*(1), 370–386. https://doi.org/10.21577/1984-6835.20170021.

Pérez-Martínez, J. I., Morillo, E., Maqueda, C., & Ginés, J. M. (2001). Ethyl cellulose polymer microspheres for controlled release of norfluazon. *Pest Management Science, 57*(8), 688–694. https://doi.org/10.1002/ps.339.

Perez, J. J., & Francois, N. J. (2016). Chitosan-starch beads prepared by ionotropic gelation as potential matrices for controlled release of fertilizers. *Carbohydrate Polymers, 148*, 134–142. https://doi.org/10.1016/j.carbpol.2016.04.054.

Pethrick, R. A. (2004). Polymer physics. Michael Rubinstein and Ralph H Colby (Eds). Oxford University Press, Oxford, 2003.. ISBN 019852059X. pp 440. Polymer International, 53(9), 1394–1395. https://doi.org/10.1002/pi.1472.

Pourjavadi, A., Harzandi, A. M., & Hosseinzadeh, H. (2004). Modified carrageenan 3. Synthesis of a novel polysaccharide-based superabsorbent hydrogel via graft copolymerization of acrylic acid onto kappa-carrageenan in air. *European Polymer Journal, 40*(7), 1363–1370. https://doi.org/10.1016/j.eurpolymj.2004.02.016.

Puoci, F., Iemma, F., Spizzirri, U. G., Cirillo, G., Curcio, M., & Picci, N. (2008). Polymer in agriculture : A review. *Review Literature And Arts Of The Americas, 3*(1), 299–314.

Ramos Campos, E. V., de Oliveira, J. L., Fraceto, L. F., & Singh, B. (2014). Polysaccharides as safer release systems for agrochemicals. *Agronomy for Sustainable Development, 35*(1), 47–66. https://doi.org/10.1007/s13593-014-0263-0.

Ramos Campos, E. V., De Oliveira, J. L., Fernandes Fraceto, L., & Singh, B. (2015). *Polysaccharides as safer release systems for agrochemicals* (Vol. 35, pp. 47–66). https://doi.org/10.1007/s13593-014-0263-0.

Rashidzadeh, A., & Olad, A. (2014). Slow-released NPK fertilizer encapsulated by NaAlg-*g*-poly(AA-*co*-AAm)/MMT superabsorbent nanocomposite. *Carbohydrate Polymers, 114*, 269–278. https://doi.org/10.1016/j.carbpol.2014.08.010.

Rashidzadeh, A., Olad, A., Salari, D., & Reyhanitabar, A. (2014). On the preparation and swelling properties of hydrogel nanocomposite based on sodium alginate-*g*-poly (acrylic acid-*co*-acrylamide)/clinoptilolite and its application as slow release fertilizer. *Journal of Polymer Research, 21*, 344. https://doi.org/10.1007/s10965-013-0344-9.

Ray, D., Gils, P. S., Mohanta, G. P., Manavalan, R., & Sahoo, P. K. (2010). Comparative delivery of diltiazem hydrochloride through synthesized polymer: Hydrogel and hydrogel microspheres. *Journal of Applied Polymer Science, 116*(2), 959–968. https://doi.org/10.1002/app.31661.

Rinaudo, M. (2006). Chitin and chitosan: Properties and applications. *Progress in Polymer Science, 31*(7), 603–632. https://doi.org/10.1016/j.progpolymsci.2006.06.001.

Rohindra, D. R., Nand, A. V., & Khurma, J. R. (2004). Advances in Computation and Intelligence: 4th International Symposium on... Cai, Z., Li, Z., Kang, Z., Liu, Y. (Eds.). Springer.

Rudzinski, W. E., Dave, A. M., Vaishnav, U. H., Kumbar, S. G., Kulkarni, A. R., & Aminabhavi, T. M. (2002). Hydrogels as controlled release devices in agriculture, *5*(1), 39–65. https://doi.org/10.1163/156855502760151580.

Sakai, S., & Kawakami, K. (2007). Synthesis and characterization of both ionically and enzymatically cross-linkable alginate. *Acta Biomaterialia, 3*(4), 495–501. https://doi.org/10.1016/j.actbio.2006.12.002.

Senna, A. M., & Botaro, V. R. (2017). Biodegradable hydrogel derived from cellulose acetate and EDTA as a reduction substrate of leaching NPK compound fertilizer and water retention in soil. *Journal of Controlled Release, 260*, 194–201. https://doi.org/10.1016/j.jconrel.2017.06.009.

Senna, A. M., Braga Do Carmo, J., Santana Da Silva, J. M., & Botaro, V. R. (2015). Synthesis, characterization and application of hydrogel derived from cellulose acetate as a substrate for slow-release NPK fertilizer and water retention in soil. *Journal of Environmental Chemical Engineering, 3*(2), 996–1002. https://doi.org/10.1016/j.jece.2015.03.008.

Shantha, K. L., & Harding, D. R. K. (2002). Synthesis and evaluation of sucrose-containing polymeric hydrogels for oral drug delivery. *Journal of Applied Polymer Science, 84*(14), 2597–2604. https://doi.org/10.1002/app.10378.

Sharma, R., Bajpai, J., Bajpai, A. K., Acharya, S., Shrivastava, R. B., & Shukla, S. K. (2014). Designing slow water-releasing alginate nanoreserviors for sustained irrigation in scanty rainfall areas. *Carbohydrate Polymers, 102*, 513–520. https://doi.org/10.1016/j.carbpol.2013.11.059.

dos Silva, M. S., Cocenza, D. S., Grillo, R., Silva de Melo, N. F., Tonello, P. S., de Oliveira, C., Lopes Cassimiro, C., Henrique, D., Rosa, A., & Fraceto, L. F. (2011). Paraquat-loaded alginate/chitosan nanoparticles: Preparation, characterization and soil sorption studies. *Journal of Hazardous Materials, 190*(1–3), 366–374. https://doi.org/10.1016/j.jhazmat.2011.03.057.

Singh, B., Sharma, D. K., Negi, S., & Dhiman, A. (2015). Synthesis and characterization of agar-starch based hydrogels for slow herbicide delivery applications. *International Journal of Plastics Technology, 19*(2), 263–274. https://doi.org/10.1007/s12588-015-9126-z.

Sopeña-Vázquez, F., Morillo, E., Villaverde, J., & Maqueda, C. (2008). Reduction of pesticide pollution in agricutural soils by using controlled release formulations: Efficacy and persistence studies. *Fresenius Environmental Bulletin, 17*(10 B), 1705–1710.

Sopeña, F., Cabrera, A., Maqueda, C., & Morillo, E. (2007). Ethylcellulose formulations for controlled release of the herbicide alachlor in a sandy soil. *Journal of Agricultural and Food Chemistry, 55*(20), 8200–8205. https://doi.org/10.1021/jf071459g.

Sopeña, F., Maqueda, C., & Morillo, E. (2009a). Controlled release formulations of herbicides based on micro-encapsulation. *Ciencia e Investigación Agraria*. scielocl.

Sopeña, F., Maqueda, C., & Morillo, E. (2009b). Formulation affecting alachlor efficacy and persistence in sandy soils. *Pest Management Science, 65*(7), 761–768. https://doi.org/10.1002/ps.1750.

Sopeña, F., Villaverde, J., Maqueda, C., & Morillo, E. (2011). Photostabilization of the herbicide norflurazon microencapsulated with ethylcellulose in the soil-water system. *Journal of Hazardous Materials, 195*, 298–305. https://doi.org/10.1016/j.jhazmat.2011.08.039.

Soppimath, K. S., Aminabhavi, T. M., Dave, A. M., Kumbar, S. G., & Rudzinski, W. E. (2002). Stimulus-responsive "smart" hydrogels as novel drug delivery systems. *Drug Development and Industrial Pharmacy, 28*(8), 957–974. https://doi.org/10.1081/ddc-120006428.

Tabata, Y. (2009). Biomaterial technology for tissue engineering applications. *Journal of the Royal Society, Interface, 6*(Suppl 3), S311–S324. https://doi.org/10.1098/rsif.2008.0448.focus.

Tanan, W., Panichpakdee, J., & Saengsuwan, S. (2018). Novel biodegradable hydrogel based on natural polymers : Synthesis, characterization, swelling / reswelling and biodegradability. *European Polymer Journal, 112*, 678–687. https://doi.org/10.1016/j.eurpolymj.2018.10.033.

Teodorescu, M., Lungu, A., Stanescu, P. O., & Neamt, C. (2009). Preparation and properties of novel slow-release NPK agrochemical formulations based on poly (acrylic acid) hydrogels and liquid fertilizers. *Industrial & Engineering Chemistry Research, 48*(14), 6527–6534. https://doi.org/10.1021/ie900254b.

Terao, K., Nagasawa, N., Nishida, H., Furusawa, K., Mori, Y., Yoshii, F., & Dobashi, T. (2003). Reagent-free crosslinking of aqueous gelatin: Manufacture and characteristics of gelatin gels irradiated with gamma-ray and electron beam. *Journal of Biomaterials Science..* Polymer Edition, *14*(11), 1197–1208.

Thombare, N., Mishra, S., Siddiqui, M. Z., Jha, U., Singh, D., & Mahajan, G. R. (2018). Design and development of guar gum based novel, superabsorbent and moisture retaining hydrogels for agricultural applications. *Carbohydrate Polymers., 185*, 169–178. https://doi.org/10.1016/j.carbpol.2018.01.018.

Vasilakoglou, I. B., & Eleftherohorinos, I. G. (2003). Persistence, efficacy, and selectivity of amide herbicides in corn. *Weed Technology, 17*(2), 381–388. https://doi.org/10.2307/3989325.

Vundavalli, R., Vundavalli, S., Nakka, M., & Rao, D. S. (2015). Biodegradable nano-hydrogels in agricultural farming - alternative source for water resources. *Procedia Materials Science, 10*, 548–554. https://doi.org/10.1016/j.mspro.2015.06.005.

Wang, T., Turhan, M., & Gunasekaran, S. (2004). Selected properties of pH-sensitive, biodegradable chitosan–poly(vinyl alcohol) hydrogel. *Polymer International, 53*(7), 911–918. https://doi.org/10.1002/pi.1461.

Wang, W., & Wang, A. (2010). Synthesis and swelling properties of pH-sensitive semi-IPN superabsorbent hydrogels based on sodium alginate-*g*-poly(sodium acrylate) and polyvinylpyrrolidone. *Carbohydrate Polymers, 80*(4), 1028–1036. https://doi.org/10.1016/j.carbpol.2010.01.020.

Wang, X.-L., Yang, K.-K., & Wang, Y.-Z. (2003). Properties of starch blends with biodegradable polymers. *Journal of Macromolecular Science, Part C, 43*(3), 385–409. https://doi.org/10.1081/mc-120023911.

Wu, Y., Zheng, Y., Yang, W., Wang, C., Hu, J., & Fu, S. (2005). Synthesis and characterization of a novel amphiphilic chitosan–polylactide graft copolymer. *Carbohydrate Polymers, 59*(2), 165–171. https://doi.org/10.1016/j.carbpol.2004.09.006.

Yegappan, R., Selvaprithiviraj, V., Amirthalingam, S., & Jayakumar, R. (2018). Carrageenan based hydrogels for drug delivery, tissue engineering and wound healing. *Carbohydrate Polymers, 198*, 385–400. https://doi.org/10.1016/j.carbpol.2018.06.086.

Zhu, Z., & Zhuo, R. (2000). Crosslinked quaternary ammonium cornstarch matrix for slow release of carboxylic groups-containing herbicides. *Starch - Stärke, 52*(2–3), 58–63. https://doi.org/10.1002/(sici)1521-379x(200004)52:2/3<58::aid-star58>3.0.co;2-3.

Chapter 8
Hydrogels: An Effective Tool to Improve Nitrogen Use Efficiency in Crops

Mariana A. Melaj and Rocío B. Gimenez

Abstract Hydrogels are hydrophilic polymeric materials with three-dimensional (3D) network structures physic and/or chemically crosslinked. They can absorb large amounts of water or aqueous solutions for a short period of time. When the hydrogel is loaded with a drug and meets an aqueous medium, the water penetrates the system and dissolves the drug. To improve the nitrogen use efficiency by plants, controlled-release fertilizers are developed with hydrogels as matrices. In particular, biodegradable polymers such as starch, xanthan, chitosan, cellulose derivatives, pectin and their mixtures have been essayed to avoid residues in soils. In addition, the superabsorbent capacity of the hydrogels improves the retention of fertilizers in the soil, reduces water losses through evaporation and decreases the frequency of irrigation. In this chapter, the behavior of different hydrogels to improve the availability of nitrogen by the plants during their crop cycle will be reviewed.

Keywords Agriculture · Fertilizers · Polymers

8.1 Introduction

The increase in world population has led to the industrialization of agriculture. Food production is a challenge related to crop yield, which can be improved through fertilization. However, conventional fertilizers have an environmental impact due to the loss of nutrients, which can cause toxicity to the crop at concentrations not tolerable in the plant roots.

Nitrogen is one of the main nutrients that are applied in high doses to crops. However, its absorption depends on the plant species of the crop, the tillage, the properties of the soil, the environmental conditions and the nitrogen source. When the fertilizer is applied, part of it is taken by the plants and the rest is lost. For this reason,

M. A. Melaj (✉) · R. B. Gimenez
ITPN. Instituto de Tecnología en Polímeros y Nanotecnología, Facultad de Ingeniería – UBA – CONICET, Buenos Aires, Argentina
e-mail: mmelaj@fi.uba.ar

T. J. Gutiérrez (ed.), *Polymers for Agri-Food Applications*,
https://doi.org/10.1007/978-3-030-19416-1_8

the nutrient use efficiency is a key point in sustainable agriculture. The nutrient use efficiency is defined as the amount of nutrients taken by the plant from soil over a period of time, related to the available nutrient present in the soil (including applied fertilizer). The nitrogen use efficiency (NUE) usually reaches 50% (Sempeho et al. 2014). Depending on the method of application and the type of fertilizer, the losses can reach 90% of the applied fertilizer (Bajpai and Giri 2002). The main processes involved in the loss of nitrogen in the crops are listed and described below:

1. Nitrogen lixiviation

Mineral nitrogen in most soils can be oxidized to NO_3^- by the action of microorganisms. Oxidized nitrogenous compounds can be leached from the root zone of the plants and reach the surface and deep waters. Large accumulations of nitrogen caused by fertilizers can lead to:

- Metahemoglobinemia in babies because of the reduction of NO_3^- to NO_2^- and the oxidation of hemoglobin.
- Gastric cancers related to the presence of NO_2^- or nitrosamines: the excess NO_3^- ions are transformed into NO_2^- in the stomach, thus reacting with the amines and leading to the production of nitrosamines, which are carcinogenic compounds.
- Eutrophication of the water surface.

According to the World Health Organization (WHO 2003), the maximum allowed limit of NO_3^- in drinking water is 50 mg/L.

2. Ammonium volatilization and nitrous oxide emission

Urea or ammonium salts when applied as fertilizers in calcareous soils can be potentially volatized. In particular, NH_4^+ can be accumulated in ecosystems, thus causing damage to vegetation. In addition, NO_3^- can be oxidized into N_2O and NO (greenhouse gases), and them transformed into nitric acid generating acid rain and lakes (Asgedom et al. 2014; Savci 2012; Shaviv 2000; Smith and Siciliano 2015).

The aforementioned adverse effects of the use of fertilizers have led to the need to reduce the application rate of fertilizers and improve their application efficiency, with the aim of improving nitrogen fixation from crops, reducing costs by their application and avoiding environmental problems and health caused by them, without placing food production at risk (Davidson and Gu 2012).

Controlled-release fertilizers (CRF) have had a great boom because they complement the desired requirements for the new generation of fertilizer: fertilizer use efficiency increases and reduces nitrogen losses through better synchronization between nitrogen availability and nutritional demand of the plant.

Figure 8.1 shows the difference between CRF and conventional fertilizers related to the nutrient demand of the plant. A nitrogen fertilizer such as urea when applied at planting, the absorption of fertilizer is low and the nutrient is vulnerable to being lost due to volatilization, denitrification or leaching. In this sense, CRFs have a better synchronization with crop demand, minimizing fertilizer losses (Chalk et al. 2015).

The CRFs also have the following advantages:

- The supply of applied nutrients is protected.
- The controlled and sustained release of the fertilizer is achieved.

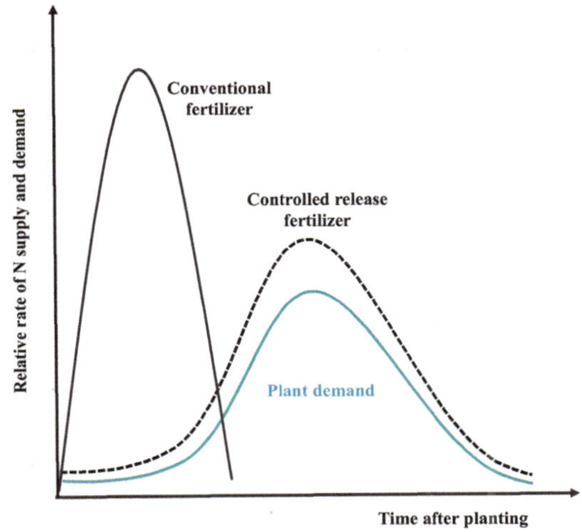

Fig. 8.1 Supply of conventional *vs* controlled release fertilizers and its synchrony with the demand for nutrients from crops. Adapted from Chalk et al. (2015)

- The concentration of nutrients between the optimal limits during a given period of time is maintained.

As a result of this:

- The availability of fertilizers lasts longer.
- It is not necessary to apply the fertilizer several times during a crop cycle.
- The cost of tillage is reduced.
- Environmental pollution is reduced (Geng et al. 2015; Anitha and Bindu 2016; Azeem et al. 2016; Geicu-Cristea et al. 2016; Trinh and KuShaari 2016).

CRFs can be tablets or granules coated with polymers, or matrices containing the dispersed nutrient, thereby delaying the release of fertilizer. The matrices and the coatings can be prepared from hydrophobic materials (polyolefins, rubber, etc.) or hydrogels (gel-forming polymers). Hydrogels loaded with fertilizers reduce their dissolution due to their water retention (swelling) (Trenkel 2010). In addition, hydrogels increase the water retention capacity of soils, reducing irrigation, increasing nutrient absorption and improving soil permeability, and this is reflected in an increase in crop yields (Elbarbary and Ghobashy 2017). In this regard, Hendrawan et al. (2016) developed fertilizer-loaded polymer composite hydrogels to achieve water and nutrient supply simultaneously.

8.2 Hydrogels

Hydrogels were first reported by Wichterle and Lím (1960). Hydrogels are hydrophilic polymers originated by the crosslinking of polymer chains, which generate macromolecular networks. These structures can absorb and retain a large amount of water or aqueous solutions within their structure (Chang and Zhang 2011).

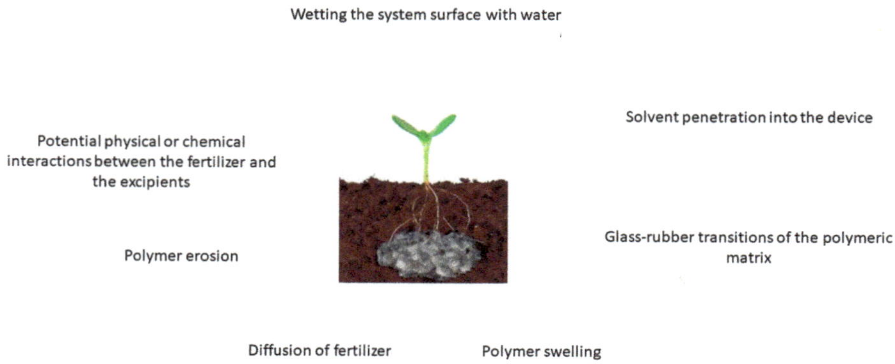

Fig. 8.2 Phenomena given during the release of fertilizers from hydrophilic matrices

The presence of hydrophilic functional groups such as – OH, – CONH, – $CONH_2$ and – SO_3H linked to the polymer backbone are responsible for this property (Mahinroosta et al. 2018; Nagam et al. 2016). There are different types of classifications of hydrogels based on their charge (anionic, cationic, nonionic, amphoteric or zwitterionic charges), method of cross-linking (chemical and physical hydrogels) and shape (matrix, film or microsphere) (Ahmed 2015; Mahinroosta et al. 2018). In particular, physically cross-linked hydrogels are obtained when the 3D network of the polymer is held together by van der Waals (dipole-dipole or hydrogen bonds) or coulombic (electrostatic) type interactions, as well as by overlapping polymer chains. In contrast, the polymer chains are covalently linked in chemically cross-linked hydrogels (Silva et al. 2009).

On the other hand, the type and quantity of the fertilizer, as well as the type of excipients, the preparation technique, the environmental conditions during the release of the fertilizer, and the geometry and dimensions of the hydrogel can affect one or more of the phenomena shown in Fig. 8.2 (Siepmann and Siepmann 2008):

8.3 Devices

Controlled release systems can be reservoir and matrix or monolithic type (Fig. 8.3). In the reservoir system, the fertilizer to be released is surrounded by a polymer layer, and this may be in a solid state or in a diluted or concentrated solution. In general, the time of dissolution of the coated fertilizer can be delayed, increasing the thickness of the coating (Du et al. 2006).

On the contrary, the fertilizer is distributed homogeneously in the polymer network when the system is constituted by a matrix and can be dissolved (in smaller amounts than its solubility allows) or dispersed (part of the fertilizer is in the form solid).

8.4 Fertilizer Release Mechanisms

In general, diffusion, swelling and erosion are the most important mechanisms that control the release rate of fertilizer from hydrophilic matrices to the soil, and these mechanisms can act in different steps or simultaneously (Siepmann and Siepmann 2012). These main fertilizer release mechanisms are described below:

8.4.1 *Diffusion*

Diffusion is a process of spontaneous mass transfer of a substance from a concentrated solution to a diluted solution, thus creating a gradient or profile concentration. Once the steady state is reached, the concentration gradient disappears, and equilibrium is achieved. The diffusion mechanism depends on the free volume of the polymer chains, and this in turn depends on the structure (matrix – Fig. 8.3a or reservoir – Fig. 8.3b) and composition of the hydrogel polymer matrix. A homogeneous hydrogel increases its free volume when subjected to temperatures around

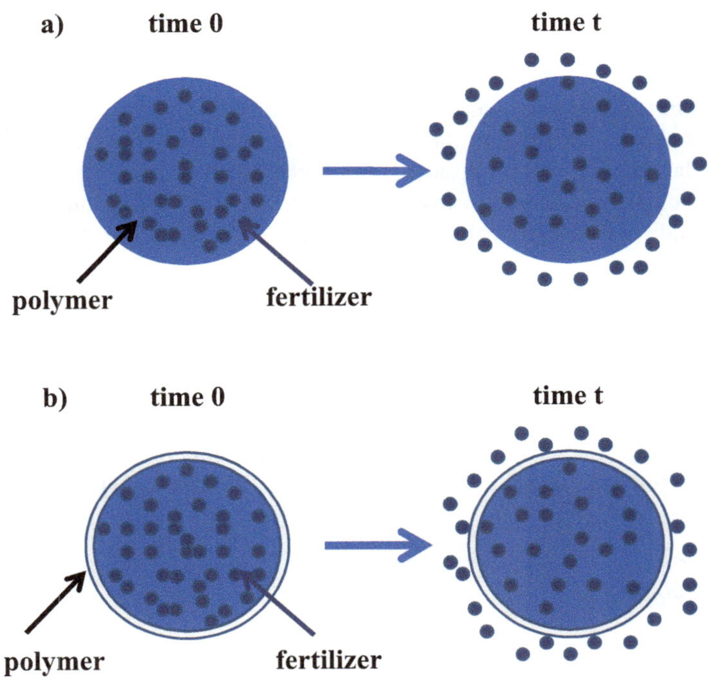

Fig. 8.3 (**a**) Matrix fertilizer delivery device and (**b**) Reservoir fertilizer delivery device. Adapted from https://www.sigmaaldrich.com/technical-documents/articles/materials-science/biomaterials/tutorial.html

or above the glass transition temperature (T_g). The free volume of the hydrogel can also be increased by copolymerization or by using of polymer blends. The plasticizing effect is another way that allows the increase of the free volume of the hydrogel, by means of using compounds with small structures such as water, which allows to lubricate the polymer chains, thus having greater mobility and free volume in the polymer network.

As indicated before, the type of structure of the hydrogel affects the release mechanism of the fertilizer. In the case of the **reservoir** type system, through the outer layer, the fertilizer must diffuse, dissolve or divide between the carrier and the membrane, diffuse through the membrane, be distributed between the membrane and the external elution medium and, finally, be transported from the surface of the membrane system. The reservoir when saturated, the concentration gradient of the fertilizer is constant in the membrane, as well as the flow rate of the compound, therefore a zero-order release kinetic is achieved (Siegel and Rathbone 2012). Zero order release kinetics is an advantage, but these systems are not easy to elaborate. The dissolution of the fertilizer in the polymeric matrix plays a decisive role in **matrix or monolithic** systems. In a first stage, diffusion occurs at the level of the surface when the fertilizer dissolves in the matrix. Then, the partition of the fertilizer between the matrix and the release medium and transport is given from the surface. In case the fertilizer is dispersed, first this compound must dissolve in the matrix before being released. Typically, matrix systems show a burst release from the surface and then release rate decreases with time.

Burst release is the release of a certain amount of fertilizer immediately after the system comes into contact with the receiving medium, after this stage the release profile is stable. Depending on the application of the release system, this phenomenon can be considered as a disadvantage, since it can provide toxicity at certain concentrations of the fertilizer, loss of active agent, decrease in the release time, which would entail the need for higher doses, and in other cases this could be an advantage such as certain treatments for wounds, encapsulated flavors, pulsed releases, etc. The impossibility of predicting the amount of fertilizer that will be released is the disadvantage of this process.

The burst release in monolithic matrices can be attributed to (1) certain conditions of synthesis and elaboration, which results in a quantity of fertilizer trapped in the surface of the device, causing a heterogeneous distribution of the substance, (2) heterogeneous polymeric matrices associated with different pore sizes and (3) fertilizer properties (molecular weight, solubility) (Huang and Brazel 2001).

8.4.2 Swelling

Materials with a hydrophilic behavior when they come into contact with water tend to be swollen. First, the polymer network is surrounded by water and expands, thus increasing its thickness and the volume of the polymer device. Once this process has occurred, chemical and physical bonds are formed. Subsequently, the water enters the polymer relatively quickly and the polymer chains are disentangled. This expansion of volume causes greater spaces between the polymer chains, which can be used to control the release of the fertilizer, regardless of the type of system used: matrix or reservoir (Fig. 8.3). Finally, the matrix disappears when the polymer is completely swollen. This type of system is called a soluble inflatable matrix (Colombo 1993). However, in some cases, the swelling of the material is limited, and the matrix is not totally disintegrated. This occurs when: (1) the water and polymer are not fully compatible, (2) the polymer chains are very long and (3) when a crosslinking agent is added. Finally, the swelling kinetics will be responsible for the controlled release of the fertilizer. Therefore, the swelling can be modulated by the hydrophilic or hydrophobic nature of the polymer, the type of fertilizer to be released, the temperature and pH of the medium and the concentration of the crosslinking agent (degree of crosslinking) (Siegel and Rathbone 2012).

The matrices showing swelling can have two different behaviors:

Classic hydrophilic matrices (unlimited swelling), diffusion is the main mechanism that controls the release. When the matrix is immersed in the release medium a small fraction of the fertilizer can be dissolved followed by hydration and the progressive gelling of the macromolecules. A highly viscous layer, increasing the thickness of the polymer over time is formed. This layer controls the entry of water into the center of the polymer device and constitutes a barrier to rapid release. The release process and rate are dependent on the physicochemical characteristics of the fertilizer, the device and its hydration.

Matrices with limited swelling (100% of the initial weight) are those materials in which the zero-order diffusion kinetics governs the release of the fertilizer, except for systems with a high concentration of fertilizer. The determining step is the rate of penetration of the release medium into the matrix, which is given by the activation of solvent on the polymer matrix.

8.4.3 Erosion

This mechanism the release of fertilizer is mainly governed by the degradation of the matrix. There are two types of erodible systems: surface erosion and erosion within the matrix. The nature of the erosion mechanism depends on the hydrophilicity and morphology of the polymer. Hydrophobic polymers tend to exclude water from the interior of the system and are eroded from the outside to the matrix. On the other hand, the hydrophilic polymers tend to be homogeneously eroded, although the amorphous regions are more susceptible to water than the crystalline regions, therefore the degree of crystallinity of the polymer affects this release mechanism of the fertilizer.

From the dissolution and diffusion of the fertilizer and the erosion of the gel layer different fronts of movements have been identified: the erosion front, the diffusion front, the transition or swelling front and the penetration front (Maderuelo et al. 2011).

These frontal movements could be adjusted by selecting different types of polymers, resulting in different release kinetics such as Higuchi, which predicts a dependence of the fertilizer mass released with the square root of time, and zero-order and first-order kinetics (Li et al. 2014; Marinich et al. 2012; Siepmann and Siepmann 2012).

8.5 Hydrogels as Constituents of CRF

The use of hydrogels as polymeric matrices of CRF, aims to improve the behavior of the coatings and reduce the environmental impact caused by the waste generated by them once the nutrient is completely released. Initially the urea granules were coated with sulfur, but this coating in contact with the microorganisms found in the soil can suffer cracks or fissures, thus releasing the fertilizer faster than required. In this context, polymer coatings using synthetic materials such as polyethylene, polystyrene and polyesters have been developed. However, most of them are non-biodegradable polymers, so once the fertilizer is released, the non-compostable material remains in the soil. In view of the above, the study of environmentally friendly coating materials have been carried out. Most of them are derived from natural materials such as chitosan (Cs), sodium alginate, starch – cellulose and its derivatives, lignin, xanthan, pectin, agricultural residues, Arabic, carrageenan, gellan and guar gum, as well as scleroglucana, etc.

An additional advantage of CRF hydrogels is that they improve soil water retention capacity and nutrient retention. In addition, its use improves the activity of soil microorganisms and reduces the frequency of irrigation. Several works on the behavior of different CRFs composed of hydrogels as polymer matrices are listed in Table 8.1.

Table 8.1 Controlled release fertilizers made from hydrogels using different polymer matrices

Fertilizer	Polymeric matrix	Release media	Device	Reference
KNO_3	Chitosan (Cs)	Water	Microespheres	Perez and Francois (2016)
	Potato starch			
KNO_3	Cs	Water	Monolayer and multilayer	Melaj and Daraio (2013)
	Xanthan		Tablet	
KNO_3	Cs	Water	Multilayer	Melaj and Daraio (2014)
	Xanthan		Tablet	
	Hydropropylmethylcellulose			
KNO_3	Cs/MMT	Water	Microespheres	França et al. (2018)
	Cs crosslinked (TPP)/MMT	Soil	Microcapsules	
Urea	Tamarind gum, Guar gum	Soil	Trilayer	Mukerabigwi et al. (2015)
	Xanthan, diatomite		Coating	
	Poly(methacrylic acid)			

(continued)

Table 8.1 (continued)

Fertilizer	Polymeric matrix	Release media	Device	Reference
Urea	Ethyl cellulose	Soil	Dual layer	Qiao et al. (2016)
	Starch based superabsoebent polymer			
Urea	Bio- epoxy	Water	Coating	Li et al. (2018)
	(bagasse and bisphenol-A diglycidyl ether)			
Urea	Starch based hydrogel	Water	Powder	Xiao et al. (2017)
Urea	Polyvinil alcohol	Water	Coating	Azeem et al. (2016)
	Modified starch			
Urea	Starch	Water	Pellets	Giroto et al. (2019)
	Melamine	Soil		
Urea	Bio-based polyurethane	Water	Coating	Liu et al. (2019)
	(waste palm oil)			
Urea	Peat Humin	Citric acid (aq)	Microespheres	Araújo et al. (2017)
	Humic acids	Soil		
Urea	Sulfonated-carboxymethyl-cellulose	Water	Dehydrated hydrogel	Olad et al. (2018)
$NH_4H_2PO_4$	Acrylic acid	Soil		
KH_2PO_4	Polyvinylpyrrolidone			
	Silica nanoparticles			
Urea	Alginate-*g*-poly(acrylic acid-*co*-acrylamide)	Water	Capsules	Rashidzadeh & Olad (2014)
$NH_4H_2PO_4$	Montmorillonite	Soil		
KH_2PO_4				
Urea	Polyvinil alcohol	Water	Powder	Li et al. (2016)
K_2HPO_4	Wheat straw cellulose-*g*-poly(acrylic acid)	pH solutions		
		Salt solutions		
Urea	Pectin	Water	Cylindrical	Guilherme et al. (2010)
Phosphate	Acrylamide		Hydrogel	
Potassium	Sodium acrylate			
KCl	Cellulose acetate and EDTAD	Soil	Powder	Senna et al. (2015)
$NH_4H_2PO_4$				
$(NH_4)_2SO_4$	PVA/crosslinked-Cs/poly(AA-*co*-AM)	Water	Trilayer	Noppakundilograt et al. (2014)
P_2O_5; KCl			Coating	

AA: Acrylic acid; AM: Acrylamide; EDTAD: Ethylenediaminetetraacetic dianhydride; MMT: Montmorillonite

The objective of CRF is to control the release of nutrients to increase the time of their availability to be used by crops. One of the tools to improve this behavior is to obtain cross-linked systems. Noppakundilograt et al. (2014) synthesized a trilayer-coated NPK fertilizer hydrogel constituted by PVA, crosslinked-Cs and an outer layer of poly(acrylic acid-*co*-acrylamide) formed *in situ*. The authors found that the level of water penetration was increased by the crosslinking of the Cs layer. Regarding the behavior of nitrogen release, the trilayer-coated NPK fertilizer hydrogel followed the Korsmeyer-Peppas & Ritger-Peppas models with pseudo-Fickian characteristics. Perez and François (2016) studied the influence of the crosslinking time in the release of KNO_3 and the degree of swelling of the microspheres prepared from Cs and starch and using sodium tripolyphosphate (TPP) as the crosslinking agent. Their results showed that the degree of swelling decreases by ~12% when the crosslinking time is longer. In another investigation carried out by Mukerabigwi et al. (2015) using different hydrocolloids (tamarind gum, guar gum and xanthan gum) in the presence of epichlorohydrin as a crosslinking agent and containing diatomite as a nanofiller were used in the coating of urea granules. These polysaccharides constituted the inner layer and the outer layer was clay, which improved the water absorption. The results showed a more beneficial water retention capacity and a higher nitrogen slow-release efficiency using xanthan gum than that prepared with the other polysaccharides. In addition, the release efficiency was improved by the authors using an outermost layer of poly (methacrylic acid). When this layer was added, nitrogen release was 43% after the first day and 79.5% after 28 days, instead of 68% and 92%, respectively, without the use of this coating. Senna et al. (2015) esterified cellulose acetate and using ethylene diamine tetrahydric dianhydride crosslinked the polymer matrix. Xiao et al. (2017) found that the content of the crosslinking agent in starch-based hydrogels could control water absorption capacity and gel strength.

Another efficient strategy to produce a CRF is to incorporate additives such as nanoparticles or clays that regulate the porosity and modify the swelling and release of the fertilizer. Olad et al. (2018) studied the effect of the addition of silica nanoparticles to a hydrogel formed by the polymerization of sulfonated-carboxymethyl cellulose with acrylic acid in the presence of polyvinylpyrrolidone. In this study, urea, potassium dihydrogen phosphate and ammonium dihydrogen phosphate were used as fertilizers. The authors analyzed the release kinetics of Korsmeyer & Peppas form in water and soil as release media and found that release was controlled by Fickian diffusion. The incorporation of silica into the matrix increased the water absorption capacity and delayed the release of the fertilizer, because the hydroxyl groups of the silica formed additional physical crosslinking points with the hydrogel. França et al. (2018) compared microcapsule and microsphere structures based on Cs crosslinked or not with TPP and adding montmorillonite. The microcapsules had higher percentages of swelling (~500%) and the release of nutrients was slower (less than 70% in the first measurements) using the Cs without crosslinking. Both phenomena are associated with a core-shell structure: the outer layer of Cs present in the microcapsule trapped in the nutrient. The swelling was also observed for both systems in water and soil release media. Rashidzadeh and Olad (2014) found that the water absorbency was higher when montmorillonite was present in the CRF, increasing porosity, and the release of the nutrient was extended in time.

On the other hand, barriers such as multilayer systems or coatings with different thickness sizes have also helped the permanence of the fertilizer in the soil. Melaj and Daraio (2013) investigated the behavior of KNO_3 release in three-layer systems made from Cs, xanthan and their mixtures. The results obtained showed a non-Fickian diffusion kinetics and the release process was controlled by the swelling, which was reflected in the linear relationship between the fraction of the nutrient release and the percentage of swelling. The Cs-xanthan matrix showed 300% swelling before starting the fertilizer release, which occurred after 14 min. In another work Melaj and Daraio (2014) observed that the addition of hydroxypropylmethylcellulose (HPMC) to xanthan matrices lead to a reduction in the release rate of fertilizer compared to the matrix without HPMC. According to the authors, HPMC formed a strong layer of gelatinous barrier on the surface, which generated a decrease in the penetration of solvent molecules. The HPMC and Cs-HPMC tablets had the same release behavior. On the other hand, the presence of HPMC allowed to obtain release systems with almost zero order release kinetics (Melaj and Daraio 2014). Azeem et al. (2016) obtained a longer release time by increasing the thickness of the CRF coating. The researchers emphasized in this type of systems the importance of achieving a homogeneous coating and the integrity of the film to obtain a promising CRF (Azeem et al. 2016; Liu et al. 2019).

Some researchers have studied the use of waste to obtain environmentally friendly polymers as an alternative to polymers obtained from non-renewable sources. Li et al. (2018) synthesized a bio-based epoxy using bagasse from sugar cane and bisphenol-A diglycidyl ether. Li et al. (2016) used wheat straw to obtain a semi-interpenetrating polymer network hydrogel. Liu et al. (2019) prepared a bio-based elastic polyurethane-coated urea fertilizer using waste palm oil. Araújo et al. (2017) also obtained humic acid and humin from peat to study the interaction with Cs to achieve the urea release. Other polymers such as pectin were also investigated as CRF matrices (Guilherme et al. 2010).

Giroto et al. (2016) developed a simple system to obtain a CRF based on the extrusion of starch, melamine and urea. The addition of melamine helped to control the release of the fertilizer and increased the total N content depending on the microorganisms present in the soil. The starch was responsible for delaying the solubilization of the urea and, therefore, its hydrolysis.

8.6 Conclusions

Good agricultural practices is one of the objectives of sustainable agriculture. Among the good agricultural practices can be highlighted the efficiency in the use of nitrogen-based fertilizers. This is the nutrient mainly applied at high concentrations to increase the yield of crops. Decreasing nitrogen losses can reduce environmental pollution. The application of controlled release fertilizers is one of the practices that contribute to the reduction of losses of this nutrient.

Hydrogels as polymeric matrices of CRF are being studied and compared with different materials in order to determine their true potential in the agroindustry. These systems, as well as the superabsorbent hydrogels, have the capacity to be swollen, and to retain the moisture and nutrients of the soil in the environment of the root zone, thus reducing the frequency of irrigation and increasing the activity of benefit microorganisms for the crops.

There are some aspects to consider in preparing a CRF based on the facts highlighted throughout this chapter:

- The crosslinking of the polymer can control the water absorption capacity and, as a consequence, the degree of swelling and the release of the fertilizer can be modulated.
- The presence of additives such as clays or silica, improve the water absorption capacity, delaying the release of fertilizers. Cationic clays can exchange anions, being suitable for transporting, e.g. nitrates. On the other hand, nutrients can be intercalated between clay layers.
- Multilayer systems delay the fertilizer delivery and improve the water retention capacity.
- Some hydrogels depending on the type of polymer used for their development need to be initially swollen before starting the fertilizer release.

In general, hydrogels are an advantageous option to improve the nitrogen use efficiency by crops and contribute to the sustainability of "green" agricultural practices.

Acknowledgments The authors would like to thank the Consejo Nacional de Investigaciones Científicas y Técnicas (CONICET) for financial support, and Dr. Analía Vazquez and Dr. Laura Foresti for their valuable technical support.

Conflicts of Interest The authors declare no conflict of interest.

References

Ahmed, E. M. (2015). Hydrogel: Preparation, characterization, and applications: A review. *Journal of Advanced Research, 6*(2), 105–121. https://doi.org/10.1016/j.jare.2013.07.006.

Anitha, K., & Bindu, G. (2016). Effect of controlled-release nitrogen fertilizer on methane emission from paddy field soil. *Procedia Technology, 24*, 196–202. https://doi.org/10.1016/j.protcy.2016.05.027.

Araújo, B. R., Romão, L. P. C., Doumer, M. E., & Mangrich, A. S. (2017). Evaluation of the interactions between chitosan and humics in media for the controlled release of nitrogen fertilizer. *Journal of Environmental Management, 190*, 122–131. https://doi.org/10.1016/j.jenvman.2016.12.059.

Asgedom, H., Tenuta, M., Flaten, D., Gao, X., & Kebreab, E. (2014). Nitrous oxide emissions from a clay soil receiving granular urea formulations and dairy manure. *Agronomy Journal, 106*(2), 732–744. https://doi.org/10.2134/agronj2013.0096.

Azeem, B., KuShaari, K., & Man, Z. (2016). Effect of coating thickness on release characteristics of controlled release urea produced in fluidized bed using waterborne starch biopolymer as coating material. *Procedia Engineering, 148*, 282–289. https://doi.org/10.1016/j.proeng.2016.06.615.

Bajpai, A., & Giri, A. (2002). Swelling dynamics of a macromolecular hydrophilic network and evaluation of its potential for controlled release of agrochemicals. *Reactive and Functional Polymers, 53*(2–3), 125–141. https://doi.org/10.1016/s1381-5148(02)00168-2.

Chalk, P., Craswell, E. T., Polidoro, J. C., & Chen, D. (2015). Fate and efficiency of ^{15}N-labelled slow- and controlled release fertilizers. *Nutrient Cycling Agroecosystems, 102*(2), 167–178. https://doi.org/10.1007/s10705-015-9697-2.

Chang, C., & Zhang, L. (2011). Cellulose-based hydrogels: Present status and application prospects. *Carbohydrate Polymers, 84*(1), 40–53. https://doi.org/10.1016/j.carbpol.2010.12.023.

Colombo, P. (1993). Swelling-controlled release in hydrogel matrices for oral route. *Advanced Drug Delivery Reviews, 11*(1–2), 37–57. https://doi.org/10.1016/0169-409x(93)90026-z.

Davidson, D., & Gu, F. X. (2012). Materials for sustained and controlled release of nutrients and molecules to support plant growth. *Journal of Agricultural and Food Chemistry, 60*(4), 870–876. https://doi.org/10.1021/jf204092h.

Du, C., Zhou, J., & Shaviv, A. (2006). Release characteristics of nutrients from polymer-coated compound controlled release fertilizers. *Journal of Polymers and the Environment, 14*(3), 223–230. https://doi.org/10.1007/s10924-006-0025-4.

Elbarbary, A. M., & Ghobashy, M. (2017). Controlled release fertilizers using superabsorbent hydrogel prepared by gamma radiation. *Radiochimica Acta, 105*(10), 865–876. https://doi.org/10.1515/ract-2016-2679.

França, D., Medina Fracon, A., Messa, L. L., Souza, C. F., & Faez, R. (2018). Chitosan spray-dried microcapsule and microsphere as fertilizer host for swellable–controlled release materials. *Carbohydrate Polymers, 196*, 47–65. https://doi.org/10.1016/j.carbpol.2018.05.014.

Geicu-Cristea, M., Popa, M., Miteluț, A., Dănăilă-Guidea, S., Neață, G., Burnichi, R., Tănase, E., Drăghici, M., & Popescu, P. (2016). Research on agriculture applications of some newly developed biobased products. *Agriculture and Agricultural Science Procedia, 10*, 485–493. https://doi.org/10.1016/j.aaspro.2016.09.021.

Geng, J., Ma, Q., Zhang, M., Li, C., Liu, Z., Lyu, X., & Zheng, W. (2015). Synchronized relationships between nitrogen release of controlled release nitrogen fertilizers and nitrogen requirements of cotton. *Field Crops Research, 184*, 9–16. https://doi.org/10.1016/j.fcr.2015.09.001.

Giroto, A. S., Guimarães, G. G., Colnago, L. A., Klamczynski, A., Glenn, G., & Ribeiro, C. (2019). Controlled release of nitrogen using urea-melamine-starch composites. *Journal of Cleaner Production, 217*, 448–455. https://doi.org/10.1016/j.jclepro.2019.01.275.

Guilherme, M. R., Reis, A. V., Paulino, A. T., Moia, T. A., Mattoso, L. H. C., & Tambourgi, E. B. (2010). Pectin-based polymer hydrogel as a carrier for release of agricultural nutrients and removal of heavy metals from wastewater. *Journal of Applied Polymer Science, 117*(6), 3146–3154. https://doi.org/10.1002/app.32123.

Hendrawan, Khoerunnisa, F., Sonjaya, Y., & Chotimah, N. (2016). Physical and chemical characteristics of alginate-poly (vinyl alcohol) based controlled release hydrogel. *Journal of Environmental Chemical Engineering, 4*(4), 4863–4869. https://doi.org/10.1016/j.jece.2016.03.043. https://www.sigmaaldrich.com/technical-documents/articles/materials-science/biomaterials/tutorial.html.

Huang, X., & Brazel, C. (2001). On the importance and mechanisms of burst release in matrix-controlled drug delivery systems. *Journal of Controlled Release, 73*(2–3), 121–136. https://doi.org/10.1016/s0168-3659(01)00248-6.

Li, L., Wang, L., Li, J., Jiang, S., Wang, Y., Zhang, X., Ding, J., Yu, T., & Mao, S. (2014). Insights into the mechanisms of chitosan–anionic polymers-based matrix tablets for extended drug release. *International Journal of Pharmaceutics, 476*(1–2), 253–265. https://doi.org/10.1016/j.ijpharm.2014.09.057.

Li, X., Li, Q., Xu, X., Su, Y., Yue, Q., & Gao, B. (2016). Characterization, swelling and slow-release properties of a new controlled release fertilizer based on wheat straw cellulose hydrogel. *Journal of the Taiwan Institute of Chemical Engineers, 60*, 564–572. https://doi.org/10.1016/j.jtice.2015.10.027.

Li, Y., Jia, C., Zhang, X., Jiang, Y., Zhang, M., Lu, P., & Chen, H. (2018). Synthesis and performance of bio-based epoxy coated urea as controlled release fertilizer. *Progress in Organic Coatings, 119*, 50–56. https://doi.org/10.1016/j.porgcoat.2018.02.013.

Liu, J., Yang, Y., Gao, B., Li, Y. C., & Xie, J. (2019). Bio-based elastic polyurethane for controlled-release urea fertilizer: Fabrication, properties, swelling and nitrogen release characteristics. *Journal of Cleaner Production, 209*, 528–537. https://doi.org/10.1016/j.jclepro.2018.10.263.

Maderuelo, C., Zarzuelo, A., & Lanao, J. M. (2011). Critical factors in the release of drugs from sustained release hydrophilic matrices. *Journal of Controlled Release, 154*(1), 2–19. https://doi.org/10.1016/j.jconrel.2011.04.002.

Mahinroosta, M., Farsangi, Z. J., Allahverdi, A., & Shakoori, Z. (2018). Hydrogels as intelligent materials: A brief review of synthesis, properties and applications. *Materials Today Chemistry, 8*, 42–55. https://doi.org/10.1016/j.mtchem.2018.02.004.

Marinich, J. A., Ferrero, C., & Jimenez-Castellanos, M. R. (2012). Graft copolymers of ethyl methacrylate on waxy maize starch derivatives as novel excipients for matrix tablets: Drug release and fronts movement kinetics. *European Journal Pharmaceutics and Biopharmaceutics, 80*(3), 674–681. https://doi.org/10.1016/j.ejpb.2011.12.005.

Melaj, M. A., & Daraio, M. E. (2013). Preparation and characterization of potassium nitrate controlled release fertilizers based on chitosan and xanthan layered tablets. *Journal of Applied Polymer Science, 130*(4), 2422–2428. https://doi.org/10.1002/app.39452.

Melaj, M. A., & Daraio, M. E. (2014). HPMC layered tablets modified with chitosan and xanthan as matrices for controlled-release fertilizers. *Journal of Applied Polymer Science, 131*(9), 40839 (1–40839 (7. https://doi.org/10.1002/app.40839.

Mukerabigwi, J. F., Wang, Q., Ma, X., Liu, M., Lei, S., Wei, H., Huang, X., & Cao, Y. (2015). Urea fertilizer coated with biodegradable polymers and diatomite for slow release and water retention. *Journal of Coatings Technology and Research, 12*(6), 1085–1094. https://doi.org/10.1007/s11998-015-9703-2.

Nagam, S., Jyothi, A., Poojitha, J., Aruna, S., & Nadendla, R. (2016). A comprehensive review on hydrogels. *International Journal of Current Pharmaceutical Research, 8*(1), 19–23.

Noppakundilograt, S., Pheatcharat, N., & Kiatkamjornwong, S. (2014). Multilayer coated NPK compound fertilizer hydrogel with controlled nutrient release and water absorbency. *Journal of Applied Polymer Science, 132*(2), 1–11. https://doi.org/10.1002/app.41249.

Olad, A., Zebhi, H., Salari, D., Mirmohseni, A., & Tabar, A. R. (2018). Slow-release NPK fertilizer encapsulated by carboxymethyl cellulose-based nanocomposite with the function of water retention in soil. *Materials Science & Engineering C, 90*, 333–340. https://doi.org/10.1016/j.msec.2018.04.083.

Perez, J. J., & François, N. J. (2016). Chitosan-starch beads prepared by ionotropic gelation as potential matrices for controlled release of fertilizers. *Carbohydrate Polymers, 148*, 134–142. https://doi.org/10.1016/j.carbpol.2016.04.054.

Qiao, D., Liu, H., Yu, L., Bao, X., Simon, G. P., Petinakis, E., & Chen, L. (2016). Preparation and characterization of slow-release fertilizer encapsulated by starch-based superabsorbent polymer. *Carbohydrate Polymers, 147*, 146–154. https://doi.org/10.1016/j.carbpol.2016.04.010.

Rashidzadeh, A., & Olad, A. (2014). Slow-released NPK fertilizer encapsulated by NaAlg-*g*-poly(AA-*co*-AAm)/MMT superabsorbent nanocomposite. *Carbohydrate Polymers, 114*, 269–278. https://doi.org/10.1016/j.carbpol.2014.08.010.

Savci, S. (2012). Investigation of effect of chemical fertilizers on environment. *APCBEE Procedia, 1*, 287–292. https://doi.org/10.1016/j.apcbee.2012.03.047.

Sempeho, S., Kim, H. T., Mubofu, E., & Hilonga, A. (2014). Meticulous overview on the controlled release fertilizers. *Advances in Chemistry, 2014*, 1–16. https://doi.org/10.1155/2014/363071.

Senna, A. M., Braga do Carmo, J., Santana da Silva, J. M., & Botaro, V. R. (2015). Synthesis, characterization and application of hydrogel derived from cellulose acetate as a substrate for slow-release NPK fertilizer and water retention in soil. *Journal of Environmental Chemical Engineering, 3*(2), 996–1002. https://doi.org/10.1016/j.jece.2015.03.008.

Shaviv, A. (2000). Advances in controlled release of fertilizers. *Advances in Agronomy, 71*, 1–49. https://doi.org/10.1016/s0065-2113(01)71011-5.

Siegel, R. A., & Rathbone, M. J. (2012). Overview of controlled release mechanisms. In R. A. S. Siepmann & M. J. Rathbone (Eds.), *Fundamentals and applications of controlled release drug delivery* (pp. 19–46). New York: Springer. https://doi.org/10.1007/978-1-4614-0881-9_2.

Siepmann, J., & Siepmann, F. (2008). Mathematical modeling of drug delivery. *International Journal of Pharmaceutics, 364*(3), 328–343. https://doi.org/10.1016/j.ijpharm.2008.09.004.

Siepmann, J., & Siepmann, F. (2012). Modeling of diffusion controlled drug delivery. *Journal of Controlled Release, 161*(2), 351–362. https://doi.org/10.1016/j.jconrel.2011.10.006.

Silva, A., Richard, C., Bessodes, M., Scherman, D., & Merten, O. (2009). Growth factor delivery approaches in hydrogels. *Biomacromolecules, 10*(1), 9–18. https://doi.org/10.1021/bm801103c.

Smith, L., & Siciliano, G. (2015). A comprehensive review of constraints to improved management of fertilizers in China and mitigation of diffuse water pollution from agriculture. *Agriculture, Ecosystems and Environment, 209*, 15–25. https://doi.org/10.1016/j.agee.2015.02.016.

Trenkel, M. E. (2010). *Slow and controlled-release and stabilized fertilizers: An option for enhancing nutrient use efficiency in agriculture*. Paris: International Fertilizer Association.

Trinh, T. H., & KuShaari, K. (2016). Dynamic of water absorption in controlled release fertilizer and its relationship with the release of nutrient. *Procedia Engineering, 148*, 319–326. https://doi.org/10.1016/j.proeng.2016.06.444.

WHO. (2003). *Nitrate and nitrite in drinking-water. Background document for preparation of WHO guidelines for drinking-water quality*. Geneva: World Health Organization. (WHO/SDE/WSH/03.04/56).

Wichterle, O., & Lím, D. (1960). Hydrophilic gels for biological use. *Nature, 185*, 117–118. https://doi.org/10.1038/185117a0.

Xiao, X., Yu, L., Xie, F., Bao, X., Liu, H., Ji, Z., & Chen, L. (2017). One-step method to prepare starch-based superabsorbent polymer for slow release of fertilizer. *Chemical Engineering Journal, 309*, 607–616. https://doi.org/10.1016/j.cej.2016.10.101.

Chapter 9
Classification and Uses of Emulsions in Food and Agro Applications

Antony Allwyn Sundarraj and Thottiam Vasudevan Ranganathan

Abstract Emulsions are partially produced in the structures of many natural and artificial (processed) foods. In general, "emulsion" is defined as a structure formed through the dispersion of one or two immiscible liquids within the other in the form of small drops. Emulsion-based delivery systems can offer many possible benefits for incorporating essential oils such as omega-3 (ω-3) into foods and beverages. There are also many terms available to describe the different types of emulsions and it is very important to define and clarify these terms as "oil/water", "micro", "macro" and "nano" and "suspo" emulsions. This review can provide a general description of the general terms of the types of emulsion, the role of various emulsifying agents and the possible application of emulsions in various sectors of the food and agricultural industry.

Keywords Active substance carries · Colloids · Controlled release · Prolonged and sustained release · Sustainable polymers

9.1 Introduction

The processing, conservation, safety and quality of food are the most important concerns in all sectors of the food industry today (Rasooli 2007; Amaral and Bhargava 2015). The process of preparing the emulsion is the determination of two different phases, such as oil droplets are formed in an aqueous phase and water droplets are formed in an oil phase. The emulsions are not determined by the ratio between the two phases that can be clearly observed, e.g. 80% oil and 20% water

A. A. Sundarraj (✉)
Department of Food Technology, Sri Shakthi Institute of Engineering and Technology, Coimbatore, India
e-mail: allwynsundarrajft@siet.ac.in

T. V. Ranganathan
Department of Food Processing Technology, Karunya Institute of Technology and Sciences, Coimbatore, India

T. J. Gutiérrez (ed.), *Polymers for Agri-Food Applications*,
https://doi.org/10.1007/978-3-030-19416-1_9

(Bergenstahl and Claesson 1997). An emulsion protects the emulsion droplets distributed uniformly throughout the storage days and consumption periods (Chanamai and McClements 2002). This has led the food manufacturing sectors and many scientists' to investigate the capability of flocculation, coalescence, edible polymers (proteins and polysaccharides) to stabilize emulsion droplets against the creation process, and evaluate their potential applications.

Agrochemical companies are increasing the development of chemical emulsions at the nanoscale and replacing the pure active ingredients with the active compounds encapsulated in their nanometric equivalent with the effort of providing a series of benefits for applications of nanoscience to insecticides (Gogos et al. 2012; Kumar et al. 2014), and other agro-chemicals such as fertilizers (Agrawal and Rathore 2014) and plant growth devices (Mura et al. 2015). The stability of a food emulsion is the most important thing to consider in relation to its industrial applications, and depends on the rheology of the continuous phase and the interactions between the particles of the dispersed phase (Melik and Fogler 1998; Brewer et al. 2016). The simple difference between a nanoemulsion and a conventional emulsion is that a nanoemulsion does not alter the appearance of the food product when added to it (Dreher 2004).

Emulsion is the basis of a wide variety of natural and artificial materials used in the processing of food products, pharmacological and industrial sectors cosmetics (Schramm 1992; Lin and Mei 2000; Achouri et al. 2012). Currently, the new ingredients are regularly combined in various food processing systems to improve their rheological, physico-chemical and dietary food properties. However, these ingredients can sometimes be degraded slowly and lose their activities, due to oxidation processes, reactions with various components present in processed foods or food systems, which may limit their change in taste and color of products food (Schrooyen et al. 2001), so its stabilization is necessary.

9.2 Emulsion

An *emulsion* is a combination of two or more liquids (such as oil and water) that are not easily combined (GuhanNath et al. 2014).

9.2.1 Oil/Water (O/W) Emulsions

O/W emulsions are now receiving a significant attention, because of the needs to eliminate or reduces the volatile organic solvent for safer handling (Hazra et al. 2017) purposes. An O/W emulsion is a system in which the active ingredient must have very a low water solubility and to avoid the crystallization problems (Tadros 1995). O/W emulsion is a type of emulsion and is the most common used in the food processing industrial sectors, such as mayonnaises, sauces and dairy products. Even though, the processed food products such as sausages and liver pates can also be considered as O/W emulsions (Knightly 1989; Gutiérrez and Álvarez 2017; Gutiérrez 2018a). The functional activity of the O/W emulsions depends to a large extent on the nature of the oil droplets (Derkach 2009; Le Reverend et al. 2010; Bai and McClements 2016). Previously, many studies have also been reported that the stability of O/W emulsions depends on their concentration and types of ingredients used in the emulsion, as well as the processing and storage conditions (McClements 1999). O/W emulsions can be classified into two different forms, such as diluted form and concentrated form. The emulsions are a single class of O/W emulsion because they can be highly consumed in diluted forms, such as fruit juice beverages and milk-based products and in their emulsification of concentrated form such as butter and creams (Lorenzo et al. 2018). The emulsion might thus have a significant degree of stability in both diluted and concentrated forms (Tan 1990).

9.2.1.1 Uses of O/W Emulsions in Food and Agro Applications

Gum arabic is one of the most commonly used biopolymers in food and beverage applications (Garti and Reichman 1993). Starches are now generally used in milk-based food products, because of their good emulsion properties, and also as promising substitutes for gum arabic (Trubiano 1995; Achouri et al. 2012; Zheng 2018). Other alternative gums include tragacanth (Imeson 1992) and λ-carrageenan (Huang et al. 2001) have also been reported as effective stabilizing and thickening agents. Proteins, containing soy and whey, are better emulsifications that can modify both the emulsion droplet charges (Singh and Ye 2000) and the interfacial membrane thickness in O/W emulsions (Achouri et al. 2012) and forms the stabilizing films against oxidation process (Djordjevic et al. 2008; McClements 2004). In addition, factors that could contribute to the emulsion stabilization or destabilization process, such as the type of hydrocolloid used, the possible combinations of hydrocolloid with protein, and the oil absorption to be dispersed have not been reported (Achouri et al. 2012). There is a growing interest in the formulation of value-added products containing various health constituents, such as minerals, vitamins, and omega-3 (ω-3), etc., for the functional food products in the market. Such emulsion concentration can be useful as feedstock emulsions and for other emulsions in suspension, as well as for organizing stability in O/W emulsion distributions. Hence, the consideration and also the monitoring the several factors that influence the stabilization and the shelf-life of prepared emulsion matrices is an acute for its continued success in the food markets.

9.2.2 W/O Emulsion

The W/O emulsion is the well-known area of industrial food processing sectors in the field of 'yellow fat' food products, such as spreads and margarines (Knightly 1989; Muschiolik and Dickinson 2017). The lipophilic emulsification which has been mainly used to prepare the primary W/O emulsions of food grade water-oil-water (W/O/W) formulations is a poly-glycerol poly-ricinoleate (Takahashi et al. 1986). In connection with the development of various functional food products that can be improved with the addition of probiotics, there is a growing interest in improving the viability of the bacterial cells through an encapsulation of W/O/W emulsions (Jimenez-Colmenero 2013). The W/O/W emulsions in polyunsaturated fatty acids through a determined selection of edible oil phases, such as linseed oil (Poyato et al. 2013), chia oil (Cofrades et al. 2014) or perilla oil (Flaiz et al. 2016) can also allow the development of improved functional foods in terms of stability and nutritional aspects. In a recent study by Altuntas et al. (2017) demonstrated that an optimal amount of poly-glycerol poly-ricinoleate is necessary so that the stability of the W/O droplets can be significantly reduced by merging the poly-glycerol polyricinoleate emulsifier with an equal amount of the lecithin (natural emulsifier).

9.2.3 *Micro-Emulsions*

According to Danielsson & Lindman, "a micro-emulsion is a system of water, oil and an amphiphile which is a single optically isotropic and thermodynamically stable liquid solution" (Serdaroglu et al. 2015).

Micro-emulsions can have different characteristic properties, such as large interfacial area, oil-soluble compounds and ultra-low interfacial tension (Friberg and Bothorel 1987; Paul and Moulik 2001). Micro-emulsions can be prepared by controlling the addition of lower alkanol's to the cloudy emulsions to produce the transparent solutions comprising W/O or O/W dispersions in colloidal systems (Shah 1998). Micro-emulsions are thermodynamically stable and have a spontaneous shape in different diameters ranging from 5 to 100 nm (Zhang 2011). In general, a scientist considers that the nano-emulsions are very similar to micro-emulsions, since both includes typically oil, water and surface-active components (surfactant) (Serdaroglu et al. 2015). The nano-emulsions are nevertheless formed by a mechanical force, while the micro-emulsion phases are formed by self-assembly (Fanun 2009; Zhang 2011). Therefore, the main differences between the micro- and micro emulsions are derived from thermodynamic characteristics of emulsions instead than compositions (Flanagan and Singh 2006; Whitesides and Grzybowski 2002; Zhang 2011).

The micro-emulsions as a functional state of lipids have been used in the preparation of processing of food products (El-Nokaly et al. 1991). The main differences between food processing and other micro-emulsions are found in the composition of oil components in food-grade surfactants (Paul and Moulik 2001). Candau and Anquetil (1998) reviewed information on O/W and W/O micro-emulsions that are formed using food grade ingredients to potentially incorporate various food ingredients such as preservatives, flavor and vitamins into the micro in emulsions. Paul and Moulik (2001) also reviewed the various micro-emulsions as drug delivery systems for the use of different types of drugs, such as peptide drugs, steroids and anti-inflammatory drugs.

9.2.4 *Macro-Emulsions*

Macro-emulsions are typical droplet diameters of various conventional emulsions, called as "macro emulsions", and vary between 100 nm and 100 μm. Macro-emulsions are the most common type of emulsification used in many different forms such as beverages, mayonnaise, desserts, sauces and milk (Serdaroglu et al. 2015). Macro-emulsions tend to be physically unstable, especially when they can undergo environmental stresses such as gravitational separation, coagulation and aggregation (Aoki et al. 2005; Dickinson 2009; Zhang 2011).

9.2.5 *Suspo-Emulsions*

Suspo-emulsions can be considered as mixtures of various suspension concentrates and emulsifications (Gasic and Tanovic 2013). The product is very difficult to formulate, because it is necessary to develop homogeneous emulsion components simultaneously while they are stable in the final formulation of the product (Tadros 2005). In addition, it is also necessary to overcome the problem of hetero-flocculation between oil droplets and solid particles (Vernner and Bauer 2007). The extensive storage stability testing of this type of formulation is very necessary (Knowles 2008). Despite the complexity of this formulation, the use and the importance of suspo-emulsions has been extraordinary and will continue to increase.

9.2.6 *Nanotechnology*

In today's competitive market, a new frontier technology is essential to maintain its leadership in the industrial food processing sector (Raj et al. 2013). Nanotechnology or nanoscience has become one of the most innovative scientific fields in the coming decades. Nanotechnology can involve in the use of diverse materials in an extremely small scale, usually 0.1–200 nano-meters (Sozer and Kokin 2008), due to the different characteristic properties of many processed foods on nano-meter sized components, such as nano-emulsions and foams (Chaudhry et al. 2008). In this sense, nanotechnology is increasingly used in the various areas of food processing, production and packaging systems (Sanguansri and Augustin 2006; Wesley et al. 2014).

9.2.6.1 Nanotechnology – Science and Technology

Nano-technology has the potential to influences the various aspects of food and agricultural systems. The numerous applications of nano-technology in food and agriculture are innovative compared to their use in pharmaceutical and medicinal products

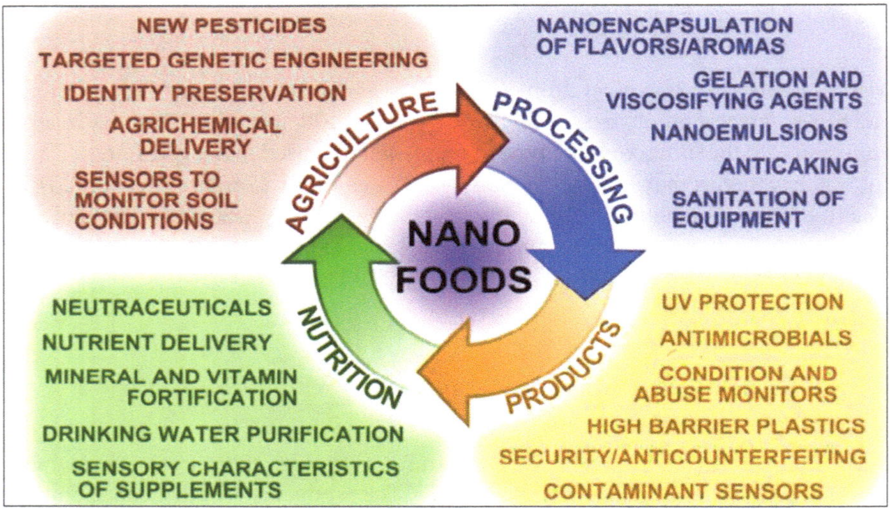

Fig. 9.1 Nanotechnology has various applications in all the areas of food science/technology, from agriculture to food processing to quality/safety to packaging to nutrition and nutraceuticals/pharmaceutical supplements (Raj et al. 2013)

(GuhanNath et al. 2014). However, scientists and industrial stakeholders have already identified the potential uses of nanotechnology in virtually all segments of the food industry (Fig. 9.1) from agriculture as fertilizer, pesticides and plant pathogen devices to food processing as encapsulation. Other applications have also been used in the food field to improve the quality and safety of food, as well as for the food packaging secto (Brody 2007; Bracone et al. 2016; Gutiérrez 2018b).

9.2.6.2 Nanotechnology in Agricultural Production

During primary production, nanoformulated agrochemicals are used to increase the efficacy of agrochemicals linked to conventional formulations. Only a few insecticides containing nanoformulates and nanometric-sized agrochemicals have been identified as commercially available. In addition, nanoparticles are used to clean the soil and water (Bouwmeester 2009). The nanostructured materials can exhibit unique properties that will open many windows of opportunity for the creation of innovative and high-performance materials, which will have an acute impact on the food processing, packaging and storage. Many reports have considered that the agricultural industry is a potential sector for nanoscience and nanotechnology applications. In the future, nanostructured catalysts will be available, which will expand the potential of nanoemulsions as pest controllers, using lower amounts of agrochemicals (IRGC 2008).

9.2.6.3 Nanotechnology in Food Processing

The numerous nanotechnologies are applied in the sectors of food processing and packaging such as the use of nano-sieves (Bouwmeester 2009). Current applications of nanotechnology in agri-food production chain is focused on the development of various nano-sized food components, additives and innovative food packaging (Chaudhry et al. 2008).

9.2.6.4 Nano-Emulsions

Nanoemulsion is one of the technologies in development, especially in the food and pharmaceutical industry, as a new system for the release of healthy substances (Chen and Wagner 2004; Wooster et al. 2008; Mao et al. 2009). The possible benefit of nano-emulsions includes good separation stability, coalescence and flocculation. The bioavailability of functional compounds can also be improved by the use of nanoemulsions (Shafiq et al. 2007; Wang et al. 2008; Shakeel et al. 2008, Dingman 2008). Nanoemulsions can be used to encapsulate oils and increase their water-dispersibility (McClements 2011). According to McClements and Decker (2000), different types of nano-emulsions with more complex properties, such as nanostructured multilayer emulsions, can offer multiple encapsulation capabilities in a single administration system. This interfacial engineering technology is used to achieve the homogenous mixture of food grade ingredients (such as phospholipids, polysaccharides and proteins) with different natural chemicals, thus expanding the use of emulsions in food (Weiss et al. 2006), e.g. Nestlé and Unilever have developed ice creams based on nano-sized emulsions with a lower fat content (Renton 2006; McClements 2004). Nano-emulsions are formulated using two different methods (Solans et al. 2005):

I. High energy approaches
High energy approaches can be used to break oil drops.
II. Low energy approaches
Low energy approaches are based on the spontaneous formation of several drops of oil within the mixed systems of oil, water and surfactant.

Nanoemulsions have great potential to overcome the various challenges associated with the development of processed food products and beverages enriched with ω-3 (Waraho et al. 2011; Li et al. 2013). It can be considered that nanoemulsions have good kinetic stability and high optical clarity, which is important for the main applications in many processed foods and beverages (Mason et al. 2006). Nanoemulsion-based delivery systems must still be carefully deliberated to ensure high bioavailability and good physical and chemical stability (Walker and Rebecca 2015). Surfactants are very critical, since emulsions must quickly cover the many new surfaces that are forming (Dickinson 2003).

In general, in food processing or food emulsions can be classified into two classes of surfactants (surface-active) as follows:

1. Small-molecule surfactants, such as sucrose esters and mono-glycerides, and
2. Macro-molecular emulsifications such as modified starches or proteins (Amaral and Bhargava 2015).

9.2.6.5 Properties of Nanoemulsions

Nanoemulsions have a low viscosity and high kinetic stability. Nanoemulsions are a core of attention for many industrial applications and some of them can be summarized below (Tadros 2004; Gutierrez 2008):

- No sedimentation during storage period: very small drop sizes can cause a large reduction in the gravitational forces.
- No amalgamation: this is due to the droplets of small size and also to the main thicknesses of the surfactant film that prevent coalescence/amalgamation, and the interruption between the O/W droplets (Sadurni 2005).
- Nanoemulsions can be applied for the delivery of fragrances and the development of formulated alcohol-free products (Baran and Maibach 1997).

9.2.6.6 Uses of Nano-Emulsions in Food and Agricultural Applications

The production of nanoemulsion to encapsulate various functional compounds is one of the fields in the development of nanoscience/technology applied to the sectors of the food industry (Silva et al. 2012). It also offers one of the most promising systems to improve solubility, bioavailability and functionality of hydrophobic compounds (Walker and Rebecca 2015; Nethaji and Parambil 2017). Some companies working on nanoemulsions for food applications can be listed below:

- Nutra-Lease, is a technology start-up company established by a scientific group, and is working to improve the bioavailability of various functional compounds. Mainly in beverages containing encapsulated functional compounds such as lycopene, ω-3, β-carotene and phytosterols (NutraLease 2011).
- Unilever has made an ice cream healthier without compromising flavor in the application of nanoemulsions. This ice cream has a low-fat content varying between 16–1% (Martins et al. 2007; Unilever 2011).
- Nestle has a patent on W/O emulsions, whose objective is to achieve a simpler and faster thawing by the addition of micelle-forming substances and poly sorbates. It is claimed that they contribute to a uniform thawing of frozen food products in the microwave process (Möller et al. 2009). This type of W/O emulsions is the drug delivery system most commonly used to incorporate ω-3 oils in processed food products and beverages such as bulk oils and powders (spray-drying emulsions) (Jacobsen 2010).

Microencapsulation has proven to be a popular way to create ω-3 powder that can be incorporated into a variety of food products, such as baked goods, spreads and fruit drinks (Drusch 2012). Nanoemulsions offer a convenient means to fortify many water-based food and beverage products with ω-3 oils. Fortified nanoemulsions could be introduced into food systems such as beverages, sauces and desserts (Velikov and Pelan 2008; Yang et al. 2012). Current liquid or semisolid food products that have been enriched with ω-3 fatty acids using emulsion-based delivery systems include table spreads, yogurts and milk (Chee et al. 2005; Sharma 2005; O' Dwyer et al. 2013; Lane et al. 2014). Nanoemulsions in the form of proteins (e.g. egg and vegetable protein) or carbohydrates (e.g. starch, pectin and guar gum) help to improve the texture and lead to the uniformity of the ice cream (Hogan et al. 2001). Brominated vegetable oil, ester gum and sucrose-acetate iso-butyrate are used as the weighting agent (Gilbert 1985). Biomolecules such as milk proteins and carbohydrates (e.g. dextrin) can prove to be potential carrier of nutrients with the help of encapsulation (Fernandez et al. 2009).

Hydrolyzed milk proteins, such as α-lactalbumin, have evolved to be a potential carrier of drugs, nutrients and supplements (Yoksan et al. 2010). Functional compounds that are generally encapsulated by self-assembled nanoemulsions are lutein, β-carotene, lycopene, vitamins A, D, E3 and Q10 and iso-flavones (Brody 2007). For example, the effectiveness of carvacrol nanoemulsions against *E. coli* on broccoli (Severino et al. 2014), radish seed, *Salmonella* enterica (Landry et al. 2015) and alfalfa seeds has been demonstrated in the literature. Recently Donsi et al. (2015) have also reported that the antibacterial and physical properties of modified chitosan-based coatings containing nanoemulsion of mandarin essential oil on green beans. In another study carried out by Kim et al. (2013) reported the application of lemon grass oil nanoemulsions on plums. These authors found an antimicrobial effect from the nanoemulsion and the physical and chemical properties of the plums were improved.

Nanoemulsions containing active nutraceutical agents may be used for the production of functional and medicinal foods (Solans et al. 2005). Nanoemulsions are acceptable because coalescence is not observed unlike macroemulsions (Gadhave 2014). Limited studies are carried out on food model and there are several challenges in application of this system in complex food matrices such as meat products (Amaral and Bhargava 2015).

9.3 Conclusions

Emulsions have a long history of use and are widely used in many foods and agrochemical inputs. The production of nanoemulsions to encapsulate and release active food and agrochemical substances is an area of knowledge in full development. However, more research at a pilot or industrial scale is required in terms of new equipment and nanoemulsion techniques for agri-food applications. The stability of the O/W emulsions are strongly influenced by the size of the drop, the charge and the concentration. In the modern period, multiple emulsions can be produced at different scales such as micro, macro, suspo or nano. The development of new food products every day is more related to the development of this type of polymer systems. The application of nanoemulsions in food processing technology still poses challenges that must be addressed both in terms of the production process and costs, but more continuous investment is required to finance research and development (R&D), in order to better understand the advantages and disadvantages that are offered from nanoemulsions systems. This chapter could thus be useful to help improve the understanding of colloidal instabilities in emulsions and how they could be improved with the objectives of their application in the agri-food industry.

Acknowledgements The authors would like to thank the management, Sri Shakthi Institute of Engineering and Technology, Coimbatore, TamilNadu, India.

Conflict of Interest The authors declare no conflicts of interest.

References

Achouri, A., Zamani, Y., & Boyne, J. I. (2012). Stability and physical properties of emulsions prepared with and without soy proteins. *Journal of Food Research, 1*(1), 254–267. https://doi.org/10.5539/jfr.v1n1p254.

Agrawal, S., & Rathore, P. (2014). Nanotechnology pros and cons to agriculture: A review. *International Journal of Current Microbiology and Applied Sciences, 3*(3), 43–55.

Altuntas, O. Y., Sumnu, G., & Sahin, S. (2017). Preparation and characterization of W/O/W type double emulsion containing PGPR–lecithin mixture as lipophilic surfactant. *Journal of Dispersion Science and Technology, 38*(4), 486–493. https://doi.org/10.1080/01932691.2016.1179121.

Amaral, D. M. F., & Bhargava, K. (2015). Essential oil nano-emulsions and food applications. *Advances in Food Technology Nutritional Sciences Open Journal, 1*(4), 84–87. https://doi.org/10.17140/aftnsoj-1-115.

Aoki, T., Decker, E. A., & McClements, D. J. (2005). Influence of environmental stresses on stability of O/W emulsions containing droplets stabilized by multi-layered membranes produced by a layer-by-layer electrostatic deposition technique. *Food Hydrocolloids, 19*(2), 209–220. https://doi.org/10.1016/j.foodhyd.2004.05.006.

Bai, L., & McClements, D. J. (2016). Extending emulsion functionality: Post-homogenization modification of droplet properties. *Process, 4*(17), 1–18. https://doi.org/10.3390/pr4020017.

Baran, R., & Maibach, H. (1997). *Textbook of cosmetic dermatology* (4th ed.). Boca Raton: CRC Press.

Bergenstahl, B. A., & Claesson, P. M. (1997). Surface forces in emulsions. In S. E. Friberg & K. Larsson (Eds.), *Food emulsions* (3rd ed.). New York/Basel: Marcel Dekker.

Bouwmeester, H. (2009). Review of health safety aspects of nanotechnologies in food production. *Regulatory Toxicology and Pharmacology, 53*(1), 52–62. https://doi.org/10.1016/j.yrtph.2008.10.008.

Bracone, M., Merino, D., González, J., Alvarez, V. A., & Gutiérrez, T. J. (2016). Chapter 6. Nanopackaging from natural fillers and biopolymers for the development of active and intelligent films. In S. Ikram & S. Ahmed (Eds.), *Natural polymers: Derivatives, blends and composites* (pp. 119–155). New York. EE.UU. ISBN: 978-1-63485-831-1: Editorial Nova Science Publishers, Inc.

Brewer, D. R., Franco, J. M., & Garcia-Zapateiro, L. A. (2016). Rheological properties of oil-in-water emulsions prepared with oil and protein isolates from sesame (*Sesamum indicum*). *Food Science and Technology, Campinas, 36*(1), 64–69. https://doi.org/10.1590/1678-457x.6761.

Brody, A. L. (2007). Case studies on nanotechnologies for food packaging. *Food Technology, 07*, 102–107.

Candau, F., & Anquetil, J. (1998). In D. O. Shah (Ed.), *Micelles, micro-emulsions and monolayers: Science and technology* (p. 193). New York: Marcel Dekker.

Chanamai, R., & McClements, D. J. (2002). Comparison of gum Arabic, modified starch and whey protein isolate as emulsifiers: Influence of pH, $CaCl_2$ and temperature. *Journal of Food Science, 67*(1), 120–125. https://doi.org/10.1111/j.1365-2621.2002.tb11370.x.

Chaudhry, Q., Scotter, M., & Blackburn, J. (2008). Applications and implications of nanotechnologies for the food sectors. *Food Additives and Contaminants, 25*(3), 241–258. https://doi.org/10.1080/02652030701744538.

Chee, C. P., Gallaher, J. J., Djordjevic, D., Faraji, H., McClements, D. J., & Decker, E. A. (2005). Chemical and sensory analysis of strawberry flavoured yogurt supplemented with an algae oil emulsion. *Journal of Dairy Research, 72*(3), 311–316. https://doi.org/10.1017/S0022029905001068.

Chen, C., & Wagner, G. (2004). Vitamin E nanoparticle for beverage applications. *Chemical Engineering Research Desiccation's, 82*(11), 1432–1437. https://doi.org/10.1205/cerd.82.11.1432.52034.

Cofrades, S., Santos-Lopez, J. A., Freire, M., Bened, J., Sanchez-Muniz, F. J., & Jimenez-Colmenero, F. (2014). Oxidative stability of meat systems made with $W_1/O/W_2$ emulsions prepared with hydroxytyrosol and chia oil as lipid phase. *LWT-Food Science and Technology, 59*(2), 941–947. https://doi.org/10.1016/j.lwt.2014.06.051.

Derkach, S. R. (2009). Rheology of emulsions. *Advances in Colloid Interface Sciences, 151*(1–2), 1–23. https://doi.org/10.1016/j.cis.2009.07.001.

Dickinson, E. (2003). Hydrocolloids at interfaces and the influence on the properties of dispersed systems. *Food Hydrocolloids, 17*(1), 25–39. https://doi.org/10.1016/S0268-005X(01)00120-5.

Dickinson, E. (2009). Hydrocolloids as emulsifiers and emulsion stabilizers. *Food Hydrocolloids, 23*(6), 1473–1482. https://doi.org/10.1016/j.foodhyd.2008.08.005.

Dingman, J. (2008). Nanotechnology: Its impact on food safety. *Journal of Environmental Health, 1*(3), 5–8.

Djordjevic, D., Cercaci, L., Alamed, J., McClements, D. J., & Decker, E. A. (2008). Stability of citral in protein and gum arabic-stabilized oil-in-water emulsions. *Food Chemistry, 106*(2), 698–705. https://doi.org/10.1016/j.foodchem.2007.06.033.

Donsi, F., Marchese, E., & Maresca, P. (2015). Green beans preservation by combination of a modified chitosan based-coating containing nano-emulsion of mandarin essential oil with high pressure or pulsed light processing. *Postharvest Biology and Technology, 106*, 21–32. https://doi.org/10.1016/j.postharvbio.2015.02.006.

Dreher, K. L. (2004). Health and environmental impact of nanotechnology: Toxicological assessment of manufactured nanoparticles. *Toxicological Sciences, 77*(1), 3–5. https://doi.org/10.1093/toxsci/kfh041.

Drusch, S. (2012). An industry perspective on the advantages and disadvantages of different fish oil delivery systems. In N. Garti & D. J. McClements (Eds.), *Encapsulation technologies and delivery systems for food ingredients and nutraceuticals* (pp. 488–504). https://doi.org/10.1533/9780857095909.4.488.

El-Nokaly, M., Hiler, G., & McGrady, J. (1991). In M. El-Nokaly & D. Cornell (Eds.), *Micro-emulsions and emulsions in foods* (pp. 26–43). Washington, DC: American Chemical Society.

Fanun, M. (2009). Properties of micro-emulsions with sugar surfactants and peppermint oil. *Colloid Polymer Science, 287*(8), 899–910. https://doi.org/10.1007/s00396-009-2043-y.

Fernandez, A., Torres-Giner, S., & Lagaron, J. M. (2009). Novel route to stabilization of bioactive antioxidants by encapsulation in electrospun fibers of zein prolamine. *Food Hydrocolloids, 23*(5), 1427–1432. https://doi.org/10.1016/j.foodhyd.2008.10.011.

Flaiz, L., Freire, M., Cofrades, S., Mateos, R., Weiss, J., Jimenez-Colmenero, F., & Bou, R. (2016). Comparison of simple, double and gelled double emulsions as hydroxytyrosol and *n*-3 fatty acid delivery systems. *Food Chemistry, 213*(15), 49–57. https://doi.org/10.1016/j.foodchem.2016.06.005.

Flanagan, J., & Singh, H. (2006). Micro-emulsions: a potential delivery systems for bioactives in food. *Critical Review in Food Science and Nutrition, 46*(3), 221–237. https://doi.org/10.1080/10408690590956710.

Friberg, S. E., & Bothorel, P. (1987). *Micro-emulsions: Structure and dynamics* (p. 1987). Boca Raton: CRC Press.

Gadhave, A. D. (2014). Nano-emulsions: Formation, stability and applications. *International Journal for Research in Sciences and Advanced Technologies, 3*(2), 38–43.

Garti, N., & Reichman, D. (1993). Hydrocolloids as food emulsifiers and stabilizers. *Food Structure, 12*, 411–426.

Gasic, S., & Tanovic, B. (2013). Biopesticide formulations, possibility of application and future trends. *Pesticides Phytomedicines (Belgrade), 28*(2), 97–102. https://doi.org/10.2298/PIF1302097g.

Gilbert, S. G. (1985). Food/package compatibility. *Food Technology, 39*(12), 54–56.

Gogos, A., Knauer, K., & Bucheli, T. D. (2012). Nanomaterials in plant protection and fertilization: current state, foreseen applications, and research priorities. *Journal of Agricultural Food Chemistry, 60*(39), 9781–9792. https://doi.org/10.1021/jf302154y.

GuhanNath, S., Aaron, S. I., Raj, A. A. S., & Ranganathan, T. V. (2014). Recent innovations in nano-technology in food processing and its various applications – A review. *International Journal of Pharmaceutical, Sciences Review and Research, 29*(2), 116–124.

Gutierrez, J. M. (2008). Nano-emulsions: New applications and optimization of their preparation. *Current Opinion in Colloid & Interface Science, 13*(4), 245–251. https://doi.org/10.1016/j.cocis.2008.01.005.

Gutiérrez, T. J. (2018a). Chapter 55. Processing nano- and microcapsules for industrial applications. In: Handbook of nanomaterials for industrial applications. Chaudhery Mustansar Hussain (Ed). Editorial Elsevier. EE.UU. ISBN: 978–0–12-813351-4. pp. 989–1011. https://doi.org/10.1016/B978-0-12-813351-4.00057-2

Gutiérrez, T. J. (2018b). Chapter 9. Biodegradability and compostability of food nanopackaging materials. In: Composite materials for food packaging. Giuseppe Cirillo, Marek A. Kozlowski, and Umile Gianfranco Spizzirri (Eds). WILEY-Scrivener Publisher. EE.UU. ISBN: 978–1–119-16020-5. pp. 269–296. https://doi.org/10.1002/9781119160243.ch9

Gutiérrez, T. J., & Álvarez, K. (2017). Chapter 6. Biopolymers as microencapsulation materials in the food industry. In M. Masuelli & D. Renard (Eds.), *Advances in physicochemical properties of biopolymers: Part 2* (pp. 296–322). Bentham Science Publishers. EE.UU. ISBN: 978–1–68108–545–6. eISBN: 978–1–68108–544–9. https://doi.org/10.2174/9781681085449117010009.

Hazra, D. K., Karmakar, R., Poi, R., Bhattacharya, S., & Mondal, W. (2017). Recent advances in pesticide formulations for eco-friendly and sustainable vegetable post management: A review. *Achieves of Agriculture and Environmental Sciences, 2*(3), 232–237.

Hogan, H. A., McNamee, B. F., O'Riordan, E. D., & O'Sullivan, M. (2001). Microencapsulating properties of sodium caseinate. *Journal of Agricultural and Food Chemistry, 49*(4), 1934–1938. https://doi.org/10.1021/jf000276q.

Huang, X., Kakuda, Y., & Cui, W. (2001). Hydrocolloids in emulsions: Particle size distribution and interfacial activity. *Food Hydrocolloids, 15*(4–6), 533–542. https://doi.org/10.1016/s0268-005x(01)00091-1.

Imeson, A.P. (1992). Exudate gums. In A. Blackie & P. Glasgow A. P. Imeson). Thickening and gelling agents for food (pp. 66–97). London: Blackie Academic and Professional.

IRGC. (2008). *International risk governance council, towards an integrative approach*. Geneva: Available at: http://www.irgc.org.

Jacobsen, C. (2010). Enrichment of foods with omega-3 fatty acids: A multidisciplinary challenge: Omega-3-enriched foods. *Annals of the New York Academy of Sciences, 1190*(1), 141–150. https://doi.org/10.1111/j.1749-6632.2009.05263.x.

Jimenez-Colmenero, F. (2013). Potential applications of multiple emulsions in the development of healthy and functional foods. *Food Research International, 52*(1), 64–74. https://doi.org/10.1016/j.foodres.2013.02.040.

Kim, I. H., Lee, H., & Kim, J. E. (2013). Plum coatings of lemongrass oil-incorporating carnauba wax-based nanoemulsion. *Journal of Food Science, 78*(10), 1551–1559. https://doi.org/10.1111/1750-3841.12244.

Knightly, W. H. (1989). Lecithin in baking. In B. F. Szuhaj (Ed.), *Lecithins: Sources, manufacture and uses*. Chicago: American Oil Chemists Society.

Knowles, A. (2008). Recent developments of safer formulations of agrochemicals. *Environmentalist, 28*(1), 35–44. https://doi.org/10.1007/s10669-007-9045-4.

Kumar, A., Negi, Y. S., Choudhary, V., & Bhardwaj, N. K. (2014). Characterization of cellulose nanocrystals produced by acid-hydrolysis from sugarcane bagasse as agro-waste. *Journal of Materials Physics and Chemistry, 2*(1), 1–8. https://doi.org/10.12691/jmpc-2-1-1.

Landry, K. S., Micheli, S., McClements, D. J., & McLandsborough, L. (2015). Effectiveness of a spontaneous carvacrol nanoemulsion against *Salmonella* enterica Enteritidis and *Escherichia coli* O157:H7 on contaminated broccoli and radish seeds. *Food Microbiology, 51*, 10–17. https://doi.org/10.1016/j.fm.2015.04.006.

Lane, K. E., Li, W., Smith, C., & Derbyshire, E. (2014). The bioavailability of an omega-3-rich algal oil is improved by nano-emulsion technology using yogurt as a food vehicle. *International Journal of Food Science & Technology, 49*(5), 1264–1271. https://doi.org/10.1111/ijfs.12455.

Le Reverend, B. J. D., Norton, I. T., Cox, P. W., & Spyropoulos, F. (2010). Colloidal aspects of eating. *Current Opinion in Colloids Interface Sciences, 15*(1–2), 84–89. https://doi.org/10.1016/j.cocis.2009.11.009.

Li, X., Anton, N., & Vandamme, T. (2013). Nano-emulsions: Overview and applications. Liang XJ, editor, pp. 21–48.

Lin, K. W., & Mei, M. Y. (2000). Influences of gums, soy protein isolate, and heating temperatures on reduced-fat meat batters in a model system. *Journal of Food Science, 65*(1), 48–52. https://doi.org/10.1111/j.1365-2621.2000.tb15954.x.

Lorenzo, G., Zaritzky, N., & Califano, A. (2018). Food gel emulsions: Structural characteristics and viscoelastic behavior. In T. J. Gutiérrez (Ed.), *Polymers for food applications* (pp. 481–507). Cham: Springer. https://doi.org/10.1007/978-3-319-94625-2_18.

Mao, L., Xu, D., Yang, J., Yuan, F., Gao, Y., & Zhao, J. (2009). Effects of small and large molecules emulsifiers on the characteristics of β-carotene nano-emulsions prepared by high pressure homogenization. *Food Technological Biotechnology, 47*(3), 336–342.

Martins, P., Dulley, R., Ramos, S., Barbosa, M., Assumpçao, R., Junior, S., & Lacerda, A. (2007). Nanotecnologias na indústria de alimentos. Available at: http://www.pucsp.br/eitt/downloads/vi_ciclo_paulomartins_ marisabarbosa_nano_puc.pdf.

Mason, T. G., Wilking, J. N., Meleson, K., Chang, C. B., & Graves, S. M. (2006). Nanoemulsions: formation, structure, and physical properties. *Journal of Physics: Condensed Matter, 18*(41), 635–666.

McClements, D. J. (1999). *Food emulsions: Principles, practice and techniques* (1st ed.). Boca Raton: CRC Press.

McClements, D. J. (2004). Role of hydrocolloids as emulsifiers in foods. In P. A. Phillips & G. O. Williams (Eds.), *Gums and stabilizers for the food industry* (pp. 381–393). Cambridge, UK: Royal Society of Chemistry. Special Publication No. 294.

McClements, D. J. (2011). Edible nano-emulsions: Fabrication, properties, and functional performance. *Soft Matter, 7*(6), 2297–2316. https://doi.org/10.1039/c0sm00549e.

McClements, D. J., & Decker, E. A. (2000). Lipid oxidation in oil-in-water emulsions: Impact of molecular environment on chemical reactions in heterogeneous food systems. *Journal of Food Science, 65*(8), 1270–1282. https://doi.org/10.1111/j.1365-2621.2000.tb10596.x.

Melik, D. H., & Fogler, H. S. (1998). Fundamentals of colloidal stability in quiscent media. In R. Beccher (Ed.), *Encyclopaedia of emulsion technology* (Vol. III., Chap. 1, pp. 3–78). New York: Marcel Dekker.

Möller, M., Eberle, U., Hermann, A., Moch, K., & Stratmann, B. (2009). *Nanotechnology in the food sector*. Zürich: TA-SWISS.

Mura, S., Greppi, G. F., & Irudayaraj, J. (2015). Latest developments of nanotoxicology in plants. In M. H. Siddiqui, Al-Whaibi, M. H., & Mohammad, F. (Eds.), *Nanotechnology and plant sciences* (pp. 125–151). Switzerland: Springer International Publishing. https://doi.org/10.1007/978-3-319-14502-0_7.

Muschiolik, G., & Dickinson, E. (2017). Double emulsions relevant to food systems: Preparation, stability and applications. *Comprehensive Reviews in Food Science and Food Safety, 16*(3), 532–555. https://doi.org/10.1111/1541-4337.12261.

Nethaji, D. K., & Parambil, K. A. (2017). Development and applications of nano emulsion in food technology. *International Journal of Science, Engineering and Management, 2*(12), 60–63.

NutraLease. (2011). Available at: http://www.nutralease.com/Nutra/Templates/showpage.asp?DBID=1&LNGID=1&TMID=84&FID=767.

O' Dwyer, S. P., O' Beirne, D., Ni Eidhin, D., Hennessy, A. A., & O' Kennedy, B. T. (2013). Formation, rheology and susceptibility to lipid oxidation of multiple emulsions (O/W/O) in table spreads containing omega-3 rich oils. *LWT – Food Science and Technology, 51*(2), 484–491. https://doi.org/10.1016/j.lwt.2012.12.008.

Paul, B. K., & Moulik, S. P. (2001). Uses and application of micro-emulsions. *Current Science, 80*(8), 990–1001.

Poyato, C., Navarro-Blasco, I., Calvo, M. I., Cavero, R. Y., Astiasaran, I., & Ansorena, D. (2013). Oxidative stability of O/W and W/O/W emulsions: effect of lipid composition and antioxidant polarity. *Food Research International, 51*(1), 132–140. https://doi.org/10.1016/j.foodres.2012.11.032.

Raj, A. A. S., Jayaraman, R., Rubila, S., & Tiroutchelvamae, D. (2013). Recent trends in nanotechnology applications in foods. *International Journal of Engineering Research and Technology, 2*(10), 956–961.

Rasooli, I. (2007). Food preservation-a bio-preservative approach. *Food, 1*, 111–136.

Renton, A., (2006). Welcome to the world of nano-foods. Guardian Unlimited UK 12. Available at: http://observer.guardian.co.uk/foodmonthly/futureoffo od/story/.

Sadurni, N. (2005). Studies on the formation of O/W nano-emulsions, by low-energy emulsification methods, suitable for pharmaceutical applications. *European Journal of Pharmaceutical Sciences, 26*(5), 438–445. https://doi.org/10.1016/j.ejps.2005.08.001.

Sanguansri, P., & Augustin, M. A. (2006). Nanoscale materials, development of food industry perspective. *Trends in Food Science & Technology, 17*(10), 547–556. https://doi.org/10.1016/j.tifs.2006.04.010.

Schramm, L. L. (1992). *Emulsions: Fundamentals and applications in the petroleum industry*. Washington, D.C: American Chemical Society.

Schrooyen, P. M. M., van der Meer, R., & De Kruif, C. G. (2001). Microencapsulation: Its application in nutrition. *Proceedings of the Nutrition Society, 60*(4), 475–479. https://doi.org/10.1079/pns2001112.

Serdaroglu, M., Ozturk, B., & Kara, A. (2015). An overview of food emulsions: Description, classification and recent potential applications. *Turkish Journal of Agriculture – Food Science and Technology, 3*(6), 430–438. https://doi.org/10.24925/turjaf.v3i6.430-438.336.

Severino, R., Vu, K. D., Donsi, F., Salmieri, S., Ferrari, G., & Lacroix, M. (2014). Antibacterial and physical effects of modified chitosan based-coating containing nano-emulsion of mandarin essential oil and three non-thermal treatments against Listeria innocua in green beans. *International Journal of Food Microbiology, 191*, 82–88. https://doi.org/10.1016/j.ijfoodmicro.2014.09.007.

Shafiq, S., Shakeel, F., Talegaonkar, S., Ali, J., Baboota, S., Ahuja, A., Khar, R. K., & Ali, M. (2007). Formation development and optimization using nano-emulsion technique: A technical note. *Pharmaceutical Sciences and Technology, 8*(2), 1–6. https://doi.org/10.1208/pt0802028.

Shah, D. O. (1998). In D. O. Shah (Ed.), *Micelles, micro-emulsions and monolayers: Science and technology* (p. 1). New York: Marcel Dekker.

Shakeel, F., Baboota, S., Ahuja, A., Ali, J., & Shafiq, S. (2008). Accelerated stability testing of celecoxib nano-emulsion containing cremophor-EL. *African Journal of Pharmaceutical Pharmacologists, 2*, 179–183.
Sharma, R. (2005). Market trends and opportunities for functional dairy beverages. *Australian Journal of Dairy Technology, 60*(2), 195–198.
Silva, H. D., Cerqueira, M. A., & Vicente, A. A. (2012). Nano-emulsions for food applications: Development and characterization. *Food and Bioprocess Technology, 5*(3), 854–867. https://doi.org/10.1007/s11947-011-0683-7.
Singh, H., & Ye, A. (2000). Influence of calcium chloride addition on the properties of emulsions stabilized by whey protein concentrate. *Food Hydrocolloids, 14*(4), 337–346. https://doi.org/10.1016/s0268-005x(00)00010–2.
Solans, T., Izquierdo, P., Nolla, J., Azemar, N., & Garcia, C. M. J. (2005). Nano-emulsions. *Current Opinion in Colloid Interface Science, 10*(3–4), 102–110. https://doi.org/10.1016/j.cocis.2005.06.004.
Sozer, N., & Kokin, J. L. (2008). Nanotechnology and its applications in the food sector. *Trends in Biotechnology, 27*(2), 82–89. https://doi.org/10.1016/j.tibtech.2008.10.010.
Tadros, T. F. (1995). *Surfactants in Agrochemicals*. New York: Marcel Dekker.
Tadros, T. (2004). Formation and stability of nano-emulsions. *Advances in Colloid and Interface Science, 108–109*, 303–318. https://doi.org/10.1016/j.cis.2003.10.023.
Tadros, F. (2005). *Applied surfactants, principles and applications* (pp. 187–256). Weinheim: Wiley-VCH Verlag GmbH and KGaA.
Takahashi, Y., Yoshida, T., & Takahashi, T., inventors. 1986. Process for the production of a water-in-oil-in-water emulsion for foods. U.S. Patent 4590086.
Tan, C. T. (1990). Beverage emulsions. In K. Larsson & S. E. Friberg (Eds.), *Food emulsions* (2nd ed., pp. 445–478). New York: Marcel Dekker.
Trubiano, P.C. (1995). *The role of specialty food starches in flavor emulsions. Flavor Technology*. Washington, D. C: ACS Symposium Series, 610, American Chemical Society.
Unilever. (2011). Available at: http://www.unilever.com/innovation/productinnovations/coolicecreaminnovations/?WT.LHNAV=Cool_ice_cream_innovations.
Velikov, K., & Pelan, E. (2008). Colloidal delivery systems for micronutrients and nutraceuticals. *Soft Matter, 4*(10), 1964–1980. https://doi.org/10.1039/b804863k.
Vernner, R., & Bauer, P. (2007). O-TEQ®, a formulation concept that overcomes the in-compability between water and oil. *Pfalzenschutz-Nachrichten Bayer, 60*(1), 7–26.
Walker, & Rebecca, M. (2015). Fish oil Nano-emulsions: Optimization of physical and chemical stability for food system applications. Masters Theses. 313, Available at: https://scholarworks.umass.edu/masters_theses_2/313.
Wang, X., Jiang, Y., Wang, Y., Huang, M., Ho, C. T., & Huang, Q. (2008). Enhancing anti-inflammation activity of curcumin through O/W nano-emulsions. *Food Chemistry, 108*(2), 419–424. https://doi.org/10.1016/j.foodchem.2007.10.086.
Waraho, T., McClements, D. J., & Decker, E. A. (2011). Mechanisms of lipid oxidation in food dispersions. *Trends in Food Science & Technology, 22*(1), 3–13. https://doi.org/10.1016/j.tifs.2010.11.003.
Weiss, J., Takhistov, P., & McClements, D. J. (2006). Functional materials in food nanotechnology. *Journal of Food Science, 71*(9), 107–116. https://doi.org/10.1111/j.1750-3841.2006.00195.x.
Wesley, S. J., Raja, P., Raj, A. A. S., & Tiroutchelvamae, D. (2014). Review on – Nanotechnology applications in food packaging and safety. *International Journal of Engineering Research, 3*(11), 645–651.
Whitesides, G. M., & Grzybowski, B. (2002). Self-assembly at all scales. *Science, 295*(5564), 2418–2421. https://doi.org/10.1126/science.1070821.
Wooster, T. J., Golding, M., & Sanguansri, P. (2008). Impact of oil type on nano-emulsion formulation and Ostwald ripening stability. *Langmuir, 24*(22), 12758–12765. https://doi.org/10.1021/la801685v.
Yang, Y., Marshall-Breton, C., Leser, M. E., Sher, A. A., & McClements, D. J. (2012). Fabrication of ultrafine edible emulsions: Comparison of high-energy and low-energy homogenization methods. *Food Hydrocolloids, 29*(2), 398–406. https://doi.org/10.1016/j.foodhyd.2012.04.009.

Yoksan, R., Jirawutthiwongchai, J., & Arpo, K. (2010). Encapsulation of ascorbyl palmitate in chitosan nanoparticles by oil-in-water emulsion and ionic gelation processes. *Colloids and Surfaces B: Biointerfaces, 76*(1), 292–297. https://doi.org/10.1016/j.colsurfb.2009.11.007.

Zhang, J. (2011). Novel emulsion-based delivery systems. PhD Dissertation, University of Minnesota, USA.

Zheng, H. (2018). Polymers for structure design of dairy foods. In T. J. Gutiérrez (Ed.), *Polymers for food applications* (pp. 509–528). Cham: Springer. https://doi.org/10.1007/978-3-319-94625-2_19.

Chapter 10
Nanoemulsions: Industrial Production and Food-Grade Applications

Sonal Agarwal, Swathika Vivekanandan, Trisha David, Mahima Mitra, Jeyanthi Palanivelu, and Ramalingam Chidambaram

Abstract The heterogeneous dispersion of two immiscible liquids forming a metastable system, with sizes ranging from 20 to 1000 nm generate nanoemulsions. Nanoemulsions are capable of efficiently delivering bioactive and flavoring molecules by crossing biological obstruction. These polymer systems in the food industry are designed to maintain the functionality of the active ingredients such as digestibility and antioxidant, anti-inflammatory and antimicrobial properties. A nanoemulsion maintains profitable concerns over conventional systems due to its high optical clarity, surface area, surface reactivity, good physical stability and better bioavailability of the substances that are encapsulated. Nanoemulsions can be produced by low and high energy methods, but there is a substantial interest towards low energy methods, since the latter are suitable to produce beverages such as soft drinks or fortified water, since the final product contains relatively less amount of surfactant. These systems can also serve as an effective platform for the incorporation of nutrients, nutraceuticals and vitamins. The objective of this chapter was to analyze the different methods of production of nanoemulsions, in particular the low energy methods and the applications of food grade nanoemulsions.

Keywords Bioactive molecule · Bioavailability · Low energy method · Nutraceuticals

10.1 Introduction

The introduction of nanotechnology into food industry was marked in late 1800's during the discovery of pasteurization process by Louis Pasteur. The first revolutionary step made in food industry was to improve the stability of the packed and un-packed food by its bactericidal action, to prevent food spoilage and maintain

S. Agarwal · S. Vivekanandan · T. David · M. Mitra · J. Palanivelu · R. Chidambaram (✉)
Department of Biotechnology, School of Bioscience & Technology,
Vellore Institute of Technology (VIT), Vellore, Tamilnadu, India
e-mail: cramalingam@vit.ac.in

T. J. Gutiérrez (ed.), *Polymers for Agri-Food Applications*,
https://doi.org/10.1007/978-3-030-19416-1_10

food quality (Chellaram et al. 2014). Recent survey suggests researcher's interest towards the advancements in nanotechnology due to its potential applications in various domains/industries. The various ways in which nanotechnology is applicable in the food industry includes agriculture, food processing and packaging (Gutiérrez and Alvarez 2018; Gutiérrez 2018a; Toro-Márquez et al. 2018; Merino et al. 2018; Gutiérrez et al. 2019; Merino et al. 2019). The derivatives of nanotechnology which found its application in food industry are nanocapsules, nanotubes, nanoparticles, nanosensors, nanofilms, nanoclays and nanoemulsions (Gutiérrez et al. 2017). Of these, there is a substantial interest in the application of nanoemulsions in food processing, since it is a potential tool for formulating new food products (Odriozola-Serrano et al. 2014; Gutiérrez and Álvarez 2017). They ensure the stability of food products during manufacturing, processing, packaging and shipping; deliver functional ingredients with beneficial activities to specific sites of action; improve food security without disturbing the environment and; are new and effective material for pathogen detection (Abdullaeva 2017). Nanoemulsions are kinetically stable and thermodynamically unstable colloidal dispersions (Komaiko and McClements 2014; Guttoff et al. 2015; Karthik et al. 2017; Walia et al. 2017). They are optically clear and can increase the chemical stability, improve bioavailability as well as increase the absorption of bioactive agents that are encapsulated (Odriozola-Serrano et al. 2014; Komaiko and McClements 2014). Nanoemulsions can be broadly categorized into oil-in-water (O/W) nanoemulsions or water-in-oil (W/O) nanoemulsions (Fig. 10.1 and Fig. 10.2), of which O/W nanoemulsions are

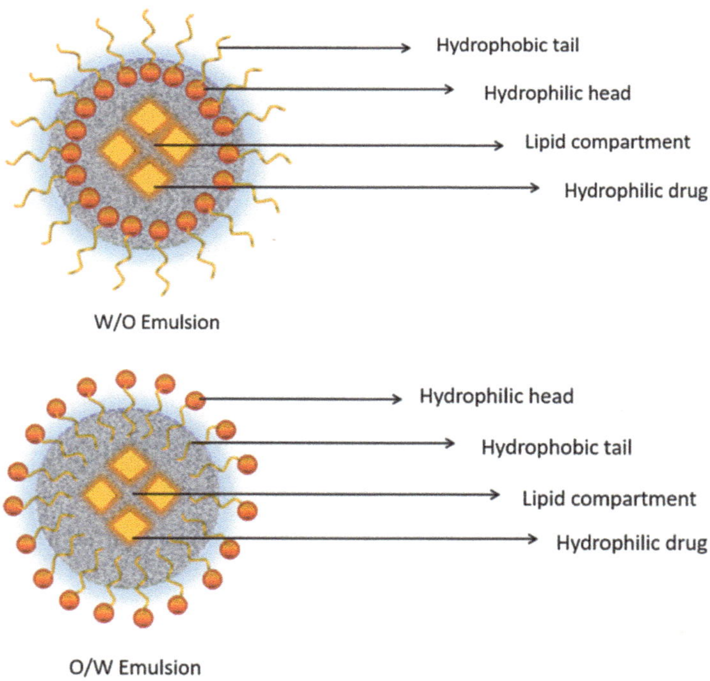

Fig. 10.1 Structure and composition of water-in-oil (W/O) and oil-in-water (O/W) nanoemulsions

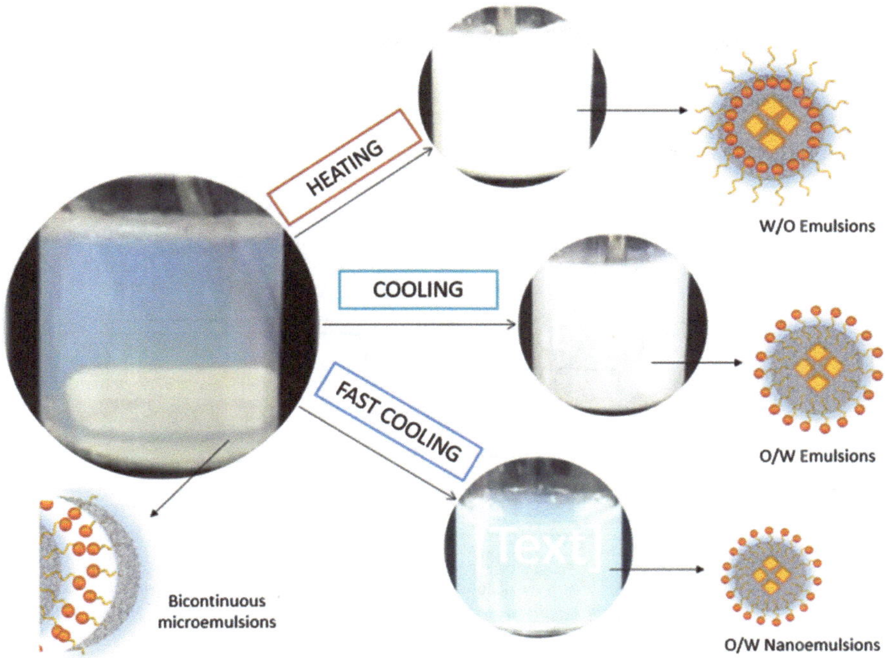

Fig. 10.2 Fabrication of nanoemulsions

commonly used in food industry (Walia et al. 2017). Nanoemulsions are produced by both low and high energy methods (Fig. 10.3) (Gutiérrez 2018b). However, food-grade nanoemulsions are fabricated through low energy techniques (Homs et al. 2018). Nanoemulsions will be selected for applications in food industry based on their performance, safety, commercial viability, robustness, and food-matrix compatibility (McClements 2015; Abdullaeva 2017). The objective of this chapter was thus to analyze the various production methods, in particular low energy methods and applications of food-grade nanoemulsions.

10.2 Properties and Characteristics of Nanoemulsions

10.2.1 Properties of Nanoemulsions

Nanoemulsion droplet sizes ranges from 50 to 1000 nm, but droplets of 20–200 nm forms stable liquid-in-liquid dispersions and 10–100 nm sized droplets are known to have applications in food industry (Lorenzo et al. 2018). They are heterogeneous dispersion mixture of two immiscible liquids, i.e. W/O or O/W (McClements 2012; Komaiko and McClements 2016). Nanoemulsions appear transparent as their sizes are significantly smaller than the wavelength of visible light, thus acting as the

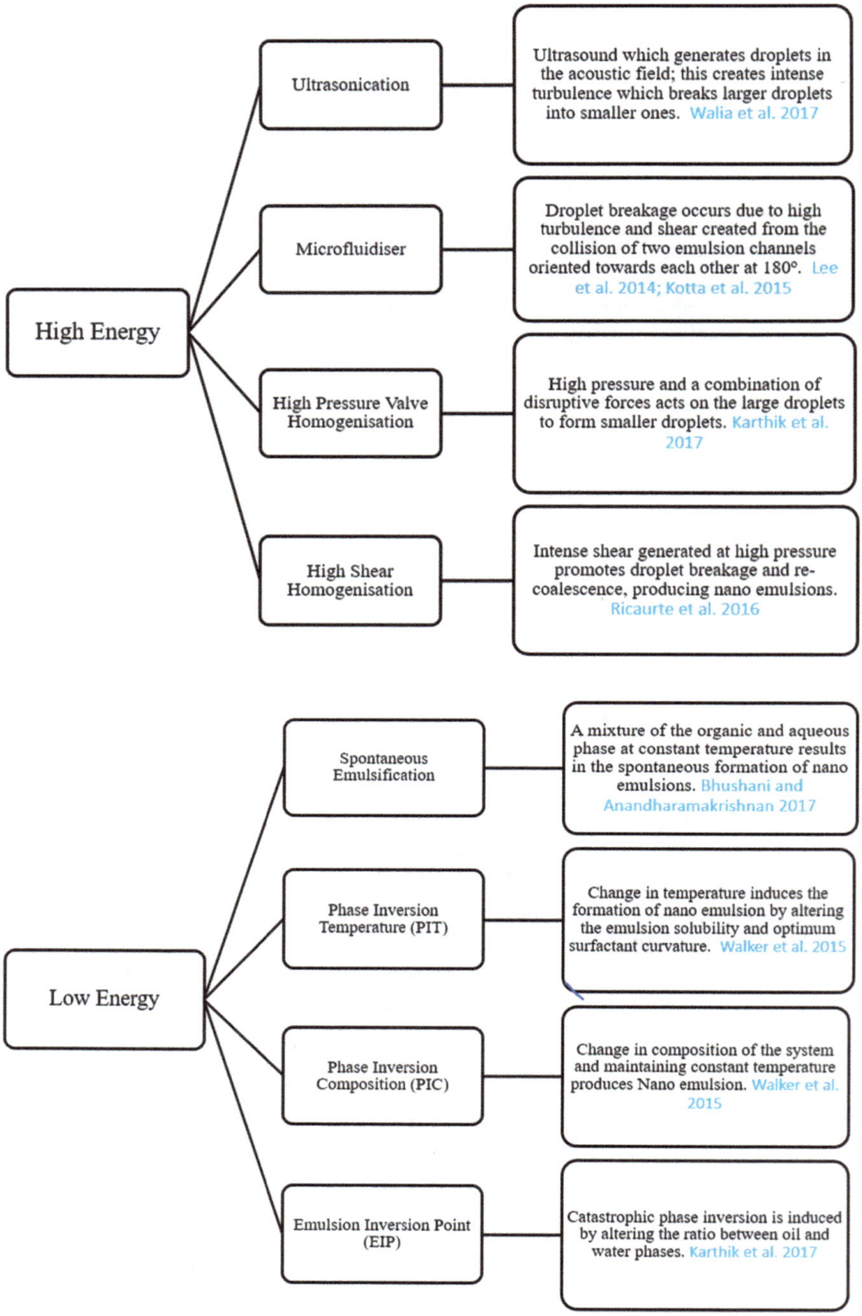

Fig. 10.3 Classification of nanoemulsion production techniques

ideal component to be opted in the food industry (Gupta et al. 2016). The size plays an important role, since it affects properties such as biological activity, encapsulation and release characteristics and optical clarity (Joye et al. 2014).

The small size property of nanoemulsions has many advantages such as extreme resistance to deformation, resulting partly in kinetic stability, the high surface-to-volume ratio of nanoemulsions improves its encapsulated component bioavailability and reactivity, increased surface area eases transportation through the plasma membrane (Helgeson 2016) and increases the enzyme activity at the oil-water interface. Small particles have the ability to cross the mucous layer and get absorbed by the epithelium cells. This leads to faster lipid digestion (Salvia-Trujillo et al. 2013; Walker et al. 2015).

The properties of nanoemulsions include optically transparent appearance, high surface area per unit volume, tunable rheology and robust stability. Nanoemulsions have lower sensitivity towards dilution, changes in pH and temperature when compared to other emulsions (Gupta et al. 2016). They can be designed to have different optical, rheological and stability properties by controlling their compositions and structures.

Optical properties: Optical properties refer to the visual appearance of the nanoemulsion. It is important that the optical properties of nanoemulsions are controlled because food industry incorporates lipophilic bioactive compounds (vitamins and nutraceuticals) into optically transparent delivery systems (Ricaurte et al. 2016). If nanoemulsions are optically opaque, then this attribute is less important. Optical properties can be altered by droplet size, droplet concentration and refractive index contrast. The optical property of nanoemulsions is inversely proportional to its radius size. Example, droplets with radius less than 40 nm appear transparent and translucent whereas when the droplet radius is approximately between 40 and 60 nm, they appear turbid or opaque (McClements 2013).

Rheological properties: From the rheological perspective, nanoemulsions exhibit stronger elasticity compared to other emulsions, thus providing better healing behavior (Gupta et al. 2016). Nanoemulsions intended to deliver lipophilic bioactive compounds in fortified water or soft drinks should not increase the viscosity ideally. However, for some functional food products such as yogurts, sauces and so on it is desirable to have a highly viscous or gel-like material (McClements 2013).

Physical stability: Nanoemulsion stability depends on its composition and structure, and is influenced by various environmental conditions it experiences within a product including mechanical agitation, thermal processing, transportation, freezing, dehydration and storage conditions (Gupta et al. 2016; Dizaj et al. 2016). Most nanoemulsions are physically stable and ensure the dominance of Brownian motion over gravitational force, thus forming a homogenized suspension (Helgeson 2016). The destabilization mechanisms which can break the metastable system are mainly creaming, flocculation, Ostwald ripening and coalescence (McClements 2012). In creaming, droplets rise up, leading to phase separation. If the droplets are denser, they tend to move downwards leading to sedimentation. In flocculation, the droplets approach each other due to attractive interactions, and move as a single entity.

Ostwald ripening, more prevalent in nanoemulsions, occurs due to the difference in chemical potential of solute within droplets of different size. Coalescence involves droplets merging together and forming a larger drop. It is often difficult to differentiate between flocculation and coalescence in an emulsion (Gupta et al. 2016).

Chemical stability: Light-catalyzed reactions may occur more rapidly in transparent nanoemulsions. Surface-catalyzed reactions, such as lipid oxidation or lipase digestion tend to occur more rapidly. Nanoemulsions are isotropic, so they retain a relatively high kinetic stability for a longer duration and, therefore, ideal nanoemulsion systems must retard any potential chemical degradation reactions (McClements 2013). Due to the ability of the solidified lipid phase to inhibit molecular diffusion processes, solid liquid nanoemulsions may be able to protect encapsulated lipophilic components from chemical degradation by lipid crystallization of lipophilic component (Joye et al. 2014).

The polarity property defines the interaction of the nanoemulsion or nanoparticles in the biological system. Another property of nanoemulsions is digestibility, which depends on the core shell structure composition (McClements 2013). As the degree of unsaturation decreases, the digestion becomes easy (McClements 2013).

10.2.2 Characteristics of Nanoemulsions

Nanoemulsions can be characterized based on their size, pH, components compatibility, content uniformity, density, conductivity and surface tension (Dizaj et al. 2016). Nanoemulsions are nanoscale delivery systems which are formulated to ensure their safety, economic viability and effectiveness. For application of nanoemulsion based delivery system in food industry, number of characteristics is to be considered specifically (McClements 2015).

10.2.3 Various Characteristic Features of a Nanoemulsion to be Used in Food

The robustness of a nanoemulsion is important to maintain its chemical and physical stability even when exposed to external factors such as environmental stress. It must also preserve its desirable functional attributes.

Nanoemulsions must be commercially available to manufacture the food grade ingredients with these delivery systems and also the processing methods that are economically viable. High loading capacity is desired to be capable for encapsulating a significant quantity of the bioactive ingredient within the system.

Active compound retention and improved bioavailability: They should be able to encapsulate the active form of the compound until its delivery to site of action. It should be able to increase the chemical stability of the encapsulated compounds during its processing, digestion in gastrointestinal tract or storage.

Food matrix compatibility: The encapsulated bioactive agent should be capable of being homogenized into the final food product with no damage to its distinctive characteristics. It should mask the off-flavor, off-odor and after taste of the encapsulated bioactive compound. The encapsulated bioactive agent should also be compatible with food matrix or beverage in which it has been fortified and it should not affect the sensory characteristic and overall acceptance of the food (Bhushani and Anandharamakrishnan 2017; McClements 2015).

Safety

Safety of nanoemulsions is an important characteristic required for its use in the food industry. If nanoemulsions are completely digested inside the gastrointestinal tract (GIT) as a result giving products of digestion that are same as those which are formed by larger particles and are formed at a similar site, then they are expected to be more toxic. However, if nanoemulsions are not digested completely, or if they are forming products of digestion that are very dissimilar, or are being digested in different regions of the GIT as compared to the bigger particles, then there might be few concerns relating to toxicity. Therefore, potential toxicity testing of the food-grade nanoemulsions stands great importance (McClements 2013).

10.3 Delivery Systems

The physical stability of food-grade nanoemulsions as it passes through the path that includes the mouth, stomach and small intestine depends upon the emulsifier used (Shu et al. 2018). The four mechanisms through which emulsifiers stabilize nanoemulsions are electrostatic, steric, electrosteric and electrostatic-steric mechanisms. Emulsifiers adsorb onto the surface of the droplets acting as a barrier for the protection of nano-sized droplets from coalescence and aggregation during emulsification process and storage (Shariffa et al. 2016). An ideal emulsifier must exhibit the following properties: surface activity, adsorption kinetics, interfacial tension reduction, stabilization and surface coverage to produce nanoemulsions with smaller droplet size during homogenization. Structure of emulsifiers influences the emulsifying and stabilizing mechanisms (Ozturk and McClements 2016). Based on the type of emulsifiers used, food-grade nanoemulsion delivery systems can be divided into three groups: lipid-based nanosystems, surfactant-based nanosystems and biopolymer-based nanosystems (Table 10.1) (Bhushani and Anandharamakrishnan 2017).

10.3.1 Lipid-Based Nanosystems

Lipid-based nanoemulsions incorporated with bioactive compounds improves the solubility property of active compounds, improves bioavailability, chemical and gastrointestinal stability (Yang and McClements 2013; Bhushani and Anandharamakrishnan 2017). Lipid-based emulsifier such as lecithin is commonly

Table 10.1 Different types of delivery systems: Properties and characteristics

Types of emulsifier	Optical characteristics	Factors that affect stability	Emulsifier properties	Examples	Reference
Surfactant	Optically clear.	Temperature.	Thermodynamically stable, have low toxicity and lack of irritability.	Tween 20, 40, 60, 80, 85 and Polysorbate 20, 28, 80.	Sugumar et al. (2013); Dizaj et al. (2016)
Lipid	Ranges from clear to opaque depending upon their size.	pH, ionic strength and temperature.	Have low solubility, low toxicity, high intestinal permeability, high encapsulation efficiency and strong electrostatic repulsion.	Lecithin, phosphated mono- and diglycerides.	Ozturk and McClements (2016)
Biopolymer	Ranges from clear to opaque depending upon their particle refractive index and size.	pH, temperature, ionic strength and metal chelating agents.	Exhibit specific interfacial behavior, high loading capability and response to environmental stresses.	Nanofibers, starch, cellulose, citrus pectin, guar gum, alginate, chitosan and dextran. Amphiphilic proteins such as casein, lactoferrin, β-lactoglobulin, whey proteins and protein isolates. Polysaccharides such as gum Arabic and modified starch.	Joye and McClements (2013); Donsì and Ferrari (2016); Bhushani and Anandharamakrishnan (2017)
Protein	Ranges from clear to opaque depending upon their size.	pH and temperature.	Produce larger droplet sizes due to their low absorption kinetics.	Corn protein, zein ultrafine fibers, whey protein and milk protein.	Bhushani and Anandharamakrishnan (2017); Shu et al. (2018)

used in the production of nanoemulsions. Surface active lecithin is extracted from milk, soybeans, rapeseeds, sunflower kernels and egg yolk. Despite possessing the surface-active property, phospholipids are not really good emulsifiers and are prone to coalescence. They are often used in combination with other protein-based emulsifiers.

10.3.2 Surfactant-Based Nanosystems

Non-ionic food-grade surfactants are relatively stable against coalescence and flocculation when compared to protein-based and lipid-based emulsifiers (Silva et al. 2015; Chang and McClements 2016; Ozturk and McClements 2016). Negatively charged ionic surfactants are also used in food grade delivery system. Hydrophilic lipophilic balance (HLB) value contributes significantly during the selection of surfactant in nanoemulsion fabrication (Sugumar et al. 2013; Dizaj et al. 2016). HLB number indicates affinity of surfactant for oil phase or aqueous phase. Small molecule surfactant used in low energy method differs by their HLB number and molecular geometry. Molecular geometry of surfactant influences whether the assembly of droplets will be into either O/W or W/O nanoemulsions. HLB >10 induces formation of O/W nanoemulsion with small droplet size (Ozturk and McClements 2016). Propylene glycol, ethanol and glycerol are co-solvents surfactants which change the properties of aqueous phase and the changes particle size in nanoemulsions (Ozturk and McClements 2016).

10.3.3 Biopolymer-Based Nanosystems

When nanoemulsions are consumed, the biopolymer emulsifiers prevent the flocculation of oil droplets under acidic conditions in stomach (Zou et al. 2015). Amphiphilic proteins and polysaccharides are biopolymers with high molecular weight which act as good emulsifiers (Ozturk and McClements 2016). When two or more biopolymer emulsifiers are combined, the resistance of nanoemulsions toward droplet growth increases (Shu et al. 2018).

10.3.4 Natural Emulsifiers

Natural emulsifiers are currently used due to the demand for "clean-label" products in food market. They exhibit good emulsifying properties, owing to their structure of non-polar proteins being attached to their hydrophilic carbohydrate chains. Some natural surface-active polysaccharides are used as emulsifiers. The most prominent natural emulsifiers available in the market that exhibit surface activity are Casein and Whey proteins, which are derived from bovine milk (de Oca-Ávalos et al. 2017). Gelatins also exhibit surface activity, but cannot be used as an emulsifier as they do not contribute significantly to stabilizing emulsions. Saponin based emulsifiers such as Quillaja Saponin is commonly used to form emulsions with smaller droplets that must be stable at wide range of environmental stresses (Yang and McClements 2013; Chen et al. 2016; Ozturk and McClements 2016).

10.4 In-Product and In-Body Behavior

Nanoemulsion digestion occurs mainly in the small intestine. Pancreatic lipases in small intestine hydrolyze triglyceride or triacylglycerol (TAG) molecules. Exogenous and endogenous surface-active molecules in bioactive compounds after getting hydrolyzed by lipases are solubilized by intestinal fluids (Rao et al. 2013). The formulation and processing of nanoemulsion in food matrix are done based on functional design principles in order to deliver encapsulated compounds to the targeted sites maintaining their functionality in product (*in-product*) and after consumption (*in-body*). *In-product* behavior depends on the dispersion efficiency and compatibility of nanoemulsion with food matrix of encapsulated compounds. Stability of the absorbed encapsulated compounds within complex environment of food matrices is decided by its reactivity with the food matrix as that environment can be in aqueous or lipid phase. Food treatments processing, intense shearing, preservation and storage at high or low temperature affects the property and reactivity of the bioactive compounds. Temperature changes can result in fragmentation of the compounds making them inactive.

In-body behavior is affected by environment changes such as temperature (high temperature during cooking, body temperature), addition of moisture (chewing), pH (oral, gastrointestinal tract), mechanical shear (chewing, grinding and mastication) and enzymes (gastrointestinal tract), which the encapsulated compounds undergoes with the food matrix (Sessa and Donsi 2015). The bioavailability of encapsulated products is affected by breaking or dissolving process of food matrix in order to promote their absorption by GIT into blood streams, epithelial cell absorption transporting them target sites.

To control and regulate in-body and in-product behaviors suitable delivery system must be used. Solidified lipid phase delivery system can be used to promote efficient dispersion; protecting the encapsulated compound during processing, preservation, preparation and also controlling their release during mastication and gastrointestinal digestion; enhancing taste, bio accessibility and bioavailability. Adequate selection of stabilization layer will maintain electrostatic repulsion and steric hindrance between nanoemulsion droplets. Surface composition and charge plays an important role on nanoemulsion's interaction with external environment (Pinheiro et al. 2013; Majeed et al. 2016). Regulating mean particle size, formulation, composition, concentration, number of lipid/polymer layers over of lipophilic core can influence the rate of lipid digestion and its kinetic stability (Donsì et al. 2013).

Lipase concentration present per unit surface area decreases with a decrease in droplet size increasing lipid digestion rate (Salvia-Trujillo et al. 2013; Walker et al. 2015; Majeed et al. 2016). As the concentration of digestible TAG molecules decrease in oil phase, the bioaccessibility of β-carotene-enriched nanoemulsions also decreased (Rao et al. 2013). A contradictory phenomenon to that proposed by Salvia-Trujillo et al. (2013) was reported by Yi et al. (2015) observing an increase in bioavailability by decreasing the particle size of β-carotene. Compared to medium chain triglyceride and non-digestible oil carrier, long chain triglyceride fatty acid oil carriers of β-carotene were found to have higher bioaccesibility (Bhushani and Anandharamakrishnan 2017). Curcumin incorporation within nanoemulsion has increased its oral bioavailability up to nine folds and the anti-inflammatory activity was also increased compared to unformulated curcumin (Dizaj et al. 2016). It was found that the higher the lipid carrier chain length, the higher is the bioaccesibility of curcumin (Salvia-Trujillo et al. 2017). Curcumin sustained release can also be achieved using multilayer nanoemulsions (Sari et al. 2015; Pinheiro et al. 2016).

10.5 Production

The fabrication of nanoemulsion requires an organic phase, aqueous phase, surfactant and energy. The formation of nanoemulsion is not a spontaneous process as either an internal or external energy input is necessary (Karthik et al. 2017). Production approaches of food-grade nanoemulsions are broadly categorized as either 'top-down' or 'bottom-up' approach. Top-down approach of nanoemulsion fabrication involves mechanical size reduction of the nanoemulsions, such as colloid milling and grinding. It is an energy intensive method, which is commonly referred as a high energy method (Donsì and Ferrari 2016; Abdullaeva 2017; Prakash et al. 2018). Bottom-up approach of nanoemulsion fabrication refers to the synthesis of large structures by the controlled assembly of smaller particles such as atoms and molecules (Abdullaeva 2017; McClements 2015). Bottom-up approach requires low energy to produce very fine droplets using low cost equipment. It is thus commonly known as a low energy method (Prakash et al. 2018). These methods applied in the fabrication of nanoemulsion have different emulsion composition and operating conditions, which regulates its droplet size (Table 10.2) (Karthik et al. 2017).

10.5.1 High Energy Methods

Top-down high energy method involves the use of mechanical instruments such as ultra sound generators, high-shear stirrer, and high-pressure homogenizer to produce nanoemulsions by generating powerful disruptive forces (Solans and Solé 2012). Intense energy generates high intensity disruptive forces which can disrupt the oil

Table 10.2 Difference between low energy and high energy production methods

Basis	High energy method	Low energy method	References
Quantity	Less quantity of nanoemulsions is formed.	A relatively large quantity of nanoemulsions is produced.	Jasmina et al. (2017)
Solvent-oil ratio (SOR)	Lower SOR is required to produce small droplets.	Higher SOR is required for the formation of smaller droplets.	Yang et al. (2012); Jasmina et al. (2017)
Energy requirement	Requires sophisticated instruments which takes up large amount of energy.	Involves user-friendly equipment, with low energy in-take. Thus, it is more acceptable in market.	Yang et al. (2012); Jasmina et al. (2017)
Cost	Expensive procedure.	Inexpensive procedure.	Jasmina et al. (2017)
Temperature	Large energy and high temperature promote component degradation.	Temperature remains constant, degradation does not occur.	Jasmina et al. (2017)
Ingredients incorporated	Thermo-labile active ingredients such as proteins and retinoids cannot be incorporated.	Any ingredient can be incorporated.	Yang et al. (2012); Jasmina et al. (2017)
Surfactant	Flexible in the choice of the internal structure and surfactant, as well as possibility of preparing nanoemulsions within a short duration.	Only non-ionic surfactant can be used, mostly all ingredients can be added.	Jasmina et al. (2017)

and water phase forming tiny oil droplets into aqueous phase (Yang et al. 2012; Silva et al. 2015; Walker et al. 2015). The disruptive forces generated are larger than the restoring forces, which helps to maintain the spherical shape of the droplets. In high energy methods, the nanoemulsion droplet size depends on interfacial tension, duration of energy input, energy intensity, relative viscosities of the phases, types of emulsifier, concentration of emulsifiers and surfactant-to-oil ratio (SOR < 0.1) (McClements 2012; Yang et al. 2012; Karthik et al. 2017). Smaller droplet sizes are preferred by the food industry for the fabrication of food-grade nanoemulsions which can be obtained by controlling the viscosity ratio, increasing the homogenization duration, or by increasing the emulsifier concentration (Karthik et al. 2017). The high energy techniques widely used to produce food-grade nanoemulsions are ultrasonication, high pressure valve homogenization and microfluidization (Yang et al. 2012; Abdullaeva 2017).

High pressure valve homogenizer applies the principle of using multiple passes and exceptional high pressure in the fabrication of nanoemulsion with the required droplet size (Lee et al. 2014). The homogenizer generates small droplets through the disintegration of large droplets by amalgamating the disruptive forces such as shear stress, cavitation and turbulent flow. Nanoemulsions enriched with thyme oil, curcumin and β-carotene are produced using high energy method with varying operating conditions (Karthik et al. 2017). Microfluidizer has a similar design, but differs in the flow of emulsion channels. The principle is to divide an emulsion flowing in a channel into two separate streams. Each stream then passes through a distinct channel. These two streams are then directed towards each other in the interaction chamber. Interaction between the two fast-moving streams creates intense disruptive forces in the interaction chamber, which results in droplet distribution (Lee et al. 2014; Komaiko et al. 2015; Ricaurte et al. 2016). In this method, the factors affecting homogenization efficiency are emulsifier concentration, emulsifier type, oil-water viscosity ratio and oil-water interfacial tension. Food-grade nanoemulsions from food ingredients can be produced using this method (Yang et al. 2012; Karthik et al. 2017).

Nanoemulsions enriched with varieties of bioactive compounds are produced through microfluidizer technique (Salvia-Trujillo et al. 2013; Karthik et al. 2017). The flaw of microfluidization is that, in low surfactant concentration sufficient surfactant quantity is not present to envelope the droplet surface, which causes the droplets to come in close proximity resulting in coalescence. High surfactant concentration does not affect droplet size during homogenization (Yang et al. 2012). Ultrasonication principle states that high ultrasonic waves with a frequency greater than 20 kHz can form nano-sized droplets of approximately 70 nm. The droplet size decreases with increase in the emulsifier concentration, sonication time and power level (Karthik et al. 2017). High energy methods can be applied to fabricate nanoemulsions in large scale, however there are several limitations associated with this method. The limitations of high energy methods are the high cost of operation, the high initial equipment required, the high-power requirement, the high probability of equipment break down and the difficulty in producing fine droplets with food ingredients such as highly viscous oils and slowly adsorbing emulsifiers (Yang et al. 2012).

10.5.2 Low Energy Methods

Energy efficient, bottom-up low energy method, utilizes the system's internal chemical energy and simple stirring to produce nanoemulsions with smaller droplets sizes and low polydispersity than high energy methods (Solans and Solé 2012; Karthik et al. 2017). In low energy mode, nanoemulsions are produced due to the chemical energy released from phase transitions which occurs during emulsification process (Karthik et al. 2017). Fabrication of nanoemulsions through low energy method requires internal energy of oil phase, aqueous phase and emulsifier system (Abdullaeva 2017). There is substantial interest in this field as low energy techniques are energy efficient, relatively inexpensive, facile, capable of producing smaller particle size, and prevent encapsulated compound from being degraded. In this method, the spontaneous formation of smaller oil droplet size is dependent upon the system composition (ionic strength, surfactant type and surfactant-oil-water ratio), and the environmental conditions (stirring speed, duration and temperature) (Karthik et al. 2017; Walker et al. 2015). This phenomenon transpires through the alteration of environmental conditions such as temperature and composition (Silva et al. 2015; Karthik et al. 2017). Solans and Solé (2012) classified low energy methods as spontaneous emulsification or self-emulsification (SE) and phase inversion method (Bhushani and Anandharamakrishnan 2017). Emulsion phase inversion (EPI) and SE are the two main isothermal methods involved in the production of food-grade nanoemulsions. Other isothermal method includes emulsion inversion point (EIP), also referred as phase inversion composition (PIC) and direct emulsification inversion (DEI) method. Thermal method requires changes in temperature to fabricate nanoemulsions, thus the main thermal method used is phase inversion temperature (PIT) method (Komaiko and McClements 2016; Karthik et al. 2017). Low energy methods can be applied to produce nanoemulsions incorporated with nutrients, nutraceuticals and vitamins (Komaiko and McClements 2016). Flavored nanoemulsions can also be prepared through this method (Gupta et al. 2016). However, food products with relatively high levels of fat are not formed by low energy methods due to the presence of excess levels of surfactant in product (de Oca-Ávalos et al. 2017).

10.5.2.1 Spontaneous Emulsification

Spontaneous emulsification (SE) is the spontaneous formation of nanoemulsions when aqueous and organic phase are mixed at constant temperature (Yang et al. 2012; Solans and Solé 2012; Komaiko and McClements 2014; Karthik et al. 2017). The organic phase of emulsion is concocted by lipophilic surfactant, water miscible solvent and oil. The aqueous phase is composed of hydrophilic surfactant and water. When both aqueous and organic phases are mixed, a non-equilibrium stage is achieved. At this stage, rapid diffusion of the water miscible solvent occurs without any phase transition between organic and aqueous phase which increases the interfacial area, resulting in a metastable system also the surfactant spontaneous curvature remains unvaried (Bhushani and Anandharamakrishnan 2017). Mercuri et al. (2011) proposed that when a surfactant/oil (S/O) mixture is in contact with the aqueous phase, a boundary initially forms between the S/O and aqueous phase. Water then penetrates the S/O layer causing swelling to occur, which initially results in a W/O microemulsion, then forms liquid crystalline phases. A liquid crystalline fragment crosses the boundary and enters into aqueous phase. Here the fragments further break down to form nano-sized emulsions (Yang et al. 2012). The turbulence caused during solvent diffusion aids in the formation of nanoemulsion (Solans and Solé 2012). The mechanisms having a key role in nanoemulsion fabrication are interfacial turbulence, negative interfacial tension, diffusion, stranding and dispersion (Bhushani and Anandharamakrishnan 2017). SE forms smaller droplet sizes; however, it requires high surfactant concentration when compared to high energy methods. SOR (surfactant-oil-ratio) for microfluidization is SOR < 0.1, whereas SOR for SE is SOR >1 (Yang et al. 2012). Spontaneous formation of nanoemulsions can be carried out by varying the composition of both phases, environmental conditions such as pH, ionic strength and temperature and mixing conditions which includes order of addition, agitation speed and rate of addition.

10.5.2.2 Phase Inversion Method

Phase inversion methods are followed if surfactant spontaneous curvature changes from negative to positive producing O/W nanoemulsions or from positive to negative producing W/O nanoemulsions during emulsification (Solans and Solé 2012). Phase inversion methods are categorized into two: PIT and PIC (Solans and Solé 2012; Walker et al. 2015). Phase transitions occur by changing the composition (PIC) or the temperature (PIT) (Kotta et al. 2013). PIT was pioneered in 1968 by Shinoda. Here the temperature is varied to alter the solubility and optimum curvature of non-ionic surfactants. This converts O/W emulsions into W/O emulsions or vice versa. In this method, the mixture is heated to a temperature higher than the PIT, followed by instant cooling by constant stirring which forms nanoemulsions (Walker et al. 2015). PIT method is applicable only for temperature sensitive surfactants such as polyoxyethylene-type nonionic surfactants that undergo hydration in poly (oxyethylene) chain and change its curvature as temperature changes. When the system's lipophilic and hydrophilic properties are balanced, the mean spontaneous curvature of surfactant molecule is zero. In PIC method phase transitions are introduced by changing the composition at a steady temperature during emulsification. This method can be applied to any type of surfactant (Solans and Solé 2012).

Recent methods used to produce nanoemulsions are evaporative ripening and bubble bursting methods at oil/water interface. Bubble bursting method involves the bubbling of gas through an aqueous phase containing surfactant. Once the bubble arrives at the interface, the oil film formed between the water phase and interface drains slowly, and a dimple is created in the water-air interface which generates and nucleates spatters of oil droplets in the water phase. In Evaporative Ripening method, smaller emulsion droplets are prepared using high viscosity oils. The initial droplets produced from HPH are a mixture of non-volatile (high molecular weight) and volatile (low molecular weight) oils. During heating, the volatile components present in the droplets evaporate causing the droplets to shrink and thus allows the formation of rich oil phase in the high molecular weight oil (Gupta et al. 2016).

10.6 Applications of Nanoemulsions

Nanotechnology has found ways for its application in the food industry by using nanoparticles in packaging and active/intelligent packaging. The latter can extend the shelf-life of the food, as well as maintaining the quality and safety of them (Bracone et al. 2016). The components involved here are indicators such as antioxidants, mineral oils, sugars, methylene blue and nanosensors for contamination detection (Abdullaeva 2017). Nanoemulsions are used to design smart foods with ingredients that are usually difficult to incorporate due to low water stability improving nanodelivery of nutrients (Gupta et al. 2016). Their various other applications in the food industry includes protection, encapsulation and delivery of food bioactives, nutrients, vitamins and controlled release of flavors and incorporation of antimicrobial and anti-oxidant compounds (Bhushani and Anandharamakrishnan 2017).

Catechins, nonpolar components such as triacylglycerol (corn, soybean, algae, fish oil), free fatty acids, mineral oils, waxes, acids are some of the compounds when encapsulated in nanoemulsions possess properties such as high stability, high retention time, improved solubility and increased absorption due to their direct absorption from gastro intestinal tract (Dizaj et al. 2016; Bhushani and Anandharamakrishnan 2017; Gadkari et al. 2017). Work has been done by several scientists in development of soya protein based nanoemulsion with improved chemical stability and bioaccessibility, docasahexaenoic acid nanoemulsion with improved physical and chemical stability and monolayer permeability of Caco-2 cells of green tea catechins (Bhushani et al. 2016; Karthik and Anandharamakrishnan 2016).

Nanoemulsions can fortify foods, such as milk cereal products or bread and can also be incorporated into beverages (Öztürk 2017). These are also used as nutraceuticals (quercetin, vitamins as D3 carotenoids) and food supplements (Bhushani and Anandharamakrishnan 2017). For a modest preservative in food industry especially for minimally processed fruits and vegetables (MPFV) essential oils are used. Due to their small size and presence of emulsifiers it gets easily transported through porin proteins of the outer membrane in gram negative bacteria, causing the leakage of cytoplasmic constituents which in turn results in cell viability (Donsì and Ferrari 2016). Some of the other applications of nanoemulsion in food industries are shown in Table 10.3.

10.7 Limitations

Nanoemulsions have great benefits over conventional emulsions, especially in the food and beverage industry. But along with its advantages it has some challenges such as the risks associated with its intake, digestion and potential toxicity to the human body (McClements 2013). There are apprehensions regarding the toxicity associated with its applications in food. Some concerns were raised regarding the incorporation of nanoparticles into foods including the delivery systems for colors, flavors, preservatives, nutraceuticals and those used to alter the properties of food or food packaging. The prospective toxicity mechanisms of different food grade nanoparticles include physiological and physiochemical mechanisms. Most of them may not have adverse effects on human health but some could have. There has been a considerable increase in the study of the potential toxic effects of food-grade nanoparticles in the past few years. These studies concluded that food grade nanoparticles may have unfavorable effects on health. But there is still some uncertainty in this area (McClements and Xiao 2017).

ADME (absorption, distribution, metabolism, excretion) is also affected in a negative way due to the potential toxicity of nanoemulsions (McClements 2013). Due to small size of nanoemulsions, their behavior within the human body may differ from conventionally utilized larger particles or bulk materials ingredients. Depending on the nature of the nanoemulsions and the properties of the food matrix they are dispersed. The safety of nanoemulsions should be judged on a case-by-case basis (McClements and Xiao 2017). To test the toxicity under reproducible and

Table 10.3 Bioactive compounds with their benefits when encapsulated in nanoemulsions and limitations as unformulated compounds

Bioactives used	Usefulness of nanoemulsion	Benefits of bioactive compound added to food	Limitations overcame	References
Omega-3 fatty acids	Increased water-dispersibility, stability, absorption and oral bioavailability, mask undesirable off-flavors	Improves brain development, reduces the risk of cardiovascular disease, mental disorders as well as diseases related to immune response disorders.	Susceptible to chemical degradation in visible or ultraviolet radiation, poor oxidative stability, low water solubility and variable bioavailability	Walker et al. (2015); Dizaj et al. (2016); Bush et al. (2017)
Carotenoids(β-carotenoid)	Improved bioavailability and bioaccessibility, improves solubility, stability, digestibility and controlled release.	Used as natural colorant, act as scavengers of active oxygen species, exhibit anti-oxidative effect, reduces risk of heart disease, cancer and macular degeneration.	Low bioavailability, very low water solubility, poor thermal and photo stability.	Rao et al. (2013); Salvia-Trujillo et al. (2013); McClements et al. (2017); Sotomayor-Gerding et al. (2016); Dizaj et al. (2016)
Essential oils	Exhibit physicochemical stability, higher bioaccessibility, absorption, improved antimicrobial activity, increased hepatoprotective property, minimized incorporation impact and reduced organoleptic characteristics.	Exhibit strong anti-bacterial, antiviral, and antifungal activities, decreased compound dosage and increased homogeneity, reduced interaction with other matrix compounds and increased compound stability under stress.	Highly volatile, susceptible to environmental conditions, have strong organoleptic characteristics, low stability, poor water solubility and fast oxidation.	Odriozola-Serrano et al. (2014); Severino et al. (2014); Donsì et al. (2014); Mostafa et al. (2015); Donsì and Ferrari (2016); Dizaj et al. (2016); Bhushani and Anandharamakrishnan (2017)
Phytosterols	Improves the solubility, intestinal absorption and bioavailability.	Act as anti-cholesterol agents.	Low solubility in both water and fat that causes their poor absorption.	Dizaj et al. (2016)
Coenzyme Q 10	Improve digestion, solubility and oral bioavailability.	Exhibit bioenergetic, antioxidant, anti-antherogenic effect, inhibits protein and DNA oxidation.	Insoluble in both water and fat.	Littarru and Tiano (2007); Cho et al. (2014); Bhushani and Anandharamakrishnan (2017)
Oil soluble vitamins (vitamins A, D, E and K)	Improve bioavailability, increased transparency, thermal and storage stability.	Used as a health supplement in the beverage industry, exhibit antioxidant, anti-inflammatory, anti-cholesterol and antimicrobial activity thus increases the shelf life of fruit juice.	Water insoluble and thermally unstable.	Hategekimana et al. (2015); Dasgupta et al. (2016); Öztürk (2017); Bhushani and Anandharamakrishnan (2017)

(continued)

Table 10.3 (continued)

Bioactives used	Usefulness of nanoemulsion	Benefits of bioactive compound added to food	Limitations overcame	References
Resveratrol	Increased stability in alkaline medium and on exposure to UV up to 88%, control release rate which aids in targeted delivery and improved transport through cell monolayer.	Antioxidant, anti-inflammatory, suppresses cancer and heart disease.	Susceptible to chemical degradation under exposure to visible light or ultraviolet radiation, instable in water and chemicals resulting in its poor bioavailability.	Joye et al. (2014); McClements (2015); Davidov-Pardo and McClements (2014); Bhushani and Anandharamakrishnan (2017)
Curcumin	Ease digestion process, increase permeation; bioaccessibility and physiochemical stability at high temperature, pH and varying ionic strength.	Exhibit anti-inflammatory, antioxidant, anticancer and antimicrobial activity.	Susceptible to chemical degradation upon exposure to visible light and ultraviolet radiation.	Li et al. (2015); Sari et al. (2015); Salvia-Trujillo et al. (2016) Pinheiro et al. (2016); Bhushani and Anandharamakrishnan (2017)
Bovine lactoferrin	Prevent its denaturation from proteolysis and dilution effects, retained its biological activity and 3D structure, inhibit the iron absorption of pathogenic bacteria.	Have anti-inflammatory, anti-oxidant, anti-viral, anti-bacterial, anti-fungal effect against pathogenic bacteria and candida albicans and mediates immunomodulatory activities.	Being a protein can be denature on high temperature, pH changes in GIT.	Balcão et al. (2013)
Quercetin	Bioaccessibility increased two-folds.	Shows antioxidant, anti-inflammatory, anti-diabetic and anti-hyperlipidemic effect, and reduce risk of cardiovascular disease, metabolic disorder, and certain types of cancer.		Joye et al. (2014); Aditya et al. (2014); David et al. (2016)
Flavors (D-limonene)	Used in baked foods and beverages, proven to improve physical and temperature stability (at 25 °C).	Act as food flavorants, exhibit antioxidant, antimicrobial and poses therapeutic properties.	Poor chemical stability to light, presence of oxidants and temperature, wherein at high temperature it forms isoprene molecules which is prone to oxidation.	Zahi et al. (2015); Li and Lu (2016); Dizaj et al. (2016); Abdullaeva (2017); Bhushani and Anandharamakrishnan (2017)

realistic conditions, there is need to standardize and develop systematic methods (McClements and Xiao 2017).

Scarcity of natural food compatible emulsifiers limits its use in large scale industry. Also, the capital cost of homogenization system required for high-throughput is typically high for food industry. Regulatory bodies have concerns about the acceptance of nanotechnology in foods as there are hardly any predictable interactions available till date between nanoemulsions and its complex food system. All these reasons limit the use of nanoemulsions as bioactive compounds delivery through food (Donsì 2018).

10.8 Conclusion and Future Aspects

Nanoemulsions with a droplet size of 20–1000 nm are used for the controlled and targeted release of lipophilic, amphiphilic and hydrophilic bioactive compounds, and nutraceuticals, as they are chemically unstable. The O/W nanoemulsions can effectively deliver bioactive compounds, antimicrobial agents, dyes, flavors, etc., and at the same time maintain and improve their chemical stability, thus increasing the bioavailability of the loaded substances. These compounds improve the nutritional content, the aroma, the shelf life, the pharmacokinetics of foods and control the release, digestion and absorption of products encapsulated in the gastrointestinal tract, making the healthier food product. Due to its potential advantage over conventional emulsions, it is receiving great attention from the food industry, although they are still not widely used today. The emulsifying properties using natural surfactants (sugar esters and lecithin), lipids and biopolymers (vegetable and animal proteins, modified starch and natural gums) are improved using nanometric emulsion methods. Nanoemulsions can also impart functional properties through the encapsulation of active substances, which can be benefits on the food quality and safety. A reduction in equipment costs and the development of new production methods for nanoemulsions will give rise to a greater boom. Since in the future functional, healthy and medicinal foods will be increasingly demanded by society, nanoemulsion systems being applied to active food ingredients will be a valuable tool.

Acknowledgments The authors wish to acknowledge Vellore Institute of Technology, Vellore, India for providing the opportunity to write this book chapter.

Conflicts of Interest The authors declare no conflict of interest.

References

Abdullaeva, Z. (2017). Nanomaterials in food industry and packaging. In *Nanomaterials in daily life* (pp. 23–46). Cham: Springer. https://doi.org/10.1007/978-3-319-57216-1_2.

Aditya, N. P., Macedo, A. S., Doktorovova, S., Souto, E. B., Kim, S., Chang, P. S., & Ko, S. (2014). Development and evaluation of lipid nanocarriers for quercetin delivery: A comparative study of solid lipid nanoparticles (SLN), nanostructured lipid carriers (NLC), and lipid nanoemulsions (LNE). *LWT-Food Science and Technology, 59*(1), 115–121. https://doi.org/10.1016/j.lwt.2014.04.058.

Balcão, V. M., Costa, C. I., Matos, C. M., Moutinho, C. G., Amorim, M., Pintado, M. E., Gomes, A. P., Vila, M. M., & Teixeira, J. A. (2013). Nanoencapsulation of bovine lactoferrin for food and biopharmaceutical applications. *Food Hydrocolloids, 32*(2), 425–431. https://doi.org/10.1016/j.foodhyd.2013.02.004.

Bhushani, A., & Anandharamakrishnan, C. (2017). Food-grade nanoemulsions for protection and delivery of nutrients. In *Nanoscience in food and agriculture* (Vol. 4, pp. 99–139). Cham: Springer. https://doi.org/10.1007/978-3-319-53112-0_3.

Bhushani, J. A., Karthik, P., & Anandharamakrishnan, C. (2016). Nanoemulsion based delivery system for improved bioaccessibility and Caco-2 cell monolayer permeability of green tea catechins. *Food Hydrocolloids, 56*, 372–382. https://doi.org/10.1016/j.foodhyd.2015.12.035.

Bracone, M., Merino, D., González, J., Alvarez, V. A., & Gutiérrez, T. J. (2016). Chapter 6. Nanopackaging from natural fillers and biopolymers for the development of active and intelligent films. In S. Ikram & S. Ahmed (Eds.), *Natural polymers: Derivatives, blends and composites* (pp. 119–155). New York. EE.UU. ISBN: 978-1-63485-831-1: Editorial Nova Science Publishers, Inc.

Bush, L., Stevenson, L., & Lane, K. E. (2017). The oxidative stability of omega-3 oil-in-water nanoemulsion systems suitable for functional food enrichment: A systematic review of the literature. *Critical Reviews in Food Science and Nutrition*, 1–15. https://doi.org/10.1080/10408398.2017.1394268.

Chang, Y., & McClements, D. J. (2016). Influence of emulsifier type on the in vitro digestion of fish oil-in-water emulsions in the presence of an anionic marine polysaccharide (fucoidan): Caseinate, whey protein, lecithin, or Tween 80. *Food Hydrocolloids, 61*, 92–101. https://doi.org/10.1016/j.foodhyd.2016.04.047.

Chellaram, C., Murugaboopathi, G., John, A. A., Sivakumar, R., Ganesan, S., Krithika, S., & Priya, G. (2014). Significance of nanotechnology in food industry. *APCBEE Procedia, 8*, 109–113. https://doi.org/10.1016/j.apcbee.2014.03.010.

Chen, X. W., Chen, Y. J., Wang, J. M., Guo, J., Yin, S. W., & Yang, X. Q. (2016). Phytosterol structured algae oil nanoemulsions and powders: Improving antioxidant and flavor properties. *Food and Function, 7*(9), 3694–3702. https://doi.org/10.1039/c6fo00449k.

Cho, H. T., Salvia-Trujillo, L., Kim, J., Park, Y., Xiao, H., & McClements, D. J. (2014). Droplet size and composition of nutraceutical nanoemulsions influences bioavailability of long chain fatty acids and coenzyme Q10. *Food Chemistry, 156*, 117–122. https://doi.org/10.1016/j.foodchem.2014.01.084.

Dasgupta, N., Ranjan, S., Mundra, S., Ramalingam, C., & Kumar, A. (2016). Fabrication of food grade vitamin E nanoemulsion by low energy approach, characterization and its application. *International Journal of Food Properties, 19*(3), 700–708. https://doi.org/10.1080/10942912.2015.1042587.

David, A. V. A., Arulmoli, R., & Parasuraman, S. (2016). Overviews of biological importance of quercetin: A bioactive flavonoid. *Pharmacognosy Reviews, 10*(20), 84–89. https://doi.org/10.4103/0973-7847.194044.

Davidov-Pardo, G., & McClements, D. J. (2014). Resveratrol encapsulation: Designing delivery systems to overcome solubility, stability and bioavailability issues. *Trends in Food Science and Technology, 38*(2), 88–103. https://doi.org/10.1016/j.tifs.2014.05.003.

de Oca-Ávalos, J. M. M., Candal, R. J., & Herrera, M. L. (2017). Nanoemulsions: Stability and physical properties. *Current Opinion in Food Science, 16*, 1–6. https://doi.org/10.1016/j.cofs.2017.06.003.

Dizaj, S. M., Yaqoubi, S., Adibkia, K., & Lotfipour, F. (2016). Nanoemulsion-based delivery systems: Preparation and application in the food industry. In *Emulsions* (pp. 293–328). https://doi.org/10.1016/b978-0-12-804306-6.00009-x.

Donsì, F. (2018). Applications of Nanoemulsions in foods. In *Nanoemulsions* (pp. 349–377). https://doi.org/10.1016/b978-0-12-811838-2.00011-4.

Donsì, F., & Ferrari, G. (2016). Essential oil nanoemulsions as antimicrobial agents in food. *Journal of Biotechnology, 233*, 106–120. https://doi.org/10.1016/j.jbiotec.2016.07.005.

Donsì, F., Sessa, M., & Ferrari, G. (2013). Nanometric-size delivery Systems for Bioactive Compounds for the nutraceutical and food industries. In *Bio-nanotechnology: A revolution in food, biomedical and health sciences* (pp. 619–666). https://doi.org/10.1002/9781118451915.ch37.

Donsì, F., Cuomo, A., Marchese, E., & Ferrari, G. (2014). Infusion of essential oils for food stabilization: Unraveling the role of nanoemulsion-based delivery systems on mass transfer and antimicrobial activity. *Innovative Food Science and Emerging Technologies, 22*, 212–220. https://doi.org/10.1016/j.ifset.2014.01.008.

Gadkari, P. V., Shashidhar, M. G., & Balaraman, M. (2017). Delivery of green tea catechins through oil-in-water (O/W) nanoemulsion and assessment of storage stability. *Journal of Food Engineering, 199*, 65–76. https://doi.org/10.1016/j.jfoodeng.2016.12.009.

Gupta, A., Eral, H. B., Hatton, T. A., & Doyle, P. S. (2016). Nanoemulsions: Formation, properties and applications. *Soft Matter, 12*(11), 2826–2841. https://doi.org/10.1039/c5sm02958a.

Gutiérrez, T. J. (2018a). Are modified pumpkin flour/plum flour nanocomposite films biodegradable and compostable? *Food Hydrocolloids, 83*, 397–410. https://doi.org/10.1016/j.foodhyd.2018.05.035.

Gutiérrez, T. J. (2018b). Chapter 55. Processing nano- and microcapsules for industrial applications. In C. M. Hussain (Ed.). Editorial Elsevier. EE.UU. ISBN: 978–0–12-813351-4. *Handbook of nanomaterials for industrial applications* (pp. 989–1011). https://doi.org/10.1016/B978-0-12-813351-4.00057-2.

Gutiérrez, T. J., & Álvarez, K. (2017). Chapter 6. Biopolymers as microencapsulation materials in the food industry. In M. Masuelli & D. Renard (Eds.), *Advances in physicochemical properties of biopolymers: Part 2* (pp. 296–322). Bentham Science Publishers. EE.UU. ISBN: 978–1–68108-545-6. eISBN: 978–1–68108-544-9,. https://doi.org/10.2174/97816810854491170100009.

Gutiérrez, T. J., & Alvarez, V. A. (2018). Bionanocomposite films developed from corn starch and natural and modified nano-clays with or without added blueberry extract. *Food Hydrocolloids, 77*, 407–420. https://doi.org/10.1016/j.foodhyd.2017.10.017.

Gutiérrez, T. J., Ponce, A. G., & Alvarez, V. A. (2017). Nano-clays from natural and modified montmorillonite with and without added blueberry extract for active and intelligent food nanopackaging materials. *Materials Chemistry and Physics, 194*, 283–292. https://doi.org/10.1016/j.matchemphys.2017.03.052.

Gutiérrez, T. J., Toro-Márquez, L. A., Merino, D., & Mendieta, J. R. (2019). Hydrogen-bonding interactions and compostability of bionanocomposite films prepared from corn starch and nano-fillers with and without added Jamaica flower extract. *Food Hydrocolloids, 89*, 283–293. https://doi.org/10.1016/j.foodhyd.2018.10.058.

Guttoff, M., Saberi, A. H., & McClements, D. J. (2015). Formation of vitamin D nanoemulsion-based delivery systems by spontaneous emulsification: Factors affecting particle size and stability. *Food Chemistry, 171*, 117–122. https://doi.org/10.1016/j.foodchem.2014.08.087.

Hategekimana, J., Chamba, M. V., Shoemaker, C. F., Majeed, H., & Zhong, F. (2015). Vitamin E nanoemulsions by emulsion phase inversion: Effect of environmental stress and long-term storage on stability and degradation in different carrier oil types. *Colloids and Surfaces A: Physicochemical and Engineering Aspects, 483*, 70–80. https://doi.org/10.1016/j.colsurfa.2015.03.020.

Helgeson, M. E. (2016). Colloidal behavior of nanoemulsions: Interactions, structure, and rheology. *Current Opinion in Colloid and Interface Science, 25*, 39–50. https://doi.org/10.1016/j.cocis.2016.06.006.

Homs, M., Calderó, G., Monge, M., Morales, D., & Solans, C. (2018). Influence of polymer concentration on the properties of nano-emulsions and nanoparticles obtained by a low-energy method. *Colloids and Surfaces A: Physicochemical and Engineering Aspects, 536*, 204–212. https://doi.org/10.1016/j.colsurfa.2017.06.009.

Jasmina, H., Džana, O., Alisa, E., Edina, V., & Ognjenka, R. (2017). Preparation of nanoemulsions by high-energy and lowenergy emulsification methods. In *cmbebih* (pp. 317–322). Singapore: Springer. https://doi.org/10.1007/978-981-10-4166-2_48.

Joye, I. J., & McClements, D. J. (2013). Production of nanoparticles by anti-solvent precipitation for use in food systems. *Trends in Food Science & Technology, 34*(2), 109–123. https://doi.org/10.1016/j.tifs.2013.10.002.
Joye, I. J., Davidov-Pardo, G., & McClements, D. J. (2014). Nanotechnology for increased micronutrient bioavailability. *Trends in Food Science and Technology, 40*(2), 168–182. https://doi.org/10.1016/j.tifs.2014.08.006.
Karthik, P., & Anandharamakrishnan, C. (2016). Fabrication of a nutrient delivery system of docosahexaenoic acid nanoemulsions via high energy techniques. *RSC Advances, 6*(5), 3501–3513. https://doi.org/10.1039/C5RA12876E.
Karthik, P., Ezhilarasi, P. N., & Anandharamakrishnan, C. (2017). Challenges associated in stability of food grade nanoemulsions. *Critical Reviews in Food Science and Nutrition, 57*(7), 1435–1450. https://doi.org/10.1080/10408398.2015.1006767.
Komaiko, J., & McClements, D. J. (2014). Optimization of isothermal low-energy nanoemulsion formation: Hydrocarbon oil, non-ionic surfactant, and water systems. *Journal of Colloid and Interface Science, 425*, 59–66. https://doi.org/10.1016/j.jcis.2014.03.035.
Komaiko, J. S., & McClements, D. J. (2016). Formation of food-grade nanoemulsions using low-energy preparation methods: A review of available methods. *Comprehensive Reviews in Food Science and Food Safety, 15*(2), 331–352. https://doi.org/10.1111/1541-4337.12189.
Komaiko, J., Sastrosubroto, A., & McClements, D. J. (2015). Formation of oil-in-water emulsions from natural emulsifiers using spontaneous emulsification: Sunflower phospholipids. *Journal of Agricultural and Food Chemistry, 63*(45), 10078–10088. https://doi.org/10.1021/acs.jafc.5b03824.
Kotta, S., Khan, A. W., Ansari, S. H., Sharma, R. K., & Ali, J. (2013). Formulation of nanoemulsion: A comparison between phase inversion composition method and high-pressure homogenization method. *Drug Delivery, 22*(4), 455–466. https://doi.org/10.3109/10717544.2013.866992.
Lee, L., Hancocks, R., Noble, I., & Norton, I. T. (2014). Production of water-in-oil nanoemulsions using high pressure homogenisation: A study on droplet break-up. *Journal of Food Engineering, 131*, 33–37. https://doi.org/10.1016/j.jfoodeng.2014.01.024.
Li, P. H., & Lu, W. C. (2016). Effects of storage conditions on the physical stability of d-limonene nanoemulsion. *Food Hydrocolloids, 53*, 218–224. https://doi.org/10.1016/j.foodhyd.2015.01.031.
Li, M., Cui, J., Ngadi, M. O., & Ma, Y. (2015). Absorption mechanism of whey-protein-delivered curcumin using Caco-2 cell monolayers. *Food Chemistry, 180*, 48–54. https://doi.org/10.1016/j.foodchem.2015.01.132.
Littarru, G. P., & Tiano, L. (2007). Bioenergetic and antioxidant properties of coenzyme Q 10: Recent developments. *Molecular Biotechnology, 37*(1), 31–37. https://doi.org/10.1007/s12033-007-0052-y.
Lorenzo, G., Zaritzky, N., & Califano, A. (2018). Food gel emulsions: Structural characteristics and viscoelastic behavior. In T. J. Gutiérrez (Ed.), *Polymers for food applications* (pp. 481–507). Cham: Springer. https://doi.org/10.1007/978-3-319-94625-2_18.
Majeed, H., Antoniou, J., Hategekimana, J., Sharif, H. R., Haider, J., Liu, F., Ali, B., Liang Rong, L., Ma, J., & Zhong, F. (2016). Influence of carrier oil type, particle size on in vitro lipid digestion and eugenol release in emulsion and nanoemulsions. *Food Hydrocolloids, 52*, 415–422. https://doi.org/10.1016/j.foodhyd.2015.07.009.
McClements, D. J. (2012). Advances in fabrication of emulsions with enhanced functionality using structural design principles. *Current Opinion in Colloid and Interface Science, 17*(5), 235–245. https://doi.org/10.1016/j.cocis.2012.06.002.
McClements, D. J. (2013). Edible lipid nanoparticles: Digestion, absorption, and potential toxicity. *Progress in Lipid Research, 52*(4), 409–423. https://doi.org/10.1016/j.plipres.2013.04.008.
McClements, D. J. (2015). Nanoscale nutrient delivery systems for food applications: Improving bioactive dispersibility, stability, and bioavailability. *Journal of Food Science, 80*(7), N1602–N1611. https://doi.org/10.1111/1750-3841.12919.

McClements, D. J., & Xiao, H. (2017). Is nano safe in foods? Establishing the factors impacting the gastrointestinal fate and toxicity of organic and inorganic food-grade nanoparticles. *npj Science of Food, 1*(1), 6. https://doi.org/10.1038/s41538-017-0005-1.

Merino, D., Gutiérrez, T. J., Mansilla, A. Y., Casalongué, C., & Alvarez, V. A. (2018). Critical evaluation of starch-based antibacterial nanocomposites as agricultural mulch films: Study on their interactions with water and light. *ACS Sustainable Chemistry & Engineering, 6*(11), 15662–15672. https://doi.org/10.1021/acssuschemeng.8b04162.

Mercuri, A., Passalacqua, A., Wickham, M. S., Faulks, R. M., Craig, D. Q., & Barker, S. A. (2011). The effect of composition and gastric conditions on the self-emulsification process of ibuprofen-loaded self-emulsifying drug delivery systems: A microscopic and dynamic gastric model study. Pharmaceutical Research, 28(7), 1540-1551. https://doi.org/10.1007/s11095-011-0387-8

Merino, D., Gutiérrez, T. J., & Alvarez, V. A. (2019). Structural and thermal properties of agricultural mulch films based on native and oxidized corn starch nanocomposites. *Starch-Stärke. 71*(7–8), 1800341. https://doi.org/10.1002/star.201800341.

Mostafa, D. M., Kassem, A. A., Asfour, M. H., Al Okbi, S. Y., Mohamed, D. A., & Hamed, T. E. S. (2015). Transdermal cumin essential oil nanoemulsions with potent antioxidant and hepatoprotective activities: *in-vitro* and *in-vivo* evaluation. *Journal of Molecular Liquids, 212*, 6–15. https://doi.org/10.1016/j.molliq.2015.08.047.

Odriozola-Serrano, I., Oms-Oliu, G., & Martín-Belloso, O. (2014). Nanoemulsion-based delivery systems to improve functionality of lipophilic components. *Frontiers in Nutrition, 1*, 24. https://doi.org/10.3389/fnut.2014.00024.

Öztürk, B. (2017). Nanoemulsions for food fortification with lipophilic vitamins: Production challenges, stability, and bioavailability. *European Journal of Lipid Science and Technology, 119*(7), 1500539. https://doi.org/10.1002/ejlt.201500539.

Ozturk, B., & McClements, D. J. (2016). Progress in natural emulsifiers for utilization in food emulsions. *Current Opinion in Food Science, 7*, 1–6. https://doi.org/10.1016/j.cofs.2015.07.008.

Pinheiro, A. C., Lad, M., Silva, H. D., Coimbra, M. A., Boland, M., & Vicente, A. A. (2013). Unravelling the behaviour of curcumin nanoemulsions during *in vitro* digestion: Effect of the surface charge. *Soft Matter, 9*(11), 3147–3154. https://doi.org/10.1039/c3sm27527b.

Pinheiro, A. C., Coimbra, M. A., & Vicente, A. A. (2016). *In vitro* behaviour of curcumin nanoemulsions stabilized by biopolymer emulsifiers–effect of interfacial composition. *Food Hydrocolloids, 52*, 460–467. https://doi.org/10.1016/j.foodhyd.2015.07.025.

Prakash, A., Baskaran, R., Paramasivam, N., & Vadivel, V. (2018). Essential oil based nanoemulsions to improve the microbial quality of minimally processed fruits and vegetables: A review. *Food Research International, 111*, 509–523. https://doi.org/10.1016/j.foodres.2018.05.066.

Rao, J., Decker, E. A., Xiao, H., & McClements, D. J. (2013). Nutraceutical nanoemulsions: Influence of carrier oil composition (digestible versus indigestible oil) on β-carotene bioavailability. *Journal of the Science of Food and Agriculture, 93*(13), 3175–3183. https://doi.org/10.1002/jsfa.6215.

Ricaurte, L., de Jesús Perea-Flores, M., Martinez, A., & Quintanilla-Carvajal, M. X. (2016). Production of high-oleic palm oil nanoemulsions by high-shear homogenization (microfluidization). *Innovative Food Science and Emerging Technologies, 35*, 75–85. https://doi.org/10.1016/j.ifset.2016.04.004.

Salvia-Trujillo, L., Qian, C., Martín-Belloso, O., & McClements, D. J. (2013). Influence of particle size on lipid digestion and β-carotene bioaccessibility in emulsions and nanoemulsions. *Food Chemistry, 141*(2), 1472–1480. https://doi.org/10.1016/j.foodchem.2013.03.050.

Salvia-Trujillo, L., Martín-Belloso, O., & McClements, D. J. (2016). Excipient nanoemulsions for improving oral bioavailability of bioactives. *Nanomaterials, 6*(1), 17. https://doi.org/10.3390/nano6010017.

Salvia-Trujillo, L., Soliva-Fortuny, R., Rojas-Graü, M. A., McClements, D. J., & Martín-Belloso, O. (2017). Edible nanoemulsions as carriers of active ingredients: A review. *Annual Review of Food Science and Technology, 8*, 439–466. https://doi.org/10.1146/annurev-food-030216-025908.

Sari, T. P., Mann, B., Kumar, R., Singh, R. R. B., Sharma, R., Bhardwaj, M., & Athira, S. (2015). Preparation and characterization of nanoemulsion encapsulating curcumin. *Food Hydrocolloids, 43*, 540–546. https://doi.org/10.1016/j.foodhyd.2014.07.011.

Sessa, M., & Donsì, F. (2015). Nanoemulsion-based delivery systems. In *Microencapsulation and microspheres for food applications* (pp. 79–94). https://doi.org/10.1016/b978-0-12-800350-3.00007-8.

Severino, R., Vu, K. D., Donsì, F., Salmieri, S., Ferrari, G., & Lacroix, M. (2014). Antibacterial and physical effects of modified chitosan based-coating containing nanoemulsion of mandarin essential oil and three non-thermal treatments against Listeria innocua in green beans. *International Journal of Food Microbiology, 191*, 82–88. https://doi.org/10.1016/j.ijfoodmicro.2014.09.007.

Shariffa, Y. N., Tan, T. B., Abas, F., Mirhosseini, H., Nehdi, I. A., & Tan, C. P. (2016). Producing a lycopene nanodispersion: The effects of emulsifiers. *Food and Bioproducts Processing, 98*, 210–216. https://doi.org/10.1016/j.fbp.2016.01.014.

Shu, G., Khalid, N., Tan, T. B., Zhao, Y., Neves, M. A., Kobayashi, I., & Nakajima, M. (2018). *In vitro* bioaccessibility of ergocalciferol in nanoemulsion-based delivery system: The influence of food-grade emulsifiers with different stabilising mechanisms. *International Journal of Food Science and Technology, 53*(2), 430–440. https://doi.org/10.1111/ijfs.13601.

Silva, H. D., Cerqueira, M. A., & Vicente, A. A. (2015). Influence of surfactant and processing conditions in the stability of oil-in-water nanoemulsions. *Journal of Food Engineering, 167*, 89–98. https://doi.org/10.1016/j.jfoodeng.2015.07.037.

Solans, C., & Solé, I. (2012). Nano-emulsions: Formation by low-energy methods. *Current Opinion in Colloid and Interface Science, 17*(5), 246–254. https://doi.org/10.1016/j.cocis.2012.07.003.

Sotomayor-Gerding, D., Oomah, B. D., Acevedo, F., Morales, E., Bustamante, M., Shene, C., & Rubilar, M. (2016). High carotenoid bioaccessibility through linseed oil nanoemulsions with enhanced physical and oxidative stability. *Food Chemistry, 199*, 463–470. https://doi.org/10.1016/j.foodchem.2015.12.004.

Sugumar, S., Nirmala, J., Ghosh, V., Anjali, H., Mukherjee, A., & Chandrasekaran, N. (2013). Bio-based nanoemulsion formulation, characterization and antibacterial activity against food-borne pathogens. *Journal of Basic Microbiology, 53*(8), 677–685. https://doi.org/10.1002/jobm.201200060.

Toro-Márquez, L. A., Merino, D., & Gutiérrez, T. J. (2018). Bionanocomposite films prepared from corn starch with and without nanopackaged Jamaica (*Hibiscus sabdariffa*) flower extract. *Food and Bioprocess Technology, 11*(11), 1955–1973. https://doi.org/10.1007/s11947-018-2160-z.

Walia, N., Dasgupta, N., Ranjan, S., Chen, L., & Ramalingam, C. (2017). Fish oil based vitamin D nanoencapsulation by ultrasonication and bioaccessibility analysis in simulated gastro-intestinal tract. *Ultrasonics Sonochemistry, 39*, 623–635. https://doi.org/10.1016/j.ultsonch.2017.05.021.

Walker, R., Decker, E. A., & McClements, D. J. (2015). Development of food-grade nanoemulsions and emulsions for delivery of omega-3 fatty acids: Opportunities and obstacles in the food industry. *Food and Function, 6*(1), 41–54. https://doi.org/10.1039/c4fo00723a.

Yang, Y., & McClements, D. J. (2013). Vitamin E bioaccessibility: Influence of carrier oil type on digestion and release of emulsified α-tocopherol acetate. *Food Chemistry, 141*(1), 473–481. https://doi.org/10.1016/j.foodchem.2013.03.033.

Yang, Y., Marshall-Breton, C., Leser, M. E., Sher, A. A., & McClements, D. J. (2012). Fabrication of ultrafine edible emulsions: Comparison of high-energy and low-energy homogenization methods. *Food Hydrocolloids, 29*(2), 398–406. https://doi.org/10.1016/j.foodhyd.2012.04.009.

Yi, J., Lam, T. I., Yokoyama, W., Cheng, L. W., & Zhong, F. (2015). Beta-carotene encapsulated in food protein nanoparticles reduces peroxyl radical oxidation in Caco-2 cells. *Food Hydrocolloids, 43*, 31–40. https://doi.org/10.1016/j.foodhyd.2014.04.028.

Zahi, M. R., Liang, H., & Yuan, Q. (2015). Improving the antimicrobial activity of d-limonene using a novel organogel-based nanoemulsion. *Food Control, 50*, 554–559. https://doi.org/10.1016/j.foodcont.2014.10.001.

Chapter 11
Biodegradable Plastic Mulch Films for Sustainable Specialty Crop Production

Douglas G. Hayes, Marife B. Anunciado, Jennifer M. DeBruyn, Sreejata Bandopadhyay, Sean Schaeffer, Marie English, Shuresh Ghimire, Carol Miles, Markus Flury, and Henry Y. Sintim

Abstract Plastic mulch films are employed in the production of vegetables and other specialty crops worldwide due to the benefits they provide, such as reduction of weeds and water loss by evaporation, and control of soil temperature. The benefits can lead to better product quality and yield, and to a more efficient utilization of agricultural inputs such as water. Unfortunately, polyethylene (PE), the most commonly employed constituent of plastic mulches, is poorly biodegradable, thereby requiring the mulch's' expensive and laborious retrieval after harvest. The opportunities for recycling and landfilling of PE mulches are not readily available or are impractical. Residual PE fragments are readily dispersed in soil-related ecosystems and watersheds, where they can harm micro- and macro-organisms. Biodegradable plastic mulches (BDMs) have been developed to address the disposal-related deficiencies. Although the purchase costs of BDMs are over two-fold higher than PE mulches, BDMs are inexpensively plowed into the soil after harvest. Despite the environmental benefits of replacing PE plastic mulches with BDMs, and potential savings of labor costs at harvest, the long-term impact of multiyear BDM employment on soil health and specialty crop productivity is still a concern. This chapter provides a review of BDMs in specialty crop production, including commonly

D. G. Hayes (✉) · M. B. Anunciado · J. M. DeBruyn
S. Bandopadhyay · S. Schaeffer · M. English
Department of Biosystems Engineering and Soil Science, University of Tennessee, Knoxville, TN, USA
e-mail: dhayes1@utk.edu

S. Ghimire
Department of Extension, Tolland County Extension Center, University of Connecticut, Vernon, CT, USA

C. Miles
Department of Horticulture, Northwestern Washington Research and Extension Center, Washington State University, Mount Vernon, WA, USA

M. Flury · H. Y. Sintim
Department of Crop and Soil Science, Washington State University, Puyallup, WA, USA

T. J. Gutiérrez (ed.), *Polymers for Agri-Food Applications*,
https://doi.org/10.1007/978-3-030-19416-1_11

employed polymeric constituents. The authors' recent interdisciplinary research on the long-term impacts of BDMs on specialty crop production and soil fertility will also be discussed.

Keywords Agro-food industry · Agroindustrial inputs · Plasticulture · Polymers

11.1 Plastic Mulches in Specialty Crop Production

Polyethylene (PE) mulch film has been used in agriculture for more than 60 years as it improves crop yield and quality by reducing weed pressure and the use of herbicides, moderating soil temperature, conserving soil moisture and preventing nutrient leaching (Emmert 1957; Ibarra-Jimenez et al. 2006; Kasirajan and Ngouajio 2012; Lamont 2005). Despite these benefits, PE mulch poses economic and environmental challenges at the end of the cropping season as it should be removed from the field and properly disposed. PE mulch is poorly biodegradable and is not easily recyclable. Most plastic recyclers in the U.S. do not accept used PE mulch because it is contaminated with soil and crop debris up to 50% by weight (Kasirajan and Ngouajio 2012). Thus, < 10% of PE mulch is currently recycled in the U.S. (Miles et al. 2017). In some areas (e.g. China and southern Spain), some farmers have incorporated PE mulch into soil annually, and PE mulch accumulation in field soil is so significant that soil water retention and crop yield are reduced (Liu et al. 2014; Steinmetz et al. 2016). Many farmers stockpile PE mulch due to the absence of disposal options, leading to its slow deterioration (Staub 2018). PE mulch fragments from these fields are dispersed into the environment by wind and water erosion, thereby causing further pollution.

11.2 Biodegradable Plastic Mulches: Motivation and Advantages Compared to Conventional PE Mulches

Biodegradable plastic mulch films, or biodegradable mulches (BDMs; see Fig. 11.1) address many of the concerns that arise from the use of PE mulches due to their lower impact on the environment. BDMs, prepared from biodegradable polymers, are designed to be tilled into the soil after their service life, after which they will be subjected to aerobic biodegradation by soil microorganisms, producing CO_2, water and biomass (Kasirajan and Ngouajio 2012; Steinmetz et al. 2016). A new standard pertaining to BDMs in the European Union, EN 10733, specifies that the BDMs must undergo 90% conversion into CO_2 within 2 years using a standardized laboratory test for assessment (discussed below). Soil incorporation is attractive to farmers, not only because of the reduction of environmental impact compared to the use

Fig. 11.1 Biodegradable mulch films (BDMs). (**a**) after machine-laying the BDM and then transplanting; (**b**) during specialty crop production; (**c**) BDM microplastic fragments recovered from the field after crop harvest and BDM soil incorporation

of PE mulch, but also for the labor cost savings involved with the retrieval of PE mulch from the fields after harvest. It is worth noting that BDMs are not designed to be used for fumigation (European Committee for Standardization 2018).

Yet, there is not widespread adoption of BDMs as replacements for PE mulches. Sociological studies on farmers' perceptions identified several barriers to the adoption of BDMs by farmers, among which are the higher purchase cost of BDMs relative to PE mulches and the "unpredictable breakdown" of the BDMs during their deployment in the field (e.g. premature tearing and ripping) and in the soil (uncertainty of achieving full biodegradation in a reasonable period of time, leading to persistent residues) (Goldberger et al. 2013; Cowan et al. 2015). Another barrier to adoption in the USA is that BDMs have not been approved for certified organic agriculture. The research community in academia, industry, and government laboratories have addressed these concerns during the last decade, including the authors' research. Results demonstrate that BDM performance in specialty crop production is generally equal to that of PE mulches, and that soil incorporation of BDMs does not affect soil quality or soil microbial communities to a significant extent. These results partially address concerns of certified organic growers and the USDA National Organic Program (NOP); but, other criteria specified by the NOP, such as fully biobased content for BDMs and the exclusion of genetically modified (GM) organisms in their manufacture, have not yet been fully addressed.

The remainder of this chapter provides an overview of BDMs: their composition and manufacture, their physicochemical properties, and their life cycle, from derivation from feedstocks (raw materials) to deployment in the field (during which their physicochemical properties may change due to agricultural weathering; see Sect. 11.4.2) to their end-of-life. Subsequently, recent research on the performance of BDMs in specialized crop production and impacts on soil ecosystems (soil quality, carbon pools, and microbial communities) will be described, which is partially based on the authors' research.

11.3 Mechanical and Physicochemical Properties of BDMs

11.3.1 Preparation of Biodegradable Mulch Films

Plastic films are processed by extrusion, which starts with melting of pellets (resins) of a thermoplastic polymer till they become pliable. The major components of the pellets are polymeric constituents and several additional minor components occur, such as fillers, plasticizers and processing aids. The molten plastic is pushed (extruded) from a circular die to form a continuous tube of plastic called the bubble. The bubble is inflated with air to reduce its thickness to the desired diameter and then it is drawn vertically upward in a tower, in which it is allowed to cool before it is flattened to its lay flat width (blowing extrusion methodology). The speed at which film is pulled from the die determines the film's thickness whereas the width of the film is controlled by the amount of air inserted in the bubble. Film color can be changed during extrusion by adding resin pellets that contain colored pigment, referred to as a "master batch" (Plastics Technology 2019). Master batches are also employed to add other minor components such as fillers and plasticizers (Kijchavengkul et al. 2008b).

Typically, mulch films are prepared by a local manufacturer, often referred to as a "converter", who will purchase polymeric feedstocks from bioplastics manufacturers, and blend the polymers with the other components (*via* master batches), such as processing aids, plasticizers and colorants, and then perform the film extrusion and processing (U.S. Department Agriculture 2015). For example, in North America, agricultural film manufacturers such as BioBag Americas (Dunevin, FL, USA), Organix Solutions (Maple Grove, MN, USA) and Dubois Agrinovation (Saint-Remi, QC, Canada) prepare BDMs from commercially available polymeric feedstocks such as Mater-Bi® (Novamont, Novarra, Italy) or ecovio® (BASF, Germany) and other components described above.

11.3.2 Physical and Mechanical Properties

The ideal polymer for BDMs would possess similar mechanical properties as PE, such as a high tensile strength (peak loads of ~8–20 Pa in the machine direction) and high % elongation (~500%). Most commercially available BDMs have 1.5–2-fold lower elongation (~200–400%); but, this is adequate for mulch function (Hayes et al. 2017; Touchaleaume et al. 2016). Because biodegradable polymers are more expensive than PE, the thickness of commercially available BDMs typically is lower than those of PE (Hayes et al. 2017). As a result, the machine-laying of BDMs on an agricultural field must be performed carefully, with minimal applied tension, so that the BDM does not tear. EN 17033 is a new EU standard for biodegradable mulch films, where several specifications for the mechanical properties of the mulches are listed (European Committee for Standardization 2018).

11.3.3 Polymeric Constituents of Biodegradable Plastic Mulches

Commonly used biopolymers and biopolymeric feedstocks for preparation of BDMs are given in Table 11.1 and the polymers' molecular structures are provided in Fig. 11.2. Many of the polymers are polyesters that are biodegradable. One of the most frequently employed polymers in BDMs is polybutylene adipate-*co*-terephthalate (PBAT), a random block copolymer consisting of the terephthalate (T) and adipic acid (A) diacid groups and 1, 4-butanediol (B). It possesses good tensile strength, an elongation near ~200–300% (which is lower than the ~500% elongation possessed by PE), and is certified by the Biodegradable Products Institute (BPI) as "compostable". Ester bonds enhance PBAT's biodegradability (*via* hydrolysis) and its T groups contribute to its stability and mechanical properties (Coltelli et al. 2008; Shah et al. 2008). Ester groups in soft aliphatic BA-rich regions are more susceptible to hydrolysis than those in the hard BT-rich regions, consequently making the BA-rich regions more readily biodegradable (Herrera et al. 2002). Other commonly employed polymers include polybutylene succinate (PBS) and PBS-*co*-adipate (PBSA). Polypropylene carbonate (PPC), prepared from propylene oxide and CO_2 *via* a cycloaddition reaction (Adeleye et al. 2014), has also been recently employed to prepare BDMs (Touchaleaume et al. 2016). PPC is therefore partially biobased, and is readily biodegradable, especially when in contact with water (Touchaleaume et al. 2016).

Table 11.1 Commercially available polymers and polymer blends employed in biodegradable agricultural mulches[a,b]

Product Name	Polymer[c,d]	Manufacturer
Biocycle®	*Sucrose / PHA blend*	*PHB Industrial (Brazil)*
Bio-Flex	PLA / copolyester blend	FKUR, Willich (Germany)
Biomax TPS	*Starch + thermoplastic starch*	*DuPont (USA)*
Biomer L	*PHA*	*Biomer (Germany)*
Biopar	*Starch co-polyester*	*United Biopolymers (Portugal)*
BioPBS	*PBS*	MCPP (Division of Mitsubishi Chemicals, Japan)
Biosafe™	PBAT / starch blend; PBS; PBSA	Xinfu Pharmaceutical Co (China)
ecoflex®	PBAT / starch blend	BASF (Germany)
ecovio®	Ecoflex® + PLA	BASF (Germany)
EnPol	PBS	IRE Chemical (Korea)
Envio®	Ecoflex® + PLA + starch blend	BASF (Germany)
GreenBio	*PHA*	*Tianjin GreenBio Materials (China)*
Ingeo®	Starch + PLA; PBS + PLA	NatureWorks (USA)
Mater-Bi®	PBAT + starch blend	Novamont (Italy)
Naturecycle	*Starch-copolyester blend*	*Custom Bioplastics (USA)*
ReNew	*Polyhydroxy alkanoate (PHA)*	*Danimer Scientific (USA)*
Succinity	*PBS*	*Succinity (Germany)*

[a]Information taken from (Martin-Closas and Pelacho 2011; Kijchavengkul and Auras 2008; Tullo 2012; Hayes et al. 2012)
[b]Italicized entries are products that are almost entirely biobased
[c]Abbreviations: PBAT: polybytylene adipate terephthalate; PHA: polyhydroxyalkanoate; PLA: polylactic acid; PBS: polybutylene succinate; PBSA: PBS-co-adipic acid
[d]The molecular structure for several of the polymers is given in Fig. 11.2

PBAT (Siegenthaler et al. 2012), PBS and PBSA are often blended with copolymers that are highly biodegradable, such as starch [at 10–30%; e.g. Mater-Bi® (Elfehri Borchani et al. 2015)], to enhance overall biodegradability (Table 11.1). To enhance their materials-related properties and their biodegradability, starch is often blended with plasticizers and other polymers, thereby producing thermoplastic starch (Dufresne et al. 2000, Zhang et al. 2014; Gutiérrez 2018a; Gutiérrez et al. 2019; Herniou--Julien et al. 2019). In addition, the blend can be used to overcome deficiencies exhibited by a single homopolymer as a polymeric material, such as embrittlement (Gutiérrez and Alvarez 2017a, b). Several examples of the commercially available blends are given in Table 11.1. It is important to note that companies frequently produce several different varieties or grades of the blends, which may vary in molecular weight and composition. Therefore, it is important for the manufacturer's identification code for the grade employed in the preparation of the BDMs to be included in any publications or presentations.

Polylactic acid (PLA) serves as another polymer that is commonly blended, because, although it possesses good strength (30–70 MPa) and is one of the least expensive biopolymers (Sodergard and Stolt 2010; Inkinen et al. 2011; Madhavan

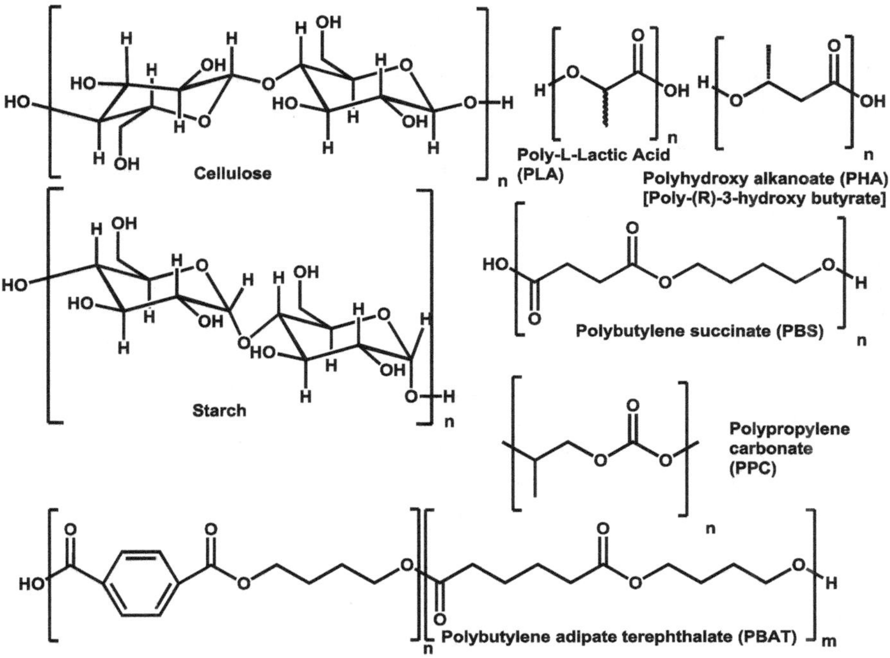

Fig. 11.2 Molecular structures of polymers and biopolymers employed in biodegradable plastic mulches

Nampoothiri et al. 2010; Shen et al. 2010), it is very crystalline (observed by a high glass transition temperature of 55–60 °C) and brittle as a homopolymer (Sodergard and Stolt 2010). Its highly crystalline nature reduces % elongation and biodegradability in soil (Shi and Palfery 2012; Rudnik and Briassoulis 2011) [but not in compost, due to the latter's higher temperature, ~40–60 °C (Kawai 2010; Hakkarainen et al. 2000; Hablot et al. 2014; Karamanlioglu and Robson 2013)]. To address the deficiencies, PLA is commonly blended with plasticizers such as citrate derivatives and other polymers such as polyhydroxyalkanoate (PHA) (Detyothin et al. 2010; Hu et al. 2008; Weng et al. 2013; Zhang and Thomas 2011; Dharmalingam et al. 2015) and PBAT (Coltelli et al. 2008). PLA is typically employed in polymeric blends for BDMs in relatively small proportions (Detyothin et al. 2010). PLA, derived from polymerization of lactides, dimers of lactic acid, a commonly produced metabolite during fermentation. The properties of PLA are controlled through molecular weight and enantiomeric distribution of L- and D- monomeric units. Microorganisms produce L-lactic acid; however, D-lactic acid can be produced by the former's racemization. An increase of D-lactic acid groups increases mechanical strength and reduces the crystallinity (Sodergard and Stolt 2010).

PHAs are produced by fermentation using genetically-modified bacteria, and serve as a organisms' long-term energy storage source (Posada et al. 2011; Roohi et al. 2018). They have several applications in biomedicine, injection molding and packaging. PHAs frequently occur as a random block copolymer, with commonly

Table 11.2 Effect of plastic mulch colors on light and weed control[a]

Color	Soil Temperature Increase (°C)[b]	Light Reflectivity	Light Absorptivity	Light Transmission	Weed Suppression
Black	1.7–2.8	Low	High	Low	Excellent
Clear	3.3–7.8	Low	Low	Very high	Poor
White/ silver	−1.1–0.4	High	Low	Low	Excellent
IRT[c]	1.7–2.8	Low	High	High	Excellent

[a]Source: (Maughan and Frost 2016)
[b]At 5–35 cm depth
[c]Infrared transmitting

encountered monomeric units including 3- and 4-hydroxybutyric acid and 3-hydroxyvaleric acid. PHA is readily biodegradable in soil and compost, but is thermally unstable, highly crystalline and therefore brittle in its homopolymeric form (Vroman and Tighzert 2009; Zhang and Thomas 2011). It thus is most commonly encountered as a component of polymeric blends (Detyothin et al. 2010).

11.3.4 Color

Polymers are blended with chemicals, “additives” (e.g. plasticizers, fillers, lubricants, nucleating agents, stabilizers, colorants), to improve their mechanical and other functional properties (Mormile et al. 2017). One of the additives used is a colorant that controls the optical and radiometric properties of plastic mulches (ISO 9050:2003). Colorants modify the absorption and reflection of light radiation that reaches the soil, thereby affecting the soil temperature (Kasirajan and Ngouajio 2012). The color of mulch is an important parameter of the microclimate around a crop plant (Mitchell et al. 2004). Depending on the color, soil microclimate conditions can vary and affect the mulch performance of soil (Table 11.2).

Carbon black is one of the most commonly used colorants and may or may not be naturally-derived (U.S. Environmental Protection Agency 2012; Miles et al. 2017). Black mulch prevents the weed growth by limiting the light transmittance to the soil and will warm the soil between 1–4 °C more than the temperature on bare soil. Carbon black derived from fossil fuel resources are extremely fine black fluffy particles (10–500 nm in diameter), which is the portion of unburned carbon produced by the reaction of a hydrocarbon fuel (e.g. oil or gas) during combustion (1320–1540 °C) (U.S. Environmental Protection Agency 2012). Biobased carbonaceous black powder from bioethanol is now being examined as a potential substitute for fossil fuel-derived carbon black (Snowdon et al. 2014). Carbon black is not only used as a colorant but also a photo-stabilizer that prevents the polymer breakdown by ultraviolet (UV) light through selective screening of UV radiation. The addition of carbon black prevents light absorption to the polymer-bonded chromophores in the film, thus stabilizing the polymer of the potential photodegradation reactions (Kijchavengkul et al. 2008b).

For some geographical regions and cropping systems, clear and white plastics are chosen over black mulches. Clear plastic mulch has the highest soil warming capability (5–8 °C higher than bare soil); but, weed growth underneath them can be substantial and herbicide may be necessary (Melek and Atilla 2009). Clear mulches effectively retain much of the heat otherwise lost to the atmosphere, and have the greatest soil warming potential among the various mulch colors (Lamont 2001). Clear plastic BDMs can also be employed for soil solarization (Mitchell et al. 2004).

Titanium dioxide (TiO_2) is traditionally used as a white colorant (Kijchavengkul et al. 2008b). However, concerns have been raised about its potential hazards to the ecosystem, including the marine environment. Direct evidence of toxicological impact of TiO_2 nanoparticles on marine microalgae was observed through cell wall damage derived from algae entrapment, positively correlated with the growth inhibition of algae (Wang et al. 2016a). In addition, TiO_2 represents a threat to human health. The inhalation of TiO_2 nanoparticles increases the potential pulmonary health risks by inducing oxidative stress, genotoxicity and apoptosis in human lung cancer cell line, A549 (Srivastava et al. 2012). Some biopolymer producers, such as Heritage Plastics, now use talc and calcium carbonate ($CaCO_3$) as a white colorant and as a filler for PHA to modify the films' stiffness and toughness (Sherman 2008). Unlike carbon black, TiO_2 in white plastic mulch catalyzes photodegradation (Kijchavengkul et al. 2008b). White or coextruded white-on-black mulches can reduce surface soil temperature relative to bare soil due to most incoming solar radiation is reflected on the soil surface and atmosphere. It is reported that the silver mulches have activity against insects (e.g. aphids) and microorganisms (e.g. fungi), mainly due to the reflected sunlight (Lamont 1999). Other colorants have also been used in plastic mulch films (Lamont 1999).

Research shows that green or brown infrared transmitting (IRT) plastics allow IR light waves to selectively penetrate through the plastic and absorb UV radiation in the photosynthetic spectral regions, allowing for soil warming without any weed growth (Lamont 1999; Maughan and Frost 2016). IRT mulches possess the weed control properties of black mulch; but, they increase the soil temperature at an intermediate level between black and clear mulch. The color of IRT mulches can be blue-green (IRT-76, AEP Industries Inc.) or brown (Lamont 1999). Recently, plastic mulch color is becoming a factor in the aesthetics of farms due to negative perceptions of visitors when there are fragments of residual black plastic on the farms, as they connote unclean agricultural practices. Therefore, growers have shown interest in brown and green mulches.

11.3.5 Minor Components of Biodegradable Plastic Mulches

"Additives" are chemicals that are blended with the polymer to modify the properties of the products *via* master batches, and include fillers to provide mechanical strength (e.g. clays and $CaCO_3$), plasticizers, processing aids, colorants, antioxidants, UV stabilizers, infrared absorbers or reflectors, anti-drip and anti-mist agents, and biocides, colorants, and others (Mormile et al. 2017). Some additives frequently added to mulch films are described below:

Anti-bacterial additives are used to create a resistance to biofouling by microorganisms, so that polymeric materials are protected from bacterial growth. The additives operate by interfering with the metabolism of microorganisms. To be effective, additives must be able to migrate to the surface of the material so that it can interact with the microorganisms (Singh et al. 2012). Thermoplastic elastomers containing zinc pyrithione and silver nanoparticles have been effective in inhibiting the growth of microorganisms, almost completely eliminating *Escherichia coli* and *Staphylococcus aureus* bacteria, and presenting an inhibition zone on fungal assay surfaces (Pittol et al. 2017).

Antioxidants are added to protect the mulch material from oxidation during and after processing, thus prolonging the material's longevity. Oxidation can result in loss of strength, flexibility, thermal stability and color of mulch films (Ram 1997).

Fillers are added to polymers mainly to reduce costs and improve the properties of the material. Fillers can improve processing, abrasion resistance, density control, dimensional and thermal stability and optical effects. Common fillers added are wood, silica, glass, clay and other polymers (Callister and Rethwisch 2015).

Lubricants are used to improve flow and processing properties. An internal lubricant is a type of additive that acts by modifying material viscosity (Murphy 2001). *Slip additives* are a type of internal lubricant that create better processability by reducing the internal friction and tackiness of polymers. Slip additives reduce the surface friction of polymers, which allows the film layers to slide over each other (Soroka 2002). Conventional plastics are mostly resistant to degradation. The use of lubricant leads to degradation of polymeric chains, which further decreases the molecular weight (Pfister and Labowsky 2003).

Nucleating or clarifying agents are chemical structures that increase the overall crystallization rate of polymers. They can be used to increase stiffness, hardness, impact properties, tensile strength, and to control the size and distribution of pores (Murphy 2001). The increase in the rate of crystallization causes more crystals to be produced in the same volume of material; therefore, crystals tend to be smaller (Soroka 2002). The low nucleation density of PHA allows it to produce spherulites that exhibit inter-spherulitic cracks. Nucleating agents are added to transform embrittled large spherulites (e.g. PHA) into small fine spherulites (Pfister and Labowsky 2003).

Plasticizers are commonly added to brittle polymers to improve their flexibility and processing properties by reducing rigidity and fracture stress. Plasticizers enhance molecular motion, lower glass transition temperature (T_g) and increase elongation and impact resistance, which affects the adhesion at the matrix/filler interface (Pfister and Labowsky 2003). Several common phthalate-based plasticizers have recently been linked to a wide range of toxicities for development, reproduction and neurotoxicity in mammals, and in some countries they have been banned (Jamarani et al. 2017; Ghosh 2017; Erickson 2015; Oehlmann et al. 2009). Therefore, the selection of plasticizers should be carefully considered. The development of newer, biobased and nontoxic plasticizers has been a topic of major interest in the research community (Howell and Lazar 2019; Howell and Sun 2018).

Several low molecular weight plasticizers such as glycerol, sorbitol, triethyl citrate and oligomers have been employed in the preparation of BDMs to reduce brittleness (Dufresne et al. 2000; Coltelli et al. 2008; Andersson et al. 2010; Shah et al. 2008; Jiang et al. 2015).

Stabilizers are additives that prevent photodegradation or photo-crosslinking caused by UV radiation when absorbed by the bulk polymer. The amount of an absorber required to provide environmental protection in a plastic is governed by factors such as thickness of the plastic and compatibility of the absorber with the polymers. The most common UV stabilizers are UV light screeners (e.g. carbon black, ZnO, TiO_2, MgO, $CaCO_3$, $BaSO_4$, Fe_2O_3), quenchers, hydroperoxide decomposers, radical scavengers and singlet oxygen ($^1O_2$), which are classified according to their mechanisms of action in the photo-stabilization process (Yousif and Haddad 2013).

11.3.6 Alternatives to Biodegradable Mulch Films

Although BDMs mostly consist of thin films prepared by extrusion, they can also be nonwovens, which consist of randomly oriented fibers of micron- and submicron widths (Shi and Palfery 2010; Hablot et al. 2014; Dharmalingam et al. 2015). Nonwovens can be considered as low-weight geotextiles possessing a great mechanical strength. Nonwovens are particularly attractive choices as a mulch for multi-season use, and can be used as landscape fabrics and row covers. Because of their strength, nonwovens can be readily retrieved by hand from a field after harvest, and transported to a nearby composting facility. A disadvantage of nonwovens is that the polymers are in a highly crystalline form, which reduces their biodegradability.

Paper mulches are another alternative, and possess the advantage of an inexpensive feedstock with high biodegradability (Haapala et al. 2014; Shogren 2000; Shogren and Hochmuth 2004; Li et al. 2014b). It has also been shown that paper mulch minimizes purple nutsedge (*Cyperus rotundus* – a persistent weed that exists in many geographic regions) compared to BDMs and PE mulches (Pereira et al. 1987; Webster 2017; Ghimire et al. 2018). However, paper mulches tend to break down too rapidly during the growing season, especially along the sides of the bed where there is direct contact with the ground (Merfield 2000; Miles et al. 2012; Martín-Closas et al. 2016). It has also been shown that brown paper mulch reduces soil temperature by 0.5 °C and 1 °C compared to bare soil in open field and high tunnel in tomato production systems, respectively, which can result in lower yields compared to black PE mulch (Cowan et al. 2014). Due to the inherently porous and hygroscopic characteristics of paper mulch, a special treatment (e.g. creping or calendaring) and additional components (e.g. mineral fillers, dry and wet strengthening agents, colorants and water repellents) are required to extend its service life. Degradation of paper mulch is variable and occurs unevenly. In general, the degradation process is strongly dependent on the quality of paper, soil characteristics and weather conditions (Haapala et al. 2014). Since paper mulches are heavier than PE mulches and plastic BDMs, their transportation costs are higher.

Other alternatives to biodegradable mulch films include "sprayable mulches." Aqueous solutions of mainly polysaccharides (e.g. chitosan, alginic acid salts, guar gum and locust tree gum, the latter two rich in galactomannan), or possibly latex (Mahmoudpour and Stapleton 1997), form hydrogels that can be applied onto the soil. The hydrophilicity and resistance to dissolution enables hydrogel to be applied onto topsoil throughout the crop cycle, where they cover the soil with a protective thin "geo-membrane" (Santagata et al. 2017; Vox et al. 2013; Immirzi et al. 2009). Sprayable biodegradable coatings need less labor for manual operations (e.g. cutting and laying-out the plastic films). Similar to mulch films, colorants are frequently added to sprayable mulches (Mahmoudpour and Stapleton 1997). Biodegradable spray mulching coatings remain in the experimental stage and costs are not yet competitive with PE films (Immirzi et al. 2009). Future research on sprayable mulches should address improving the mechanical performances and lengthening mulch lifetime upon exposure to climactic agents, to allow its use in open fields on a large-scale (Vox et al. 2013).

11.3.7 Photo- and Oxodegradables: Poor Substitutes for BDMs

Mulch products containing "oxodegradable" polymers have been misunderstood as being BDMs. Such materials frequently consist of poorly biodegradable polymers such as PE along with triggerable groups that are susceptible to UV radiation ("photodegradables"), or heat and oxidation ("oxodegradables"), which are grafted onto the polymeric backbones or are compounds that are blended with the polymers. These materials typically undergo an accelerated macroscopic degradation upon the application of the trigger. However, the resulting fragments are not likely to be completely biodegraded, and will persist in the soil and in watersheds indefinitely, and are capable of forming microplastics that can adsorb toxins such as pesticides and carry them into the food chain (Steinmetz et al. 2016; European Commission 2018). The oxidation reagents themselves may be ecotoxic (European Commission 2018). The misrepresentation of photo- and oxodegradables as being biodegradable has led to mistrust of plastic manufacturers by specialty crop growers. The EU is considering a restriction on the use of oxodegradable products (European Commission 2018), and in February 2019, France banned the use of oxodegradable plastics for agricultural applications. PDQ-H (ACCRAbond Inc., Olive Branch, MS, USA) is an additive used for either PE or polypropylene mulches, and it is made from proprietary ingredients to disintegrate plastic by oxidation and photodegradation. TDPA (EPI Environmental Products Inc., Vancouver, Canada) is an additive to PE that assists in the oxidation, fragmentation and biodegradation of the plastic and its by-products that are formed from the action of heat, mechanical stress or sunlight (Mohee et al. 2008). Both PDQ-H and TDPA are non-biodegradable.

11.4 Life Cycle of BDMs

The life cycle of BDMs typically consists of the following stages, chronologically listed: their derivation from feedstocks, their manufacture, their service life (on top of the soil) to enhance specialty crop production, and their disposal (typically soil incorporation, but possibly retrieval followed by composting). As described below, there are direct linkages between the events that occur during mulches' service life (agricultural weathering) and biodegradability; however, the start-of-life (biobased *vs* petroleum-derived) has no impact on the BDM's performance in its service life and during biodegradation.

11.4.1 Biobased BDMs

The ideal plastic mulch film for sustainable agriculture should possess a favorable beginning- and end-of-life, i.e. be biobased and highly biodegradable, respectively. "Biobased" materials are defined by U.S. Senate Committee on Agriculture Nutrition and Forestry (2006) as derived "…in whole or significant part of biological products or renewable domestic agricultural materials (including plant, animal, and marine materials) or forestry materials". Biobased content is typically measured by measuring the stable distribution of carbon isotopes *via* standardized tests such as ASTM D6866 (ASTM International 2012c). Most of the polyesters used in BDMs possess a minimal or partial biobased content; moreover, most commercially available BDMs possess only ~20% biobased content (U.S. Department Agriculture 2015). Unlike fossil fuel-derived feedstocks, the use of renewable starting materials does not increase atmospheric levels of CO_2, a greenhouse gas associated with climate change. PBAT, the polymeric constituent most commonly used in BDMs, is mainly not biobased; however, Novamont has recently established a facility in Italy to produce biobased butanediol by recombinant *E. coli*. In addition, adipic acid can be replaced at least partially by azaelic acid, which can be derived from seed oils such as castor or safflower oil. Terephthalic acid can also be produced from renewable resources, such as furan and furfural, which are derived from lignocellulosic biomass, although the synthetic approaches are expensive (Collias et al. 2014; Tachibana et al. 2015; Smith 2015). Fully and partially biobased PBS are commercially available (Table 11.1).

11.4.2 Impact of Agricultural Weathering on BDMs

Exposure to the environment leads to significant physicochemical changes in the structure of BDMs. Sunlight, particularly its UV component, is the most significant factor. Once the canopy of a plant is formed to block solar radiation exposure from

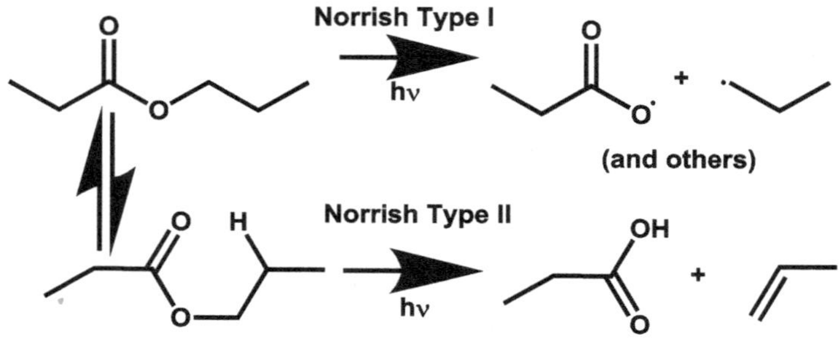

Fig. 11.3 Norrish type photochemical reactions

the BDMs, the degree of degradation is reduced. It is known that biopolymers such as PLA and PBAT undergo photodegradative reactions such as Norrish Type I and II reactions (Fig. 11.3) (Kijchavengkul et al. 2010; Hayes et al. 2017). The former involves the formation of free radicals, and can lead to cross-linking, which can reduce the biodegradability of the plastic (Kijchavengkul et al. 2008b). It is known that colorants, particularly TiO_2 (white), enhance the Norrish Type I reaction (Kijchavengkul et al. 2008a). Norrish Type II reaction entails chain scission, thus leading to depolymerization, which can enhance biodegradation (Hablot et al. 2014). In addition, due to the presence of moisture and warming from sunlight, hydrolysis of ester bonds can also occur. FTIR spectroscopy is an effective analytical tool to differentiate between the different weathering mechanisms. In addition to sunlight, other important agricultural weathering factors include moisture and precipitation, heat, wind, technique used for BDM machine-laying and making perforations for plant growth, soil type and preparation, and BDM storage conditions (European Committee for Standardization 2018). Mulch manufacturers typically recommend BDMs be stored no more than 1.5 years under cool and dark conditions.

It is also believed that minor components in the air, groundwater and soil, such as pesticides, fumigants, air and water pollutants, and low-molecular weight oligomers, plasticizers or other components that are leached out of the plastic films by water play a catalytic role in environmental weathering. Moreover, physicochemical changes of BDMs observed under environmental conditions cannot always be simulated by artificial weathering, which uses specialized weatherometry equipment to simulate solar radiation and moisture exposure (Hayes et al. 2017). The most significant changes in the BDMs under environmental weathering conditions are: (1) embrittlement, observed by a significant loss of % elongation and tensile strength, and (2) depolymerization (Hayes et al. 2017; Briassoulis et al. 2004; Briassoulis 2006, 2007). These changes weaken the mulch films, leading to a faster macroscopic deterioration (Ghimire et al. 2018).

11.4.3 Assessment of BDMs' Biodegradability Using Standards

Biodegradability is an often-misused term (Gutiérrez 2018b). This term must be referenced to (1) a specific set of environmental conditions, (2) a specific benchmark for the degree of bioconversion (e.g. 90% conversion of BDMs' carbon atoms into CO_2 for aerobic biodegradation), and (3) a duration. Frequently, it has been miscommunicated or misconstrued that achievement of the criteria laid out in ASTM D6400 (ASTM International 2012a) (or equivalent standards from other organizations) for biodegradability under industrial composting conditions is equivalent to the biodegradability of the plastics under ambient soil conditions. The conditions of industrial composting involve higher temperatures (~58 °C) and different microbial communities compared to the environmental soil. A second area of miscommunication involves unsubstantiated claims of full biodegradation of oxodegradable materials in soil (described in Sect. 11.3.7). Reviews of standards related to biodegradable plastics have been published (Briassoulis and Dejean 2010; Briassoulis et al. 2010).

The development of standards specifically related to biodegradation of plastics in soils has been a recent achievement (Hayes and Flury 2018). EN 17033 is a standard that was recently adopted by the European Union and is specific for BDMs and their employment in agriculture (European Committee for Standardization 2018). The standard includes criteria for constituents (threshold levels for heavy metals and hazardous substances), inherent biodegradability, ecotoxicity and mechanical properties (e.g. tensile strength ≥18 MPa in the machine direction), as well as "best practices" for BDMs' deployment. The inherent biodegradability criterion consists of a 90% conversion of the BDM's carbon atoms into CO_2 within 2 years using a standardized laboratory test conducted with soil at 25–28 °C, such as ASTM D5988 (ASTM International 2012b). Alternatively, the achievement may consist of reaching 90% conversion compared to the conversion of the cellulosic positive control. Or, 90% biodegradation in two years can be achieved for the film's base material, or each organic constituent that is contained at >1% in the BDM. The ecotoxicity criteria use standardized tests for germination of plants, ecotoxicity to earthworms and nitrification inhibition assays for bacteria. Although ecotoxocity test is more thorough in EN 17033 compared to any biodegradable plastic standards issued previously, the development of an additional test is necessary, e.g. ecotoxicological effects against plant germination and plant growth should be differentiated (Martin-Closas et al. 2014). As described in Sect. 11.5, the incorporation of BDMs into the soil results in the formation of microplastics and likely nanoplastics. It has recently been shown that reducing the size of the BDMs to 50–75 μm does not have any adverse effect on the inherent biodegradability (Chinaglia et al. 2018). However, recent reports have pointed to the ecotoxicity of microplastics on earthworms and collembolans (hexapods), as well as in fish and other marine-based organisms (Huerta Lwanga et al. 2017; Zhu et al. 2018; Alimi et al. 2018). The ecotoxicity criteria should thus include testing with micro- and nanoplastics that are formed from the BDMs.

11.5 Impacts of BDMs on Specialty Crop Production and Soil Fertility

As described above, one of the barriers to increasing adoption of BDMs as replacements to PE mulches in sustainable agriculture has been the uncertainty about the long-term impacts of incorporating BDM into the soil on the production of specialty crops and soil health. The authors of this chapter as well as other research groups worldwide have been investigating this topic in recent years. A summary of the impacts of BDMs on specialty crop production and soil fertility are given in this section.

11.5.1 Impacts on Specialty Crop Production

In the last decade, many studies have reported crop yields that are similar for plastic BDMs and PE mulch in crops of: lettuce (Brault et al. 2002), melon (Filippi et al. 2011; Iapichino et al. 2014; Shogren and Hochmuth 2004); tomato (Cirujeda et al. 2012; Cowan et al. 2014; Martín-Closas et al. 2008; Moreno and Moreno 2008), cucumber (Wortman et al. 2016) and pumpkin (Ghimire et al. 2018). Some of these studies also reported higher quality of products grown in PE and BDMs compared to bare soil. A review on this topic was recently published by Martín-Closas et al. (2017).

The authors of this chapter participated in a 4 year study, where the effect of mulch treatments on the production of vegetable crops was examined in two places that differed greatly in climate and soil type: Mount Vernon, WA (Pacific Northwest region of the USA, mild and humid climate, poorly drained Skagit silt loam soil) and Knoxville, TN (Southeastern USA, subtropical climate, moderately well drained Shady-Whitwell complex soil). The study compared the performance of three commercially available BDMs (one of which was prepared from Mater-Bi® and another from ecovio®), an experimental BDM film prepared from a PLA/PHA blend, PE mulch, and paper mulch, and bare soil as a control. The treatments were organized in a randomized complete split-split-plot design with four replications. Pie pumpkin (*Cucurbita pepo* L.) was employed as the test crop for the first two years (2015 and 2016) in both locations, and green pepper in Knoxville and sweet corn in Mount Vernon during the last two years (2017 and 2018). Raised beds were formed from the soil and the mulches were machine-laid. After crop harvest each year, the BDMs were plowed into the soil. The parameters measured included crop quality and yield, weed amount, macroscopic deterioration of the BDMs, and soil quality indices. Environmental data were also collected. The results for the 2015 and 2016 seasons were published by our research group (Ghimire et al. 2018). All mulching treatments were effective in suppressing weeds compared to bare soil, and mulches generally remained intact during the growing season (June–September) with the exception of paper mulch in Knoxville, which likely was weakened by high moisture exposure. The latter was similar to results reported by Martín-Closas et al. (2016). Fruit yield was higher for all mulching treatments relative to bare soil, and was similar between the BDMs and PE mulch in Mount Vernon; however, yields did not differ between mulching treatments and bare soil in Knoxville. The difference in fruit yield trends between sites is probably due to higher soil temperatures under mulches in relation to bare soil in Mount Vernon (by ~1 °C), with no effect on soil temperature in Knoxville. BDM fragments adhered to pumpkins in Mount Vernon, which reduces fruit commercialization, but this did not occur using PE mulch. When the growing fruit (e.g. pumpkin or watermelon) rests on the BDM, problems with mulch durability may arise due to the weight and pressure imposed by the fruit (Limpus 2012).

Another recent study carried out by Touchaleaume et al. (2016) compared several different BDMs, PE mulch and bare soil in vineyards in the southern France over a period of 2 years. Six months after planting, BDMs and PE mulch produced a greater vine biomass density and a fruit yield compared to bare soil; but, there were no significant differences between mulch treatments. For the same period, BDMs suffered a macroscopic deterioration, but not PE mulch. During the subsequent fruit harvests, BDMs and PE mulch performed equally well, despite the macroscopic deterioration of the former. The authors hypothesized that during the first six months, BDMs (similar to PE) enhanced soil moisture relative to bare soil due to increased soil moisture at the upper levels of soil. After six months, the root system of the plants grew deeper into the soil to find new sources of water (since water evaporation occurred at the upper levels due to the deterioration of the BDMs), leading to more vigorous plants and a better distribution of the roots compared to PE mulch, which contained a high upper-level soil moisture and shallower root systems.

11.5.2 Assessing Effects of Biodegradable Mulches on Soil Health

As discussed in the previous section, many studies have indicated that BDMs are equivalent to PE mulches in terms of agronomic performance. However, less is known about the impact of BDMs on soil health. According to the U.S. Department of Agriculture (2019) soil health is defined as the capacity of a soil to perform its essential functions as a living ecosystem. This is an important aspect for the sustainable use of soil resources in agriculture. The vitality and productivity of soils is of utmost importance to sustain agriculture and food production. It is therefore essential that new management practices, such as BDMs, are rigorously evaluated for their impact on soil health.

Soil processes usually occur on time scales of years, decades and centuries. As such, changes in soil health also occur at these time-scales, and long-term assessments of soil health are needed to evaluate whether new management practices are impacting soils in a positive or negative manner. Soil health is assessed by measuring selected soil properties or processes and comparing the measurements with benchmark values. While there is no standard set of measurements used by the scientific community, soil health measurements usually include soil physical, chemical and biological properties. Examples of such properties typically include bulk density, soil water retention characteristics, infiltration rates, aggregate stability, pH, electrical conductivity, organic matter and respiration (U.S. Department of Agriculture 2008). Efforts are currently underway to standardize measurement parameters and methods for assessing soil health (Soil Health Institute 2019).

Measures in the change of the soil properties before and after the application of the BDMs, and having a control as reference (bare soil) have been used to evaluate the impact of the BDMs on the health of the soils (Sintim et al. 2019). While some studies have reported on the effects of BDMs on soil health, the results have often been inconsistent, probably because of differences in edaphic factors, management practices, such as cropping system and the type of BDM tested.

Moore-Kucera et al. (2014) did not observe significant effects of the BDMs on the structures of soil bacteria and fungi communities. Nonetheless, Koitabashi et al. (2012) observed changes in the composition of the population of soil fungi and *Acantamoeba* spp. after 4 weeks of incubation of BDMs in soil. Li et al. (2014a) did not observe a definite trend of BDM effects on soil pH, electrical conductivity and total organic carbon, after 18 months of incorporation of the BDMs into the soil, in three different locations and two cropping systems. Sintim et al. (2019) reported significant effects of some BDMs on aggregate stability, infiltration, soil pH, electrical conductivity, nitrate-N, and exchangeable potassium over a 24 month period. However, the effects were not consistent at two sites and five assessments (three spring and two fall dates) during the 24 month period. The nitrate-N levels in fall, e.g. were generally lower but similar in spring compared to no-mulch treatment. This was attributed to better plant uptake of nitrate-N in the BDM plots, whereas rainfall leached the residual nitrate-N in the bare soil plots during winter and early spring. Sintim et al. (2019) also noted that soil disturbance caused by tillage, which is a usual operation for the production of vegetable crops, restores many of the soil properties and recommended long-term studies to better elucidate the impact of BDMs on soil health. Despite inconsistent reports in the literature, some scientists and growers believe that BDMs probably have minimal impacts on soil health and are a better alternative to PE mulch. Actually, BDMs were included in the 2018 revision of EU Fertilizers Regulation, citing that they provide agronomical, economic and environmental benefits in agriculture (European Committee for Standardization 2018).

EN 17033 describes that BDMs should be completely decomposed into CO_2 and water, or incorporated into microbial biomass (European Committee for Standardization 2018). We also think that, theoretically, BDMs should not have an adverse impact on soil health. However, our ongoing studies in Mount Vernon, WA and Knoxville, TN (Section 11.5.1) showed that most BDMs did not undergo complete degradation after 36 months (unpublished results), which raises environmental concerns.

11.5.3 *Fate and Pathways of Carbon from Biodegradable Mulch in Soils*

Mulches have the potential ability to alter soil organic carbon (SOC) content. SOC is the carbon portion of soil organic matter in the soil and is also an important indicator of healthy soil. Thousands of different C compounds of varying degrees of recalcitrance make up SOC (Davidson and Janssens 2006; Baldock and Skjemstad 2000; Lehmann and Kleber 2015). As SOC constantly undergoes the decomposition by microorganisms, its replacement during the growth of the plants is achieved to maintain the C balance of the soil. Equilibrium is reached when soils have the same amount of carbon accumulation as is lost to the atmosphere by the respiration (Paul et al. 1997; Haynes 2005). Soil C with a turnover time of days to months is readily accessible to microorganisms and also includes dissolved organic C and plant and microbial biomass (e.g. proteins, nucleic acids and polysaccharides). Soil C with a turnover time of decades to millennia is much more stable and contains more alkyl and aromatic compounds, as well as SOC that is otherwise inaccessible to microbes due to organic matter-mineral interactions and physical occlusion in soil aggregates (Sollins et al. 1996; Krull et al. 2003; Jastrow et al. 2007).

By altering the soil temperature and moisture, conventional plastic film mulching can increase decomposition rates in the soil, leading to a loss of SOC (Steinmetz et al. 2016). Mulching has led to increases in above-ground carbon (C) inputs (Cuello et al. 2015; Ghimire et al. 2018) from higher yields and increases in below-ground C inputs (Wang et al. 2016b; Li et al. 2007) from increased development. One study by Zhang et al. (2017a) modelled the effects of conventional mulching for 30 years and found a significant increase in decomposition of SOC, but also an increase plant biomass (above and below ground) was obtained, thus leading to no net change in SOC concentration. In contrast, Steinmetz et al. (2016) considered it unlikely that increases in net primary production in mulch treatments would compensate the C losses due to the increase in decomposition.

BDMs, in addition to affecting soil temperature and moisture, could have a positive effect on C pools as they are an additional C source once they enter the soil. There is an incomplete understanding of soil C dynamics under mulch (Luo et al. 2015). It has been shown that the use of BDM in the short term does not have a net effect on total SOC (Li et al. 2014a) nor organic matter (Sintim et al. 2019), or a slight positive effect on soil microbial biomass C (Moreno and Moreno 2008). After the use of conventional short-term mulch, SOC can decrease (Li et al. 2004; Li et al. 2007; Cuello et al. 2015), increase (Muñoz et al. 2017), or show no net change (Zhang et al. 2017b; Wang et al. 2016b). Interestingly, the use of mulch can increase the potential for lengthening the turnover time for SOC by promoting the formation of larger water-stable aggregates (Domagała-Świątkiewicz and Siwek 2013; Zhang et al. 2013; Siwek et al. 2015), which increases the stability of SOC (Jastrow et al. 2007). The effect of mulching on the destabilization (through increased mineralization) and stabilization (through increased aggregation) of SOC requires more research (Steinmetz et al. 2016).

11.5.4 Impacts of Biodegradable Mulch on Soil Microbial Communities

There is a small but growing body of research reporting the effects of plastic mulch on soil microbial communities (Bandopadhyay et al. 2018). As a physical barrier on the soil surface, plastic mulches (both conventional and biodegradable) modify the soil microclimate, which can indirectly control the composition and function of the microbial communities. In particular, plastic mulches regulate soil moisture *via* reduced evaporation, and moderate soil temperature *via* warming (black plastic) or cooling (white plastic). These modifications to soil temperature and moisture can also affect soil physical properties, e.g. by increasing aggregation (Siwek et al. 2015; Sintim et al. 2019), and plant root development and exudation (Wang et al. 2016b). Together these effects on the soil environment result in changes in the composition and function of soil microbial communities. Studies of soil mulched with conventional PE mulches reveal changes in taxonomic composition of soil microbial communities, e.g. increases in the bacterial phyla *Proteobacteria* and *Actinobacteria* (Farmer et al. 2017) and decrease pressure in the oomycete plant pathogen, *Phytophthera capsici* (Núñez-Zofío et al. 2011). Microbial activity is also affected by plastic mulches, and most studies report an increase in soil respiration under PE mulch (Mu et al. 2014; Zhang et al. 2015; Mu et al. 2016). It should be noted that in some cases where mulches increase soil temperatures above the microbial growth optimum, microbial activity is diminished (Moreno and Moreno 2008). While these studies were performed on PE mulches, not BDMs, it is expected that as a surface barrier, BDMs would have a similar environmental modification that would affect microbial communities.

A difference between BDMs and PE mulches is that the former are ultimately incorporated into the soil to be degraded by microorganisms. BDM fragments are a physical and biogeochemical (C) input that can be expected to have some effect on soil microbial communities. Although the amount of carbon introduced from mulches is very small compared to the amount of SOC already present, studies of incorporation of BDM into the soil have shown changes in the structure of the microbial communities (Li et al. 2014b; Muroi et al. 2016), and increases in fungal abundances (Rychter et al. 2006; Li et al. 2014b; Muroi et al. 2016); though the responses are not necessarily consistent between locations (Li et al. 2014b). BDM incorporation can also cause increases in microbial biomass and enzyme activities (Li et al. 2014b; Yamamoto-Tamura et al. 2015). Studies directly comparing PE mulch with BDMs have generally shown enhanced microbial activity under BDMs, suggesting that it is a response to the incorporation of the BDMs into soil, and not only an indirect effect of soil environmental modifications (Moreno and Moreno 2008; Li et al. 2014b; Yamamoto-Tamura et al. 2015; Barragan et al. 2016). While this body of knowledge has provided some insight, there are still some key gaps in our understanding of the impact of BDMs on soil microbial communities, including the long-term effects of tilling BDM into soils, and the relative importance of BDM composition in soil microbial responses (Bandopadhyay et al. 2018).

11.6 Conclusions

In recent years, important advances in research and development related to BDMs have been achieved. It has been shown that BDMs are just as effective as PE mulch in improving the production of specialty crops compared to bare soil for several different cropping systems and different environmental conditions. Environmental studies have not shown any adverse impacts associated with the incorporation of BDMs into the soil to date. More research is needed to ensure that the production of micro and nanoplastics from BDMs does not adversely affect ecotoxicity, soil ecosystems and crop quality, and to better adapt the BDM physicochemical properties with the needs of specific cropping systems and climates. BDMs may be formulated to deliver macro and micro nutrients to the crop as they are biodegraded, or pesticides that are released directly into the soil at the base of the plant, thereby reducing their applications to adjacent areas and reducing overall applications. A major use of mulches overall is for weed control; but, plastic mulches (PE and BDMs) are ineffective against weeds such as nutsedge, horsetail and other persistent weeds that can puncture plastic mulch films. Perhaps embedding herbicides specific for these types of weeds into the mulch may provide control. BDMs may also be formulated with additives that when released can increase the nutraceutical content of crops. Furthermore, advances in materials chemistry are needed to enable cost-effective production of biobased BDMs that completely biodegrade in soil. Finally, as scientists, we must continue to educate farmers, government regulators and the general public on the importance of replacing the use of PE mulch with that of BDMs.

Acknowledgements The authors acknowledge financial support for their research pertaining to biodegradable plastic mulches from the USDA Specialty Crops Research Initiative program (Award 2014-51181-22382) and NIFA Hatch projects 1017286 and 1014527 for Carol Miles and Markus Flury, respectively.

Conflicts of Interest The authors declare no conflict of interest.

References

Adeleye, A. I., Patel, D., Niyogi, D., & Saha, B. (2014). Efficient and greener synthesis of propylene carbonate from carbon dioxide and propylene oxide. *Industrial & Engineering Chemistry Research, 53*(49), 18647–18657. https://doi.org/10.1021/ie500345z.

Alimi, O. S., Farner Budarz, J., Hernandez, L. M., & Tufenkji, N. (2018). Microplastics and nanoplastics in aquatic environments: Aggregation, deposition, and enhanced contaminant transport. *Environmental Science & Technology, 52*(4), 1704–1724. https://doi.org/10.1021/acs.est.7b05559.

Andersson, S. R., Hakkarainen, M., & Albertsson, A.-C. (2010). Tuning the polylactide hydrolysis rate by plasticizer architecture and hydrophilicity without introducing new migrants. *Biomacromolecules, 11*(12), 3617–3623. https://doi.org/10.1021/bm101075p.

ASTM International. (2012a). *Standard specification for labeling of plastics designed to be aerobically composted in municipal or industrial facilities (ASTM D6400)*. West Conshohocken.

ASTM International. (2012b). *Standard test method for determining aerobic biodegradation in soil of plastic materials or residual plastic materials after composting (ASTM D5988)*. West Conshohocken.

ASTM International. (2012c). *Standard test methods for determining the biobased content of solid, liquid, and gaseous samples using radiocarbon analysis (ASTM D6866)*. West Conshohocken.

Baldock, J. A., & Skjemstad, J. O. (2000). Role of the soil matrix and minerals in protecting natural organic materials against biological attack. *Organic Geochemistry, 31*(7), 697–710. https://doi.org/10.1016/S0146-6380(00)00049-8.

Bandopadhyay, S., Martin-Closas, L., Pelacho, A. M., & DeBruyn, J. M. (2018). Biodegradable plastic mulch films: Impacts on soil microbial communities and ecosystem functions. *Frontiers in Microbiology, 9*, 819. https://doi.org/10.3389/fmicb.2018.00819.

Barragan, D. H., Pelacho, A. M., & Martin-Closas, L. (2016). Degradation of agricultural biodegradable plastics in the soil under laboratory conditions. *Soil Research, 54*(2), 216–224. https://doi.org/10.1071/sr15034.

Brault, D., Stewart, K. A., & Jenni, S. (2002). Growth, development, and yield of head lettuce cultivated on paper and polyethylene mulch. *HortScience, 37*(1), 92–94. https://doi.org/10.21273/hortsci.37.1.92.

Briassoulis, D. (2006). Mechanical behavior of biodegradable agricultural films under real field conditions. *Polymer Degradation and Stability, 91*(6), 1256–1272. https://doi.org/10.1016/j.polymdegradstab.2005.09.016.

Briassoulis, D. (2007). Analysis of the mechanical and degradation performances of optimized agricultural biodegradable films. *Polymer Degradation and Stability, 92*(6), 1115–1132. https://doi.org/10.1016/j.polymdegradstab.2007.01.024.

Briassoulis, D., & Dejean, C. (2010). Critical review of norms and standards for biodegradable agricultural plastics part I: Biodegradation in soil. *Journal of Polymers and the Environment, 18*(3), 384–400. https://doi.org/10.1007/s10924-010-0168-1.

Briassoulis, D., Aristopoulou, A., Bonora, M., & Verlodt, I. (2004). Degradation characterisation of agricultural low-density polyethylene films. *Biosystems Engineering, 88*(2), 131–143. https://doi.org/10.1016/j.biosystemseng.2004.02.010.

Briassoulis, D., Dejean, C., & Picuno, P. (2010). Critical review of norms and standards for biodegradable agricultural plastics part II: Composting. *Journal of Polymers and the Environment, 18*(3), 364–383. https://doi.org/10.1007/s10924-010-0222-z.

Callister, W. D., & Rethwisch, D. G. (2015). *Fundamentals of materials science and engineering: An integrated approach* (4th ed.). Hoboken: Wiley.

Chinaglia, S., Tosin, M., & Degli-Innocenti, F. (2018). Biodegradation rate of biodegradable plastics at molecular level. *Polymer Degradation and Stability, 147*, 237–244. https://doi.org/10.1016/j.polymdegradstab.2017.12.011.

Cirujeda, A., Aibar, J., Anzalone, Á., Martín-Closas, L., Meco, R., Moreno, M. M., Pardo, A., Pelacho, A. M., Rojo, F., Royo-Esnal, A., Suso, M. L., & Zaragoza, C. (2012). Biodegradable mulch instead of polyethylene for weed control of processing tomato production. *Agronomy for Sustainable Development, 32*(4), 889–897. https://doi.org/10.1007/s13593-012-0084-y.

Collias, D. I., Harris, A. M., Nagpal, V., Cottrell, I. W., & Schultheis, M. W. (2014). Biobased terephthalic acid technologies: A literature review. *Industrial Biotechnology, 10*(2), 91–105. https://doi.org/10.1089/ind.2014.0002.

Coltelli, M.-B., Maggiore, I. D., Bertoldo, M., Signori, F., Bronco, S., & Ciardelli, F. (2008). Poly(lactic acid) properties as a consequence of poly(butylene adipate-*co*-terephthalate) blending and acetyl tributyl citrate plasticization. *Journal of Applied Polymer Science, 110*(2), 1250–1262. https://doi.org/10.1002/app.28512.

Cowan, J. S., Miles, C., Andrews, P. K., & Inglis Debra, A. (2014). Biodegradable mulch performed comparable to polyethylene in high tunnel tomato (*Solanum lycopersicum L.*) production. *Journal of the Science of Food and Agriculture, 94*(9), 1854–1864. https://doi.org/10.1002/jsfa.6504.

Cowan, J. S., Goldberger, J. R., Miles, C. A., & Inglis, D. A. (2015). Creating tactile space during a university extension field day event: The case of a sustainable agriculture innovation. *Rural Sociology, 80*(4), 456–482. https://doi.org/10.1111/ruso.12073.

Cuello, J. P., Hwang, H. Y., Gutierrez, J., Kim, S. Y., & Kim, P. J. (2015). Impact of plastic film mulching on increasing greenhouse gas emissions in temperate upland soil during maize cultivation. *Applied Soil Ecology, 91*, 48–57. https://doi.org/10.1016/j.apsoil.2015.02.007.

Davidson, E. A., Janssens, I. A. (2006). Temperature sensitivity of soil carbon decomposition and feedbacks to climate change. Nature, 440(7081), 165–173. https://doi.org/10.1038/nature04514

Detyothin, S., Kathuria, A., Jaruwattanayon, W., Selke, S. E. M., & Auras, R. (2010). In R. Auras, L. T. Lim, S. E. M. Selke, & H. Tsuji (Eds.), *Poly(lactic acid) blends in poly(Lactic Acid): Synthesis, structures, properties, processing, and applications* (pp. 227–271). Hoboken: Wiley. https://doi.org/10.1002/9780470649848.ch16.

Dharmalingam, S., Hayes, D. G., Wadsworth, L. C., Dunlap, R. N., DeBruyn, J. M., Lee, J., & Wszelaki, A. L. (2015). Soil degradation of polylactic acid/polyhydroxyalkanoate-based nonwoven mulches. *Journal of Polymers and the Environment, 23*(3), 302–315. https://doi.org/10.1007/s10924-015-0716-9.

Domagała-Świątkiewicz, I., & Siwek, P. (2013). The effect of direct covering with biodegradable nonwoven film on the physical and chemical properties of soil. *Polish Journal of Environmental Studies, 22*(3), 667–674.

Dufresne, A., Dupeyre, D., & Vignon, M. R. (2000). Cellulose microfibrils from potato tuber cells: Processing and characterization of starch–cellulose microfibril composites. *Journal of Applied Polymer Science, 76*(14), 2080–2092. https://doi.org/10.1002/(sici)1097-4628(20000628)76:14<2080::aid-app12>3.0.co;2-u.

Elfehri Borchani, K., Carrot, C., & Jaziri, M. (2015). Biocomposites of alfa fibers dispersed in the mater-bi® type bioplastic: Morphology, mechanical and thermal properties. *Composites Part A: Applied Science and Manufacturing, 78*, 371–379. https://doi.org/10.1016/j.compositesa.2015.08.023.

Emmert, E. M. (1957). Black polyethylene for mulching vegetables. *Proceedings of the American Society for Horticultural Science, 69*, 464–469.

Erickson, B. E. (2015). Regulators and retailers raise pressure on phthalates. *Chemical and Engineering News, 95*(25), 11–15.

European Commission. (2018). *Report from the commission to the European Parliament and the council on the impact of the use of oxo-degradable plastic, including* [Online]. Brussels: European Union. Available: http://ec.europa.eu/environment/circular-economy/pdf/oxo-plastics.pdf. Accessed 13 Feb 2019.

European Committee for Standardization. (2018). EN 17033: Plastics-biodegradable mulch films for use in agriculture and horticulture- requirements and test methods. Brussels.

Farmer, J., Zhang, B., Jin, X., Zhang, P., & Wang, J. (2017). Long-term effect of plastic film mulching and fertilization on bacterial communities in a brown soil revealed by high through-put sequencing. *Archives of Agronomy and Soil Science, 63*(2), 230–241. https://doi.org/10.1080/03650340.2016.1193667.

Filippi, F., Magnani, G., Guerrini, S., & Ranghino, F. (2011). Agronomic evaluation of green biodegradable mulch on melon. *Italian Journal of Agronomy, 6*(2), 111–116. https://doi.org/10.4081/ija.2011.e18.

Ghimire, S., Wszelaki, A. L., Moore, J. C., Inglis, D. A., & Miles, C. (2018). The use of biodegradable mulches in pie pumpkin crop production in two diverse climates. *HortScience, 53*(3), 288–294. https://doi.org/10.21273/hortsci12630-17.

Ghosh, A. (2017). Phthalate puzzle. *Resonance, 22*(7), 691–696. https://doi.org/10.1007/s12045-017-0512-z.

Goldberger, J. R., Jones, R. E., Miles, C. A., Wallace, R. W., & Inglis, D. A. (2013). Barriers and bridges to the adoption of biodegradable plastic mulches for US specialty crop production. *Renewable Agriculture and Food Systems, 30*(2), 143–153. https://doi.org/10.1017/S1742170513000276.

Gutiérrez, T. J. (2018a). Are modified pumpkin flour/plum flour nanocomposite films biodegradable and compostable? *Food Hydrocolloids, 83*, 397–410. https://doi.org/10.1016/j.foodhyd.2018.05.035.

Gutiérrez, T. J. (2018b). Chapter 9. Biodegradability and compostability of food nanopackaging materials. In: Composite materials for food packaging. Giuseppe Cirillo, Marek A. Kozlowski, and Umile Gianfranco Spizzirri (Eds). WILEY-Scrivener Publisher. EE.UU. ISBN: 978-1-119-16020-5. pp. 269-296. doi:https://doi.org/10.1002/9781119160243.ch9.
Gutiérrez, T. J., & Alvarez, V. A. (2017a). Eco-friendly films prepared from plantain flour/PCL blends under reactive extrusion conditions using zirconium octanoate as a catalyst. *Carbohydrate Polymers, 178*, 260–269. https://doi.org/10.1016/j.carbpol.2017.09.026.
Gutiérrez, T. J., & Alvarez, V. A. (2017b). Data on physicochemical properties of active films derived from plantain flour/PCL blends developed under reactive extrusion conditions. *Data in Brief, 15*, 445–448. https://doi.org/10.1016/j.dib.2017.09.071.
Gutiérrez, T. J., Toro-Márquez, L. A., Merino, D., & Mendieta, J. R. (2019). Hydrogen-bonding interactions and compostability of bionanocomposite films prepared from corn starch and nano-fillers with and without added Jamaica flower extract. *Food Hydrocolloids, 89*, 283–293. https://doi.org/10.1016/j.foodhyd.2018.10.058.
Haapala, T., Palonen, P., Korpela, A., & Ahokas, J. (2014). Feasibility of paper mulches in crop production — A review. *Agricultural and Food Science, 23*(1), 60–79. https://doi.org/10.23986/afsci.8542.
Hablot, E., Dharmalingam, S., Hayes, D. G., Wadsworth, L. C., Blazy, C., & Narayan, R. (2014). Effect of simulated weathering on physico-chemical properties and inherent biodegradation of PLA/PHA non-woven-based agricultural mulches. *Journal of Polymers and the Environment, 22*(4), 417–429. https://doi.org/10.1007/s10924-014-0697-0.
Hakkarainen, M., Karlsson, S., & Albertsson, A. C. (2000). Rapid (bio)degradation of polylactide by mixed culture of compost microorganisms-low molecular weight products and matrix changes. *Polymer, 41*(7), 2331–2338. https://doi.org/10.1016/s0032-3861(99)00393-6.
Hayes, D. G. & Flury, M. (2018). *Summary and assessment of EN 17033:2018. a new standard for biodegradable plastic mulch films* [Online]. Available: https://ag.tennessee.edu/biodegradablemulch/Pages/factsheets.aspx. Accessed 11 Feb 2019.
Hayes, D. G., Dharmalingam, S., Wadsworth, L. C., Leonas, K. K., Miles, C., & Inglis, D. A. (2012). Biodegradable agricultural mulches derived from biopolymers. In K. C. Khemani & C. Scholz (Eds.), *Degradable polymers and materials: Principles and practice* (ACS Symposium Series) (Vol. 1114, 2nd ed., pp. 201–223). Washington, DC: American Chemical Society. https://doi.org/10.1021/bk-2012-1114.ch013.
Hayes, D. G., Wadsworth, L. C., Sintim, H. Y., Flury, M., English, M., Schaeffer, S., & Saxton, A. M. (2017). Effect of diverse weathering conditions on the physicochemical properties of biodegradable plastic mulches. *Polymer Testing, 62*, 454–467. https://doi.org/10.1016/j.polymertesting.2017.07.027.
Haynes, R. J. (2005). Labile organic matter fractions as central components of the quality of agricultural soils: An overview. In *Advances in agronomy* (pp. 221–268). Cambridge: Academic. https://doi.org/10.1016/S0065-2113(04)85005-3.
Herniou--Julien, C., Mendieta, J. R., & Gutiérrez, T. J. (2019). Characterization of biodegradable/non-compostable films made from cellulose acetate/corn starch blends processed under reactive extrusion conditions. *Food Hydrocolloids, 89*, 67–79. https://doi.org/10.1016/j.foodhyd.2018.10.024.
Herrera, R., Franco, L., Rodríguez-Galán, A., & Puiggalí, J. (2002). Characterization and degradation behavior of poly(butylene adipate-*co*-terephthalate)s. *Journal of Polymer Science Part A: Polymer Chemistry, 40*(23), 4141–4157. https://doi.org/10.1002/pola.10501.
Howell, B. A., & Lazar, S. T. (2019). Biobased plasticizers from carbohydrate-derived 2, 5-bis(hydroxymethyl)furan. *Industrial & Engineering Chemistry Research, 58*(3), 1222–1228. https://doi.org/10.1021/acs.iecr.8b05442.
Howell, B. A., & Sun, W. (2018). Biobased plasticizers from tartaric acid, an abundantly available, renewable material. *Industrial & Engineering Chemistry Research, 57*(45), 15234–15242. https://doi.org/10.1021/acs.iecr.8b03486.
Hu, Y., Sato, H., Zhang, J., Noda, I., & Ozaki, Y. (2008). Crystallization behavior of poly(L-lactic acid) affected by the addition of a small amount of poly(3-hydroxybutyrate). *Polymer, 49*(19), 4204–4210. https://doi.org/10.1016/j.polymer.2008.07.031.

Huerta Lwanga, E., Gertsen, H., Gooren, H., Peters, P., Salánki, T., van der Ploeg, M., Besseling, E., Koelmans, A. A., & Geissen, V. (2017). Incorporation of microplastics from litter into burrows of Lumbricus terrestris. *Environmental Pollution, 220.* (Part A),, 523–531. https://doi.org/10.1016/j.envpol.2016.09.096.

Iapichino, G., Mustazza, G., Sabatino, L., & D'Anna, F. (2014). Polyethylene and biodegradable starch-based mulching films positively affect winter melon production in Sicily. *Acta Horticulturae, 1015*(1), 225–231. https://doi.org/10.17660/ActaHortic.2014.1015.25.

Ibarra-Jimenez, L. R., Quezada-Martin, R., Cedeno-Rubalcava, B., Lozano, A. L., & de la Rosa-Ibarra, M. (2006). Watermelon response to plastic mulch and row covers. *European Journal of Horticultural Science, 71*(6), 262–266.

Immirzi, B., Santagata, G., Vox, G., & Schettini, E. (2009). Preparation, characterisation and field-testing of a biodegradable sodium alginate-based spray mulch. *Biosystems Engineering, 102*(4), 461–472. https://doi.org/10.1016/j.biosystemseng.2008.12.008.

Inkinen, S., Hakkarainen, M., Albertsson, A.-C., & Sodergard, A. (2011). From lactic acid to poly(lactic acid) (PLA): Characterization and analysis of PLA and its precursors. *Biomacromolecules, 12*(3), 523–532. https://doi.org/10.1021/bm101302t.

International Organization for Standardization. (2003). ISO 9050:2003. Glass in building — determination of light transmittance, solar direct transmittance, total solar energy transmittance, ultraviolet transmittance and related glazing factors.

Jamarani, R., Erythropel, C. H., Burkat, D., Nicell, A. J., Leask, L. R., & Maric, M. (2017). Rheology of green plasticizer/poly(vinyl chloride) blends via time–temperature superposition. *Processes, 5*(3), 43. https://doi.org/10.3390/pr5030043.

Jastrow, J. D., Amonette, J. E., & Bailey, V. L. (2007). Mechanisms controlling soil carbon turnover and their potential application for enhancing carbon sequestration. *Climatic Change, 80*(1), 5–23. https://doi.org/10.1007/s10584-006-9178-3.

Jiang, X., Han, J., Han, Q., Zhou, X., & Ma, J. (2015). Preparation and characteristics of paper-based biodegradable plastics. *BioResources, 10*(2), 2982–2994.

Karamanlioglu, M., & Robson, G. D. (2013). The influence of biotic and abiotic factors on the rate of degradation of poly(lactic) acid (PLA) coupons buried in compost and soil. *Polymer Degradation and Stability, 98*(10), 2063–2071. https://doi.org/10.1016/j.polymdegradstab.2013.07.004.

Kasirajan, S., & Ngouajio, M. (2012). Polyethylene and biodegradable mulches for agricultural applications: A review. *Agronomy for Sustainable Development, 32*(2), 501–529. https://doi.org/10.1007/s13593-011-0068-3.

Kawai, F. (2010). Polylactic acid (PLA)-degrading microorganisms and PLA depolymerases. In Green polymer chemistry: Biocatalysis and biomaterials (ACS Symposium Series, Vol. 1043), eds Cheng, HN & Gross, RA, American Chemical Society, Washington, DC, pp. 405-414. doi:https://doi.org/10.1021/bk-2010-1043.ch027.

Kijchavengkul, T., & Auras, R. (2008). Compostability of polymers. *Polymer International, 57*(6), 793–804. https://doi.org/10.1002/pi.2420.

Kijchavengkul, T., Auras, R., Rubino, M., Ngouajio, M., & Fernandez, R. T. (2008a). Assessment of aliphatic-aromatic copolyester biodegradable mulch films. Part I: Field study. *Chemosphere, 71*(5), 942–953. https://doi.org/10.1016/j.chemosphere.2007.10.074.

Kijchavengkul, T., Auras, R., Rubino, M., Ngouajio, M., & Fernandez, R. T. (2008b). Assessment of aliphatic-aromatic copolyester biodegradable mulch films. Part II: Laboratory simulated conditions. *Chemosphere, 71*(9), 1607–1616. https://doi.org/10.1016/j.chemosphere.2008.01.037.

Kijchavengkul, T., Auras, R., Rubino, M., Alvarado, E., Camacho Montero, J. R., & Rosales, J. M. (2010). Atmospheric and soil degradation of aliphatic-aromatic polyester films. *Polymer Degradation and Stability, 95*(2), 99–107. https://doi.org/10.1016/j.polymdegradstab.2009.11.048.

Koitabashi, M., Noguchi, M. T., Sameshima-Yamashita, Y., Hiradate, S., Suzuki, K., Yoshida, S., Watanabe, T., Shinozaki, Y., Tsushima, S., & Kitamoto, H. K. (2012). Degradation of biodegradable plastic mulch films in soil environment by phyllopland fungi isolated from framineous plants. *AMB Express, 2*(1), 40. https://doi.org/10.1186/2191-0855-2-40.

Krull, E. S., Baldock, J. A., & Skjemstad, J. O. (2003). Importance of mechanisms and processes of the stabilisation of soil organic matter for modelling carbon turnover. *Functional Plant Biology, 30*(2), 207–222. https://doi.org/10.1071/FP02085.

Lamont, W. J. (1999). The use of different colored mulches for yield and earliness. Proceedings New England Vegetable and Berry Growers Conference and Trade Show, Sturbridge. p. 299–302 [Online]. Available: http://ipm.uconn.edu/root/publications.php. Accessed 11 Feb 2019.

Lamont, W. J. (2001). Vegetable production using agriculture [Online]. Available: http://www.fftc.agnet.org/htmlarea_file/library/20110808093747/eb476.pdf. Accessed 5 Feb 2019.

Lamont, W. J. (2005). Plastics: Modifying the microclimate for the production of vegetable crops. *HortTechnology, 15*(3), 477–481. https://doi.org/10.21273/horttech.15.3.0477.

Lehmann, J., & Kleber, M. (2015). The contentious nature of soil organic matter. *Nature, 528*, 60–68. https://doi.org/10.1038/nature16069.

Li, F.-M., Song, Q.-H., Jjemba, P. K., & Shi, Y.-C. (2004). Dynamics of soil microbial biomass C and soil fertility in cropland mulched with plastic film in a semiarid agro-ecosystem. *Soil Biology and Biochemistry, 36*(11), 1893–1902. https://doi.org/10.1016/j.soilbio.2004.04.040.

Li, Y.-S., Wu, L.-H., Zhao, L.-M., Lu, X.-H., Fan, Q.-L., & Zhang, F.-S. (2007). Influence of continuous plastic film mulching on yield, water use efficiency and soil properties of rice fields under non-flooding condition. *Soil and Tillage Research, 93*(2), 370–378. https://doi.org/10.1016/j.still.2006.05.010.

Li, C., Moore-Kucera, J., Lee, J., Corbin, A., Brodhagen, M., Miles, C., & Inglis, D. (2014a). Effects of biodegradable mulch on soil quality. *Applied Soil Ecology, 79*, 59–69. https://doi.org/10.1016/j.apsoil.2014.02.012.

Li, C., Moore-Kucera, J., Miles, C., Leonas, K., Lee, J., Corbin, A., & Inglis, D. (2014b). Degradation of potentially biodegradable plastic mulch films at three diverse U.S. locations. *Agroecology and Sustainable Food Systems, 38*(8), 861–889. https://doi.org/10.1080/21683565.2014.884515.

Limpus, S. (2012). Comparison of biodegradable mulch products to polyethylene in irrigated vegetable, tomato and melon crops. Final report MT09068. Queensland Department of Agriculture, Fisheries and Forestry, Queensland.

Liu, E. K., He, W. Q., & Yan, C. R. (2014). 'White revolution' to 'white pollution'—Agricultural plastic film mulch in China. *Environmental Research Letters, 9*(9), 091001. https://doi.org/10.1088/1748-9326/9/9/091001.

Luo, S., Zhu, L., Liu, J., Bu, L., Yue, S., Shen, Y., & Li, S. (2015). Sensitivity of soil organic carbon stocks and fractions to soil surface mulching in semiarid farmland. *European Journal of Soil Biology, 67*, 35–42. https://doi.org/10.1016/j.ejsobi.2015.01.004.

Madhavan Nampoothiri, K., Nair, N. R., & John, R. P. (2010). An overview of the recent developments in polylactide (PLA) research. *Bioresource Technology, 101*(22), 8493–8501. https://doi.org/10.1016/j.biortech.2010.05.092.

Mahmoudpour, M. A., & Stapleton, J. J. (1997). Influence of sprayable mulch colour on yield of eggplant (*Solanum melongena* L. cv. Millionaire). *Scientia Horticulturae, 70*(4), 331–338. https://doi.org/10.1016/s0304-4238(97)00039-3.

Martin-Closas, L., & Pelacho, A. M. (2011). Agronomic potential of biopolymer films. In L. Martín-Closas & A. M. Pelacho (Eds.), *Biopolymers – New materials for sustainable films and coatings* (pp. 277–299). https://doi.org/10.1002/9781119994312.ch13.

Martín-Closas, L., Bach, M. A., & Pelacho, A. M. (2008). Biodegradable mulching in an organic tomato production system. *Acta Horticulturae, 767*(1), 267–274. https://doi.org/10.17660/ActaHortic.2008.767.28.

Martin-Closas, L., Botet, R., & Pelacho, A. M. (2014). An in vitro crop plant ecotoxicity test for agricultural bioplastic constituents. *Polymer Degradation and Stability, 108*, 250–256. https://doi.org/10.1016/j.polymdegradstab.2014.03.037.

Martín-Closas, L., Costa, J., Cirujeda, A., Aibar, J., Zaragoza, C., Pardo, A., Suso, M. L., Moreno, M. M., Moreno, C., Lahoz, I., Mácua, J. I., & Pelacho, A. M. (2016). Above-soil and in-soil degradation of oxo- and bio-degradable mulches: A qualitative approach. *Soil Research, 54*(2), 225–236. https://doi.org/10.1071/SR15133.

Martín-Closas, L., Costa, J., & Pelacho, A. M. (2017). Agronomic effects of biodegradable films on crop and field environment. In M. Malinconico (Ed.), *Soil degradable bioplastics for a sustainable modern agriculture. Green chemistry and sustainable technology* (pp. 67–104). Heidelberg: Springer. https://doi.org/10.1007/978-3-662-54130-2_4.

Maughan, T. & Frost, D. (2016). Use of plastic mulch for vegetable production. *Horticulture Extension, Utah State University* [Online]. Available: https://digitalcommons.usu.edu/extension_curall/786/. Accessed 5 Feb 2019.

Melek, E., & Atilla, D. (2009). Effect of different mulch materials on the plant growth, some quality parameters and yield of melon (*Cucumis melo* L.) cultivars in high altitude environmental condition. *Pakistan Journal of Botany, 41*(4), 1891–1901.

Merfield, C. (2000). Organic weed management: A practical guide. [Online]. Available: http://www.merfield.com/research/2003/organic-weed-management-2003-merfield.pdf. Accessed 11 Feb 2019.

Merino, D., Gutiérrez, T. J., Mansilla, A. Y., Casalongué, C., & Alvarez, V. A. (2018a). Critical evaluation of starch-based antibacterial nanocomposites as agricultural mulch films: Study on their interactions with water and light. *ACS Sustainable Chemistry & Engineering, 6*(11), 15662–15672. https://doi.org/10.1021/acssuschemeng.8b04162.

Merino, D., Mansilla, A. Y., Gutiérrez, T. J., Casalongué, C. A., & Alvarez, V. A. (2018b). Chitosan coated-phosphorylated starch films: Water interaction, transparency and antibacterial properties. *Reactive and Functional Polymers, 131*, 445–453. https://doi.org/10.1016/j.reactfunctpolym.2018.08.012.

Merino, D., Gutiérrez, T. J., & Alvarez, V. A. (2019a). Potential agricultural mulch films based on native and phosphorylated corn starch with and without surface functionalization with chitosan. *Journal of Polymers and the Environment, 27*(1), 97–105. https://doi.org/10.1007/s10924-018-1325-1.

Merino, D., Gutiérrez, T. J., & Alvarez, V. A. (2019b). Structural and thermal properties of agricultural mulch films based on native and oxidized corn starch nanocomposites. *Starch-Stärke, 71*(7–8), 1800341. https://doi.org/10.1002/star.201800341.

Miles, C., Wallace, R., Wszelaki, A., Martin, J., Cowan, J., Walters, T., & Inglis, D. A. (2012). Deterioration of potentially biodegradable alternative to black plastic mulch in three tomato production regions. *HortScience, 47*(9), 1270–1277. https://doi.org/10.21273/hortsci.47.9.1270.

Miles, C., De Vetter, L., Ghimire, S., & Hayes, D. G. (2017). Suitability of biodegradable plastic mulches for organic and sustainable agricultural production systems. *HortScience, 52*(1), 10–15. https://doi.org/10.21273/hortsci11249-16.

Mitchell, J., Summers, C., McGiffen, M., Aguiar, J., Aslan, S. & Stapleton, J. (2004). Mulches in california vegetable crop production. Agriculture and natural resources, University of California, Publication No. 8129.

Mohee, R., Unmar, G. D., Mudhoo, A., & Khadoo, P. (2008). Biodegradability of biodegradable/degradable plastic materials under aerobic and anaerobic conditions. *Waste Management, 28*(9), 1624–1629. https://doi.org/10.1016/j.wasman.2007.07.003.

Moore-Kucera, J., Cox, S. B., Peyron, M., Bailes, G., Kinloch, K., Karich, K., Miles, C., Inglis, D. A., & Brodhagen, M. (2014). Native soil fungi associated with compostable plastics in three contrasting agricultural settings. *Applied Microbiology and Biotechnology, 98*(14), 6467–6485. https://doi.org/10.1007/s00253-014-5711-x.

Moreno, M. M., & Moreno, A. (2008). Effect of different biodegradable and polyethylene mulches on soil properties and production in a tomato crop. *Scientia Horticulturae, 116*(3), 256–263. https://doi.org/10.1016/j.scienta.2008.01.007.

Mormile, P., Stahl, N., & Malinconico, M. (2017). The world of plasticulture. In M. Malinconico (Ed.), *Soil degradable bioplastics for a sustainable modern agriculture* (pp. 1–21). Berlin/Heidelberg: Springer. https://doi.org/10.1007/978-3-662-54130-2_1.

Mu, L., Liang, Y., Zhang, C., Wang, K., & Shi, G. (2014). Soil respiration of hot pepper (*Capsicum annuum* L.) under different mulching practices in a greenhouse, including controlling factors in China. *Acta Agriculturae Scandinavica, Section B — Soil & Plant Science, 64*(1), 85–95. https://doi.org/10.1080/09064710.2014.887141.

Mu, L., Fang, L., & Liang, Y. (2016). Temporal and spatial variation of soil respiration under mulching in a greenhouse cucumber cultivation. *Pesquisa Agropecuária Brasileira, 51*(7), 869–879. https://doi.org/10.1590/S0100-204x2016000700010.

Muñoz, K., Buchmann, C., Meyer, M., Schmidt-Heydt, M., Steinmetz, Z., Diehl, D., Thiele-Bruhn, S., & Schaumann, G. E. (2017). Physicochemical and microbial soil quality indicators as affected by the agricultural management system in strawberry cultivation using straw or black polyethylene mulching. *Applied Soil Ecology, 113*, 36–44. https://doi.org/10.1016/j.apsoil.2017.01.014.

Muroi, F., Tachibana, Y., Kobayashi, Y., Sakurai, T., & Kasuya, K. (2016). Influences of poly(butylene adipate-*co*-terephthalate) on soil microbiota and plant growth. *Polymer Degradation and Stability, 129*, 338–346. https://doi.org/10.1016/j.polymdegradstab.2016.05.018.

Murphy, J. (2001). *Additives for plastics handbooks*. New York: Elsevier Science Ltd.

Natural Resources Conservation Service (U.S. Department of Agriculture). (2019). *Soil Health* [Online]. Available: http://www.nrcs.usda.gov/wps/portal/nrcs/main/soils/health/. Accessed 31 Jan 2019.

Núñez-Zofío, M., Larregla, S., & Garbisu, C. (2011). Application of organic amendments followed by soil plastic mulching reduces the incidence of *Phytophthora capsici* in pepper crops under temperate climate. *Crop Protection, 30*(12), 1563–1572. https://doi.org/10.1016/j.cropro.2011.08.020.

Oehlmann, J., Schulte-Oehlmann, U., Kloas, W., Jagnytsch, O., Lutz, I., Kusk, K. O., Wollenberger, L., Santos, E. M., Paull, G. C., Van Look, K. J., & Tyler, C. R. (2009). A critical analysis of the biological impacts of plasticizers on wildlife. *Philosophical Transactions of the Royal Society B, Biological Sciences, 364*(1526), 2047–2062. https://doi.org/10.1098/rstb.2008.0242.

Paul, E. A., Paustian, K., Elliott, E. T., & Cole, C. V. (1997). *Soil organic matter in temperate agroecosystems: Long-term experiments in North America*. Boca Raton: Cambridge University Press.

Pereira, W., Crabtree, G., & William, R. D. (1987). Herbicide action on purple and yellow nutsedge (*Cyperus rotundus* and *C. esculentus*). *Weed Technology, 1*(1), 92–98. https://doi.org/10.1017/s0890037x00029201.

Pfister, B. & Labowsky, H. J. (2003). Biodegradable materials and natural fibre composites in agriculture and horticulture: International symposium, June 2–4, 2002 in Hanover, Darmstadt, KTBL.

Pittol, M., Tomacheski, D., Simões, D. N., Ribeiro, V. F., & Santana, R. M. C. (2017). Antimicrobial performance of thermoplastic elastomers containing zinc pyrithione and silver nanoparticles. *Materials Research, 20*(5), 1266–1273. https://doi.org/10.1590/1980-5373-mr-2017-0137.

Plastics Technology. (2019). *Overview of the extrusion process* [Online]. Available: https://www.ptonline.com/knowledgecenter/Profile-Extrusion/profile-extrusion-fundamentals. Accessed 8 Feb 2019.

Posada, J. A., Naranjo, J. M., Lopez, J. A., Higuita, J. C., & Cardona, C. A. (2011). Design and analysis of poly-3-hydroxybutyrate production processes from crude glycerol. *Process Biochemistry (Amsterdam, Neth.), 46*(1), 310–317. https://doi.org/10.1016/j.procbio.2010.09.003.

Ram, A. (1997). *Fundamentals of polymer engineering*. Boston: Springer.

Roohi, Zaheer, M. R., & Kuddus, M. (2018). PHB (poly-β-hydroxybutyrate) and its enzymatic degradation. *Polymers for Advanced Technologies, 29*(1), 30–40. https://doi.org/10.1002/pat.4126.

Rudnik, E., & Briassoulis, D. (2011). Degradation behaviour of poly(lactic acid) films and fibres in soil under Mediterranean field conditions and laboratory simulations testing. *Industrial Crops and Products, 33*(3), 648–658. https://doi.org/10.1016/j.indcrop.2010.12.031.

Rychter, P., Biczak, R., Herman, B., Smylla, A., Kurcok, P., Adamus, G., & Kowalczuk, M. (2006). Environmental degradation of polyester blends containing atactic poly(3-hydroxybutyrate). Biodegradation in soil and ecotoxicological impact. *Biomacromolecules, 7*(11), 3125–3131. https://doi.org/10.1021/bm060708r.

Santagata, G., Schettini, E., Vox, G., Immirzi, B., Mugnozza, G. S., & Malinconico, M. (2017). Biodegradable spray mulching and nursery pots: New frontiers for research. In *Soil degradable bioplastics for a sustainable modern agriculture* (pp. 105–137). https://doi.org/10.1007/978-3-662-54130-2_5.

Shah, A. A., Hasan, F., Hameed, A., & Ahmed, S. (2008). Biological degradation of plastics: A comprehensive review. *Biotechnology Advances, 26*(3), 246–265. https://doi.org/10.1016/j.biotechadv.2007.12.005.

Shen, L., Worrell, E., & Patel, M. (2010). Present and future development in plastics from biomass. *Biofuels, Bioproducts & Biorefining, 4*(1), 25–40. https://doi.org/10.1002/bbb.189.

Sherman, L. M. (2008). Additives are needed for toughness, heat resistance & processability. *Plastics Technology, 54*(7), 58–63.

Shi, B., & Palfery, D. (2010). Enhanced mineralization of PLA meltblown materials due to plasticization. *Journal of Polymers and the Environment, 18*(2), 122–127. https://doi.org/10.1007/s10924-010-0190-3.

Shi, B., & Palfery, D. (2012). Temperature-dependent polylactic acid (PLA) anaerobic biodegradability. *International Journal of Environment and Waste Management, 10*(2/3), 297–306. https://doi.org/10.1504/ijewm.2012.048324.

Shogren, R. L. (2000). Biodegradable mulches from renewable resources. *Journal of Sustainable Agriculture, 16*(4), 33–47. https://doi.org/10.1300/J064v16n04_05.

Shogren, R. L., & Hochmuth, R. C. (2004). Field evaluation of watermelon grown on paperpolymerized vegetable oil mulches. *HortScience, 39*(7), 1588–1591. https://doi.org/10.21273/hortsci.39.7.1588.

Siegenthaler, K. O., Kuenkel, A., Skupin, G., & Yamamoto, M. (2012). Ecoflex and Ecovio: Biodegradable, performance-enabling plastics. In B. Rieger, A. Künkel, G. Coates, R. Reichardt, E. Dinjus, & T. Zevaco (Eds.), *Synthetic biodegradable polymers* (pp. 91–136). Berlin/Heidelberg: Springer. https://doi.org/10.1007/12_2010_106.

Singh, P., Saengerlaub, S., Abas Wani, A., & Langowski, H.-C. (2012). Role of plastics additives for food packaging. *Pigment & Resin Technology, 41*(6), 368–379. https://doi.org/10.1108/03699421211274306.

Sintim, H. Y., Bandopadhyay, S., English, M. E., Bary, A. I., DeBruyn, J. M., Schaeffer, S. M., Miles, C. A., Reganold, J. P., & Flury, M. (2019). Impacts of biodegradable plastic mulches on soil health. *Agriculture, Ecosystems & Environment, 273*, 36–49. https://doi.org/10.1016/j.agee.2018.12.002.

Siwek, P., Domagala-Swiatkiewicz, I., & Kalisz, A. (2015). The influence of degradable polymer mulches on soil properties and cucumber yield. *Agrochimica, 59*(2), 108–123. https://doi.org/10.12871/0021857201522.

Smith, P. B. (2015). Bio-based sources for terephthalic acid. In H. N. Cheng, R. A. Gross, & P. B. Smith (Eds.), *Green polymer chemistry: Biobased materials and biocatalysis* (ACS Symposium Series) (Vol. 1192, pp. 453–469). American Chemical Society. https://doi.org/10.1021/bk-2015-1192.ch027.

Snowdon, M. R., Mohanty, A. K., & Misra, M. (2014). A study of carbonized lignin as an alternative to carbon black. *ACS Sustainable Chemistry & Engineering, 2*(5), 1257–1263. https://doi.org/10.1021/sc500086v.

Sodergard, A., & Stolt, M. (2010). Industrial production of high molecular weight poly(lactic acid). In R. Auras, L. T. Lim, S. E. M. Selke, & H. Tsuji (Eds.), *Poly(Lactic Acid): Synthesis, structures, properties, processing, and applications* (pp. 27–41). Hoboken: Wiley. https://doi.org/10.1002/9780470649848.ch3.

Soil Health Institute. (2019). *North American project to evaluate soil health measurements.* [Online]. The Soil Health Institute, Morrisville. Available: https://soilhealthinstitute.org/north-american-project-to-evaluate-soil-health-measurements. Accessed 5 Feb 2019.

Sollins, P., Homann, P., & Caldwell, B. A. (1996). Stabilization and destabilization of soil organic matter: Mechanisms and controls. *Geoderma, 74*(1), 65–105. https://doi.org/10.1016/S0016-7061(96)00036-5.

Soroka, W. (2002). *Fundamentals of packaging technology*. Naperville: Institute of Packaging Professionals.

Srivastava, R. K., Rahman, Q., Kashyap, M. P., Singh, A. K., Jain, G., Jahan, S., Lohani, M., Lantow, M., & Pant, A. B. (2012). Nano-titanium dioxide induces genotoxicity and apoptosis in human lung cancer cell line, A549. *Human & Experimental Toxicology, 32*(2), 153–166. https://doi.org/10.1177/0960327112462725.

Staub, C. (2018). Ag plastics recycling operation branches out [Online]. Available: https://resource-recycling.com/recycling/2018/05/30/details-on-a-growing-ag-plastics-recycling-operation/. Accessed 9 Feb 2019.

Steinmetz, Z., Wollmann, C., Schaefer, M., Buchmann, C., David, J., Tröger, J., Muñoz, K., Frör, O., & Schaumann, G. E. (2016). Plastic mulching in agriculture. Trading short-term agronomic benefits for long-term soil degradation? *Science of the Total Environment, 550*, 690–705. https://doi.org/10.1016/j.scitotenv.2016.01.153.

Tachibana, Y., Kimura, S., & Kasuya, K.-i. (2015). Synthesis and verification of biobased terephthalic acid from furfural. *Scientific Reports, 5*, 8249. https://doi.org/10.1038/srep08249.

Touchaleaume, F., Martin-Closas, L., Angellier-Coussy, H., Chevillard, A., Cesar, G., Gontard, N., & Gastaldi, E. (2016). Performance and environmental impact of biodegradable polymers as agricultural mulching films. *Chemosphere, 144*, 433–439. https://doi.org/10.1016/j.chemosphere.2015.09.006.

Tullo, A. H. (2012). Old plastics, fresh dirt. *Chemical and Engineering News, 90*(12), 12–18.

U.S. Department Agriculture. (2015). *Report on biodegradable biobased mulch films* [Online]. Available: https://www.ams.usda.gov/sites/default/files/media/Biobased%20mulches%20report.pdf. Accessed 10 Feb 2019.

U.S. Department of Agriculture. (2008). *Soil quality test kit guide* [Online]. Available: https://www.nrcs.usda.gov/wps/portal/nrcs/detail/soils/health/assessment/?cid=nrcs142p2_053873. Accessed 10 Feb 2019.

U.S. Environmental Protection Agency. (2012). Report to congress on black carbon (EPA-450/R-12-001) [Online]. Available: https://nepis.epa.gov/. Accessed 10 Feb 2019.

U.S. Senate Committee on Agriculture Nutrition and Forestry. (2006). *Energy title (Title IX) of the farm security and rural investment act of 2002* [Online]. Available: http://www.agriculture.senate.gov/imo/media/doc/107-171%20-%20Farm%20Security%20And%20Rural%20Investment%20Act%20Of%202002.pdf. Accessed 10 Sept 2018.

Vox, G., Santagata, G., Malinconico, M. I., Mmirzi, B., Scarascia Mugnozza, G., & Schettini, E. (2013). Biodegradable films and spray coatings as eco-friendly alternative to petro-chemical derived mulching films. *Journal of Agricultural Engineering, 44*(2s), 221–225. https://doi.org/10.4081/jae.2013.286.

Vroman, I., & Tighzert, L. (2009). Biodegradable polymers. *Materials, 2*(2), 307–344. https://doi.org/10.3390/ma2020307.

Wang, Y., Zhu, X., Lao, Y., Lv, X., Tao, Y., Huang, B., Wang, J., Zhou, J., & Cai, Z. (2016a). TiO_2 nanoparticles in the marine environment: Physical effects responsible for the toxicity on algae. *Phaeodactylum tricornutum, Science of The Total Environment, 565*(1), 818–826. https://doi.org/10.1016/j.scitotenv.2016.03.164.

Wang, Y. P., Li, X. G., Fu, T. T., Wang, L., Turner, N. C., Siddique, K. H. M., & Li, F. M. (2016b). Multi-site assessment of the effects of plastic-film mulch on the soil organic carbon balance in semiarid areas of China. *Agricultural and Forest Meteorology, 228*(1), 42–51. https://doi.org/10.1016/j.agrformet.2016.06.016.

Webster, T. M. (2017). Patch expansion of purple nutsedge (*Cyperus rotundus*) and yellow nutsedge (*Cyperus esculentus*) with and without polyethylene mulch. *Weed Science, 53*(6), 839–845. https://doi.org/10.1614/WS-05-045R.1.

Weng, Y.-X., Wang, L., Zhang, M., Wang, X.-L., & Wang, Y.-Z. (2013). Biodegradation behavior of P(3HB,4HB)/PLA blends in real soil environments. *Polymer Testing, 32*(1), 60–70. https://doi.org/10.1016/j.polymertesting.2012.09.014.

Wortman, S. E., Kadoma, I., & Crandall, M. D. (2016). Biodegradable plastic and fabric mulch performance in field and high tunnel cucumber production. *HortTechnology, 26*(2), 148–155. https://doi.org/10.21273/horttech.26.2.148.

Yamamoto-Tamura, K., Hiradate, S., Watanabe, T., Koitabashi, M., Sameshima-Yamashita, Y., Yarimizu, T., & Kitamoto, H. (2015). Contribution of soil esterase to biodegradation of aliphatic polyester agricultural mulch film in cultivated soils. *AMB Express, 5*(1), 10. https://doi.org/10.1186/s13568-014-0088-x.

Yousif, E., & Haddad, R. (2013). Photodegradation and photostabilization of polymers, especially polystyrene: Review. *SpringerPlus, 2*(1), 398. https://doi.org/10.1186/2193-1801-2-398.

Zhang, M., & Thomas, N. L. (2011). Blending polylactic acid with polyhydroxybutyrate: The effect on thermal, mechanical, and biodegradation properties. *Advances in Polymer Technology, 30*(2), 67–79. https://doi.org/10.1002/adv.20235.

Zhang, G., Zhang, X., & Hu, X. (2013). Runoff and soil erosion as affected by plastic mulch patterns in vegetable field at Dianchi lake's catchment, China. *Agricultural Water Management, 122*, 20–27. https://doi.org/10.1016/j.agwat.2013.02.004.

Zhang, Y., Rempel, C., & McLaren, D. (2014). In J. H. Han (Ed.), *Chapter 16 – thermoplastic starch in innovations in food packaging* (2nd ed., pp. 391–412). San Diego: Academic. https://doi.org/10.1016/b978-0-12-394601-0.00016-3.

Zhang, F., Li, M., Qi, J., Li, F., & Sun, G. (2015). Plastic film mulching increases soil respiration in ridge-furrow maize management. *Arid Land Research and Management, 29*(4), 432–453. https://doi.org/10.1080/15324982.2015.1018456.

Zhang, F., Zhang, W., Li, M., Yang, Y., & Li, F.-M. (2017a). Does long-term plastic film mulching really decrease sequestration of organic carbon in soil in the loess plateau? *European Journal of Agronomy, 89*, 53–60. https://doi.org/10.1016/j.eja.2017.06.007.

Zhang, F., Zhang, W., Li, M., Zhang, Y., Li, F., & Li, C. (2017b). Is crop biomass and soil carbon storage sustainable with long-term application of full plastic film mulching under future climate change? *Agricultural Systems, 150*, 67–77. https://doi.org/10.1016/j.agsy.2016.10.011.

Zhu, D., Chen, Q.-L., An, X.-L., Yang, X.-R., Christie, P., Ke, X., Wu, L.-H., & Zhu, Y.-G. (2018). Exposure of soil collembolans to microplastics perturbs their gut microbiota and alters their isotopic composition. *Soil Biology and Biochemistry, 116*, 302–310. https://doi.org/10.1016/j.soilbio.2017.10.027.

Chapter 12
Performance of Bio-Based Polymeric Agricultural Mulch Films

Danila Merino, Andrea Y. Mansilla, Claudia A. Casalongué, and Vera A. Alvarez

Abstract In order to improve yields and crop traits, polymeric films are commonly used to cover soils like agricultural mulches. Its use provides important benefits to farmers and the environment, as they maintain humidity, temperature and prevent the growth of weeds. However, the most widely used mulches are mainly made of low-density polyethylene. The use of PE films or other non-biodegradable polymers present great environmental and economic disadvantages for responsible farmers, who must remove them before the next crop period. The objective of this chapter is to discuss the main properties presented by the biodegradable and bio-based materials that have recently been proposed for this application, what is their effect on the yields of different crops in which they have been tested, as well as their biodegradation time and their effect on the soil quality will also be analyzed. Finally, based on these findings and the recently approved new standard for biodegradable agricultural mulches, the materials that will set trends in the future will be remarked.

Keywords Biodegradable Agricultural Mulch (BDM) · Biodegradation · Biopolymers · Soil quality

12.1 Introduction

Polymeric materials began to be used in agriculture from the 1940s, after the Second World War. Until then, the crops were not produced out of season, but the demanding situation of fresh products of high quality led to the development of new techniques

D. Merino (✉) · V. A. Alvarez
Grupo de Materiales Compuestos Termoplásticos (CoMP), Instituto de Investigaciones en Ciencia y Tecnología de Materiales (INTEMA), Facultad de Ingeniería, Universidad Nacional de Mar del Plata (UNMdP) y Consejo Nacional de Investigaciones Científicas y Técnicas (CONICET), Mar del Plata, Argentina
e-mail: danila.merino@fi.mdp.edu.ar

A. Y. Mansilla · C. A. Casalongué
Instituto de Investigaciones Biológicas, Facultad de Ciencias Exactas y Naturales, UE UNMdP y CONICET, Mar del Plata, Argentina

T. J. Gutiérrez (ed.), *Polymers for Agri-Food Applications*,
https://doi.org/10.1007/978-3-030-19416-1_12

that allow obtaining food throughout the year. This situation led to the gradual introduction of plastics to replace the inflexible, heavy and expensive glass in the greenhouses. Since then, the use of plastics in agriculture has been increasing (Robinson and Brae 1991). Today, they are used in soil fertilization, as controlled release systems for agrochemicals and nutrients, in seed coatings, as well as agricultural mulches and tunnels for the protection of plants (Mitrus et al. 2010). However, the polymers traditionally used are derived from petroleum and do not biodegrade, i.e. are accumulated in the environment, so, for several decades, researchers have tried to develop new formulations that can replace them, using polymers of natural origin that can be biodegraded and in turn improve the state of soils (Mitrus et al. 2010).

The word "mulch" has German origin and means soft, flexible or begins to crumble or deteriorate. The mulches are materials that are applied on the surface of the soil, while mulching is a technique used to conserve water, since an increase in the amount of water retained by the soil causes a reduction in its erosion (Kader et al. 2017). In this way, agricultural mulches are materials that are placed on the ground before sowing or transplanting to maintain their humidity and temperature and prevent the growth of weeds (Mitrus et al. 2010). The use of agricultural mulches provides important benefits for farmers and the environment, since it allows reducing the use of water for irrigation and the use of fertilizers and herbicides, helps shorten harvest time and increase yields, thus contributing to the development of farmers around the world (Mitrus et al. 2010; Miles et al. 2012).

The choice of suitable mulch for a given crop depends on weather conditions, crop management practices and, of course, the type of crop (Kader et al. 2017). Because of the multiple advantages of the mulching technique, it is widely used all over the world to improve yields and crop conditions. The materials used as mulches are classified mainly into three large groups (Kader et al. 2017):

- Organic: agricultural waste such as straw, stems, husks and manure,
- Inorganic: plastic films based on petroleum, as well as biodegradable or photodegradable synthetic polymers, and
- Special: concrete and sand.

Each of them has particular characteristics and their choice will depend on factors such as climate and the cost-benefit ratio (Kader et al. 2017). In general, plastic mulches have the highest capacity for soil moisture retention compared to organic ones. The special ones, on the other hand, present other types of disadvantages, such as the loss of nutrients when sand is used, and the high price involved in its construction. In addition, organic mulches may contain seeds that end up germinating and acting as weeds and are not suitable for preventing the growth of weeds compared to plastic mulches (Kader et al. 2017). Organic mulches, being more permeable than inorganic ones, can increase soil moisture during periods of abundant rainfall (Wortman et al. 2015). An additional problem presented by these mulches is that they do not prevent the growth of weeds unless the applied layer is sufficiently thick. However, this generates another drawback that is a low temperature of the soil and, consequently, a reduction in crop yield (Wortman et al. 2015). Although paper mulches have given yields like those obtained with polyethylene (PE) for vegetable crops, they degrade rapidly during the growth stage of the plant and mainly in the buried areas of the mulch, so that the growth of weeds and the yield of crops decreases. Temperatures are also lower under this type of mulches (Wortman et al. 2015).

In this way, farmers around the world prefer polymeric mulches. In 2011, 2, 9 million tons were used, of which 545,000 tons were used only in Europe (mainly for applications in horticultural crops) (D'Avino et al. 2015) and 120,000 tons were used in China (Gu et al. 2017).

The use of PE films or other non-biodegradable polymers presents significant environmental and economic disadvantages for responsible farmers. It is necessary after the harvests to remove the mulch films to avoid problems with the crops of the next period, which entails great costs for the farmers who must use labor, equipment and infrastructure in their harvesting (Mitrus et al. 2010). According to Miles et al. (2012), in 2004 in the United States, the use of mulch films cost USD 250 *per* hectare (Miles et al. 2012). In addition, many farmers do not know what to do with this waste, so some of them eliminate their plastic waste in local landfills, others openly burn them, causing inconveniences to the environment and the health of the local residents (Miles et al. 2012), and others incorporate them in the soil in the tillage, which entails a serious risk to the environment, since PE is accumulated in the soil, which interferes with the root growth of the crop in the next period (Briassoulis and Dejean 2010). Additionally, plastic waste from agricultural mulches is usually polluted with traces of soil and plants and, mainly, with toxic chemicals that are sprayed during use. These materials under these conditions and considering that they have suffered prolonged exposure to sunlight, present a significant decline in their properties and makes recycling difficult, mainly due to the high costs involved in this process and the low quality of the product that would be obtained (Martín-Closas et al. 2017).

In this context, the use of biodegradable agricultural mulch (BDM) films or with short useful life would be highly desirable (Mitrus et al. 2010). It has been initially attempted to reduce the disadvantages associated with this practice by reducing the shelf life of the PE films by means of different strategies including the addition of photodegradation catalyzing substances such as nickel and ferric dibutylthiocarbamates, or a combination of substituted benzophenones and titanium or zirconium chelates, or the mixture with biodegradable polymers such as starch (Imam et al. 2012). In 1973, the aggregate of granular starch in low percentages (6–7% by weight) was made for the first time (Griffin 1974) and, in 1977 the first mixtures were made using plasticized starch (Otey et al. 1977). However, the studies carried out in these mixtures showed biodegradation, due to the fact that the starch left the PE matrix weakened (Imam et al. 2012; Halley et al. 2001). On the other hand, photodegradable films have the same drawback. After the fragmentation of the material, small non-compostable plastic residues are obtained and accumulated, thus making their disposal even more difficult (Halley et al. 2001).

Several authors have reported that the benefits of mulching in the short term could be transformed into long-term environmental problems (Wortman et al. 2015). The disadvantages would be related to the quality of the soil that would be obtained after years of carrying out this type of agricultural practices. In particular,

Steinmetz et al. (2016) reported that although the long-term environmental effects are not known, this practice could produce an accelerated degradation of soil organic matter, which in turn would be directly related to the loss of nutrients (Mg, K, N) and, particularly, of soil carbon. In addition, this practice could contribute to the development of water repellency of the soil (loss of water retention capacity), together with an increase in the runoff of pesticides and other agrochemicals, mainly in furrows. It has also been reported that plants grown under these conditions show a significant increase in the accumulation of heavy metals such as Zn, Cd, Cu and Pb. Another alarming impact includes changes in the microbial community of the soil that could involve increases in the production of mycotoxins. These effects on the soil microbiota could be seen, in turn, reflected in the diversity of higher organisms such as insects and earthworms (Steinmetz et al. 2016).

Starting in the 1980s, new fully or partially biodegradable polymers began to be developed from renewable sources, however, their use in agriculture did not begin until the late 1990s (Martín-Closas et al. 2017). In this way, during the last 20 years approximately, numerous formulations have been proposed and evaluated for this purpose and some of them have even reached the commercial market. Some of the formulations include polymers such as polybutylene adipate terephthalate (PBAT), polyhydroxyalkanoates (PHA), polylactic acid (PLA), thermoplastic starch (TPS) and their mixtures (Martín-Closas et al. 2017). This type of materials has ester-type bonds or are polysaccharides that can be easily hydrolyzed by microorganisms and converted into biomass, carbon dioxide (CO_2) and water (Bandopadhyay et al. 2018). In this way, BDMs have two fundamental advantages that are: (1) the possibility of being adjusted to the shelf life to the crop cycle and (2) the biodegradation is produced in the place where they were used, therefore it is not necessary to remove them from the soil after use (Martín-Closas et al. 2017). However, biodegradation in acceptable times is not easily controllable and is often unpredictable and incomplete (Bandopadhyay et al. 2018).

BDMs also have radiometric properties similar to those of PE, in fact they produce a significant increases in soil temperature relative to bare soil and this produces higher yields and growth rates of the crop compared with the bare soil (Wortman et al. 2015). However, they present a lower performance, including a lower mechanical properties compared to PE, making them more susceptible to breakage, tears and holes and, therefore, allow the growth of weeds (Wortman et al. 2015) and, in addition, they produce a lower increase in soil temperature and presents a lower water vapor barrier capacity (Bandopadhyay et al. 2018).

The replacement of PE by other biodegradable polymers such as polycaprolactone (PCL) or PLA and its combination with starch, allow to obtain biodegradable films, although these have some disadvantages. Although the PLA is a brittle polymer from renewable sources, PCL is a synthetic polymer derived from petroleum and both materials require future research to reduce its cost and develop formulations to optimize its properties (Halley et al. 2001).

Bio-based polymers are those polymers obtained from renewable sources, which can be acquired directly from natural materials such as plants, e.g. starch and cellulose; produced by the polymerization of monomers obtained from nature, e.g. PLA;

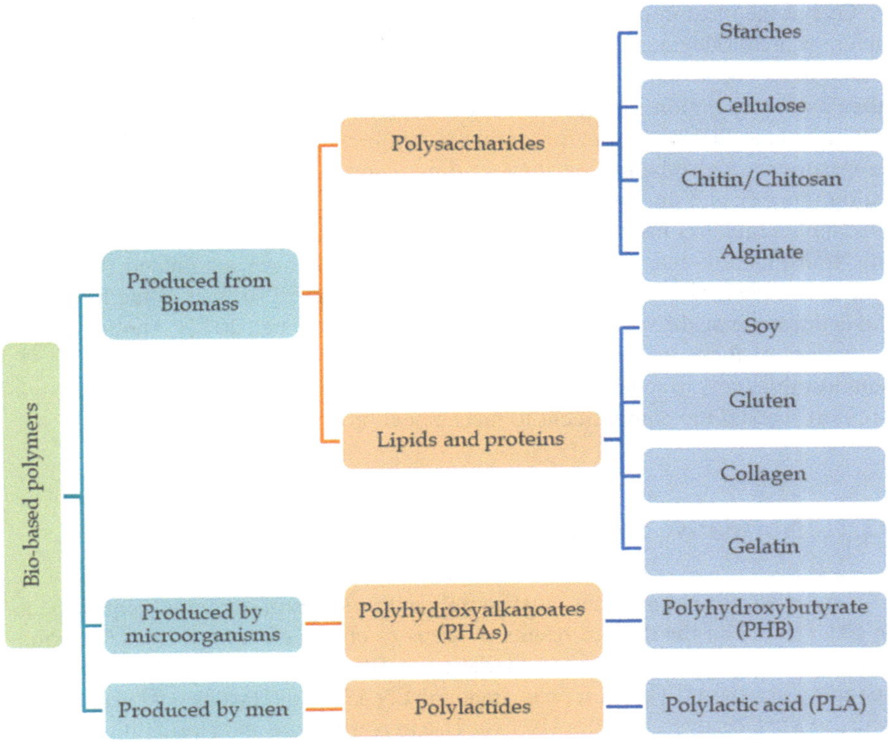

Fig. 12.1 Classification of bio-based polymers according to their method of production

or by microorganisms or bacteria, e.g. PHAs (Fig. 12.1). The most sustainable bio-based polymers are those extracted from plants such as starch, cellulose, pectins and proteins (Mitrus et al. 2010). It is important to mention at this point that the production of bio-based polymers does not represent a threat to food production. According to the 2016 bioplastics market data report carried out by "European Bioplastics", the land used in the production of bio-plastics during 2004 represented 0.01% of the total global agricultural area, while by 2019 this value will doubled, i.e. it will only represent 0.02% (European Bioplastics 2016).

Polymers such as PLA and PHAs are the main responsible for the market growth of biodegradable and bio-based polymers. These polymers are very versatile and offer a wide range of physicochemical and mechanical properties depending on their chemical composition. It is expected that PHAs production could be tripled by 2022, while that of PLA could be increased by 50% up to the same year, compared to 2017 (European Bioplastics e.V 2018a). However, these polymers have deficiencies in their mechanical properties that can be overcome, e.g. by the formation of blends with other polymers (Ghimire et al. 2018). According to Ghimire et al. (2018), the bio-based polymers that are most used for the formulation of BDMs are starch, PLA and PHAs, the former is exceptional because of its low cost.

One of the most important design criteria as mentioned above for the mulches is that they are biodegradable (Briassoulis 2006). It is ideally waiting for its biodegradation to be 100% before the cultivation of the next period begins and that it occurs after its incorporation into the soil in the tillage, without the need to be removed (Ghimire et al. 2018). A series of minimum requirements related to their mechanical properties must also be met, as they must maintain their integrity during the installation process, and during the aging of the film (Briassoulis 2006). The process of placing the mulches has been completely mechanized. In fact, the process of planting, transplanting and placing the irrigation pipes has been automated (Robinson and Brae 1991). This process involves at least mechanical stresses in the longitudinal and transverse direction of the films (Martín-Closas et al. 2017). Another design parameter of these materials is their thickness. The idea is to be able to use the smallest thickness to reduce costs, increase the rate of biodegradation, and also taking into account the commitment to mechanical properties (Briassoulis 2006).

12.2 Normative and Regulations

The materials used as mulches must comply with a series of minimum requirements in order to obtain the desired results in terms of crop yield and quality. A thermoplastic film ready to be used and marketed as BDM must comply with the corresponding standards. The European standard EN 13655: "Plastics – Thermoplastic Mulch Films Recoverable after Use, For Use in Agriculture and Horticulture" applicable to PE mulches or their copolymers, specifies the requirements for the optical and mechanical properties of films used in agriculture and horticulture: transparent, transparent and diffuse (thermal), white and black films with thicknesses between 10 μm and up to 250 μm (Briassoulis and Giannoulis 2018). In January 2018, a new standard was approved in Europe, EN 17033: "Plastics: biodegradable mulch films for use in agriculture and horticulture: requirements and test methods", which is specific for agricultural mulches based on biodegradable materials. The standard was designed to be a clear reference for farmers, distributors and stakeholders, and to be the basis for further certification and according labels for biodegradable mulch films. EN 17033 is likely to replace other pre-existing national standards in Europe. The development of this new standard began in 2012 as a result of the revision of the previous standard EN 13655 (PRI/75 2002) for conventional mulch films, which had to be divided into two standards in order to consider for the most appropriate end-of-life option: (a) for mulch films that must be recovered after use (revised EN 13655) and (b) for mulch films that can be incorporated into the soil due to their property of biodegradation (new EN 17033). The standard regulates its composition, biodegradability in soil, ecotoxicity, mechanical and optical properties, and test procedures for each of the categories listed (Hayes and Flury 2018; European Bioplastics 2018). The responsible European Committee CEN/TC 249/WG 7 (Thermoplastic films for use in agriculture) decided to use the pre-existing and well-established certification "OK Biodegradable soil" as basis for the new standard, which requires a 90% CO_2 conversion within 24 months in a soil biodegradation test. The standard also includes a new and more comprehensive ecotoxicity testing and evaluation scheme considering relevant terrestrial organism groups such as plants, invertebrates (e.g. earthworm) and microorganisms (e.g. nitrification inhibition test). In addition, the standard strictly defines the restrictions of use regarding different potentially harmful constituents, such as regulated metals and substance of very high concern. Specifically, the materials used for the preparation of agricultural mulches should be tested in the laboratory to establish the concentration of heavy metals (EN 17294–2), which should not exceed the amounts of 150 mg of Zn/kg, 50 mg of Cr, Cu and Pb/kg, 25 mg of Ni/kg and 0.5 mg of Cd and Hg/kg and must have less than 0.1% of dangerous substances. With respect to ecotoxicity, testing must ensure more than 90% of seed germination and plant growth compared to bare soil, less than 10% difference in earthworm mortality rate and the amount of biomass compared to uncovered soil and the nitrification of the bacteria should be ≥80% of those reached for free soil (Hayes and Flury 2018).

The standard also specifies the dimensional evaluation of the films, as well as the optical properties for weed control and mechanical properties. Although biodegradable mulch films only must be strong enough to be placed in the field, they must also have the mechanical properties necessary for their recovery. Accordingly, a recommendation to only use conventional mulch film thicker than 25 μm was added into the scope of the relevant standard EN 13655 when revised to make sure they can be collected after use. In addition, the standard recommends to clearly mark biodegradable mulch films and their packaging and to include a reference to the standard, in order to allow farmers to be certain the mulch films will disintegrate and biodegrade without leaving harmful residue in the soil. The standard EN 17033 also contains nine annexes that include detailed specifications on how to perform the different tests mentioned in the standard, and provides relevant information on the classification of films depending on the crops, as well as relevant recommendations for farmers on the application of biodegradable mulch films (CEN/TC 249 – Plastics EN 17033 2018).

The BDMs must have the capacity to withstand the stresses generated during the laying process, resist the propagation of cracks during the drilling process and, at the same time, be as thin as possible to reduce their cost. The standard EN 17033, classifies to the quilted materials according to their thickness, as follows: < 10 μm, between 10 and 15 μm and > 15 μm (Briassoulis and Giannoulis 2018). Another important factor is the aging of these films, since this produces a reduction in their mechanical properties and the material must maintain its integrity in a specific period of time according to the crop (Hayes et al. 2017). Other important properties are the penetration resistance, impact resistance and propagation of tears (Briassoulis and Giannoulis 2018). However, in the field of scientific research, such complete studies are not always carried out. Many times, only the mechanical properties are reported and, sometimes, these properties are reported along with the water vapor barrier or radiometric properties. The search for the most suitable materials for this application can be difficult because, although these may have adequate properties, in most cases, the results of the tests of composition, ecotoxicity and biodegradability should still be considered (Gutiérrez 2018).

12.3 State of the Art of Mechanical Properties of Biodegradable Mulches. What We Have Achieved?

A quite complete analysis and carried out within the framework of the current standards to evaluate the performance of the commercial polymers Mater-Bi®, Ecovio and linear low density polyethylene (LLDPE) was reported by Briassoulis (Briassoulis and Giannoulis 2018). Mater-Bi® is a bio-based material composed of starch and other fully biodegradable and compostable aliphatic/aromatic polyesters developed by Novamont. It has properties very similar to those of the PE and is recommended in a thickness of 15 μm with addition of carbon black for its application as mulch. Ecovio is a mixture of biodegradable polymers, Ecoflex (copolyester) and PLA produced by BASF. The control was carried out with 20 μm linear low-density PE from Plastika Kritis. The analyzed materials are commercial and have already passed the tests and adaptation to standards, although prior to EN 17033. According to the reported results for the tensile tests in the machine direction (MD) and the transverse direction (TD), Mater-Bi® showed a tensile strength of MD/TD = 41.0/34.2 MPa, Ecovio of 31.2/22.5 MPa and LLDPE of 33.6/27.3 MPa, while the elongation at break was in the MD/TD = 400/550% for Mater-Bi®, 100/500% for Ecovio and 400/630% for LLDPE. All the materials presented a tensile strength higher than 16/9 MPa, thus satisfying EN 17033, while the LLDPE was adjusted, on the other hand, to the standard EN 13655, for films with thicknesses less than 30 μm (Briassoulis and Giannoulis 2018). Regarding the elongation at break, 180% is expected for black conventional mulches with thickness less than 30 μm (according to EN 13655) and for biodegradable mulches of 15 μm, an elongation at break in MD/TD = 150/300%.

Other materials evaluated in bibliography, however, have not achieved the required mechanical properties. PLA composites (Dow Cargill, 148,000 Da) with 10% or 25% w/w Osage Orange (OO) wood designed to decrease the cost of PLA and act as a matrix for the controlled release of naturally occurring phytoactive agents within the wood had tensile strengths less than 50 MPa (do not specify the direction of the test) for experiments with different OO particle sizes and conditioning: dry (50% RH) and wet (95% RH). The elongation at break was in all cases less than 20% (Finkenstadt and Tisserat 2010). On the other hand, starch is the main renewable source used for the preparation of biodegradable and economical thermoplastic films (European Bioplastics e.V 2018b). However, its use alone is not recommended due to the fragility and high moisture sensitivity. The effect of chemical modifications (Merino et al. 2018a; b) and the formation of nanocomposites (Merino et al. 2018c; 2019) on the resulting properties has been studied, but even so they have not been sufficient. It is generally recommended that starch be combined with other polymers, such as PLA and PHAs (bio-based) or PBAT, polybutylene succinate (PBS) and PCL (petrochemical base) (Ghimire et al. 2018; Briassoulis 2004). Mixtures of 30% cassava TPS (Indemil) with PBAT (Ecoflex 7011, BASF) with or without carbon black aggregate and thicknesses of 123 and 218 μm, respectively, showed a tensile strength of 8.4 MPa for white and 6.1 MPa for black films, with elongations in all cases exceeding 300%. Rapid decay in the properties was also reported after the first week with the appearance of cracks from the second week onwards for the white films that allowed the growth of weeds (Bilck et al. 2010). On the other hand, TPS/PLA blends have been studied widely and several compatibilization strategies have been approached, obtaining in general very fragile materials, with elongations at break of less than 10%, such as for the mixture of PLA with grafted TPS with maleic anhydride or with epoxidized cardanol (Yang et al. 2015), or PLA with TPS grafted with maleic anhydride or PE grafted with glycol (Akrami et al. 2016), for the ternary mixture of PLA with TPS and PCL (Carmona et al. 2015), or for PLA/TPS with epoxidized soybean oil (Przybytek et al. 2018). Good results were found for PLA mixtures with 30% TPS and compatible with maleinized linseed oil up to the limit of that required by EN 17033 (Ferri et al. 2016).

PHAs are natural bacterial polymers produced by the fermentation of sugars (Bugnicourt et al. 2014; Kasirajan and Ngouajio 2012). Poly (3-hydroxybutyrate) (PHB) is the most commonly used and best characterized member of the PHA family (Bugnicourt et al. 2014). However, its use alone is not recommended, as it has a high brittleness characterized by an elastic modulus of 1.7 GPa, a tensile strength of 35 MPa and an elongation at break of only 10% (Briassoulis 2004). The pure copolymer polyhydroxybutyrate/valerate (PHB/V), with an elongation at break of less than 15%, a modulus of 1.2 GPa and a tensile strength of 25 MPa, is also fragile, which is why it is generally used in mixtures with other polymers or with the appropriate processing additives (Bugnicourt et al. 2014). For example, commercial PHAs such as Biomer have a tensile strength of 15–20 MPa and an elongation at break of 600–1200% (Bugnicourt et al. 2014), adapting to the mechanical stress requirements established in EN 17033. The suitable properties were also obtained for PLA/PHB mixtures plasticized with a lactic acid oligomer (Armentano et al. 2015) and PLA/PHB mixtures compatible with maleic anhydride (Jandas et al. 2014), while PHA/TPS mixtures, although showed improvements in their mechanical properties, do not reach the elongation required by the standard (Shogren 2009).

Greater efforts must be made in order to improve and adjust the properties of bio-based polymer films and their blends to the new standards. Economic estimates maintain that the world market of agricultural mulches will reach USD 4070 million in 2020 and that this will be dominated by LLDPE followed by low-density polyethylene (LDPE), being the main use of these mulches for horticultural crops (Briassoulis and Giannoulis 2018). This growth in the use of non-biodegradable mulches could be discouraging for ecosystems. The world market for BDMs will increase to USD 52.43 million by 2021 (Briassoulis and Giannoulis 2018). However, these materials still include petrochemical raw materials in percentages of around 80%. The main drawback of the use of biological raw materials lies in their price that is often up to three times higher than that of fossil resources (Ghimire et al. 2018), in that sense, economic and abundant starch could represent an alternative (Research 2015).

12.4 About the Biodegradation of Agricultural Mulch Films

Just like any natural organic material, such as leaves or straw, soil biodegradable polymers are assimilated as food and energy source by the microbial population present in soil. Bacteria and fungi use the organic carbon present in the biodegradable polymer to extract biochemical energy for driving their life processes by aerobic oxidation of readily utilizable organic carbon (Boelens 2017). The only (scientifically correct) parameter to quantify biodegradation is the conversion of organic carbon into CO_2. In terms of biodegradation, 'complete' does not mean '100% conversion to CO_2' as part of the organic carbon is also used for microbial growth (which cannot be accurately quantified). This biomass yield typically ranges from 10% to 40%, depending on the substrate.

$$C_{Polymer} + O_2 \rightarrow CO_2 + H_2O + C_{Residual} + C_{Biomass} \quad (12.1)$$

It is important to note that only the amount of CO_2 can be quantified in Eq. 12.1. $C_{Residual}$ and $C_{Biomass}$ cannot be quantified, since technically and analytically it is not possible to carry out this precisely.

Biodegradation of a polymer is not a uniform process. Instead, both the rate and the maximum level of biodegradation of a specific polymer are very much determined by the environmental conditions at the micro level. These environmental conditions can differ in terms of moisture content, oxygen availability, pH, temperature, types of microorganisms (bacteria, fungi and/or actinomycetes), etc. While the temperature and presence or absence of fungi can have a large impact on the level of biodegradation, other factors mainly influence on the rate of biodegradation (Bastoli 2014).

It is also worth mentioning that conventional PE mulch films are not biodegraded in the soil, thus they are accumulated over time, resulting in an average accumulation of 462.5 kg of mulch film (fragments) *per* hectare *per* decade, assuming a crop period *per* year and a film thickness of 20 μm (Boelens 2017).

12.4.1 Key Definitions

The literature associated with biodegradable polymer degradation and biodegradation is inconsistent with respect to the terms used to describe different stages and aspects of degradation (Mitrus et al. 2010). Some important definitions can be extracted from CEN/TR 15351:2006 (Kader et al. 2017) and ASTM D883–12 (Miles et al. 2012):

- ***Biodegradable***: status of a polymeric item that can be biodegraded.
- ***Biodegradable plastic***: a degradable plastic in which the degradation is due to the action of natural microorganisms such as bacteria, fungi and algae.

- ***Biodegradation***: degradation of a polymeric item due to cell-mediated phenomena.
- ***Biodisintegration***: disintegration resulting from the action of cells.
- ***Bioerosion***: faster degradation on the surface than in the interior as a result of biodegradation.
- ***Biofragmentation***: fragmentation of a polymeric item due to the action of cells.
- ***Bulk degradation***: faster degradation in the interior than in the surface of a polymeric item.
- ***Compostable plastic***: a plastic that undergoes biological degradation during composting to produce CO_2, water, inorganic compounds and biomass at a rate consistent with other known compostable materials and leaves no toxic or visually distinguishable residues.
- ***Degradable***: status of a polymeric item that can undergo degradation.
- ***Degradable plastic***: a plastic designed to undergo a significant change in its chemical structure under specific environmental conditions resulting in a loss of some properties that may vary as measured by standard test methods appropriate for the plastic and the application over a period of time that determines its classification.
- ***Degradation***: a detrimental change in the chemical structure, physical properties, or appearance of a polymer, which can result from chemical cleavage of the macromolecules forming a polymeric item, regardless of the mechanism of chain cleavage.
- ***Disintegration***: fragmentation to particles of an acceptable size (depending on the application).
- ***Erosion***: faster alteration at the surface than in the interior.
- ***Fragmentation***: breakdown of a polymeric item into particles regardless of the mechanism.
- ***Maximum degree of biodegradation***: maximum value of the degree of biodegradation that can be achieved under selected experimental conditions.
- ***Mineralization***: conversion of an organic compound to methane or CO_2 and water and other minerals.
- ***Theoretical degree of biodegradation***: theoretical value of the degree of biodegradation corresponding to total conversion of the organic matter present in an original polymer-based item to minerals and biomass.

All definitions are commonly used, and several times misused in the literature. It is important on the other hand to define how the biodegradation of a material can be measured.

By declaring that a plastic is biodegradable, it is necessary to establish the environment under which it is biodegradable. The ASTM International standard D6400 (Wortman et al. 2015), and a similar standard from the International Organization for Standardization (ISO 17088) (D'Avino et al. 2015), e.g. specify criteria for the biodegradability of plastics under industrial composting conditions based on standardized laboratory tests. ASTM D6400 uses ASTM D5338, a standardized test that measures the aerobic biodegradation of plastic materials under controlled composting conditions and specifies that 90% of carbon atoms must be mineralized in 180 days to CO_2 and water. The capacity of biodegradable mulch to meet those standards is the first and critical biodegradability test, since if mulch is not compostable, it is expected that it will not be biodegraded under field conditions (Gu et al. 2017).

The best-known test method to determine the biodegradation of a polymer in the soil is the international standard ISO 17556 Plastics – Determination of the ultimate aerobic biodegradability of plastic materials in soil by measuring the oxygen demand in a respirometer or the amount of CO_2 evolved (Briassoulis and Dejean 2010). This method prescribes that tests are considered valid only if the degree of biodegradation (based on carbon to CO_2 conversion) of the cellulose reference material has reached 60% or more at the plateau phase or at the end of the test.

12.4.2 Biodegradation of Bio-Based Polymeric Agricultural Mulch Films

Biodegradable materials can be integrated directly into the soil at the end of their useful life where the microflora converts them into methane or CO_2, water and biomass (European Bioplastics 2016; European Bioplastics e.V 2018a; Ghimire et al. 2018; Briassoulis 2006). A typical biodegradable mulch film has a thickness of 15 μm and a density of 1250 kg/m^3. As such, 187.5 kg of mulch film *per* crop *per* hectare is used. If complete biodegradation in soil has been obtained under laboratory conditions, all organic carbon within the polymer ($C_{Polymer}$) will be converted to CO_2 or used for biomass production. Under laboratory conditions (25 °C), this means that at least 90% of organic carbon has been converted to CO_2 in a maximum period of 2 years. In fact, the rate of biodegradation will depend on the environmental conditions (temperature, moisture content, etc.). It must, however, be noted that one of the reasons of using the mulching film is to increase the temperature of the soil. Nevertheless, even at temperatures below 25 °C, complete biodegradation is expected. The biodegradation profile of biodegradable polymers varies from one to the other, which makes it impossible to calculate the level of temporary accumulation in the soil. Accurately biodegradable material should be destroyed by microorganisms present in the soil, mineralized or bioassimilated (Briassoulis and Giannoulis 2018; PRI/75 2002; Hayes and Flury 2018).

Biodegradation is governed by several factors, including polymer characteristics (crystallinity, molecular weight, tacticity, functional groups, plasticizers, other additives, etc) and exposure conditions (Biotic: enzymes, biosurfactants, hydrofilycity and abiotic: pH, moisture, UV radiation and temperature) (European Bioplastics 2018; CEN/TC 249 – Plastics EN 17033 2018; Hayes et al. 2017; Finkenstadt and Tisserat 2010).

Several biopolymers are designed to be waste in landfills, compost or soil. These materials will be decomposed, as long as the required microorganisms are present, i.e. the original biota of the soil, water and everything necessary (European

Bioplastics e.V 2018b). To be biodegradable, some parts of the polymer backbone need to have similar natural substances, consequently, microorganisms will be able to use their enzymes to break the polymer chains at specific locations and use them as an energy source. Thus, in the case of starches, the microorganisms decompose the amylose and amylopectin macromolecules to glucose which will then be broken down into two molecules of pyruvic acid, which can be fermented more into lactic acid or converted aerobically into CO_2 to generate energy. Portions of sufficiently small polymers can be conveyed into microbial cells and the consumed as a food source. In the case of biodegradable polyesters, such as PLA, PHAs and PBAT, the ester bonds make them susceptible hydrolysis, which is a chemical degradation process. The hydrolysis produces a random scission of the main chain, causing a rapid reduction of the molecular weight, accelerating the biodegradation process because smaller molecules are more susceptible to the enzymatic reactions. Consequently, the chemical structure of the polymer is the key factor in determining whether polymers can be biodegraded, but also in the determination of their mechanism of biodegradation.

Biodegradation in the soil or compost depends on a complex synergy of biological and abiotic degradative processes. Some examples of degradation in soil of biodegradable mulches are summarized in the next paragraphs.

Cowan et al. (2013) studied the biodegradability of three potentially biodegradable plastic mulch products, Mater-Bi®-based black film (BioAgri: Corn starch and non-disclosed biopolymers; biodegradable and compostable: BIO group USA, Palm Harbor, FL), experimental polyhydroxy alkanoate film (Crown 1: Experimental polyhydroxy alkanoate film; biodegradable and compostable: Crown Films, Burlington, WA) and experimental spun bonded PLA fabric (SB-PLA-11: Experimental nonwoven spun bond, translucent: Feedstock from NatureWorks, Blair, NE; fabricated by Saxony Textiles, Chemnitz, Germany) finding that two of the mulch systems evaluated: BioAgri and Crown 1, deteriorated enough to comply with the recommendations of the National Organic Standards Board (NOSB) on biodegradation for certified organic production.

Jandas et al. (2013) completely prepared the BDM film from modified PLA by incorporating PLA and nanoclays (OMMT and C30B). The PLA mulch film was degraded in a period of time of 60 days in specific inoculums, but a reduction in more than 50% in the blends and nanocomposites was observed.

Hablot et al. (2014) prepared two mulches made from PLA using spun bond processing, one naturally white and the other black (SB-W and SB-B, respectively), and two *via* melt blown (MB) processing, from 100% PLA and a 75/25% w/w blend of PLA and polyhydroxy alkanoate [PHA; poly (3-hydroxybutyrate-*co*-4-hydroxyb utyrate); MB-PLA and MB-PLA + PHA, respectively]. Weathering under accelerated conditions were carried out, finding that biodegradation (91–93% in 90 days) almost coincides with the value obtained for the positive cellulose control.

The commercially available Ecovio mulch from BASF (BASF 2018) is a blend of Ecoflex and PLA which is optimized for agricultural applications. It can be used with a thickness of 10 μm without experiencing tears and perforations when placed in the field with the machine and offers biodegradable and compostable plastics that comply with European standard EN 13432 with regard to biodegradability, compostability, compost quality and compatibility with the plant.

All the examples described above of mulch based on different bio-based polymers meet the requirements of the standard EN 17033 in the terms of biodegradability. Alvarez et al. (2006) studied the degradation in soil of short sisal fibers/Mater-Bi®-Y biocomposites during indoor burial and showed that composites absorb less water than the matrix due to the presence of fiber-fiber and fiber-matrix interactions. These authors also shown that, because the matrix is mainly amorphous, the preferential removal of the starch was favored, making the material more bio-susceptible. However, this formulation was only degraded by 40% in a year of assay buried in the soil, thus being outside the standard EN 17033. The chemical composition can thus severally influence the biodegradation process of mulches.

Biodegradation in compost or in soil depends on a complex synergy of several factors: abiotic and biological degradative processes. The existence of standards that allow to determine the biodegradability of mulches is a crucial tool in the design of polymeric films for such agricultural application. It is also important to point out that compost and soil are not the only alternative for the end-of-life of mulches based on biodegradable polymers. PCL, PHB and PHBV have been successfully discarded as electron donors and solid substrates for bacterial nitrification of wastewater (Hiraishi and Khan 2003), as is PLA (Takahashi et al. 2011). Yoshida et al. (2013) shown that starch-based polymers provided the required electron donor, H_2, for polychlorinated polyaromatic hydrocarbon by microbial reductive dechlorination. The studies are mainly relevant to agriculture because abundant wastewater is produced from animal husbandry and because although fragments from agricultural mulch accumulate chemicals, these materials can simplify the decomposition of toxins.

12.5 Impact of Mulch on Soil Quality

Soil quality has been defined as the capacity of a specific kind of soil to function, within the limits of natural or managed ecosystems, to sustain the productivity of plants and animals, maintain or improve water and air quality, and support housing and human health (Karlen et al. 1997). The structure of the soil microbial community is one of the important parameters for the evaluation of soil quality (Gil-Sotres et al. 2005). Soil microorganisms participate in the mineralization of nutrients, the formation of soil structure and the capture and transport of nutrient to crops (Ma et al. 2016). Determining factors that influence microbial community structure has a significant impact on the understanding how management affects the quality of the crop, the ecology of the diseases and the biogeochemical cycle (Maul et al. 2014) and has become a fundamental aspect of sustainable agriculture (Dong et al. 2017).

According to Bandopadhyay et al. (2018), BDMs can influence soil microbial communities indirectly and directly. Indirectly, as a surface barrier prior to the incorporation of the soil, affecting the microclimate and the atmosphere of the soil, and directly, after the incorporation of the soil, as a direct input of physical fragments and a source of carbon, microorganisms, additives and products adhering chemicals (Kasirajan and Ngouajio 2012).

The application of mulching practices reduces the water vapor diffusivity, conserves soil moisture, controls soil structure and temperature, among other parameters, and therefore affects soil microbial communities (Kader et al. 2017). Ma et al. (2016) reported that in the early stage of mulching, the excellent soil temperature preservation and moisture maintenance functions of mulching film provided a good environment for growth and reproduction for microorganisms, causing an increase in the amount of microbial in the soil. It was also shown that the amount of fungi, bacteria and actinomycetes was higher in soils covered with biodegradable mulch film based on citric acid fermentation wastes compared to PE films. In this research work a high fungal/bacterial ratio was also reported, which can promote the organic matter decomposition and N mineralization in soil, thus being beneficial for the growth of crops according to De Vries et al. (2006). Farmer et al. (2017) investigated the effect of long-term fertilization and film mulching on the soil properties and structure of the bacterial community in a corn-growing system. It was reported that film mulching and manure fertilization significantly increased the bacterial diversity and richness during long-term fertilization. The results also showed that the predominant groups in the bacterial community were *Proteobacteria* and *Actinobacteria. Proteobacteria.* It is worth remembering that both bacterial communities play a key role in the soil carbon, sulfur and nitrogen cycles, including free-living aerobic nitrogen fixers (*Azotobacter*, *Azomonas*, *Azospirillum* and *Beyerinckia*) (Nosheen et al. 2016), nitrifying bacteria (*Nitrobacter*, *Nitrococcus* and *Nitrospira*) (Saijai et al. 2016) and the presence of them is related to healthy soils (Wang et al. 2017). Most of *Actinobacteria* can produce antibiotics that inhibit plant pathogens and control plant diseases (Barka et al. 2016). Chen et al. (2014) also reported that bacterial communities in different mulching treatments (grass, cornstalk and PE mulches) were significantly different compared to the control without mulch, and these changes were correlated with plant performance. However, other authors have reported a decrease in microbial activity under mulches (Moreno and Moreno 2008). This difference can most probably be explained by the amount of heating under the mulches: during the warmer seasons or in places with high temperatures, the mulches can raise temperatures above the optimum ones, whith limits the microbial activity of the soil; the opposite occurs during the winter seasons (Moreno and Moreno 2008). Interestingly Muñoz et al. (2017) reported that the amount of the filamentous fungi in the soil was drastically reduced in PE mulching and this decrease was accompanied by increased production of mycotoxins. The authors relate this mycotoxin production to a stress-mediated response of mycotoxigenic fungi species against unfavorable soil environmental conditions.

Changes in soil microbiota also affect nutrient cycling and storage. The change in soil organic carbon (SOC) is the result of inputs from organic sources derived from plants and microbial decomposition and loss of CO_2 (Nan et al. 2016). It has been reported that continuous use of plastic film mulch maintains the SOC level in temperature- and rainfall-limited semiarid regions by balancing the increase in

SOC mineralization with the increase in carbon input from the root (Wang et al. 2016). However, other authors have reported changes in SOC levels under the mulches (An et al. 2015; Cuello et al. 2015; Zhang et al. 2017). This difference may be due to the times in which the mulching experiment was carried out, usually in the short-term, and these long-term dynamics are not captured. A recent study, which compared four BDMs at two locations in the United States, showed that nutrient cycle was the only soil function affected by mulch treatment and using PLA/PHA was reduced compared to no-mulch, in only one of the localities studied (Sintim et al. 2019).

The microbial communities in the soil are also directly affected by the degradation products of the films. Muroi et al. (2016) investigated changes in soil microbiota associated with PBAT degradation. The surfaces of the PBAT film were enriched in fungi belonging to the phylum *Ascomycota* and affected the growth of specific fungal species in the bulk soil. On the other hand, the compositions of the bacterial flora were not strongly influenced by the presence of PBAT. The abundances of plant growth-promoting rhizobacteria, genera *Azospirillum* and *Mesorhizobium*, were approximately constant in all soil types during the course of 7 month incubation. A recent study on the impact of biodegradable seedling trays (BST) on microbial communities in rice soils showed that the use of BSTs increases microbial activity and reduces the community functional diversity in early stages in certain soils (Meng et al. 2019). It should be noted that the microbial response to the BDMs can be affected by the type of soil and/or location. Li et al. (2014) e.g. found that the degradation of the cellulose paper mulch produces enrichment of fungi in one locality and of Gram positive bacteria in another. Therefore, the application of BDMs should be evaluated in different regions and studies on degradation and long-term effects should be carried out, since the impacts of its application in the microbial community of the soil may be even greater.

Finally, there are not many studies carried out on the effect of the degradation of BDMs in the transformation of soil nutrients. A study conducted with biodegradable plastic material based on corn starch and copolyesters showed a total biodegradability and did not inhibit the nitrification potential of the soil (Bettas Ardisson et al. 2014). However, effects on other nutrients remain unknown.

12.6 Are Bio-based Polymeric Agricultural Mulch Films Improving the Quality and Yield of Crops?

Plastic mulch technology has played an important role in agriculture worldwide by helping counteract effects on soil, moisture and weed control. It was initially used to maintain soil temperature and humidity, thus facilitating seed germination and seedling growth by preventing evaporation and also reducing pressure and competition from weeds (Steinmetz et al. 2016). Therefore, throughout cultivation time, plants grown under mulch films result in an improvement in crop yields and faster crop cycle times. The plastic film can be pre-punched, and the seedlings are planted through holes or, alternatively, seeds are sown on the ridges and the holes are cut at the position of the seedlings. The first studies carried out by Rongsen (1994) showed that, depending on the implementation of plastic film technology, the production of chilli can be increased by 74%, tomato production by 52% and the production of garden peas by 31%. Nevertheless, most of plastic films included the PE remain in the field produce negative consequences in the environment (Subrahmaniyan et al. 2006; Gao et al. 2019). Thus, the properties of short-term agronomic benefits and long-term soil degradation are currently being discussed (Steinmetz et al. 2016; Kumar et al. 2018; Lamont 2005). Current studies are thus addressed to the development mulch films made from biodegradable materials composed mainly of natural polymers such as cellulose, starch, PLA and PHAs. Fortunately, compared to PE, in addition to the ability to directly enter the soil and become microbial biomass, CO_2 and water (Bandopadhyay et al. 2018; Kijchavengkul et al. 2008; Briassoulis 2007; César et al. 2009), a large number of biodegradable mulch films have been shown to improve yield and quality of crops in different vegetables, including cereal as spring wheat (Li et al. 2004) and horticultural species such as tomato and muskmelon, among others (Moreno and Moreno 2008; Saraiva et al. 2012).

The color of mulch films can also have an impact on the plant productivity (Scaringelli et al. 2016). Below, some examples of studies will be interpreted in which different types of biodegradable and colored mulch films have been used with the intention of identifying successful case studies that could facilitate the implementation of crops grown under biopolymer cover. Since black mulches provide less soil warming than clear mulches are suitable in environments where harvests of warm season vegetables are grown in a short, cool growing season (Waterer 2010). Therefore, during three growing seasons were evaluated and compared by Waterer (Waterer 2010) different types of mulch: clear, black and selective wavelength, including biodegradable starch-based mulch, a non-biodegradable LDPE and a non-mulch treatment. The clear and selective wavelength types produced beneficial effects on yields and crop development rates, probably by increasing soil temperatures, especially at the start of the growing season of different plant species such as sweet corn, zucchini, cantaloupe, pepper and eggplant. In the third growing season, statically significant results indicate higher yields for crops grown under biodegradable mulching.

In the case of tomato crop, a comparative study between black Mater-Bi® biodegradable mulch film (15 μm) and black PE film (15 μm) for weed control was carried out during a 2 year study in open fields in Spain (Cirujeda et al. 2012). Both mulches had similar benefits in crop yields and weed control, ranging between 80% and 100% for all of them. However, from the economic point of view, PE is the most recommended. A previous and similar 2 year study (2004–2005) was conducted by Moreno and Moreno (2008) on tomato (*Lycopersicon sculentum* Mill.) cv Mina using seven mulch films (four non-specified biodegradable and three PE films) including different types of thicknesses (range 15–30 μm), width (1.20–1.35 m) and color (green, black, brown, blue/yellow), found that PE mulch films increased soil temperatures relative to non-mulching soil. In addition, the temperatures under PE films were always higher than under biodegradable films, which represents a disadvantage especially in hot climates and an advantage in cold conditions. During 2005, average mean soil temperatures changed between 26.6–14.8 °C under biodegradable mulch films while the average mean soil temperature varied between 27.4–15.1 °C under PE mulches. Probably for this increase of temperature, the lowest values of soil microbial biomass C and the mineralization of soil organic matter were detected under PE films. Throughout the growing season, biodegradable mulch films were progressively degraded, which caused the temperature differences in relation to the non-mulching soil were undetectable. Contrary to the previous study, Locher et al. (2005) using color mulch films did not detect any noticeable changes in tomato yield, suggesting that the differences in soil temperature between mulches are mainly due to their compositions. Similarly, to crop yield, no significant differences in tomato fruit shape, firmness, juice and solid soluble solid contents between all tested treatments were obtained, except for one of the biodegradable mulch films which showed the lowest yield value apparently associated to the earliest degradation of this material.

A study done by Morra et al. (2016) evaluating the qualitative and quantitative action of Mater-Bi® biodegradable mulch film in two cultivars of strawberry cv Fortuna and cv Sabrina revealed that both cultivars had a different behavior compared to PE; meanwhile, the yield of cv Sabrina was 18% higher in PE, while the yield of cv Fortuna was 10% higher in Mater-Bi® biodegradable film (Morra et al. 2016). In general, the quality of strawberry fruits was much more beneficial from the nutritional point of view for plants grown under Mater-Bi® biodegradable since, they had a higher level of antioxidant compounds (Morra et al. 2016). In this same sense, the action of biodegradable mulch sheets made from PLA, Ecoflex (from BASF) and modified starch were studied by Tachibana et al. (2009) on the growth and quality of mandarin orange groves cultivated in the field during 2004 and 2005. The inclusion of the modified starch in the film was probably appropriate due to its adequate degradability (Tachibana et al. 2009). Compared to unmulched oranges, the fruit quality measured by the Brix degree, the acidity index and peel color was better than that of plants grown under mulch films (Tachibana et al. 2009). However, in this study made by Tachibana et al. (2009) antioxidative metabolites were not analyzed.

More recently, Sun et al. (2018) investigated the use of peanuts as a model crop to test whether the biodegradable film can promote soil benefits throughout the growth stages of this crop. The authors evaluated biodegradable films with different degradation capacity based on four PBAT/starch ratios changing starch from 0 to 20% and compared them with PE and whitout mulching as controls during 2013 and 2014. In 2013, the degradation rates increased according to the increase in starch content in the film. However, degradation rates of the films in 2014 were slower than in 2013. In 2014, similarly to the trend measured for the net photosynthetic rate, the leaf area index was higher (~25% at 120 days after sowing) under mulch films compared with non-mulch treatment. The application of the biodegradable film with PBAT/20% starch triggered the highest photosynthesis capacity and peanut yield because of adequate temperature and moisture in the soil. In the last stages of growth, the soil mulched with PBAT/15% starch maintained more water in the soil and the highest peanut leaf area, the chlorophyll content and the net photosynthetic rate compared to the PE mulch films and the other biodegradable mulch films. However, the peanut yields under the biodegradable films were slightly lower (in average 4991.75 kg/ha) than PE (5304 kg/ha) mulch film; the difference between PBAT/15% starch and PE films was not significant.

Nowadays, the development of highly functional bio-based polymeric mulch films is emerging at the level of fundamental and applied scientific knowledge. In this sense, a new trend for bio-based polymeric mulch films is their uses as physical support for the controlled release of bioactive compounds. This is the case of a biodegradable PBAT/PLA blend with the herbicide 2-methyl-4-chlorophenoxyacetic acid (MCPA)-PHBV conjugate which suppresses the growth of weeds and results in a good candidate for field applications. The biological targeting of this MCPA-PHBV conjugate in the mulch film was assayed by Kwiecien et al. (2018) in *Vicia faba* as a model crop species. This blend specifically suppressed the growth of broadleaf weed species under greenhouse conditions (Kwiecien et al. 2018). However, the effectiveness of this conjugate pesticide has not yet been proven in field bioassays.

Iwata (2015) warned that new biopolymeric-based mulch films could be prepared from a strong knowledge of different types of biomass, including diverse sources of lignocellulosic compounds. Research on bio-based plastics, including the development of new materials and processing techniques is still required. Thus, this same author also emphasizes that this kind of study must be interdisciplinary. One of the main goals of the bio-based polymeric mulch industry is to achieve the highest possible crop yield and supply the right materials for sustainable agriculture. Therefore, blends between different bio-based and biodegradable materials are studied more and more.

Despite the positive potential of the new bio-based materials, the application of biopolymer films has remained little expanded throughout the world, probably because they are more expensive and their breakdown is generally premature in the field (Kasirajan and Ngouajio 2012). In general, studies on the useful life of the material required for long-term crops mulching are also very scarce (Scaringelli et al. 2016; Touchaleaume et al. 2016).

Finally, in addition to bio-based polymeric mulch films, a recent technique based on spraying water solutions onto soil forming a biodegradable mulching could be another alternative and disruptive tool for modern and organic agriculture (Sartore et al. 2018).

12.7 Conclusions

We in this chapter interpreted how bio-based polymeric agricultural mulch films can overcome the traditional PE plastic film technology. However, more basic research on bio-based polymeric films, including the development of new formulations and material processing techniques that can be used for their production as films is still required. It is well known that plastic film technology can increase the yield of some crops compared to conventional agriculture. However, this new technology based on biopolymeric sources is still relatively new. It will be necessary to continue sharing experiences and promote awareness from different specialized fields of knowledge.

Despite the advantages of degradable films, several challenges still remain to be solved, especially some properties related to the quality of biodegradation products, the temporal control of degradation and its incidence on soil temperature. In addition, the type of mulch films used is a key issue for the effect on the soil, the yield and the quality of fruits. The fact that specific environmental conditions and each growth season have marked influences on the production and yield of the plants requires that the type and manner of use of mulch films should be optimized individually for each crop species that is planned cultivate. More studies should be conducted with a more diverse set of crop varieties on a wider range of types of cultivated areas before concluding on yield and on each type of biodegradable mulch films.

In addition, before agricultural mulches can reach the market, its properties must be adjusted to the corresponding standards. The European standard EN 13655:2002, which was used for many years as a reference for producers and researchers was divided into two standards, the revised version of EN 13655 for removable mulches (non-biodegradable materials based on PE) and the standard EN 17033 for materials to be incorporated into the soil after the cultivation cycle. The new standard regulates not only the mechanical, optical and barrier properties as in standard EN 13655, but also regulates the chemical composition, biodegradation in soil and ecotoxicity.

Biodegradation in several media (compost or soil) is a complex process that depends on different factors: abiotic and biological degradative processes. It is important to remark that the existence of standards that allow determining the biodegradability of the mulches is a crucial tool in the design of polymeric films for such agricultural application. The biodegradability and mechanic requirements demanded by this standard can be achieved by combining bio-based polymers such as PLA, PHAs and starch, although not without efforts. In addition, all commercially available biodegradable mulches comply with EN 17033.

Bio-based plastics produced from natural compounds have particular and valuable properties. The use of additives and bioactive agents that could be released from bio-based polymeric mulch films should also be considered precisely because of their influence on the environment. With respect to its impact on soil, BDMs offer an environmentally sustainable alternative to PE plastic mulches. However, there is still a need to better understand the impact of BDM on nutrient biogeochemistry, microbial response and long-term effects in agricultural applications.

Acknowledgements The authors would like to thank Agencia Nacional de Promoción Científica y Tecnológica (ANPCyT), Consejo Nacional de Investigaciones Científicas y Técnicas (CONICET), and Universidad Nacional de Mar del Plata (UNMdP), Argentina. DM, AYM, CAC and VAA are members of CONICET research staff.

Conflicts of Interest The authors declare no conflict of interest.

References

Akrami, M., Ghasemi, I., Azizi, H., Karrabi, M., & Seyedabadi, M. (2016). A new approach in compatibilization of the poly(lactic acid)/thermoplastic starch (PLA/TPS) blends. *Carbohydrate Polymers, 144*, 254–262. https://doi.org/10.1016/j.carbpol.2016.02.035.

Alvarez, V. A., Ruseckaite, R. A., & Vázquez, A. (2006). Degradation of sisal fibre/Mater Bi-Y biocomposites buried in soil. *Polymer Degradation and Stability, 91*(12), 3156–3162. https://doi.org/10.1016/j.polymdegradstab.2006.07.011.

An, T., Schaeffer, S., Li, S., Fu, S., Pei, J., Li, H., Zhuang, J., Radosevich, M., & Wang, J. (2015). Carbon fluxes from plants to soil and dynamics of microbial immobilization under plastic film mulching and fertilizer application using ^{13}C pulse-labeling. *Soil Biology and Biochemistry, 80*, 53–61. https://doi.org/10.1016/j.soilbio.2014.09.024.

Armentano, I., Fortunati, E., Burgos, N., Dominici, F., Luzi, F., Fiori, S., Jiménez, A., Yoon, K., Ahn, J., Kang, S., & Kenny, J. M. (2015). Processing and characterization of plasticized PLA/PHB blends for biodegradable multiphase systems. *Express Polymer Letters, 9*(7), 583–596. https://doi.org/10.3144/expresspolymlett.2015.55.

Bandopadhyay, S., Martin-Closas, L., Pelacho, A. M., & DeBruyn, J. M. (2018). Biodegradable plastic mulch films: Impacts on Soil microbial communities and ecosystem functions. *Frontiers in Microbiology, 9*, 819. https://doi.org/10.3389/fmicb.2018.00819.

Barka, E. A., Vatsa, P., Sanchez, L., Gaveau-Vaillant, N., Jacquard, C., Klenk, H.-P., Clément, C., Ouhdouch, Y., & van Wezel, G. P. (2016). Taxonomy, physiology, and natural products of *Actinobacteria. Microbiology and Molecular Biology Reviews, 80*(1), 1–43. https://doi.org/10.1128/mmbr.00019-15.

BASF. (2018). Biodegradable polymers – ecovio®.

Bastoli, C. (Ed.). (2014). *Handbook of biodegradable polymers, a smithers group company*. Shawbury.

Bettas Ardisson, G., Tosin, M., Barbale, M., & Degli-Innocenti, F. (2014). Biodegradation of plastics in soil and effects on nitrification activity. A laboratory approach. *Frontiers in Microbiology, 5*, 710. https://doi.org/10.3389/fmicb.2014.00710.

Bilck, A. P., Grossmann, M. V. E., & Yamashita, F. (2010). Biodegradable mulch films for strawberry production. *Polymer Testing, 29*(4), 471–476. https://doi.org/10.1016/j.polymertesting.2010.02.007.

Boelens, J. (2017). Expert statement (bio)degradable mulching films.

Briassoulis, D. (2004). An overview on the mechanical behaviour of biodegradable agricultural films. *Journal of Polymers and the Environment, 12*(2), 65–81. https://doi.org/10.1023/b:jooe.0000010052.86786.ef.

Briassoulis, D. (2006). Mechanical behaviour of biodegradable agricultural films under real field conditions. *Polymer Degradation and Stability, 91*(6), 1256–1272. https://doi.org/10.1016/j.polymdegradstab.2005.09.016.

Briassoulis, D. (2007). Analysis of the mechanical and degradation performances of optimised agricultural biodegradable films. *Polymer Degradation and Stability, 92*(6), 1115–1132. https://doi.org/10.1016/j.polymdegradstab.2007.01.024.

Briassoulis, D., & Dejean, C. (2010). Critical review of norms and standards for biodegradable agricultural plastics Part I. Biodegradation in soil. *Journal of Polymers and the Environment, 18*(3), 384–400. https://doi.org/10.1007/s10924-010-0168-1.

Briassoulis, D., & Giannoulis, A. (2018). Evaluation of the functionality of bio-based plastic mulching films. *Polymer Testing, 67*, 99–109. https://doi.org/10.1016/j.polymertesting.2018.02.019.

Bugnicourt, E., Cinelli, P., Lazzeri, A., & Alvarez, V. (2014). Polyhydroxyalkanoate (PHA): Review of synthesis, characteristics, processing and potential applications in packaging. *Express Polymer Letters, 8*(11), 791–808. https://doi.org/10.3144/expresspolymlett.2014.82.

Carmona, V. B., Corrêa, A. C., Marconcini, J. M., & Mattoso, L. H. C. (2015). Properties of a biodegradable ternary blend of thermoplastic starch (TPS), poly(ε-caprolactone) (PCL) and poly(lactic acid) (PLA). *Journal of Polymers and the Environment, 23*(1), 83–89. https://doi.org/10.1007/s10924-014-0666-7.

CEN/TC 249 – Plastics EN 17033. (2018). Plastics – Biodegradable mulch films for use in agriculture and horticulture – Requirements and test methods.

César, M. E. F., Mariani, P. D. S. C., Innocentini-Mei, L. H., & Cardoso, E. J. B. N. (2009). Particle size and concentration of poly(ε-caprolactone) and adipate modified starch blend on mineralization in soils with differing textures. *Polymer Testing, 28*(7), 680–687. https://doi.org/10.1016/j.polymertesting.2009.05.002.

Chen, Y., Wen, X., Sun, Y., Zhang, J., Wu, W., & Liao, Y. (2014). Mulching practices altered soil bacterial community structure and improved orchard productivity and apple quality after five growing seasons. *Scientia Horticulturae, 172*, 248–257. https://doi.org/10.1016/j.scienta.2014.04.010.

Cirujeda, A., Aibar, J., Anzalone, Á., Martín-Closas, L., Meco, R., Moreno, M. M., Pardo, A., Pelacho, A. M., Rojo, F., Royo-Esnal, A., Suso, M. L., & Zaragoza, C. (2012). Biodegradable mulch instead of polyethylene for weed control of processing tomato production. *Agronomy for Sustainable Development, 32*(4), 889–897. https://doi.org/10.1007/s13593-012-0084-y.

Cowan, J. S., Inglis, D. A., & Miles, C. A. (2013). Deterioration of three potentially biodegradable plastic mulches before and after soil incorporation in a broccoli field production system in northwestern washington. *HortTechnology, 23*(6), 849–858. https://doi.org/10.21273/horttech.23.6.849.

Cuello, J. P., Hwang, H. Y., Gutierrez, J., Kim, S. Y., & Kim, P. J. (2015). Impact of plastic film mulching on increasing greenhouse gas emissions in temperate upland soil during maize cultivation. *Applied Soil Ecology, 91*, 48–57. https://doi.org/10.1016/j.apsoil.2015.02.007.

D'Avino, L., Rizzuto, G., Guerrini, S., Sciaccaluga, M., & Pagnotta, E. (2015). Environmental implications of crude glycerin used in special products for the metalworking industry and in biodegradable mulching films. *Industrial Crops and Products, 75*(Part A), 29–35. https://doi.org/10.1016/j.indcrop.2015.02.043.

de Vries, F. T., Hoffland, E., van Eekeren, N., Brussaard, L., & Bloem, J. (2006). Fungal/bacterial ratios in grasslands with contrasting nitrogen management. *Soil Biology and Biochemistry, 38*(8), 2092–2103. https://doi.org/10.1016/j.soilbio.2006.01.008.

Dong, W., Si, P., Liu, E., Yan, C., Zhang, Z., & Zhang, Y. (2017). Influence of film mulching on soil microbial community in a rainfed region of northeastern China. *Scientific Reports, 7*(1), 8468. https://doi.org/10.1038/s41598-017-08575-w.

European Bioplastics. (2016). Bioplastic market data 2016. Global production capacities of bioplastics 2016–2021.

European Bioplastics. (2018). New EU standard for biodegradable mulch films in agriculture published – European Bioplastics e.V.

European Bioplastics e.V. (2018a). Bioplastics market data.

European Bioplastics e.V. (2018b). Biodegradable plastics.

Farmer, J., Zhang, B., Jin, X., Zhang, P., & Wang, J. (2017). Long-term effect of plastic film mulching and fertilization on bacterial communities in a brown soil revealed by high through-put sequencing. *Archives of Agronomy and Soil Science, 63*(2), 230–241. https://doi.org/10.1080/03650340.2016.1193667.

Ferri, J. M., Garcia-Garcia, D., Sánchez-Nacher, L., Fenollar, O., & Balart, R. (2016). The effect of maleinized linseed oil (MLO) on mechanical performance of poly(lactic acid)-thermoplastic starch (PLA-TPS) blends. *Carbohydrate Polymers, 147*, 60–68. https://doi.org/10.1016/j.carbpol.2016.03.082.

Finkenstadt, V. L., & Tisserat, B. (2010). Poly(lactic acid) and Osage Orange wood fiber composites for agricultural mulch films. *Industrial Crops and Products, 31*(2), 316–320. https://doi.org/10.1016/j.indcrop.2009.11.012.

Gao, H., Yan, C., Liu, Q., Ding, W., Chen, B., & Li, Z. (2019). Effects of plastic mulching and plastic residue on agricultural production: A meta-analysis. *The Science of the Total Environment, 651*. (Part 1, 484–492. https://doi.org/10.1016/j.scitotenv.2018.09.105.

Ghimire, S., Hayes, D., Cowan, J. S., Inglis, D., DeVetter, L. W., & Miles, C. A. (2018). *Biodegradable plastic mulch and suitability for sustainable and organic agriculture*. Washington: Washington State University Extension.

Gil-Sotres, F., Trasar-Cepeda, C., Leirós, M. C., & Seoane, S. (2005). Different approaches to evaluating soil quality using biochemical properties. *Soil Biology and Biochemistry, 37*(5), 877–887. https://doi.org/10.1016/j.soilbio.2004.10.003.

Griffin, G. J. L. (1974). Biodegradable fillers in thermoplastics. In R. D. Deanin & N. R. Schott (Eds.), *Fillers and reinforcements for plastics* (Vol. 134, pp. 159–170). American Chemical Society. https://doi.org/10.1021/ba-1974-0134.ch016.

Gu, X.-B., Li, Y.-N., & Du, Y.-D. (2017). Biodegradable film mulching improves soil temperature, moisture and seed yield of winter oilseed rape (*Brassica napus* L.). *Soil and Tillage Research, 171*, 42–50. https://doi.org/10.1016/j.still.2017.04.008.

Gutiérrez, T. J. (2018). Chapter 9. Biodegradability and compostability of food nanopackaging materials. In G. Cirillo, M. A. Kozlowski, & U. G. Spizzirri (Eds.), *Composite materials for food packaging* (pp. 269–296). WILEY-Scrivener Publisher. EE.UU. ISBN: 978-1-119-16020-5. https://doi.org/10.1002/9781119160243.ch9.

Hablot, E., Dharmalingam, S., Hayes, D. G., Wadsworth, L. C., Blazy, C., & Narayan, R. (2014). Effect of simulated weathering on physicochemical properties and inherent biodegradation of PLA/PHA nonwoven mulches. *Journal of Polymers and the Environment, 22*(4), 417–429. https://doi.org/10.1007/s10924-014-0697-0.

Halley, P., Rutgers, R., Coombs, S., Kettels, J., Gralton, J., Christie, G., Jenkins, M., Beh, H., Griffin, K., Jayasekara, R., & Lonergan, G. (2001). Developing biodegradable mulch films from starch-based polymers. *Starch – Stärke, 53*(8), 362–367. https://doi.org/10.1002/1521-379x(200108)53:8<362::aid-star362>3.0.co;2-j.

Hayes, D.G., & Flury, M. (2018). A new standard for biodegradable plastic mulch films.

Hayes, D. G., Wadsworth, L. C., Sintim, H. Y., Flury, M., English, M., Schaeffer, S., & Saxton, A. M. (2017). Effect of diverse weathering conditions on the physicochemical properties of biodegradable plastic mulches. *Polymer Testing, 62*, 454–467. https://doi.org/10.1016/j.polymertesting.2017.07.027.

Hiraishi, A., & Khan, S. T. (2003). Application of polyhydroxyalkanoates for denitrification in water and wastewater treatment. *Applied Microbiology and Biotechnology, 61*(2), 103–109. https://doi.org/10.1007/s00253-002-1198-y.

Imam, S. H., Wood, D. F., Abdelwahab, M. A., Chiou, B.-S., Williams, T. G., Glenn, G. M., & Orts, W. J. (2012). Starch chemistry, microstructure, processing, and enzymatic degradation. In J. Ahmed, B. Tiwari, S. Imam, & M. Rao (Eds.), *Starch-based polymeric materials and nanocomposites: Chemistry, processing, and applications* (pp. 5–32). Boca Raton: CRC Press.

Iwata, T. (2015). Biodegradable and bio-based polymers: future prospects of eco-friendly plastics. *Angewandte Chemie, 54*(11), 3210–3215. https://doi.org/10.1021/acs.biomac.7b01636.

Jandas, P. J., Mohanty, S., & Nayak, S. K. (2013). Sustainability, compostability, and specific microbial activity on agricultural mulch films prepared from poly(lactic acid). *Industrial and Engineering Chemistry Research, 52*(50), 17714–17724. https://doi.org/10.1021/ie4023429.

Jandas, P. J., Mohanty, S., & Nayak, S. K. (2014). Morphology and thermal properties of renewable resource-based polymer blend nanocomposites influenced by a reactive compatibilizer. *ACS Sustainable Chemistry & Engineering, 2*(3), 377–386. https://doi.org/10.1021/sc400395s.

Kader, M. A., Senge, M., Mojid, M. A., & Ito, K. (2017). Recent advances in mulching materials and methods for modifying soil environment. *Soil and Tillage Research, 168*, 155–166. https://doi.org/10.1016/j.still.2017.01.001.

Karlen, D. L., Mausbach, M. J., Doran, J. W., Cline, R. G., Harris, R. F., & Schuman, G. E. (1997). Soil quality: A concept, definition, and framework for evaluation (a guest editorial). *Soil Science Society of America Journal, 61*(1), 4–10. https://doi.org/10.2136/sssaj1997.03615995006100010001x.

Kasirajan, S., & Ngouajio, M. (2012). Polyethylene and biodegradable mulches for agricultural applications: A review. *Agronomy for Sustainable Development, 32*(2), 501–529. https://doi.org/10.1007/s13593-011-0068-3.

Kijchavengkul, T., Auras, R., Rubino, M., Ngouajio, M., & Fernandez, R. T. (2008). Assessment of aliphatic–aromatic copolyester biodegradable mulch films. Part I: Field study. *Chemosphere, 71*(5), 942–953. https://doi.org/10.1016/j.chemosphere.2007.10.074.

Kumar, T., Haque, M. A., Islam, M. S., Hoque, M. F., & Jodder, R. (2018). Effect of polythene mulch on growth and yield of Sunflower (Helianthus annuus). *Archives Crop Science, 2*(1), 38–46.

Kwiecien, I., Adamus, G., Jiang, G., Radecka, I., Baldwin, T. C., Khan, H. R., Johnston, B., Pennetta, V., Hill, D., Bretz, I., & Kowalczuk, M. (2018). Biodegradable PBAT/PLA blend with bioactive MCPA-PHBV conjugate suppresses weed growth. *Biomacromolecules, 19*(2), 511–520. https://doi.org/10.1002/anie.201410770.

Lamont, W. J. (2005). Plastics: Modifying the Microclimate for the Production of Vegetable Crops. *HortTechnology, 15*(3), 477–481.

Li, F.-M., Wang, J., Xu, J.-Z., & Xu, H.-L. (2004). Productivity and soil response to plastic film mulching durations for spring wheat on entisols in the semiarid Loess Plateau of China. *Soil and Tillage Research, 78*(1), 9–20. https://doi.org/10.1016/j.still.2003.12.009.

Li, C., Moore-Kucera, J., Miles, C., Leonas, K., Lee, J., Corbin, A., & Inglis, D. (2014). Degradation of potentially biodegradable plastic mulch films at three diverse U.S. locations. *Agroecology and Sustainable Food Systems, 38*(8), 861–889. https://doi.org/10.1080/21683565.2014.884515.

Locher, J., Ombódi, A., Kassai, T., & Dimény, J. (2005) Influence of coloured mulches on soil temperature and yield of sweet pepper. European Journal of Horticultural Science, 70 (3). Available in: https://www.pubhort.org/ejhs/2005/file_22635.pdf.

Ma, Z., Ma, Y., Qin, L., Liu, J., & Su, H. (2016). Preparation and characteristics of biodegradable mulching films based on fermentation industry wastes. *International Biodeterioration & Biodegradation, 111*, 54–61. https://doi.org/10.1016/j.ibiod.2016.04.024.

Martín-Closas, L., Costa, J., & Pelacho, A. M. (2017). Agronomic effects of biodegradable films on crop and field environment. In M. Malinconico (Ed.), *Soil degradable bioplastics for a sustainable modern agriculture. Green chemistry and sustainable technology* (pp. 67–104). Berlin/Heidelberg: Springer. https://doi.org/10.1007/978-3-662-54130-2_4.

Maul, J. E., Buyer, J. S., Lehman, R. M., Culman, S., Blackwood, C. B., Roberts, D. P., Zasada, I. A., & Teasdale, J. R. (2014). Microbial community structure and abundance in the rhizosphere and bulk soil of a tomato cropping system that includes cover crops. *Applied Soil Ecology, 77*, 42–50. https://doi.org/10.1016/j.apsoil.2014.01.002.

Meng, K., Ren, W., Teng, Y., Wang, B., Han, Y., Christie, P., & Luo, Y. (2019). Application of biodegradable seedling trays in paddy fields: Impacts on the microbial community. *The Science of the Total Environment, 656*, 750–759. https://doi.org/10.1016/j.scitotenv.2018.11.438.

Merino, D., Gutiérrez, T. J., & Alvarez, V. A. (2018a). Potential agricultural mulch films based on native and phosphorylated corn starch with and without surface functionalization with chitosan. *Journal of Polymers and the Environment, 27*(1), 97–105. https://doi.org/10.1007/s10924-018-1325-1.

Merino, D., Mansilla, A. Y., Gutiérrez, T. J., Casalongué, C. A., & Alvarez, V. A. (2018b). Chitosan coated-phosphorylated starch films: Water interaction, transparency and antibacterial properties. *Reactive and Functional Polymers, 131*, 445–453. https://doi.org/10.1016/j.reactfunctpolym.2018.08.012.

Merino, D., Gutiérrez, T. J., Mansilla, A. Y., Casalongué, C., & Alvarez, V. A. (2018c). Critical evaluation of starch-based antibacterial nanocomposites as agricultural mulch films: Study on their interactions with water and light. *ACS Sustainable Chemistry & Engineering, 6*(11), 15662–15672. https://doi.org/10.1021/acssuschemeng.8b04162.

Merino, D., Gutiérrez, T. J., & Alvarez, V. A. (2019). Structural and thermal properties of agricultural mulch films based on native and oxidized corn starch nanocomposites. *Starch-Stärke. 71*(7–8), 1800341. https://doi.org/10.1002/star.201800341.

Miles, C., Wallace, R., Wszelaki, A., Martin, J., Cowan, J., Walters, T., & Inglis, D. (2012). Deterioration of potentially biodegradable alternatives to black plastic mulch in three tomato production regions. *HortScience, 47*(9), 1270–1277. https://doi.org/10.21273/horttech.23.6.849.

Mitrus, M., Wojtowicz, A., & Moscicki, L. (2010). Biodegradable polymers and their practical utility. In L. P. B. M. Janssen & L. Moscicki (Eds.), *Thermoplastic starch* (pp. 1–33). Weinheim: Wiley-VCH Verlag GmbH & KGaA. https://doi.org/10.1002/9783527628216.ch1.

Moreno, M. M., & Moreno, A. (2008). Effect of different biodegradable and polyethylene mulches on soil properties and production in a tomato crop. *Scientia Horticulturae, 116*(3), 256–263. https://doi.org/10.1016/j.scienta.2008.01.007.

Morra, L., Bilotto, M., Cerrato, D., Coppola, R., Leone, V., Mignoli, E., Pasquariello, M. S., Petriccione, M., & Cozzolino, E. (2016). The Mater-Bi® biodegradable film for strawberry (*Fragaria* x *ananassa* Duch.) mulching: effects on fruit yield and quality. *Italian Journal of Agronomy, 11*(3), 203. https://doi.org/10.4081/ija.2016.731.

Muñoz, K., Buchmann, C., Meyer, M., Schmidt-Heydt, M., Steinmetz, Z., Diehl, D., Thiele-Bruhn, S., & Schaumann, G. E. (2017). Physicochemical and microbial soil quality indicators as affected by the agricultural management system in strawberry cultivation using straw or black polyethylene mulching. *Applied Soil Ecology, 113*, 36–44. https://doi.org/10.1016/j.apsoil.2017.01.014.

Muroi, F., Tachibana, Y., Kobayashi, Y., Sakurai, T., & Kasuya, K. (2016). Influences of poly(butylene adipate-co-terephthalate) on soil microbiota and plant growth. *Polymer Degradation and Stability, 129*, 338–346. https://doi.org/10.1016/j.polymdegradstab.2016.05.018.

Nan, W., Yue, S., Huang, H., Li, S., & Shen, Y. (2016). Effects of plastic film mulching on soil greenhouse gases (CO_2, CH_4 and N_2O) concentration within soil profiles in maize fields on the Loess Plateau, China. *Journal of Integrative Agriculture, 15*(2), 451–464. https://doi.org/10.1016/S2095-3119(15)61106-6.

Nosheen, A., Bano, A., Yasmin, H., Keyani, R., Habib, R., Shah, S. T. A., & Naz, R. (2016). Protein quantity and quality of safflower seed improved by NP fertilizer and rhizobacteria (*Azospirillum* and *Azotobacter* spp.). *Frontiers in Plant Science, 7*, 104. https://doi.org/10.3389/fpls.2016.00104.

Otey, F. H., Westhoff, R. P., & Russell, C. R. (1977). Biodegradable films from starch and ethylene-acrylic acid copolymer. *Industrial and Engineering Chemistry Product Research and Development, 16*(4), 305–308. https://doi.org/10.1021/i360064a009.

PRI/75. (2002). EN 13655 Plastics – Mulching thermoplastic films for use in agriculture and horticulture.

Przybytek, A., Sienkiewicz, M., Kucińska-Lipka, J., & Janik, H. (2018). Preparation and characterization of biodegradable and compostable PLA/TPS/ESO compositions. *Industrial Crops and Products, 122*, 375–383. https://doi.org/10.1016/j.indcrop.2018.06.016.

Research, G.V. (2015). Biodegradable mulch films market analysis by raw material (thermoplastic starch (TPS), polylactic acid (PLA), polyhydroxyalkanoates (PHA), aliphatic-aromatic copolyesters (AAC)), By Crop Type (Fruits& Vegetables, Grains & Oilseeds, Flowers & Plants) And S.

Robinson, D., & Brae, B. (1991) Developments in plastic structures and materials for horticultural crops. ASPAC, Food Fertilizer Technology Center, Taipei city. Available in: http://en.fftc.org.tw/htmlarea_file/library/20110729171201/eb331.pdf.

Rongsen, L. (1994). The application of plastic film technology in China. *agris.fao.org*.

Saijai, S., Ando, A., Inukai, R., Shinohara, M., & Ogawa, J. (2016). Analysis of microbial community and nitrogen transition with enriched nitrifying soil microbes for organic hydroponics. *Bioscience, Biotechnology, and Biochemistry, 80*(11), 2247–2254. https://doi.org/10.1080/09168451.2016.1200459.

Saraiva, A., Costa, R., Carvalho, L., & Duarte, E. (2012). The use of biodegradable mulch films in muskmelon crop production. *Basic Research Journal of Agricultural Science and Review, 1*(4), 88–95.

Sartore, L., Schettini, E., de Palma, L., Brunetti, G., Cocozza, C., & Vox, G. (2018). Effect of hydrolyzed protein-based mulching coatings on the soil properties and productivity in a tunnel greenhouse crop system. *The Science of the Total Environment, 645*, 1221–1229. https://doi.org/10.1016/j.scitotenv.2018.07.259.

Scaringelli, M. A., Giannoccaro, G., Prosperi, M., & Lopolito, A. (2016). Adoption of biodegradable mulching films in agriculture: is there a negative prejudice towards materials derived from organic wastes? *Italian Journal of Agronomy, 11*(2), 92. https://doi.org/10.4081/ija.2016.716.

Shogren, R. (2009). Starch-poly(hydroxyalkanoate) composites and blends. In L. Yu (Ed.), *Biodegradable polymer blends and composites from renewable resources* (pp. 209–226). Hoboken: Wiley. https://doi.org/10.1002/9780470391501.ch9.

Sintim, H. Y., Bandopadhyay, S., English, M. E., Bary, A. I., DeBruyn, J. M., Schaeffer, S. M., Miles, C. A., Reganold, J. P., & Flury, M. (2019). Impacts of biodegradable plastic mulches on soil health. *Agriculture, Ecosystems and Environment, 273*, 36–49. https://doi.org/10.1016/j.agee.2018.12.002.

Steinmetz, Z., Wollmann, C., Schaefer, M., Buchmann, C., David, J., Tröger, J., Muñoz, K., Frör, O., & Schaumann, G. E. (2016). Plastic mulching in agriculture. Trading short-term agronomic benefits for long-term soil degradation? *The Science of the Total Environment, 550*, 690–705. https://doi.org/10.1016/j.scitotenv.2016.01.153.

Subrahmaniyan, K., Kalaiselvan, P., Balasubramanian, T. N., & Zhou, W. (2006). Crop productivity and soil properties as affected by polyethylene film mulch and land configurations in groundnut (*Arachis hypogaea* L.). *Archives of Agronomy and Soil Science, 52*(1), 79–103. https://doi.org/10.1080/03650340500421786.

Sun, T., Li, G., Ning, T.-Y., Zhang, Z.-M., Mi, Q.-H., & Lal, R. (2018). Suitability of mulching with biodegradable film to moderate soil temperature and moisture and to increase photosynthesis and yield in peanut. *Agricultural Water Management, 208*, 214–223. https://doi.org/10.1016/j.agwat.2018.06.027.

Tachibana, Y., Maeda, T., Ito, O., Maeda, Y., & Kunioka, M. (2009). Utilization of a biodegradable mulch sheet produced from poly(lactic acid)/ecoflex/modified starch in mandarin orange groves. *International Journal of Molecular Sciences, 10*(8), 3599–3615. https://doi.org/10.3390/ijms10083599.

Takahashi, M., Yamada, T., Tanno, M., Tsuji, H., & Hiraishi, A. (2011). Nitrate removal efficiency and bacterial community dynamics in denitrification processes using poly (L-lactic acid) as the solid substrate. *Microbes and Environments, 26*(3), 212–219. https://doi.org/10.1264/jsme2.me11107.

Touchaleaume, F., Martin-Closas, L., Angellier-Coussy, H., Chevillard, A., Cesar, G., Gontard, N., & Gastaldi, E. (2016). Performance and environmental impact of biodegradable polymers as agricultural mulching films. *Chemosphere, 144*, 433–439. https://doi.org/10.1016/j.chemosphere.2015.09.006.

Wang, Y. P., Li, X. G., Fu, T., Wang, L., Turner, N. C., Siddique, K. H. M., & Li, F.-M. (2016). Multi-site assessment of the effects of plastic-film mulch on the soil organic carbon balance in semiarid areas of China. *Agricultural and Forest Meteorology, 228–229*, 42–51. https://doi.org/10.1016/j.agrformet.2016.06.016.

Wang, R., Zhang, H., Sun, L., Qi, G., Chen, S., & Zhao, X. (2017). Microbial community composition is related to soil biological and chemical properties and bacterial wilt outbreak. *Scientific Reports, 7*(1), 343. https://doi.org/10.1038/s41598-017-00472-6.

Waterer, D. (2010). Evaluation of biodegradable mulches for production of warm-season vegetable crops. *Canadian Journal of Plant Science, 90*(5), 737–743. https://doi.org/10.4141/cjps10031.

Wortman, S. E., Kadoma, I., & Crandall, M. D. (2015). Assessing the potential for spunbond, nonwoven biodegradable fabric as mulches for tomato and bell pepper crops. *Scientia Horticulturae (Amsterdam), 193*, 209–217. https://doi.org/10.1016/j.scienta.2015.07.019.

Yang, Y., Tang, Z., Xiong, Z., & Zhu, J. (2015). Preparation and characterization of thermoplastic starches and their blends with poly(lactic acid). *International Journal of Biological Macromolecules, 77*, 273–279. https://doi.org/10.1016/j.ijbiomac.2015.03.053.

Yoshida, N., Ye, L., Liu, F., Li, Z., & Katayama, A. (2013). Evaluation of biodegradable plastics as solid hydrogen donors for the reductive dechlorination of fthalide by *Dehalobacter* species. *Bioresource Technology, 130*, 478–485. https://doi.org/10.1016/j.biortech.2012.11.139.

Zhang, F., Zhang, W., Li, M., Yang, Y., & Li, F.-M. (2017). Does long-term plastic film mulching really decrease sequestration of organic carbon in soil in the Loess Plateau. *European Journal of Agronomy, 89*, 53–60. https://doi.org/10.1016/j.eja.2017.06.007.

Chapter 13
Agronomical Overview of Mulch Film Systems

S. Guerrini, C. Yan, M. Malinconico, and P. Mormile

Abstract Mulching is intended to reduce or eliminate the aggression of weeds, and has been used for more than 10,000 years using different materials and techniques, increasingly advanced. A significant advance was obtained with the development of plastic materials for agricultural applications. Plastic mulch films for more than half a century have helped to control weeds, improving crop development and achieving greater yield in terms of quantity and quality, as well as reducing phytosanitary risks and irrigation water consumption, due to the reduction of soil evapotranspiration rate. Agronomic techniques and materials have been developed in recent years, providing new options and opportunities for growers. A brief introduction to the history of the mulch, the state of the art, and its market will be reported in this chapter. The recent evolution of mulch films will also be analyzed, in particular, the development and use of photo-selective and biodegradable mulch films in soil. This innovative solution aims to overcome the simple concept of mulch with the purpose of improving the agronomical performances by using photo-selective films, while biodegradable films can contribute to environmental protection by eliminating the problems of collecting post-harvest plastic films. In this chapter different photo-selective mulch films, from their physical characteristics to the agronomic performances are discussed. Photo-selective mulch films for soil solarization are also highlighted.

S. Guerrini (✉)
Novamont Spa, Novara, Italy
e-mail: sara.guerrini@novamont.com

C. Yan
Institute of Environment and Sustainable Development in Agriculture,
Chinese Academy of Agricultural Sciences (CAAS), Beijing, China
e-mail: yanchangrong@caas.cn

M. Malinconico
Institute of Polymers, Composites and Biomaterials, National Council of Research (IPCB-CNR), Pozzuoli (NA), Italy
e-mail: mario.malinconico@ipcb.cnr.it

P. Mormile
Institute of Applied Sciences and Intelligent Systems, National Council of Research (ISASI-CNR), Pozzuoli (NA), Italy
e-mail: p.mormile@isasi.cnr.it

T. J. Gutiérrez (ed.), *Polymers for Agri-Food Applications*,
https://doi.org/10.1007/978-3-030-19416-1_13

The use of photo-selective mulch films in orchards will complete the overview on mulching. The current challenge and future steps in research and development is to perform photo-selective BMFs to combine the advantages of photo-selective effect with respect for the environment. Finally, the recent essays on the performances of photo-selective mulch films in China are presented.

Keywords Biodegradable mulch films in soil · Photo-selective mulches · Plastic mulch films

13.1 Historical Background and State of the Art

Mulching is a common and very old agronomic practice designed to cover the soil with different materials (organic and inorganic) to block the growth of weeds that adversely affect crops.

No one knows how and exactly when this practice was used for the first time, but it is true that our ancestors used stones more than 10,000 years ago in gardening applications as a method to inhibit the growth of weeds.

When Neolithic man approached agriculture, he discovered very early on that plants in their natural environment (e.g. in the forest), where there was an even distribution of dry leaves and branches in the soil, grew better than similar plants in open field, where competition with weeds was strong and more aggressive. So, he just simulated the weed; the dry leaves and the wood residues, the sediments in the soil, prevent the growth of weeds, due to the combined effect of the block of solar radiation and the biochemical action of the decomposition of the organic materials (Fig. 13.1).

Fig. 13.1 Agriculture and mulch in the past

When the Neolithic man understood that the presence of weeds was an obstacle to growth and learned to use straw, shrubs, leaves, twigs or stones that covered the soil to inhibit the growth of weeds, he invented the practice of mulch.

Several peoples (Egyptians, Etruscans, Persians, Romans) adopted the technique of mulching during 2500–3000 years before Christ. This practice continued for centuries and was improved using alternative elements such as manure, pets' dung, waste and seaweeds.

For more than 2000 years, seaweed has been a source of food for many cultures and countries bordering the oceans. It has also been used as animal feed, medicinal herbs and even dyes. The pioneers of the use of seaweed as mulch have been the Scottish and Irish farmers.

Many different materials were used for centuries for mulching such as: straw, dried leaves, bark of trees, cardboard, gravel, lapilli, jute, cocoa husks, natural fiber (mainly coconut and hemp), mulch and organic waste. The real revolution occurred with the advent of plastics in the 1960s. Films based on black polyethylene, with thicknesses of 50–60 μm, were widely introduced in agriculture as mulch.

Agricultural plastic films (greenhouse, mulch and silage films) are growing, along with other plastic items (e.g. irrigation pipes) for agricultural uses have increased their demand to almost double compared to 10 years ago. The areas that drive this increase are mainly Asia and Australia, especially China, which accounts for 70% of regional demand (as discussed in depth in Sect. 13.6). There has also been strong growth in Latin America, North Africa, the Levant and the Middle East (Reynolds 2018).

The state of the art of the use of agricultural films and the growth by region (year 2016) was reported in the literature (Fig. 13.2). Worldwide, the market size

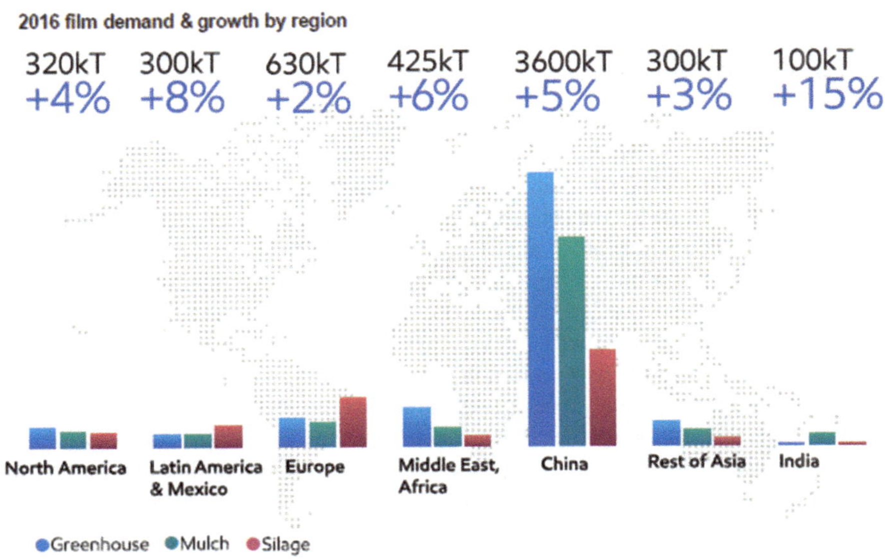

Fig. 13.2 Overview of the agricultural film market: 2016 demand and growth of films by region. Source EM estimates

is 5,6 M tons, with an average annual growth of 7% (different in different areas), with a selection of growth in Asian areas, such as India, where the demand for increase in the food production is stronger.

13.2 Mulch Films

Plastic mulch films have invaded the agricultural world in recent decades and so far, is a common practice to use black films to cover the soil (Garnaud 1988; Brown 2004). These films are completely opaque to solar radiation, mainly in the visible (Vis) part of the solar radiation, and therefore, fully guarantee the "mulch effect", which means non-germination of the weeds (Merino et al. 2018a). Recently, in this sector there was an interesting evolution due to new materials and concepts that overcome the concept of simple traditional mulching.

A film of plastic mulch has the following advantages:

- Soil protection against weeds, thus reducing reduces the use of herbicides.
- Protection of the soil against the weather, in particular cold and strong rains, which determines the erosion of the upper soil.
- Protection of plants against some parasites (Integrated Pest Management), reducing the use of chemical products.
- Preservation of soil moisture and temperature, reducing the need for irrigation (water saving).
- The increase in the formation of humus in the soil and, as a consequence, improves its organic and nutritional characteristics and this affects the quality and the quantity of the crop.
- Possibility of extending the agronomic cycle of crops.

Polyethylene-based mulch films played a leading role from the beginning and today plastic films, such as black, transparent and semitransparent (smoke), are widely used (Espi et al. 2006; Ramakrishna et al. 2006). Recently, innovative mulch films such as photo-selective, biodegradable and functionalized films are gaining much consideration (Mormile et al. 2012; Cerny et al. 2003).

The most common mulch films are monolayer, produced by extrusion of raw materials (sometimes recycled plastic). The most sophisticated mulches are multilayer films that are manufactured using two methods: the co-extrusion of multilayer blown films and the cast extrusion (flat die) (Merino et al. 2018b; 2019a, b). These films, which also contain additives (antioxidants) for ultraviolet (UV) protection, are available on the market in the relief version (cast) or smooth (blown) and in a wide range of thicknesses, widths and holes (with different distribution on the surface) according to the specific needs of the growers, the specifications (duration, soil characteristics and transplant times) and the type of crop.

The cast production method involves embossing while the material is hot, thus changing the surface of the sheet giving rise to a material of greater flexibility, softness and strength. This allows a better adherence of the sheets to the surface of the

Fig. 13.3 Some of different types of plastic mulch films

soil and is particularly suitable for crops grown in high beds. It is possible to process raw materials that guarantee the highest level of mechanical strength and flexibility, and compatibility with all standards of mechanical testing due to multilayer co-extrusion technology (Mormile et al. 2017). Figure 13.3 shows different types of mulch films. This new type of mulch films will be covered in the following paragraphs.

13.3 Biodegradable Much Films

Biodegradable mulch films (BMF) first appeared commercially on the market about 20 years ago, and have since been used more extensively in a wide range of crops, due to their biodegradability in the soil and its characteristics of use. Nowadays, a wide literature is available, and all the aspects related to the functional characteristics such as the mode of use, the biodegradability and the ecotoxicological aspects, together with the agronomic results in different areas and climates have been analyzed. The following pages will summarize these different topics, with the aim of helping to better understand how these materials can be an innovative and useful solution to reduce the end of the useful life related to plastic mulch, the possible leakage of plastics in the environment, the global production of unprofitable plastic waste and labor.

Conventional plastic films must be removed, collected and disposed of properly from the environment at the end of their use. The practice of burning plastic waste after use is banned in many countries, while the removal of plastic films from the field leads to accumulation in the soil (so-called "white pollution"), which can interfere with growth and production of the crops.

It has been estimated that only in Europe in 2014 were produced 1326 thousand tons of agricultural plastic waste, and only 28% was recycled, 30% were sent to energy recovery, while 42% is still sent to landfills. This, as explained above, is due to the characteristics of this waste: heavily contaminated by soil, water and crop residues. In particular, the plastic mulch film can accumulate large amounts of contaminants, reaching up to 4 times its initial weight, which makes mechanical recycling economically unfeasible (European Commission 2018).

One of the main and oldest markets for BMF is Europe. The European Parliamentary Association (APE) has estimated that the size of the market for mulch film in the United States is 80,000 t, and European Bioplastics suggested that 5% of this market is represented by biodegradable mulches (APE Europe, http://www.apeeurope.eu/statistiques). However, this market is growing and is estimated to reach USD 52.3 million in 2021 and USD 64.3 million in 2024 (Briassoulis and Giannoulis 2018).

13.4 Definition and Characteristics of Biodegradable Much Films

When talking about biodegradable materials, clear definitions become essential to correctly use the term "biodegradability" according to the medium where the material will end its life (soil for mulch films) and also to fully benefit from this property (Oranic Waste System (OWS) 2013). A complete BMF can be left in the soil, where it must be buried to put it in direct contact with soil microorganisms: the agents of biodegradation. A clear definition, supported by existing standards will also help to differentiate between BMFs and polyolefin plastics containing additives, which are not degraded, but simply segregate in the environment due to the effect of UV light (light) and/or the temperature (Gutiérrez 2018a). These materials eventually create plastic waste (and microplastics) that remain in the soil for a long time, since they have not proven to comply with the requirements of biodegradability and compostability according to official standards (Oranic Waste System (OWS) 2013). It is worth noting in this regard that biodegradable materials are not necessarily compostable (Gutiérrez 2018b; Herniou--Julien et al. 2019; Gutiérrez et al. 2019).

13.4.1 What Does Biodegradability Means? And What Are Biodegradable Plastics?

Biodegradation is a biochemical process during which the different aerobic or anaerobic microorganisms available in the environment convert plastic materials (produced directly or indirectly from photosynthesis) into natural substances such as water, carbon dioxide (CO_2), methane (CH_4) and compost (no artificial additives are required) (European Commission 2018). The biodegradation process depends on the surrounding environmental conditions (e.g. location or temperature) of the material and the application (European Bioplastic Association, n.d.).

Microorganisms (fungi, bacteria, protozoa) are the agents of biodegradation and are the transforming agents of organic macromolecules so that they can be used by plants.

To evaluate the feasibility of a material being biodegradable and not having negative effects in the end-of-life environment, national and international standards have been developed (e.g. ASTM, UNI, EN, ISO). The biodegradability standards of plastics are based on the biochemical principles of the use of polymer carbon by microorganisms and their transformation into CO_2 under aerobic conditions or CO_2 and methane (CH_4) under anaerobic conditions. Standards provide the specification for an approved/failed test and also provide the basis for making claims about biodegradability in a specific environment (e.g. compost, soil, liquid). Since the biodegradability tests are always defined in relation to the specific end-of-life environment, the test schemes are configured according to the different conditions to which the test material is exposed. However, the different standards share the same test approach, based on:

- The measurement of the biodegradation of the material as a percentage of mineralization using the respirometric test. This means that the CO_2 consumption or evolution at a given moment and temperature (depending on the temperature of the actual end of the useful life) and testing in parallel a reference material (usually cellulose) allows to define the approved/failed criterion. The minimum percentage for a material to be considered biodegradable is 90% in relation to the control.
- Ecotoxicology assessment: the polymer will not induce ecotoxicity effects on the main target organisms and will not increase the level of heavy metals (in accordance with international guidelines and standards) in the final environment. In this case the approved/failed criterion are also given: it is important to evaluate a minimum required level of fragmentation in a composting cycle so as not to spoil the quality of the final compost. Studies have been published in the last 15 years evaluating the impact of BMFs on soils in terms of ecotoxicity. One of the first works published in this line was made by (Kapanen et al. 2008), evaluating the microbial population in the soils in which BMF was buried, demonstrating that no differences were observed. A more complete article published by (Sforzini et al. 2016) analyzing the quality of the soils, after the biodegradation of a biodegradable mulch, using a great variety of biomasses based on the representative organism of the different trophic levels in the food chains of the edaphic and aquatic ecosystems. In this study no negative effects on the soil were detected by using biodegradable mulches (Sforzini et al. 2016).

The biodegradation test carried out to verify the conformity of a plastic material with the biodegradation criterion of 90%, measures the "ultimate biodegradation". Aerobic biodegradation is the transformation of a polymer (plastic material) into CO_2, water and biomass. The evolution of CO_2 is measured to quantify biodegradation. Currently there are no standardized methods available to measure the biomass produced during the biodegradation process. For this reason, the most reliable way to measure biodegradation is by measuring the CO_2 evolution or O_2 consumption, leaving out the reaction of the biomass produced. According to the scientific community, the values of biodegradation of a plastic material corresponding to or greater than 90% mean that complete biodegradation has been achieved. Recent studies analyzed in detail the steps of the biodegradation of a polymeric mulch film in the soil, showing how the carbon in the polymer is used by soil microorganisms and their fate in the soil (Sander 2019).

The European Committee for Standardization (CEN) in 2018 published the first standard on BMF: EN 17033 – "*Plastics - Biodegradable mulch films for use in agriculture and horticulture - Requirements and test methods*". This standard defines the requirements for biodegradable films, in terms of biodegradation in the soil (90% conversion of CO_2 at 25 °C for a maximum time of 24 months, according to ISO 17556 and ASTM D5988), more complete ecotoxicity tests (earthworms, nitrification inhibition test, plants) and, finally, restrictions regarding different potentially harmful components, such as regulated metals and highly worrisome substances. In addition, the standards define the minimum mechanical and optical properties that a mulch film must have in order to have a good performance in the field.

Certification schemes based on the EN (TŪV Austria) and private label systems (OK Biodegradable SOIL) are available and used by the main producers of mulch films to certify the total biodegradability of their materials (TŪV Austria, OK Biodegradable Soil Programme n.d.). The main biodegradable polymers in the soil used for the application of mulch films are shown in Table 13.1. In this table the different agricultural applications of biodegradable and compostable materials are summarized as reported by Martín-Closas et al. (2017). Mulch films represent 50% of the applications for biodegradable and compostable materials in agriculture (Martín-Closas et al. 2017).

Table 13.2 presents the main raw materials and films of BDF producers (Martín-Closas et al. 2017). Biodegradable certified mulch films are commercially available in the market with a thickness of 15 to 10 μm, mainly black (the master black batch is added in percentages of 5–7% to ensure that no light passes through the film). In addition, clear biodegradable films are present in the market and are used mainly for early production in crops such as sweet corn, corn or melon. The BMF can be mechanically placed in the field with the same equipment that is used for the conventional plastic mulch films (mainly low-density polyethylene - LDPE), with no particular differences if not to release the tension in the roll, in order to avoid extra-stretching.

Briassoulis (Briassoulis and Giannoulis 2018) has reported an exhaustive study on the mechanical and radiometric properties of two of the main certified BMFs

Table 13.1 Main application of type of biodegradable polymers and share in agriculture (Martín-Closas et al. 2017)

Type of product	Bioplastic material	Share (%)
Mulching films	PBAT, PBSA, BS-LA copolymer, PBLDA, TP Starch, PCL, PLA	51
Films for fumigation	PBAT	3
Sheet/forestry films	PBAT, PCL	12
Bands, tapes, ties	PBAT, PBSA, PBS, PLA	12
Floating covers	PLA	1
Nets	PBA, PLA	4
Yarns, ropes	PBS, PBA, PLA	3
Pots/planters	PBSA, PBS, PBLDA	7
Other products (clips…)	PBSA, PBS, BS-LA, PLA	10

Table 13.2 Main biodegradable raw materials used for mulch films: company's name, and commercial name (Martín-Closas et al. 2017)

Product	Raw material	Company
(a) *Granulates*		
Biolice®/Biofilm® (BF)	PBAT-cereal flour	Limagrain Céréales Ingrédients
DaniMer®	PHA	DaniMer Scientific
Ecovio® (EcV)	PBAT-PLA	Basf Se
Mater-Bio® (MB)	PBAT-TPS-Veg.oils	Novamont S.p.a.
Meredian®	PHA	Meredian Inc.
Mirel®	PHA	Metabolix Inc.
So Green®	P (3, 4 HB)	Tianjin Greenbio Material Co. Ltd.
(b) *Finishead garden, horticultural and agricultural products as films*		
BioBag Agri Film	Mater-Bio®	BioBag International AS
Biofilm Sylva	Mater-Bio®	GroenCreatie
Mater-Bio EF 04P	Mater-Bio®	Novamont S.p.a.

used (Ecovio® by BASF and Mater-Bi® by Novamont) in accordance with the requirements of EN 17033, to show its functionality. It is important to emphasize that BMFs have properties and behavior in the field that do not completely resemble conventional LDPE coverages, as they are different materials, with their own specificity and a different end of life. In general, BMF shows a greater permeability to water vapor and this aspect, in relation to the dynamics of water in the soil and its availability for plants has been studied. The first field study was conducted by the University of Pisa in 2005. BMF, LDPE and bare soil were compared with different irrigation systems in broccoli crops, evaluating the amount of water needed for different materials and the yields obtained (Magnani et al. 2005). More recently, (Saglam et al. 2017) analyzed and proposed models for soil water dynamics under a raised bed cake pumpkin growing system covered with different types of mulches: BMF, LDPE and paper mulch. The results have shown that biodegradable mulches (paper and BMF) reduced evaporation by restricting water and vapor flow through the soil-atmosphere interface and therefore conserved water in the root zone. These materials proved to be a valid alternative to LDPE films. It was however stressed that a premature degradation of BMF could interfere with the soil water dynamics, allowing for greater water transpiration, where the soil is not adequately covered (Magnani et al. 2005). The differences do not change the effect of overall mulching, and in certain cases and for certain crops, the use of BMF is allowed when LDPE mulch films cannot be used.

13.4.2 Applications

BMF guarantees the same agronomic functions as a LDPE film without having to be removed from the soil. Thus, biodegradability becomes a functional property of a material that is important and a tangible advantage in those cases where there are risks of environmental contamination due to the leakage of plastic materials that are not properly disposed.

In general, BMFs are used to produce vegetables such as lettuce, zucchini, melon, strawberry and pepper. More than 15 years of studies highlighting the main aspects of BMF and different types of biodegradable polymers are available in terms of yield and quality of crops, precocity, weed control and, finally, in the degradation dynamics in different climates.

An extensive study has been carried out by (Martín-Closas et al. 2017), collecting the main results in horticultural crops. In particular, traditionally mulched crops (e.g. table tomatoes and melons) from Mater-Bi® and Ecovio® biodegradable films have shown that the yield and quality of the crops are comparable to crops mulched from LDPE films. When black films are not used early, the effect is remarkable (Magnani et al. 2005; di Cozzolino et al. 2015; Limpus et al. 2012; Candido et al. 2004).

In melon crops, 6 years of data are available at the Po di Tramontana Italian Research Center, demonstrating that black BMF films of 12 and 15 μm (Mater-Bi®) are comparable to 40 μm black LDPE films in terms of yield and quality of the fruits and the BMF films covered the soil for a sufficient period of time for the crop cycle (Guerrini et al. 2009). In melon crops, some studies have also been carried out usinf green photoselective BMF to increase the soil temperature and obtain an early production, as well as to prevent the growth of weeds. The results obtained in a 2 year trial in Italy showed an increase in the yield both in number and weight of the fruits, since the green color improved the environmental conditions (soil temperature). In addition, according to the results of this work (Filippi et al. 2011), the green BMF has shown a better commercial quality than the black BMF and the LDPE films, especially for the sugar content and the degree of maturation. At the other extreme, green BMF has been shown to be more prone to early degradation in the exposed part and, therefore, commercial optimization is required, provided that these films are used in a crop with a more vertical canopy and, for therefore, offer less shade to the BMF (Filippi et al. 2011).

Some crops that do not normally use plastic mulches such as rice and wine grapes have been shown to benefit by using a BMF. In both cases, a plastic layer from the field at the end of the crop would be difficult to remove, with the consequence of remaining in the environment.

In rice, the first trials in Europe began 4 years ago, in the field of low inputs, where the objective was to reduce the amount of chemicals introduced into the environment. Black BMF (Mater-Bi® and Ecovio®) have been used, which shows good control of weeds and an "initiating effect" on the crop, which germinated 4–5 days earlier than plants that grow on bare soil and have a performance comparable to environmentally sustainable techniques such as "green mulching" or mechanical weeding (Guerrini S. Personal communication).

The use of BMF in French and Italian vineyards has shown the benefit of this technique compared to the use of LDPE films or bare soil where the weeds are controlled the first year with herbicides or with mechanical control. In some areas, the first few years of planting plastic mulch films are used to control weeds and improve crop growth, but these films are generally not removed from the field, where they will remain. Tarricone (2011), Gastaldi et al. (2012) and Gastaldi (2018) have shown that in southern Italy and France, respectively, BMFs are a valid technique to accelerate the growth of young grape, cutting in the first years of the fruits, with the results to obtain a higher yield in the first two years compared to bare soil, as well as a more developed rooting system Gastaldi (2018). Gastaldi (2018) also showed that this effect can be obtained in the southern France with a soil cover of approx. 6 months from the plantation.

13.5 Photo-Selective Mulch Films: Physical and Chemical Characterization

The concept of selecting the solar radiation that passes through a plastic film for agricultural use in order to determine the best physical and agronomic conditions for the growth and defense of crops is known since some decades as a "photo-selective effect" and, recently, it is gaining great interest and consideration both from plastic film producers and farmers (Torres-Olivar et al. 2018; Canul-Tun et al. 2017).

The selection of the solar spectrum contributes significantly to the success of the current agronomic practice and to the concrete improvement of production in terms of quality and quantity of crops. The management of sunlight is possible due to the use of special plastics, called photo-selective because their optical properties can select the part of solar radiation necessary according to the farmer's requests in terms of the type of plant, the time of transplant and the period of culture. Since photo-selective films allow the soil to be heated or cooled at convenience, this can improve the climatic conditions in the greenhouse by selecting solar energy to obtain a high sterilization effect of the soil, thus reducing the presence of harmful insects or obtaining a high water saving.

In the last two decades, photo-selective plastic films have gained increasing consideration, and today they play an important role for some high-quality crops. Growers of horticulture, fruits (strawberry, melon, peach, cherry), flowers, pots and nurseries have understood that due to the use of photo-selective films, they have superior advantages that traditional plastics cannot give.

The intellectual approach of agricultural plastic films, which is finally spreading, is that plastic is not only a cover, but could be an essential tool to exploit solar radiation at best. Films with this target (mulching, solarization, low tunnel and greenhouse covering) are gaining, year after year, important market shares.

In this scenario, the contribution of scientific research from chemistry, physics, materials engineering and agronomy is increasingly determining for the development of new materials for agricultural applications. New proposals for advanced materials with special properties are continuously offered by the leading industries in the agricultural film sector. The agricultural films can heat or cool the soil, thus improving the growth of the plant and resulting in the best quality and quantity of crops, as well as help the integrated management of pests. In this section, the physical properties based on experimental data obtained in recent trials, using some photo-selective films will be presented with the objective of evaluating the contribution of photo-selective mulch films on the growth of plants. The experimental results of the photo-selective films have been compared with those obtained with the traditional ones.

The physical characteristics in term of optical and thermal properties that are found in photoelective mulch films, determine the agronomic performance of the films used for mulching and, in general, for the solarization of the soil or to cover greenhouses. The experimental results show the high potential of this type of plastic film for agriculture and confirm the considerable potential that is still the subject of active research, whose objective is to demonstrate the advantages and peculiarities of photo-selective films and their contribution to the achievement of healthier products and more environmentally friendly.

Some photo-selective films that are among the most innovative in the international market have been considered. In particular, a brown mulch film (for transplants in winter) and a yellow-brown film (for transplants in summer), both have a thickness of 30 μm were selected.

As a first approach to the study of the considered films, the transmissivity as a function of the incident light wavelength ranging from 0.2 μm and 2.5 μm (UV-Vis-infrared regions of the solar spectrum) and in the IR region was measured. Up to 25 μm (medium infrared and far infrared regions) were obtained the spectra of the films considered using a two-module spectrophotometer, Jasco model 560. This preliminary information allowed to foresee the behavior of the films in their operative functions (mulching, heating or cooling and the blocking of IR radiation from the soil).

Figure 13.4 shows two transmittance spectra *versus* the wavelength of the brown film, in two regions 200–2500 nm (Fig. 13.4a) and 2500–12,000 nm (Fig. 13.4b). In the same way, the two spectra of the yellow-brown film are shown in Figs. 13.5a and 13.5b.

The spectrum of a traditional 30 μm thick black film is shown in Fig. 13.6 as a comparison. In this case, only the region 200–2500 nm was only shown, because also in the middle infrared region (MIR) the transmittivity is constantly zero.

It is worth noting that the spectra of the two mulch films (brown and yellow), ensuring the mulching function due to very low transmission capacity (< 6%) in the Vis region. For the brown film (see Fig. 13.4a) the transmissivity increases in the near infrared region (NIR) up to 80%. The high transmissivity in the NIR region is the necessary condition for the thermal effect, while the absorption peaks in the MIR are sufficient (Fig. 13.4b), because it means that this film blocks the heat emitted from the soil, associated with the two peaks of wavelength around 3500 nm and 6800 nm.

The low transmittance in the NIR region was noted for the yellow film, which corresponds to a high reflectivity that ensures the cooling function of the soil. The spectrum analysis of the black mulch film confirms that the film behaves only as a mulch film. On the contrary, the photo-selective films (brown and yellow), according to their spectra, from a theoretical point of view, could ensure the heating or cooling of the soil, respectively. To confirm this behavior, some of the essays were organized, which will be presented in the next section.

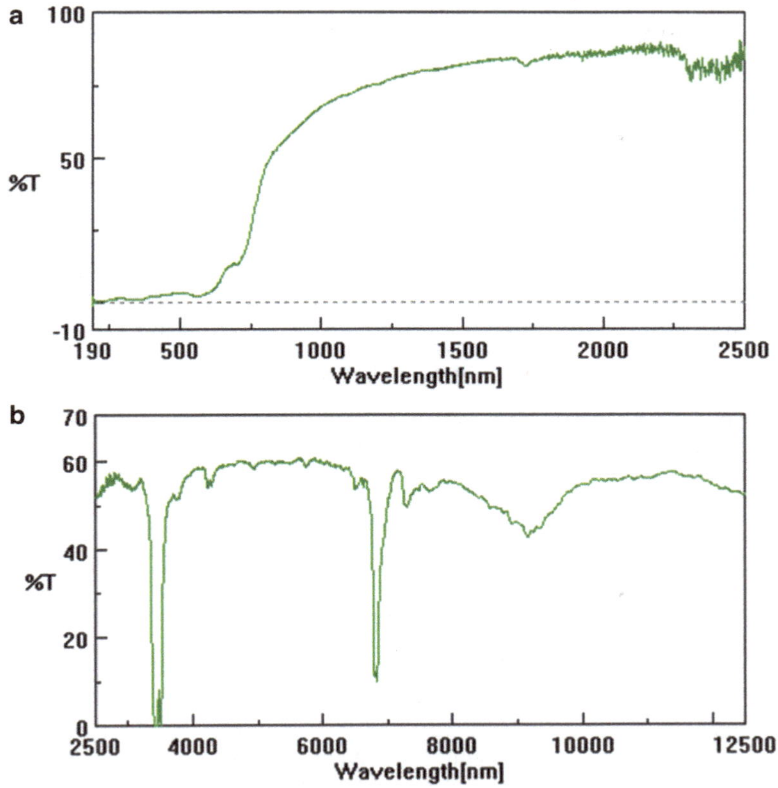

Fig. 13.4 Optical spectra of a thermal mulch film used for winter transplants in UV-Vis-NIR region (**a**) and IR region (**b**)

13.5.1 Experimental Results

The results obtained from the spectral analysis of the analyzed films were considered to organize some subsequent tests under real conditions with the in order to confirm the potentialities of the photo-selective films and being compared with the traditional ones. The objective of the first test was to verify the thermal capacity of the brown mulch film and measure the eventual increase in temperature compared to a traditional film (Mormile et al. 2012). In this experiment, under greenhouse conditions two lines with brown films and two lines with black films both with the same thickness of 30 μm were used to cover the soil. Each line had three thermocouples at different depths (2, 5 and 10 cm) in the soil and temperatures were recorded for 12 hours every half hour. The temperatures as a function of time at 5 cm and 10 cm depth in the soil for the two films were reported in Fig. 13.7, while the difference in temperature (ΔT) between the two films was shown in Fig. 13.8.

The results shown in Figs. 13.7 and 13.8 clearly confirm that the brown film has very good thermal properties and can guarantee higher temperatures at the root of

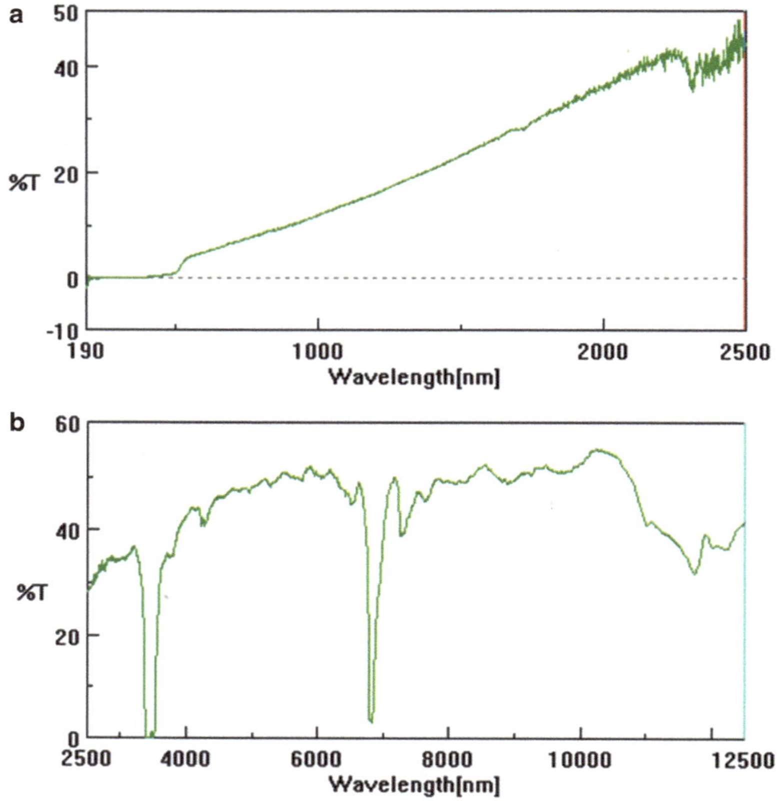

Fig. 13.5 Optical spectra of a photo-reflective mulch film used for spring and summer transplants in UV-Vis-NIR region (**a**) and IR region (**b**)

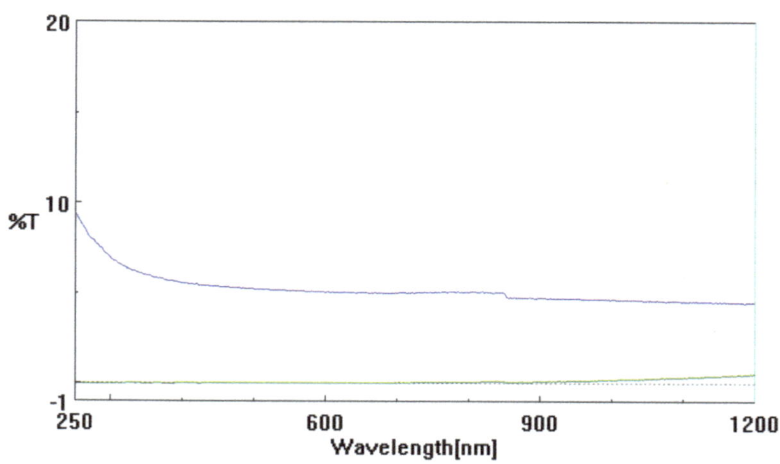

Fig. 13.6 Total transmittivity (curve a) and reflectivity (curve b) *vs* wavelength of a common black mulch film in the region 250–1200 nm

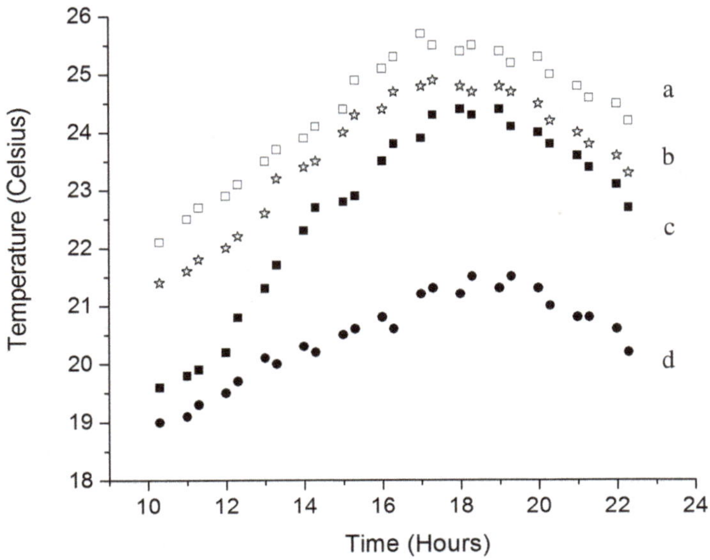

Fig. 13.7 Temperatures as a function of the time at 5 cm and 10 cm depth in the soil for the two compared mulches (brown film and black film): (**a**) brown mulch at 5 cm in the soil, (**b**) brown mulch at 10 cm in the soil, (**c**) black mulch at 5 cm in the soil and (**d**) black mulch at 10 cm in the soil

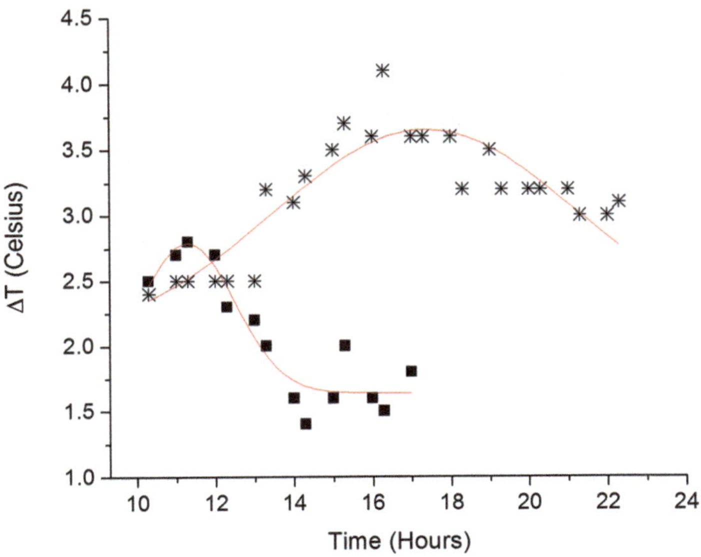

Fig. 13.8 Difference in temperature (ΔT) calculated between the two mulches at different depths. Cross-point curve: ΔT at 10 cm in the soil, square black point curve: ΔT at 5 cm in the soil. The continuous curve represents the Gaussian adjustment of the experimental data

the plant. This behavior, especially for winter transplants, offers many agronomic benefits, since it provides the plant with better conditions, without stress due to low temperatures, and helps the plant to better form its root system. These aspects give rise not only to a better plant state of the plant, but also determine the anticipation of the crop, a better quality of the products and, finally, an increase in terms of quantity. The yields described are confirmed by many agronomic tests carried out in recent years, which compare different types of mulch films for several crops, both in open field and under greenhouse.

An additional open field test was carried out in southern Italy (Foggia nearby), comparing the thermal contribution of the same mulch film to increase soil temperature in the area of the root plants. Four film systems: brown, black, transparent and yellow were compared in terms of temperatures at different depths in the soil (soil-film gap, 5, 10 and 20 cm). Table 13.3 reports the experimental data for each mulch film and the temperatures recorded in the soil.

As expected in the soil-film gap the highest temperature was found in the case of the black film, due to heat transfer by contact. The surface of the black film heats up more than others, and the heat is transferred directly to the contact layer, which in this case, is the soil-film gap. The yellow film, due to its high reflectivity, as expected, was not able to heat the soil. The brown film, because of its thermal properties, heats the part of the soil where the roots develop, and this, as is well known, is very important for the growth of the plants.

In order to evaluate the agronomic yields of the photo-selective mulch films, tests were carried out with four different types of mulch films for the cultivation of melon (cultivar Macigno/Clause-Tezier) under low tunnel conditions being compared with the one of traditional mulch films. A brown film, a black film and a transparent film was considered.

Some agronomic aspects related to the crop cultivation considered can be summarized in Table 13.4. According to the experimental data reported in Table 13.4, the photo-selective brown film confirms its good performance compared to other traditional films.

Table 13.3 Temperatures at different depths in the soil using several mulch films

Film	Soil-film gap (°C)	5 cm depth (°C)	10 cm depth (°C)	20 cm depth (°C)
Brown	30.7	26.6	23.5	21.7
Black	33	24.5	22	20.5
Transparent	30.5	25.5	22.3	19.5
Yellow	27.5	21	18	15

Table 13.4 Agronomical results

Plastic film	First week production (Kg/m^2)	Average weight of the fruit (gr)	Brix grade	Mulch effect
Brown	6.2	1513	11.6	Yes
Black	5.8	1493	10.8	Yes
Transparent	6.1	1393	10.8	No

13.5.2 Functionalized Mulch Films for Saving Water

As described above, there is a type of film, belonging to the photo-selective, that works to cool the soil due to its high reflectivity to solar radiation, in addition to the mulch (Mormile et al. 2015; Rippa et al. 2018). This property allows to avoid damages caused by the high temperatures during the warm period of the transplants. In this section, the optical properties of photo-reflective mulch (PR) films will be considered, as well as some experimental results of trials organized in different geographical conditions will be presented with the aim of demonstrating that it is possible to save water in agriculture and, finally, an estimate of the percentage of water saved compared to traditional mulches will be reported.

The main objective of the tests was to demonstrate that PR films (yellow, silver and white) offer the possibility of water saving in agriculture. The test was carried out in the greenhouse and several films (two PR mulches and one black as reference) were tested using an automatic irrigation system in each parcel, according to the water content in the soil. This system also allows the acquisition of data, in order to calculate day by day the water consumption for each line covered with different mulch films. It has been shown in accordance with the experimental results that the use of the PR mulch film could contribute a valuable contribution to the water problem in agriculture, which is the massive water consumption.

The critical phase of the transplant, due to the stress induced by the edge conditions such as the type of soil and its temperature and the temperature around the baby-leafs can be overcome due to the high reflecting action of the solar radiation that reduces the temperature in the surface of the mulch film. In addition, the temperature variation between the film and the soil results in the water condensation, which is the basis for water saving. This means that after the irrigation, due to the effect of condensation, the water remains in the upper layer of the soil so that it is fully exploited by the root of the plant. In the case of black mulch, the situation is completely opposite. This means that there is no water condensation that descends by gravity in the soil, away from the roots of the plants. A schematic representation of a mechanism that mainly affects water saving is shown in Fig. 13.9. An approach to solving the problem of water sources was presented here, demonstrating through experimental results of the trials that it is possible to save water with this simple method.

13.5.3 Materials and Methods

The optical characteristics of the yellow mulch film considered for the tests were shown in Sect. 13.5.1 (Figs. 13.4a and 13.4b). For each test, the same experimental configuration was adopted (Fig. 13.9), based on the use of tensiometers to control soil moisture near the roots.

Three different parcels were considered: one is bare earth (**a**), another is covered with a traditional black film (**b**) and the last is covered with a yellow PR film (**c**). The automatic irrigation system is independent for each parcel, which means that

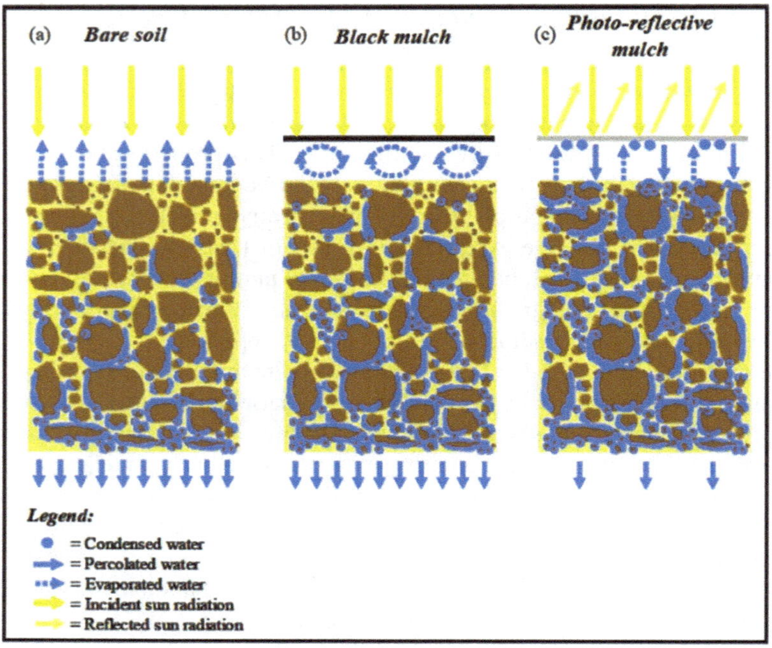

Fig. 13.9 Schematic representation of the state of the water in the soil according to the different situations

Fig. 13.10 Experimental configuration adopted for the tests. (**A**) bare soil, (**B**) black mulch, (**C**) PR yellow mulch, (**D**) driver, ADS) acquisition data system, AIS) automatic irrigation system, WD) water distributor and WM) water meter

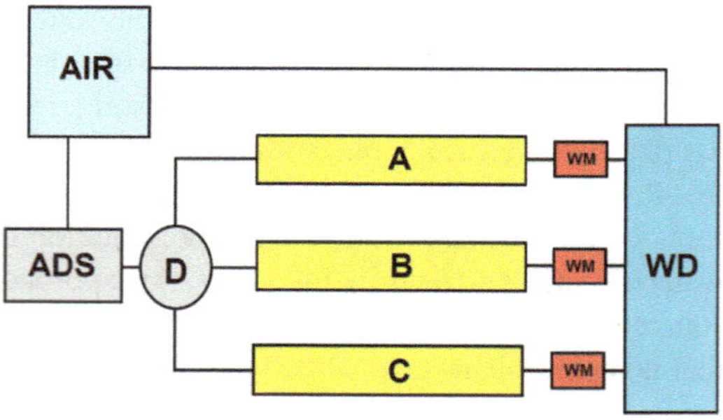

the irrigation process is executed entirely by the output data of the tensiometers placed in each parcel. In other words, the water needs for each parcel are monitored through tensiometers (Fig. 13.10).

When the humidity in the soil decreases below a threshold value, the output tensiometer signal (soil water voltage) enables the irrigation system for each parcel, and the water meter measures the amount of water supplied until the soil moisture tension returns to the reference value. At this stage, the water pump is switched off and the amount of water is automatically recorded. The objective of the experimental activity was to evaluate and quantitatively demonstrate that PR films (yellow, silver or black) offer the possibility of water saving on crops. Tomato

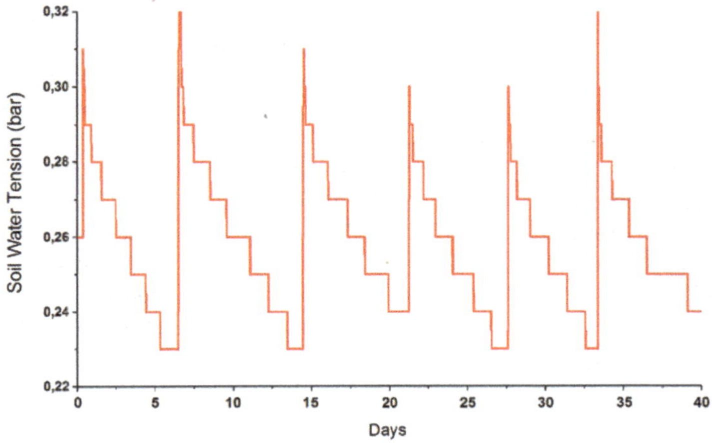

Fig. 13.11 Soil water tension as a function of the time

Table 13.5 Experimental data of water consumption (m³/ha) for each parcel and water saved (%)

Product	BS (m³/ha)	YM (m³/ha)	BM (m³/ha)	S (%)
Tomato	3720	1120	1471	23.8
Melon	2231	1022	1412	27.6

BS: Bare Soil, YM: Yellow Mulch, BM: Black Mulch and S: Saved Water (%)

and melon crops in two different farms were tested and monitored in terms of water consumption for each parcel with the system described for the entire plant cycle. The water consumption data were stored day by day for A, B and C parcels mulched with PR yellow films (**C**), with black (**B**) and bare soil (**A**) (Fig. 13.10).

The soil water tension as a function of time for a few days during a trial is shown as an example in Fig. 13.11. The minimums of the curve indicate the threshold value at which the irrigation system (including the system for measuring the amount of water used) begins to work. In this way, it is possible to monitor hour by hour the water consumption for each parcel.

The data of each parcel was collected and used to calculate the theoretical water consumption for 1 ha in the case of bare soil, yellow and black film. This estimate is presented in Table 13.5.

The percentage of water saved is referred to YM with respect to BM. It indicates that with a very simple adjustments, as the use of a different mulch film, in the traditional agronomical practice, it is possible to save water up to 30%.

13.6 State of the Art in China

Plasplastic film mulching was introduced for the first time in the late 1970s (1979) in China, and since the early 1980s it has been used commercially to produce vegetables. It has also been further developed in different types of mulching, including flat mulching, mulching with ridges, partial and total mulching of the land surface, etc. In recent years, plastic mulch films are playing an important role in Chinese agriculture, due to the advantages for the growth of plants, which affect both the quality and the quantity of the crop. In the last three decades, the areas dedicated to agriculture in the open field increased drastically and in parallel the consumption of mulch film was also exploded. Two graphs are shown in Fig. 13.12, where trends and forecast can be observed until 2025 for the use of mulch films in China. In addition, the introduction of special mulch films, in the Chinese agronomic system, represents a qualitative step forward for the improvement of agricultural production.

In 2012, around 13% of China's farmland was mulched, representing 60% of the plastic film used for agricultural crops in the world. Previous studies have focused mainly on improving crop yields and water use efficiency (WUE) of the main crops,

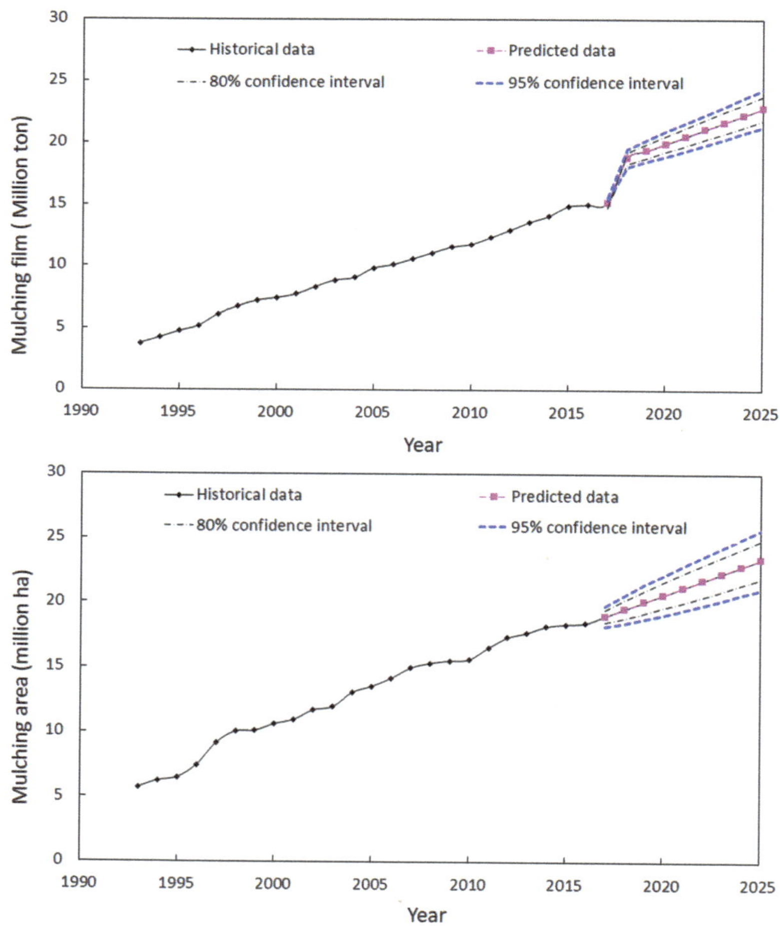

Fig. 13.12 Trends in the areas for open field agriculture in China and the relative consumption of mulch films, with a forecast of up to 2025

i.e. corn, peanuts, cotton wheat and potatoes by mulching in specific sites or regions. However, a rational use of plastic films, respecting the environment is also required. No comprehensive studies are available to evaluate the impact of film mulching on the yield and WUE of all crops in China.

The growing world population demands a 110% increase in crop production by 2050 to ensure food security and adapt to changes in diet. However, since 1990, crop yields have stagnated in many parts of the world, representing a major challenge for global food security (Alston et al. 2009; Chen et al. 2014). Innovative solutions are needed to increase efficiency in the use of crop system resources to produce more grains *per* unit area. In China, the country with 1/5 of the world's population, a simple plastic mulching film technology has helped increase crop productivity by up to 180% in Gansu Province (Yan et al. 2015). It can be seen here that plastic mulch has helped Chinese farmers increase their crop productivity by more than 40%. We used data, including 3570 pairs of crop yields and 1327 pairs of field evapotranspiration measurements collected from 1095 publications and 6 long-term experiments across China, to conduct a meta-analysis (Rippa et al. 2018). Our results showed that the plastic film mulching improved crop yield and WUE by 45.5% and 58.0%, respectively. The highest yield increase occurred in the dry (annual precipitation <400 mm) and relatively warm areas (total thermal time 3000-4000 °C, base ≥0 °C). In 2012, the estimated contribution to yield increase from plastic film mulching was 3.0×10^7 metric tons for the three main wheat, corn and rice crops in China. In summary, plastic film mulching increased crop yields more than any other technique in semiarid and arid areas of China. New and degradable films are being developed to minimize negative impacts on the soil and the wider environment. This provides an example to use the technology in other regions of the world (e.g. Central Asia) to advance the productivity of crops for global food security.

Acknowledgments The authors would like to thank Ginegar Plastic Products, Israeli company supplier of photo-selective mulch films used for the trials.

Conflicts of Interest The authors declare no conflict of interest.

References

Alston, J. M., Beddow, J. M., & Pardey, P. G. (2009). Agricultural research, productivity, and food prices in the long run. *Science, 325*(5945), 1209–1210. https://doi.org/10.1126/science.1170451.

Briassoulis, D., & Giannoulis, A. (2018). Evaluation of the functionality of bio-based plastic mulching films. *Polymer Testing, 67*, 99–109. https://doi.org/10.1016/j.polymertesting.2018.02.019.

Brown, R. P. (2004). *Polymers in agriculture and horticulture*. Oxford: Rapra Technology Ltd..

Candido, V., Miccolis, V., Castronuovo, D., Manera, C., & Margiotta, S. (2004). Mulching studies in greenhouse by using eco-compatible plastic films on fresh tomato crop. *Acta Horticultuturae., 710*, 415–420. https://doi.org/10.17660/actahortic.2006.710.50.

Canul-Tun, C. E., Ibarra-Jiménez, L., Valdez-Aguilar, L. A., Lozano-del Río, A. J., Cárdenas-Flores, A., Zermeño-González, A., Lozano-Cavazos, C. J., Valenzuela-Soto, J. H., & Torres-Olivar, V. (2017). Influence of colored plastic mulch on soil temperature, growth, nutrimental status, and yield of bell pepper under shade house conditions. *Journal of Plant Nutrition, 40*(8), 1083–1090. https://doi.org/10.1080/01904167.2016.1263331.

Cerny, T. A., Faust, J. E., Layne, D. R., & Rajapakse, N. C. (2003). Influence of photoselective films and growing season on stem growth and flowering of six plant species. *Journal of the American Society for Horticultural Science, 128*(4), 486–491. https://doi.org/10.21273/jashs.128.4.0486.

Chen, X., Cui, Z., Fan, M., Vitousek, P., Zhao, M., Ma, W., et al. (2014). Producing more grain with lower environmental costs. *Nature, 514*(7523), 486–489. https://doi.org/10.1038/nature13609.

di Cozzolino, E., Bilotto, M., Leone, V., Zampella, L., Petriccione, M., Cerrato, D., & Morra, L. (2015). Produzione e qualità di melone retato su pacciamatura in Mater-Bi®. *Colture Protette, 6*, 66–71.

Espi, E., Salmeron, A., Fontecha, A., García, Y., & Real, A. I. (2006). Plastic films for agricultural applications. *Journal of Plastic Film & Sheeting, 22*(2), 85–102. https://doi.org/10.1177/8756087906064220.

European Bioplastic Association.(n.d.). Available in: https://www.european-bioplastics.org/bioplastics/

European Commission. (2018). "A European strategy for plastic in a circular economy", COMM, 16.1.2018.

Filippi, F., Magnani, G., Guerrini, S., & Ranghino, F. (2011). Agronomic evaluation of green biodegradable mulch on melon. *Italian Journal of Agronomy, 6*(2), 111–116. https://doi.org/10.4081/ija.2011.e18.

Garnaud, J. C. (1988). *Agricultural and horticultural application of polymers*. Shawbury: Rapra Technology Ltd.

Gastaldi, E. (2018). "Agronomical performances of biodegradable films as alternative to polyethylene mulches in vineyard", CIPA Congress 2018, Agriculture, Plastics and Environment, XXI CIPA Congress, Bordeaux, Arcachon, France, May 29th, 30th and 31st 2018; Presentation.

Gastaldi, E., Cesar, G., Angellier, H., Berger, F., & Chevillard, A. (2012). "Impact of bio-mulch film on a French Vineyars.", Agrobiofilm Workshop, Almeiring, June 2012.

Guerrini, S., Ranghino, F., Chiarini, F., & Mantoan, C. (2009). "Biodegradable mulch film: A review of performance and agronomical results of six years trials in the Veneto Region, 2009, Proceeding of the 35th Agricultural Plastic Congress, State College, Pennsylvania, July 13–16.

Gutiérrez, T. J. (2018a). Chapter 9. Biodegradability and compostability of food nanopackaging materials. In G. Cirillo, M. A. Kozlowski, & U. G. Spizzirri (Eds.), *Composite materials for food packaging* (pp. 269–296). WILEY-Scrivener Publisher.. EE.UU. ISBN: 978-1-119-16020-5. https://doi.org/10.1002/9781119160243.ch9.

Gutiérrez, T. J. (2018b). Are modified pumpkin flour/plum flour nanocomposite films biodegradable and compostable? *Food Hydrocolloids, 83*, 397–410. https://doi.org/10.1016/j.foodhyd.2018.05.035.

Gutiérrez, T. J., Toro-Márquez, L. A., Merino, D., & Mendieta, J. R. (2019). Hydrogen-bonding interactions and compostability of bionanocomposite films prepared from corn starch and nano-fillers with and without added Jamaica flower extract. *Food Hydrocolloids, 89*, 283–293. https://doi.org/10.1016/j.foodhyd.2018.10.058.

Herniou--Julien, C., Mendieta, J. R., & Gutiérrez, T. J. (2019). Characterization of biodegradable/non-compostable films made from cellulose acetate/corn starch blends processed under reactive extrusion conditions. *Food Hydrocolloids, 89*, 67–79. https://doi.org/10.1016/j.foodhyd.2018.10.024.

Kapanen, A., Schettini, E., Vox, G., & Itävaara, M. (2008). Performance and environmental impact of biodegradable films in agriculture: a field study on protected cultivation. *Journal of Polymers and the Environment, 16*(2), 109–122. https://doi.org/10.1007/s10924-008-0091-x.

Limpus, S., Heisswolf, S., Kreymborg, D., Wright, R., Hall, W., & Guerrini, S. (2012). Comparison of biodegradable mulch products to polyethylene in irrigated vegetable, tomato and melon crops. Department of agriculture, fishery and forestry, final report for horticulture Australia Limited Project MT09068. Sydney.

Magnani, G., Filippi, F., Graifenberg, A., & Bertolacci, M. (2005). Valutazione agronomica di film biodegradabili per la pacciamatura. *Colture Protette, 1*, 49–68.

Martín-Closas, L., Costa, J., & Pelacho, A. M. (2017). Agronomic effects of biodegradable films on crop and field environment. In M. Malinconico (Ed.), *Soil degradable bioplastics for a sustainable modern agriculture* (pp. 67–104). Berlin: Springer. https://doi.org/10.1007/978-3-662-54130-2_4.

Merino, D., Gutiérrez, T. J., Mansilla, A. Y., Casalongué, C., & Alvarez, V. A. (2018a). Critical evaluation of starch-based antibacterial nanocomposites as agricultural mulch films: Study on their interactions with water and light. *ACS Sustainable Chemistry & Engineering, 6*(11), 15662–15672. https://doi.org/10.1021/acssuschemeng.8b04162.

Merino, D., Mansilla, A. Y., Gutiérrez, T. J., Casalongué, C., & Alvarez, V. A. (2018b). Chitosan coated-phosphorylated starch films: Water interaction, transparency and antibacterial properties. *Reactive and Functional Polymers, 131*, 445–453. https://doi.org/10.1016/j.reactfunctpolym.2018.08.012.

Merino, D., Gutiérrez, T. J., & Alvarez, V. A. (2019a). Potential agricultural mulch films based on native and phosphorylated corn starch with and without surface functionalization with chitosan. *Journal Polymers and the Environment, 27*(1), 97–105. https://doi.org/10.1007/s10924-018-1325-1.

Merino, D., Gutiérrez, T. J., & Alvarez, V. A. (2019b). Structural and thermal properties of agricultural mulch films based on native and oxidized corn starch nanocomposites. *Starch-Stärke, 71*(7–8), 1800341.https://doi.org/10.1002/star.201800341.

Mormile, P., Capasso, R., Rippa, M., & Petti, L. (2012). Light filtering by innovative plastic films for mulching and soil solarization. *Acta Horticulturae, 1015*, 113–121. https://doi.org/10.17660/actahortic.2014.1015.12.

Mormile, P., Rippa, M., Bonanomi, G., Scala, F., Yan, C., & Petti, L. (2015). Photo-reflective mulches for saving water. In agriculture, world academy of science, engineering and technology, agricultural and biosystems engineering 2(5).

Mormile, P., Stahl, N., & Malinconico, M. (2017). The world of plasticulture. In M. Malinconico (Ed.), *Soil degradable bioplastics for a sustainable modern agriculture. Green chemistry and sustainable technology*. Berlin: Springer. https://doi.org/10.1007/978-3-662-54130-2_1.

Oranic Waste System (OWS). (2013). Benefits and challenges of bio and oxo-degradable plastics, Available in: http://www.ows.be/wp-content/uploads/2013/10/Executive-summary1.pdf

Ramakrishna, A., Tam, H. M., Wani, S. P., & Long, T. D. (2006). Effect of mulch on soil temperature, moisture, weed infestation and yield of groundnut in northern Vietnam. *Field Crops Research, 95*(2–3), 115–125. https://doi.org/10.1016/j.fcr.2005.01.030.

Reynolds, A. (2018). Trends in the agricultural film business, AMI Agricultural Film 2018, Presentation.

Rippa, M., Yan, C., Petti, L., & Mormile, P. (2018). A simple method for saving water in agriculture based on the use of photo-selective mulch film. *International Journal of Agriculture and Environmental Research, 4*(1).

Saglam, M., Sintim, H. Y., Bary, A. I., Miles, C. A., Ghimire, S., Inglis, D. A., & Flury, M. (2017). Modeling the effect of biodegradable paper and plastic mulch on soil moisture dynamics. *Agricultural Water Management, 193*, 240–250. https://doi.org/10.1016/j.agwat.2017.08.011.

Sander, M. (2019). Biodegradation of polymeric mulch films in agricultural soils: Concepts, Knowledge gaps, and future research directions. *Environmental Science & Technology., 53*(5), 2304–2315. https://doi.org/10.1021/acs.est.8b05208.

Sforzini, S., Oliveri, L., Chinaglia, S., & Viarengo, A. (2016). Application of biotests for the determination of soil ecotoxicity after exposure to biodegradable plastics. *Frontiers in Environmental Science, 4*(68). https://doi.org/10.3389/fenvs.2016.00068.

Tarricone, L. (2011). Evaluation of the effect of biodegradable mulching on grapevine nursery soil management, Conference Paper, 17th GIESCO Meeting 2011, pg. 139–142.

Torres-Olivar, V., Ibarra-Jiménez, L., Cárdenas-Flores, A., Lira-Saldivar, R. H., Valenzuela-Soto, J. H., & Castillo-Campohermoso, M. A. (2018). Changes induced by plastic film mulches on soil temperature and their relevance in growth and fruit yield of pickling cucumber. *Acta Agriculturae Scandinavica, Section B—Soil & Plant Science, 68*(2), 97–103. https://doi.org/10.1080/09064710.2017.1367836.

TŪV Austria, OK Biodegradable Soil Programme.(n.d.) Available in: http://www.tuv-at.be/it/certifications/ok-biodegradable/

Yan, C. R., He, W. Q., Liu, S., & Cao, S. L. (2015). *Application of mulch films and prevention of its residual pollution in China*. Beijing: Science Press.

Chapter 14
Mulch Plastic Systems: Recent Advances and Applications

Verónica Manzano, Nancy L. García, Carlos Rodríguez Ramírez, Norma D'Accorso, and Silvia Goyanes

Abstract In the field of agriculture, the introduction of mulches revolutionized the commercial production of some vegetables, fruits and grain crops due to its instant economic benefits such as higher yields and improved crop quality. While the agricultural surface assisted with mulching films shows growing demand throughout the world, the sustainability of these films remains vague in terms of both an environmental and agronomic perspective. This chapter describes the evolution of the films used with special attention to the associated problems of undesirable end-of-life of non-biodegradable plastic mulch (non-BDM) films as compared with BDM designed to degrade without hazard to biota. Current trends in future, new production techniques for these films such as electrospinning as well as standardized laboratory testing processes are also discussed.

Keywords Biodegradable mulch · Electrospinning · Polyethylene mulch · Plastic mulches

V. Manzano · N. D'Accorso (✉)
Facultad de Ciencias Exactas y Naturales (FCEyN), Departamento de Química Orgánica, Universidad de Buenos Aires, Buenos Aires, Argentina

Centro de Investigaciones en Hidratos de Carbono (CIHIDECAR), CONICET- Universidad de Buenos Aires, Buenos Aires, Argentina
e-mail: norma@qo.fcen.uba.ar

N. L. García · C. Rodríguez Ramírez
Centro de Investigaciones en Hidratos de Carbono (CIHIDECAR), CONICET- Universidad de Buenos Aires, Buenos Aires, Argentina

S. Goyanes (✉)
Facultad de Ciencias Exactas y Naturales (FCEyN), Departamento de Física, Laboratorio de Polímeros y Materiales Compuestos (LP&MC), Universidad de Buenos Aires, Buenos Aires, Argentina
e-mail: goyanes@df.uba.ar

T. J. Gutiérrez (ed.), *Polymers for Agri-Food Applications*,
https://doi.org/10.1007/978-3-030-19416-1_14

14.1 Introduction

In many fields especially in agriculture, the introduction of mulches revolutionized the commercial production of some vegetables, fruits and grain crops. Mulching is an essential cultural technique that can reduce the amount of work inherent in gardening; helping to produce healthier plants and potentially increasing vegetable yields. Mulching is one of the most beneficial practices for soil preservation in plant production (Merino et al. 2019a). It also benefits for growth, annual and perennial yield and health of crops, weeds and plants and its low costs make them a widespread technique.

Mulch is defined as any material, synthetic or natural, that is spread or laid as a cover over the surface of the soil. Its main purpose is to retain moisture in the soil, increase or decrease soil temperature, suppress weeds, prevent nutrient leaching, and more efficient use of fertilizers and soil nutrients (Kasirajan and Ngouajio 2012). The vantage of mulches includes early seeding and transplanting of certain crops and encourages faster growth. In this sense, it forms a layer between the soil and the atmosphere which prevents sunlight to reach the soil surface, diminishing the evaporation. It is worth mentioning that mulches also provide a physical barrier to soil pathogens and increases water infiltration into the soil. The type and depth of mulch as well as the time of the year in which is applied, will depend on the function, purpose and aesthetic value of the crop or landscape, as reported by many researchers (Cushman et al. 2005). It is usually applied in spring and early summer but there are several cases in which they are used all along the year. There are also some problems related to the use of them, e.g. the slower-maturing of crops, the specialized management and the cost. Table 14.1 summarizes the benefits and disadvantages of the use of mulches.

In this chapter, we will review the use of mulches in agriculture, with special reference to biodegradable mulches and the new developments in the area. We will also focus on the structure and the possibility of charging them with additives, relating them to its application. It is also an aim of this chapter the review of the new advances in this field, especially the development of woven mulches made by electrospinning and also the multi layers mulches.

Table 14.1 Advantages and disadvantages of mulching

Advantages	Disadvantages
Improves nutrients	Slower-maturing costs
Weed suppression	Retards the heating of the soil by the sun
Water retention in the soil	Cooler temperatures above and below mulch
Encourage favorable microbial activity	Cover crop mulch re-growth
Reduced soil erosion	Cover small slugs
Increased soil quality	Removal and disposal (for plastic mulches)
Enhanced nitrogen availability	High production cost (for plastic mulches)
Reduced fertilizer leaching	Specialized management
Earlier and higher overall yields	Special equipment required

As mentioned above, there are many types of mulches which goes from the conventional mulches, organic and inorganic mulches, to the special materials which include all the bio- and photodegradable plastics that have recently been developed and studied.

14.2 Conventional Mulches

The organic and inorganic mulches can be mentioned among the conventional mulches. *Organic mulches* proceed from living materials, such as bark, plant and animal waste. They are usually obtained as a byproduct of other industries such as lumber and paper. The advantages of them rely in the low cost, the natural appearance and the fact that they allow the infiltration of water and oxygen. Its main characteristic is that they readily decompose, improving soil structure and quality, and returning nutrients to the soil and maintain the tilth (Tindall et al. 1991). Because of decomposing they decay over time and are temporary. The replacement frequency can be mentioned as a main disadvantage. The fact that it moves around and becomes displaced and that, sometimes, pests can grow if it is too close to the tree trunks and that its quality is inconsistent, and more labor is required to spread it.

Inorganic *mulch* is understood as any material derived from not living sources such as rocks or gravel. These types of materials also known as mulches, and are generally used to create barriers to weeds and for landscape fabric (decorative purposes). Inorganic mulch, such as rocks or gravel, does not readily decompose. Rocks absorb and reflect heat which can be detrimental during hot, dry weather. Because inorganic mulches do not decompose rapidly they do not improve the quality of the soil.

Thus, the use of plastic mulches in agriculture increased dramatically throughout the world in the last few years (Lamont 2017). Plastic mulches are often manufactured from synthetic materials such as polyethylene (PE) sheets and petroleum based products. The advantages of these types of mulches, relies in the low cost of maintenance, the low frequency of replacement, the versatility and liability. The difficulty of installation together with the cost of them is one of its negative points.

While bio-based mulches are biodegradable and compostable, i.e. they are decomposed in a relatively short time, the synthetic plastic mulches derived from petroleum are not easily decomposed, thus generating pollution since they remain in the environment. Plastic mulching has become popular in agriculture due to its instantaneous economic benefits, which results in higher yields, earlier harvests, better fruit quality and greater efficiency in water use (Steinmetz et al. 2016). Synthetic mulches are easy processing, have excellent chemical and mechanical properties, are cleaner and impermeable to water but also to nutrients (Lamont 1993). With the time, they deteriorate with the exposure to sunlight. It is important to highlight that an ideal plastic mulch must be adjusted to certain characteristics such as flexibility and rigidity enough to be able to eliminate it from environments (Kasirajan and Ngouajio 2012).

We will focus our attention in the plastic derived mulches and the advances in the next section. We will also discuss the additives and characteristics of these mulches.

14.3 Plastic Mulches

The term plasticulture refers to the practice in which plastic materials have their performance for agricultural applications (Merino et al. 2019b). Generally, these plastic materials are also known as "ag plastics" and most of them are based on non-renewable petroleum-based polymers. In this category (plasticulture or ag plastics) are included soil fumigation film, irrigation drip tape/tubing, nursery pots and silage bag, but this term is often used to describe all kinds of plastic plant/soil coverings. There is a wide range of coverings that goes from plastic mulch film and row coverings to plastic greenhouses.

In recent years, the use of plastic mulches has increased due to the fact that: (1) better production yields, (2) earlier maturation of the products, (3) higher quality of the products, (4) control of insect/disease and (5) not least the low cost (Kasirajan and Ngouajio 2012). It is interesting to note that these polymers mulches can be modified to incorporate different additives (dyes, plasticizers, antiblock agents, antioxidants, flame retardants, and photodegradable additives) with the aim of modifying, improving or minimizing properties according to the use or the application area (Wright 1968). It is well known that these mulches directly affect the microclimate around the plant by modifying the radiation budget of the surface and decreasing the soil water loss and the heat gain by the soil (Liakatas et al. 1986). It is worth noting also that the control of insects or diseases, usually incorporates an antiblocking agent in the plastic film. This is a substance that inhibits the adhesion of the surface of the films.

As mention before, the main aim for using plastic mulches is the protection of seedlings and shoots by maintaining or slightly increasing the temperature and humidity of the soil (Tarara 2000). All these issues made plastic films an upcoming technology which governs the majority of the covered agricultural surface. A clear advantage is also the fact that these mulches are placed in the field and can be plowed on them. On the other hand, mulches modify the microclimatic conditions and, in some cases, increases biological degradation of litter and soil organic matter. A great limitation of these mulches constitutes the non-biodegradability, the null compostability and the great problem of the final disposal of the films, which generally cause pollution and contamination, which makes their elimination necessary (Kasirajan and Ngouajio 2012). Thus, degradation of PE mulches needs synergistic action of photo- and thermo-oxidative degradation and biological activity (Hakkarainen and Albertsson 2004). This decomposition takes more than 300 years in soil, could lead to degradation harmful products such as aldehydes and ketones, which are not completely eliminated, so that remains of 5–10% of the mulch residues (Ghimire et al. 2018). Another undesirable byproduct formed when the mulch

Table 14.2 The LPLE mass remains in soil after harvesting of non-biodegradable inert mulch film (fragments) *per* hectare of crop

Thickness (μm)	Average collection rate (%)	Mass in kg *per* ha of crop
10	25	62.9
20	32	46.25
25	90	23.13

is burned is 1,4-dioxane, which is released into the air causing pollution. It is estimated that around 60% on average is not removed from the field, which implies that a significant amount of money, up to $288 *per* acre must be assigned to the final disposal (Galinato et al. 2012).

In this sense, it is important to note that the thickness of mulch influences the final residues to be released. Thus, a thicker mulch film has a lower degradation rate, which results in a greater elimination or contamination problem. To see the influence of the thickness of mulch films used, Table 14.2 shows the mass remain in soil after harvesting of non-biodegradable inert mulching film (fragments) *per* hectare of crop, taking for example a low density PE (LDPE) as the most used material in this type of mulch, with an average density of 925 kg/m^3 for LDPE and with the known data: 185.0 kg of mulch film *per* hectare of crop is needed (Bioplastics.org):

The most used mulches are made mainly from PE, polypropylene (PP), polyvinyl chloride and ethylene vinylacetate. One of the first plastic mulches used for this purpose was made from PE, its development dates from 1939 and its industrial use was made at the beginning of 1950s. The PE resin is available in the form of pellets, which when are treated by slot casting or blown bubble, films can be obtained (Clarke 1987). PE mulch films can be obtained as linear or branched polymers (Fig. 14.1), which influences directly the properties of the final product. High-density PE (HDPE) and ultrahigh-molecular-weight PE (UHMWPE) are the linear versions, while LDPE or linear LDPE (LLDPE) are branched. PE also comes in an increasing variety of kinds such as very low-density PE (VLDPE), ultra-low-density PE (ULDPE), PE elastomer, ethylene vinyl acetate (EVA) copolymer and ethylene-methacrylic acid (EMAA) copolymer.

HDPE mulch films have a higher density and rigidity compared to LDPE mulch films as might be expected. This due to the minimal branching of the polymer chain which leads to good packing of the polymer chains. The most common PE mulches are made from LDPE (density values ranging between 0.91–0.94 g/cm^3). LLDPE and HDPE are also used, but less for these purposes. As mention before, LDPE is a white material with high degree of ramification which results in a low density polymer with excellent properties such as good thermic and chemical resistance, excellent tensile strength, good resistance to the impacts and high durability and flexibility (Clarke 1987; Lamont 1993; 2017). There are many reports on the uses of PE mulches in vegetable production. The advantages of its use in agriculture are well documented for crops such as cabbages and vegetable production (Emmert 1957; Bryan 1966) of muskmelons, watermelons, cucumbers, tomatoes, peppers, eggplant, sweet corn, to mention some (Lament 1993).

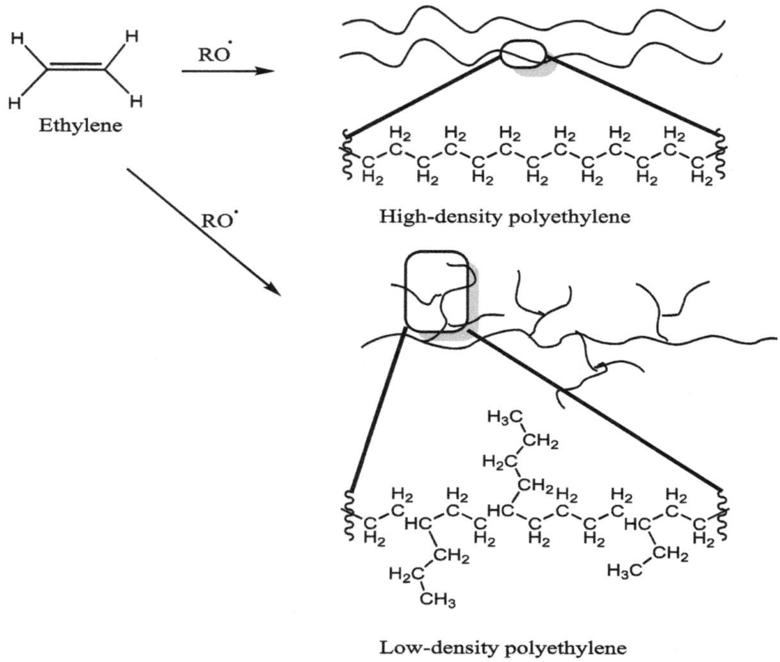

Fig. 14.1 Linear and branched polyethylene

14.4 Effect and Influence of the Color on PE Mulches

The color parameters of an agricultural mulch film together with the optical properties are directly related to its radiometric properties and will influence the temperature of the surface and the microclimate around the vegetable plant (Tarara 2000). This is because parameters such as reflectivity, absorption capacity or transmittance depend on incoming solar radiation (Lamont 2017). Mulches will prevent the growth of weed species as a result of the reduced transmission of light by the mulch. Ngouajio and Ernest (2004) studied the influence of color on certain properties of PE mulch, such as light transmission, weed populations and their relation to light transmission. They concluded that there were differences in the ability to transmit light through the colored mulches, and that this affected the germination and growth of weed seeds.

The most common colors used in plastic mulches are black, red, green, clear, beige and white, as well as combinations of them are also found, e.g. white on black. A black plastic mulch will absorb most of the wavelengths, which will then be lost as thermal energy, thus increasing the soil temperature up to 10 cm depth. This is the type of mulch chosen to produce vegetables in USA (Lamont 2017).

On the other hand, the transparent plastic mulches do not absorb many solar radiation, but in contrast, they do transmit the solar radiation. These transmissions depend on the thickness and degree of opacity of the PE. As a consequence, the inner part of the mulch is covered by condensed water that will retain the heat produced by the transmission of the radiation. These mulches are found in cold regions where heat is needed to preserve, but strict control of the weeds is needed. Finally, if a slight decrease in temperature is desired, then the correct choice is white, silver or white on black. This is because they reflect most incoming solar radiation (Lamont 2017).

With this in mind, many scientists have based their research on this issue (mulch color) and its influence on the growth of different vegetable crops (Gordon et al. 2008; Streck et al. 1995; Brault and Stewart 2002; Ban et al. 2009; Abdalla et al. 2016). In these sense, Streck et al. (1995) related the color of PE films with soil temperature and yields of tomato production. They observed that transparent mulches allowed higher temperatures in the soil and that the yield of the tomato was not affected by the color even though a higher yield was observed in the opaque mulches, especially in the whites. Brault and Stewart (2002) also studied the effect of optical properties of paper mulches as compared to PE-colored mulches for weed control in lettuce crops. They concluded that all the PE-mulches used maintained stable their optical properties and were able to control weeds. In the same way, a research on watermelon production was carried out by Ban et al. (2009) using five different colors (black, white, clear, green and brown) for the PE films. They observed that the best results in terms of plant growth and early yield were obtained for transparent PE mulches with respect to the results obtained for white films. No differences were observed for brown and green mulches. Later, Abdalla et al. (2016) reported on a study of four plastic mulches of different colors in the quality of potato crop, and showed that the lowest rate of germinated tuber was obtained when transparent mulch was used. It was interesting to note that when covering a field with black plastic mulch, a low rate of field emergence and the highest percentage of foliage dry weight was obtained, while white sheets resulted in the highest number of stems.

14.5 Plastic Mulches Together with Row Covers

Other approaches uses row covers along with colored plastic mulches to produce a crop, as demonstrated by Gordon et al. (2008). This research group studied the effect of different mulch color with and without the row and concluded that the coverage of the row affected the plant height and the stem diameter, but the colored mulch did not increase early fruit yield. It was interesting to note that the mulch color and the year significantly affected air and soil temperatures and row cover significantly affected the air temperature.

14.6 Mulches as Repellent of Pests

Plagues are a serious problem with regards to vegetable production and in this sense; many researches focused their attention in the utilization of PE mulches as repellents of pests, fungus and viruses. Since depending on the color of the plastic mulch, the amount of light transmission through the mulch and the weed of the soil can be controlled, this directly affects the development of several pests and plagues (Decoteau et al. 1989). One of the first pioneers in this field was Adlerz and Evereit (1968), who tested white PE mulches as repellent of aphid vectors of watermelon mosaic virus. The results were that these PE mulch trapped more aphids than over the bare ground. Moreno and Moreno (2008) demonstrated that the use of PE films resulted low values of soil microbial biomass and soil organic matter mineralization compared to biodegradable mulches, which will be explained later.

14.7 Photodegradable PE Mulches

In order to solve the problem of the non-degradation PE mulches, photodegradable plastics are considered for degradable mulch applications. Usually these plastics contain antioxidants and photodegradation catalyst. The main disadvantage is the continual use of non-renewable resources. There have been several attempts to use materials that are ultimately degraded for the reasons explained above. In this way, Romic et al. (2003) evaluated the effects of different mulch materials, including transparent photodegradable PE film and paper biodegradable cellulose mulch, and irrigation systems in the culture of watermelons. They concluded that the PE mulches allowed an earlier harvest and that the better results were obtained using the black film (104.5 t *per* ha in 1995 and 79 t *per* ha in 1996). It was remarkable that clear mulches led to early growth, but did not give higher yields, maybe because of an earlier photodegradation of materials.

14.8 Other Alternatives to PE Plastic Mulches

Although, the most used plastic mulches are the one made from PE, other approaches have also been tested for this purpose. For instance, the well-known paper mulches have been uses since the early 1931 for potato crops (Smith 1931). It was thought that these mulches were a good alternative due to the easy and fast biodegradation, but they have the limitation of the initial cost of paper and the need to take care of the crop to avoid paper injuries. An interesting approach was made by Sanchez et al. (2008) who used mulches made from shredded newspaper and newspaper sheets as weed suppressors in cucumber crops. Another alternative for PE mulches were the PP mulches. These mulches are usually restricted to perennial crops. Some stimulating advances have used black PP woven mulch in potato crops, which shows that these PP mulches had a positive effect on soil temperature and early emergence of potatoes (Dvořák et al. 2001). PP woven fabrics were also applied as agricultural mulch materials in rubber plantations, dry leaves and coconut tree mats (George et al. 2015). The results were an improvement in the retention of soil moisture, as well as suppression of weed growth.

14.9 Biodegradable Mulches

In order to introduce eco-friendly alternatives to replace synthetic polymers as agricultural mulch materials, it is first necessary to evaluate their biodegradability and compostability. Biodegradation in soils is a process of assimilation of organic material by the microbial population present in the soil (bacteria and fungi) as a food and energy source. These microbial populations use organic carbon to extract biochemical energy to boost their life processes through the aerobic oxidation of usable organic carbon (de Wilde and Deconinck 2017). In polymers, this summary process can be expressed by Eq. 14.1:

$$C_{polymer} + O_2 \rightarrow CO_2 + H_2O + C_{biomas} + C_{residual} \quad \text{(Eq. 14.1)}$$

Equation 14.1 shows that most of this organic carbon is converted into carbon dioxide (CO_2), while a minor part is converted into microbial carbon (the so-called biomass yield or $C_{biomass}$ in the equation) and $C_{residual}$.

Biomass yield is typically between 10% and 40% depending on the substrate which is converted. The $C_{residual}$ consists of (partially) undegraded polymer but can also be metabolites (to be considered as in-between degradation products). The CO_2 is the only one that can be quantified technically and analytically, for this reason speak of 100% complete biodegradation of a polymer is not correct, even though this expression is found in multiple references. It should also be noted that additionally, the rate and level of biodegradation of a polymer is determined by their environmental conditions. These conditions vary, since they depend on moisture content, oxygen availability, pH value, temperature, types of microorganisms.

One of the methods used to determine the biodegradation of a polymer in soil is the international standard ISO 17556 Plastics: *Determination of the ultimate aerobic biodegradability of plastic materials in soil International Standardization Organization (ISO) 17,556 Plastics, by measuring the oxygen demand in a respirometer or the amount of carbon dioxide evolved* (2012). This test is considered valid only if the degree of biodegradation (based on carbon to carbon dioxide conversion) of the reference material (cellulose), has reached 60% or more at the plateau phase or at the end of the protocol.

Although there is no standard specification at the level (ISO) for biodegradation of polymers, but by agreement at the Committee European of Normalization (CEN) level, a polymer is considered completely biodegradable when at least 90%

(absolute or relative to cellulose) of the organic carbon present in the polymer (C polymer) is converted to CO_2 within a maximum period of 2 years. However, since January 2018, the European Standard EN 17033 was released; this novelty is presented after this section.

The soil depth where these plastic fragments typically end after use is presumed to be 20 cm, which corresponds to the normal depth of soil tillage. In other words, at the end of each cultivation period, for 20 μm of thickness, 46.25 kg of (non-biodegradable) PE film fragments will end up in a volume of soil equal to 2000 m^3.

The level of accumulation when using 20 μm PE mulch films and one cultivation period *per* year increases to 462.5 kg mulch film (fragments) *per* hectare after 10 years, 2.3 t *per* ha after 50 years and even 4.6 t *per* ha after 100 years (Fig. 14.2). It must, however, be noted that these figures are average figures and the actual accumulation load depends on the thickness and density of the PE mulch film used, and the number of crops cultivated and, therefore, number of mulch films used *per* year. Figure 14.2 shows the increase in accumulation for 3 scenarios: one crop *per* year and a 20 μm film thickness, two crops *per* year and a 20 μm film thickness and two crops *per* year and a 10 μm film thickness.

The demand for high quality crops and the need for (more) optimization of food production due to the increase in population is becoming more intense, so the market for quilting films is expected to grow. In the early 2000s, 'only' 2–3 million t of mulch film were used on an annual basis, while in 2012 this was already 4.4 million t. According to Markets and markets, Global Biodegradable Mulch Films Market size

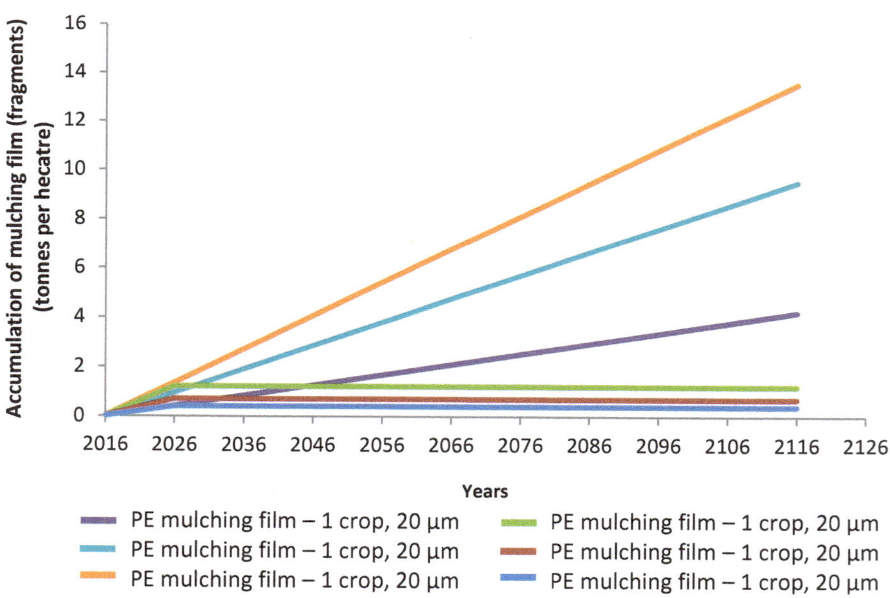

Fig. 14.2 Overview of the increase in accumulation of PE and biodegradable plastic fragments in soil. Adapted from de Wilde and Deconinck (2017)

to grow from USD 35.76 million in 2016 to USD 52.43 million by 2021, at a Compound Annual Growth Rate of 7.95% from 2016 to 2021. Another study predicts that the global biodegradable mulching films market will reach USD 64.3 million by 2024 (Grand View Research 2016).

In order to increase the sustainability of agricultural practices and overcome the disposal problems associated with conventional plastic films, the development and use of mulch films based on biodegradable materials it became essential. In this context, several biodegradable polymers enter the scene as a desirable alternative to PE and all those materials that do not respond to the care of the environment.

Mulch films produced from biodegradable polymers and which comply with current soil biodegradation standards are a good alternative, because among other advantages, they can be ploughed into the soil at the end of the growing season without causing any plastic accumulation on long term.

Today, several films capable of being degraded by microorganisms in the soil have been developed, including oxo-degradable plastic, paper film, and biodegradable film. The main idea has always been the same, the search for non-harmful substances resulting from the degradation of biodegradable materials used as mulches, in order not to cause undesirable effects on the yield of agricultural land and non-cumulative effects on the soil (Briassoulis and Giannoulis 2018).

Some biodegradable polymers such as starch, poly(vinyl alcohol) (PVA), poly(butylene adipate-co-terephthalate) (PBAT), poly(butylene succinate-co-adipate) (PBSA), and poly(lacticacid)(PLA) have been used to prepare biodegradable agricultural films (Scarascia-Mugnozza et al. 2006; Miles et al. 2017; Merino et al. 2018a; Merino et al. 2018b). In some cases, blends and composites have been used with fibrous materials derived from these crops or from byproducts or agroindustry wastes. However, some natural fibers such as ramie, flax, hemp, and cotton, are abundant but, the waste fiber has not been suitably recycled after industrial production (Miles et al. 2017).

Most biodegradable plastic mulches (BDMs) that are currently commercially available are made from plant starch. The agricultural films made of starch are prepared using conventional plastics processing technology, and in some cases blended with other polymers and/or plasticizers (Ghimire et al. 2018). Products currently on the market that are derived from or blended with plant starch include BiosafeTM (Xinfu Phar-maceutical Co., China), Eastar BioTM (Novamont, Italy), Eco-Flex® (BASF, Germany), Ingeo® (NatureWorks, USA), Mater-Bi® (Novamont, Italy) and Paragon (Avebe, Nether-lands) (Ghimire et al. 2018). Two polymers that may have a role in BPMs in the future are polylactic acid (PLA) and poly-hydroxyalkanoate (PHA).

Table 14.3 presents the constituents of the main brands of the market, where PBAT — polybutylene adipate terephthalate; PBS — polybutylene succinate; PBSA — PBS-co-adipic acid; PHA — polyhydroxyalkanoate; PLA; TPS — thermoplastic starch (Miles et al. 2017).

PLA is a highly versatile, biodegradable polyester derived from 100% renewable resources such as corn and sugar beet starch, and offers great promise in a wide range of commodity applications (Drumright et al. 2000). Starch is converted by

Table 14.3 Commercial biodegradable mulches

Trade name	Polymer/Polymer Blend	Trade name	Polymer/Polymer Blend
Bio 360	Mater-Bi(Starch)+PBAT	Ecovio	PLA + PBAT
BioAgri	Mater-Bi(Starch)+PBAT	Envio	PBAT + PLA + Starch Blend
Biocycle	Sucrose/PHA blend	GreenBio	PHA
Bio-Flex	PLA/co-polyester	Ingeo	Starcha + PLA; PBS + PLA
BiomaxTPS	Starch + TPS	Mater Bi	Starch blend
Biomer L	PHA	Naturecycle	Starch
Biopar	Starch co-polyester	Paragon	Starch + TPS
Biosafe	PBAT/Starch blend; PBS; PSA	Renew	PHA
Ecoflex	PBAT		

microorganisms into lactic acid through fermentation. Lactic acid molecules are then linked together into long chains called polymers. PLA is a relatively inexpensive biopolymer to manufacture and can be produced in large quantities. The PLA polymer is very attractive for biological and medical applications because it can be spun into filaments which can then be used to make textiles or films (Gupta et al. 2006).

PHAs are promising biodegradable plastics that have been highlighted as "green" polymers because they are made from renewable resources in a one-step process through bacterial fermentation of sugars and/or lipids (Kaihara et al. 2005; Posada et al. 2011; Ghimire et al. 2018). PHA polymers can be produced from microbes or plants, but currently, microbes are the main source (Keshavarz and Roy 2010).

Mater-Bi® is one of the most commercially available biodegradable and compostable films in Europe (Martin-Closas et al. 2008). Mater-Bi® is a range of fully biodegradable and compostable thermoplastic materials, designed to provide a low impact environmental solution and solve specific problems in different sectors such as separate collection of organic waste, packaging, catering, hygiene, agriculture and many other areas. Mater-Bi® is produced using renewable resources made from plant material such as starches from different crops (e.g. corn, potato as well as other cereals and tubers) and vegetable oils. Specific types of Mater-Bi® have been developed for a range of agricultural applications, including mulch films. It is a completely compostable film, which reduces the consumption of energy and non-renewable sources, thus completing a virtuous circle: raw materials of agricultural origin return to the earth through biodegradation and composting processes, without releasing pollutants (Novamont 2013). The Mater-Bi® mulch film fully complies with the requirements of national and international standards on biodegradation and environmental impact such as EN13432/EN14995 Compostability mark, ASTM D 6400 Compostability mark and EN13432 Compostability mark.

Briassoulis (2006) investigated the mechanical performance of several experimental biodegradable films made of different grades of Mater-Bi® material (starch based polymers complexed with biodegradable polyesters) and additives; different thickness and processing parameters, and compared these results against the corresponding behavior of conventional agricultural films under real cultivation conditions. The investigation of the selected critical mechanical properties describing the mechanical behavior of these experimental mulches and low tunnel biodegradable films, exposed to four different locations in Europe, confirms that a rather good mechanical behavior is possible for these films, comparable to the behavior of conventional agricultural films in terms of strength, inferior, however, to the conventional films in terms of elongation at break. The most important points of this work are summarized below:

The performance of the biodegradable mulches and low tunnel films during their useful lifetime can be comparable to that of conventional (thicker) films in terms of tensile strength in the parallel direction; the tensile strength of thin biodegradable films can reach values below the initial stress at yield in the transverse direction.

Biodegradable mulches and low tunnel films, especially thin films in the transverse direction, may exhibit a very low elongation at break values within the first week of exposure.

The evolution of the initial to tearing of the Mater-Bi® based biodegradable films follows in general follows the corresponding behavior of the elongation at break as far as the displacement at maximum load and the tensile strength is concerned.

Two different commercial processing schemes (Scheme A and B) were employed in manufacturing the experimental biodegradable film concerns the size of the commercial extruder (a rather large machine was used with Scheme B). Carbon black was used in most cases (carbon black masterbatch based on Mater-Bi for the processing Scheme A or carbon black master-batch based on polyethylene for the processing Scheme B). Any possible stabilizing effects of carbon black on the degradation of Mater-Bi based biodegradable films are yet unknown (act as a strong stabilizer for the conventional film during its useful lifetime). In some other cases of biodegradable mulching films, special additives that might be of interest for specific cultivations like colors (especially green) or stabilizers for transparent mulching films were tested. As far as the biodegradable low-tunnel films are concerned, the effect of the UV stabilizers belonging to two main groups was investigated: (a) HALS based UV stabilizers and (b) UV adsorbed. Briassoulis (2006) concluded that particular stabilization schemes used with the biodegradable films do not appear to affect significantly the tensile strength and the tear resistance of the film, and certainly do not improve the weak behavior of the films produced with processing schemes that have not been optimized for the specific biodegradable films production. Processing problems appear to be more critical and controllable at this moment, than stabilization. Basic research is needed though in the direction of developing new additives systems, appropriate for biodegradable Mater-Bi® based low tunnel and mulch films. In general, the performance of the thin biodegradable films in the transverse direction remains the weak point compared with the conventional films (thicker films).

The Mater-Bi grade NF 803/P was found to be best suited for blow extrusion of thin biodegradable agricultural films. It has been shown that it is possible to develop very thin biodegradable films made of this grade (with a thickness of at least up to 30 mm for low tunnels and 15 mm for mulching films) that perform satisfactory for the specific applications and may replace conventional (thicker) polyethylene films (Briassoulis 2006).

On the other hand, a no less important fact is the economic approach for mulch films. The bio-based plastics production reached 1.0% (1.7 mt) of the global plastics market in 2014 (300 mt) with the European market share estimated at 15.4% (0.2–0.5 mt) (Briassoulis and Giannoulis 2018). The participation of the agricultural sector in the world market of bio-based plastics was estimated in 2014 at 105 kt/yr. The global bio-based polymer market, which already grew by an 20% annual growth (Bioplastic 2018) and is estimated to reach 7.9 mt this year (2019). The European market for bio-based plastics in 2020 is predicted at 1–3 mt (Briassoulis and Giannoulis 2018). Among the recent technological developments in the market of innovative bio-based plastics, are new grades of mulch films designed for targeted agricultural and horticultural applications (Martin-Closas et al. 2008). According to the size of the global market for biodegradable mulch film is expected to grow to USD 52.43 million by 2021. Another study predicts that the global biodegradable mulch films market will reach USD 64.3 million by 2024 (Briassoulis and Giannoulis 2018). The predominant use of biodegradable mulch films is horticultural crops and has a 51.1% share of total market revenues in 2015, which is also expected to show the fastest growth in the next 8 years.

Many agricultural plastics advertise themselves as biodegradable. However, such claims must be carefully evaluated. For a manufacturer to use the claim: biodegradability, a series of specific standards must be met. ASTM International (formerly known as the American Society for Testing and Materials) has prepared a series of standards for compostable plastics to measure biodegradability under industrial composting conditions, called ASTM D6400. The ASTM D6400 specification employs several ASTM standardized tests, such as the measuring the inherent biodegradability of the plastic material through ASTM D5988–03. Many types of mulch claiming to be biodegradable are actually compostable, and fulfill the requirements of ASTM D6400, or related standards. Many types of mulch are said to be biodegradable when they are also compostable and meet the requirements of ASTM D6400, or related standards. There is currently no standard for measuring the biodegradability of plastics buried in the soil under field soil conditions. To meet this need to measure soil biodegradability, ASTM is developing a standard through a specification (Work Element 29,802) entitled “Aerobically Biodegradable Plastics in the Soil Environment” (Ramani Narayan, ASTM member, personal communication). In this new standard, biodegradable mulches must be decomposed into CO_2, water and environmentally benign substances within one or two years, without leaving harmful residues.

In January 2018, the European Standard EN 17033: Plastics– Biodegradable mulch films for use in agriculture and horticulture– Requirements and test methods was released (Hayes and Flury 2018). The standard was developed by the European Committee for Standardization, Technical Committee CEN/TC 249 Plastics and applies to all European Union countries plus Macedonia, Norway, Sweden,

Switzerland, Serbia, Turkey, and the United Kingdom. This standard regulates the requirements for BDM films: their composition, biodegradability in the soil, effect on the soil environment (ecotoxicity), mechanical and optical properties, and the test procedures for each of the listed categories. It does not apply to mulch films that are removed from the fields after use.

EN 17033 requires laboratory testing of BDMs for (1) chemical composition (in particular, for regulated metals and hazardous substances, (2) biodegradation in soil, (3) ecotoxicity (i.e. toxic effects on plants, invertebrates, microorganisms) and (4) selected physical characteristics (e.g. thickness, tensile stress, light transmission). The inclusion of the composition criteria (e.g. maximal concentrations of heavy metals), biodegradability and ecotoxicity for the plants are consistent with compostability, the Vinçotte 'OK Soil' standards, and ASTM WK29802. An important difference between EN 17033 and WK29802 is however that the former specifically addresses plastics before exposure to environmental factors (which can alter chemical bond), while the latter includes separate categories for unweathered and environmentally weathered plastics that meet the criteria. A major criterion of EN 17033 is the requirement of $\geq$90% biodegradation under aerobic conditions for the plastic (i.e. conversion of organic carbon into CO_2) in natural topsoil from an agricultural field or forest at 20–28 °C conditions within 2 years using a standardized test to measure CO_2 respiration.

It is also important to highlight that it is very important to study the by-products of decomposition of biodegradable mulches, thus some of this compounds may comprise the growth and development of the plant or crop. For instance, Serrano-Ruíz et al. (2018) demonstrated by the used of an *in vitro* system, that some of the extracts from biodegradable plastics affected the seed germination and growth of lettuce and tomato. In summary, the complete and full knowledge of this products is very important in order to design and develop a safe mulch material for application in the agroindustry and agriculture.

14.10 Compostable Mulches

A polymer is compostable when it has the ability to become useful compost. Compost (humus or soil improver); it is the result of the decomposition and humification of a mixture of organic material (e.g. material from the pruning of plants, kitchen scraps, manure, effluent or gardening waste) by macro and micro-organisms under specific conditions: the presence of oxygen and a balance between the chemical elements of the material being transformed.

In this sense, it is also important for a mulch not only the biodegradability but also the compostability. Many researchers have focused their attention in this topic, and have developed different methods to study the biodegradability of the mulches and whether this products of degradation are able to form compost. Gutiérrez (2018) has designed nanocomposite films from phosphated or methylated flours derived from pumpkin (*Cucurbita maxima*) and proved that both films were biodegradable but only the phosphate result compostable by means of measurement of ecotoxicity biomarkers. Similarly, Herniou--Julien et al. (2019) used food hydrocolloids to manufacture food packaging materials and studied the biodegradability and compostability of the films. In this case, this group prepared films using a reactive extruder from corn (*Zea mays*) thermoplastic starch (TPS) containing either cellulose acetate (C) or chromium octanoate (Cat – a potential food grade catalyst), or a blend of both (C + Cat). As conclusion they demonstrated that the Cat-containing films resulted biodegradable but non-compostable but showed creep and resilience which makes them potential candidates for applications as shape-memory food packaging materials.

Another interesting approach was done by Gutiérrez et al. (2019) which developed biocomposite films from corn starch (Zea mays) containing nanofillers packaged with Jamaica flower (*Hibiscus sabdariffa*) extract (JFE). They demonstrated that the interaction with the nanofiller strengthened the hydrogen (H)-bonding interactions and influenced directly the physicochemical and surface properties of the materials. As a consequence of this, all the films result to be biodegradable but were non-compostable confirming that not all the biodegradable materials are compostable.

14.11 New Developments in Mulches

In order to produce thin films composed of ultrathin polymer fiber networks, electrospinning technology has gained considerable attention. The diameters range from the submicrometer level to the nanometer range and offer large surface areas, small interfibrous pore size and high porosity. In this context, electrospinning is a good candidate to fabricate a controlled thin environmental film for agricultural uses (Toncheva et al. 2014; Nicosia et al. 2015; Ribba et al. 2017). In this method, an electrical potential to overcome the surface tension of a solution is used to produce an ultrafine jet, which is elongated, thinned and solidified as it travels through the electric field to a collector (Fig. 14.3). Despite being a relatively simple procedure to carry out in a

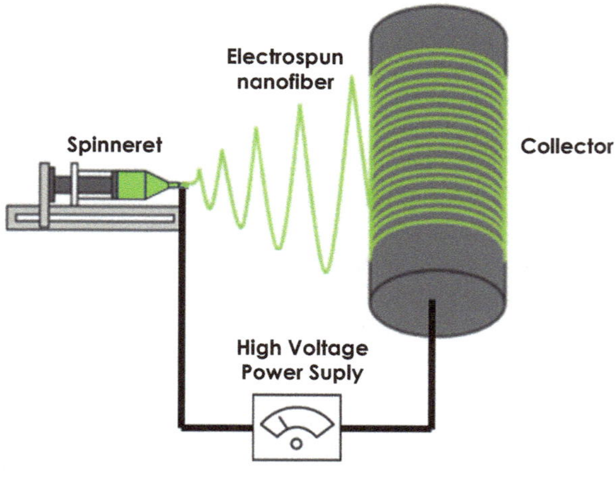

Fig. 14.3 Schematic representation of a typical electrospinning apparatus

Table 14.4 Bionanocomposite film formulations. Reprinted with permission from Arrieta et al. (2016a)

Formulations	PLA (wt%)	PHB (wt%)	ATBC (wt%)	CNC (wt%)
PLA–PHB	75	25	–	–
PLA–PHB–ATBC	63.75	21.25	15	–
PLA–PHB–ATBC–CNC1	63	21	15	1
PLA–PHB–ATBC–CNC5	60	20	15	5

laboratory, since it requires minimal equipment, the physics behind the process is complex. To obtain an insufficient description of all the variables and interactions involved in electrospinning, one should consider polymer chemistry, electric field interactions, fluid mechanics, environmental conditions and kinetics (Robb and Lennox 2011), e.g. the dependence of the opacity of electrospun materials with humidity (Cimadoro et al. 2018).

Electrospinning, for the recent years, has been acknowledged for its capacity to manufacture a micro/nanofibrous architecture from different polymers, which is otherwise difficult to obtain, this technique is widely used due to its profitability and user-friendly. Such polymer micro/nanofibers can be tailored to meet applications such as drug delivery, tissue engineering, filtration, fuel cell, food packaging, mulching, etc. (Ribba et al. 2014; Kakoria and Sinha-Ray 2018). New studies in the agricultural area have focused their efforts on the development of bio-based and biodegradable electrospun nanocomposites for sustainable flexible films whose main use is in mulch (Arrieta et al. 2016a).

Arrieta et al. (2016b) studied the effect the PLA with poly(hydroxybutyrate) (PHB) bionanocomposite fibers obtained with electrospun on different matrices that contained acetyl(tributyl citrate) (ATBC) and cellulose nanocrystals (CNC). The obtained mat formulations were designed as PLA–PHB, PLA–PHB–ATBC, PLA–PHB–ATBC–CNC1 and PLA–PHB–ATBC–CNC5 and the proportion of each component is summarized in Table 14.4.

The materials once generated were subjected to biodegradability tests, using the SEM observed the changes of the surface. Figure 14.4 shows the SEM micrographs of electrospun samples after 10 days of disintegration. In disintegrated PLA–PHB mat (Fig. 14.3a), the straight fibers were replaced by more contorted fibers, while some of them appear wrenched. More signs of disintegration were observed in the samples of plasticized mats (Fig. 14.4a, b, c), which show regions where the fibers were merged forming junction areas since they were dissolved because of the initiation of the hydrolysis process. In the case of the mixtures containing different percentages of CNC (Fig. 14.4c and d) some fibers were still straight in accordance with the higher tensile strength offered by these bionanocomposites.

Regarding the biodegradability analysis, the different films were compared in order to observe the contribution of each material in the film during the biodegradability process (Fig. 14.5).

On the other hand, an increase of the disintegration was observed when the plasticizer and nanofiller were added. The electrospun mats were broken in 10 days

Fig. 14.4 SEM observations of electrospun mats after 10 days of disintegration in compost: (**a**) PLA–PHB, (**b**) PLA–PHB–ATBC, (**c**) PLA–PHB–ATBC–CNC1 and (**d**) PLA–PHB–ATBCkCNC5. Reprinted with permission from Arrieta et al. (2016a)

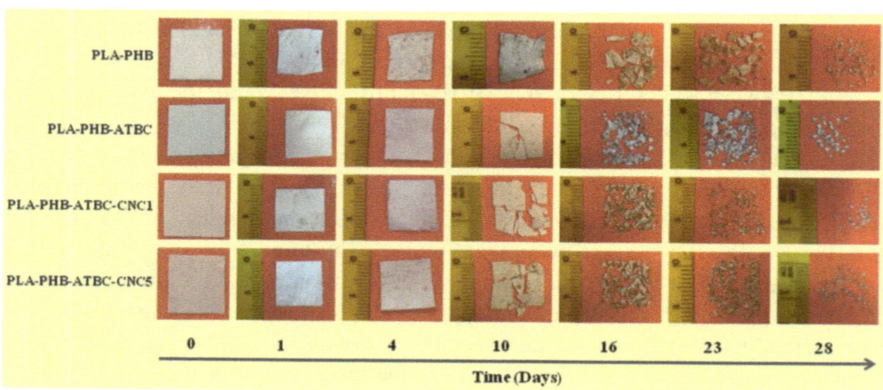

Fig. 14.5 Visual appearance of disintegrated electrospun mats under composting conditions. Reprinted with permission from Arrieta et al. (2016a)

of exposition to the compost medium. Small pieces of bionanocomposites were recovered at 16 days, while PLA–PHB sample still presented higher portions of the recovered material, and in general, a lower rate of disintegration in the materials developed can potentially allow use in agricultural coatings (Arrieta et al. 2016a).

Table 14.5 Electrospun biocomposite mat formulations. Formulations. Reprinted with permission from Arrieta et al. (2016b)

Formulations	PLA (wt%)	PHB (wt%)	ATBC (wt%)	Ch (wt%)	Cat (wt%)
PLA–PHB–ATBC	63.75	21.26	15	–	–
PLA–PHB–ATBC-Ch1%	63	21	15	1	–
PLA–PHB–ATBC-Ch5%	60	20	15	5	–
PLA–PHB–ATBC-Cat1%	63	21	15	–	1
PLA–PHB–ATBC-Cat5%	60	20	15	–	5

Similar materials were made by this same research group, but changing the loads used in the formulations. Arrieta et al. (2016a) studied the processing and properties of flexible electrospun biocomposites based on PLA, PHB, plasticized with ATBC and further loaded with 1 wt% and 5 wt% of chitosan (CS) or catechin (Cat) microparticles. Both fillers (CS and Cat) present a high content of hydroxyl groups on their surfaces (Arrieta et al. 2016b). The biocomposite mat formulations and the proportion of each component as well as their designation are summarized in Table 14.5.

The morphological investigations (Fig. 14.6) of the biocomposites obtained revealed that, while 1 wt% of CS or Cat produced some structural changes in the fiber such as smaller diameters and bead-free electros-pun fibers, when increasing the charge percentage to a 5 wt% of CS and Cat produced some structural defects. Cat induced a decrease of the average fiber diameter, but no significant effects on the average fiber diameters were observed in CS based biocomposites. On the other hand, the wettability of the biocomposites was increased by increasing amount of polar particles. The CS based formulations showed more hydrophilic character (Arrieta et al. 2016b).

The dynamic and isothermal TGAs revealed that the incorporation of the CS or Cat increases the decomposition temperatures of the different formations. The addition of CS and Cat always resulted in biocomposites with higher tensile strength at the expense of the reduction of the elongation at break. Finally, all biocomposites disintegrated under composting conditions as a sustainable end life around 16 days. This study showed that the electrospun plasticized PLA-PHB biocomposites loaded with both CS and Cat were promising materials for biodegradable film applications in several fields such as agricultural mulch films or films for food packaging (Arrieta et al. 2016b).

Other studies carried out by Chizhov et al. (2018) found water absorption and thermo-oxidative destruction as biodegradation processes of films and nonwoven fibrous materials based on the natural polymers (polylactide, polyhydroxybutyrate and their mixtures). Fibrous materials obtained by electrospinning, as compared with film ones, showed a high water absorption and a high oxidation rate, which significantly accelerates biodegradation processes under environmental conditions (Chizhov et al. 2018).

The kinetic oxidation curves of the nonwoven fibrous material and film from the PHB/PLA (1:1) blend showed no degradation by oxidation. PLA/PHB blend with acts as an inert component with respect to the oxidation process. On the other hand, in the area of the ratio of polymers of (1:1) in fiber and film, both the polymer

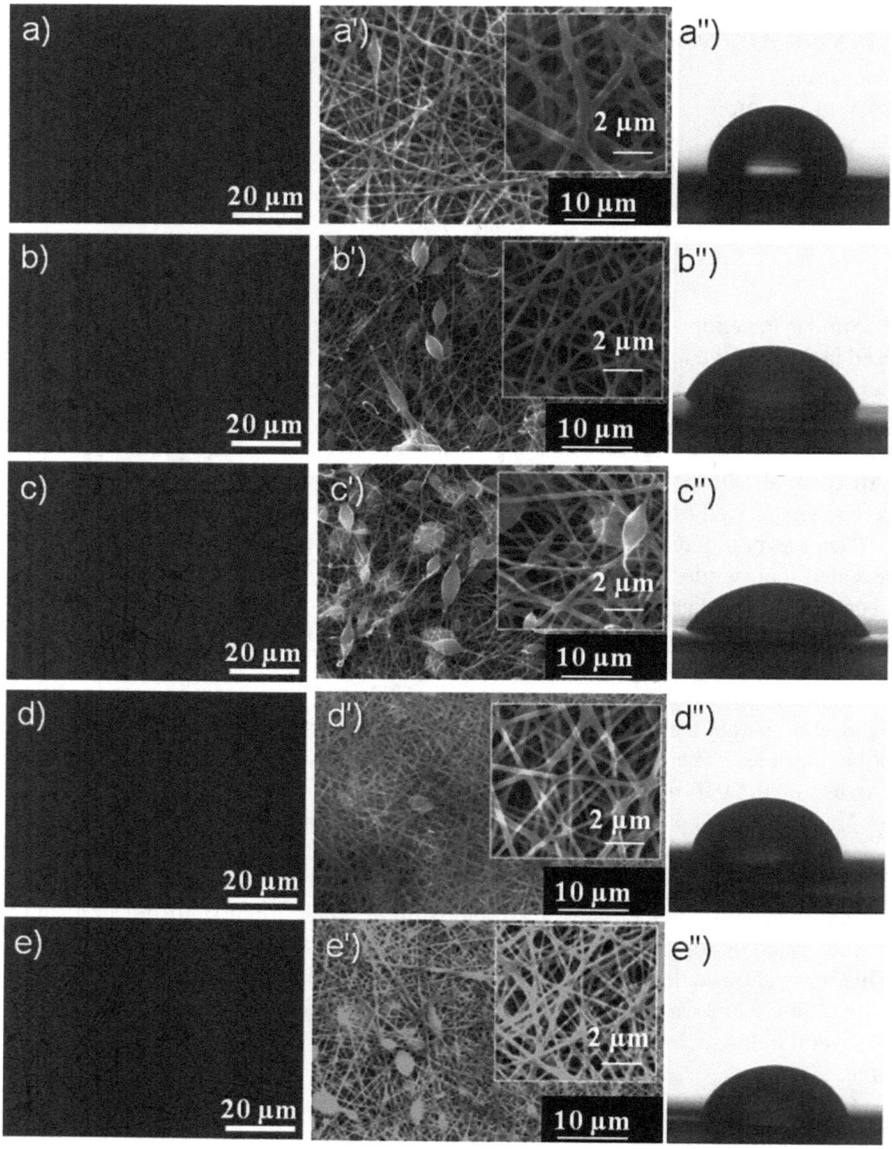

Fig. 14.6 Electrospun PLA-PHB-ATBC and the biocomposites Optical (left column), SEM observation (middle column) and water contact angles (right column). (**a**) PLA-PHB-ATBC, (**b**) PLA-PHB-ATBC-Ch1%, (**c**) PLA-PHB-ATBC-Ch5%, (**d**) PLA-PHB-ATBC-Cat1% and (**e**) PLA-PHB-ATBC-Cat5%. Reprinted with permission from Arrieta et al. (2016b)

phases are continuous, PLA significantly hinders the interaction of PHB with oxygen (Chizhov et al. 2018).

The samples were subjected to biodegradability tests carried out under soil conditions. The tests were carried out for the nonwoven fibrous material and films

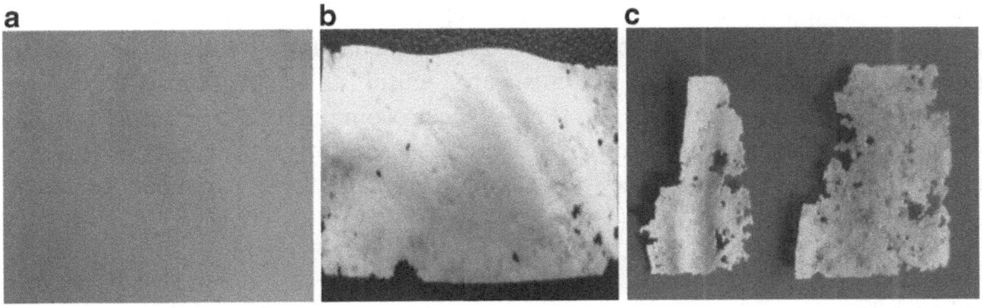

Fig. 14.7 Photographs of PHB films at various exposures in soil: (**a**) original film, (**b**) after 45 days, and (**c**) after 70 days. Reprinted with permission from Chizhov et al. (2018)

from PHB. Figure 14.7 shows photos of film samples during the experience. The onset of biodegradation in the material is easier to detail after 45 days (Fig. 14.4b). The exhibition has separate focal traces of biodegradation but retains its original integrity. A film sample after 70 days of exposure in the soil (Fig. 14.4c) loses its geometric integrity, and its fragments show typical defects in the form of through holes and shells (Chizhov et al. 2018).

The results obtained by Chizhov et al. (2018) emphasized that this material can be used as a disposable mulch material for the soil and as a cover material, with a useful life of less than 45 days under natural conditions. In addition, this product is harmless and quickly decomposes and will be easy to apply (Chizhov et al. 2018).

14.12 Perspectives

In the last years, there is a widespread trend in the use of synthetic and natural polymers in the production of different materials for daily uses. Currently in the agroindustry, these materials have been introduced in the production of coatings, such as mulches for the growth of crops. These materials increase soil temperature, reduce weed pressure, conserve moisture, reduce certain insect pests and increase the crop yields by making a more efficient use of soil nutrients. However, the use of conventional petroleum-derived polymers have generated threats to the environment with the concomitant problem of non-biodegradation and elimination. In this scenario, renewable and biodegradable materials have received great attention from the scientific and industrial community. The main disadvantage of natural polymers lies in the limited properties due to their poor physical properties. In this context, modifications of natural polymers by generation of composite biomaterials are an excellent alternative for the creation of renewable materials that can equal or exceed the performance of commonly used petroleum-based engineering polymers. Likewise, the combination of "bio-based" and renewable polymers with reinforcements or renewable charges would be an alternative to address the property gap that corresponds to the performance between renewable and traditional synthetic polymers. There are many attempts in this way, to produce a material bio-based, biodegradable and sustainable from biopolymers, e.g. Mater-Bi®, is a good example of this, and nowadays new mulches of this material are being used in the agricultural sector with good results and performance. Scientists are focusing their attention on the design of new processing methodologies, such as electrospinning or 3D printing to make the mulches. Although good results at laboratory scale have been obtained, it is a priority to work on a large scale for its massive use in industrial applications.

Annexes

Common certifications in bioplastics are specified, such as the following:

UNI EN 13432, "Requirements for packaging recoverable by composting and biodegradation – Test scheme and evaluation criteria for the final acceptance of packaging": defines the compostability characteristics and requirements of a material. According to this standard, compostable materials must have the following characteristics: – Biodegradability, meaning the capacity of a material to be converted into CO_2 by the action of micro-organisms. In order to present complete biodegradability, the material must be at least 90% biodegraded in less than 6 months; – Disintegratability, namely fragmentation and invisibility in the final compost. After 3 months in compost the fragments of the compostable material must have disappeared, leaving no more than 10% of the substance with a diameter of over 2 mm; – Absence of negative effects on the composting process; – Low levels of heavy metals (in relation to predefined maximum amounts) and the absence of negative effects on the final compost (e.g. reduction in the agronomic value and presence of phytotoxic effects on plant growth). Standard UNI EN 13432 is a harmonized standard, meaning it has been published in the Official Gazette of the European Community and has been implemented in Europe at a national level. Moreover, it grants presumption of conformity with European Directive 94/62 EC on packaging and packaging waste.

UNI EN 14995:2006, "Plastics – evaluation of compostability – test scheme and specifications": extends the scope of UNI EN 13432 to plastic materials not used in packaging, such as plastics used in agriculture and bags for the collection of waste. The requirements and test scheme are identical to those of UNI EN 13432.

ASTM D6400:2004, "Standard Specification for Compostable Plastics" – relating to specific techniques employed in order to define a plastic material as compostable. Very similar to the methods used in UNI EN 13432:2002 and UNI EN 14995:2006.

NF U52–001, "Matériaux biodégradables pour l'agriculture et l'horticulture – Produits de paillage – Exigences et méthodes d'essai": a French standard relating to mulch film. Alongside methods and tests to determine the biodegradation of the material, it also defines the duration of mulch films in the field in different environments (average life of the product). Based on their duration, biodegradable mulch films are divided into different classes.

UNI 11183:2006, "Plastic materials biodegradable at ambient temperature – Requirements and test methods"; this standard defines the biodegradability requirements that must be met by plastics used to make items that can be disposed of through aerobic biodegradation at ambient temperature. Ambient temperature means the temperature range of temperate regions excluding the high temperatures typical of industrial composting. As a rule, the ecotoxicity requirements of the plastic materials are also taken into consideration.

OK Biodegradable Soil: a conformity mark issued by the Belgian certification organization Vinçotte for products and materials which are specifically biodegradable in a typical soil. This mark guarantees the product/material is completely biodegradable in the soil, without the need for additional treatments, during a period determined by international criteria, and also guarantees it will not have any negative effect on the soil.

"Compostable DIN CERTCO" certification: DIN CERTCO is the German certification organisation which issues a range of certificates, including those for biodegradable materials. DIN CERTCO has prepared a certification scheme for compostable products made from biodegradable materials. The certification process is conducted in accordance with the principal international standards: DIN EN 13432 and ASTM D6400.

References

Abdalla, M. M. A., El-Dein, M. S., Badawy, A. S., & Hassaan, A. (2016). Effect of plastic mulch and drought periods on growth and yield of potato (solanum tuberosum L). I- vegetative growth. *Assiut Journal of Agricultural Sciences, 47*, 615–632. https://doi.org/10.21608/ajas.1999.3830.

Adlerz, W. C., & Evereit, P. H. (1968). Aluminum foil and white polyethylene mulches to repel aphids and control watermelon mosaic 1. *Journal of Economic Entomology, 61*, 1276–1279. https://doi.org/10.1093/jee/61.5.1276.

Arrieta, M. P., López, J., López, D., Kenny, J. M., & Peponi, L. (2016a). Biodegradable electrospun bionanocomposite fibers based on plasticized PLA–PHB blends reinforced with cellulose nanocrystals. *Industrial Crops and Products, 93*, 290–301. https://doi.org/10.1016/j.indcrop.2015.12.058.

Arrieta, M. P., López, J., López, D., Kenny, J. M., & Peponi, L. (2016b). Effect of chitosan and catechin addition on the structural, thermal, mechanical and disintegration properties of plasticized electrospun PLA-PHB biocomposites. *Polymer Degradation and Stability, 132*, 145–156. https://doi.org/10.1016/j.polymdegradstab.2016.02.027.

Ban, D., Žanić, K., Dumičić, G., Čuljak, T. G., & Ban, S. G. (2009). The type of polyethylene mulch impacts vegetative growth, yield, and aphid. *Journal of Food Agriculture & Environment, 7*, 543–550.

Brault, D., & Stewart, K. A. (2002). Optical properties of paper and polyethylene mulches used for weed control in lettuce. *Hort Science, 37*(1), 87–91.

Briassoulis, D. (2006). Mechanical behaviour of biodegradable agricultural films under real field conditions. *Polymer Degradation and Stability, 91*, 1256–1272. https://doi.org/10.1016/j.polymdegradstab.2005.09.016.

Briassoulis D, Giannoulis A (2018) Evaluation of the functionality of bio-based plastic mulching films. Polym Test 67:99–109. https://doi.org/10.1016/j.polymertesting.2018.02.019.

Bryan, H. H. (1966). Effect of plastic mulch on the yield of several vegetable crops in Noth Florida. Florida Sate Hortic Soc 139–146.
Clarke, A. D. (1987). Some plastic industry development; their impact on plastic films for agricultural application. *Plasticulture, 74*, 15–26.
Chizhov, A. S., Ol'khov, A. A., Monakhova, T. V., Shibryaeva, L., & Iordanskii, A. (2018). Thermooxidation and biodegradation of nonwoven biopolymer fibrous materials. *Polymer Science Series D, 11*, 339–343. https://doi.org/10.1134/S1995421218030036.
Cimadoro, J., Ribba, L., Ledesma, S., & Goyanes, S. (2018). Electrospun Mats: From white to transparent with a drop. *Macromolecular Materials and Engineering, 1800237*, 1800237. https://doi.org/10.1002/mame.201800237.
Cushman, K. E., Maqbool, M., & Gerard, P. D. (2005). Mulch type, mulch depth, and rhizome planting depth for field-grown American mayapple. *Hort Science, 40*(3), 635–639.
Decoteau, D. R., Kasperbauer, M. J., & Hunt, P. G. (1989). Mulch surface color affects yield of fresh market tomatoes. *Journal of the American Society for Horticultural Science, 114*(2), 216–219.
Dvořák, P., Hajšlová, J., Hamouz, K., Schulzová, V., Kuchtova, P., & Tomasek, J. (2001). Black polypropulene mulch textile in organic Agriculture. *Lucr Stiint, 52*, 2–6.
Emmert, E. M. (1957). Black polyethylene for mulching vegetables. *Proceedings of the American Society for Horticultural Science, 69*, 464–469.
Galinato, S. P., Miles, C., & Ponnaluru S. (2012). 2011 Cost of producing fresh market field-grown tomatoes in Western Washington. Washington State University Extension Publisher FS080E 1–6.
George, S., Idicula, S. P., & Joseph, K. (2015). *Polypropylene woven fabric – A good mulch material for young rubber plants, 43*, 171–177. https://doi.org/10.19071/jpc.2015.v43.i3.2850.
Ghimire, S., Hayes, D., DeVetter, L. W., Inglis, D., Cowan, J. S., & Miles, C. A. (2018). Biodegradable plastic mulch and suitability for sustainable and organic agriculture. *WSU Ext*, 1–11.
Gordon, G. G., Foshee, W. G., Reed, S. T., Brown, J. E., Vinson, E., & Woods, F. M. (2008). Plastic mulches and row covers on growth and production of summer squash. *International journal of vegetable science, 14*(14), 322–338. https://doi.org/10.1080/19315260802215830.
Grand View Research. (2016). Biodegradable mulch films market worth $64.3 million by 2024, Grand View Research, Inc, https://www.grandviewresearch.com. Accessed 18 Dec 2018.
Gupta, B., Revagade, N., Anjum, N., Atthoff, B., & Hilborn, J. (2006). Preparation of poly(lactic acid) fiber by dry-jet-wet-spinning. I. Influence of draw ratio on fiber properties. *Journal of Applied Polymer Science, 100*(2), 1239–1246. https://doi.org/10.1002/app.23497.
Gutiérrez, T. J. (2018). Are modified pumpkin flour/plum flour nanocomposite films biodegradable and compostable? *Food Hydrocolloids, 83*, 397–410. https://doi.org/10.1016/j.foodhyd.2018.05.035.
Gutiérrez, T. J., Toro-Márquez, L. A., Merino, D., & Mendieta, J. R. (2019). Hydrogen-bonding interactions and compostability of bionanocomposite films prepared from corn starch and nano-fillers with and without added Jamaica flower extract. *Food Hydrocolloids, 89*, 283–293. https://doi.org/10.1016/j.foodhyd.2018.10.058.
Hakkarainen, M., & Albertsson, A.-C. (2004). In A.-C. Albertsson (Ed.), *Environmental degradation of polyethylene BT – long term properties of polyolefins* (pp. 177–200). Berlin/Heidelberg: Springer. https://doi.org/10.1007/b13523.
Hayes, D. G., & Flury, M. (2018). Summary and assessment of EN 17033 : 2018, a new standard for biodegradable plastic mulch films. 1–7.
Herniou--Julien, C., Mendieta, J. R., & Gutiérrez, T. J. (2019). Characterization of biodegradable/non-compostable films made from cellulose acetate/corn starch blends processed under reactive extrusion conditions. *Food Hydrocolloids, 89*, 67–79. https://doi.org/10.1016/j.foodhyd.2018.10.024.
Kaihara, S., Osanai, Y., Nishikawa, K., Toshima, K., Doi, Y., & Matsumura, S. (2005). Enzymatic transformation of bacterial polyhydroxyalkanoates into repolymerizable oligomers directed

towards chemical recycling. *Macromolecular Bioscience, 5*(7), 644–652. https://doi.org/10.1002/mabi.200500030.

Kakoria, A., & Sinha-Ray, S. (2018). A review on biopolymer-based fibers via electrospinning and solution blowing and their applications. *Fibers, 6*, 45. https://doi.org/10.3390/fib6030045.

Kasirajan S., & Ngouajio, M. (2012). Polyethylene and biodegradable mulches for agricultural applications: a review. 501–529. https://doi.org/10.1007/s13593-011-0068-3.

Keshavarz, T., & Roy, I. (2010). Polyhydroxyalkanoates: Bioplastics with a green agenda. *Current Opinion in Microbiology, 13*, 321–326. https://doi.org/10.1016/j.mib.2010.02.006.

Lament, W. J. Jr. (1993) Plastic Mulches for the Production of Vegetable Crops. Hort. *Technology 3*(1):35–39.

Lamont, W. J. (2017). Plastic mulches for the production of vegetable crops. In M. D. Orzolek (Ed.), *A guide to the manufacture, performance, and potential of plastics in agriculture* (pp. 45–60., ISBN 9780081021705). Elsevier Ltd. https://doi.org/10.1016/B978-0-08-102170-5.00003-8.

Liakatas, A., Clark, J. A., & Monteith, J. L. (1986). Measurements of the heat balance under plastic mulches. Part I. radiation balance and soil heat flux. *Agricultural and Forest Meteorology, 36*, 227–239. https://doi.org/10.1016/0168-1923(86)90037-7.

Martin-Closas, L., Pelacho, A. M., Picuno, P., & Rodríguez, D. (2008). Properties of new biodegradable plastics for mulching, and characterization of their degradation in the laboratory and in the field. Acta Horticulturae 801 PART 1: 275–282. https://doi.org/10.17660/ActaHortic.2008.801.27

Merino, D., Gutiérrez, T. J., & Alvarez, V. A. (2019a). Structural and thermal properties of agricultural mulch films based on native and oxidized corn starch nanocomposites. *Starch-Stärke, 71*(7–8), 1800341. https://doi.org/10.1002/star.201800341.

Merino, D., Gutiérrez, T. J., & Alvarez, V. A. (2019b). Potential agricultural mulch films based on native and phosphorylated corn starch with and without surface functionalization with chitosan. *Journal of Polymers and the Environment, 27*(1), 97–105. https://doi.org/10.1007/s10924-018-1325-1.

Merino, D., Gutiérrez, T. J., Mansilla, A. Y., Casalongué, C., & Alvarez, V. A. (2018a). Critical evaluation of starch-based antibacterial nanocomposites as agricultural mulch films: Study on their interactions with water and light. *ACS Sustainable Chemistry & Engineering, 6*(11), 15662–15672. https://doi.org/10.1021/acssuschemeng.8b04162.

Merino, D., Mansilla, A. Y., Gutiérrez, T. J., Casalongué, C. A., & Alvarez, V. A. (2018b). Chitosan coated-phosphorylated starch films: Water interaction, transparency and antibacterial properties. *Reactive and Functional Polymers, 131*, 445–453. https://doi.org/10.1016/j.reactfunctpolym.2018.08.012.

Miles, C., DeVetter, L., & Ghimire, S. (2017). Suitability of biodegradable plastic mulches for organic and sustainable agricultural production systems. *HortScience, 52*(1), 10–15. https://doi.org/10.21273/hortsci11249-16.

Moreno, M. M., & Moreno, A. (2008). Effect of different biodegradable and polyethylene mulches on soil properties and production in a tomato crop. *Scientia Horticulturae, 116*(3), 256–263. https://doi.org/10.1016/j.scienta.2008.01.007.

Ngouajio, M., & Ernest, J. (2004). Light transmission through colored polyethylene mulches affects weed populations. *HortScience, 39*, 1302–1304.

Nicosia, A., Gieparda, W., Foksowicz-Flaczyk, J., Wesołek, D., Vazquez, B., Prodi, F., & Belosi, F. (2015). Air filtration and antimicrobial capabilities of electrospun PLA/PHB containing ionic liquid. *Separation and Purification Technology, 154*, 154–160. https://doi.org/10.1016/j.seppur.2015.09.037.

Novamont (2013) Biodegradable and compostable definitions. In: Biodegrad. Compost. Defin. www.novamont.com. Accessed 13 Dec 2018.

Posada, J. A., Naranjo, J. M., López, J. A., Higuita, J. C., & Cardona, C. A. (2011). Design and analysis of poly-3-hydroxybutyrate production processes from crude glycerol. *Process Biochemistry, 46*(1), 310–317. https://doi.org/10.1016/j.procbio.2010.09.003.

Drumright, R. E., Gruber, P. R., & Henton, D. E. (2000). Polylactic acid technology: A review. *Advanced Materials, 12*, 1841–1846. https://doi.org/10.1201/9780203508206.ch16.

Ribba, L. G., Cimadoro, J. D., D'Accorso, N. B., & Goyanes, S. N. (2017). Removal of pollutants using electrospun nanofiber membranes. In S. N. Goyanes & N. B. D'Accorso (Eds.), *Industrial applications of renewable biomass products: Past, present and future* (pp. 301–324). Cham: Springer International Publishing. https://doi.org/10.1007/978-3-319-61288-1_12.

Ribba, L., Parisi, M., D'Accorso, N. B., & Goyanes, S. (2014). Electrospun nanofibrous mats: From vascular repair to osteointegration. *Journal of Biomedical Nanotechnology, 10*(12), 3508–3535. https://doi.org/10.1166/jbn.2014.2046.

Robb, B., & Lennox, B. (2011). The electrospinning process, conditions and control. In L. A. Bosworth & S. Downes (Eds.), *Electrospinning tissue Regen/Electrospinning for tissue Regen* (pp. 51–66). United Kingdom: Elsevier. https://doi.org/10.1016/B978-1-84569-741-9.50003-3.

Romic, D., Borosic, J., Poljak, M., & Romic, M. (2003). Polyethylene mulches and drip irrigation increase growth and yield in watermelon (Citrullus lanatus L.). *European Journal of Horticultural Science, 68*(4), 192–198. ISSN 1611-4426.

Sanchez, E., Lamont, W. J., & Orzolek, M. D. (2008). Newspaper mulches for suppressing weeds for organic high-tunnel cucumber production. *HortTechnology, 18*(1), 154–157. https://doi.org/10.21273/horttech.18.1.154.

Scarascia-Mugnozza, G., Schettini, E., Vox, G., Malinconico, M., Immirzi, B., & Pagliara, S. (2006). Mechanical properties decay and morphological behaviour of biodegradable films for agricultural mulching in real scale experiment. *Polymer Degradation and Stability, 91*, 2801–2808. https://doi.org/10.1016/j.polymdegradstab.2006.02.017.

Serrano-Ruíz, H., Martín-Closas, L., & Pelacho, A. M. (2018). Application of an in vitro plant ecotoxicity test to unused biodegradable mulches. *Polymer Degradation and Stability, 158*, 102–110. https://doi.org/10.1016/j.polymdegradstab.2018.10.016.

Smith, A. (1931). Effect of paper mulches on soil temperature, soil moisture, and yields of certain crops. *Hilgardia : A journal of agricultural science, 6*(6), 159–201. https://doi.org/10.3733/hilg.v06n06p159.

Steinmetz, Z., Wollmann, C., Schaefer, M., Buchmann, C., David, J., Tröger, J., Muñoz, K., Frör, O., & Schaumann, G. E. (2016). Science of the total environment plastic mulching in agriculture. Trading short-term agronomic benefits for long-term soil degradation? *Science of the Total Environment, 550*, 690–705. https://doi.org/10.1016/j.scitotenv.2016.01.153.

Streck, N. A., Schneider, F. M., Buriol, G. A., & Heldwein, A. B. (1995). Effect of polyethylene mulches on soil temperature and tomato yield in plastic greenhouse. *Scientia Agricola, 52*, 587–593. https://doi.org/10.1590/S0103-90161995000300028.

Toncheva, A., Spasova, M., Paneva, D., Manolova, N., & Rashkov, I. (2014). Polylactide (PLA)-based electrospun fibrous materials containing ionic drugs as wound dressing materials: A review. *International Journal of Polymeric Materials and Polymeric Biomaterials, 63*(13), 657–671. https://doi.org/10.1080/00914037.2013.854240.

Tarara, J. M. (2000). Microclimate modification with plastic mulch. *HortScience, 35*, 169–180.

Tindall, J. A., Beverly, R. B., & Radcliffe, D. E. (1991). Mulch effect on soil properties and tomato growth using micro-irrigation. *Agronomy Journal, 83*, 1028–1034. https://doi.org/10.2134/agronj1991.00021962008300060019x.

De Wilde, B., Deconinck, S (2017) (Bio) Degradable Mulching Films. Expert statement FORM A.097 Rev 00, pp 10. http://www.ows.be/wp-content/uploads/2017/08/Expert-statement-mulching-films.pdf.

Wright, J. (1968). Production of polyethylene film. In: Proc. 8th Natl. Agr. Plastics Conf. pp 72–79.

Chapter 15
Biodegradable Polymer Nanofibers Applied in Slow Release Systems for Agri-Food Applications

Danilo Martins, Vanessa P. Scagion, Rodrigo Schneider, Ana Carolina C. Lemos, Juliano Oliveira, and Daniel S. Correa

Abstract Polymer nanofibers possess remarkable properties including high aspect ratio, nanoscale diameters, mechanical flexibility, varied morphology and porosity and large surface area for bulk or surface modification. These features make them highly potential for technological applications, where active components can be added to the nanofibers depending on the specific target application. Polymer nanofibers can be produced by several techniques, but electrospinning (ES) and solution blow-spinning (SB-Spinning) are two techniques of outmost importance, once they allow controlling the nanofibers features, including diameter and morphology, combined at a high production rate. Such polymer nanofibers have been applied in biotechnological and medical applications, sensors, catalysts and pollutant adsorption platforms. In agriculture, specifically, nanofibers can be used in the design of slow release systems using biodegradable micro- and nanofibers containing active compounds such as antibiotics, fertilizers and pesticides. Such systems can increase the application efficiency by using a smaller amount of the active material, reducing costs and toxicity-related problems. In this context, in this book chapter the recent developments of biodegradable nanofibers produced by ES and SB-Spinning

D. Martins
Embrapa Instrumentação, Nanotechnology National Laboratory for Agriculture (LNNA), São Carlos, SP, Brazil

V. P. Scagion · R. Schneider · D. S. Correa (✉)
Embrapa Instrumentação, Nanotechnology National Laboratory for Agriculture (LNNA), São Carlos, SP, Brazil

PPGQ, Department of Chemistry, Center for Exact Sciences and Technology, Federal University of São Carlos (UFSCar), São Carlos, SP, Brazil
e-mail: daniel.correa@embrapa.br

A. C. C. Lemos
PPGBiomat, Federal University of Lavras (UFLA), Lavras, MG, Brazil

J. Oliveira
Department of Engineering, Federal University of Lavras (UFLA), Lavras, MG, Brazil

T. J. Gutiérrez (ed.), *Polymers for Agri-Food Applications*,
https://doi.org/10.1007/978-3-030-19416-1_15

containing active compounds for slow release system aiming at agricultural and food applications are reviewed and presented.

Keywords Electrospinning · Solution blow-spinning

15.1 Introduction

Nanotechnology has emerged as a powerful tool capable of developing and transforming the agri-food sector, with the potential to increase agricultural productivity, food security and economic growth for industries (Villena de Francisco and García-Estepa 2018). Among one-dimensional nanostructures potentially applied in agri-food sector, polymer nanofibers have attracted great attention because of their exceptional properties, including high surface-to-volume ratio and porosity, mechanical flexibility, possibility of chemical functionalization and ease of production process (Noruzi 2016; Yang et al. 2018).

Electrospinning (ES) and solution blow spinning (SB-Spinning) are currently two of the most investigated techniques for producing polymer fibers because they are cost-effective, simple and highly versatile, and allow the manufacture of micro- and nanoscale fibers from various polymers, polymer blends and polymer composites (Daristotle et al. 2016; Medeiros et al. 2009; Wang et al. 2013).

Polymer nanofibers, mainly those prepared by electrospining have been widely studied to encapsulate active compounds applied in agriculture and food industry, such as pheromones, fertilizers, pesticides and nutraceuticals, protecting them from adverse environmental conditions, maintaining their properties and achieving their controlled release (Daristotle et al. 2016; Noruzi 2016; Wen et al. 2017). These systems can increase the application efficiency by using a smaller amount of the active compound, thus promoting sustainable practices and reducing costs (Noruzi 2016).

A wide range of biodegradable polymers have already been successfully spun into fibers, including synthetic polymers such as poly(ε-caprolactone) (PCL), poly(lactic acid) (PLA), poly(lactic-*co*-glycolic acid) (PLGA), poly(ethylene glycol)-*co*-poly(d,l-lactide) (PELA) and poly(ethylene glycol)-PCL (PEG-PCL) (Yang et al. 2018). Natural polymers such as alginate, starch, cellulose and chitosan have also been utilized in ES and SB-Spinning (Mele 2016). However, some biodegradable synthetic polymers, including PCL and PLA are often required to be added as carrier polymers due to low spinnability of aforementioned natural polymers (dos Santos et al. 2018; Mele 2016; Yang et al. 2018). Overall, the application of biodegradable polymers in the development of novel materials by using ES or SB-Spinning is in accordance with current demands for sustainable developments and environmental protection (Laycock et al. 2017).

In this context, this chapter provides an overview on the design and fabrication through electrospinning and SB-Spinning of biodegradable polymeric nanofibers as vehicles for sustained and controlled release of active compounds used in agri-food sector. First, an introduction to the principle of ES and SB-Spinning is presented. Then, recent developments on the ES and SB-Spinning of biodegradable polymers and their potential application in the encapsulation of antibiotics, fertilizers, pesticides, pheromones, biocontrol agents and nutraceuticals are reviewed and discussed. In addition, the future scopes and existing limitations of these approaches are also highlighted.

15.2 Methods of Fiber Fabrication

15.2.1 Electrospinning

15.2.1.1 Introduction and History

The first record of the electrostatic attraction of a liquid was reported by William Gilbert in 1600 (Tucker et al. 2012). Then, in 1914, John Zeleny published his work on the behavior of fluid droplets at the end of metallic capillaries (Zeleny 1917), which initiated the study of mathematically modeling the liquids under electrostatic forces. In 1934, Anton Formhals filled a patent describing the experimental apparatus for the production of synthetic fibers using electrical field (Formhals 1934). The works published by Taylor between 1964 and 1969 contributed enormously to electrospinning area by mathematically modelling the cone shape that the fluid droplet assumes under the effect of an electric field (Lübbert and Peukert 2018). Nowadays, this characteristic droplet shape observed in electrospinning is known as the Taylor cone (Sui et al. 2016). In 1995, the study reported by Renecker revived electrospinning by demonstrating the fabrication of nanostructured fibers from various polymers (Doshi and Reneker 1995).

This technique is used today to fabricate nonwoven nanofibers displaying remarkable properties, such as high surface area to volume ratio, permeability, stability and porosity (Mori et al. 2015; Wendorff et al. 2012a). Nanofibers can be applied in a wide range of applications including sensors (Bagchi et al. 2017; Hidayat et al. 2017; Park et al. 2012; Zamora-Pérez et al. 2012), biotechnology (Al-Enizi et al. 2018; Ashraf et al. 2018; Ramakrishna et al. 2006), food industry (Torres-Giner et al. 2018; Weiss et al. 2012; Wen et al. 2017), healthcare (Venugopal et al. 2012), agriculture (Damasceno et al. 2013; Krishnamoorthy and Rajiv 2017b; Peršin et al. 2017) and environmental engineering (Agarwal and Greiner 2011; Yoon et al. 2008).

15.2.1.2 Electrospinning Process

Electrospinning is a process in which electrostatic forces are employed to produce fibers from polymer solutions or melts (Feltz et al. 2017; Wendorff et al. 2012b; Yousefzadeh 2017). Fig. 15.1a illustrates the general set-up of an electrospinning apparatus used to prepare polymeric nanofibers. The equipment consists of a high voltage power supply, a syringe and a syringe pump, a spinneret connected to the syringe and a metal collector. In this technique, an electric field is applied to a solution or melt polymer droplet. As a consequence, an electrostatic charge accumulates

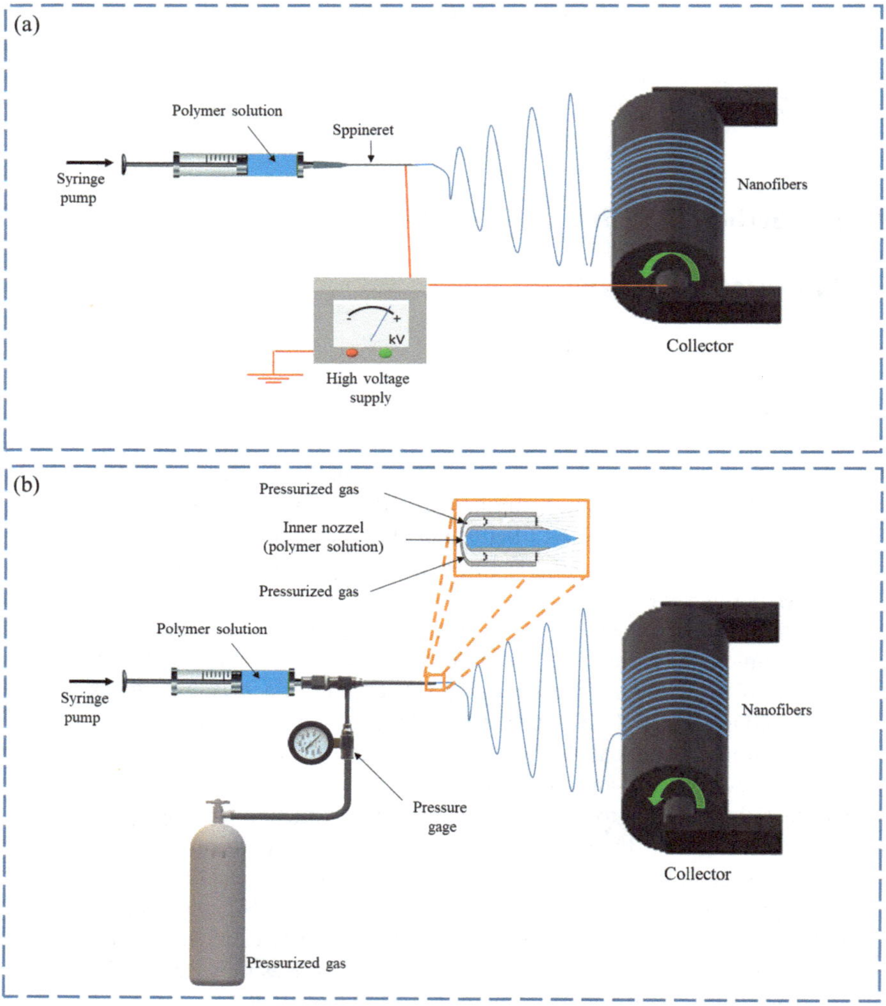

Fig. 15.1 Schematic representation of (**a**) conventional electrospinning (**a**) and (**b**) solution blow spinning apparatus

at the tip of polymeric droplet (Mercante et al. 2017). By force balance, charge repulsion works against the surface tension, causing the shape of droplet to change from a spherical surface to an elongated cone shape (Taylor cone) (Roque et al. 2014). Other important factor is the rheological behavior of solution (Kong and Ziegler 2014; Pontrelli et al. 2014). If the polymer entanglements are sufficiently large, the concentration of polymer in the solution will oppose the droplet jet break (Plateau–Rayleigh instability) (Kong and Ziegler 2012; Manasco et al. 2012). This produces a jet of polymer solution which will be stretched and whipped by electric forces at a complex path to metallic collector (Liu et al. 2009; Oliveira et al. 2013). In this process, the solvent evaporates, and a network of fibers is collected on the ground target.

15.2.1.3 Parameters of Electrospinning Process

The range of parameters known to affect the electrospinning of the nanofibers and the corresponding nonwovens are classified into solution parameters (viscosity, concentration, molecular weight, surface tension, conductivity); processing parameters (polymer solution feed rate, needle (tip) shape, applied voltage, distance between spinneret and collector, collector composition and geometry) and environmental parameters (temperature and humidity).

Each of these parameters can affect the quality of the fibers, and by proper control of these parameters, electrospun fibers with desired morphologies can be prepared, such as porous nanofibers (Thorvaldsson et al. 2008), core-shell nanofibers (Qin 2017), nanotubes (Li et al. 2007), nanoribbons (Li et al. 2010), helicoidal nanofibers (Chang and Shen 2011), aligned nanofibers (Ismar and Sarac 2018), honeycomb nanofiber (Chen et al. 2018c) and nanofiber yarns (Chang et al. 2016).

15.2.1.3.1 Solution Parameters

Solution parameters are related to the physical-chemical properties of polymers, solvents and interactions between polymer and solvent (Li and Xia 2004; Picciani et al. 2009). A polymer solution with appropriate viscosity, surface tension, and conductivity is essential to fabricate nanofibers without beads or beads-on-a-string appearance.

The polymer concentration influences the solution rheological behavior and plays a crucial role in the electrospinning (Peer et al. 2018; Valipour et al. 2018). To obtain homogeneous nanofibers without beads, a high degree of macromolecular entanglement is required. The diameter of the fibers can be increased by increasing the concentration of the polymer solution, while maintaining the other parameters of the process unchanged (Cramariuc et al. 2013; Fridrikh et al. 2003; Thompson et al. 2007). When the polymer solution concentration is too low, the degree of entanglement of the polymer chains is small, causing capillary instability at the end of the jet. This instability can result in an electrostatic scattering (electrospraying) instead of electrospinning. In this case, spherical particles or nanofibers with beads are obtained (Shariatpanahi et al. 2016; Yarin et al. 2001).

The surface tension is directly related to the formation of the nanofibers, since the formation of polymer jets occurs when the applied voltage is sufficiently high to cause the electrostatic forces to overcome the surface tension force acting on the drop (Liu et al. 2010). From this voltage value, called critical voltage, the spinning process is started. Several authors have reported that the decrease in surface tension can favor the formation of fibers without beads (Dong et al. 2011; Nagarajan et al. 2007).

The electrical conductivity of the solution and the dielectric constant of solvent also influence the morphology of the produced fibers. The ions in the polymeric solution can be oriented by the electric field, which allows the jet to undergo a greater stretching, resulting in thinner segments and thus reducing the fiber diameter (Chayad et al. 2016; Rezaei et al. 2018).

15.2.1.3.2 Processing Parameters

A minimum applied voltage is required to cause the deformation of the solution drop and the jet formation (Mercante et al. 2017). An increase in applied voltage can result in a decrease of the nanofiber diameters. Nonetheless, a further increase can also lead to a faster acceleration of jet polymers towards the substrate collector, which reduces the flight time for the jet to stretch prior to deposition and results in a larger fiber diameter (Ray et al. 2016). In addition, an increase in the applied voltage beyond the critical value can result in the formation of nanofibers with beads (Deitzel et al. 2001).

The distance between the needle tip and the collector is an important processing parameter as it influences the electric field, jet flying time and the solvent evaporation rate (Agarwal et al. 2013). In this sense, a proper distance is required to give the fibers sufficient time to solvent evaporation before reaching the collector, otherwise with distances that are either too close or too far, no fiber formation or beads can be observed (Bhardwaj and Kundu 2010).

Flow rate of polymer solution is related to the jet velocity and the material transfer rate, which are factors that affects the solvent evaporation rate and consequently the fiber morphology and size (Bhardwaj and Kundu 2010). High flow rates can result in beaded or flatter fibers due to unavailability of proper time to solvent evaporation prior to reaching the collector, while lower feed rate can hinder the ejection of polymer solution from the needle tip due to the drying of droplet (Haider et al. 2015; Rogina 2014).

The collector composition and geometry affect the morphology of electrospun fibers. Aluminium foil is commonly used as a collector, but aiming to prepare nanofibers with complex geometries, distinct collecting apparatuses have been proposed, such as, wire mesh, parallel or grided bar, rotating rod, rotating wheel and liquid non solvent as coagulation bath (He et al. 2014; Ki et al. 2007; Rogina 2014).

15.2.1.3.3 Environment Parameters

The electrospinning process and the physical characteristics of the fibers are affected by the temperature and humidity of the environment, since these parameters may influence the solvent evaporation rate of the polymer jet. Investigations have revealed that an increase in relative humidity results in decrease of fiber diameter (De Vrieze et al. 2009; Pelipenko et al. 2013). It has also been demonstrated that humidity increase causes an increase in the number, diameter, shape, and distribution of the pores on surface of the fibers (Casper et al. 2004). It is important to emphasize that this behavior is dependent on the physico-chemical properties of polymer and solvent. A decrease of the temperature has been found to decrease fiber diameters, which was attributed to the decrease of solution viscosity and the increase of solvent evaporation rate (De Vrieze et al. 2009).

15.2.2 Solution Blow Spinning (SB-Spinning)

15.2.2.1 Introduction and History

Solution blow spinning (SB-Spinning) technique was developed from the combination of two other techniques, namely electrospinning and melt blowing (Ellison et al. 2007; Farias et al. 2015; Medeiros et al. 2009; Rezende 2004; Silva et al. 2015; Taylor 1969; Zeleny 1914). SB-Spinning produces fibers with similar features to those obtained by electrospinning, but with a much higher production rate and no need of high electrical voltages (Costa et al. 2016; Santos et al. 2016; Zhang et al. 2003). Additionally, SB-Spinning allows the deposition of fibers directly on any type of surfaces, including living systems, as in the case of surgical scaffolds, sealants or tissue adhesives. SB-Spinning has been used for the production of fibers applied in agroindustry, catalysts, energy storage, among others (Costa et al. 2016; Paschoalin et al. 2017; Tomecka et al. 2017).

15.2.2.2 SB-Spinning Process

SB-Spinning is a fiber manufacturing process that requires two parallel concentric fluid flows (Fig. 15.1). The polymer solution produced from a polymer dissolved in a volatile solvent flows at the same time as a pressurized gas (Daristotle et al. 2016). The geometry of the needle creates a low pressure region around the inner nozzle, and the external concentric nozzle (where the gas passes through) makes use of the Bernoulli principle (Eisberg and Lerner 1982) for fluid flow, in which changes in pressure are converted into kinetic energy. At the contact, the polymer solution is drawn by pressurized air, where the drop in air pressure in the center of the jet creates a driving force responsible for the acceleration of the jet, forming a cone shape (da Silva Parize et al. 2016; Tomecka et al. 2017). The polymer solution feed rate is controlled by an injection pump and the air is pressurized by a pressure regulator. The produced nanofibers are collected in collector in the direction of the gas flow, forming a nonwoven web or fiber mat (da Silva Parize et al. 2016; Tomecka et al. 2017). The collector can be flat and static or cylindrical and rotational. The rotation of the collector can cause a stretch in the fibers, so they are deposited in a preferential direction, making them aligned (Daristotle et al. 2016; Rotta et al. 2016).

15.2.2.3 Parameters of SB-Spinning Process

The final features of fibers produced by SB-Spinning depend on several parameters, which include the molecular weight of the polymer, the type of solvent, the concentration and the viscosity of the polymer solution. In addition, other process variables, such as gas pressure, injection rate, and working distance are also important. Other important factors are the distance between nozzles, the nozzle geometry, the humidity and the ambient temperature (da Silva Parize et al. 2016; Daristotle et al. 2016; Li et al. 2017). These parameters can directly interfere in the diameters and the morphology of the fiber (Oliveira et al. 2013), droplets, porosity and the formation of film zones (Venturelli and Gripa 2017).

Some authors observed that the increase in fiber diameter is directly linked to the increase in the concentration of the solution and, consequently, increase in viscosity (Benito et al. 2017; da Silva Parize et al. 2016; Kuk et al. 2016; Paschoalin et al. 2017). The velocity of fluid transport through the nozzle is higher at low viscosities, resulting in more stretched fibers and, consequently finer ones (Li et al. 2017). A solution with high surface tension (more viscous/concentrated) hinders the solvent evaporation, which upon reaching the collector can generate undesirable film formation zones (Benito et al. 2017; Bonan et al. 2017). The choice of solvent can be made from solubility parameters of Hansen (Oliveira et al. 2014; Vay et al. 2011).

Although the effect of pressure on the fiber diameter is small, the effect of the combined pressure on the solution injection rate is expressive. Very low pressures do not allow the fibers to reach the collector correctly, causing obstruction (Daristotle et al. 2016; Kuk et al. 2016; Silva et al. 2015). Insufficient pressure can generate discontinuous jets, which increases the diameter of the fibers, causing greater porosity, formation of granules and interconnected fibers (Bonan et al. 2017; Kuk et al. 2016; Li et al. 2017).

The working distance is a processing variable that influences the formation of fibers, but with little effect on the diameter. The effective working range should be the minimum distance required for the solvent evaporation before reaching the collector and a maximum distance that prevents excessive loss of the aligned fibers (Benito et al. 2017; da Silva Parize et al. 2016; Daristotle et al. 2016; Li et al. 2017).

15.3 Applications Involving ES and SB-Spinning for Encapsulation

15.3.1 Antibiotics Encapsulation

Antibiotics, including human therapeutic drugs, have been widely used in food animals to treat and prevent bacterial diseases or to promote growth (Scott et al. 2018). However, it is recognized nowadays that the widespread use of antibiotics in food animals could contribute to the development of antibiotic-resistant bacteria, which threatens human and animal health (Huang et al. 2018). In this regard, the development of technologies that allow the control of antibiotic dose is imperative to maximize the therapeutic effect, while minimizing side effects on animals, costs and the risks associated with the problem of antibiotic resistance.

Polymer micro- and nanofibers prepared by electrospinning and SB-spinning offer different features including a large surface area-to-volume ratio, high porosity, uniformity in fiber size, diversity in the composition, flexibility in assembled structure, ease of drug encapsulation and high versatility in drug formulations, making them promising candidates as drug delivery systems (Chen et al. 2018a; Daristotle et al. 2016).

Mastitis is a inflammation of the mammary glands in dairy cows, and can be caused by a wide spectrum of pathogens, including *Staphylococcus aureus, Escherichia coli, Pseudomonas aeruginosa* and *Streptococcus agalactiae* (Erskine et al. 2003). This inflammation is responsible for financial losses to dairies by reducing milk yield and quality while increasing additional treatment costs (Huang et al. 2018). Antimicrobials have been applied in high doses to treat mastitis and the design of platforms that allows the control over drug dosage and its release profile is required. For this purpose, Schneider et al. (2018) encapsulated cloxacillin benzatine (CLOX), a penicillin derivate, poorly water soluble, and commonly used to treat mastitis, into a blend of PLA/poly (butylene adipate-*co*-terephthalate) (PBAT) (namely Ecovio (EC)) nanofibers prepared by electrospinning (Fig. 15.2a). One of effects presented by this infection is the pH decrease, so the authors demonstrated that CLOX release was modulated by pH, which included a change of the release

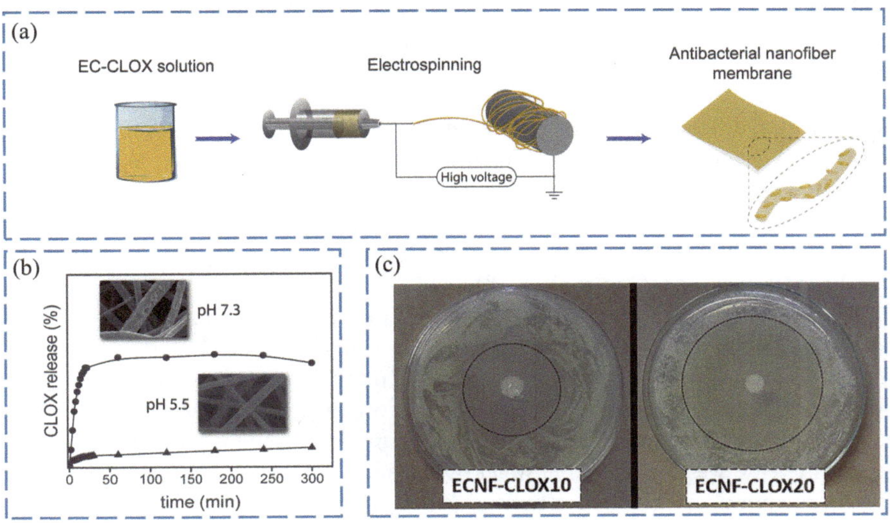

Fig. 15.2 Schematic representation of preparation of CLOX-encapsulated EC nanofibrous membranes (**a**), CLOX release profiles at different pHs (**b**) and antibacterial properties of ECNF-CLOX nanofibers against *S. aureus* after 24 h of incubation. The dashed black line is a guide to the eye to visualize the inhibition zone. Reprinted with permission from Schneider et al. (2018). Copyright 2018 Elsevier

mechanism from a Fickian diffusion at pH = 7.3 (with a burst release), to a Non-Fickian mechanism at pH = 5.5, with a sustainable release (Fig. 15.2b). *In vitro studies* revealed that the developed drug-loaded nanofibrous membrane containing 10% and 20% w/w of CLOX with respect to the polymer exhibited activity against *S. aureus* (Fig. 15.2c), thus demonstrating that electrospun nanofibers has a great potential to treat mastitis.

Giram et al. (2018) successfully incorporated moxifloxacin hydrochloride (MOX-HCL), a widely used antibacterial drug with activity against various type of Gram-positive and Gram-negative bacteria, into Eudragit nanofibers prepared by electrospinning. For this purpose, three concentrations of MOX-HCL (1%, 5% and 15% w/w, drug weight to polymer weight) were incorporated into Eudragit nanofibers. The authors evaluated the release prolife of MOX-HCL at pH 1.2 and 6.8 demonstrating that Eudragit nanofibers showed pH-dependent drug release profiles, with slow release at pH 1.2 for 60 h and burst release at pH 6.8 (around 30 s). The *in vitro* quantitative and qualitative antimicrobial assay showed that nanofibers containing different MOX-HCL concentrations exhibited antibacterial activity against Gram positive (*S. aureus*) and Gram negative (*E. coli*) bacteria, which are pathogens related to bovine mastitis, indicating its potential to treat this inflammation.

15.3.2 Fertilizers Encapsulation

Current agricultural practices depend to a large extent on the use of fertilizers, which plays an important role in maintaining soil fertility, increasing crop yields and improving harvest quality (Chen et al. 2018b). An important limitation of conventional fertilizers is associated with the fact that significant portions of them are lost during the field application, resulting not only in large economic losses but in serious environmental pollution (Calabi-Floody et al. 2018; Zhang et al. 2014). An important approach to overcome these deficiencies is the development of new technologies aiming to deliver nutrients to crops in a controlled manner, thus minimizing the negative effects associated with overdose and reducing the frequency of fertilizer applications (Guo et al. 2018).

Recent studies have reported the use of electrospinning to encapsulate fertilizers into polymeric nanofibers (Azarian et al. 2018; Hassounah et al. 2014; Kampeerapappun and Phanomkate 2013; Krishnamoorthy et al. 2016; Krishnamoorthy and Rajiv 2017a). Krishnamoorthy et al. (2016) e.g. prepared electrospun nanofibers of polyvinylpyrrolidone (PVP) incorporating urea and cobalt nanoparticles (CoNPs). The study of *in vitro* nutrient release was carried out in water by means of conductivity measurements and the results showed a controlled release for 200 h. In another study, Krishnamoorthy and Rajiv (2017) demonstrated that electrospun nanofibers of PVP containing urea and cobalt nanoparticles have a great potential as seed coating and improve the germination of cowpea seeds (*Vigna unguiculata*).

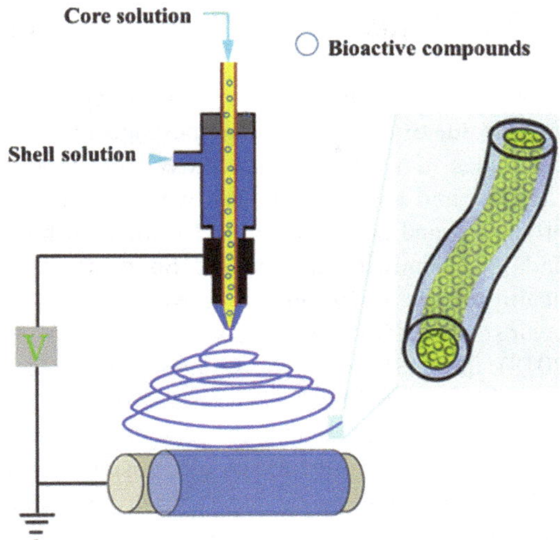

Fig. 15.3 Schematic representation of coaxial electrospinning. Reprinted with permission from Jiang et al. (2014). Copyright 2018 Elsevier

Coaxial electrospinning is a modification of the conventional electrospinning process in which two concentrically aligned capillaries are used to produce fibers with a core–shell structure (Qin 2017). This approach has emerged as an efficient alternative to encapsulate a high load of bioactive compounds without compromising their activity, as well as to achieve a sustained release profile (Isik et al. 2018; Torkamani et al. 2018; Yao et al. 2016). In this process, shell solution is injected through the outer needle, whereas the core solution containing the bioactive compound is injected through the inner needle, as illustrated in Fig. 15.3 (Jiang et al. 2014).

Kampeerapappun and Phanomkate (2013) encapsulated NPK fertilizer into core–shell fibers using polyhydroxybutyrate (PHB) as the shell material and NPK-PLA as the core content. The *in vitro* release of the NPK fertilizer from nanofibers in water at 30 °C was monitored by UV-vis absorption spectroscopy. The results showed that the release profile of NPK fertilizer from coaxial fibers was modulated by the feed rate of core solution and NPK concentration, where the increase of feed rate and NPK concentration resulted in a faster release. The release took 1 month to complete and was controlled by the diffusion mechanism, indicating the potential of the system to be applied in slow-release fertilizer application.

Although electrospinning has shown great potential to encapsulate fertilizers, its actual application in crops still requires considerable advances. SB-Spinning also has great potential to encapsulate fertilizers due to the fact that SB-Spinning preserves the active compound and the nanofibers can be deposit onto any target *in-situ*. In addition, this technique is cheaper and faster to generate nanofibers compared to electrospinning. Nonetheless, the potential of SB-Spinning to encapsulate fertilizers remains unexplored and more research is needed.

15.3.3 Pesticides Encapsulation

Global agricultural production suffers considerable losses during production and storage due to insects and other pest species (Luiz de Oliveira et al. 2018). Synthetic pesticides of several classes have been widely applied for pest control in crops (Mfarrej and Rara 2018). However, conventional pesticide formulations have low efficiency and are responsible for the continuous use of harmful organic solvents and indiscriminate application. Such effects cause negative impacts on human health and the environment, as well as an increase in the resistance of pathogens against pesticides (Lamichhane et al. 2016; Mfarrej and Rara 2018; Patil et al. 2018). In this sense, the development of a safe and effective application, as well as pesticide formulations that are harmless to the environment are required (Nuruzzaman et al. 2016).

Polymer fibers prepared by electrospinning and SB-Spinning have a great potential to encapsulate pesticides and deliver them at a sustained rate, which increases their effectiveness in controlling pests for a longer period while minimizing the amount of pesticide needed to reach the target site, with lower environmental impacts to ecosystem. Despite this, few studies have reported the use of electrospinning to encapsulate pesticides (Roshani et al. 2017; Thitiwongsawet et al. 2011; Xiang et al. 2013), whereas, to the best of our knowledge, no work involving the use of SB-Spinning technique to this purpose has been described in the literature.

Roshani et al. (2017) prepared PLA electrospun nanofibers loaded with Thiram pesticide in different concentrations (5, 10, 15, 20, and 25% (w/w), in respect to the polymer weight). The release of Thiram from PLA nanofibers in deionized water was evaluated by UV absorption spectrophotometry, revealing that the release was extended in a sustained manner during the 40 days of measurement for all Thiram concentrations. Additionaly, both diffusion and degradation played a role in the controlled release during long-term periods.

In another study, Thitiwongsawet et al. (2011) encapsulated the fungicide, 2, 6-dichloro-4-nitroaniline (DCNA) at three different concentrations (5, 10 and 20%) into electrospun cellulose acetate (CA) nanofibers. The study of *in vitro* release of DCNA from nanofibers was carried out in deionized water for 60 day using UV–vis spectroscopy to quantify the DCNA released. The results showed that the nanofibers exhibited sustained-release profile that could be modulated according to the DCNA concentration, showing the potential to control the release of pesticides in agriculture.

Xiang et al. (2013) used the electrospinning technique to encapsulate Columbia Blue, a nonionic hydrophobic dye with a molecular weight and partition coefficient that simulates a systemic pesticide, into PLA electrospun fibers containing 0, 1 and 10% cellulose nanocrystals. *In vitro* release study was investigated for 48 h in water by liquid chromatography/mass spectroscopy (LC/MS) demonstrating that the release of Columbia Blue displayed a minimal burst release. It was also found that the Columbia Blue release by diffusion-controlled mechanism followed a Fickian diffusion mechanism. In addition, the increasing cellulose nanocrystal content in the fibers increased the fiber degradation rate and the Columbia Blue release rate. The authors also evaluated the application of the electrospun PLA system in a greenhouse trial and verified a high whitefly mortality, thus confirming the efficacy of the system developed against these insects.

15.3.4 Pheromones and Biocontrol Agents

An alternative approach to the use of pesticides for plant protection against insect pests is the application of pheromones, which are volatile organic molecules that trigger a behavioral response in individuals of the same species (Hellmann et al. 2011). Among the different types of pheromones, sex pheromones have been most exploited in the protection of plants, once they are species-specific, effective in very small amounts and non-toxic for animals (Bell 2014; Ujváry 2010). For use in pest management, pheromone components require formulation in appropriate dispensing systems that protect the pheromone and can provide constant release over prolonged periods (Bansal et al. 2012). In this sense, polymeric nanofibers can be good candidates for this purpose. Hellmann et al. (2011) e.g. successfully incorporated (Z)-9-dodecenyl acetate, the sex pheromone of European grape berry moth (*Eupoecilia ambiguella*) into different electrospun nanofibers such as polyamide (PA) and CA. The authors demonstrated that the pheromone could be incorporated in high concentrations (30% w/w, pheromone weight to polymer weight) in the different polymer nanofiber systems. In addition, the release of pheromones from nanofibers was evaluated by a thermogravimetric analysis (TGA), revealing that the release extends for more than 50 days with a almost linear behavior.

Bansal et al. (2012) developed a hybrid encapsulation structure containing sex pheromone of *Lobesia botrana* through electrospinning. In this study, pheromone was encapsulated in the synthesized oligolactide particles, and then, pheromone-loaded nanofibers were prepared by electrospinning the mixture of the encapsulated pheromone with the suspension of a block copolymer (polyhexyleneadipate-*b*-methoxypolyethylene glycol). The release of pheromone from nanofibers was evaluated by TGA at 30 °C, showing that the release delay time was 7 h. Despite the slower release of encapsulated pheromone compared to the free one, the authors concluded that a longer delay in pheromone release is required for an adequate practical application in control of *Lobesia botrana* population.

Kikionis et al. (2017) developed electrospun micro/nanofiber matrices of different biodegradable polymers such as PCL and PHB for the controlled release of 1,7-dioxaspiro[5.5]undecane and (Z)-7-tetradecenal, the sex pheromones of olive fruit fly, *Bactrocera oleae*, and olive moth, *Prays oleae*, respectively. In this study, pheromones were incorporated into nanofibers in three concentrations (5%, 10% and 20% (w/w), in respect to polymer weight) and the amount released from different matrices was determined by gas chromatography mass (GC/MS) spectrometry. The results revealed that the release of pheromones was extended during the 16 weeks of measurement in all concentrations, regardless of the matrix composition.

Laboratory bioassays and field trapping tests showed that the fiber mats obtained by electrospinning of PHB solution containing 5% w/w 1,7-dioxaspiro[5.5]undecane and PCL solution containing 5% w/w (Z)-7-tetradecenal were almost twice as effective to attract the *B. oleae* and *P. oleae* males, respectively, compared to the positive controls used.

Another approach for the protection of agricultural crops consists in the use of biocontrol agents, including bioinsecticides (*Bacillus thuringiensis*), biofungicides (*Trichoderma*), bioherbicides (*Phytopthora*) (Ram et al. 2018). A possible strategy to increase the efficiency of biocontrol agents is their encapsulation in an appropriate polymer carrier, which facilitates the handling of the microorganisms, increases their protection against environmental factors and increases their viability during storage (Noruzi 2016).

Spasova et al. (2011) encapsulated *Trichoderma viride* spores within chitosan-based nanofibers. In this study, solutions containing different ratios of chitosan–poly(ethylene oxide) (PEO) and chitosan–polyacrylamide (PAAm) were blended with fungal *Trichoderma* spores and electrospun to form nanofibers. PEO and PAAm were used as facilitating polymers in the electrospinning process. The presence of the fungal spores in the nanofiber mats was confirmed by scanning electron microscopy. In addition, the microbiological tests confirmed that fungal spores immobilized in nanofibers maintained viability and were effective to inhibit the growth of pathogenic soil microorganisms: *Alternaria* and *Fusarium*. The authors also demonstrated that fibers containing fungal spores can be directly electrospun on the plant root system or onto the plant leaves of different crops, without damaging the plant.

Despite the wide interest and progress in electrospinning, there is a great need for additional research studies in the field of encapsulation of pheromones and biocontrol agents to fully explore this potential. To best of our knowledge, there is no paper describing the application of SB-Spinning to obtain nanofibers as carrier systems for the sustained and controlled release of pheromones and biocontrol agents, thus research studies are required to explore the application of this technique in the field of protection of agricultural crops.

15.3.5 Food Industry

A growing interest in the use of ES and SB-Spinning is attributed to the fact that these techniques do not involve severe temperature conditions, which are compatible with food grade polymers and biopolymers, and allow efficient encapsulation by reducing denaturation and bioactive stability compared to other conventional encapsulation methods, such as spray drying (Daristotle et al. 2016; Medeiros et al. 2009; Wen et al. 2017).

Essential oils (EOs) are lipophilic compounds extracted from plants that exhibit antimicrobial and antioxidant properties. EOs are unstable and fragile volatile compounds that can be easily degraded by oxidation, volatilization, heating, light, which

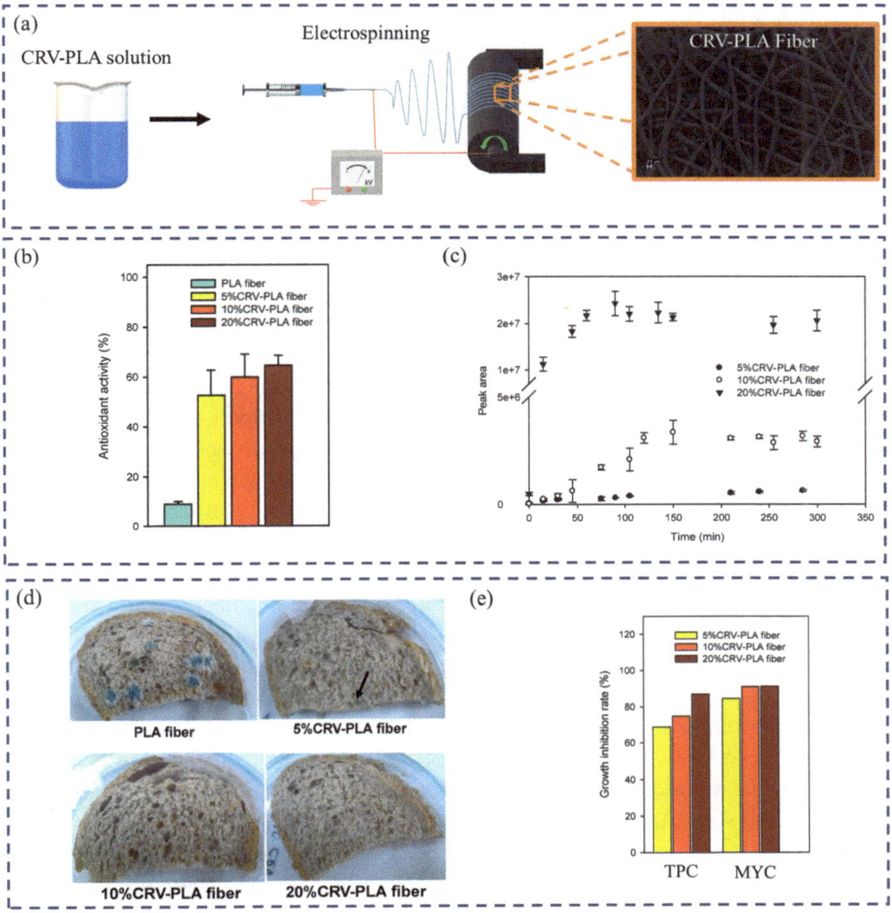

Fig. 15.4 Schematic representation of preparation of CRV-loaded PLA fibers (**a**), antioxidant activity (%) of electrospun fibers of PLA with different CRV content (**b**), cumulative CRV release from electrospun PLA fibers with different CRV content (**c**), images of whole wheat bread samples packed with electrospun PLA fibers with different CRV content stored at 25 °C for 7 days (**d**), and growth inhibition rate (%) of aerobic bacteria (TPC) and mold and yeast (MYC) (**e**). Reprinted with permission from Altan et al. (2018). Copyright 2018 Elsevier

limits their applications in food industry. Numerous studies have shown that electrospun fibers are effective means to increase EO stability and bioactivity (Prakash et al. 2018).

Altan et al. (2018) incorporated carvacrol (CRV), a main component of oregano and thyme oil, at three different concentrations (5, 10 and 20%) into PLA electrospun nanofibers (Fig. 15.4a). The results showed that the antioxidant activity of CRV-loaded PLA nanofibers ranged from 53 to 65% for 5–20% CRV content (Fig. 15.4b). The amount of CRV released from PLA nanofibers were determined by CG/MS spectrometry and indicated that a sustained low release of CRV at 5 and 10% CRV level occurred, while the release profile of PLA fibers at 20% CRV showed the initial burst effect followed by sustained release (Fig. 15.4c). The results revealed that CRV release from PLA fibers at 5 and 10% CRV level shows a non-Fickian diffusion mechanism. On the other hand, the CRV release was controlled by the Fickian diffusion in 20% PLA nanofibers. The authors also demonstrated that CRV-loaded electrospun PLA nanofibers are able to preserve whole wheat bread samples from degradation by aerobic bacteria (TPC) and mold and yeast (MYC) (Fig. 15.4d, e), which indicates that are good candidates for active food packaging applications to extend the shelf life of whole wheat.

Souza et al. (2015) encapsulated linalool, a monoterpene alcohol widely present as a major constituent of plant EOs, into PLA nanofibers prepared by electrospinning and SB-spinning. In this study, the effect of linalool concentration (10, 15 and 20 wt. %) on the morphology of the fibers, structural properties and the release behavior was evaluated. The results showed that both processes were efficient to encapsulate linalool into polymeric nanofibers. In addition, the authors verified that linalool concentration did not influence the fiber morphology, but changed physical properties, such as crystallinity degree, interplanar distance, crystals sizes and surface energy. The *in vitro* release rate of the linalool from nanofibers was evaluated by TGA at 35 °C. The curves of linalool release were non-linear, demonstrating a biphasic pattern consistent with one or more Fickian release components. It was also found that the time required to release half the amount of encapsulated linalool ($T_{1/2}$) decreased with increasing linalool concentration and that $t_{1/2}$ values for SB-Spinning nanofibers were higher (291–1645s) than the $t_{1/2}$ values for electrospun fibers (76–575 s). Since linalool is a well-known antimicrobial and insect repellant, the results reported by the authors indicate the potential of linalool-loaded spun PLA nanofibers in active food packaging.

Vitamin A is an essential micronutrient that plays a vital biological role in human body, such as in vision health and immune function. The low water-solubility and the poor stability of vitamin A, as well as its unfavorable effects on the flavor, odor and transparency of beverages, limits theirs applications in commercial products, such as functional foods. Therefore, efficient strategies are needed to increase the vitamin A stability and mask the unpleasant off-flavor (Katouzian and Jafari 2016; Sauvant et al. 2012). For this purpose, Fahami and Fathi (2018) incorporated Vitamin A into polyvinyl alcohol (PVA)-based nanofibers. The encapsulation efficiency and loading capacity were calculated and the results showed that nanofibers with 30% w/w vitamin A could be recognized as the optimum sample, displaying an encapsulation efficiency of 97% and a load capacity of 29.51%. The release behavior of vitamin A was studied in simulated gastric (2 h) and intestinal (6 h) fluids and burst release was not observed. A small amount of vitamin was released in the SGF (5.27%), while 15.43% of the vitamin release was observed in the simulated intestinal fluid (SIF), thus demonstrating the ability of controlled release in different conditions. The results also showed that the main release mechanism in SGF was diffusional transport, whereas for simulated intestinal media, a whole release time cooperation of Fickian diffusion and dissolution were observed. The lower release rate of vitamin A in simulated gastric fluids indicated that the produced nanofibers are appropriate carriers to increase the stability of food bioactives against acidic conditions.

Yang et al. (2017) encapsulated fish oil into core–shell fibers with PVP as the shell material and fish oil-zein constituting the core content. The oxidative stability, the shelf life and the release behavior of encapsulated fish oil were investigated and compared to single nanofibers prepared by conventional electrospinning. The results showed that the oxidative stability of encapsulated fish oil in the coaxial nanofibers was significantly enhanced compared to that of the single nanofibers. It was also found that the shelf life of encapsulated oil in the single nanofibers and coaxial nanofibers were 374 and 439 days, respectively, thus demonstrating that coaxial electrospinning is a more recommended option to slow down the oxidation process. The release of fish oil was evaluated in SGF (HCl solution, pH 1.2) and SIF (phosphate buffer, pH 6.8), revealing that core-shell nanofibers showed sustained release in both media during a period of 300 min. In addition, the release mechanism was controlled by a combination of diffusion and macromolecular chain relaxation. Overall, the results demonstrated that coaxial electrospinning provides a platform for the encapsulation of fish oil without compromising its stability and consists in a promising technique to produce fish oil functional foods.

Although electrospinning has emerged as a potential process to encapsulate bioactive compounds, the application of electrospinning in food industry is limited due to its relatively low productivity, which restricts its large-scale commercial exploitation. In this sense, advanced research is still needed to modify the structural aspects of the electrospinning setup (e.g. multi-needle arrangement setup or multiaxial technologies) to allow the industrial application in food science. On the other hand, SB-Spinning has shown great potential to develop innovative products with new specific properties on a large scale, nevertheless, its application in food science is still in the early stages of development, and studies are required to explore this potential in the food industry.

15.4 Final Remarks

In this book chapter we reviewed how the electrospinning (ES) and SB-Spinning techniques can be used to fabricate suitable composite polymer nanofibers to encapsulate agricultural inputs (pesticides, hormones, fertilizers, etc.) aiming at slow release systems. Both techniques are highly suitable for producing nanometric platforms with a high surface area/volume ratio, varied diameters and morphologies, with the possibility of chemical functionalization for the intended agricultural applications. The chapter reported the successful use of such fiber forming techniques to design novel slow release systems based on biodegradable micro- and nanofibers containing different antibiotics to treat cattle diseases, fertilizers to improve productivity of crops and pesticides to mitigate or avoid pest infestations in varied crops. Such systems greatly increase the application efficiency of agri-food inputs, with simultaneous effect on increasing productivity while requiring a smaller amount of the active inputs. Although some advances are still critical and necessary to scale-up the production of nanofibers at a reasonable cost, the recent advances pave the way for sustainable agricultural practices.

Acknowledgments The authors thank the financial support from FAPESP (Fundação de Amparo à Pesquisa do Estado de São Paulo – Project numbers: 2017/12174-4, 2017/20973-4), CNPQ (Conselho Nacional de Desenvolvimento Científico e Tecnológico), MCTI-SisNano (CNPq/402.287/2013-4), Coordenação de Aperfeiçoamento de Pessoal de Nível Superior – Brasil (CAPES) – Código de Financiamento 001, FAPEMIG (Fundação de Amparo a Pesquisa do Estado de Minas Gerais - Project Numbers: APQ-01505-15; APQ-00906-17) and Rede Agronano-Embrapa from Brazil.

References

Agarwal, S., & Greiner, A. (2011). On the way to clean and safe electrospinning-green electrospinning: Emulsion and suspension electrospinning. *Polymers for Advanced Technologies, 22*(3), 372–378. https://doi.org/10.1002/pat.1883.

Agarwal, S., Greiner, A., & Wendorff, J. H. (2013). Functional materials by electrospinning of polymers. *Progress in Polymer Science, 38*(6), 963–991. https://doi.org/10.1016/j.progpolymsci.2013.02.001.

Al-Enizi, A., Zagho, M., & Elzatahry, A. (2018). Polymer-based electrospun nanofibers for biomedical applications. *Nanomaterials, 8*(4), 259. https://doi.org/10.3390/nano8040259.

Altan, A., Aytac, Z., & Uyar, T. (2018). Carvacrol loaded electrospun fibrous films from zein and poly(lactic acid) for active food packaging. *Food Hydrocolloids, 81*, 48–59. https://doi.org/10.1016/j.foodhyd.2018.02.028.

Ashraf, R., Sofi, H. S., Malik, A., Beigh, M. A., Hamid, R., & Sheikh, F. A. (2018). Recent trends in the fabrication of starch nanofibers: Electrospinning and non-electrospinning routes and their applications in biotechnology. *Applied Biochemistry and Biotechnology, 187*(1), 47–74. https://doi.org/10.1007/s12010-018-2797-0.

Azarian, M. H., Kamil Mahmood, W. A., Kwok, E., Bt Wan Fathilah, W. F., & Binti Ibrahim, N. F. (2018). Nanoencapsulation of intercalated montmorillonite-urea within PVA nanofibers: Hydrogel fertilizer nanocomposite. *Journal of Applied Polymer Science, 135*(10), 45957. https://doi.org/10.1002/app.45957.

Bagchi, S., Achla, R., & Mondal, S. K. (2017). Electrospun polypyrrole-polyethylene oxide coated optical fiber sensor probe for detection of volatile compounds. *Sensors and Actuators, B: Chemical, 250*, 52–60. https://doi.org/10.1016/j.snb.2017.04.146.

Bansal, P., Bubel, K., Agarwal, S., & Greiner, A. (2012). Water-stable all-biodegradable microparticles in nanofibers by electrospinning of aqueous dispersions for biotechnical plant protection. *Biomacromolecules, 13*(2), 439–444. https://doi.org/10.1021/bm2014679.

Bell, C. H. (2014). Chapter 29 – Pest management. In Y. Motarjemi & H. Lelieveld (Eds.), *Food safety management* (pp. 799–820). San Diego: Academic. https://doi.org/10.1016/b978-0-12-381504-0.00029-9.

Benito, J. G., Teno, J., Torres, D., & Diaz, M. (2017). Solution blow spinning and obtaining submicrometric fibers of different polymers. *International Journal of Nanoparticles and Nanotechnology, 3*(007), 1–10.

Bhardwaj, N., & Kundu, S. C. (2010). Electrospinning: A fascinating fiber fabrication technique. *Biotechnology Advances, 28*(3), 325–347. https://doi.org/10.1016/j.biotechadv.2010.01.004.

Bonan, R. F., Bonan, P. R. F., Batista, A. U. D., Perez, D. E. C., Castellano, L. R. C., Oliveira, J. E., & Medeiros, E. S. (2017). Poly(lactic acid)/poly(vinyl pyrrolidone) membranes produced by solution blow spinning: Structure, thermal, spectroscopic, and microbial barrier properties. *Journal of Applied Polymer Science, 134*(19). https://doi.org/10.1002/app.44802.

Calabi-Floody, M., Medina, J., Rumpel, C., Condron, L. M., Hernandez, M., Dumont, M., & Mora, M. d. l. L. (2018). Smart fertilizers as a strategy for sustainable agriculture. *Advances in Agronomy, 147*, 119–157. https://doi.org/10.1016/bs.agron.2017.10.003.

Casper, C. L., Stephens, J. S., Tassi, N. G., Chase, D. B., & Rabolt, J. F. (2004). Controlling surface morphology of electrospun polystyrene fibers: Effect of humidity and molecular weight in the electrospinning process. *Macromolecules, 37*(2), 573–578. https://doi.org/10.1021/ma0351975.

Chang, G., & Shen, J. (2011). Helical nanoribbons fabricated by electrospinning. *Macromolecular Materials and Engineering, 296*(12), 1071–1074. https://doi.org/10.1002/mame.201100060.

Chang, G., Li, A., Xu, X., Wang, X., & Xue, G. (2016). Twisted polymer microfiber/nanofiber yarns prepared via direct fabrication. *Industrial and Engineering Chemistry Research, 55*(25), 7048–7051. https://doi.org/10.1021/acs.iecr.6b00686.

Chayad, F. A., Jabur, A. R., & Jalal, N. M. (2016). Effect of NaCl solution addition on improving some of the physical properties of Nylon 6 solutions used for electro spinning purpose. *Engineering and Technology Journal, 34*(7), 1265–1274.

Chen, S., Li, R., Li, X., & Xie, J. (2018a). Electrospinning: An enabling nanotechnology platform for drug delivery and regenerative medicine. *Advanced Drug Delivery Reviews, 132*, 188–213. https://doi.org/10.1016/j.addr.2018.05.001.

Chen, J., Lü, S., Zhang, Z., Zhao, X., Li, X., Ning, P., & Liu, M. (2018b). Environmentally friendly fertilizers: A review of materials used and their effects on the environment. *Science of the Total Environment, 613–614*, 829–839. https://doi.org/10.1016/j.scitotenv.2017.09.186.

Chen, X., Xu, Y., Liang, M., Ke, Q., Fang, Y., Xu, H., Jin, X., & Huang, C. (2018c). Honeycomb-like polysulphone/polyurethane nanofiber filter for the removal of organic/inorganic species from air streams. *Journal of Hazardous Materials, 347*, 325–333. https://doi.org/10.1016/j.jhazmat.2018.01.012.

Costa, D. L., Leite, R. S., Neves, G. A., Santana, L. N. d. L., Medeiros, E. S., & Menezes, R. R. (2016). Synthesis of TiO_2 and ZnO nano and submicrometric fibers by solution blow spinning. *Materials Letters, 183*, 109–113. https://doi.org/10.1016/j.matlet.2016.07.073.

Cramariuc, B., Cramariuc, R., Scarlet, R., Manea, L. R., Lupu, I. G., & Cramariuc, O. (2013). Fiber diameter in electrospinning process. *Journal of Electrostatics, 71*(3), 189–198. https://doi.org/10.1016/j.elstat.2012.12.018.

da Silva Parize, D. D., de Oliveira, J. E., Foschini, M. M., Marconcini, J. M., & Mattoso, L. H. C. (2016). Poly(lactic acid) fibers obtained by solution blow spinning: Effect of a greener solvent on the fiber diameter. *Journal of Applied Polymer Science, 133*(18). https://doi.org/10.1002/app.43379.

Damasceno, R., Roggia, I., Pereira, C., & de Sá, E. (2013). Rhizobia survival in seeds coated with polyvinyl alcohol (PVA) electrospun nanofibres. *Canadian Journal of Microbiology, 59*(11), 716–719. https://doi.org/10.1139/cjm-2013-0508.

Daristotle, J. L., Behrens, A. M., Sandler, A. D., & Kofinas, P. (2016). A Review of the fundamental principles and applications of solution blow spinning. *ACS Applied Materials & Interfaces, 8*(51), 34951–34963. https://doi.org/10.1021/acsami.6b12994.

De Vrieze, S., Van Camp, T., Nelvig, A., Hagström, B., Westbroek, P., & De Clerck, K. (2009). The effect of temperature and humidity on electrospinning. *Journal of Materials Science, 44*(5), 1357–1362. https://doi.org/10.1007/s10853-008-3010-6.

Deitzel, J., Kleinmeyer, J., Harris, D., & Beck Tan, N. (2001). The effect of processing variables on the morphology of electrospun nanofibers and textiles. *Polymer, 42*(1), 261–272. https://doi.org/10.1016/s0032-3861(00)00250-0.

Dong, Z., Wu, Y., & Clark, R. L. (2011). Thermodynamic modeling and investigation of the formation of electrospun collagen fibers. *Langmuir, 27*(20), 12417–12422. https://doi.org/10.1021/la201859c.

dos Santos, D. M., Souza Leite, I., de Lacerda Bukzem, A., de Oliveira Santos, R. P., Frollini, E., Inada, N. M., & Campana-Filho, S. P. (2018). Nanostructured electrospun nonwovens of poly(ε-caprolactone)/quaternized chitosan for potential biomedical applications. *Carbohydrate Polymers, 186*, 110–121. https://doi.org/10.1016/j.carbpol.2018.01.045.

Doshi, J., & Reneker, D. H. (1995). Electrospinning process and applications of electrospun fibers. *Journal of Electrostatics, 35*, 151–160. https://doi.org/10.1109/ias.1993.299067.

Eisberg, R. M., & Lerner, L. S. (1982). *Física fundamentos e aplicações*. São Paulo: McGrawHill.

Ellison, C. J., Phatak, A., Giles, D. W., Macosko, C. W., & Bates, F. S. (2007). Melt blown nanofibers: Fiber diameter distributions and onset of fiber breakup. *Polymer, 48*(11), 3306–3316. https://doi.org/10.1016/j.polymer.2007.04.005.

Erskine, R. J., Wagner, S., & DeGraves, F. J. (2003). Mastitis therapy and pharmacology. *Veterinary Clinics of North America – Food Animal Practice, 19*(1), 109–138. https://doi.org/10.1016/s0749-0720(02)00067-1.

Fahami, A., & Fathi, M. (2018). Development of cress seed mucilage/PVA nanofibers as a novel carrier for vitamin A delivery. *Food Hydrocolloids, 81*, 31–38. https://doi.org/10.1016/j.foodhyd.2018.02.008.

Farias, R. M. D. C., Menezes, R. R., Oliveira, J. E., & De Medeiros, E. S. (2015). Production of submicrometric fibers of mullite by solution blow spinning (SBS). *Materials Letters, 149*(3), 47–49. https://doi.org/10.1016/j.matlet.2015.02.111.

Feltz, K. P., Kalaf, E. A. G., Chen, C., Martin, R. S., & Sell, S. A. (2017). A review of electrospinning manipulation techniques to direct fiber deposition and maximize pore size. *Electrospinning, 1*(1), 46–61. https://doi.org/10.1515/esp-2017-0002.

Formhals, A. (1934). Apparatus for producing artificial filaments from material such as cellulose acetate. *Section Title: Cellulose and Paper*.

Fridrikh, S. V., Yu, J. H., Brenner, M. P., & Rutledge, G. C. (2003). Controlling the fiber diameter during electrospinning. *Physical Review Letters, 90*(14), 4. https://doi.org/10.1103/physrevlett.90.144502.

Giram, P. S., Shitole, A., Nande, S. S., Sharma, N., & Garnaik, B. (2018). Fast dissolving moxifloxacin hydrochloride antibiotic drug from electrospun Eudragit L-100 nonwoven nanofibrous Mats. *Materials Science and Engineering C, 92*, 526–539. https://doi.org/10.1016/j.msec.2018.06.031.

Guo, H., White, J. C., Wang, Z., & Xing, B. (2018). Nano-enabled fertilizers to control the release and use efficiency of nutrients. *Current Opinion in Environmental Science & Health, 6*, 77–83. https://doi.org/10.1016/j.coesh.2018.07.009.

Haider, A., Haider, S., & Kang, I. K. (2015, December 12). A comprehensive review summarizing the effect of electrospinning parameters and potential applications of nanofibers in biomedical and biotechnology. *Arabian Journal of Chemistry, 11*(8), 1165–1188. https://doi.org/10.1016/j.arabjc.2015.11.015.

Hassounah, I., Shehata, N., Hudson, A., Orler, B., & Meehan, K. (2014). Characteristics and 3D formation of PVA and PEO electrospun nanofibers with embedded urea. *Journal of Applied Polymer Science, 131*(3). https://doi.org/10.1002/app.39840.

He, C., Nie, W., & Feng, W. (2014). Engineering of biomimetic nanofibrous matrices for drug delivery and tissue engineering. *Journal of Materials Chemistry B, 2*(45), 7828–7848. https://doi.org/10.1039/c4tb01464b.

Hellmann, C., Greiner, A., & Wendorff, J. H. (2011). Design of pheromone releasing nanofibers for plant protection. *Polymers for Advanced Technologies, 22*(4), 407–413. https://doi.org/10.1002/pat.1532.

Hidayat, S. N., Julian, T., Rianjanu, A., Kusumaatmadja, A., Triyana, K., & Roto. (2017). Quartz crystal microbalance coated by PAN nanofibers and PEDOT: PSS for humidity sensor. *International Seminar on Sensors, Instrumentation, Measurement and Metrology (ISSIMM), 2017–Janua*, 119–123. https://doi.org/10.1109/issimm.2017.8124274.

Huang, Q., Liu, X., Zhao, G., Hu, T., & Wang, Y. (2018). Potential and challenges of tannins as an alternative to in-feed antibiotics for farm animal production. *Animal Nutrition, 4*(2), 137–150. https://doi.org/10.1016/j.aninu.2017.09.004.

Isik, B. S., Altay, F., & Capanoglu, E. (2018). The uniaxial and coaxial encapsulations of sour cherry (*Prunus cerasus* L.) concentrate by electrospinning and their in vitro bioaccessibility. *Food Chemistry, 265*, 260–273. https://doi.org/10.1016/j.foodchem.2018.05.064.

Ismar, E., & Sarac, A. S. (2018). Oxidation of polyacrylonitrile nanofiber webs as a precursor for carbon nanofiber: Aligned and non-aligned nanofibers. *Polymer Bulletin, 75*(2), 485–499. https://doi.org/10.1007/s00289-017-2043-x.

Jiang, H., Wang, L., & Zhu, K. (2014). Coaxial electrospinning for encapsulation and controlled release of fragile water-soluble bioactive agents. *Journal of Controlled Release, 193*, 296–303. https://doi.org/10.1016/j.jconrel.2014.04.025.

Kampeerapappun, P., & Phanomkate, N. (2013). Slow release fertilizer from core-shell electrospun fibers. *Chiang Mai Journal of Science, 40*(4), 775–782.

Katouzian, I., & Jafari, S. M. (2016). Nano-encapsulation as a promising approach for targeted delivery and controlled release of vitamins. *Trends in Food Science & Technology, 53*, 34–48. https://doi.org/10.1016/j.tifs.2016.05.002.

Ki, C. S., Kim, J. W., Hyun, J. H., Lee, K. H., Hattori, M., Rah, D. K., & Park, Y. H. (2007). Electrospun three-dimensional silk fibroin nanofibrous scaffold. *Journal of Applied Polymer Science, 106*(6), 3922–3928. https://doi.org/10.1002/app.26914.

Kikionis, S., Ioannou, E., Konstantopoulou, M., & Roussis, V. (2017). Electrospun micro/nanofibers as controlled release systems for pheromones of bactrocera oleae and prays oleae. *Journal of Chemical Ecology, 43*(3), 254–262. https://doi.org/10.1007/s10886-017-0831-2.

Kong, L., & Ziegler, G. R. (2012). Role of molecular entanglements in starch fiber formation by electrospinning. *Biomacromolecules, 13*(8), 2247–2253. https://doi.org/10.1021/bm300396j.

Kong, L., & Ziegler, G. R. (2014). Molecular entanglement and electrospinnability of biopolymers. *Journal of Visualized Experiments*, (91). https://doi.org/10.3791/51933.

Krishnamoorthy, V., & Rajiv, S. (2017a). An eco-friendly top down approach to nutrient incorporated electrospun seed coating for superior germination potential. *Journal of Advanced Applied Scientific Research, 1*(7).

Krishnamoorthy, V., & Rajiv, S. (2017b). Potential seed coatings fabricated from electrospinning hexaaminocyclotriphosphazene and cobalt nanoparticles incorporated polyvinylpyrrolidone for sustainable agriculture. *ACS Sustainable Chemistry and Engineering, 5*(1), 146–152. https://doi.org/10.1021/acssuschemeng.6b01088.

Krishnamoorthy, V., Elumalai, G., & Rajiv, S. (2016). Environment friendly synthesis of polyvinylpyrrolidone nanofibers and their potential use as seed coats. *New Journal of Chemistry, 40*(4), 3268–3276. https://doi.org/10.1039/c5nj03008k.

Kuk, E., Ha, Y.-M., Yu, J., Im, I.-T., Kim, Y., & Jung, Y. C. (2016). Robust and flexible polyurethane composite nanofibers incorporating multi-walled carbon nanotubes produced by solution blow spinning. *Macromolecular Materials and Engineering, 301*(4), 364–370. https://doi.org/10.1002/mame.201500298.

Lamichhane, J. R., Dachbrodt-Saaydeh, S., Kudsk, P., & Messéan, A. (2016). Toward a reduced reliance on conventional pesticides in European agriculture. *Plant Disease, 100*(1), 10–24. https://doi.org/10.1094/pdis-05-15-0574-fe.

Laycock, B., Nikolić, M., Colwell, J. M., Gauthier, E., Halley, P., Bottle, S., & George, G. (2017). Lifetime prediction of biodegradable polymers. *Progress in Polymer Science, 71*, 144–189. https://doi.org/10.1016/j.progpolymsci.2017.02.004.

Li, D., & Xia, Y. (2004). Electrospinning of nanofibers: Reinventing the wheel? *Advanced Materials, 16*(14), 1151–1170. https://doi.org/10.1002/adma.200400719.

Li, X. H., Shao, C. L., & Liu, Y. C. (2007). A simple method for controllable preparation of polymer nanotubes via a single capillary electrospinning. *Langmuir, 23*(22), 10920–10923. https://doi.org/10.1021/la701806f.

Li, Z., Fan, Y., & Zhan, J. (2010). In_2O_3 nanofibers and nanoribbons: Preparation by electrospinning and their formaldehyde gas-sensing properties. *European Journal of Inorganic Chemistry, 2010*(21), 3348–3353. https://doi.org/10.1002/ejic.201000313.

Li, J., Song, G., Yu, J., Wang, Y., Zhu, J., & Hu, Z. (2017). Preparation of solution blown polyamic acid nanofibers and their imidization into polyimide nanofiber mats. *Nanomaterials, 7*(11), 395. https://doi.org/10.3390/nano7110395.

Liu, K., Ertley, C., & Reneker, D. (2009). *The path and motion of an electrospinning jet observed with videography and stereography* (Vol. 54, 19011). American Physical Society.

Liu, Y., Ren, Z. F., & Wang, S. Q. (2010). Theoretical analysis and experimental verification of the effect of surface tension on morphology of nanofibers in bubble electrospinning. *International Journal of Nonlinear Sciences and Numerical Simulation, 11*(8), 625–630. https://doi.org/10.1515/ijnsns.2010.11.8.625.

Lübbert, C., & Peukert, W. (2018). The mass transfer at Taylor cones. *Journal of Aerosol Science, 123*, 39–51. https://doi.org/10.1016/j.jaerosci.2018.05.014.

Luiz de Oliveira, J., Ramos Campos, E. V., & Fraceto, L. F. (2018). Recent developments and challenges for nanoscale formulation of botanical pesticides for use in sustainable agriculture. *Journal of Agricultural and Food Chemistry, 66*(34), 8898–8913. https://doi.org/10.1021/acs.jafc.8b03183.

Manasco, J. L., Saquing, C. D., Tang, C., & Khan, S. A. (2012). Cyclodextrin fibers *via* polymer-free electrospinning. *RSC Advances, 2*(9), 3778–3784. https://doi.org/10.1039/c2ra00004k.

Medeiros, E. S., Glenn, G. M., Klamczynski, A. P., Orts, W. J., & Mattoso, L. H. C. (2009). Solution blow spinning: A new method to produce micro- and nanofibers from polymer solutions. *Journal of Applied Polymer Science, 113*(4), 2322–2330. https://doi.org/10.1002/app.30275.

Mele, E. (2016). Electrospinning of natural polymers for advanced wound care: towards responsive and adaptive dressings. *Journal of Materials Chemistry B, 4*(28), 4801–4812. https://doi.org/10.1039/c6tb00804f.

Mercante, L. A., Scagion, V. P., Migliorini, F. L., Mattoso, L. H. C., & Correa, D. S. (2017). Electrospinning-based (bio)sensors for food and agricultural applications: A review. *TrAC Trends in Analytical Chemistry, 91*, 91–103. https://doi.org/10.1016/j.trac.2017.04.004.

Mfarrej, M. F. B., & Rara, F. M. (2018). Competitive, sustainable natural pesticides. *Acta Ecologica Sinica., 39*, 145–151. https://doi.org/10.1016/j.chnaes.2018.08.005.

Mori, C. L. S. D. O., Dos Passos, N. A., Oliveira, J. E., Altoé, T. F., Mori, F. A., Mattoso, L. H. C., Scolforo, J. R., & Tonoli, G. H. D. (2015). Nanostructured polylactic acid/candeia essential oil mats obtained by electrospinning. *Journal of Nanomaterials, 439253*, 1–9. https://doi.org/10.1155/2015/439253.

Nagarajan, R., Drew, C., & Mello, C. M. (2007). Polymer-micelle complex as an aid to electrospinning nanofibers from aqueous solutions. *Journal of Physical Chemistry C, 111*(44), 16105–16108. https://doi.org/10.1021/jp077421o.

Noruzi, M. (2016). Electrospun nanofibres in agriculture and the food industry: A review. *Journal of the Science of Food and Agriculture, 96*(14), 4663–4678. https://doi.org/10.1002/jsfa.7737.

Nuruzzaman, M., Rahman, M. M., Liu, Y., & Naidu, R. (2016). Nanoencapsulation, nano-guard for pesticides: A new window for safe application. *Journal of Agricultural and Food Chemistry, 64*(7), 1447–1483. https://doi.org/10.1021/acs.jafc.5b05214.

Oliveira, J. E., Mattoso, L. H. C., Orts, W. J., & Medeiros, E. S. (2013). Structural and morphological characterization of micro and nanofibers produced by electrospinning and solution blow spinning: A comparative study. *Advances in Materials Science and Engineering, 2013*, 1–14. https://doi.org/10.1155/2013/409572.

Oliveira, J., Brichi, G. S., Marconcini, J. M., Mattoso, L. H. C., Glenn, G. M., & Medeiros, E. S. (2014). Effect of solvent on the physical and morphological properties of poly(lactic acid) nanofibers obtained by solution blow spinning. *Journal of Engineered Fibers and Fabrics, 9*(4), 117–125. https://doi.org/10.1177/155892501400900414.

Park, J. Y., Choi, S.-W., Jung, S.-H., & Kim, S. S. (2012). Synthesis of NiO nanofibers and their gas sensing properties. *Journal of Nanoscience and Nanotechnology, 12*(2), 1288–1291. https://doi.org/10.1166/jnn.2012.4602.

Paschoalin, R. T., Traldi, B., Aydin, G., Oliveira, J. E., Rütten, S., Mattoso, L. H. C., Zenke, M., & Sechi, A. (2017). Solution blow spinning fibres: New immunologically inert substrates for the analysis of cell adhesion and motility. *Acta Biomaterialia, 51*, 161–174. https://doi.org/10.1016/j.actbio.2017.01.020.

Patil, M. D., Patil, V. D., Sapre, A. A., Ambone, T. S., Torris, A. T., Shukla, P. G., & Shanmuganathan, K. (2018). Tuning controlled release behavior of starch granules using nanofibrillated cellulose derived from waste sugarcane bagasse. *ACS Sustainable Chemistry & Engineering, 6*(7), 9208–9217. https://doi.org/10.1021/acssuschemeng.8b01545.

Peer, P., Polaskova, M., & Suly, P. (2018). Rheology of poly(vinyl butyral) solution containing fumed silica in correlation with electrospinning. *Chinese Journal of Polymer Science, 36*(6), 742–748. https://doi.org/10.1007/s10118-018-2077-z.

Pelipenko, J., Kristl, J., Jankoví, B., Baumgartner, S., & Kocbek, P. (2013). The impact of relative humidity during electrospinning on the morphology and mechanical properties of nanofibers. *International Journal of Pharmaceutics, 456*(1), 125–134. https://doi.org/10.1016/j.ijpharm.2013.07.078.

Peršin, Z., Ravber, M., Stana Kleinschek, K., Knez, Ž., Škerget, M., & Kurečič, M. (2017). Bio-nanofibrous mats as potential delivering systems of natural substances. *Textile Research Journal, 87*(4), 444–459. https://doi.org/10.1177/0040517516631323.

Picciani, P. H. S., Medeiros, E. S., Pan, Z., Orts, W. J., Mattoso, L. H. C., & Soares, B. G. (2009). Development of conducting polyaniline/Poly(lactic acid) nanofibers by electrospinning. *Journal of Applied Polymer Science, 112*(2), 744–753. https://doi.org/10.1002/app.29447.

Pontrelli, G., Gentili, D., Coluzza, I., Pisignano, D., & Succi, S. (2014). Effects of non-linear rheology on electrospinning process: A model study. *Mechanics Research Communications, 61*, 41–46. https://doi.org/10.1016/j.mechrescom.2014.07.003.

Prakash, B., Kujur, A., Yadav, A., Kumar, A., Singh, P. P., & Dubey, N. K. (2018). Nanoencapsulation: An efficient technology to boost the antimicrobial potential of plant essential oils in food system. *Food Control, 89*, 1–11. https://doi.org/10.1016/j.foodcont.2018.01.018.

Qin, X. (2017). Coaxial electrospinning of nanofibers. In *Electrospun Nanofibers* (pp. 41–71). Elsevier. https://doi.org/10.1016/b978-0-08-100907-9.00003-9.

Ram, R. M., Keswani, C., Bisen, K., Tripathi, R., Singh, S. P., & Singh, H. B. (2018). Chapter 10 – Biocontrol technology: Eco-friendly approaches for sustainable agriculture. In D. Barh & V. Azevedo (Eds.), *Omics technologies and bio-engineering* (pp. 177–190). Academic. https://doi.org/10.1016/b978-0-12-815870-8.00010-3.

Ramakrishna, S., Fujihara, K., Teo, W. E., Yong, T., Ma, Z., & Ramaseshan, R. (2006). Electrospun nanofibers: Solving global issues. *Materials Today, 9*(3), 40–50. https://doi.org/10.1016/s1369-7021(06)71389-x.

Ray, S. S., Chen, S.-S., Li, C.-W., Nguyen, N. C., & Nguyen, H. T. (2016). A comprehensive review: Electrospinning technique for fabrication and surface modification of membranes for water treatment application. *RSC Advances, 6*(88), 85495–85514. https://doi.org/10.1039/c6ra14952a.

Rezaei, B., Shoushtari, A. M., Rabiee, M., Uzun, L., Turner, A. P. F., & Cheung Mak, W. (2018). Multifactorial modeling and optimization of solution and electrospinning parameters to generate superfine polystyrene nanofibers. *Advances in Polymer Technology, 37*(8), 2743–2755. https://doi.org/10.1002/adv.21947.

Rezende, S. M. (2004). *Materiais e Dispositivos Eletrônicos*. Editora Livraria da Física.

Rogina, A. (2014). Electrospinning process: Versatile preparation method for biodegradable and natural polymers and biocomposite systems applied in tissue engineering and drug delivery. *Applied Surface Science, 296*, 221–230. https://doi.org/10.1016/j.apsusc.2014.01.098.

Roque, A. P., Mercante, L. A., Scagion, V. P., Oliveira, J. E., Mattoso, L. H. C., De Boni, L., Mendonca, C. R., & Correa, D. S. (2014). Fluorescent PMMA/MEH-PPV electrospun nanofibers: Investigation of morphology, solvent, and surfactant effect. *Journal of Polymer Science, Part B: Polymer Physics, 52*(21), 1388–1394. https://doi.org/10.1002/polb.23574.

Roshani, B., Tavanai, H., Morshed, M., & Khajehali, J. (2017). Controlled release of thiram pesticide from poly (L-lactic acid) nanofibers. *The Journal of The Textile Institute, 108*(9), 1504–1509. https://doi.org/10.1080/00405000.2016.1258950.

Rotta, M., Zadorosny, L., Carvalho, C. L., Malmonge, J. A., Malmonge, L. F., & Zadorosny, R. (2016). YBCO ceramic nanofibers obtained by the new technique of solution blow spinning. *Ceramics International, 42*(14), 16230–16234. https://doi.org/10.1016/j.ceramint.2016.07.152.

Santos, A. M. C., Medeiros, E. L. G., Blaker, J. J., & Medeiros, E. S. (2016). Aqueous solution blow spinning of poly(vinyl alcohol) micro- and nanofibers. *Materials Letters, 176*, 122–126. https://doi.org/10.1016/j.matlet.2016.04.101.

Sauvant, P., Cansell, M., Sassi, A. H., & Atgié, C. (2012). Vitamin A enrichment: Caution with encapsulation strategies used for food applications. *Food Research International, 46*(2), 469–479. https://doi.org/10.1016/j.foodres.2011.09.025.

Schneider, R., Mercante, L. A., Andre, R. S., Brandão, H. d. M., Mattoso, L. H. C., & Correa, D. S. (2018). Biocompatible electrospun nanofibers containing cloxacillin: Antibacterial activity and effect of pH on the release profile. *Reactive and Functional Polymers, 132*, 26–35. https://doi.org/10.1016/j.reactfunctpolym.2018.09.001.

Scott, A. M., Beller, E., Glasziou, P., Clark, J., Ranakusuma, R. W., Byambasuren, O., Bakhit, M., Page, S. W., Trott, D., & Del Mar, C. (2018). Is antimicrobial administration to food animals a direct threat to human health? A rapid systematic review. *International Journal of Antimicrobial Agents, 52*(3), 316–323. https://doi.org/10.1016/j.ijantimicag.2018.04.005.

Shariatpanahi, S. P., Bonn, D., Ejtehadi, M. R., & Zad, A. I. (2016). Electrical bending instability in electrospinning visco-elastic solutions. *Journal of Polymer Science, Part B: Polymer Physics, 54*(11), 1036–1042. https://doi.org/10.1002/polb.24029.

Silva, T. H., de Oliveira, J. E., & Medeiros, E. S. d. (2015). Obtenção de micro e nanofibras de PVC pela técnica de Fiação por Sopro em Solução. *Polímeros, 25*(2), 229–235. https://doi.org/10.1590/0104-1428.1694.

Souza, M. A., Oliveira, J. E., Medeiros, E. S., Glenn, G. M., & Mattoso, L. H. C. (2015). Controlled release of linalool using nanofibrous membranes of poly(lactic acid) obtained by electrospinning and solution blow spinning: A comparative study. *Journal of Nanoscience and Nanotechnology, 15*(8), 5628–5636. https://doi.org/10.1166/jnn.2015.9692.

Spasova, M., Manolova, N., Naydenov, M., Kuzmanova, J., & Rashkov, I. (2011). Electrospun biohybrid materials for plant biocontrol containing chitosan and Trichoderma viride spores. *Journal of Bioactive and Compatible Polymers, 26*(1), 48–55. https://doi.org/10.1177/0883911510391446.

Sui, T., Ying, S., Titov, K., Dolbnya, I. P., Tan, J. C., & Korsunsky, A. M. (2016). Operando observation of the Taylor cone during electrospinning by multiple synchrotron X-ray techniques. *Materials and Design, 110*, 933–934. https://doi.org/10.1016/j.matdes.2016.08.097.

Taylor, G. (1969). Electrically driven jets. *Proceedings of the Royal Society A: Mathematical, Physical and Engineering Sciences, 313*(1515), 453–475. https://doi.org/10.1098/rspa.1969.0205.

Thitiwongsawet, P., Ouykul, P., & Khaoroppan, A. (2011). 2, 6-dichloro-4-nitroaniline –loaded electrospun cellulose acetate fiber mats and their release characteristics. *ASEAN Journal of Chemical Engineering, 2*(0), 40–47. Retrieved from http://aseanjche.ugm.ac.id/ojs/index.php/jce/article/view/292.

Thompson, C. J., Chase, G. G., Yarin, A. L., & Reneker, D. H. (2007). Effects of parameters on nanofiber diameter determined from electrospinning model. *Polymer, 48*(23), 6913–6922. https://doi.org/10.1016/j.polymer.2007.09.017.

Thorvaldsson, A., Stenhamre, H., Gatenholm, P., & Walkenstrom, P. (2008). Electrospinning of highly porous scaffolds for cartilage regeneration. *Biomacromolecules, 9*(3), 1044–1049. https://doi.org/10.1021/bm701225a.

Tomecka, E., Wojasinski, M., Jastrzebska, E., Chudy, M., Ciach, T., & Brzozka, Z. (2017). Poly(L-lactic acid) and polyurethane nanofibers fabricated by solution blow spinning as potential substrates for cardiac cell culture. *Materials Science and Engineering: C, 75*, 305–316. https://doi.org/10.1016/j.msec.2017.02.055.

Torkamani, A. E., Syahariza, Z. A., Norziah, M. H., Wan, A. K. M., & Juliano, P. (2018). Encapsulation of polyphenolic antioxidants obtained from *Momordica charantia* fruit within zein/gelatin shell core fibers *via* coaxial electrospinning. *Food Bioscience, 21*, 60–71. https://doi.org/10.1016/j.fbio.2017.12.001.

Torres-Giner, S., Busolo, M., Cherpinski, A., & Lagaron, J. M. (2018). Electrospinning in the packaging industry. *Electrospinning: From Basic Research to Commercialization.*

Tucker, N., Stanger, J., Staiger, M., Razzaq, H., & Hofman, K. (2012). The History of the Science and Technology of Electrospinning from 1600 to 1995. *Journal of Engineered Fibers and Fabrics, 7*(SPECIAL ISSUE), 63–73. https://doi.org/10.1177/155892501200702s10.

Ujváry, I. (2010). Chapter 3 – Pest control agents from natural products. In R. Krieger (Ed.), *Hayes' handbook of pesticide toxicology* (3rd ed., pp. 119–229). New York: Academic. https://doi.org/10.1016/b978-0-12-374367-1.00003-3.

Valipour, P., Tahvili, A., Mohsenian, S., Atouei, S. A., & Khosravi, M. (2018). Investigation of electrospun nanofibers with an electrified non-Newtonian jet using differential quadrature method. *International Journal of Hydrogen Energy, 43*(14), 6789–6794. https://doi.org/10.1016/j.ijhydene.2018.02.084.

Vay, K., Scheler, S., & Frieß, W. (2011). Application of Hansen solubility parameters for understanding and prediction of drug distribution in microspheres. *International Journal of Pharmaceutics, 416*(1), 202–209. https://doi.org/10.1016/j.ijpharm.2011.06.047.

Venturelli, R. B., & Gripa, S. (2017). Eletrofiação: uma alternativa para a produção de não-tecidos. Revista Da UNIFEBE.

Venugopal, J., Prabhakaran, M. P., Zhang, Y. Z., Deepika, G., Giri Dev, V. R., Low, S., Choon, A. T., & Ramakrishna, S. (2012). Nanotechnology: A global challenge in healthcare. *Journal of Characterization and Development of Novel Materials, 4*(3), 241–264.

Villena de Francisco, E., & García-Estepa, R. M. (2018). Nanotechnology in the agrofood industry. *Journal of Food Engineering, 238*, 1–11. https://doi.org/10.1016/j.jfoodeng.2018.05.024.

Wang, X., Ding, B., Sun, G., Wang, M., & Yu, J. (2013). Electro-spinning/netting: A strategy for the fabrication of three-dimensional polymer nano-fiber/nets. *Progress in Materials Science, 58*(8), 1173–1243. https://doi.org/10.1016/j.pmatsci.2013.05.001.

Weiss, J., Kanjanapongkul, K., Wongsasulak, S., & Yoovidhya, T. (2012). Electrospun fibers: Fabrication, functionalities and potential food industry applications. In *Nanotechnology in the food, beverage and nutraceutical industries* (pp. 362–397). https://doi.org/10.1016/b978-1-84569-739-6.50013-7.

Wen, P., Wen, Y., Zong, M.-H., Linhardt, R. J., & Wu, H. (2017). Encapsulation of bioactive compound in electrospun fibers and its potential application. *Journal of Agricultural and Food Chemistry, 65*(42), 9161–9179. https://doi.org/10.1021/acs.jafc.7b02956.

Wendorff, J. H., Agarwal, S., & Greiner, A. (2012a). Nanofiber properties. In *Electrospinning* (pp. 69–104). https://doi.org/10.1002/9783527647705.ch3.

Wendorff, J. H., Agarwal, S., & Greiner, A. (2012b). Nature of the electrospinning process – experimental observations and theoretical analysis. In *Electrospinning* (pp. 29–68). https://doi.org/10.1002/9783527647705.ch2.

Xiang, C., Taylor, A. G., Hinestroza, J. P., & Frey, M. W. (2013). Controlled release of nonionic compounds from poly(lactic acid)/cellulose nanocrystal nanocomposite fibers. *Journal of Applied Polymer Science, 127*(1), 79–86. https://doi.org/10.1002/app.36943.

Yang, H., Wen, P., Feng, K., Zong, M. H., Lou, W. Y., & Wu, H. (2017). Encapsulation of fish oil in a coaxial electrospun nanofibrous mat and its properties. *RSC Advances, 7*(24), 14939–14946. https://doi.org/10.1039/c7ra00051k.

Yang, G., Li, X., He, Y., Ma, J., Ni, G., & Zhou, S. (2018). From nano to micro to macro: Electrospun hierarchically structured polymeric fibers for biomedical applications. *Progress in Polymer Science, 81*, 80–113. https://doi.org/10.1016/j.progpolymsci.2017.12.003.

Yao, Z.-C., Chang, M.-W., Ahmad, Z., & Li, J.-S. (2016). Encapsulation of rose hip seed oil into fibrous zein films for ambient and on demand food preservation via coaxial electrospinning. *Journal of Food Engineering, 191*, 115–123. https://doi.org/10.1016/j.jfoodeng.2016.07.012.

Yarin, A. L., Koombhongse, S., & Reneker, D. H. (2001). Bending instability in electrospinning of nanofibers. *Journal of Applied Physics, 89*(5), 3018–3026. https://doi.org/10.1063/1.1333035.

Yoon, K., Hsiao, B. S., & Chu, B. (2008). Functional nanofibers for environmental applications. *Journal of Materials Chemistry, 18*(44), 5326–5334. https://doi.org/10.1039/b804128h.

Yousefzadeh, M. (2017). Modeling and simulation of the electrospinning process. In *Electrospun nanofibers* (pp. 277–301). https://doi.org/10.1016/B978-0-08-100907-9.00012-X.

Zamora-Pérez, G. A., González-Morán, C. O., & Suaste-Gómez, E. (2012). Humidity sensor for biological applications via electrospinning. In *2012 Pan American Health Care Exchanges, PAHCE 2012 – Conference, workshops, and exhibits. Cooperation / linkages: An independent forum for patient aare and technology support* (pp. 97–100). https://doi.org/10.1109/pahce.2012.6233449.

Zeleny, J. (1914). The electrical discharge from liquid points, and a hydrostatic method of measuring the electric intensity at their surfaces. *Physical Review, 3*(2), 69–91. https://doi.org/10.1103/physrev.3.69.

Zeleny, J. (1917). Instability of electrified liquid surfaces. *Physical Review, 10*(1), 1–6. https://doi.org/10.1103/physrev.10.1.

Zhang, X., Min, B. G., & Kumar, S. (2003). Solution spinning and characterization of poly(vinyl alcohol)/soybean protein blend fibers. *Journal of Applied Polymer Science, 90*(3), 716–721. https://doi.org/10.1002/app.12699.

Zhang, M., Gao, B., Chen, J., Li, Y., Creamer, A. E., & Chen, H. (2014). Slow-release fertilizer encapsulated by graphene oxide films. *Chemical Engineering Journal, 255*, 107–113. https://doi.org/10.1016/j.cej.2014.06.023.

Chapter 16
Synthesis Methods of Starch-Based Polymer Foams and Its Comparison With Conventional Polymer Foams for Food Packaging Applications

Shatakshi Pachori, Anushka Sarkar, Anjama Dutta, Jeyanthi Palanivelu, and Ramalingam Chidambaram

Abstract Due to growing concerns in the environmental and economic perspectives of traditional petroleum-based packaging materials, biodegradable polymers have gained increasing attention. The complete or partial natural sources of packaging materials based on biopolymers make them potentially biodegradable. Polyurethane (PU) foams are polymers that consist of urethane bonds that bind to several organic units and are widely used for food packaging applications. The conventional PU foams usually cost more than other alternatives, release toxic fumes when burned and have problems associated with disposal, since they are not biodegradable. At present, the only biodegradable plastics that have been close to competing with conventional petrochemical plastics are those based on starch, either from potatoes, corn or other cereal crops. Starch foams with different cell structures and properties can be produced using many techniques such as extrusion, hot-mol baking/compression, microwave heating, freeze-drying and superficial fluid extrusion. Starch is being blended with other materials such as cellulose, polyvinyl alcohol to improve its mechanical properties and water sensitivity. The purpose of this chapter was to investigate the production methods of starch-based polymer foams and compare their application with PU and establish how the former is the better alternative in terms of environmental concerns.

Keywords Biodegradable · Biopolymer · Extrusion · Polyurethane

---

S. Pachori · A. Sarkar · A. Dutta · J. Palanivelu · R. Chidambaram (✉)
Department of Industrial Biotechnology, School of Bio-Sciences and Technology, Vellore Institute of Technology, Vellore, Tamil Nadu, India
e-mail: cramalingam@vit.ac.in

T. J. Gutiérrez (ed.), *Polymers for Agri-Food Applications*,
https://doi.org/10.1007/978-3-030-19416-1_16

16.1 Introduction

Food packaging consists of long-term storage and protection of all types of foods and their raw materials from oxidative and microbial contamination (Ashjari et al. 2018). Synthetic polymers such as polyethylene (PE), polypropylene, polystyrene (PS) and polyvinylchloride (PVC) have revolutionized packaging in the form of films, packages and foams (Suárez and Gutiérrez 2017). Due to their availability in large quantities at low cost and favorable characteristics such as high tensile and tear stress, high barrier strength against oxygen and volatile compounds and make them thermally stable petrochemical-based polymers such as polyolefins, polyesters, polyamides, polyurethane (PU) etc. have been conventionally used in food packaging (Pornsuksomboon et al. 2016; Araque et al. 2018). However, their non-biodegradability has raised environmental concerns about disposal (Soykeabkaew et al. 2015). Keeping in mind the environmental threats of these non-biodegradable materials, there is a search for bio renewable polymer-based environmentally friendly materials (Prasanth et al. 2015). Many types of biobased polymers such as poly (lactic acid) (PLA), poly-3-hydroxybutyrate, starch and cellulose have been introduced (Hu 2014). Currently, the only biodegradable plastics that have come close to competing with conventional petrochemical plastics in terms of production costs are those based on starch extracted from potatoes, corn or other cereals (Herniou--Julien et al. 2019; Gutiérrez et al. 2019; Merino et al. 2019a, b). Biodegradability includes the easy-to-use aspect as well as eco-friendly quality of raw materials, which are mandatorily acquired from industries of renewable agricultural products or sea-food waste processing (Razza et al. 2015).

The botanical source from which the starch has been isolated determines its physic-chemical properties (Gutiérrez et al. 2014; 2015a, b; Gutiérrez 2018a). Corn, potatoes, wheat, cassava and waxy corn are the major commercial sources of starch. Starch, as a natural polymer, can be processed by techniques such as extrusion into films/sheets, foams, shaped articles, hot-mol baking/compression, microwave heating, freeze-drying and superficial fluid extrusion. However, due to some limitations, such as weakness and water sensitivity, very low abrasion and moisture resistance, starch is being blended with other materials such as cellulose and polyvinyl alcohol (PVA) to improve its mechanical properties and water sensitivity (Debiagi et al. 2011).

Different methods have been developed for the characterization of starch-based polymers. Some of them are: determination of viscosity, measurement of water absorption capacity (Pornsuksomboon et al. 2016), compression test (Priya et al. 2014), scanning electron microscopy (Lopez-Gil et al. 2015), biological evaluation, cytotoxic assay, alamar blue assay and statistical analysis (Kaisangsri et al. 2014; Liming Ge, Xinying Li et al. 2015).

16.2 Polymer Foams

Polymer foams are found in various objects used in our daily life. These are porous microstructures that are made from synthetic polymers for many applications, especially food packaging. Polymeric foam is dispersion of a gas in a polymer matrix.

These are formed when small packs of gases are trapped inside a solid. The foam has a polymer matrix with air bubbles or built-in air tunnels, which is known as a closed cell or open cell structure. Closed-cell aluminum foams are a class of novel materials with continuous metallic matrix being separated by equally distributed gas bubbles. They have many useful properties, such as good energy absorption property, ultra-low density, high electromagnetic shielding effect, good damping performance and excellent sound-absorbing property (Xia et al. 2014). Open cell foams are solid foams that present excellent electrical, thermal and acoustic properties. The open cell foams also improve the mixture and have excellent specific mechanical properties (Kumar et al. 2014).

Production of polymer foams involves several steps (Fig. 16.1). Firstly, a gas created by chemical reaction of molten polymer is dissolved under high pressure (chemical blowing agent) or the gas is dissolved in a supersaturated state in the molten polymer (physical blowing agent). Secondly, a population of gas clusters is nucleated in the supersaturated solution when pressure is released to the ambient pressure. Finally, the nucleated bubbles in the polymer grow to their final equilibrium size (Ruiz et al. 2017).

Polymer foams can be produced by several methods. Some of the examples include extrusion, slab-stock by pouring and different forms of molding. The most common of these is extrusion foaming. The word "extrusion" refers to forcing a material through a restricted orifice. An extruder is basically composed of one or two rotating Archimedes screws fitted in a barrel in order to progressively increase the pressure and push forward the ingredients through a die where expansion takes place (Sauceau et al. 2011).

Polymer foams can be divided into thermoplastics and thermosets. Thermoplastic polymers are made up of linear molecular chains which soften on heating and harden on cooling. These are further subdivided into three types: crystalline thermoplastics, amorphous thermoplastics and semi-crystalline polymers. Thermoplastics can be easily recycled. Thermoset polymers refer to the irreversible polymerization and this type of polymer is cured by chemical reaction or heat and becomes infusible and insoluble material. Thermosets are further divided into rigid and flexible foams (Grigore 2017). It is harder to recycle thermosets.

There are various types of polymer foams, e.g. PU and synthetic and natural biopolymers. PUs are a class of polymer with urethane linkages joining several monomeric units. Other groups such as ether, ester and amide may also be found in the polymer molecule. PUs have a high demand due to their flexibility for long and short

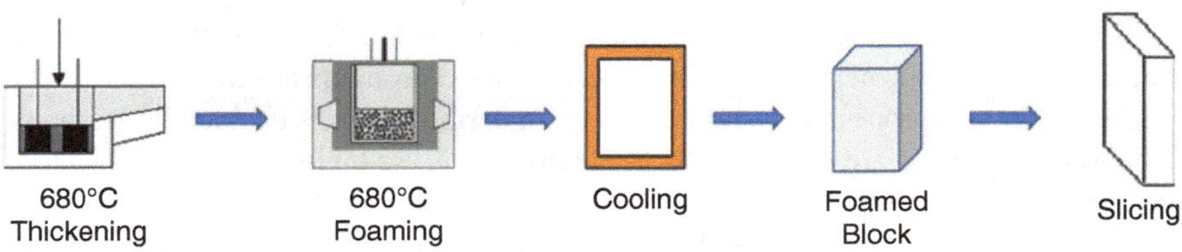

Fig. 16.1 Synthesis of polymer foams

production runs, relatively low tooling and manpower costs. PU materials tend to have a very high abrasion resistance. Due to these advantages, PUs are being increasingly used in the automotive, furniture, construction, insulation, coating, adhesives, fibers, footwear components, and biomaterials industries (Nayani et al. 2013). Biodegradable foams are organic polymers that are generated from renewable natural sources, are often biodegradable and non-toxic. They can be produced by biological systems (i.e. micro-organisms, plants and animals), or chemically synthesized from biological starting materials (Pawar and Purwar 2013). Starch-based foams are one of the possible alternative materials for making biodegradable foams. The starch is abundant and economical, which makes it a substance of choice to produce foams. These are produced mainly by the extrusion method.

16.3 Conventional Polymers and Their Limitations

Food packaging in general has a crucial role in improving the shelf life of the product (Álvarez et al. 2017; 2018). The selection of the correct kind of package shape and material combined with automation will give a more efficient and eco-friendlier package. Food packaging consists of long-term storage and the protection of all types of foods and its raw materials against oxidative and microbial contamination (Abreu et al. 2015; Ashjari et al. 2018). Petrochemical based polymers such as polyolefins, polyesters, polyamides, PU etc. have been conventionally used in food packaging for their availability in large quantities at low cost and favorable characteristics such as high tensile and tear stress, high barrier strength against oxygen and volatile compounds and make them thermally stable (Pornsuksomboon et al. 2016). Contrarily they have a very low water vapor transmission rate and are totally non-biodegradable, which is a big factor and therefore a big cause of environmental pollution posing serious threat to the environment. Therefore, its use in any form or shape should be restricted and may even be abandoned altogether to eliminate waste disposal problems (Bergel et al. 2018).

Recently, due to the growing environmental awareness, the search for packaging films made from biodegradable and compostable materials has started, since they are compatible with the environment (Scarfato et al. 2015; Gutiérrez 2018b). Biodegradability is not only an essential functional requirement but also an important environmental hallmark. Therefore, it exploits the conservation of natural resources, reinforcing respect for the environment and the protected atmosphere. The additional advantage of biodegradable packaging materials is that in decomposition and composting they can act as fertilizer and condition soil, improving the yield. Biopackaging is expensive, however, there lies the future of the packaging industry, especially when selective value-added food products are treated.

Food, either in processed form or as a raw material, which is decided by its water retention activity and are based on storage temperature, are highly decomposable and, therefore, need intelligent technological intervention to preserve them during longer periods. The conservation of food quality is an important task in the current food processing industry (Biji et al. 2015). Post-harvest losses of the farm produce are significant enough to have an impact on the food industry. These losses are majorly caused by improper handling and unsuitable post-harvest technologies that are being performed. Transportation from the production site to distant places for commercialization leads to additional losses due to pollution. The accessibility of agricultural products with freshness, long shelf life, as well as a better aroma, texture and flavor and having a higher nutritional value is the ideal case that is needed (Han 2014).

16.4 Biodegradable Polymer Foams for Food Packaging

Due to the large number of negative effects of petroleum-based foams on the environment, attempts are being made at developing biodegradable foams (Mali 2018). These foams made from renewable sources provide solutions to various ecological problems, such as the amount of waste and the disposal of waste. The attribute of compostability is preferred due to the high cost of recycling. Also, degradation of biomass in the soil leads to the production of water, carbon dioxide (CO_2) and other inorganic substance. No toxic substances are produced. Agricultural sources can be used as alternative raw materials. They can also serve as fertilizers and soil conditioners. Examples of such resources are starch and blends of starch with other compounds and lignin.

Lignin is found in the cell wall of plants, and is obtained as a by-product in several industries related to papermaking, the production of ethanol from biomass, etc. Lignin has several properties which make it suitable for packaging materials, such as high abundance, low weight, antioxidant and antimicrobial and biodegradability (Gutiérrez and Alvarez 2017). It also has CO_2 neutrality and reinforcing capability (Thakur et al. 2014).

Starch is present in most plant. Starch-based polymers are the most commonly used natural raw materials for manufacture of foams. It can be obtained from natural sources, it is inexpensive and biodegradable. Starch can be easily converted into water, CO_2, without the production of any toxic substances (Soykeabkaew et al. 2015). Cereal grains such as corn (82%) and wheat (8%) and tuber crops such as potato (5%) and cassava (5%) are the primary sources for commercial starch (Glenn et al. 2014). However, the properties of starch need to be improved. For example, the mechanical properties of starch have been improved by blending the starch with the fibers, such as the vegetable fibers of the pulp. The tensile strength and flexibility of the starch have been improved, since the fiber gets dispersed throughout the matrix. Improvement in the moisture resistance capacity of starch has been reported for its composites with acrylic acid, PVA, ethylene-vinyl acetate copolymer, poly(caprolactone) and PLA (Teixeira et al. 2014).

16.5 Starch Properties

Starch is a very important natural resource found in the roots, tubers and seeds of plants. Commercial starch is obtained from crops such as corn, wheat, potatoes and cassava (Table 16.1). It is one of the most abundant polymers in plants and they are the

Table 16.1 Properties of starch from different sources

Source	Appearance/Texture	Flavor	Clarity
Potato	Pulpy	Nil	Transparent
Corn	Smooth/slightly gelatinous	Cereal like	Opaque
Waxy corn	Smooth	Cereal like	Opaque
Cassava	Very smooth	Nil	Transparent
Wheat	Smooth/slightly gelatinous	Cereal like	Opaque
Barley ap	Very smooth	Cereal like	Opaque

carbohydrates for energy storage at present. The biological source determines the size, shape and morphology of the starch (Robyt 2008). The two polymers consisting of D-glucose repeating units, amylose and amylopectin constitute starch. The tight packing of amylopectin molecules makes the starch impervious to large molecules, resulting in a relatively smooth surface. However, the granules are hydrophilic and begin to hydrate and swell when heated in water. The viscosity of starch pastes increases as granules swell and hydrate (Glenn et al. 2014). The extrusion and viscous behavior of the starch is determined by temperature, moisture and thermo-mechanical treatment (Garg and Jana 2007). The hydrophilic nature of starch and its tendency to embrittle with age do not make it very suitable as a replacement for plastics. Despite its shortcomings, starch continues to remain a viable alternative to petroleum-based plastics primarily because of its low cost and abundance and because it is a renewable material that degrades readily in composting conditions and in many landscape and aquatic environments.

16.6 Starch Based Polymer Foams

The development and research of polymer foams based on starch has increased in recent decades due to its high biodegradability and its ability to be environment-friendly. The methods of synthesis and modification of starch matrix will be discussed in the following sections:

16.6.1 Synthesis of Starch-Based Foams

Many methods have been developed for the manufacture of starch-based foams. Starch foams with different cellular structures and properties can be produced using many techniques such as extrusion, hot-mold baking/compression, microwave heating, freeze-drying and superficial fluid extrusion (Fig. 16.2).

Extrusion can be used to produce film and foams shaped articles. This includes conversion from the natural granular form to homogenous gelatinous form. This can be further improved by shear treatment. The phase transition and rheology, which

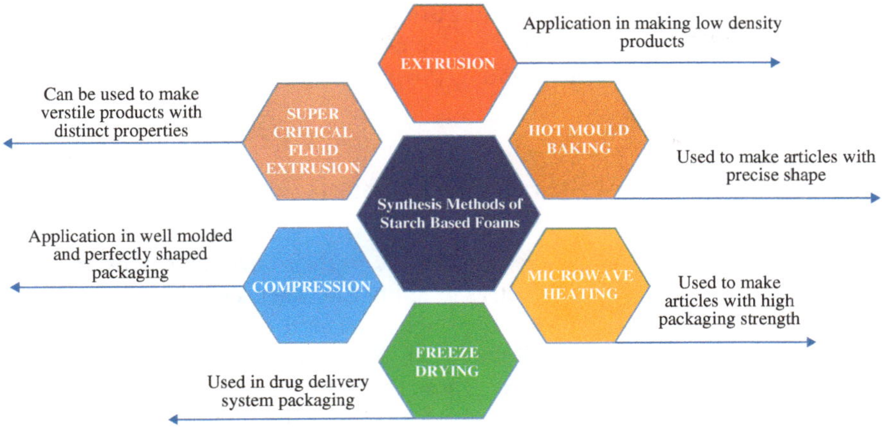

Fig. 16.2 Different methods of production of starch-based foams and their applications

play an important role in starch processing, can be largely influenced by plasticizer type and content and starch type and structure (amylose/amylopectin ratios) (Xie et al. 2014).

A co-rotating twin-screw extruder machine was used to develop cushion foam sheets by blending wheat and starch. The extruded materials were suitable for cushioning since they had comparable physical properties with the commercial plastic foams. Their dynamic cushion curve test indicated that the starch-based foam sheets provided good shock absorption properties. They deceleration peak was lower and larger than the expandable PE foams (Abinader et al. 2014).

Biodegradable foam trays have been developed by Kaisangsri et al. (2012) blending cassava starch with fibers and chitosan (Cs). This was achieved by hot-mold baking in an oven with a controlled temperature at 250 °C for 5 min. The water solubility and water absorption indexes of these foams were better than those of PS foams.

Starch-based foams combined with natural resources such as grape waste, thistle waste, barley and straw fibers have been developed by Lopez-Gil et al. (2015) using a microwave foaming process. The product formed could be used for proving structural support, due to its high stiffness and strength. It can also be used for packing, since it is biodegradable.

Cellulose nanofibrils (CNF) are also used along with starch to produce biodegradable foams. This involved freeze-drying with CNF/S water suspensions within a range of 1–7.5 wt% solids content. The mechanical and thermal properties of the composite foams were higher than that of the foams produced from control starch. The increase in the solid content of starch increases the viscosity of the paste. This in turn decreases the rate of bubble steam expansion and results in higher densities and mechanical properties. These CNF/starch composite foams are cheaper than the conventional PE and PS foams (Yildirim et al. 2014).

Lastly, there is a novel technology of extrusion of supercritical fluids that can be more versatile and controllable, thus, a wide range of foam structures with desired properties can be designed better for uses (Soykeabkaew et al. 2015).

16.7 Modification of Starch-Based Foams

Starch based foams, although very popular due to their environmentally friendly nature, have some disadvantages such as: fragility and water sensitive. Native starch foams have very low abrasion and moisture resistance, due to their hydrophilic nature generated by the presence of the free hydroxyl groups in the glucopyranosyl unit of the starch polymer molecules. These hydroxyl groups can form hydrogen bonds with the surrounding water molecules, so they are hygroscopic in nature. The starch foams can collapse when in contact with water molecules or have low dimension stability in an atmosphere with high relative humidity. Another important problem with starch-based foams is their lack of enough strength and flexibility. All these drawbacks associated with starch-based foams have made its use very limited (Soykeabkaew et al. 2015). To improve its microstructure, water resistance, mechanical and thermal characteristics, moldability, abrasion, lightness and other properties of starch-based foams, starch has thus been blended with various biodegradable polymers (Table 16.2).

16.7.1 Modified Starch-Based Foams

Converting hydrophilic hydroxyl groups to larger and less polar groups, such as acetyl groups, can be an approach to overcome the low water resistance of starch. This increases its hydrophobicity and decreases its moisture sensitivity. This comes

Table 16.2 Polymer foams based on starch blends and their advantages

Blends	Advantages
Starch acetate blended with substances such as corn cob, cellulose and PVA	When cellulose content is increased, the density increases and so does the compressive strength of the foam.
	This can thus be used for a better protection of goods during shipment.
Addition of PVA	Improve the tensile strength, hardness, crystallinity and porosity of the foam.
Starch blended with naturally hydrophobic polymers such as latex and Cs	Shows noticeable reduction in water sensitivity
Addition of Kraft, zein and gluten	Improve flexural and compressive strength of the CSF trays
Natural fibers incorporated in starch-based biocomposite systems	A significant improvement in mechanical properties

under the class of chemical modification of starch-based foams. Some examples of these types of foams are acetylated and hydroxypropylated. Acetylated starch when mixed with substances such as cellulose and PVA, improved properties are observed. When cellulose content is increased, the density increases and so does the compressive strength of the foam. Thus, this can be used for a better protection of goods during shipment. Another way of increasing hydrophobicity is adding crosslinking agent to create some network structures within the medium. This will prevent starch-water interaction when it is exposed to moisture and pressure (Soykeabkaew et al. 2015). Addition of PVA can improve the tensile strength, hardness, crystallinity and porosity of the foam. This is due their structural similarity which results in good adhesion between starch and PVA (Jiang et al. 2014).

16.7.2 Starch Polymer Blend Foams

When the starch is mixed with natural hydrophobic polymers such as latex and Cs, a marked reduction in water sensitivity can be observed. Insoluble composite foams can be produced by cross-linking potato starch and Cs by microwave treatment. The electrostatic interactions between phosphate groups of potato starch and amine groups of Cs prevent disintegration of the foam while in water for a long period of time. Cassava starch foams (CSFs) have a high tensile strength but low flexibility. Foam based on mixtures of starch with compounds such as limestone and eucalyptus cellulose fibers also show similar properties (Schmidt and Laurindo 2010). An improvement in the properties is observed when biodegradable polymers are added into the baked foam systems. According to Kaisangsri et al. (2014) when baked CSFs were blended with natural polymers including proteins, fiber and vegetable oil, its water resistance and mechanical properties were improved. It was also found that the addition of kraft, zein and gluten could improve flexural and compressive strength of the CSF trays. In addition, the water absorption and water solubility indexes of CSFs blended with zein and gluten proteins were low. Although adding palm oil into CSFs increased the water resistance, their flexural and compressive strength decreased (Kaisangsri et al. 2014).

16.7.3 Thermoplastic Starch/Natural Fiber Composite Foams

Various natural fibers have been incorporated in starch-based biocomposite systems, which generally provided a significant improvement in mechanical properties of the foams at appropriate fiber content. One main reason is because both starch and natural fiber (cellulose) are similar in their chemistry, therefore, good compatibility and strong interaction between them are obtained, leading to good stress transfer capability in the composite system.

16.7.4 Thermoplastic Starch/Polyolefin Composite Foams

Cerestech Inc. is a company in Montreal, Canada that produces thermoplastic starch (TPS) blends and commodity thermoplastics in a one-step extrusion process. The starch preparation is fed into a twin-screw extruder to form a TPS melt. A second single-screw extruder is attached to the twin-screw extruder in a perpendicular position to allow a thermoplastic polyolefin such as high-density PE (HDPE) to be melted and injected directly into the TPS melt. The melt blend is compounded further using high shear to form a blend of the two incompatible resins. Blends containing up to 50% starch have been produced with excellent mechanical properties and moisture resistance. The carbon footprint of these blends is significantly reduced compared to the neat polymer due to the starch content.

16.8 Characterization of Starch Based Polymers

Starch based foams can be characterized based on different parameters (Fig. 16.3). Some of them are discussed below:

Fig. 16.3 Different methods of characterization of starch-based foams

16.8.1 Viscosity Determination

The viscosity of various concentrations can be evaluated using of a rotational rheometer (Ares, Rheometric Scientific) that has parallel plates with 50 mm diameter. The viscosity can be measured on three independent samples resulting from a stock solution of each concentration. Average of the three values determines the final values. The Carreau–Yasuda model can be used for the experimental data in this case, which is as follows:

$$\eta = \eta\infty + (\eta\infty - \eta 0)\left[1 + (\lambda \times \gamma)^a\right]^{(n-1)/a} \tag{16.1}$$

η is the shear rate, $\eta\infty$ is the shear rate viscosity at infinite shear rates, $\eta 0$ is the zero-shear rate viscosity and λ the relaxation time in seconds. Parameters 'a' and 'n' are variable. Parameter 'a' dictates the curvature at the top of the curve. Higher value of 'a' means a sharper transition of viscosity into the power-law rule (Priya et al. 2014; Kaisangsri et al. 2014).

16.8.2 Measurement of Water Absorption Capacity

The water absorption was calculated using equation:

$$\text{Water absorption} = \frac{\left(W_{swell} - W_{dry}\right)}{W_{dry}} \tag{16.2}$$

W_A is the water absorption capacity value; $\mathbf{W}_{swell}$ is the swollen foam weight and; $\mathbf{W}_{dry}$ is the dried foam weight (Pornsuksomboon et al. 2016).

16.8.3 Compression Test

Uniaxial compression test till specific percent of deformation at a constant deformation using Ares rheometer is the ideal method for this test. Before starting the test, the diameter and weight of all the samples are measured to calculate the stress and to standardize the Young's modulus based on the average weight. Stress and strain can be calculated based on the following Eqs. 16.3 and 16.4:

$$\sigma = \frac{F}{Ao} \tag{16.3}$$

$$\varepsilon = \frac{\Delta l}{lo} \tag{16.4}$$

σ and ϵ are the stress and strain, respectively F is the force; Δl is the deformation; Ao and lo are the original cross-sectional area of the sample and the weight, respectively.

The average of Young's modulus can be calculated by taking linear range of each compression curve composition and is tested in triplicates.

16.8.4 Scanning Electron Microscopy

Morphology of cross section of each sample can be determined by using environmental scanning electron microscopy (ESEM, XL30 FEI) in low vacuum technique and by applying a tension of specific kW. The dried samples were cut into slices by using a surgical blade. The dried and cut samples are placed on the SEM sample holder using double-sided adhesive tape and observed (Lopez-Gil et al. 2015; Kaisangsri et al. 2014).

16.8.5 Biological Evaluation

The biocompatibility of different scaffold materials can be investigated using the MG-63 human osteosarcoma cell line, widely used to study the biocompatibility of orthopaedic and dental materials (LONZA, Milan, Italy). Since these cells are at an early stage in the osteoblastic lineage, they represent a good model for studying the initial stages of biomaterial induced cell osteogenic differentiation.

16.8.6 Cytotoxic Assay

Elution studies can be conducted by adding specific volumes of DMEM solution to each of material samples and placed on a rotating mixer for a period of 3 days; the eluent from each sample has to be removed at different times and 500 μL has to be pipetted out into sterile 24-well cell culture plates (Corning) earlier seeded with 1×10^3 cell per scaffold. The negative, i.e. non-toxic control is taken as tissue culture plastic with DMEM solution eluent. The plates are then incubated at 37 °C, 5% CO_2 concentration and 95% of air humidity for 24-hour duration. Cell viability can be evaluated using the Alamar Blue assay.

16.8.7 Open Cell Content Measurements

The percentage of open cells (OC) can be measured using a gas pycnometer.

$$\mathrm{OC} = \left[\mathrm{V} - \frac{2V1 - V2}{V(vf)} \right] \times 100 \quad (16.5)$$

V represents the geometric volume of the sample; V1 is the volume of the cubic samples measured by the gas pycnometer and; V2 is the volume of the same cubic samples after being cut in three planes; Vf is the porosity of the foamed sample; ϵs and ϵf are the density of the solid and that of the foam, respectively (Lopez-Gil et al. 2015).

$$Vf = 1 - \frac{\sigma \mathrm{f}}{\sigma \mathrm{s}} \quad (16.6)$$

16.8.8 Statistical Analysis

Single-way analysis of variance (ANOVA) can be performed to inspect consequential effects among treatments using the post-Bonferroni *t*-test. Duncan's multiple range test can be used to identify the differences between physical and mechanical properties, water resistance and oil resistance of starch foam trays obtained with different formulations (Kaisangsri et al. 2014; Ge et al. 2015).

16.8.9 Nanoindentation Analysis

The mechanical properties of foams can be measured with a UBI1 Nanomechanical Test Instrument (HYSITRON, Inc., Minneapolis, MN, USA). The continuous stiffness measurement technique explains the contact stiffness, hardness, elastic modulus and creep resistance (Obradovic et al. 2017). Nanoindentation reduced modulus (E) and hardness (H) are defined by the following equations:

$$\mathrm{E} = \frac{\mathbf{1dP}}{\mathbf{2\,dh}} \sqrt{\frac{\pi}{\mathbf{A}}} \quad (16.7)$$

$$\mathrm{H} = \frac{\mathbf{Pmax}}{\mathbf{A}} \quad (16.8)$$

16.9 Applications of Starch-Based Foam in Field of Food Biotech

Nearly 500 billion USD are constituted by packaging and containers. This makes up about 29.5% of municipal solid wastes in United States. The development of renewable/sustainable plastics is perceived by the industry as a hedge against the prospect that traditional plastics will be much more costly in the future as the petroleum prices go on increasing. The sustainability movement is further seen as a positive development for plastic processors since it will drive further innovation and a new generation of materials with properties more comparable to commodity plastics. In the last few years starch-based polymers were one of the most common biopolymers discussed in reference to sustainability. Broadening the spectrum of sustainability by increasing renewable content and decreasing the overall weight of traditional plastic used has been the main concern in the recent years. Broadening the scope of sustainability has helped balance the need for providing affordable packaging today while stimulating investment in research to develop more sustainable alternatives for tomorrow (Glenn et al. 2014).

The field of application of starch-based polymers in food-contact articles includes plates, drinking cups, salad cups, disposable cutlery, straws, stirrers, lids, overwraps and containers for food at fast food establishments. These articles will be in contact with aqueous, acidic and fatty foods that may be maintained at or below room temperature. These may also be heated at temperature as high as 60 °C and brought down to room temperature or below. Biopolymers, especially starch-based polymers have garnered increasing attention in the recent years due to their recyclable and environment-friendly nature. Also, their physical, chemical, optical and mechanical properties can be relatively easily tailored through polymer architecture. Therefore, as a consequence, starch-based polymers can be compared to other synthetic polymers used in food packaging such as PU, oriented PS (OPS) and polyethylene terephthalate (PET).

The most commonly used starch-based polymers in food packaging is TPS (Gutiérrez and González 2016, 2017; Gutiérrez et al. 2016a, b; Gutiérrez et al. 2016a, b; Medina Jaramillo et al. 2016; Gutiérrez and Álvarez 2016). Castillo et al. (2017) developed a prototype package at laboratory scale from active biodegradable films based on corn TPS and Cs oligomers (CsO). It was demonstrated that the antimicrobial agent migrated from active films to aqueous simulant media due to the intrinsic characteristics of CsO (low molecular weight and degree of polymerization), the nature of the matrix and the microstructure, and the physicochemical properties of the media. It was also shown that CsO diffusion was not conditioned by the acidity of the simulant media in the pH studied range, highlighting the potentiality of using this antimicrobial agent for a wide variety of foods. Cannarsi et al. (2005) also demonstrated that two biodegradable films based on starch (one polyester/starch blend and one blend of 3 biodegradable/biobased polyesters, Novamont, Italy) could be used to replace PVC films to pack fresh cut beef steaks.

Starch can be transformed also into a foamed material using water steam, replacing the PS foam as packaging material. It can be pressed into trays or disposable dishes and dissolves in water leaving a non-toxic solution, consumed by microbic environment in about 10 days giving only water and CO_2 as by-products. Avella et al. (2005) prepared starch/clay nanocomposite films which could be used in food packaging sector.

16.10 Future of Starch-Based Foams: Nanotechnology in Food Packaging Applications

Major issues associated with biodegradable polymers are including the following: performance, processing and low cost and problems due to performance and processing. Specifically, brittleness, low heat distortion temperature, high gas, high vapor permeability, poor resistance to protracted processing operations has altogether strongly limited the application of biodegradable polymers. The application of nanotechnology to these biodegradable polymers can offer new opportunities to improve the cost-effectiveness and properties of these materials. Nanotechnology can be used to improve the taste and texture of food and to produce packaging that keeps fresh product (Bratovčić et al. 2015).

Nanotechnology is generally defined as the development of structures with at least one dimension in nanometer length scale (10^{-9} m) and their utilization. Such structures are called nanocomposites and could exhibit changes in the properties of the materials or develop new properties and exhibit novel phenomena. To achieve that, a good interaction between the polymer matrix (continuous phase) and the nanofiller (discontinuous phase) is desirable (Ghanbarzadeh et al. 2015). When bioplastics are mixed with clay nanoparticles (NPs), the resulting nanocomposites exhibit excellent barrier properties as compared with the pure bioplastics, which after their functional life can be composted (Bracone et al. 2016; Gutiérrez et al. 2017). Other nanomaterials can also be utilized including NPs, nanofibers and nanowhiskers. The clay has a distinct morphology that contains at least one dimension on the nanoscale. Clay nanocomposites have been developed for use in a variety of food packaging applications and have achieved success in the market (Malathi et al. 2014; Gutiérrez and Alvarez 2018; Merino et al. 2018; Toro-Márquez et al. 2018).

Nanotechnology-enabled food packaging can be divided into two different key points: improved packaging, where nanomaterials are mixed into the polymer matrix to improve the gas barrier properties such as polymer/clay nanocomposites and "active packaging", where the NPs interact directly with the food or the environment to allow a better protection of the food, such as silver NPs as potent antimicrobial agents (Carbone et al. 2016).

Polymer nanocomposite food packaging material with antimicrobial properties is particularly useful because of the high surface-to-volume ratio of nanofillers. This property also improves the surface reactivity of nanosized antimicrobial agents compared to their bulk counterparts, which allows them able to inactivate or kill microorganisms. The properties such as mechanical, barrier, optical, thermal, biodegradation, antimicrobial and other functional properties are found in polymer nanocomposites for the packaging applications (Shankar and Rhim 2016).

Nanotechnology can result in production of novel products. Nanoscale fillers have been integrated with the polymer matrix to form "nanocomposite materials". As far as starch-based foams are concerned, their methods involving nanofillers or nano-reinforcement, i.e. layered silicates (clays), nanocelluloses and nanohydroxy-apatite have been applied to form nanocomposite foams. It has been reported that addition of organoclays (Cloisite) results in the formation of extruded starch-based foams with improved cell density, smaller cell size and better uniformity of cells (Insert reference). This has shown that the organoclays acted as nucleation sites for cell formation during extrusion of the foam.

It has also been reported that addition of kaolin can potentially improve about five times the impact strength of the biomaterial and reduces the water absorption ability of the starch-based foams. In addition, the thermal stability of the clay-filled foams can be improved by an increase in the initial degradation temperatures and a reduction in the percentage of mass loss.

The choice of the type of clay or the step of modification is fundamental for a better performance of the nanocomposite foam. It has been demonstrated, however, that with the appropriate processing conditions, the production of clay structures intercalated and exfoliated in the biocomposite foams could be obtained, regardless of the type of clay. The foams with adequate clay dispersions reveal a substantial increase in the foam strength, around ten times higher than the control starch-based foam, without any reduction in the elongation. In the case of foam incorporating microfibrillated cellulose (MFC), it has been found that the addition of 40% by weight of MFC can increase the compressibility and water resistance of the amylopectin-based foam (Dufresne and Castaño 2017). This work also revealed that the strong interactions between bacterial cellulose (BC) and the starch matrix give rise to an exceptional dispersion of BC within the matrix. It is evident that the addition of BC could decrease the water absorption capacities of the starch-based foams. Surprisingly, the foams integrated with the BC films revealed considerably higher elongations (30–37%) compared to the control foam.

The main advantages of NP, such as CuO, which are very sensitive to copper compounds in human tissues compared to other metals and the antimicrobial effect of NPs, even at very low concentrations, allow its possible use in hospital environments (Ashjari et al. 2018).

It has also been found that the cellulose nanofibers integrated in starch scaffolds can potentially increase cell attachment and cell proliferation for use in tissue engineering application involving cartilage (Zheng et al. 2015). Conclusively, there is an optimum range of nanofiller addition in every foam system, relying on the nature of each composite and its components.

Montmorillonite is hydrated alumina-silicate layered clay. This clay has a structure that restricts the gas permeation due to the large aspect ratio. In order to obtain the polymer composite improvements, small percentage of clay can be integrated in the polymer matrix. This process is called solid layer dispersion in polymers and involves two main steps: intercalation and exfoliation.

In exfoliation, the clay particles release from the system and are dispersed in the matrix polymer with no significant particle interactions. This results in layers of nanoclay in the form of polymer structural matrix. Integration of the dispersed clay layers into the polymer matrix improves substantially the overall mechanical strength and barrier properties of the (Rhim and Kim 2014). In some cases, these nanopolymer composites have superior properties to the traditionally cast polymer, including higher heat resistance. Polymer nanocomposites are developed by dispersing a filler material into NPs that form flat platelets. These platelets are furthermore distributed into a polymer matrix forming multiple parallel layers forcing gases to flow through the polymer in a turbulent path, creating a complex. More tortuosity in the polymer structure will allow much lower load levels than traditional fillers to obtain a favorable performance (Gutiérrez and Álvarez 2017). Generally, the addition levels of nanofillers are <5%, having a significant impact on the weight reduction of nanocomposite films. This dispersion process results in higher aspect ratio and surface area resulting in higher performance polymers than the conventional ones. The incorporation of nanosized fiber also increases the matrix filler interaction that adds to the mechanical reinforcement of the film (Salehudina et al. 2013). Such films can thus potentially be converted into an effective packaging material to improve the food quality, safety and performance.

In addition to nanotechnology, in recent years a new technology known as cold plasma technology has emerged. This has enormous potential in food packaging. Although, it was originally developed to increase the surface energy of polymers, improving adhesion and printability, it has recently emerged as a powerful tool for decontaminating the surface of foodstuffs and food packaging materials (Pankaj et al. 2014).

16.11 Conclusion

As environmental concerns about global waste problems increase, this motivates global interests in the production of ecological products from renewable resources. Starch based foams have emerged as PU replacements and the development in this field aims to improve the drawbacks of the starch itself. In this chapter, conventional and biopolymeric foams, and especially starch-based foams and their mixtures, were discussed and how the latter is a more sustainable alternative compared to PU and PS foams for food packaging applications. This work showed detailed information on the various methods of polymer processing based on starch. One of the most common foaming production methods is extrusion, which is suitable for film and foam-shaped articles. This can be useful for the application where weight is of high priority, such as packaging for transport. The hot mold baking/compression technique can be used to produce articles that are preferred in well-molded shapes, such as disposable containers and foam trays. Thick cell walls with high strength and rigidity can be manufactured by a microwave foaming process. The freeze-drying and solvent exchange method can be used when uniform cells with cell connectivity are required as a scaffolding of drug delivery systems. Finally, the newly designed fluid extrusion technology can be used to make a wide range of foam structures of desired properties.

Starch has been widely used as an alternative for PU foams. But the exclusive use of starch is not usually adequate due to its numerous disadvantages, such as low water resistance, low flexibility and low resistance, as well as high brittleness and many others. Several ideas have been proposed and tested to overcome this problem.

The strength and flexibility of the starch-based polymers can be improved by using starch blends. The problem of low resistance polymers can be solved by using modified starches or mixing with more hydrophobic polymers, such as PLA, latex, soy protein, poly (3-hydroxybutyrate-hydroxyvalerate). The mechanical properties of the starch foams and products can be improved by adding different types of fillers into the foam. Several types of natural fibers have been used, which will produce desirable results. In addition, nanotechnology has opened new windows to the development of starch-based foams. Due to the nanometric size effect, various properties of the foams have been shown to be effectively improved at low filler content. The success and continuity in the development of starch-based foams will make possible many more of their new applications.

Acknowledgements We want to express our immense gratitude to our honorable chancellor G. Vishwanathan and vice president G.V. Selvam for their constant support.

Conflicts of Interest The authors declare no conflict of interest.

References

Abinader, G., Lacoste, C., Baillif, M. L., Erre, D., & Copinet, A. (2014). Effect of the formulation of starch-based foam cushions on the morphology and mechanical properties. *Journal of Cellular Plastics, 51*(1), 31–44. https://doi.org/10.1177/0021955x14527979.

Abreu, A. S., Oliveira, M., de Sá, A., Rodrigues, R. M., Cerqueira, M. A., Vicente, A. A., & Machado, A. V. (2015). Antimicrobial nanostructured starch based films for packaging. *Carbohydrate Polymer, 129*, 127–134. https://doi.org/10.1016/j.carbpol.2015.04.021.

Álvarez, K., Famá, L., & Gutiérrez, T. J. (2017). Chapter 12. Physicochemical, antimicrobial and mechanical properties of thermoplastic materials based on biopolymers with application in the food industry. In M. Masuelli & D. Renard (Eds.), *Advances in physicochemical properties of biopolymers: Part 1* (Vol. 2017, pp. 358–400). Bentham Science Publishers. EE.UU. ISBN: 978–1–68108-454-1. eISBN: 978–1–68108-453-4. https://doi.org/10.2174/97816810845341l 7010015.

Álvarez, K., Alvarez, V. A., & Gutiérrez, T. J. (2018). Chapter 3. Biopolymer composite materials with antimicrobial effects applied to the food industry. In V. K. Thakur & M. K. Thakur (Eds.), *Functional biopolymers* (pp. 57–96). Editorial Springer International Publishing.. EE.UU. ISBN: 978-3-319-66416-3. eISBN: 978-3-319-66417-0. https://doi.org/10.1007/978-3-319-66417-0_3.

Araque, L. M., Alvarez, V. A., & Gutiérrez, T. J. (2018). Chapter 13. Composite foams made from biodegradable polymers for food packaging applications. In T. J. Gutiérrez (Ed.), *Polymers for food applications* (pp. 347–355). Cham. ISBN: 978–3–319-94624-5. eISBN: 978–3–319-94625-2: Editorial Springer International Publishing. https://doi.org/10.1007/978-3-319-94625-2_13.

Ashjari, H. R., Dorraji, M. S. S., Fakhrzadeh, V., Eslami, H., Rasoulifard, M. H., Rastgouy-Houjaghan, M., Gholizadeh, P., & Kafil, H. S. (2018). Starch-based polyurethane/CuO nanocomposite foam: Antibacterial effects for infection control. *International Journal of Biological Macromolecules, 111*, 1076–1082. https://doi.org/10.1016/j.ijbiomac.2018.01.137.

Avella, M., De Vlieger, J. J., Errico, M. E., Fischer, S., Vacca, P., & Volpe, M. G. (2005). Biodegradable starch/clay nanocomposite films for food packaging applications. *Food Chemistry, 93*(3), 467–474. https://doi.org/10.1016/j.foodchem.2004.10.024.

Biji, K. B., Ravishankar, C. N., Mohan, C. O., & Gopal, T. S. (2015). Smart packaging systems for food applications: A review. *Journal of Food Science and Technology, 52*(10), 6125–6135. https://doi.org/10.1007/s13197-015-1766-7.

Bracone, M., Merino, D., González, J., Alvarez, V. A., & Gutiérrez, T. J. (2016). Chapter 6. Nanopackaging from natural fillers and biopolymers for the development of active and intelligent films. In S. Ikram & S. Ahmed (Eds.), *Natural polymers: Derivatives, blends and composites* (pp. 119–155). New York. EE.UU. ISBN: 978-1-63485-831-1: Editorial Nova Science Publishers, Inc.

Bratovčić, A., Odobašić, A., Ćatić, S., & Šestan, I. (2015). Application of polymer nanocomposite materials in food packaging. *Croatian Journal of Food Science and Technology, 7*(2), 86–94. https://doi.org/10.17508/cjfst.2015.7.2.06.

Cannarsi, M., Baiano, A., Marino, R., Sinigaglia, M., & Del Nobile, M. A. (2005). Use of biodegradable films for fresh cut beef steaks packaging. *Meat Science, 70*(2), 259–265. https://doi.org/10.1016/j.meatsci.2005.01.006.

Carbone, M., Donia, D. T., Sabbatella, G., & Antiochia, R. (2016). Silver nanoparticles in polymeric matrices for fresh food packaging. *Journal of King Saud University-Science, 28*(4), 273–279. https://doi.org/10.1016/j.jksus.2016.05.004.

Castillo, L. A., Farenzena, S., Pintos, E., Rodríguez, M. S., Villar, M. A., García, M. A., & López, O. V. (2017). Active films based on thermoplastic corn starch and chitosan oligomer for food packaging applications. *Food Packaging and Shelf Life, 14*.(Part B, 128–136. https://doi.org/10.1016/j.fpsl.2017.10.004.

Debiagi, F., Mali, S., Grossmann, M. V. E., & Yamashita, F. (2011). Biodegradable foams based on starch, polyvinyl alcohol, chitosan and sugarcane fibers obtained by extrusion. *Brazilian Archives of Biology and Technology, 54*(5), 1043–1052. https://doi.org/10.1590/s1516-89132011000500023.

Dufresne, A., & Castaño, J. (2017). Polysaccharide nanomaterial reinforced starch nanocomposites: A review. *Starch-Stärke, 69*(1–2), 1500307. https://doi.org/10.1002/star.201500307.

Garg, S., & Jana, A. K. (2007). Studies on the properties and characteristics of starch–LDPE blend films using cross-linked, glycerol modified, cross-linked and glycerol modified starch. *European Polymer Journal, 43*(9), 3976–3987. https://doi.org/10.1016/j.eurpolymj.2007.06.030.

Ghanbarzadeh, B., Oleyaei, S. A., & Almasi, H. (2015). Nanostructured materials utilized in biopolymer-based plastics for food packaging applications. *Critical Reviews in Food Science and Nutrition, 55*(12), 1699–1723. https://doi.org/10.1080/10408398.2012.731023.

Glenn, G. M., Orts, W., Imam, S., Chiou, B.-S., & Wood, D. F. (2014). Starch plastic packaging and agriculture applications. In P. J. Halley & L. Avérous (Eds.), *Starch polymers: From genetic engineering to green applications* (pp. 421–452). Amsterdam: Elsevier. https://doi.org/10.1016/b978-0-444-53730-0.00032-4.

Grigore, M. (2017). Methods of recycling, properties and applications of recycled thermoplastic polymers. *Recycling, 2*(4), 24. https://doi.org/10.3390/recycling2040024.

Gutiérrez, T. J. (2018a). Characterization and *in vitro* digestibility of non-conventional starches from Guinea arrowroot and La Armuña lentils as potential food sources for special diet regimens. *Starch-Stärke, 70*(1–2). https://doi.org/10.1002/star.201700124.

Gutiérrez, T. J. (2018b). Chapter 9. Biodegradability and compostability of food nanopackaging materials. In G. Cirillo, M. A. Kozlowski, & U. G. Spizzirri (Eds.), *Composite materials for food packaging* (pp. 269–296). Hoboken.. EE.UU. ISBN: 978–1–119-16020-5: WILEY-Scrivener Publisher. https://doi.org/10.1002/9781119160243.ch9.

Gutiérrez, T. J., & Álvarez, K. (2016). Physico-chemical properties and *in vitro* digestibility of edible films made from plantain flour with added *Aloe vera* gel. *Journal of Functional Foods, 26*, 750–762. https://doi.org/10.1016/j.jff.2016.08.054.

Gutiérrez, T. J., & Alvarez, V. A. (2017). Cellulosic materials as natural fillers in starch-containing matrix-based films: A review. *Polymer Bulletin, 74*(6), 2401–2430. https://doi.org/10.1007/s00289-016-1814-0.

Gutiérrez, T. J., & Álvarez, K. (2017). Chapter 4. Transport phenomena in biodegradable and edible films. In M. A. Masuelli (Ed.), *Biopackaging* (pp. 58–88). Miami,. EE.UU. ISBN: 978–1–4987-4968-8: Editorial CRC Press Taylor & Francis Group.

Gutiérrez, T. J., & Alvarez, V. A. (2018). Bionanocomposite films developed from corn starch and natural and modified nano-clays with or without added blueberry extract. *Food Hydrocolloids, 77*, 407–420. https://doi.org/10.1016/j.foodhyd.2017.10.017.

Gutiérrez, T. J., & González, G. (2016). Effects of exposure to pulsed light on surface and structural properties of edible films made from cassava and taro starch. *Food and Bioprocess Technology, 9*(11), 1812–1824. https://doi.org/10.1007/s11947-016-1765-3.

Gutiérrez, T. J., & González, G. (2017). Effect of cross-linking with *Aloe vera* gel on surface and physicochemical properties of edible films made from plantain flour. *Food Biophysics, 12*(1), 11–22. https://doi.org/10.1007/s11483-016-9458-z.

Gutiérrez, T. J., Pérez, E., Guzmán, R., Tapia, M. S., & Famá, L. (2014). Physicochemical and functional properties of native and modified by crosslinking, dark-cush-cush yam (*Dioscorea Trifida*) and cassava (*Manihot Esculenta*) starch. *Journal of Polymer and Biopolymer Physics Chemistry, 2*(1), 1–5. https://doi.org/10.12691/jpbpc-2-1-1.

Gutiérrez, T. J., Morales, N. J., Pérez, E., Tapia, M. S., & Famá, L. (2015a). Physico-chemical study of edible films based on native and phosphating cush-cush yam and cassava starches. *Food Packaging and Shelf Life, 3*, 1–8. https://doi.org/10.1016/j.fpsl.2014.09.002.

Gutiérrez, T. J., Tapia, M. S., Pérez, E., & Famá, L. (2015b). Structural and mechanical properties of native and modified cush-cush yam and cassava starch edible films. *Food Hydrocolloids, 45*, 211–217. https://doi.org/10.1016/j.foodhyd.2014.11.017.

Gutiérrez, T. J., Guzmán, R., Medina Jaramillo, C., & Famá, L. (2016a). Effect of beet flour on films made from biological macromolecules: Native and modified plantain flour. *International Journal of Biological Macromolecules, 82*, 395–403. https://doi.org/10.1016/j.ijbiomac.2015.10.020.

Gutiérrez, T. J., Suniaga, J., Monsalve, A., & García, N. L. (2016b). Influence of beet flour on the relationship surface-properties of edible and intelligent films made from native and modified plantain flour. *Food Hydrocolloids, 54*, 234–244. https://doi.org/10.1016/j.foodhyd.2015.10.012.

Gutiérrez, T. J., González Seligra, P., Medina Jaramillo, C., Famá, L., & Goyanes, S. (2017). Chapter 14. Effect of filler properties on the antioxidant response of thermoplastic starch composites. In V. K. Thakur, M. K. Thakur, & M. R. Kessler (Eds.), *Handbook of composites from renewable materials* (pp. 337–370). Hoboken: WILEY-Scrivener Publisher. EE.UU. ISBN: 978-1-119-22362-7. https://doi.org/10.1002/9781119441632.ch14.

Gutiérrez, T. J., Toro-Márquez, L. A., Merino, D., & Mendieta, J. R. (2019). Hydrogen-bonding interactions and compostability of bionanocomposite films prepared from corn starch and nano-fillers with and without added Jamaica flower extract. *Food Hydrocolloids, 89*, 283–293. https://doi.org/10.1016/j.foodhyd.2018.10.058.

Han, J. (2014). *Innovations in food packaging* (2nd ed.). London: Imprint: Academic Press. ISBN:9780123948359 Hardcover ISBN: 9780123946010.

Herniou--Julien, C., Mendieta, J. R., & Gutiérrez, T. J. (2019). Characterization of biodegradable/non-compostable films made from cellulose acetate/corn starch blends processed under reactive extrusion conditions. *Food Hydrocolloids, 89*, 67–79. https://doi.org/10.1016/j.foodhyd.2018.10.024.

Hu, B. (2014). Biopolymer-based lightweight materials for packaging applications. In Y. Yang, H. Xu, & X. Yu (Eds.), *Lightweight materials from biopolymers and biofibers* (Vol. 1175, pp. 239–255). https://doi.org/10.1021/bk-2014-1175.ch013.

Jiang, H., Wang, L., & Zhu, K. (2014). Coaxial electrospinning for encapsulation and controlled release of fragile water-soluble bioactive agents. Journal of Controlled Release, 193, 296–303. https://doi.org/10.1016/j.jconrel.2014.04.025.

Kaisangsri, N., Kerdchoechuen, O., & Laohakunjit, N. (2012). Biodegradable foam tray from cassava starch blended with natural fiber and chitosan. *Industrial Crops and Products, 37*(1), 542–546. https://doi.org/10.1016/j.indcrop.2011.07.034.

Kaisangsri, N., Kerdchoechuen, O., & Laohakunjit, N. (2014). Characterization of cassava starch based foam blended with plant proteins, Kraft fiber, and palm oil. *Carbohydrate Polymers, 110*, 70–77. https://doi.org/10.1016/j.carbpol.2014.03.067.

Kumar, P., Topin, F., & Vicente, J. (2014). Determination of effective thermal conductivity from geometrical properties: Application to open cell foams. *International Journal of Thermal Sciences, 81*, 13–28. https://doi.org/10.1016/j.ijthermalsci.2014.02.005.

Lopez-Gil, A., Silva-Bellucci, F., Velasco, D., Ardanuy, M., & Rodriguez-Perez, M. A. (2015). Cellular structure and mechanical properties of starch-based foamed blocks reinforced with natural fibers and produced by microwave heating. *Industrial Crops and Products, 66*, 194–205. https://doi.org/10.1016/j.indcrop.2014.12.025.

Malathi, A. N., Santhosh, K. S., & Nidoni, U. (2014). Recent trends of biodegradable polymer: Biodegradable films for food packaging and application of nanotechnology in biodegradable food packaging. Current Trends in Technology and Science, 3(2), 73–79. Available in: http://www.ctts.in/assets/upload/1071ctts-31027%20vol-3%20iss-2.pdf.

Mali, S. (2018). Biodegradable foams in the development of food packaging. In T. J. Gutiérrez (Ed.), *Polymers for food applications* (pp. 329–345). Cham: Springer. https://doi.org/10.1007/978-3-319-94625-2_12.

Medina Jaramillo, C., Gutiérrez, T. J., Goyanes, S., Bernal, C., & Famá, L. (2016). Biodegradability and plasticizing effect of *yerba mate* extract on cassava starch edible films. *Carbohydrate Polymers, 151*, 150–159. https://doi.org/10.1016/j.carbpol.2016.05.025.

Merino, D., Gutiérrez, T. J., Mansilla, A. Y., Casalongué, C., & Alvarez, V. A. (2018). Critical evaluation of starch-based antibacterial nanocomposites as agricultural mulch films: Study on their interactions with water and light. *ACS Sustainable Chemistry & Engineering, 6*(11), 15662–15672. https://doi.org/10.1021/acssuschemeng.8b04162.

Merino, D., Gutiérrez, T. J., & Alvarez, V. A. (2019a). Potential agricultural mulch films based on native and phosphorylated corn starch with and without surface functionalization with chitosan. *Journal Polymers and the Environment, 27*(1), 97–105. https://doi.org/10.1007/s10924-018-1325-1.

Merino, D., Gutiérrez, T. J., & Alvarez, V. A. (2019b). Structural and thermal properties of agricultural mulch films based on native and oxidized corn starch nanocomposites. *Starch-Stärke, 71*(7–8), 1800341. https://doi.org/10.1002/star.201800341.

Nayani, M., Gunashekar, S., & Abu-Zahra, N. (2013). Synthesis and characterization of polyurethane-nanoclay composites. *International Journal of Polymer Science*,. 717895, 1–5. https://doi.org/10.1155/2013/717895.

Obradovic, J., Voutilainen, M., Virtanen, P., Lassila, L., & Fardim, P. (2017). Cellulose fibre-reinforced biofoam for structural applications. *Materials, 10*(6), 619. https://doi.org/10.3390/ma10060619.

Pankaj, S. K., Bueno-Ferrer, C., Misra, N. N., Milosavljević, V., O'donnell, C. P., Bourke, P., Keener, K. M., & Cullen, P. J. (2014). Applications of cold plasma technology in food packaging. *Trends in Food Science & Technology, 35*(1), 5–17. https://doi.org/10.1016/j.tifs.2013.10.009.

Pawar, P. A., & Purwar, A. H. (2013). Biodegradable polymers in food packaging. *American Journal of Engineering Research, 2*(5), 151–164. Available in: https://pdfs.semanticscholar.org/aa8d/a9f18b260eeea48bdf52c74bf17373211775.pdf.

Pornsuksomboon, K., Holló, B. B., Szécsényi, K. M., & Kaewtatip, K. (2016). Properties of baked foams from citric acid modified cassava starch and native cassava starch blends. *Carbohydrate Polymers, 136*, 107–112. https://doi.org/10.1016/j.carbpol.2015.09.019.

Prasanth, R., Owuor, P. S., Shankar, R., Joyner, J., Kosolwattana, S., Jose, S. P., Dong, P., Thakur, V. K., Cho, J. H., & Shelke, M. (2015). Eco-friendly polymer-layered silicate nanocomposite–preparation, chemistry, properties, and Applications. In V. Thakur & M. Thakur (Eds.), *Eco-friendly polymer nanocomposites. Advanced structured materials* (pp. 1–42). New Delhi: Springer. https://doi.org/10.1007/978-81-322-2473-0_1.

Priya, B., Gupta, V. K., Pathania, D., & Singha, A. S. (2014). Synthesis, characterization and antibacterial activity of biodegradable starch/PVA composite films reinforced with cellulosic fibre. *Carbohydrate Polymers, 109*, 171–179. https://doi.org/10.1016/j.carbpol.2014.03.044.

Razza, F., Degli Innocenti, F., Dobon, A., Aliaga, C., Sanchez, C., & Hortal, M. (2015). Environmental profile of a bio-based and biodegradable foamed packaging prototype in comparison with the current benchmark. *Journal of Cleaner Production, 102*, 493–500. https://doi.org/10.1016/j.jclepro.2015.04.033.

Rhim, J. W., & Kim, Y.-T. (2014). Biopolymer-based composite packaging materials with nanoparticles. In J. H. Han. (Ed.), *Innovations in food packaging* (Second ed., pp. 413–442). London: Academic. https://doi.org/10.1016/b978-0-12-394601-0.00017-5.

Robyt, J. F. (2008). Starch: Structure, properties, chemistry, and enzymology. In B. O. Fraser-Reid, K. Tatsuta, & J. Thiem (Eds.), *Glycoscience: Chemistry and chemical biology* (pp. 1437–1472). Berlin: Springer. https://doi.org/10.1007/978-3-540-30429-6_35.

Sauceau, M., Fages, J., Common, A., Nikitine, C., & Rodier, E. (2011). New challenges in polymer foaming: A review of extrusion processes assisted by supercritical carbon dioxide. *Progress in Polymer Science, 36*(6), 749–766. https://doi.org/10.1016/j.progpolymsci.2010.12.004.

Shankar, S., & Rhim, J.-W. (2016). Polymer nanocomposites for food packaging applications. In A. Dasari & J. Njuguna (Eds.), *Functional and physical properties of polymer nanocomposites* (pp. 29–55). Chichester: Wiley. https://doi.org/10.1002/9781118542316.ch3.

Soykeabkaew, N., Thanomsilp, C., & Suwantong, O. (2015). A review: Starch-based composite foams. *Composites Part A: Applied Science and Manufacturing, 78*, 246–263. https://doi.org/10.1016/j.compositesa.2015.08.014.

Suárez, G., & Gutiérrez, T. J. (2017). Chapter 15. Recent advances in the development of biodegadable films and foams from cassava starch. In C. Klein (Ed.), *Handbook on cassava: Production, potential uses and recent advances* (pp. 297–312). New York.. EE.UU. ISBN: 978-1-53610-307-6: Editorial Nova Science Publishers, Inc.

Teixeira, E. D. M., De Campos, A., Marconcini, J. M., Bondancia, T. J., Wood, D., Klamczynski, A., Mattoso, L. H. C., & Glenn, G. M. (2014). Starch/fiber/poly(lactic acid) foam and compressed foam composites. *RSC Advances, 4*(13), 6616–6623. https://doi.org/10.1039/c3ra47395c.

Thakur, V. K., Thakur, M. K., Raghavan, P., & Kessler, M. R. (2014). Progress in green polymer composites from lignin for multifunctional applications: A review. *ACS Sustainable Chemistry & Engineering, 2*(5), 1072–1092. https://doi.org/10.1021/sc500087z.

Toro-Márquez, L. A., Merino, D., & Gutiérrez, T. J. (2018). Bionanocomposite films prepared from corn starch with and without nanopackaged Jamaica (*Hibiscus sabdariffa*) flower extract. *Food and Bioprocess Technology, 11*(11), 1955–1973. https://doi.org/10.1007/s11947-018-2160-z.

Xia, X., Chen, X., Zhang, Z., Chen, X., Zhao, W., Liao, B., & Hur, B. (2014). Compressive properties of closed-cell aluminum foams with different contents of ceramic microspheres. *Materials & Design (1980-2015), 56*, 353–358. https://doi.org/10.1016/j.matdes.2013.11.040.

Xie, F., Liu, P., & Yu, L. (2014). Processing of plasticized starch-based materials: state of the art and perspectives. In P. J. Halley & L. Avérous (Eds.), *Starch polymers: From genetic engineering to green applications* (pp. 257–289). Amsterdam: Elsevier. https://doi.org/10.1016/b978-0-444-53730-0.00024-5.

Yildirim, N., Shaler, S. M., Gardner, D. J., Rice, R., & Bousfield, D. W. (2014). Cellulose nanofibril (CNF) reinforced starch insulating foams. *Cellulose, 21*(6), 4337–4347. https://doi.org/10.1007/s10570-014-0450-9.

Zheng, Y., Monty, J., & Linhardt, R. J. (2015). Polysaccharide-based nanocomposites and their applications. *Carbohydrate Research, 405*, 23–32. https://doi.org/10.1016/j.carres.2014.07.016.

Chapter 17
Coatings in the Postharvest

María Roberta Ansorena and Alejandra G. Ponce

Abstract The development of new edible coatings (ECs) with improved functionality and performance for fresh and minimally processed fruits and vegetables is one of the challenges of the postharvest industry. This technique has been successful not only in reducing water loss and delaying senescence, but also in increasing the antimicrobial properties of the coated product. The new generation of ECs is specially designed to allow the incorporation and/or controlled release of antioxidants, vitamins, nutraceuticals and natural antimicrobial agents by means of the application of promising technologies such as nanoencapsulation and layer-by-layer (LbL) assembly. This review analyzes the maintenance of postharvest quality through the application of new ECs on fresh products that cover the use of nanoemulsions, nanoparticles, nanofibers and LbL as techniques to form active edible coatings and films.

Keywords Edible coatings · IV gama · Micro and nanoencapsulation

17.1 Introduction

Production and consumption of minimally processed foods is gaining popularity (Saberi and Golding 2018). In this context, fresh-cut fruits and vegetables are being welcomed by the consumers due to the desire for new and natural products and change in life style of the consumers (Yousuf and Srivastava 2015). Due to growing consumer preferences for ready-to-use foods, the economic importance of the fresh fruit industry is becoming increasingly significant. The main aim of the fresh-cut technology is to provide consumers with a convenient and fresh like product having desirable nutritional and sensory quality.

M. R. Ansorena · A. G. Ponce (✉)
Food Engineering Research Group (GIIA), Faculty of Engineering,
National University of Mar del Plata (UNMdP) and National Scientific and Technical Research Council (CONICET), Mar del Plata, Buenos Aires, Argentina
e-mail: agponce@mdp.edu.ar

T. J. Gutiérrez (ed.), *Polymers for Agri-Food Applications*,
https://doi.org/10.1007/978-3-030-19416-1_17

Fig. 17.1 General flow diagram of fresh-cut processing for fruits and vegetables with film/coating application

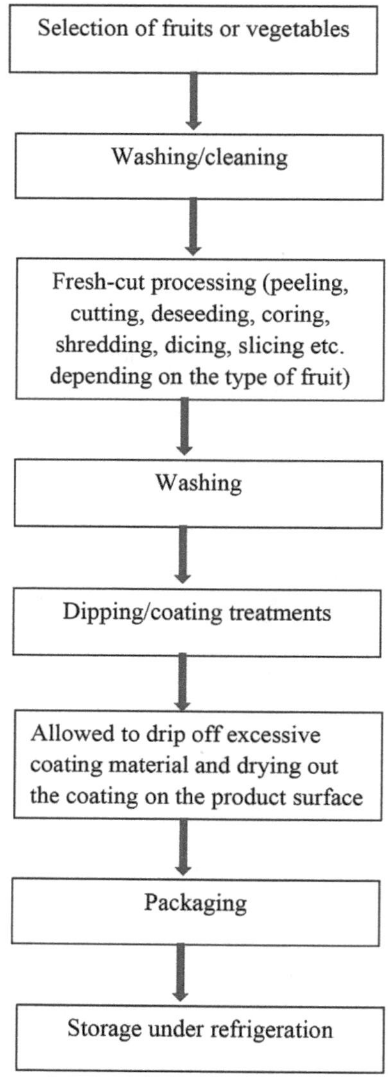

There are numerous factors which limit the quality and shelf-life of fresh-cut products. Quality of fresh-cut products can be affected both by internal factors including morphological, physiological, and biochemical defense mechanisms, type of fruit, genotype, stress-induced senescence programs and external factors representing environmental situations such as storage temperature, humidity, sharpness of cutting-knife and chemical treatments (Hodges and Toivonen 2008). Figure 17.1 represents a general flow sheet for fresh-cut processing of fruits and vegetables (Yousuf et al. 2018).

Food scientists are involved in the development of new technologies that can improve the quality and quantity of fresh products in order to meet the expectations of consumers. Different technologies including irradiation with gamma rays, surface treatments by means of coating, modified atmosphere packaging (MAP) to maintain the appropriate concentration of gas around the cut surface, controlled atmosphere to delay the softening of the fruit, chemical treatments such as calcium dives at low temperatures has been investigated to reduce harmful changes such as microbial growth, drying and discoloration (Mantilla et al. 2013; Lamikanra and Watson 2004). Denoya et al. (2015), studied the effect of high-pressure processing on preservation of fresh-cut peaches and concluded that high pressure processing in combination with vacuum packaging could preserve fresh-cut peaches. The concept of biopreservation has also been attempted for fresh-cut fruits and vegetables (Qadri et al. 2015).

Coating is considered as a convenient and safe method and is therefore, more and more concerned in food industry (Mantilla et al. 2013). It is a promising technology to preserve the quality and to prolong shelf-life of fresh fruits and vegetables.

Through this chapter, an attempt has been made to highlight, update and compile the scientific literature available on emerging technology of edible coating (EC) in context to horticultural crops.

17.2 Edible Coating Technology: In the Context of Horticultural Crops

Fruits and vegetables remain as living tissues up until the time they are consumed fresh, cooked for consumption, or processed for preservation. Controlling respiration of these living tissues would improve storability and extend shelf life of fresh products a certain level of respiration activity is required to prevent plant tissues from senescing and dying. Minimally processed products such as fresh-cut fruits and vegetables are essentially wounded tissues, leading directly to tissue softening and browning discoloration on cut surfaces (Fig. 17.1). The intensity of the wound response is affected by a number of factors, including species and variety, O_2 and carbon dioxide (CO_2) concentrations, water vapor pressure, and the presence of inhibitors (Brecht et al. 2004). The concept of using ECs to extend shelf life of fresh and minimally processed products and protect them from harmful environmental effects has been emphasized based on the need for high quality and the demand for minimal food processing and storage technologies. ECs have demonstrated the capability of improving food quality and prolonging shelf life of fresh products by regulating the transfer of moisture, oxygen, CO_2, aroma and taste compounds in a food system (Gutiérrez and Álvarez 2017).

ECs indicate the application of commercial food grade waxes or films to protect the loss of natural glossiness during the postharvest period. The best part of the EC is that these are edible, healthy and biodegradable in nature, unlike other chemical postharvest treatments which leave a residue (Brecht et al. 2004).

ECs may also be used to advantage on processed fruits and vegetables for improving structural integrity of frozen fruits and vegetables and preventing moisture absorption and oxidation of freeze-dried fruits or vegetables (Baker and Nisperos-Carriedo 1994; Baldwin and Baker 2002; Olivas et al. 2007; Park 2005). ECs can also contain functional ingredients such as antioxidants, antimicrobials, nutrients and flavors to further improve food stability, quality, functionality and safety (Krochta 1992; Krochta 1997; Debeaufort et al. 1998; Min and Krochta 2005; Gutiérrez 2017a; 2018).

Fruits or vegetables are usually coated by dipping in or spraying with a range of edible materials, so that a semipermeable membrane is formed on the surface for suppressing respiration, controlling moisture loss and providing other functions (Ukai et al. 1976; Thompson 2010). A variety of edible materials, including lipids, polysaccharides and proteins, alone or in combinations, have been formulated to product ECs (Ukai et al. 1976; Kester and Fennema 1986). Lipid-based coatings made of acetylated monoglycerides, waxes (beeswax, carnauba, candelilla, paraffin, and rice bran), and surfactants were the 1st successful ones on whole fruits and vegetables (Paredes-Lopez et al. 1974; Lawrence and Iyengar 1983; Warth 1986), used for reducing surface abrasion during handling and serving as moisture barrier (Hardenburg 1967). Colloidal suspensions of oils or waxes dispersed in water were typical early fruit-coating formulations.

Appropriately formulated ECs can be utilized for most foods to meet challenges associated with stable quality, market safety, nutritional value and economic production cost. Regarding the fresh product industry, the potential benefits of using ECs include (Fig. 17.2):

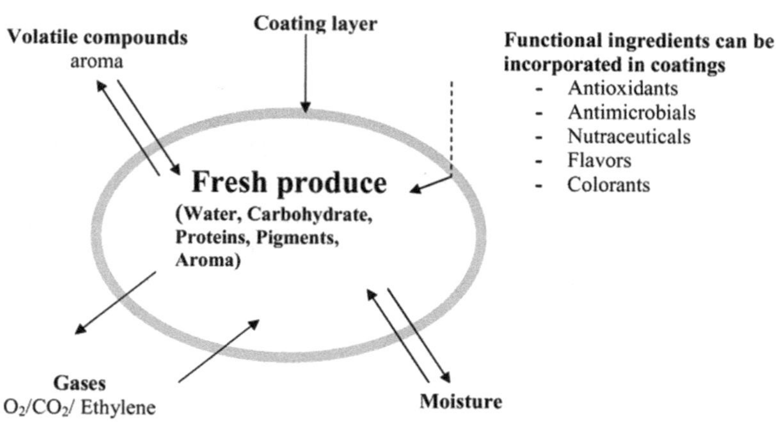

Fig. 17.2 Functional properties of an edible coating on fresh fruits and vegetables

17.3 Recent Advances in the Development of Coatings in the Postharvest

ECs also called "environment friendly coatings", are thin layers of edible substances, applied to product surface, that do not pose any harm if consumed at concentrations effective for maintaining product quality (Dhall 2013). Maintenance of horticultural products by application of ECs based on plant extracts (ECPE) is necessary (Murmu and Mishra 2018; Tesfay and Magwaza 2017). Application of animal extracts is hindered by their limited acceptance by consumers with allergens for animal food, religious beliefs that do not encourage consumption of animal products and their probability of forming impermeable layers due to a waxy nature (Oms-Oliu et al. 2010; Shiekh et al. 2013). The nature of ECPE is usually based on extracts obtained from plants regarded healthy for human consumption such as fruit, herbs, or spices with high antioxidant and antimicrobial properties (Florez-López et al. 2016). Therefore, there are probable medicinal benefits from their consumption (Gutiérrez 2017a).

In the last decade, numerous studies have been carried out to develop ECs of fruit and vegetables. Many review manuscripts have also been published in the literature (Table 17.1).

Advances in the procedures used during the development of coatings have led to the introduction of multiple approaches such as nano-coatings (coatings with their thickness in nanometers), compounds (mixture of hydrocolloids and lipids), different extraction solvents (alcohol, water or acids) and multilayer system in which the application of the coating is modified by a layer of active substances such as antimicrobials, flavoring agents, antioxidants or emulsifiers (Ponce et al. 2008; Campos et al. 2011; Florez-López et al. 2016; Valencia-Chamorro et al. 2011 Ansorena et al. 2011; Moreira et al. 2011; Tomadoni et al. 2018).

Although coatings have been used successfully for decades in the fresh plant products industry, some coatings such as Becox wax and diphenylamine are known to be harmful to consumers. Even though the coatings have the desired characteristics to control the quality and antimicrobial activities, there is a need to know the risks to the health of the consumer associated with said post-harvest treatments, therefore, the need for the study of ECs (Dhall 2013; Ncama et al. 2018). The main aim of the fresh-cut technology is to provide consumers with a convenient and fresh like product having desirable nutritional and sensory quality. Coating is considered as a convenient and safe method and is therefore, more and more concerned in food industry (Mantilla et al. 2013; Ansorena et al. 2011; Moreira et al. 2011; Tomadoni et al. 2018). It is a promising technology to preserve the quality and to prolong shelf-life of fresh fruits and vegetables.

Edible films and coatings are synthesized from substances having film forming properties. Throughout manufacturing, film and coating substances are dissolve in water, alcohol and combination of water and alcohol solvents. Various types of colorants, antimicrobial agents, flavors and plasticizers can be fused in this system. To obtain highly effective coatings, solution is synthesized and dried at a precise

Table 17.1 Previous literature reviews of edible coatings application on fresh fruits and vegetables

Topic	References
Plant-based edible coatings for managing postharvest quality of fresh horticultural produce: A review	Ncama et al. (2018)
Edible coating technology for extending market life of horticultural produce	Prasad et al. (2018)
Recent developments in shelf-life extension of fresh-cut fruits and vegetables by application of different edible coatings: A review	Yousuf et al. (2018)
Bio-based and environmentally benign coatings	Tiwari et al. (2016)
Perspectives on utilization of edible coatings and nano-laminate coatings for extension of postharvest storage of fruits and vegetables	Florez-López et al. (2016)
Advances in edible coatings for fresh fruits and vegetables: a review	Dhall (2013)
Chitosan based edible films and coatings: A review	Elsabee and Abdou (2013)
Chitosan as a novel edible coating for fresh fruits	Shiekh et al. (2013)
Application of chitosan-based coating in fruit and vegetable preservation: A review	Duan and Zhang (2013)
Development of edible films and coatings with antimicrobial activity	Campos et al. (2011)
Antimicrobial edible films and coatings for fresh and minimally processed fruits and vegetables: A review	Valencia-Chamorro et al. (2011)
Edible films and coatings: Structures, active functions and trends in their use	Falguera et al. (2011)
Use of essential oils in bioactive edible coatings: A review	Sanchez-Gonzalez (2016)
Edible coatings as carriers of food additives on fresh-cut fruits and vegetables	Raybaudi-Massilia et al. (2010)
Edible films and coatings for fruits and vegetables	Olivas and Barbosa-Cánovas (2009)
Edible coatings to incorporate active ingredients to fresh-cut fruits: A review	Rojas-Graü et al. (2009a)
Recent advances in edible coatings for fresh and minimally processed fruits	Vargas et al. (2008)
Edible films and coatings: Characteristics and properties	Bourtoom (2008)
Applications of chitosan for improvement of quality and shelf life of foods: A review	No et al. (2007)
Innovations in the development and application of edible coatings for fresh and minimally processed fruits and vegetables	Lin and Zhao (2007)

temperature and relative humidity. Adjusting the pH or heating the solutions can be accomplished for the particular polymer to enable dispersion. In food processing, film/coating solutions can be applied to food by various strategies inclusive of dipping, spraying, brushing and panning accompanied with the aid of drying (Hassan et al. 2018).

The success of an EC for meeting the specific needs of fresh and minimally processed fruits and vegetables strongly depends on its barrier property to moisture, oxygen, and CO_2, which in turn depends on the chemical composition and structure of the coating-forming polymers, the characteristics of the products and the storage conditions. Several ECs made from cellulose, casein, zein, soy protein and chitosan (Cs) have shown such desirable characteristics on fresh products: good barrier properties, odorless, tasteless and transparent (Álvarez et al. 2017; 2018). However, commercial success is still limited. Challenges in the ECs targeted for fruits and vegetables may include: (1) limited moisture-barrier properties of the hydrophilic nature of most EC materials, (2) variable oxygen and CO_2-barrier properties due to poor temperature and relative humidity control in the storage, transportation, and marketing of fruits and vegetables, (3) unfavorable storage environment induced by improper gas barrier property of some coating materials, (4) inefficient coating coverage and poor coating adhesion between coating layer and the surface of fruits and vegetables and (5) adverse sensory effect from exogenous flavor imparted by some of the coating materials. Major quality deteriorations involved in fresh products are *via* mass transfer phenomena, including moisture adsorption, oxygen invasion, flavor loss, undesirable odor absorption, and the migration of packaging components into the food. Therefore, permeation, absorption, and diffusion into water, oxygen and CO_2 are among the most important functional properties of any EC applied to fruits and vegetable (Lin and Zhao 2007).

Several studies have been reported in which antimicrobial agents are applied to a variety of ECs to extend the shelf-life of fresh-cut fruits. Brasil et al. (2012) e.g. evaluated the incorporation of cinnamaldehyde into Cs and pectin coatings on papaya. Salinas-Roca et al. (2016) evaluated the incorporation of mallic acid into alginate coatings on mango, combined with light pulses. In addition, the incorporation of essential oils into alginate EC reduced the microbial growth in fresh-cut pineapple (Azarakhsh et al. 2013) and in fresh-cut apples (Rojas-Graü et al. 2007).

Reviews focussing on the literature for the application of plant-based ECs are limited. The commercial limitation of ECs could be associated with a lack of connection between research and legislative issues affecting their acceptance in the market. Future perspectives will be drawn for increasing the accuracy in production of ECs in attempt to improve their effectiveness in the market (Ncama et al. 2018).

17.4 Nanotechnology in Edible Films and Coatings

Nanotechnology is a multidisciplinary technological and scientific field that undergoes rapid development. Size reduction to the nanometer scale (10^{-9} m) may fundamentally modify compounds' physical, chemical, and biological properties resulting in new application potential (Ayala-Zavala et al. 2014; Viacava et al. 2017). The potential benefits of nanomaterials in the food industry are tremendous and cover many aspects such as food safety, nanosensors, nutrients delivery systems, bioavailability, new materials for pathogen detection and packaging materials (Duran and Marcato 2013; Bracone et al. 2016; Pereda et al. 2017; Ponce et al. 2017; Gutiérrez et al. 2017). In the following sections are presented various nanotechnology approaches in the field of edible films and ECs.

17.4.1 Nanoemulsions

Nanoemulsions are used in the food sector, mainly as a method to disperse lipophilic active ingredients in an aqueous media (Mason et al. 2006; Lorenzo et al. 2018). They present unique physic chemical and functional characteristics like supreme kinetic stability and clear or translucent appearance that make them compatible as delivery systems mainly in drinks and beverages (Duncan 2011). Edible films and ECs based on nanoemulsion containing solutions can be utilized (Mason et al. 2006).

Essential oils are the most common active agents encapsulated into nanoemulsified edible films and ECs (Gutiérrez et al. 2008). They have proven to be effective natural antimicrobial agents and gained particular interest due to their ecofriendly properties, relative safety and acceptance by consumers (Burt 2004). Edible films and edible coatings with nanoemulsified essential oils are mainly applied on fruits and vegetables in order to inhibit or prevent microbial growth (Donsì et al. 2015; Kim et al. 2013; Salvia-Trujillo et al. 2015; Sessa et al. 2015; Severino et al. 2014, 2015). Sessa et al. (2015) e.g. encapsulated nanoemulsified lemon, mandarin, oregano, and clove essential oils to modified Cs ECs and examined their antimicrobial effect. The combined use of nanoemulsified lemon essential oil with a modified Cs showed a remarkable *in vitro* antimicrobial activity. Salvia-Trujillo et al. (2015) studied the effect of alginate ECs with several concentrations of nanoemulsified lemongrass essential oil (LEO) on the bacterial growth on fresh-cut Fuji apples. Comparison of alginate coatings with coarse-emulsified and nanoemulsified LEO showed that coatings with nanoemulsified LEO exhibited a faster and greater inactivation of *Escherichia coli.*

Incorporation of essential oils to EFs and ECs in a nanoemulsion form can also reduce their sensory impact. In addition, the nanoemulsion method of encapsulation enables to minimize the essential oil concentration required for antimicrobial activity by increasing its accessibility (Duran and Marcato 2013).

17.4.2 Layer-by-Layer Edible Films and Coatings

The Layer-by-Layer (LbL) technique is an approach that originated from materials science and is based on the alternate deposition of various polymers to products multilayers with controlled properties (Decher 1997). Due to a nanometer scale buildup, the approach facilitates a delicate and precise tuning of the materials' features.

It is especially valuable due to its potential abilities to control and manipulate material properties and incorporate a wide range of functional biomolecules without substantial loss of their biological functions (Hammond 2012).

The LbL approach has entered the field of food nanotechnology relatively recently (Weiss et al. 2006). Since then, multilayered EFs and ECs have demonstrated their abilities to improve food safety and quality (Dhall 2013). The majority of multilayered EFs and ECs developed for food products are applied on fruit and vegetables, whole or fresh-cut (Vargas et al. 2008). Frequently such LbL edible films and ECs consist of oppositely charged polyelectrolytes and are applied by utilizing the electrostatic deposition method.

The LbL films/coatings have already been reported as an effective drug delivery biomedical system that allows controlling the release of functional biomolecules without substantial loss of their bioactivity through manipulating film/coating's properties (Jiang and Li 2009; Wang and Sain 2007; Zhong et al. 2007).

Another approach for encapsulating active agents in LbL edible films and ECs involves the formation of an inclusion complex (IC) with the active agent (guest molecule) using cyclodextrins. An IC improves the solubility of an active agent, controls its volatility and protects its bioactivity. In case of active agents with dominant flavors such as essential oils, an IC may mask undesired organoleptic effects (Brasil et al. 2012; Mantilla et al. 2013; Martiñon et al. 2014; Sipahi et al. 2013). Wen et al. (2016) demonstrated that polyvinyl alcohol/cinnamon essential oil/β-cyclodextrin (PVA/CEO/β-CD) antimicrobial nanofibrous film can effectively prolong the shelf-life of strawberry, indicating its potential for the application in active food packaging.

17.4.3 Nanoparticles

The possibility to improve the performances of polymers for food packaging by adding nanoparticles (NPs) has led to the development of a variety of polymer nanomaterials (Brody 2007; De Abreu et al. 2007; Kotsilkova et al. 2007; Duncan 2011). Incorporation of nanometric-sized particles was found to improve physical, optical and mechanical properties of edible films and coatings (Acevedo-Fani et al. 2015). NPs incorporated into edible films and coatings can also serve as delivery systems for active agents (Rojas-Graü et al. 2009a, b). This has led to the development of a variety of NPs reinforced edible films and coatings, also termed as "nanocomposites" which typically contain up to 5% w/w NPs.

NPs can often be produced from food-grade biopolymers such as polysaccharides and proteins or natural bioactive compounds such as curcumin and solid lipids. The majority of biopolymeric NPs in edible films and coatings are based on polysaccharides such as Cs and cellulose derivatives. Cs is also well known for its antimicrobial activity against a wide variety of microorganisms (Gutiérrez 2017b). Several works have studied the effect of Cs NPs' addition to the mechanical, and water vapor and oxygen permeability properties of several coatings and films (Lorevice et al. 2012; Martelli et al. 2013; Pilon et al. 2014; Antoniou et al. 2015). Solid lipid NPs (SLNs) are lipid colloidal submicronic systems that have been developed to encapsulate and deliver lipophilic functional components.

17.5 Bioactive Packaging

Bioactive packaging is the novel packaging technology that alters the package or coating in a way so as to have positive effect on consumer's health (Majid et al. 2018). Various techniques known to retain characteristic properties of biopolymers and employed in this novel packaging approach include enzyme encapsulation, nanoencapsulation, microencapsulation and enzyme immobilization. Keeping in view the required functional properties of particular bioactive components, functional or bioactive packaging has the potential to maintain bioactive substances in desired proportions until their controlled or fast diffusion within the packed food during its storage or prior to its consumption (Lopez-Rubio et al. 2004). Process of bioactive packaging technology is implemented *via* (1) utilization of biodegradable packaging materials for the release of functional or bioactive components, (2) encapsulating bio active ingredients into the foods or to the packaging materials and (3) introducing packaging materials exhibiting enzyme activity and capable of transforming some food components in order to deliver health benefits. The development of such packaging systems exerting health promotion effect involves the concept of marine oils, prebiotics, probiotics, encapsulated vitamins, phytochemicals, lactose free foods, bioavailable flavonoids and many more will boost the packaging industry in near future because of growing human health consciousness (Lagaron 2005).

Acknowledgements The authors would like to thank the Consejo Nacional de Investigaciones Científicas y Técnicas (CONICET) and Universidad Nacional de Mar del Plata (UNMdP) for financial support.

Conflicts of Interest The authors declare no conflict of interest.

References

Acevedo-Fani, A., Salvia-Trujillo, L., & Rojas-Grau, M. A. (2015). Edible films from essential-oil-loaded nanoemulsions: Physicochemical characterization and antimicrobial properties. *Food Hydrocolloids, 47*, 168–177. https://doi.org/10.1016/j.foodhyd.2015.01.032.

Álvarez, K., Famá, L., & Gutiérrez, T. J. (2017). Chapter 12. Physicochemical, antimicrobial and mechanical properties of thermoplastic materials based on biopolymers with application in the food industry. In M. Masuelli & D. Renard (Eds.), *Advances in physicochemical properties of biopolymers: Part 1* (pp. 358–400). Bentham Science Publishers. EE.UU. ISBN: 978–1–68108–454–1. eISBN: 978–1–68108–453–4. https://doi.org/10.2174/9781681084534117010015.

Álvarez, K., Alvarez, V. A., & Gutiérrez, T. J. (2018). Chapter 3. Biopolymer composite materials with antimicrobial effects applied to the food industry. In V. K. Thakur & M. K. Thakur (Eds.), *Functional biopolymers* (pp. 57–96). Editorial Springer International Publishing. EE.UU. ISBN: 978-3-319-66416-3. eISBN: 978-3-319-66417-0. https://doi.org/10.1007/978-3-319-66417-0_3.

Ansorena, M. R., Marcovich, N. E., & Roura, S. I. (2011). Impact of edible coatings and mild heat shocks on quality of minimally processed broccoli (*Brassica oleracea* L.) during refrigerated storage. *Postharvest Biology and Technology, 59*(1), 53–63. https://doi.org/10.1016/j.postharvbio.2010.08.011.

Antoniou, J., Liu, F., Majeed, H., & Zhong, F. (2015). Characterization of tara gum edible films incorporated with bulk chitosan and chitosan nanoparticles: A comparative study. *Food Hydrocolloids, 44*, 309–319. https://doi.org/10.1016/j.foodhyd.2014.09.023.

Ayala-Zavala, J. F., González-Aguilar, G. A., Ansorena, M. R., Alvarez-Párrilla, E., & de la Rosa, L. (2014). Nanotechnology tools to achieve food safety. Chapter 17. In R. Bhat & V. M. Gómez-López (Eds.), *Practical food safety: Contemporary issues and future directions* (pp. 341–353). Editorial Wiley-Blackwell. ISBN: 978–1–118–47460–0. 632. https://doi.org/10.1002/9781118474563.ch17.

Azarakhsh, N., Osman, A., Ghazali, H. M., Tan, C. P., & Mohd Adzahan, N. (2013). Effect of limonene incorporation into gellan-based edible coating on the changes in microbiological and sensory characteristics of fresh-cut pineapple during cold storage. *Acta Horticulturae*, (1012), 999–1004. https://doi.org/10.17660/actahortic.2013.1012.134.

Baker, R. A., & Nisperos-Carriedo, M. O. (1994). Edible coatings and films for processed foods. In J. M. Krochta, E. A. Baldwin, & M. O. Nisperos-Carriedo (Eds.), *Edible coatings and films to improve food quality* (pp. 89–104). Lancaster: Technomic Publishing Co., Inc.

Baldwin, E. A., & Baker, R. A. (2002). Use of proteins in edible coatings for whole and minimally processed fruits and vegetables. In A. Gennadios (Ed.), *Protein based films and coatings* (pp. 501–515). Boca Raton: CRC Press.

Bourtoom, T. (2008). Edible films and coatings: Characteristics and properties. *International Food Research Journal, 15*, 237–248.

Bracone, M., Merino, D., González, J., Alvarez, V. A., & Gutiérrez, T. J. (2016). Chapter 6. Nanopackaging from natural fillers and biopolymers for the development of active and intelligent films. In: Natural polymers: Derivatives, blends and composites. Ikram, S., & Ahmed, S. (Eds). Editorial Nova Science Publishers, Inc. New York. EE.UU. ISBN: 978-1-63485-831-1. pp. 119–155.

Brasil, I. M., Gomes, C., Puerta-Gomez, A., Castell-Perez, M. E., & Moreira, R. G. (2012). Polysaccharide-based multilayered antimicrobial edible coating enhances quality of fresh-cut papaya. *LWT – Food Science and Technology, 47*(1), 39–45. https://doi.org/10.1016/j.lwt.2012.01.005.

Brecht, J. K., Felkey, K., Bartz, J. A., Schneider, K. R., Saltveit, M. E., & Talcott, S. T. (2004). Fresh-cut vegetables and fruits. In J. Janick (Ed.), *Horticultural reviews* (Vol. 30, pp. 185–251). Wiley. https://doi.org/10.1002/9780470650837.ch6.

Brody, A. L. (2007). Nanocomposites technology in food packaging. *Food Technology, 61*, 80–83.

Burt, S. (2004). Essential oils: Their antibacterial properties and potential applications in foods – A review. *International Journal of Food Microbiology, 94*(3), 223–253. https://doi.org/10.1016/j.ijfoodmicro.2004.03.022.

Campos, C. A., Gerschenson, L. N., & Flores, S. K. (2011). Development of edible films and coatings with antimicrobial activity. *Food and Bioprocess Technology, 4*(6), 849–875. https://doi.org/10.1007/s11947-010-0434-1.

De Abreu, D. A. P., Losada, P. P., Angulo, I., & Cruz, J. M. (2007). Development of new polyolefin films with nanoclays for application in food packaging. *European Polymer Journal, 43*(6), 2229–2243. https://doi.org/10.1016/j.eurpolymj.2007.01.021.

Debeaufort, F., Quezada-Gallo, J. A., & Voilley, A. (1998). Edible films and coatings: tomorrow's packagings: A review. *Critical Reviews in Food Science and Nutrition, 38*(4), 299–313. https://doi.org/10.1080/10408699891274219.

Decher, J. (1997). Fuzzy nanoassemblies: Toward layered polymeric multicomposites. *Science, 277*, 1232–1237. https://doi.org/10.1126/science.277.5330.1232.

Denoya, G. I., Vaudagna, S. R., & Polenta, G. (2015). Effect of high pressure processing and vacuum packaging on the preservation of fresh-cut peaches. *LWT-Food Science and Technology, 62*(1), 801–806. https://doi.org/10.1016/j.lwt.2014.09.036.

Dhall, R. K. (2013). Advances in edible coatings for fresh fruit and vegetables: A review. *Critical Reviews in Food Science and Nutrition, 53*(5), 435–450. https://doi.org/10.1080/10408398.2010.541568.

Donsì, F., Marchese, E., Maresca, P., Pataro, G., Vu, K. D., Salmieri, S., Lacroix, M., & Ferrari, G. (2015). Green beans preservation by combination of a modified chitosan based-coating containing nanoemulsion of mandarin essential oil with high pressure or pulsed light processing. *Postharvest Biology and Technology, 106*, 21–32. https://doi.org/10.1016/j.postharvbio.2015.02.006.

Duan, J., & Zhang, S. (2013). Application of chitosan based coating in fruit and vegetable preservation: A review. *Journal of Food Processing and Technology, 4*, 227–258. https://doi.org/10.4172/2157-7110.1000227.

Duncan, T. V. (2011). Applications of nanotechnology in food packaging and food safety: Barrier materials, antimicrobials and sensors. *Journal of Colloid and Interface Science, 363*(1), 1–24. https://doi.org/10.1016/j.jcis.2011.07.017.

Duran, N., & Marcato, P. D. (2013). Nanobiotechnology perspectives. Role of nanotechnology in the food industry: A review. *International Journal of Food Science and Technology, 48*(6), 1127–1134. https://doi.org/10.1111/ijfs.12027.

Elsabee, M. Z., & Abdou, E. S. (2013). Chitosan based edible films and coatings: A review. *Materials Science and Engineering: C, 33*(6), 1819–1841. https://doi.org/10.1016/j.msec.2013.01.010.

Falguera, V., Quintero, J. P., Jiménez, A., Muñoz, J. A., & Ibarz, A. (2011). Edible films and coatings: Structures, active functions and trends in their use. *Trends in Food Science & Technology, 22*(6), 292–303. https://doi.org/10.1016/j.tifs.2011.02.004.

Florez-López, M. L., Cerqueira, M. A., de Rodríguez, D. J., & Vicente, A. A. (2016). Perspectives on utilization of edible coatings and nano-laminate coatings for exten- Sion of postharvest storage of fruit and vegetables. *Food Engineering Reviews, 8*(3), 292–305. https://doi.org/10.1007/s12393-015-9135-x.

Gutiérrez, T. J. (2017a). Surface and nutraceutical properties of edible films made from starchy sources with and without added blackberry pulp. *Carbohydrate Polymers, 165*, 169–179. https://doi.org/10.1016/j.carbpol.2017.02.016.

Gutiérrez, T. J. (2017b). Chapter 8. Chitosan applications for the food industry. In: Chitosan: Derivatives, composites and applications. Ahmed, S., & Ikram, S. (Eds). WILEY-Scrivener Publisher. EE.UU. ISBN: 978–1–119-36350-7. pp. 183–232. https://doi.org/10.1002/9781119364849.ch8

Gutiérrez, T. J. (2018). Active and intelligent films made from starchy sources/blackberry pulp. *Journal Polymers and the Environment, 26*(6), 2374–2391. https://doi.org/10.1007/s10924-017-1134-y.

Gutiérrez, T. J., & Álvarez, K. (2017). Chapter 4. Transport phenomena in biodegradable and edible films. In M. A. Masuelli (Ed.), *Biopackaging* (pp. 58–88). Miami., EE.UU. ISBN: 978–1–4987–4968–8: Editorial CRC Press Taylor & Francis Group.

Gutiérrez, J. M., González, C., Maestro, A., Solè, I. M. P. C., & Nolla, J. (2008). Nano-emulsions: New applications and optimization of their preparation. *Current Opinion in Colloid and Interface Science, 13*(4), 245–251. https://doi.org/10.1016/j.cocis.2008.01.005.

Gutiérrez, T. J., Ponce, A. G., & Alvarez, V. A. (2017). Nano-clays from natural and modified montmorillonite with and without added blueberry extract for active and intelligent food nanopackaging materials. *Materials Chemistry and Physics, 194*, 283–292. https://doi.org/10.1016/j.matchemphys.2017.03.052.

Hammond, H. (2012). Building biomedical materials layer-by-layer. *Materials Today, 15*(5), 196–206. https://doi.org/10.1016/S1369-7021(12)70090-1.

Hardenburg, R. E. (1967). Wax and related coatings for horticultural products: A bibliography. In *Agricultural Research Service ARS 51–15* (p. 26). Washington, DC: U.S. Department of Agriculture.

Hassan, B., Ali, S., Chatha, S., Hussain, A., Zia, K. M., & Akhtar, N. (2018). Recent advances on polysaccharides, lipids and protein based edible films and coatings: A review. *International Journal of Biological Macromolecules, 109*, 1095–1107. https://doi.org/10.1016/j.ijbiomac.2017.11.097.

Hodges, D. M., & Toivonen, P. M. A. (2008). Quality of fresh-cut fruits and vegetables as affected by exposure to abiotic stress. *Postharvest Biology and Technology, 48*(2), 155–162. https://doi.org/10.1016/j.postharvbio.2007.10.016.

Jiang, B. B., & Li, B. Y. (2009). Tunable drug loading and release from polypeptide multilayer nanofilms. *International Journal of Nanomedicine, 4*(1), 37–53.

Kester, J. J., & Fennema, O. R. (1986). Edible films and coatings: A review. *Food Technology, 40*(12), 47–59.

Kim, I. H., Lee, H., Kim, J. E., Son, K. B., Lee, Y. S., Chung, D. S., & Min, S. C. (2013). Plum coatings of lemongrass oil-incorporating carnauba wax-based nanoemulsion. *Journal of Food Science, 78*(10), 1551–1559. https://doi.org/10.1111/1750-3841.12244.

Kotsilkova, R., Silvestre, C., & Cimmino, S. (2007). Thermoset nanocomposites for engineering applications. In R. Kotsilkova (Ed.), *Thermoset nanocomposites for engineering applications* (pp. 93–116). Shawbury: Rapra Technology.

Krochta, J. M. (1992). Proteins as raw materials for films and coatings: Definitions, current status, and opportunities. In R. P. Singh & M. A. Wirakartakasumah (Eds.), *Advances in food engineering* (pp. 517–538). Boca Raton: CRC Press.

Krochta, J. M. (1997). Edible protein films and coatings. In S. Damodaran & A. Paraf (Eds.), *Food proteins and their applications*. New York: Marcel Dekker.

Lagaron, J.M. (2005). Biodegradable and sustainable plastics as essential elements in novel bioactive packaging technologies. In: First conference on biodegradable polymers for packaging applications. PIRA International, Leatherhead (UK). 5–6 July 2005.

Lamikanra, O., & Watson, M. A. (2004). Effect of calcium treatment temperature on fresh-cut cantaloupe melon during storage. *Journal of Food Science, 69*(6), 468–472. https://doi.org/10.1111/j.1365-2621.2004.tb10990.x.

Lawrence, J. F., & Iyengar, J. R. (1983). Determination of paraffin wax and mineral oil on fresh fruits and vegetables by high temperature gas chromatography. *Journal of Food Safety, 5*(3), 119–129. https://doi.org/10.1111/j.1745-4565.1983.tb00464.x.

Lin, D., & Zhao, Y. (2007). Innovations in the development and application of edible coatings for fresh and minimally processed fruit and vegetables. *Comprehensive Reviews in Food Science and Food Safety, 6*(3), 60–75. https://doi.org/10.1111/j.1541-4337.2007.00018.x.

Lopez-Rubio, A., Almenar, E., Hernandez-Munoz, P., Lagaron, J. M., Catala, R., & Gavara, R. (2004). Overview of active polymer-based packaging technologies for food applications. *Food Reviews International, 20*(4), 357–387. https://doi.org/10.1081/fri-200033462.

Lorenzo, G., Zaritzky, N., & Califano, A. (2018). Food gel emulsions: Structural characteristics and viscoelastic behavior. In T. J. Gutiérrez (Ed.), *Polymers for food applications* (pp. 481–507). Cham: Springer. https://doi.org/10.1007/978-3-319-94625-2_18.

Lorevice, M. V., de Moura, M. R., Aouada, F. A., & Mattoso, S. L. H. C. (2012). Development of novel guava puree films containing chitosan nanoparticles. *Journal of Nanoscience and Nanotechnology, 11*, 1–7. https://doi.org/10.1166/jnn.2012.5716.

Majid, I., Thakur, M., & Nanda, V. (2018). Innovative and Safe Packaging Technologies for Food and Beverages: Updated Review. In Innovations in Technologies for Fermented Food and Beverage Industries (pp. 257–287). Springer, Cham.

Mantilla, N., Castell-Perez, M. E., Gomes, C., & Moreira, R. G. (2013). Multilayered antimicrobial edible coating and its effect on quality and shelf-life of fresh-cut pineapple (*Ananas comosus*). *LWT-Food Science and Technology, 51*(1), 37–43. https://doi.org/10.1016/j.lwt.2012.10.010.

Martelli, M. R., Barros, T. T., de Moura, M. R., Mattoso, L. H. C., & Assis, O. B. G. (2013). Effect of chitosan nanoparticles and pectin content on mechanical properties and water vapor permeability of banana puree films. *Journal of Food Science, 78*(1), 98–104. https://doi.org/10.1111/j.1750-3841.2012.03006.x.

Martiñon, M. E., Moreira, R. G., Castell-Perez, M. E., & Gomes, C. (2014). Development of a multilayered antimicrobial edible coating for shelf life extension of fresh-cut cantaloupe (*Cucumis melo* L.) stored at 4 °C. *LWT – Food Science and Technology, 56*(2), 341–350. https://doi.org/10.1016/j.lwt.2013.11.043.

Mason, T. G., Wilking, J. N., Meleson, K., Chang, C. B., & Graves, S. M. (2006). Nanoemulsions: Formation, structure, and physical properties. *Journal of Physics: Condensed Matter, 18*(41), R635–R666.

Min, S., & Krochta, J. M. (2005). Antimicrobial films and coatings for fresh fruit and vegetables. In W. Jongen (Ed.), *Improving the safety of fresh fruit and vegetables* (pp. 455–492). New York: CRC Press. https://doi.org/10.1533/9781845690243.3.454.

Moreira, M. D. R., Roura, S. I., & Ponce, A. (2011). Effectiveness of chitosan edible coatings to improve microbiological and sensory quality of fresh cut broccoli. *LWT-Food Science and Technology, 44*(10), 2335–2341. https://doi.org/10.1016/j.lwt.2011.04.009.

Murmu, S. B., & Mishra, H. N. (2018). The effect of edible coating based on Arabic gum, sodium caseinate and essential oil of cinnamon and lemon grass on guava. *Food Chemistry, 245*, 820–828. https://doi.org/10.1016/j.foodchem.2017.11.104.

Ncama, K., Magwaza, L., Mditshwa, A., & Tesfay, S. (2018). Plant-based edible coatings for managing postharvest quality of fresh horticultural produce: A review. *Food Packaging and Shelf Life, 16*, 157–167. https://doi.org/10.1016/j.fpsl.2018.03.011.

No, H. K., Meyers, S. P., Prinyawiwatkul, W., & Xu, Z. (2007). Applications of chitosan for improvement of quality and shelf life of foods: A review. *Journal of Food Science, 72*(5), 1750–3841. https://doi.org/10.1111/j.1750-3841.2007.00383.x.

Olivas, G. I., & Barbosa-Cánovas, G. (2009). Edible films and coatings for fruits and vegetables. In K. Huber & M. Embuscado (Eds.), *Edible films and coatings for food applications* (pp. 211–244). New York: Springer. https://doi.org/10.1007/978-0-387-92824-1_7.

Olivas, G. I., Mattinson, D. S., & Barbosa-Canovas, G. V. (2007). Alginate coatings for preservation of minimally processed 'Gala' apples. *Postharvest Biology and Technology, 45*(1), 89–96. https://doi.org/10.1016/j.postharvbio.2006.11.018.

Oms-Oliu, G., Rojas-Graü, M. A., González, L. A., Varela, P., Soliva-Fortuny, R., & Hernando, M. I. H. (2010). Recent approaches using chemical treatments to preserve quality of fresh-cut fruit: A review. *Postharvest Biology and Technology, 57*(3), 139–148. https://doi.org/10.1016/j.postharvbio.2010.04.001.

Paredes-Lopez, O., Camargo-Rubio, E., & Gallardo-Navarro, Y. (1974). Use of coatings of candelilla wax for the preservation of limes. *Journal of the Science of Food and Agriculture, 25*(10), 1207–1210. https://doi.org/10.1002/jsfa.2740251003.

Park, H. J. (2005). Edible coatings for fruit. In W. Jongen (Ed.), *Fruit and vegetable processing*. Boca Raton: CRC Press LLC.

Pereda, M., Marcovich, N., & Ansorena, M.R. (2017). Nanotechnology in food packaging applications: Barrier materials, antimicrobial agents, sensors and safety assessment. In: Dr. Leticia Myriam Torres Martínez, Dr. Oxana Vasilievna Kharissova, Dr. Boris Ildusovich Kharisov (Eds). Handbook of ecomaterials. Springer International Publishing. pp: 1–22. ISBN: 978–3–319–48281–1978–3–319–48281–1. https://doi.org/10.1007/978-3-319-48281-130-1.

Pilon, L., Spricigo, P. C., Miranda, M., de Moura, M. R., Assis, O. B. G., Mattoso, L. H. C., & Ferreira, M. D. (2014). Chitosan nanoparticle coatings reduce microbial growth on fresh-cut apples while not affecting quality attributes. *International Journal of Food Science and Technology, 50*(2), 440–448. https://doi.org/10.1111/ijfs.12616.

Ponce, A. G., Roura, S. I., del Valle, C. E., & Moreira, M. R. (2008). Antimicrobial and antioxidant activities of edible coatings enriched with natural plant extracts: In vitro and in vivo studies. *Postharvest Biology and Technology, 49*(2), 294–300. https://doi.org/10.1016/j.postharvbio.2008.02.013.

Ponce, A. G., Ayala-Zavala, J. F., Marcovich, N. E., Vázquez, F. J., & Ansorena, M. R. (2017). Nanotechnology trends in the food industry: Recent developments, risks and regulation. In A. M. Grumezescu & A. M. Holban (Eds.), *Multi-volume SET (I-XX) Handbook of food bioengineering: Volume 12: Impact of nanoscience in the food industry. Chapter 5* (pp. 113–134). Elsevier Academic Press. ISBN: 978-0-12-811441-4.

Prasad, K., Guarav, A., Preethi, P., & Neha, P. (2018). Edible coating technology for extending market life of horticultural produce. *Acta Scientific Agriculture, 2*, 55–64.

Qadri, O. S., Yousuf, B., & Srivastava, A. K. (2015). Fresh-cut fruits and vegetables: Critical factors influencing microbiology and novel approaches to prevent microbial risks – A review. *Cogent Food & Agriculture, 1*, 1–11. https://doi.org/10.1080/23311932.2015.1121606.

Raybaudi-Massilia, R. M., Mosqueda-Melgar, J., & Tapia, M. S. (2010). Edible coatings as carriers of food additives on fresh-cut fruit and vegetables. *Stewart Postharvest Review, 6*(3), 1–7. https://doi.org/10.2212/spr.2010.3.3.

Rojas-Graü, M. A., Tapia, M. S., Rodríguez, F. J., Carmona, A. J., & Martin-Belloso, O. (2007). Alginate and gellan-based edible coatings as carriers of antibrowning agents applied on fresh-cut Fuji apples. *Food Hydrocolloids, 21*(1), 118–127. https://doi.org/10.1016/j.foodhyd.2006.03.001.

Rojas-Graü, M. A., Soliva-Fortuny, R., & Martín-Belloso, O. (2009a). Edible coatings to incorporate active ingredients to fresh-cut fruit: A review. *Trends in Food Science & Technology, 20*(10), 438–447. https://doi.org/10.1016/j.tifs.2009.05.002.

Rojas-Graü, M., Oms-Oliu, G., Soliva-Fortuny, R., & Martín-Belloso, O. (2009b). The use of packaging techniques to maintain freshness in fresh-cut fruits and vegetables: A review. *International Journal of Food Science and Technology, 44*(5), 875–889. https://doi.org/10.1111/j.1365-2621.2009.01911.x.

Saberi, B., & Golding, J. B. (2018). Postharvest application of biopolymer-based edible coatings to improve the quality of fresh horticultural produce. In T. J. Gutiérrez (Ed.), *Polymers for food applications* (pp. 211–250). Cham: Springer. https://doi.org/10.1007/978-3-319-94625-2_9.

Salinas-Roca, B., Soliva-Fortuny, R., Welti-Chanes, J., & Martín-Belloso, O. (2016). Combined effect of pulsed light, edible coating and malic acid dipping to improve fresh-cut mango safety and quality. *Food Control, 66*, 190–197. https://doi.org/10.1016/j.foodcont.2016.02.005.

Salvia-Trujillo, L., Rojas-Graü, M. A., Soliva-Fortuny, R., & Martín-Belloso, O. (2015). Use of antimicrobial nanoemulsions as edible coatings: Impact onsafety and quality attributes of fresh-cut Fuji apples. *Postharvest Biology and Technology, 105*, 8–16. https://doi.org/10.1016/j.postharvbio.2015.03.009.

Sanchez-Gonzalez, L. (2016). Effect of hydroxypropyl methylcellulose and chitosan coatings with and without bergamot essential oil on quality and safety of cold-stored grapes. *Postharvest Biology and Technology, 60*(1), 57–63. https://doi.org/10.1016/j.postharvbio.2010.11.004.

Sessa, M., Ferraria, G., & Donsì, F. (2015). Novel edible coating containing essential oil nanoemulsions to prolong the shelf life of vegetable products. *Chemical Engineering Transactions, 43*, 55–60. https://doi.org/10.3303/cet1543010.

Severino, R., Vu, K. D., Donsì, F., Salmieri, S., Ferrari, G., & Lacroix, M. (2014). Antibacterial and physical effects of modified chitosan based-coating containing nanoemulsion of mandarin essential oil and three non-thermal treatments against *Listeria innocua* in green beans. *International Journal of Food Microbiology, 191*, 82–88. https://doi.org/10.1016/j.ijfoodmicro.2014.09.007.

Severino, R., Ferrari, G., Vu, K. D., Donsì, F., & Lacroix, M. (2015). Antimicrobial effects of modified chitosan based coating containing nanoemulsion of essential oils, modified atmosphere packaging and gamma irradiation against *Escherichia coli* O157: H7 and *Salmonella typhimurium* on green beans. *Food Control, 50*, 215–222. https://doi.org/10.1016/j.foodcont.2014.08.029.

Shiekh, R. A., Malik, M. A., Al-Thabaiti, S. A., & Shiekh, M. A. (2013). Chitosan as a novel edible coating for fresh fruit. *Food Science and Technology Research, 19*(2), 139–155. https://doi.org/10.3136/fstr.19.139.

Sipahi, R. E., Castell-Perez, M. E., & Moreira, R. G. (2013). Improved multilayered antimicrobial alginate-based edible coating extends the shelf life of fresh-cut watermelon (*Citrullus lanatus*). *LWT-Food Science and Technology, 51*(1), 9–15. https://doi.org/10.1016/j.lwt.2012.11.013.

Tesfay, S. Z., & Magwaza, L. S. (2017). Evaluating the efficacy of moringa leaf extract, chitosan and carboxymethyl cellulose as edible coatings for enhancing quality and extending postharvest life of avocado (*Persea americana* Mill.) fruit. *Food Packaging and Shelf Life, 11*, 40–48. https://doi.org/10.1016/j.fpsl.2016.12.001.

Thompson, A. K. (2010). *Controlled atmosphere storage of fruit and vegetables* (2nd ed.). CABI Chapter 1.

Tiwari, A., Galanis, A., & Soucek, M. D. (2016). *Biobased and environmentally benign coatings. Material degradation and failures series* (pp. 87–120). Wiley.

Tomadoni, B., Moreira, M. R., Pereda, M., & Ponce, A. G. (2018). Gellan-based coatings incorporated with natural antimicrobials in fresh-cut strawberries: Microbiological and sensory evaluation through refrigerated storage. *LWT – Food Science and Technology, 97*, 384–389. https://doi.org/10.1016/j.lwt.2018.07.029.

Ukai, Y. N., Tsutsumi, T., & Marakami, K. (1976). Preservation of agricultural products. U.S. patent 3,997, 674.

Valencia-Chamorro, S. A., Palou, L., del Río, M. A., & Pérez-Gago, M. B. (2011). Antimicrobial edible films and coatings for fresh and minimally processed fruits and vegetables: A review. *Critical Reviews in Food Science and Nutrition, 51*(9), 872–900. https://doi.org/10.1080/10408398.2010.485705.

Vargas, M., Pastor, C., Chiralt, A., McClements, D. J., & González-Martínez, C. (2008). Recent advances in edible coatings for fresh and minimally processed fruits. *Critical Reviews in Food Science and Nutrition, 48*(6), 496–511. https://doi.org/10.1080/10408390701537344.

Viacava, G. E., Ayala-Zavala, J. F., Vázquez, J., & Ansorena, M. R. (2017). Sustainability challenges involved in use of nanotechnology in agro-food sector. In R. Bhat & V. Gómez-López (Eds.), *Sustainability challenges in the agro-food sector* (p. 675). Editorial Wiley-Blackwell.

Wang, B., & Sain, M. (2007). Isolation of nanofibers from soybean source and their reinforcing capability on synthetic polymers. *Composites Science and Technology, 67*(11), 2521–2527. https://doi.org/10.1016/j.compscitech.2006.12.015.

Warth, A. H. (1986). *The chemistry and technology of waxes* (Vol. 37). New York: Reinhold Publishing Corporation.

Wen, P., Zhu, D. H., Feng, K., Liu, F. J., Lou, W. Y., Li, N., ... & Wu, H. (2016). Fabrication of electrospun polylactic acid nanofilm incorporating cinnamon essential oil/β-cyclodextrin inclusion complex for antimicrobial packaging. Food chemistry, 196, 996–1004.

Weiss, J., Takhistov, P., & Mcclements, D. J. (2006). Functional materials in food nanotechnology. *Journal of Food Science, 71*(9), 107–116. https://doi.org/10.1111/j.1750-3841.2006.00195.x.

Yousuf, B., & Srivastava, A. K. (2015). Psyllium (*Plantago*) gum as an effective edible coating to improve quality and shelf life of fresh-cut papaya (*Carica papaya*). *International Journal of Biological, Biomolecular, Agricultural, Food and Biotechnological Engineering, 9*(7), 702–707.

Yousuf, B., Magwazaa, L., & Srivastava, A. K. (2018). Recent developments in shelf-life extension of fresh-cut fruits and vegetables by application of different edible coatings: A review. *LWT – Food Science and Technology, 16*, 157–167. https://doi.org/10.1016/j.lwt.2017.10.051.

Zhong, Y., Whittington, C. F., Zhang, L., & Haynie, D. T. (2007). Controlled loading and release of a model drug from polypeptide multilayer nanofilms. *Nanomedicine: Nanotechnology. Biology and Medicine, 3*(2), 154–160. https://doi.org/10.1016/j.nano.2007.03.002.

Chapter 18
Trends in the Manufacture of Coatings in the Postharvest Conservation of Fruits and Vegetables

Loveleen Sharma, Alok Saxena, and Tanushree Maity

Abstract The deterioration of perishable commodities, such as fresh fruits and vegetables, is a major problem which results in a shorter shelf life with a compromised nutritional quality. Several techniques can be used to minimize the spoilage, such as the use of pesticides and fungicides during pre- harvest conditions, postharvest storage conditions at low temperature, the application of edible coatings along with chemical additives or natural plant extracts and different nanotechnological techniques such as the development of nanocapsules or multilayer systems, etc. Recently, the use of natural plant extracts or essential oils in edible coatings have been in trend. Polysaccharide-based edible films and coatings such as chitosan (Cs), alginate, starch; protein based edible films and coatings such as caseinate, whey protein isolates and lipid-based edible films and coatings such as beeswax, etc. have been used with the incorporation of plant extracts such as pomegranate peel extract, pineapple extract or incorporation of essential oils such as cinnamon oil, citrus oil, etc. to enhance the overall quality of the fruits or vegetables. In chapter the benefits of utilizing this technique in fresh fruits and vegetables contributing to a longer shelf life will be analyzed.

Keywords Active packaging · Edible polymers · Modified atmosphere

18.1 Introduction

The deterioration of fresh fruits and vegetables is a serious concern which results in a shorter shelf life with a compromised nutritional quality. Micro-organisms, insects, pre and post-harvest conditions during transportation and storage are responsible for the maximum losses in quality and quantity of fresh fruits and vegetables. Minimally processed fruits and vegetables are the main growing segments

---

L. Sharma · A. Saxena (✉)
Amity Institute of Food Technology, Amity University Uttar Pradesh, Noida, India

T. Maity
Defence Research and Development Organization, DRDO Bhawan, New Delhi, India

T. J. Gutiérrez (ed.), *Polymers for Agri-Food Applications*,
https://doi.org/10.1007/978-3-030-19416-1_18

in retail food market. Due to the convenience in service and consumption, the demand for fresh-cut products in the market is increasing day by day. However, metabolic activities, such as the delocalization of enzymes, the increase in respiration rate increases due to peeling and cutting, lead to deterioration of products such as browning, softening, microbial growth and off-flavour (Gutiérrez 2019). The development of edible coatings based on biodegradable materials for fruits and vegetables have exhibited favourable results in improving the shelf life of fresh products (Pavlath and Orts 2009; Álvarez et al. 2018). Edible coatings can be considered as an alternative to modified atmosphere storage to increase the postharvest life of fresh fruits and vegetables due to similar storage effects (Park 1999).

Edible coatings are a thin layer of edible material which can be applied directly on the food surface and provide a barrier against moisture, oxygen and solute movement of food (Pavlath and Orts 2009). Edible coatings can be prepared from protein, polysaccharides and lipids, alone or in combination (Álvarez et al. 2017; Garrido et al. 2018). Edible coatings have appeared as a substitute for synthetic plastic for food applications and have received appreciable attention in recent years because of their benefits over synthetic films (Herniou--Julien et al. 2019; Gutiérrez et al. 2019). Coatings are developed from renewable and edible ingredients, so it is anticipated that they will degrade more quickly than polymeric materials. Edible coatings can also be utilized to enhance the organoleptic properties of packaged foods, as long as they contain different components such as flavourings, colourings and sweetening agents (Saberi and Golding 2018). As edible coatings are generally consumed with food products, therefore, the material used in edible coatings should be Generally Recognized as Safe (GRAS) and must comply with the regulations (Bracone et al. 2016; Gutiérrez et al. 2017).

The main role of edible coatings is to restrict migration of oxygen, carbon dioxide (CO_2), moisture, or any other solute materials. The coating should also act as a carrier of different food additives such as antioxidants or antimicrobials, and decrease the decay without any change in the quality of the food product (López-Córdoba 2018).

The functional properties of coatings are improved by the addition of different additives to produce a biodegradable active packaging material (Gutiérrez 2017a, 2018a). Substantial research has been conducted to test natural active compounds as an alternative to synthetic ones, mainly due to safety concerns associated with synthetic active compounds (Gimenez et al. 2012).

18.2 Applications and Functions to Preserve Quality of Fruits and Vegetables

The most recent coating applications are for fruits and vegetables, to reduce moisture loss, prevent weight loss, and improve appearance and to reduce gas transfer rates (Shah et al. 2016; de Aquino et al. 2015; Talens et al. 2012; Riberio et al. 2007;

Gutiérrez and Álvarez 2017; Tapia-Blácido et al. 2018). Fruits and vegetables respire, where cells consume oxygen and produce CO_2. Fruits also produce ethylene responsible for ripening and senescence. The shelf life of fruits and vegetables can be improved by storage in controlled or modified atmospheres of relatively low oxygen and high CO_2 content, which slows respiration rates and reduces ethylene production and ripening process. Coatings can have a similar effect to controlled and modified storages, depending on their formulations. The coatings decrease the gas permeability, developing an internal atmosphere of increased CO_2 and decreased oxygen concentration (Ansorena et al. 2018). The organoleptic properties of packaged foods can also be modified with edible coatings by incorporating different compounds (flavourings, colourings, and sweetening agents) (Gutiérrez and Álvarez 2016). Coatings can act as carriers of many functional ingredients and enhance the shelf life of processed foods by inhibiting microbial deterioration, flavours, rancidity, enzymatic browning, off flavours etc. (Álvarez et al. 2017). These ingredients may include antimicrobial agents, antioxidants, spices, colorants and nutraceuticals (Salmieri and Lacroix 2006; Tapia et al. 2013; Gutiérrez et al. 2016a, b). The concept of bio preservation by coatings has also been adopted (Nair et al. 2018a). The use of natural plant extract such as pomegranate peel extract (PPE) with chitosan (Cs) and alginate enhanced the quality of the guava fruit and increased the shelf life up to 20 days at low temperature storage (Tables 18.1, 18.2 and 18.3).

Table 18.1 Effect of edible coatings with different antimicrobials applied on various fruits or vegetables

Functional ingredients	Coating material	Product	Major effect	Reference
Lemongrass oil	Chitosan (Cs) and Tween	Grape berries	Improved microbiological safety against *Salmonella*	Oh et al. (2017)
Lemongrass oil	Sodium alginate	Fuji apples	Exhibited a faster and greater inactivation of *Escherichia coli* during storage	Salvia-Trujillo et al. (2015)
Citral and Eugenol	Sodium alginate and pectin	Raspberry	Edible coatings enriched with citral and eugenol were effective in reducing microbial spoilage	Guerreiro et al. (2015)
Thyme essential oil	Cs and Gelatin	Black reddish	Exhibited strong antimicrobial activity on the growth of *L. monocytogenes* during storage	Jovanović et al. (2016)
Alginate	Clove oil	Apples	Inhibited growth of *E. coli* in the first week of storage	Rojas-Grau et al. (2008)
Pectin	Transglutaminase-crosslinked whey protein	Carrots	Coating decreased food weight loss and prevented the microbial growth	Marquez et al. (2017)
Lemongrass oil	Cs and Tween	Grape berries	Improved microbiological safety against *Salmonella*	Oh et al. (2017)

Table 18.2 Effect of edible coatings with anti-browning agents and antioxidants applied on various fruits and vegetables

Functional ingredients	Coating material	Commodity	Effect	Reference
Ferulic acid	Soy protein	Fresh cut apples	Coatings prevent weight loss and control firmness.	Álvarez et al. (2017)
Ascorbic acid and citric acid	Sodium alginate and pectin	Apples	Coatings and anti-browning agents maintained most quality attributes.	Guerreiro et al. (2017)
N-acetylcysteine and glutathione	Alginate and gellan	Fresh cut pear	Prevented browning for 2 weeks without affecting firmness of fruit wedges.	Oms-Oliu et al. (2008)
Potassium sorbate, sodium benzoate and nisin	Pectin	Persimmon slices	Microbial growth, colour, firmness, polyphenol oxidase (PPO) activity, visual quality and overall sensory flavour were maintained during storage.	Sanchís et al. (2016)
Calcium and ascorbic acid	Carboxymethyl cellulose	Fresh cut apples	Combination of CMC and AA controled surface browning and maintained total antioxidant, vitamin C and fruit quality attributes during storage.	Saba and Sogvar (2016)
Green tea extract	Chitosan	Fresh walnut kernel	Coating provided effective inhibition of lipid oxidation.	Sabaghi et al. (2015)
Tea polyphenols	Alginate	Chinese winter jujube	The decline of total chlorophylls content, ascorbic acid and total phenol content was delayed.	Zhang et al. (2016)

18.3 Components of Edible Coatings

The main components of edible coatings are polymers, proteins and polysaccharides (Sandhu et al. 2017). Sometimes, coatings made from waxes and resins, with inclusion of polymers to form either bilayer or composite coatings have been used (Sharma et al. 2018). Apart from polymers and waxes, other ingredients also have prime role in the development of coatings. Plasticizers such as propylene glycol, glycerol, or sorbitol are added to control the flexibility and viscosity of the coatings (Gutiérrez et al. 2015a, b). Most coatings are made up of solvent, ordinarily water or aqueous alcohol (Bai et al. 2003; Hagenmaier 2004; Riberio et al. 2007).

Table 18.3 Effect of edible coatings with different plant extracts applied on various fruits and vegetables

Functional ingredient	Coating material	Commodity	Major effect	Reference
Pomegranate peel extract	Alginate and chitosan (Cs)	Capsicum	Antifungal activity against strains of *Colletotrichum gloeosporioides*	Nair et al. (2018a)
Pomegranate peel extract	Alginate and Cs	Guava	Prolongation in shelf life of fruit up to 20 days	Nair et al. (2018b)
Pomegranate peel extract	Cs and locust bean gum	Oranges	Inhibition in growth of *P. digitatum*	Kharchoufi et al. (2018).
			Enhancement in bio-control efficacy	
Moringa plant extract	Carboxyl methycellulose	Avocado	Reduced rate of ripening and inhibition in the growth of postharvest pathogens	Tesfay et al. (2017)
Green tea extracts	Alginate-oleic acid	Strawberries and raspberries	Exhibited anti-viral activities against murine norovirus (MNV) and HAV	Falco et al. (2019)
Ginseng extract	Guar gum	Sweet cherries	Improvement in the nutrient content and quality of cherries	Dong and Wang (2018)
Lotus leaf extract	Sodium alginate and starch	Fresh goji	Exhibit inhibitory effect on microbial growth and reduced weight loss during storage	Fan et al. (2019)
			Prolong shelf life.	

18.4 Protein Based Coatings

The properties of the proteins made them an exceptional material for the development of coatings (Garrido et al. 2018). Proteins are polymers made up of distinct amino acid sequences and molecular structure. The structures (secondary, tertiary and quaternary) of proteins can be easily altered to ameliorate the protein configuration and protein interactions, which results in coating formation. Different factors affecting the interactions during coating formations, such as the interaction between proteins and small molecules, which include water, plasticizer, lipids, waxes and other additives, physical treatments and enzymatic treatments, effect the mechanical and barrier properties of the coatings (Park et al. 2002). Protein coating materials can be obtained from various plant or animal sources. The different proteins, such as corn and wheat based, casein, whey proteins are described below:

18.4.1 Zein Based Edible Coatings

Maize or corn (*Zea mays*) is one of the most important food and industrial crops. Zein is a protein and a biopolymer from corn and can be converted into biodegradable plastics due to its unique properties (Biswas et al. 2009; Tihminlioglu et al. 2010). The group of alcohol-soluble proteins are found in the corn endosperm with a high content of non-polar amino acids, i.e. leucine, alanine and proline. Zein coatings are generally developed in alcohol or aqueous solutions of alcohol. A plasticizer is required in the development of zein-based coatings since they are extremely brittle after drying on surface of product. Plasticizer commonly used for preparation of zein coatings are glycerol, polyethylene glycol, polypropylene glycol and different fatty acids. Santos et al. (2018) e.g. improved the storage stability of guava by the application of tannic acid-crosslinked zein coatings. Both the unmodified and modified coatings proved to be effective in extending the shelf life of guavas. The effect of coatings on visual appearance, weight loss, total solids and respiration rate of guava fruit stored at 23 °C was observed. The cross-linking of zein with tannic acid improved the barrier properties of the protein and is responsible for the reduction of the gas exchange rate of coated guava fruits during storage.

18.4.2 Whey Protein-Based Coating

Whey protein is the remaining portion in the milk serum after the cheese is manufactured (Ramos et al. 2012). Whey proteins are divided into two categories: whey protein concentrates and whey protein isolates, depending on their industrial use. The heat denaturation of whey protein is necessary for the formation of films. The heating improves the three-dimensional structure and exposes the internal SH and hydrophobic groups. Ramos et al. (2012) investigated the antimicrobial activity of whey protein films induced with various antimicrobial agents. Chito oligosaccharide and lactic acid exhibited an antimicrobial effect against Gram –ve and Gram +ve bacteria. The surface of the coated cheese was more resistant to bacterial growth, whereas it was less effective against yeast and mould. In addition, effect of whey protein purity and glycerol content were studied by Ramos et al. (2013). The films were developed from whey protein concentrate and isolated from whey protein and glycerol at three different levels. The films were characterized in terms of molecular structure, as well as mechanical, thermal and barrier properties. It was observed that the whey protein films had good barrier and mechanical properties compared to the soy, zein and wheat protein films. Films plasticized at 40 and 50% glycerol exhibited overall better properties. The thermal properties were determined by differential scanning calorimetry (DSC) and thermal gravimetric analysis (TGA), which revealed that whey proteins exhibited excellent thermal properties and were comparable to synthetic films. Coatings prepared from transglutaminase-crosslinked whey protein were found to be effective in enhancing the shelf life of fresh-cut apples, potatoes and carrots for up to 10 days and prevented the microbial growth (Marquez et al. 2017).

18.4.3 Soy Protein-Based Coatings

Proteins from soy protein isolate (SPI) ranged from 200 to 600 kDa. Globulins (~90%) are the main soy proteins, and are classified according to their sedimentation coefficients (Cho and Rhee 2004). A two-step process that included denaturing the protein by heat and then dehydrating the surface to prepare films from the soy protein isolate was carried out by Kokoszka et al. (2010). These authors found that the intermolecular interactions in the protein were unfolded, thus being linked by the formation of networks during drying. Alves et al. (2017) also showed that soy protein-based coatings with ferulic acid exhibited a significant improvement in the shelf-life of fresh-cut apples during seven days of storage at 10 °C. Zhang et al. (2018) studied the impact of soybean protein isolate (SPI)-Cs based edible coatings on the softening of apricot fruit during storage. Several parameters were studied, including weight loss, firmness, titrable acidity, soluble solid content (SSC) and nanostructural properties of pectin to analyse the effectiveness of coatings. The atomic force microscopy (AFM) and nanostructural properties indicated SPI-Cs coatings could inhibit pectin degradation and be effective in preserving the quality of apricoats.

18.4.4 Sesame Protein-Based Coatings

Sesame flour has a wide range of use as a protein source or as an ingredient in the food industry. Sesame flour (protein content of 40–60%) and isolated protein have particularly high contents of methionine (2.4–4.0%) and total sulphur-containing amino acids (3.8–5.5%) (Saini et al. 2018). Better barrier properties were determined by Sharma and Singh (2016) for films prepared from sesame protein as compared to soy, peanut, whey and mung bean protein films. The effectiveness of the bilayer coating was also analyzed by Sharma et al. (2018) for coatings based on native sesame protein and organic acid crosslinked-sesame protein to enhance the quality and increase the shelf-life of fresh-cut pineapple. The cut pineapple fruit was coated by using a bilayer technique with a dipping method and stored in polystyrene trays for 15 days at 5 °C. The bilayer coatings based on native/crosslinked sesame protein and calcium chloride-pineapple juice extract exhibited an effect on the drip loss, the firmness, the titratable acidity (TA) and the colour of fresh-cut pineapple fruit as compared to noncoated samples. Coatings showed a non-significant effect on the total sugars and the reducing sugars during the 15 days of storage. The coatings were effective in controlling microbial spoilage as total plate count was less for all coated samples compared to non-coated samples during storage (Sharma et al. 2018). The biocomposite edible bi-layer coatings were also developed and analysed by Sharma et al. (2019) from native/crosslinked sesame protein and mango puree with the addition of guar gum and calcium chloride to evaluate the shelf life stability of fresh-cut mango fruit for 15 days. These authors demonstrated that all the coatings evaluated had the potential to enhance the shelf life of fresh-cut mango fruit. The isolate of sesame protein cross-linked with organic acids can act as a better and cost-effective substitute for coatings obtained from other sources.

18.5 Polysaccharide Based Edible Coatings

Polysaccharide based coatings are composed of starch, alginate, cellulose ethers, Cs, carrageenan, or pectin and have film forming abilities with solvents (Gutiérrez 2018b, c). Coatings from polysaccharides impart excellent gas barrier properties due to presence of the polymer chains (Dehghani et al. 2018). The demand and use of polysaccharides have increased in recent years for the development of edible coatings, especially in the field of dairy, bakery, meat products, ready-to-eat meals, fresh and fresh-cut fruits and vegetables. The polysaccharides widely used in coating formulations are starch and cellulose derivatives, and other sources such as gums and pectin are also gaining importance (Suárez and Gutiérrez 2017).

18.5.1 Cellulose and Derivatives

Cellulose materials are biodegradable; therefore, they have gained attention in recent decades as a prospective substitute for synthetic polymers (Gutiérrez and Alvarez 2017a). Regenerated cellulose films such as cellophane have not been used for food applications as compared to their synthetic counterparts due to poor water vapour permeability (WVR). While transparent cellulose films have shown gas barrier properties in dry conditions and are comparable to low density polyethylene (LDPE) and high-density polyethylene (HDPE) films (Su et al. 2010). The main cellulose and cellulose derivatives used for the development of biodegradable/edible films are microcrystalline cellulose, nanocrystalline cellulose, microfibrillated and nanofibrillated cellulose, methylcellulose (MC), hydroxypropylcellulose (HPC), hydroxylpropyl methylcellulose (HPMC) and carboxy methyl cellulose.

The microcrystalline or nano-crystalline cellulose is developed by acid hydrolysis of native cellulose to disrupt the reactive crystalline regions like fine crystals. Microfibril and nanofibril cellulose is prepared by passing the native cellulose at high pressure and shear force. The carboxy, methoxy and hydroxypropyl groups also affect the physical and chemical properties of cellulose. Sayanjali et al. (2011) e.g. evaluated the antimicrobial and physical properties of edible coatings based on carboxymethyl cellulose containing potassium sorbate in fresh pistachios.

The antimicrobial effects were studied against *Aspergillus* species using an agar diffusion assay. The pistachios were coated with an antimicrobial film with three concentrations (1, 0.5 and 0.25/100 ml of film solution) of sorbate. All concentrations did not show mould growth. WVP values were increased with the increase in the concentration of sorbate. Boumail et al. (2016) studied the effect of coatings on microbiological, sensorial and physico-chemical properties of pre-cut cauliflowers. The coatings were prepared from methylcellulose, maltodextrin and starch with antimicrobial formulations of citral extract, lactic acid and lemongrass essential oil. Six coatings containing different ratios of polysaccharides and antimicrobials were prepared and tested *in vitro* against *Listeria monocytogenes*. All the selected coatings exhibited total *in vitro* inhibition of bacteria. Choi et al. (2016) evaluated the effect of HPMC-based edible coatings enriched with oregano and bergamot essential oils on the quality attributes of 'Formosa' plums. The coating at the 2% level was effective in decreasing the respiration rate, ethylene production, total weight loss and total cell count, compared to the plums treated with other coatings at 23 ± 2 °C. In addition, 2% coated plum was firmer and exhibited less surface colour change than the control fruit at the end of storage.

18.5.2 Starch and Derivatives

Starch is composed of amylose and amylopectin (Gutiérrez and Alvarez 2017b, c, d, e). Both are comprised of glucose but differ in size and shape. Amylose is a linear molecule with (1–4)-α-D bonds of glycosidic units. Amylopectin is a branched macromolecule with glucose units that are linked in a linear way with α (1 → 4) glycosidic bonds (Gutiérrez et al. 2014). Branching takes place with α (1 → 6) bonds that are produced every 24–30 glucose units, emerging as a soluble molecule that can be quickly degraded as it has many endpoints to which enzymes can attach (Rodriguez et al. 2006). On the other hands, amylose contains very few α (1 → 6) bonds, or even none, which causes insolubility or slow hydrolysis of amylose.

Starch is a very common food hydrocolloid due to its low cost. The native and modified form of starch can provide a wide range of functional properties in coatings. Tubers, legume, cereal grains and certain fruits contains up to 30–85% starch (dry basis) (Zobel and Stephen 2006). Starch exists as insoluble granules with a certain crystallinity and characteristic shape in its native state. The film-forming capacity and mechanical properties of aqueous native maize starches was evaluated by Palviainen et al. (2001). Saberi et al. (2016) recently optimized the physical and optical properties of coatings based on pea starch and guar gum. The effect of the processing variables (pea starch, guar gum and glycerol) on transparency, viscosity, solubility, moisture content, hunter parameters and total colour difference were determined. The linear effect of all three processing levels significantly influenced the response variables. The results showed that pea starch/guar gum films were successfully produced and can be used more for food packaging purposes.

18.5.3 Pectin and Derivative-Based Coatings

Pectins are the group of polysaccharides present as the main component of all cell walls. Pectin is composed of (1–4) α-D-galacturonic acid units with single (1–2) α-L-rhamnose residues (Ridley et al. 2001). The galactouronic acid units contain the carboxyl groups, which are partly esterified by methyl groups. The degree of esterification (DE) represents the dispersion of the methylated groups along the polysaccharide chains, as well as the number of methyl esterified-galacturonic acid residues *versus* the total number of galacturonic acid units (Solvia-Fortuny et al. 2012). Pectins to form gels with high methoxyl content require a low pH (2.5–3.5) and a minimum amount of SSC (55–88%) (Lopes da Silva and Rao 2006). Guerreiro et al. (2016) studied the effect of pectin and alginate-based coatings enriched with citral and eugenol essential oil on the capacity and the nutritional and sensory quality of the raspberry. Colour, firmness, SSC, weight loss, microbial growth, phenolic compounds (total phenolics, flavonoids, anthocyanins), sugars, organic acids, antioxidant activity (TEAC and ORAC), acetaldehyde and CO_2 production were determined during 0, 5, 10 and 15 days of storage. All the coatings were efficient to control the microbial spoilage of the food. The acceptability of the coated raspberries was good up to 14 days, while control fruit were acceptable only until 7 days.

18.5.4 Seaweed Extracts

18.5.4.1 Alginates

Alginate is an attractive film-forming compound because of its non-toxicity, biodegradability, and low price (Vu and Won 2013). Several authors have studied their functional properties, suspension, thickening, stabilization, film formation, gel-producing and emulsion-stabilizing (Dhanapal et al. 2016; Zactiti and Kieckbusch 2006). Seaweed is broken into two pieces and stirred in a hot alkali solution to extract the alginate. The alginate is dissolved in this solution and extracted as sodium alginate. Lastly, alginate is separated by precipitation as alginic acid or calcium alginate (McHugh 2003). Rojas-Grau et al. (2007) developed the apple puree-alginate coatings as carriers of antimicrobials agents to enhance the shelf life of fresh cut apples. Lemongrass, oregano oil and vanillin incorporated into apple puree alginate edible coatings as antimicrobial agents and their effect was determined. All antimicrobial coatings significantly inhibited the growth of psychrophilic aerobes, yeasts and moulds (Table 18.2). Alginate based edible coatings with antibrowning agents were developed to maintain the sensory and antioxidant properties of fresh cut pears (Oms-Oliu et al. 2008). *N*-acetylcysteine and glutathione were used as antibrowning agents. Both anti-browning agents inhibited the browning of fresh cut pears for 2 weeks and also reduced the microbial growth. An increase in the vitamin C content and polyphenols was observed in case of coated fresh cut pears as compared to the non-coated sample. Azarakhsh et al. (2012) optimized the alginate-gellan based edible coatings for fresh-cut pineapples. The influence of alginate-based (sodium alginate 0–2% (w/v), glycerol 0–2% (w/v) and sunflower oil 0.025% (w/v)) and gellan-based (gellan 0–1% (w/v), glycerol 0–1% (w/v) and sunflower oil 0.025% (w/v)) edible coatings on fresh-cut pineapple were studied by response surface methodology (RSM). The authors concluded that the models were significant ($p < 0.05$) for all responses.

18.5.4.2 Carrageenan Based Edible Coatings

Carrageenan has a great potentiality as a film-forming material because it consists of water-soluble polymers with a linear chain of partially sulphated galactans. The extraction of these sulphated polysaccharides can be carried out from the cell walls of various red seaweeds (*Rhodophyceae*) (Riberio et al. 2007). Different types of carrageenans were extracted from different seaweeds. Carrageenan as edible coatings have various applications on fresh and frozen meat, poultry and fish, sausage-casings, dry solid foods, granulation coated powders and fatty foods (Lee 2003).

18.5.5 Chitosan (Cs)

Chitosan (Cs) is developed by deacetylation of chitin, which is the main component of the exoskeleton of crustaceans, as well as a cell wall of green algae and fungi (Gutiérrez 2017b; Merino et al. 2018). Chitin is formed of β-1, 4 linked linear polymers of 2-acetamido-2-deoxy-D-glucopyranosyl residues (Zhang et al. 2018). Cs has been used substantially as coating applications due to its film-forming properties. Cs is insoluble in water, so coatings from Cs are also insoluble in water and exhibit good wet tensile properties. Jafarizadeh et al. (2011) evaluated and concluded that the ripening process of banana by applying Cs-glycerol based coatings was delayed. RSM was used as an experimental design to optimize the concentrations of Cs and glycerol for coating of Berangan banana (*Musa sapientum* cv. *Berangan*). The effects of main edible coating components, Cs (0.5–2.5%, w/w) and glycerol (0–2%, w/w) on the firmness, weight loss, total colour difference, TA and total soluble solids content (TSS) of coated banana were determined during 10 days of storage at 26 ± 2 °C and 40–50% relative humidity (Jafarizadeh et al. 2011). The results in this study exhibited that the Cs concentration appeared to be the most significant factor influencing all the variables, except for TSS. It was predicted that the optimal concentration of Cs and glycerol would be 2.02 and 0.18%, respectively. Guerra et al. (2016) developed edible composite coatings from Cs and Mentha (*piperita* L. or *x villosa huds*) essential oil and studied the effect on the appearance of mould in the postharvest and the quality of the table grape cv. *Isabella*. The effect of coatings on physicochemical and sensory characteristics of table grapes was also determined. The coatings were effective in delaying the mould growth on table grapes. Parameters for colour and firmness were enhanced in coated fruits. Coatings did not significantly affect the physicochemical properties of the table grapes. Nair et al. (2018b) developed Cs and alginate coatings with pomegranate peel extract (PPE) to evaluate the effectiveness on capsicum (*Capsicum annuum* L.). PPE contains polyphenols such as punicalagin, catechin and rutin and also exhibited antifungal activity against the fungal strain of *Colletotrichum gloeosporioides*. Cs and alginate coatings with PPE were effective to inhibit microbial growth, enhancing shelf life and maintaining the sensory properties up to 25 days at 10 °C (Nair et al. 2018b) (Table 18.3).

18.5.6 Gums

Basil seed gum (BSG) is a typical gum which can be isolated from *Ocimum basilicum* L, a member of genus Ocimum. The extracted BSG as a hetero-polysaccharide is composed of two main fractions: commonly known as glucomannan (43%) along with highly branched arabinogalactan, as the hydrophobic fraction, and xylan (24.29%), the accountable part for its hydrophilic properties. The hydrophilic portion can absorb water during soaking and swell in the mucilage due to the presence of a polysaccharide substance (Hosseini-Parvar et al. 2010). Although BSG has various advantages, such as low production cost, high availability, extraction convenience and hydrophilic properties, its applications in the food industry particularly for food packaging, are not well manifested (Hosseini-Parvar et al. 2010; Karimi and Kenari 2016). Hashemi et al. (2017) developed BSG based edible coatings containing Origanum vulgare subsp. viride essential oil for the preservation of fresh cut apricots. The chemical attributes, the microbial load and the sensory characteristics of coated fresh cut apricot during cold storage at 4 °C for 8 days were determined. Among all treatments tested, the BSG + 6% essential oil was analysed as the most effective to reduce the microbial populations of apricot cuttings.

Xanthan gum (XG), produced as an exo-polysaccharide by Xanthomonas campestris under unfavourable conditions, is a GRAS compound for use as an emulsifier or stabilizer. XG develops a highly viscous solution in hot or cold water at low concentration with excellent stability in a variable range of temperature and pH. It is stable to enzymatic degradation. In complex formulations it also accelerates particle suspension, even for a long time (Sworn and gum 2009). Sharma and Rao (2015) investigated the effect of XG-based edible coatings containing cinnamic acid on the quality of fresh cut pears. The coatings were effective to prevent the browning and extend the shelf life. The enrichment of cinnamic acid, as an antioxidant into XG-based edible coating exhibited a significant decline of ascorbic acid, retardation of oxidative browning, reduction of antioxidant capacity and degradation of total phenolic content compared to uncoated fresh-cut pears and coated with only XG. Zambrano-Zaragoza et al. (2017) evaluated the effect of β-carotene release rate from nanocapsules incorporated XG coatings on physical and physicochemical properties of fresh cut melon. Coating formulations were prepared from XG alone,

XG combined with nanocapsules (Ncs/XG), XG combined with nanospheres (Nsp/XG), nanocapsules (Ncs), and nanospheres (Nsp) and compared to untreated fresh-cut melons in order to evaluate its preservation efficiency. The incorporation of β-carotene nanocapsules into a polysaccharide matrix improves the properties of the coatings, thereby increasing storage time to 21 days at 4 °C.

18.6 Effect of Coatings on Whole and Fresh-Cut Fruits and Vegetables

The sensory, nutritional, textural and safety aspects are the basic parameters for the quality evaluation of fruits and vegetables. In this section, the effect of coatings with different plant extracts on whole and fresh-cut fruits, the effect on sensory and nutritional properties and the effect on aroma and volatile components is discussed.

18.6.1 Effect of Coatings With Different Plant Extracts Applied on Various Fruits and Vegetables

The addition of natural plant extracts in edible coatings and their application on different fruits and vegetables is gaining popularity over the use of chemical preservatives. Aloui et al. (2014) improved the overall quality of grapes with the application of alginate coatings enriched with grape seed extract. Tesfay et al. (2017) reported that Moringa leaf extract in carboxyl methyl cellulose-based coatings reduced the rate of ripening and inhibited the growth of postharvest pathogens. Nair et al. (2018a, b) concluded that pomegranate peel extract-enriched alginate and Cs-based coatings act as antifungal against strains of *Colletotrichum gloeosporioides* in the case of capsicum and prolong the shelf life of the guava fruit for up to 20 days. Kharchoufi et al. (2018) found that the pomegranate peel extract also inhibited the growth of *Penicillium digitatum* in oranges coated with Cs and locust bean gum. Foodborne diseases due to human enteric viruses, such as human norovirus (NOVs) and the hepatitis A virus (HAV) with the consumption of contaminated berries is a major concern for food safety. Falco et al. (2019) coated strawberries and raspberries with alginate-oleic acid and green tea extracts exhibiting anti-viral activities against NOVs and HAV.

18.6.2 Effect on Sensory and Nutritional Properties Applied on Various Fruits and Vegetables

The effect of edible coatings and storage conditions on the sensory quality of fresh-cut products has been scrutinized and colour is also one of the main quality attributes of fresh-cut fruits and vegetables that can change during storage. Browning is the biggest concern in extending the shelf life of fresh-cut fruits, which greatly influences consumer acceptance. Appearance and freshness are used to judge the quality of whole and fresh-cut fruits and vegetables (Kader 2002). Anti-browning agents are being incorporated into edible coatings to inhibit browning, e.g. ferulic acid has been used to prevent browning in fresh-cut apples with the use of soy protein-based coatings, which also reduced the weight loss and control firmness (Alves et al. 2017). According to Oms-oliu et al. (2008) *N*-acetyl cysteine and glutathione in alginate and gellan-based coatings also prevented browning for 2 weeks without affecting firmness of fruit wedges in fresh-cut pears. Following Sanchís et al. (2016) potassium sorbate, sodium benzoate and nisin in pectin coatings prevent microbial growth, colour, firmness, polyphenol oxidase activity, visual quality and overall sensory flavour during storage in persimmon slices.

18.6.3 Effect of Antimicrobial Coatings Applied on Various Fruits and Vegetables

The minimal processing steps and the elimination of the natural protection of fruits and vegetables are the main responsible for microbial spoilage. Cross-contamination during shredding and cutting operations, damaged tissues/plant tissues are also susceptible to pathogenic. Oh et al. (2017) using Cs and sodium alginate coatings enriched with lemongrass oil improved microbial safety against *Salmonella* and exhibited the growth of *E. coli* during storage in grape berries and fuji apples. Guerreiro et al. (2015) reduced the microbial spoilage of raspberry during storage by using edible coatings based on alginate and pectin enriched with citral and eugenol. Jovanovic et al. (2016) achieved to strongly exhibit the antimicrobial activity on growth of *L. monocytogenes* during storage using Cs-based coatings containing thyme essential oil. Rojas-Grau et al. (2008) concluded that alginate coatings enriched with clove oil were effective in inhibiting the growth of *E. coli* in the first week of storage.

18.7 Conclusions

The development of innovative technologies to improve the quality and shelf life of post-harvest products is a subject of concern to the food industries. The use of biodegradable or edible packaging materials is opening a potential economic advantage for farmers and agricultural processors. Bilayer or composite coatings should be developed in a similar way to synthetic packaging materials with excellent mechanical and barrier properties. It is necessary to adopt innovative techniques to preserve food safety, structural integrity and biodegradability of post-harvest food. The cost-effective edible materials could thus replace the synthetic polymers, as well as improve the shelf life and nutritional value of whole and fresh-cut fruit and vegetables. For this reason, much of the research conducted in the area of food is currently concentrated and focused on this topic.

Conflicts of Interest The authors declare no conflict of interest.

References

Aloui, H., Khwaldia, K., Sánchez-González, L., Muneret, L., Jeandel, C., Hamdi, M., & Desobry, S. (2014). Alginate coatings containing grapefruit essential oil or grapefruit seed extract for grapes preservation. *International Journal of Food Science & Technology, 49*(4), 952–959.

Álvarez, K., Famá, L., & Gutiérrez, T. J. (2017). Chapter 12. Physicochemical, antimicrobial and mechanical properties of thermoplastic materials based on biopolymers with application in the food industry. In M. Masuelli & D. Renard (Eds.), *Advances in physicochemical properties of biopolymers: Part 1* (pp. 358–400). Bentham Science Publishers. EE.UU. ISBN: 978-1-68108-454-1. eISBN: 978-1-68108-453-4. https://doi.org/10.2174/97816810845341170100015.

Álvarez, K., Alvarez, V. A., & Gutiérrez, T. J. (2018). Chapter 3. Biopolymer composite materials with antimicrobial effects applied to the food industry. In V. K. Thakur & M. K. Thakur (Eds.), *Functional biopolymers* (pp. 57–96). Editorial Springer International Publishing. EE.UU. ISBN: 978-3-319-66416-3. eISBN: 978-3-319-66417-0. https://doi.org/10.1007/978-3-319-66417-0_3.

Alves, M. M., Goncalves, M. P., & Rocha, C. M. (2017). Effect of ferulic acid on the performance of soy protein isolate-based edible coatings applied to fresh-cut apples. *LWT- Food Science and Technology, 80*, 409–415. https://doi.org/10.1016/j.lwt.2017.03.013.

Ansorena, M. R., Pereda, M., & Marcovich, N. E. (2018). Edible films. In T. J. Gutiérrez (Ed.), *Polymers for food applications* (pp. 5–24). Cham: Springer. https://doi.org/10.1007/978-3-319-94625-2_2.

Azarakhsh, N., Osman, A., Ghazali, H. M., Tan, C. P., & Mohd Adzahan, N. (2012). Optimization of alginate and gellan-based edible coating formulations for fresh-cut pineapples. *International Food Research Journal, 19*(1), 279–285.

Bai, J., Alleyne, V., Hagenmaier, R. D., Mattheis, J. P., & Baldwin, E. A. (2003). Formulation of zein coatings for apples (*Malus Domestica* Borkh). *Postharvest Biology and Technology, 28*(2), 259–268. https://doi.org/10.1016/s0925-5214(02)00182-5.

Biswas, A., Selling, G. W., Woods, K. K., & Evans, K. (2009). Surface modification of zein films. *Industrial Crop Production, 30*(1), 168–171. https://doi.org/10.1016/j.indcrop.2009.02.002.

Boumail, A., Salmieri, S., St-Yves, F., Lauzon, M., & Lacroix, M. (2016). Effect of antimicrobial coatings on microbiological, sensorial and physico-chemical properties of pre-cut cauliflowers. *Postharvest Biology and Technology, 116*, 1–7. https://doi.org/10.1016/j.postharvbio.2015.12.017.

Bracone, M., Merino, D., González, J., Alvarez, V.A., Gutiérrez, T.J. (2016). Chapter 6. Nanopackaging from natural fillers and biopolymers for the development of active and intelligent films. In: *Natural polymers: Derivatives, blends and composites*. Ikram S. and Ahmed S. (Eds). Editorial Nova Science Publishers, Inc. New York. EE.UU. ISBN: 978-1-63485-831-1. pp. 119–155.

Cho, S. Y., & Rhee, C. (2004). Mechanical properties and water vapor permeability of edible films made from fractionated soy proteins with ultrafiltration. *LWT - Food Science and Technology, 37*(8), 833–839.

Choi, W. S., Singh, S., & Lee, Y. S. (2016). Characterization of edible film containing essential oils in hydroxypropyl methylcellulose and its effect on quality attributes of 'Formosa' plum (*Prunus salicina* L.). *LWT- Food Science and Technology, 70*, 213–222. https://doi.org/10.1016/j.lwt.2016.02.036.

de Aquino, A. B., Blank, A. F., & de Aquino Santana, L. C. L. (2015). Impact of edible chitosan–cassava starch coatings enriched with Lippia gracilis Schauer genotype mixtures on the shelf life of guavas (*Psidium guajava* L.) during storage at room temperature. *Food Chemistry, 171*, 108–116.
Dehghani, S., Hosseini, S. V., & Regenstein, J. M. (2018). Edible films and coatings in seafood preservation: A review. *Food Chemistry, 240*, 505–513. https://doi.org/10.1016/j.foodchem.2017.07.034.
Dhanapal, A., Sasikala, P., Rajamani, L., Kavitha, V., Yazhini, G., & Shakila Banu, M. (2016). Edible films from polysaccharides. *Food Science and Quality Management, 3*, 9–18.
Dong, F., & Wang, X. (2018). Guar gum and ginseng extract coatings maintain the quality of sweet cherry. *LWT- Food Science and Technology, 89*, 117–122. https://doi.org/10.1016/j.lwt.2017.10.035.
Falco, I., Flores-Meraz, P. L., Randazzo, W., Sanchez, G., Lopez-Rubio, A., & Fabra, M. J. (2019). Antiviral activity of alginate-oleic acid based coatings incorporating green tea extract on strawberries and raspberries. *Food Hydrocolloids, 87*, 611–618. https://doi.org/10.1016/j.foodhyd.2018.08.055.
Fan, X. J., Zhang, B., Yan, H., Feng, J. T., Ma, Z. Q., & Zhang, X. (2019). Effect of lotus leaf extract incorporated composite coating on the postharvest quality of fresh goji (*Lycium barbarum* L.) fruit. *Postharvest Biology and Technology, 148*, 132–140. https://doi.org/10.1016/j.postharvbio.2018.10.020.
Garrido, T., Uranga, J., Guerrero, P., & de la Caba, K. (2018). The potential of vegetal and animal proteins to develop more sustainable food packaging. In T. J. Gutiérrez (Ed.), *Polymers for food applications* (pp. 25–59). Cham: Springer. https://doi.org/10.1007/978-3-319-94625-2_3.
Gimenez, B., Gómez-Guillén, M. C., López-Caballero, M. E., Gómez Estaca, J., & Montero, P. (2012). Role of sepiolite in the release of active compounds from gelatin-egg white films. *Food Hydrocolloids, 27*(2), 475–486. https://doi.org/10.1016/j.foodhyd.2011.09.003.
Guerra, I. C. D., Lima, P., Lima, P. D., Fernandez, M. M., Carneiro, A. S. S., Tavares, J. F., Barbosa, J. M., Madruga, M. S., & Leite, E. (2016). The effects of composite coatings containing chitosan and *Mentha* (*piperita* L. or *x villosa Huds*) essential oil on postharvest mold occurrence and quality of table grape cv. Isabella. *Innovative Food Science & Emerging Technologies, 34*, 112–121. https://doi.org/10.1016/j.ifset.2016.01.008.
Guerreiro, A. C., Gago, C., Faleiro, M., Miguel, M., & Antines, M. (2015). Raspberry fresh fruit quality as affected by pectin-and alginate-based edible coatings enriched with essential oils. *Scientia Horticulturae, 194*, 138–146. https://doi.org/10.1016/j.scienta.2015.08.004.
Guerreiro, A. C., Gago, C., Miguel, M., Faleiro, M., & Antines, M. (2016). The influence of edible coatings enriched with citral and eugenol on the raspberry storage ability, nutritional and sensory quality. *Food Packaging and Shelf Life, 9*, 20–28. https://doi.org/10.1016/j.fpsl.2016.05.004.
Guerreiro, A. C., Gago, C. M. L., Faleiro, M. L., Miguel, M. G. C., & Antunes, M. D. C. (2017). The effect of edible coatings on the nutritional quality of 'bravo de Esmolfe' fresh-cut apple through shelf-life. *LWT, 75*, 210–219.
Gutiérrez, T. J. (2017a). Surface and nutraceutical properties of edible films made from starchy sources with and without added blackberry pulp. *Carbohydrate Polymers, 165*, 169–179. https://doi.org/10.1016/j.carbpol.2017.02.016.
Gutiérrez, T. J. (2017b). Chapter 8. Chitosan applications for the food industry. In S. Shakeel Ahmed & S. Ikram (Eds.), *Chitosan: Derivatives, composites and applications* (pp. 183–232). WILEY-Scrivener Publisher. EE.UU. ISBN: 978-1-119-36350-7. https://doi.org/10.1002/9781119364849.ch8.
Gutiérrez, T. J. (2018a). Active and intelligent films made from starchy sources/blackberry pulp. *Journal Polymers and the Environment, 26*(6), 2374–2391. https://doi.org/10.1007/s10924-017-1134-y.
Gutiérrez, T. J. (2018b). Biological macromolecule composite films made from sagu starch and flour/poly(ε-caprolactone) blends processed by blending/thermo molding. *Journal Polymers and the Environment, 26*(9), 3902–3912. https://doi.org/10.1007/s10924-018-1268-6.

Gutiérrez, T. J. (2018c). Are modified pumpkin flour/plum flour nanocomposite films biodegradable and compostable? *Food Hydrocolloids, 83*, 397–410. https://doi.org/10.1016/j.foodhyd.2018.05.035.

Gutiérrez, T. J. (2019). Chapter 19. Antibiofilm enzymes as an emerging technology for food quality and safety. In M. Kuddus (Ed.), *Enzymes in food biotechnology: Production, applications, and future prospects* (pp. 321–342). Editorial Academic Press. EE.UU. ISBN: 978-0-12813-280-7. https://doi.org/10.1016/b978-0-12-813280-7.00019-0.

Gutiérrez, T. J., & Álvarez, K. (2016). Physico-chemical properties and *in vitro* digestibility of edible films made from plantain flour with added *Aloe vera* gel. *Journal of Functional Foods, 26*, 750–762. https://doi.org/10.1016/j.jff.2016.08.054.

Gutiérrez, T. J., & Álvarez, K. (2017). Chapter 4. Transport phenomena in biodegradable and edible films. In M. A. Masuelli (Ed.), *Biopackaging* (pp. 58–88). Miami., EE.UU. ISBN: 978-1-4987-4968-8: Editorial CRC Press Taylor & Francis Group.

Gutiérrez, T. J., & Alvarez, V. A. (2017a). Cellulosic materials as natural fillers in starch-containing matrix-based films: A review. *Polymer Bulletin, 74*(6), 2401–2430. https://doi.org/10.1007/s00289-016-1814-0.

Gutiérrez, T. J., & Alvarez, V. A. (2017b). Films made by blending poly(ε-caprolactone) with starch and flour from sagu rhizome grown at the Venezuelan Amazons. *Journal Polymers and the Environment, 25*(3), 701–716. https://doi.org/10.1007/s10924-016-0861-9.

Gutiérrez, T. J., & Alvarez, V. A. (2017c). Properties of native and oxidized corn starch/polystyrene blends under conditions of reactive extrusion using zinc octanoate as a catalyst. *Reactive and Functional Polymers, 112*, 33–44. https://doi.org/10.1016/j.reactfunctpolym.2017.01.002.

Gutiérrez, T. J., & Alvarez, V. A. (2017d). Eco-friendly films prepared from plantain flour/PCL blends under reactive extrusion conditions using zirconium octanoate as a catalyst. *Carbohydrate Polymers, 178*, 260–269. https://doi.org/10.1016/j.carbpol.2017.09.026.

Gutiérrez, T. J., & Alvarez, V. A. (2017e). Data on physicochemical properties of active films derived from plantain flour/PCL blends developed under reactive extrusion conditions. *Data in Brief, 15*, 445–448. https://doi.org/10.1016/j.dib.2017.09.071.

Gutiérrez, T. J., Pérez, E., Guzmán, R., Tapia, M. S., & Famá, L. (2014). Physicochemical and functional properties of native and modified by crosslinking, dark-cush-cush yam (*Dioscorea Trifida*) and cassava (*Manihot Esculenta*) starch. *Journal of Polymer and Biopolymer Physics Chemistry, 2*(1), 1–5. https://doi.org/10.12691/jpbpc-2-1-1.

Gutiérrez, T. J., Morales, N. J., Pérez, E., Tapia, M. S., & Famá, L. (2015a). Physico-chemical study of edible films based on native and phosphating cush-cush yam and cassava starches. *Food Packaging and Shelf Life, 3*, 1–8. https://doi.org/10.1016/j.fpsl.2014.09.002.

Gutiérrez, T. J., Tapia, M. S., Pérez, E., & Famá, L. (2015b). Structural and mechanical properties of native and modified cush-cush yam and cassava starch edible films. *Food Hydrocolloids, 45*, 211–217. https://doi.org/10.1016/j.foodhyd.2014.11.017.

Gutiérrez, T. J., Guzmán, R., Medina Jaramillo, C., & Famá, L. (2016a). Effect of beet flour on films made from biological macromolecules: native and modified plantain flour. *International Journal of Biological Macromolecules, 82*, 395–403. https://doi.org/10.1016/j.ijbiomac.2015.10.020.

Gutiérrez, T. J., Suniaga, J., Monsalve, A., & García, N. L. (2016b). Influence of beet flour on the relationship surface-properties of edible and intelligent films made from native and modified plantain flour. *Food Hydrocolloids, 54*, 234–244. https://doi.org/10.1016/j.foodhyd.2015.10.012.

Gutiérrez, T. J., González Seligra, P., Medina Jaramillo, C., Famá, L., & Goyanes, S. (2017). Chapter 14. Effect of filler properties on the antioxidant response of thermoplastic starch composites. In V. K. Thakur, M. K. Thakur, & M. R. Kessler (Eds.), *Handbook of composites from renewable materials* (pp. 337–370). WILEY-Scrivener Publisher. EE.UU. ISBN: 978-1-119-22362-7. https://doi.org/10.1002/9781119441632.ch14.

Gutiérrez, T. J., Toro-Márquez, L. A., Merino, D., & Mendieta, J. R. (2019). Hydrogen-bonding interactions and compostability of bionanocomposite films prepared from corn starch and nano-fillers with and without added Jamaica flower extract. *Food Hydrocolloids, 89*, 283–293. https://doi.org/10.1016/j.foodhyd.2018.10.058.

Hagenmaier, R. D. (2004). Fruit coatings containing ammonia instead of morpholine. *Proceedings of the Florida State Horticultural Society, 117*, 396–402.

Hashemi, S. M. B., Khaneghah, A. M., Ghahfarrokhi, M. G., & Eş, I. (2017). Basil-seed gum containing *Origanum vulgare* subsp. *viride* essential oil as edible coating for fresh cut apricots. *Postharvest Biology and Technology, 125*, 26–34. https://doi.org/10.1016/j.postharvbio.2016.11.003.

Herniou--Julien, C., Mendieta, J. R., & Gutiérrez, T. J. (2019). Characterization of biodegradable/non-compostable films made from cellulose acetate/corn starch blends processed under reactive extrusion conditions. *Food Hydrocolloids, 89*, 67–79. https://doi.org/10.1016/j.foodhyd.2018.10.024.

Hosseini-Parvar, S. H., Matia-Merino, L., Goh, K. K. T., Razavi, S. M. A., & Mortazavi, S. A. (2010). Steady shear flow behavior of gum extracted from *Ocimum basilicum* L. seed: Effect of concentration and temperature. *Journal of Food Engineering, 101*(3), 236–243. https://doi.org/10.1016/j.jfoodeng.2010.06.025.

Jafarizadeh, M., Osman, A., Tan, C. P., & Abdul, R. R. (2011). Development of an edible coating based on chitosan-glycerol to delay 'Berangan' Banana (*Musa Sapientum* cv. Berangan) ripening process. *International Food Research Journal, 18*(3), 989–997.

Jovanovic, G. D., Klaus, A. S., & Niksić, M. P. (2016). Antimicrobial activity of chitosan coatings and films against *Listeria monocytogenes* on black radish. *Revista Argentina de Microbiología, 48*(2), 128–136. https://doi.org/10.1016/j.ram.2016.02.003.

Kader, A. A. (2002). Quality parameters of fresh-cut fruit and vegetable products. In O. Lamikanra (Ed.), *Fresh-cut fruits and vegetables*. CRC Press.

Karimi, N., & Kenari, R. E. (2016). Functionality of coatings with salep and basil seed gum for deep fried potato strips. *Journal of the American Oil Chemists' Society, 93*(2), 243–250. https://doi.org/10.1007/s11746-015-2762-9.

Kharchoufi, S., Parafati, L., Licciardello, F., Muratore, G., Hamdi, M., Cirvilleri, G., & Restuccia, C. (2018). Edible coatings incorporating pomegranate peel extract and biocontrol yeast to reduce *Penicillium digitatum* postharvest decay of oranges. *Food Microbiology, 74*, 107–112. https://doi.org/10.1016/j.fm.2018.03.011.

Kokoszka, S., Debeaufort, F., Hambleton, A., Lenart, A., & Voilley, A. (2010). Protein and glycerol contents affect physico-chemical properties of soy protein isolate-based edible films. *Innovative Food Science and Emerging Technologies, 11*(3), 503–510. https://doi.org/10.1016/j.ifset.2010.01.006.

Lee, J. Y. (2003). Extending shelf life of minimally processed apples with edible coatings and anti-browning agents. *LWT- Food Science and Technology, 6*(3), 323–329. https://doi.org/10.1016/S0023-6438(03)00014-8.

Lopes da Silva, J., & Rao, M. A. (2006). Pectins: Structure, functionality and uses. In A. Gennadios (Ed.), *Food polysaccharides and their applications*. Boca Raton: CRC Press.

López-Córdoba, A. (2018). Antimicrobial films and coatings incorporated with food preservatives of microbial origin. In T. J. Gutiérrez (Ed.), *Polymers for food applications* (pp. 193–209). Cham: Springer. https://doi.org/10.1007/978-3-319-94625-2_8.

Marquez, G. R., Di Pierro, P., Mariniello, L., Esposito, M., Giosafatto, C. V. L., & Porta, R. (2017). Fresh-cut fruit and vegetable coatings by transglutaminase-crosslinked whey protein/pectin edible films. *LWT- Food Science and Technology, 75*, 124–130. https://doi.org/10.1016/j.lwt.2016.08.017.

McHugh, D.J. (2003). A guide to the seaweed industry. FAO fisheries technical paper, Rome.

Merino, D., Mansilla, A. Y., Gutiérrez, T. J., Casalongué, C. A., & Alvarez, V. A. (2018). Chitosan coated-phosphorylated starch films: Water interaction, transparency and antibacterial properties. *Reactive and Functional Polymers, 131*, 445–453. https://doi.org/10.1016/j.reactfunctpolym.2018.08.012.

Nair, M. S., Saxena, A., & Kaur, C. (2018a). Effect of chitosan and alginate based coatings enriched with pomegranate peel extract to extend the postharvest quality of guava (*Psidium guajava* L.). *Food Chemistry, 240*, 245–252. https://doi.org/10.1016/j.foodchem.2017.07.122.

Nair, M. S., Saxena, A., & Kaur, C. (2018b). Characterization and antifungal activity of pomegranate peel extract and its use in polysaccharide-based edible coatings to extend the shelf-life of capsicum (*Capsicum annuum* L.). *Food and Bioprocess Technology, 11*(7), 1317–1327. https://doi.org/10.1007/s11947-018-2101-x.

Oh, Y. A., Oh, Y. J., Song, A. Y., Won, J. S., Song, K. B., & Min, S. C. (2017). Comparison of effectiveness of edible coatings using emulsions containing lemongrass oil of different size droplets on grape berry safety and preservation. *LWT- Food Science and Technology, 75*, 742–750. https://doi.org/10.1016/j.lwt.2016.10.033.

Oms-Oliu, G., Soliva-Fortuny, R., & Martín-Belloso, O. (2008). Edible coatings with antibrowning agents to maintain sensory quality and antioxidant properties of fresh-cut pears. *Postharvest Biology and Technology, 50*(1), 87–94. https://doi.org/10.1016/j.postharvbio.2008.03.005.

Palviainen, P., Heinamaki, J., Myllarinen, P., Lahtinen, R., Yliruusi, J., & Forssell, P. (2001). Corn starches as film formers in aqueous-based film coating. *Pharmaceutical Development and Technology, 6*(3), 351–359. https://doi.org/10.1081/pdt-100002617.

Park, H. J. (1999). Development of advanced edible coatings for fruits. *Trends in Food Science & Technology, 10*(8), 254–260.

Park, S. K., et al. (2002). *Formation and properties of soy protein films and coatings: Protein based films and coatings*. Boca Raton: CRC press.

Pavlath, A. E., & Orts, W. (2009). Edible films and coatings: Why, What, and How? In K. Huber & M. Embuscado (Eds.), *Edible films and coatings for food applications*. New York: Springer. https://doi.org/10.1007/978-0-387-92824-1_1.

Ramos, Ó. L., Santos, A. C., Leão, M. V., Pereira, J. O., Silva, S. I., Fernandes, J. C., Franco, M. I., Pintado, M. E., & Malcata, F. X. (2012). Antimicrobial activity of edible coatings prepared from whey protein isolate and formulated with various antimicrobial agents. *International Dairy Journal, 25*(2), 132–141.

Ramos, Ó. L., Reinas, I., Silva, S. I., Fernandes, J. C., Cerqueira, M. A., Pereira, R. N., Vicente, A. A., Poças, M. F., Pintado, M. E., & Malcata, F. X. (2013). Effect of whey protein purity and glycerol content upon physical properties of edible films manufactured therefrom. *Food Hydrocolloids, 30*(1), 110–122.

Riberio, C., Vicente, A. A., Teixeira, J. A., & Miranda, C. (2007). Optimization of edible coating composition to retard strawberry fruit senescence. *Postharvest Biology and Technology, 44*(1), 63–70. https://doi.org/10.1016/j.postharvbio.2006.11.015.

Ridley, B. L., O'Neill, M. A., & Mohnen, D. (2001). Pectins: Structure, biosynthesis, and oilgogalacturonide-related signaling. *Phytochemistry, 57*(6), 929–967. https://doi.org/10.1016/s0031-9422(01)00113-3.

Rodriguez, M., Osés, J., Ziani, K., & Maté, J. I. (2006). Combined effect of plasticizers and surfactants on the physical properties of starch based edible films. *Foodservice Research International, 39*(8), 840–846. https://doi.org/10.1016/j.foodres.2006.04.002.

Rojas-Grau, M. A., Raybaudi-Massilia, R. M., Solica-Fortuny, R., Avena-Bustillos, R. J., McHugh, T. H., & Belloso, O. M. (2007). Apple puree-alginate edible coating as carrier of antimicrobial agents to prolong shelf-life of fresh-cut apples. *Postharvest Biology and Technology, 45*(2), 254–264. https://doi.org/10.1016/j.postharvbio.2007.01.017.

Rojas-Grau, M. A., Tapia, M. S., & Martín-Belloso, O. (2008). Using polysaccharide-based edible coatings to maintain quality of fresh-cut Fuji apples. *LWT- Food Science and Technology, 41*(1), 139–147. https://doi.org/10.1016/j.lwt.2007.01.009.

Saba, M. K., & Sogvar, O. B. (2016). Combination of carboxymethyl cellulose-based coatings with calcium and ascorbic acid impacts in browning and quality of fresh-cut apples. *LWT- Food Science and Technology, 66*, 165–171. https://doi.org/10.1016/j.lwt.2015.10.022.

Sabaghi, M., Maghsoudlou, Y., Khomeiri, M., & Ziaiifar, A. M. (2015). Active edible coating from chitosan incorporating green tea extract as an antioxidant and antifungal on fresh walnut kernel. *Postharvest Biology and Technology, 10*, 224–228. https://doi.org/10.1016/j.postharvbio.2015.08.025.

Saberi, B., & Golding, J. B. (2018). Postharvest application of biopolymer-based edible coatings to improve the quality of fresh horticultural produce. In T. J. Gutiérrez (Ed.), *Polymers for food applications* (pp. 211–250). Cham: Springer. https://doi.org/10.1007/978-3-319-94625-2_9.

Saberi, B., Thakur, R., Vuong, Q. V., Chockchaisawasdee, S., Golding, J. B., Scarlet, C. J., & Stathopoulos, C. E. (2016). Optimization of physical and optical properties of biodegradable edible films based on pea starch and guar gum. *Industrial Crops and Products, 86*, 342–352. https://doi.org/10.1016/j.indcrop.2016.04.015.

Saini, C. S., Sharma, H. K., & Sharma, L. (2018). Thermal, structural and rheological characterization of protein isolate from sesame meal. *Journal of Food Measurement and Characterization, 12*(1), 426–432. https://doi.org/10.1007/s11694-017-9655-6.

Salmieri, S., & Lacroix, M. (2006). Physiochemical properties of alginate/polycaprolactone-based films containing essential oils. *Journal of Agricultural and Food Chemistry, 54*(26), 0205–10214. https://doi.org/10.1021/jf062127z.

Salvia-Trujillo, L., Rojas-Grau, M. A., Soliva-Fortuny, R., & Martín-Belloso, O. (2015). Use of antimicrobial nanoemulsions as edible coatings: Impact on safety and quality attributes of fresh-cut Fuji apples. *Postharvest Biology and Technology, 105*, 8–16. https://doi.org/10.1016/j.postharvbio.2015.03.009.

Sanchís, E., González, S., Chidelli, C., Sheth, C. C., Mateos, M., Palou, L., & Pérez-Gago, M. B. (2016). Browning inhibition and microbial control in fresh-cut persimmon (*Diospyros kaki Thunb. cv.* Rojo Brillante) by apple pectin-based edible coatings. *Postharvest Biology and Technology, 112*, 186–193. https://doi.org/10.1016/j.postharvbio.2015.09.024.

Sandhu, K. S., Sharma, L., Singh, C., & Siroha, A. K. (2017). Recent advances in biodegradable films, coatings and their applications. In S. Gahlawat, R. Salar, P. Siwach, J. Duhan, S. Kumar, & P. Kaur (Eds.), *Plant biotechnology: Recent advancements and developments* (pp. 271–296). Singapore: Springer. https://doi.org/10.1007/978-981-10-4732-9_14.

Santos, T. M., Souza Filho, M. d. S. M., Silva, E. d. O., da Silveira, M. R. S., de Miranda, M. R. A., Lopes, M. M. A., & Azeredo, H. M. C. (2018). Enhancing storage stability of guava with tannic acid-crosslinked zein coatings. *Food Chemistry, 257*, 252–258.

Sayanjali, S., Ghanbarzadeh, B., & Ghiassifar, S. (2011). Evaluation of antimicrobial and physical properties of edible film based on carboxymethyl cellulose containing potassium sorbate on some mycotoxigenic Aspergillus species in fresh pistachios. *LWT- Food Science and Technology, 44*(4), 1133–1138. https://doi.org/10.1016/j.lwt.2010.12.017.

Shah, N. N., Vishwasrao, C., Singhal, R. S., & Nnrhanarayan, L. (2016). *n*-Octenyl succinylation of pullulan: Effect on its physico-mechanical and thermal properties and application as an edible coating on fruits. *Food Hydrocolloids, 55*, 178–188. https://doi.org/10.1016/j.foodhyd.2015.11.026.

Sharma, S., & Rao, T. R. (2015). Xanthan gum based edible coating enriched with cinnamic acid prevents browning and extends the shelf-life of fresh-cut pears. *LWT- Food Science and Technology, 62*(1), 791–800. https://doi.org/10.1016/j.lwt.2014.11.050.

Sharma, L., & Singh, C. (2016). Sesame protein based edible films: Development and characterization. *Food Hydrocolloids, 61*, 139–147. https://doi.org/10.1016/j.foodhyd.2016.05.007.

Sharma, L., Singh Saini, C., & Sharma, H. K. (2018). Development of crosslinked sesame protein and pineapple extract-based bilayer coatings for shelf-life extension of fresh-cut pineapple. *Journal of Food Processing and Preservation, 42*(2), 1–11. https://doi.org/10.1111/jfpp.13527.

Sharma, L., Saini, C. S., Sharma, H. K., & Sandhu, K. S. (2019). Biocomposite edible coatings based on cross linked-sesame protein and mango puree for the shelf life stability of fresh-cut mango fruit. *Journal of Food Process Engineering, 42*(1), e12938. https://doi.org/10.1111/jfpe.12938.

Solvia-Fortuny, Rojas-Grau, M. A., & Martin-Belloso, O. (2012). Polysaccharide coatings. In E. A. Baldwin, R. D. Hagenmair, & J. Bai (Eds.), *Edible coatings and films to improve food quality*. Boca Raton: Taylor and Francis.

Su, J. F., Huang, Z., Yuan, X. Y., Wang, X. Y., & Li, M. (2010). Structure and properties of carboxymethyl cellulose/soy protein isolate blend edible films crosslinked by maillard reactions. *Carbohydrate Polymers, 79*(1), 145–153. https://doi.org/10.1016/j.carbpol.2009.07.035.

Suárez, G., & Gutiérrez, T. J. (2017). Chapter 15. Recent advances in the development of biodegadable films and foams from cassava starch. In C. Klein (Ed.), *Handbook on cassava: Production, potential uses and recent advances* (pp. 297–312). New York. EE.UU. ISBN: 978-1-53610-307-6: Editorial Nova Science Publishers, Inc.

Sworn, G., & gum, X. (2009). In G. O. Phillips & P. A. Williams (Eds.), *Handbook of hydrocolloids* (2nd ed.). Boca Raton: CRC Press.

Talens, P., Pérez-Masía, R., fabra, M. J., Vargas, M., & Chiralt, A. (2012). Application of edible coatings to partially dehydrated pineapple for use in fruit–cereal products. *Journal of Food Engineering, 112*(1-2), 86–93. https://doi.org/10.1016/j.jfoodeng.2012.03.022.

Tapia, M. S., Rojas-Grau, M. A., Carmona, A., Rodríguez, F. J., Soliva-Fortuny, R., & Martin-Belloso, O. (2013). Use of alginate and gellan-based coatings for improving barrier, texture and nutritional properties of fresh-cut papaya. *Food Hydrocolloids, 22*(8), 1493–1503. https://doi.org/10.1016/j.foodhyd.2007.10.004.

Tapia-Blácido, D. R., Maniglia, B. C., & Tosi, M. M. (2018). Transport phenomena in edible films. In T. J. Gutiérrez (Ed.), *Polymers for food applications* (pp. 149–192). Cham: Springer. https://doi.org/10.1007/978-3-319-94625-2_7.

Tesfay, S. Z., Magwaza, L. S., Mbili, N., & Mditshwa, A. (2017). Carboxyl methylcellulose (CMC) containing moringa plant extracts as new postharvest organic edible coating for Avocado (*Persea americana* Mill.) fruit. *Scientia Horticulturae, 226*, 201–207. https://doi.org/10.1016/j.scienta.2017.08.047.

Tihminlioglu, F., Atik, İ. D., & Özen, B. (2010). Water vapor and oxygen-barrier performance of corn–zein coated polypropylene films. *Journal of Food Engineering, 96*(3), 342–347. https://doi.org/10.1016/j.jfoodeng.2009.08.018.

Vu, C. H. T., & Won, K. (2013). Novel water-resistant UV-activated oxygen indicator for intelligent food packaging. *Food Chemistry, 140*(1-2), 52–56. https://doi.org/10.1016/j.foodchem.2013.02.056.

Zactiti, E. M., & Kieckbusch, T. G. (2006). Potassium sorbate permeability in biodegradable alginate films: Effect of the antimicrobial agent concentration and crosslinking degree. *Journal of Food Engineering, 77*(3), 462–467. https://doi.org/10.1016/j.jfoodeng.2005.07.015.

Zambrano-Zaragoza, M. L., Quintanar-Guerrero, D., Del Real, A., Piñon-segundo, E., & Zambrano-zaragoza, J. F. (2017). The release kinetics of β-carotene nanocapsules/xanthan gum coating and quality changes in fresh-cut melon (cantaloupe). *Carbohydrate Polymers, 157*, 1874–1882. https://doi.org/10.1016/j.carbpol.2016.11.075.

Zhang, L., Li, S., Dong, Y., Zhi, H., & Zong, W. (2016). Tea polyphenols incorporated into alginate-based edible coating for quality maintenance of Chinese winter jujube under ambient temperature. *LWT- Food Science and Technology, 70*, 155–161. https://doi.org/10.1016/j.lwt.2016.02.046.

Zhang, L., Chen, F., Lai, S., Wang, H., & Yang, H. (2018). Impact of soybean protein isolate-chitosan edible coating on the softening of apricot fruit during storage. *LWT- Food Science and Technology, 96*, 604–611. https://doi.org/10.1016/j.lwt.2018.06.011.

Zobel, H. F., & Stephen, A. M. (2006). Starch: structure, analysis and application. In A. Gennadios (Ed.), *Food polysaccharides and their applications*. Boca Raton: Taylor and Francis.

Chapter 19
Development of Edible Coatings in the Preservation of Fruits and Vegetables

Sabina Galus

Abstract Edible coatings are thin layers obtained from biopolymers (proteins, polysaccharides, lipids or combinations of these components) and are applied to food products. Some of its functions are to protect the product from physical, chemical and biological deterioration during storage and distribution. Due to the problem related to losses of horticultural products caused by different factors, including postharvest spoilage due to high moisture content, edible coatings have been recently studied in order to enhance the quality of food products. The application of edible coatings on fresh fruits and vegetables has shown promising results in extending its shelf life and improving safety. This chapter aims to analyze the recent advances in the development of edible coatings in the postharvest.

Keywords Biopolymers · Edible coatings · Postharvest

19.1 Introduction

The use of edible coatings as protective layers on the food surface is not a recent approach, as they have been used for centuries in food production (Ansorena et al. 2018). The first documented applications were documented in the twelfth century in China. In the twentieth century, different coatings to prevent the moisture loss from fruits and vegetables, as well as to add the bright effect were used (Hassan et al. 2018). An edible coating is a thin layer formed directly on a food product, while the edible film is first formed as a self-supporting layer and can then be placed on a food product or between food components. Both structures are prepared from a liquid film-forming solution of edible materials which play a role as a structural matrix (polysaccharide, protein, lipid or different mixture of those components) (Saberi and Golding 2018). The main difference between them is the application technique:

S. Galus (✉)
Department of Food Engineering and Process Management, Faculty of Food Sciences, Warsaw University of Life Sciences – SGGW (WULS-SGGW), Warsaw, Poland
e-mail: sabina_galus@sggw.pl

T. J. Gutiérrez (ed.), *Polymers for Agri-Food Applications*,
https://doi.org/10.1007/978-3-030-19416-1_19

the edible coating is applied by an immersing or spraying method, while the edible film is dried or molded and then applied as a separate layer or a wrap on the food product (Falguera et al. 2011).

Fruits and vegetables are horticultural products which are characterized as highly perishable products with a short shelf life, which constitutes a continuous challenge for the food industry, particularly in the regions where refrigerated storage is limited. Despite of different postharvest technologies which have been investigated and applied, many harvested fruits and vegetables are still wasted, thus adequate and advanced postharvest processing approaches are required. Therefore, the use of edible coatings as protective layers is the most desired approach in order to avoid or limit a waste problem by improving its quality acceptance. It can be observed that recent scientific studies are focused on the development of different edible coatings and their application to extend the shelf life of fresh of fresh fruits and vegetables, both whole and fresh-cut.

19.2 Film-Forming Materials

Edible films and coatings are produced from edible polymers which are usually classified into three categories: hydrocolloids, lipids and their composites (Gutiérrez 2018a). Table 19.1 summarizes the main materials used in the preparation of edible films and coatings. Hydrocolloids (polysaccharides and proteins) are the most widely used structural materials for edible materials (Valencia and do Amaral Sobral 2018). They are more neutral, transparent and tasteless than those obtained from lipids, which are rather opaque. Edible films and coatings obtained from polysaccharides or proteins are usually good oxygen barriers and provide good mechanical strength (Garrido et al. 2018). However, these materials are hydrophilic in nature, thus they are characterized as poor water vapor barriers. On the other hand, lipids (fats and oils) are excellent hydrophobic materials and a low water vapor permeability. Nevertheless, the hydrophilic character of polysaccharide and protein-based

Table 19.1 Main materials used in the preparation of edible films and coatings

Source	Polysaccharides	Proteins	Lipids
Vegetable	Cellulose and cellulose derivatives, starch, pectin, tara gum	Soy proteins, corn zein, wheat gluten	Vegetable oils, seed oils, nut oils, candelilla wax, carnauba wax, rice bran wax
Animal	Chitosan	Gelatin, milk proteins, collagen, egg proteins, myofibrillar proteins	Beeswax, shellac, fish oil
Seaweed extracts	Alginate, agar, carrageenan	–	–
Microbial	Pullulan, gellan gum, xanthan gum	–	–

Note: – not found

materials can be improved efficiently by incorporation of lipids (Galus and Kadzińska 2015). In recent studies on edible film-forming materials, the combination of different polysaccharides, proteins and/or lipids can be observed (Merino et al. 2019). The aim of these researches is to take advantages of properties of each component in order to obtain new coating materials with improved functional properties which depend on their miscibility and compatibility. Nonetheless, the new materials must be developed, characterized and analyzed. One of the examples can be fruit or/and vegetable purees, wastes or byproduct-based films and coatings usually prepared with the addition of hydrocolloid agents. Wang et al. (2011) observed that only one kind of hydrocolloid did not solve the problem of the poor mechanical strength of carrot films. The authors combined carboxylmethyl cellulose, corn starch and gelatin in order to improve functional properties of carrot puree films which can be suitable as wrappers or edible packaging for food products.

Among polysaccharides, proteins and lipids, additional compounds are usually needed in order to provide continuous and flexible structure of films and coatings (Álvarez et al. 2017). Plasticizers are used to limit negative characteristic such as fragility, brittleness or cracking. Polyols, oligosaccharides and lipids are different types of plasticizers used in edible materials prepared from hydrocolloids (Medina Jaramillo et al. 2016). In addition, water plays a role as a plasticizer and effect many physical properties of films. Other components can also be added to enhance its functionality, including crosslinking agents, emulsifiers or reinforcers (Collazo-Bigliardi et al. 2018). However, new research and recent publications on the production of edible films and coatings are focused on different substances or compounds (antioxidants, antimicrobials, color or flavor agents, nutraceuticals) incorporated into film matrix to improve physical properties as well as to give a new function such as active and intelligent (A&I) packaging films (Bracone et al. 2016; Gutiérrez 2018b).

The use of bioactive substances and biopolymers derived from agricultural byproducts or waste promotes the development of these environmentally friendly materials which can replace or at least limit the use of conventional plastics (Salgado et al. 2015). There is also a great research effort focused in developing and investigating the new film-forming materials, as well as the new processing systems, in order to optimize the composition, costs and functional properties of films and coatings.

19.3 Factors that Affect the Practical Approaches of Edible Coatings Applications for Fruits and Vegetables

Edible coatings have been applied to fruits and vegetables in order to provide a semi-permeable barrier to water vapor and other gases, which reduces weight loss and extends the shelf life of them. Fruits and vegetables are obtained from different parts of the plant and are characterized as highly heterogeneous products in terms of tissue structure, stage of development or metabolic rate. Therefore, the application of edible coatings can be designed for fresh, minimally processed horticultural products or other processed horticultural products. In general, many factors must be considered when designing coatings for fruits and vegetables, such as shape, size, maturity stage, tissue structure, porosity, juiciness, turgidity, surface texture, firmness, metabolic activity, processing, packaging and storage conditions.

These aforementioned factors affect the selection of the adequate edible material and the application technique (Montero-Calderón et al. 2016). The applications of edible coatings for the development of horticultural products with improved quality and prolonged shelf life can be designed for different approaches, including the modification of the internal atmosphere, the control of the mass transfer properties, the inhibition of spoilage microorganisms, the reduction of oxidative reactions and the improvement of the texture properties or gloss and shine on its surface (Gutiérrez and Álvarez 2017; Tapia-Blácido et al. 2018).

In recent times, the extensive review articles have focused on edible films and coatings for fruit and vegetable applications (Vargas et al. 2008; Dhall 2013; Ghidelli and Pérez-Gago 2018; Yousuf et al. 2018). In this chapter, the most recent trends are summarized, with emphasis on the new formulations and applications of edible coatings, which are listed in Table 19.2.

19.4 Advances in Edible Coatings for Fruits and Vegetable

For a long time, biopolymers were applied more in one-component coating formulation, whose tendency is still observed. However, two-component and multicomponent edible coatings provide improved functional properties. In this context, edible coatings based on sodium alginate and pectin were used by Silva et al. (2018) to preserve the quality of the minimally processed mangoes. Chitosan (Cs) is one of the promising materials that has been applied, either as a one-component coating or as a main polymer in the component structures, for many horticultural products due to the excellent film-forming ability and characteristics antimicrobial (Gutiérrez 2017a; Paul et al. 2018). Ortiz-Duarte et al. (2019) observed that the respiration rate and the ethylene production of the fresh-cut melon were reduced after the Ag-Cs nanocomposite coating treatments. The coated samples also had better sensory quality with lower translucency and texture degradation after 13 days of storage at 5 °C. Kharchoufi et al. (2018) incorporated the by-product pomegranate peel extract and the biocontrol agent into Cs and locus bean gum coating in order to control the growth of *Penicillium digitatum* and reduce the postharvest decay of oranges. The results of antifungal effectiveness showed the potential synergistic effect between the biocontrol agents and the natural bioactive compounds. Yan et al. (2018) observed that a layer-by-layer edible coating based on Cs and carboxymethyl cellulose was significantly effective in inhibiting the loss of fruit firmness and aroma volatiles of strawberry. The authors found, after an untargeted metabolomics analysis, that the layer-by-layer application reduced the primary metabolite contents involved in the metabolism of carbohydrates, fatty acids and amino acids.

Table 19.2 Latest potential applications of edible coatings for fruits and vegetables and their main advantages

Fruit or vegetable	State	Coating material	Functional compound	Advantage	Reference
Apple	Fresh-cut	Carboxymethyl cellulose	*Aloe vera*, ascorbic acid, calcium chloride, cysteine	Inhibited browning, retained color, decreased CO_2 production and O_2 consumption	Kumar et al. (2018)
Melon	Fresh-cut	Chitosan (Cs) or carboxymethyl cellulose	Citral	Superior antimicrobial protection; improved appearance; extended storability	Arnon-Rips et al. (2019)
Strawberry	Whole	Cs	Shrimp carotenoproteins	Reduced fungal decay and weight loss, and inhibited phytopathogenic growth	Hajji et al. (2018)
Melon	Fresh-cut	Cs	Ag nanoparticles	Reduced respiration rate and ethylene production, and improved sensory quality	Ortiz-Duarte et al. (2019)
Banana	Whole	Rice starch, carrageenan	–	Reduced weight loss, maintained firmness, improved the visual appearance and delayed fruit ripening	Thakur et al. (2019)
Garambullo fruits	Whole	Gelatin	Tomato oily extract	Delayed changes in weight loss, brix degrees, titriatable acidity and pH; retained concentration of bioactive compounds	López-Palestina et al. (2018)
Plum	Whole	Rice starch, carrageenan	Sucrose fatty acid ester	Reduced weight loss and respiration rate; inhibited ethylene production	Thakur et al. (2018)
Tomato Chili pepper Eggplant	Whole	Cs	Cs nanoparticles	Decreased weight loss and improved antifungal activity	Divya et al. (2018)
Eggplant	Fresh-cut	Soy protein isolate	Cysteine	Improved quality and prolonged shelf-life	Ghidelli et al. (2018)
Apricot	Whole	Soy protein isolate, Cs	–	Decreased weight loss, retained firmness and textural properties, and inhibited pectin degradation	Zhang et al. (2018)
Orange	Whole	Cs, locus bean gum	Pomegranate peel extract, biocontrol yeast	Reduced disease incidence and inhibited of green mold	Kharchoufi et al. (2018)
Tomato	Whole	Cassava starch, Cs	Rosemary pepper essential oil, pomegranate peel extract	Delayed ripening; reduced total soluble solids content and weight loss; maintained firmness	Araújo et al. (2018)

(continued)

Table 19.2 (continued)

Fruit or vegetable	State	Coating material	Functional compound	Advantage	Reference
Strawberry	Whole	Cs	Cinnamon essential oil, extract of Roselle calyces	Reduced weight loss, remained fruit firmness and increased antioxidant capacity	Ventura-Aguilar et al. (2018)
Dates	Whole	Poly vinyl alcohol, Cs	Tannic acid	Prolonged shelf life	El-Dein et al. (2018)
Figs	Whole	–	Mucilage extract from *O puncia ficus-indica* cladodes	Maintained weight loss and fruit firmness	Allegra et al. (2018)
Strawberry	Whole	Protein isolate and gum from *Cajanus cajan* seeds	–	Reduced weight loss, total soluble content and the consumption of citric acid	Robles-Flores et al. (2018)
Capsicum	Whole	Cs, sodium alginate	Pomegranate peel extract	Inhibited microbial growth, reduced weight loss and fruit firmness, maintained color and sensory scores	Sneha Nair et al. (2018)
Carrot	Fresh-cut	Sodium alginate, sunflower oil	Probiotic culture *L. acidophilus*	Reduced carrot metabolism, maintained weight loss and color changes	Shigematsu et al. (2018)
Cherry tomatoes	Whole	Cs	Grapefuit seed extract	Delayed microorganism growth and reduced weight loss	Won et al. (2018)
Apple	Fresh-cut	Gelatin	Aloe vera, green tea extracts	Inhibited microbial growth and reduced softening trend	Amiri et al. (2018)
Guava	Whole	Arabic gum, sodium caseinate	Cinnamon and lemon grass essential oils	Improved overall quality and maintained bioactive compounds	Murmu and Mishra (2018)
Avocado	Whole	Carboxyl methylcellulose	Moringa plant extracts	Reduced weight loss, ethylene production and respiration rate, prolonged shelf life and maintained overall quality	Tesfay et al. (2017)
Kiwi	Fresh-cut	Sodium alginate	Poly-ε-lysine	Reduced CO_2 production; maintained green color, total chlorophylls content, ascorbic acid, antioxidant capacity and morphological properties	Li et al. (2017)
Mandarins	Whole	Carboxymethyl cellulose, arabic gum, persian gum	Beeswax, carnauba wax	Reduced weight loss	Khorram et al. (2017)

Emulsified films and coatings have generally been investigated to provide a better moisture barrier of the coated products. However, lipid-containing coatings, both coarse and nano emulsions, can be used for the delivery of valuable bioactive compounds, such as antimicrobial agents, antioxidants, dyes, flavors or nutraceutical agents (Galus and Kadzińska 2015). Arnon-Rips et al. (2019) analyzed the encapsulation of citral by utilizing coarse and nano emulsions using sunflower oil, which were integrated into Cs and carboxylmethyl cellulose coatings to be applied on melons. Both polysaccharides showed a good stabilizing effect in the formulation, however, the emulsions containing Cs were more stable. The nano emulsions were more stable, so the films obtained from these solutions were more organized and denser, with a greater water vapor barrier and mechanical resistance. The coatings based on nano emulsions applied to fresh-cut melons resulted in superior antimicrobial protection and improved appearance, as well as increased storage capacity. Nevertheless, the citrus aroma was significant in coated melons, which may limit their sensory acceptability, which indicates that the coating formulation analyzed would be more appropriate for citrus fruits. Thakur et al. (2019) managed to prolong the shelf life of banana for 12 days by applying starch-ι-carrageenan edible coatings containing sucrose fatty acid esters.

In the last decades, the incorporation of different natural compounds on the surface of food products in solid state began, but a more exhaustive investigation in this area is still necessary to be applied such substances in the food industry. For example, the extensive studies on the biological activity of rosemary extracts as an antioxidant agent resulted in their approval as a food additive E 392 by the Directives 2010/67/EU and 2010/69/EU (European Commission 2010a, b).

19.5 Advances in the Combination of Methods to Extend the Shelf Life of Fruits and Vegetables

The extension of the shelf life of fruits and vegetables is necessary due to several aspects, such as market demand and the limitation of a global waste problem. Recently published studies have focused on new formulations of edible coatings, as well as on the use of the combination method in order to produce functional protective layers. In this context, copolymer coatings based on hydrolyzed poly (vinyl alcohol) and Cs with the addition of tannic acid were prepared by El-Dein et al. (2018) using a triple blend to maintain the quality of shelf life of the dates during commercialization. The film-forming solutions were further exposed to the γ-irradiation before casting or using them. The results obtained showed that the γ-irradiation and the addition of tannic acid increased the mechanical properties of the films and successfully prolonged the shelf life of the dates during the marketing period, from 1 week to 1 month with an acceptable freshness and quality. Muñoz-Labrador et al. (2018) applied citrus pectin gels prepared by power ultrasound as edible coatings for fresh strawberries, resulting in a better fruit quality than control samples in terms of weight loss and color parameters.

The application of edible coatings can be done by different methods, among them soaking, or immersion are the most popular. However, Soares et al. (2018) used vacuum impregnation as an alternative method to soak techniques in the application of edible coatings of Cs and Cs/lauric acid on minimally processed pumpkin. The results showed that the vacuum impregnation method caused a greater incorporation of components, and more uniform and thicker coatings, meanwhile some important changes in pH, acidity, color and firmness indicated that this method led to larger changes in the properties of the pumpkin than the popular coating method. Todisco et al. (2018) analyzed the effect of the edible coating composition and the drying temperatures on the drying kinetics and the quality of the dried red guava. The authors obtained pectin-based coatings containing disintegrated guava by-products and applied them to guava slices prior to hot-air drying. The coatings improved nutritional quality, by high carotenoid and total phenolic retentions, without having an important impact on the drying times.

Salinas-Roca et al. (2018) evaluated the combined effect of pulsed light treatment, alginate coating and malic acid dip on quality attributes of fresh-cut mango. These treatment combinations prevented fruit browning during storage and resulted in the highest radical scavenging activity, indicating that this novel processing system can preserve bioactive compounds of fresh-cut mango for 14 days. Vivek and Subbarao (2018) observed that Cs coatings could delay the decomposition, both microbial and non-microbial, in fresh-cut kiwi treated in combination with ultrasound and sodium hypochloride, thus extending its shelf life for 10 days at 5 °C.

19.6 New Coating Materials Testing for Applications to Fruits and Vegetables

Among the known biopolymers, new coating materials are being tested. In this context, the different concentration of *Aloe vera* gel were used by Mubarak and Rao Engakanah (2017) as edible coating on to maintain the postharvest quality of wax applies. These authors observed that 100% *Aloe vera* is most effective in retaining the bioactive compounds of stored apples and also has a tendency to reduce weight loss and fruit firmness. Allegra et al. (2018) applied an edible coating extracted from cladodes of *Opuntia ficus-indica* on breba fig fruit which was effective in maintaining the fruit fresh weight and fruit firmness. In addition, metabolic study showed that fig fruits exhibited important changes in primary metabolism as a result of coating treatment, including increased amounts of carbohydrates and most of the amino acids. Robles-Flores et al. (2018) developed new edible coatings from protein isolate and gum obtained from *Cajanus cajan* seeds to be applied to strawberries. The authors observed that coatings resulted in a reduction of the weight loss, the total soluble solid and the citrus consumption, while no sensory changes were obtained.

Shigematsu et al. (2018) developed alginate-based coatings containing sunflower oil and the probiotic culture *Lactobacillus acidophilus* to coat the minimally processed carrot slices. The results showed that application of these coatings had some advantages, such as better preservation of moisture content and color. The incorporation of *L. acidophilus* to the film-forming emulsions reduced the metabolism of the minimally processed carrot slices. Alvarez et al. (2018) prepared sodium alginate and Cs coatings incorporating different dietary fibers (apple fiber, orange fiber, inulin and oligofructose) in order to improve quality attributes of fresh blueberries. The results showed that the fiber-enriched Cs coatings allow the maintenance of freshness and the improve of the quality of blueberries as an innovative product with prebiotic potential and prolonged shelf life.

19.7 Edible Coatings From Fruits and Vegetables

In recent years, different purees of fruits and vegetables, pulp, juice or extract have been used in the production of edible films and coatings. The components that are added to the biopolymers provide different bioactive compounds with nutritional value and can play an important role on factors such as color or taste, thus affecting the physical properties of the edible films or coatings (Gutiérrez 2017b). In addition, some operations applied during the processing of fruits or vegetables can generate high amounts of waste, which are not attractive and are underutilized due to low market values, and which can be used as raw material for the development of films and coatings edible. However, the reduction of losses sustained during post-harvest treatments is the subject of latest research, which are focused on the application of overproduced, by-products or residues as a new unconventional source in the preparation of edible packaging. In this context, Torres-León et al. (2018) used mango by-products (peel and seeds) for the formation of biodegradable coatings for fresh peaches. The mango films showed good properties of permeability, color, antioxidants and greater hydrophobicity. The incorporation of antioxidants from extract of mango seed into film formulation, positively affected the surface properties. Fresh peaches coated with the solution containing mango peel and mango seed extract showed reduced ethylene and carbon dioxide production, as well as lower oxygen consumption compared to uncoated fruits. This study demonstrated the great potential of mango by-products in the production of low-cost biopackaging material and the longer shelf life of peaches. Nogueira et al. (2019) developed flexible films made from arrowroot starch and blackberry pulp. The obtained films provide the typical color and flavor of blackberry, bioactive compound and antioxidant capacity which may be attractive for some food applications. Rangel-Marrón et al. (2019) used a papaya puree as a base for the production of edible coatings containing alginate, glycerol and citric acid. The authors suggested that formulation obtained could be an alternative packaging system for different applications, e.g. sushi wraps or minimally processed fruits and vegetables with compatible colors.

19.8 Conclusions

The research and development (R&D) of edible coatings for fruits and vegetables has been intensified in recent decades, but its low presence in the market shows that future studies should be carried out. The benefits of edible coatings for horticultural products are numerous, but the most important should be highlighted as follows: reduction of weight loss, respiration rate and ethylene production, as well as the inhibition of the growth of microorganisms and the maintaining the content of active compounds. Therefore, the application of edible coatings for fruits and vegetables can positively affect the prolongation of the shelf life and, consequently, impact the economy of the food processes.

Acknowledgments The work was co-financed by a statutory activity subsidy from the Polish Ministry of Science and Higher Education for the Faculty of Food Sciences of Warsaw University of Life Sciences.

Conflicts of Interest The author declares no conflict of interest.

References

Allegra, A., Gallotta, A., Carimi, F., Mercati, F., Inglese, P., & Martinelli, F. (2018). Metabolic profiling and post-harvest behavior of "Dottato" Fig (*Ficus carica* L.) fruit covered with an edible coating from *O. ficus-indica*. *Frontiers in Plant Science, 9*(1321). https://doi.org/10.3389/fpls.2018.01321.

Álvarez, K., Famá, L., & Gutiérrez, T. J. (2017). Chapter 12. Physicochemical, antimicrobial and mechanical properties of thermoplastic materials based on biopolymers with application in the food industry. In M. Masuelli & D. Renard (Eds.), *Advances in physicochemical properties of biopolymers: Part 1* (pp. 358–400). Bentham Science Publishers. EE.UU. ISBN: 978–1–68108-454-1. eISBN: 978–1–68108-453-4. https://doi.org/10.2174/97816810845341117010015.

Alvarez, M. V., Ponce, A. G., & Moreira, M. R. (2018). Influence of polysaccharide-based edible coatings as carriers of prebiotic fibers on quality attributes of ready-to-eat fresh blueberries. *Journal of the Science of Food and Agriculture, 98*(7), 2587–2597. https://doi.org/10.1002/jsfa.8751.

Amiri, S., Akhavan, H. R., Zare, N., & Radi, M. (2018). Effect of gelatin-based edible coatings incorporated with aloe vera and green tea extracts on the shelf-life of fresh-cut apple. *Italian Journal of Food Science, 30*(1), 61–74. https://doi.org/10.14674/1120-1770-ijfs699.

Ansorena, M. R., Pereda, M., & Marcovich, N. E. (2018). Edible films. In T. J. Gutiérrez (Ed.), *Polymers for food applications* (pp. 5–24). Cham: Springer. https://doi.org/10.1007/978-3-319-94625-2_2.

Araújo, J. M. S., de Siqueira, A. C. P., Blank, A. F., Narain, N., & de Aquino Santana, L. C. L. (2018). A cassava starch–chitosan edible coating enriched with *Lippia sidoides* Cham. Essential oil and pomegranate peel extract for preservation of italian tomatoes (*Lycopersicon esculentum Mill.*) stored at room temperature. *Food and Bioprocess Technology, 11*(9), 1750–1760. https://doi.org/10.1007/s11947-018-2139-9.

Arnon-Rips, H., Porat, R., & Poverenov, E. (2019). Enhancement of agricultural produce quality and storability using citral based edible coatings; the valuable effect of nano-emulsification in a solid state delivery on fresh-cut melons model. *Food Chemistry, 277*, 205–212. https://doi.org/10.1016/j.foodchem.2018.10.117.

Bracone, M., Merino, D., González, J., Alvarez, V. A., & Gutiérrez, T. J. (2016). Chapter 6. Nanopackaging from natural fillers and biopolymers for the development of active and intelligent films. In S. Ikram & S. Ahmed (Eds.), *Natural polymers: Derivatives, blends and composites* (pp. 119–155). New York. EE.UU. ISBN: 978-1-63485-831-1: Editorial Nova Science Publishers, Inc.

Collazo-Bigliardi, S., Ortega-Toro, R., & Chiralt, A. (2018). Properties of micro- and nano-reinforced biopolymers for food applications. In T. J. Gutiérrez (Ed.), *Polymers for food applications* (pp. 61–99). Cham: Springer. https://doi.org/10.1007/978-3-319-94625-2_4.

Dhall, R. K. (2013). Advances in edible coatings for fresh fruits and vegetables: A review. *Critical Review in Food and Nutrition, 53*(5), 435–450. https://doi.org/10.1080/10408398.2010.541568.

Divya, K., Smitha, V., & Jisha, M. S. (2018). Antifungal, antioxidant and cytotoxic activities of chitosan nanoparticles and its use as an edible coating on vegetables. *International Journal of Biological Macromolecules, 114*, 572–577. https://doi.org/10.1016/j.ijbiomac.2018.03.130.

El-Dein, A. E., Khozemy, E. E., Farag, S. A., El-Hamed, N. A., & Dosoukey, I. M. (2018). Effect of edible co-polymers coatings using γ-irradiation on Hyani date fruit behavior during marketing. *International Journal of Biological Macromolecules, 117*, 851–857. https://doi.org/10.1016/j.ijbiomac.2018.05.170.

European Commission. (2010a). Commission directive 2010/67/EU of 22 October. Official Journal of the European Union, L277, 17–26.

European Commission. (2010b). Commission Directive 2010/69/EU of 22 October 2010. Official Journal of the European Union, L279, 22–31.

Falguera, V., Quintero, J. P., Jiménez, A., Muñoz, A., & Ibarz, A. (2011). Edible films and coatings: Structures, active functions and trends in their use. *Trends Food Science and Technology, 22*(6), 291–303. https://doi.org/10.1016/j.tifs.2011.02.004.

Galus, S., & Kadzińska, J. (2015). Food applications of emulsion-based edible films and coatings. *Trends in Food Science and Technology, 45*(2), 273–283. https://doi.org/10.1016/j.tifs.2015.07.011.

Garrido, T., Uranga, J., Guerrero, P., & de la Caba, K. (2018) The potential of vegetal and animal proteins to develop more sustainable food packaging. In: Gutiérrez T. J. (Ed.). Polymers for food applications. Springer, Cham. pp 25–59. doi:https://doi.org/10.1007/978-3-319-94625-2_3.

Ghidelli, C., & Pérez-Gago, M. B. (2018). Recent advances in modified atmosphere packaging and edible coatings to maintain quality of fresh-cut fruits and vegetables. *Critical Review in Food and Nutrition, 58*(4), 622–679. https://doi.org/10.1080/10408398.2016.1211087.

Ghidelli, C., Sanchís, E., Rojas-Argudo, C., Pérez-Gago, M. B., & Mateos, M. (2018). Controlling enzymatic browning of fresh-cut eggplant by application of an edible coating and modified atmosphere packaging. *Acta Horticulturae, 1209*(1209), 239–245. https://doi.org/10.17660/actahortic.2018.1209.34.

Gutiérrez, T. J. (2017a). Chapter 8. Chitosan applications for the food industry. In S. Ahmed & S. Ikram (Eds.), *Chitosan: Derivatives, composites and applications* (pp. 183–232). WILEY-Scrivener Publisher. EE.UU. ISBN: 978-1-119-36350-7. https://doi.org/10.1002/9781119364849.ch8.

Gutiérrez, T. J. (2017b). Surface and nutraceutical properties of edible films made from starchy sources with and without added blackberry pulp. *Carbohydrate Polymers, 165*, 169–179. https://doi.org/10.1016/j.carbpol.2017.02.016.

Gutiérrez, T. J. (2018a). Polymers for food applications: News. In T. J. Gutiérrez (Ed.), *Polymers for food applications* (pp. 1–4). Cham: Springer. https://doi.org/10.1007/978-3-319-94625-2_1.

Gutiérrez, T. J. (2018b). Active and intelligent films made from starchy sources/blackberry pulp. *Journal of Polymers and the Environment, 26*(6), 2374–2391. https://doi.org/10.1007/s10924-017-1134-y.

Gutiérrez, T. J., & Álvarez, K. (2017). Chapter 4. Transport phenomena in biodegradable and edible films. In: *Biopackaging*. Masuelli, M. A. (Ed). Editorial CRC Press Taylor & Francis Group. Miami, EE.UU. ISBN: 978–1–4987–4968–8. pp. 58–88.

Hajji, S., Younes, I., Affes, S., Boufi, S., & Nasri, M. (2018). Optimization of the formulation of chitosan edible coatings supplemented with carotenoproteins and their use for extending strawberries postharvest life. *Food Hydrocolloids, 83*, 375–392. https://doi.org/10.1016/j.foodhyd.2018.05.013.

Hassan, B., Chatha, S. A. S., Hussain, A. I., Zia, K. M., & Akhtar, N. (2018). Recent advances on polysaccharides, lipids and protein based edible films and coatings: A review. *International Journal of Biological Macromolecules, 109*, 1095–1107. https://doi.org/10.1016/j.ijbiomac.2017.11.097.

Kharchoufi, S., Parafati, L., Licciardello, F., Muratore, G., Hamdi, M., Cirvilleri, G., & Restuccia, C. (2018). Edible coatings incorporating pomegranate peel extract and biocontrol yeast to reduce *Penicillium digitatum* postharvest decay of oranges. *Food Microbiology, 74*, 107–112. https://doi.org/10.1016/j.fm.2018.03.011.

Khorram, F., Ramezanian, A., & Hosseini, S. M. H. (2017). Effect of different edible coatings on postharvest quality of 'Kinnow' mandarin. *Journal of Food Measurement and Characterization, 11*(4), 1827–1833. https://doi.org/10.1007/s11694-017-9564-8.

Kumar, P., Sethi, S., & Sharma, R. R. (2018). Inhibition of browning in fresh-cut apple wedges through edible coatings and anti-browning agents. *Indian Journal of Horticulture, 75*(3), 517–522. https://doi.org/10.5958/0974-0112.2018.00087.7.

Li, S., Zhang, L., Liu, M., Wang, X., Zhao, G., & Zong, W. (2017). Effect of poly-ε-lysine incorporated into alginate-based edible coatings on microbial and physicochemical properties of fresh-cut kiwifruit. *Postharvest Biology and Technology, 134*, 114–121. https://doi.org/10.1016/j.postharvbio.2017.08.014.

López-Palestina, C. U., Aguirre-Mancilla, C. L., Raya-Pérez, J. C., Ramínez-Pimentel, J. G., Gutiérrez-Tlahque, J., & Hernández-Fuentes, A. D. (2018). The effect of an edible coating with tomato oily extract on the physicochemical and antioxidant properties of garambullo (*Myrtillocactus geometrizans*) fruits. *Agronomy, 8*(11), 248. https://doi.org/10.3390/agronomy8110248.

Medina Jaramillo, C., Gutiérrez, T. J., Goyanes, S., Bernal, C., & Famá, L. (2016). Biodegradability and plasticizing effect of yerba mate extract on cassava starch edible films. *Carbohydrate Polymers, 151*, 150–159. https://doi.org/10.1016/j.carbpol.2016.05.025.

Merino, D., Gutiérrez, T. J., & Alvarez, V. A. (2019). Potential agricultural mulch films based on native and phosphorylated corn starch with and without surface functionalization with chitosan. *Journal of Polymers and the Environment, 27*(1), 97–105. https://doi.org/10.1007/s10924-018-1325-1.

Montero-Calderón, M., Soliva-Fortuny, R., & Martín-Belloso, O. (2016). In M. Â. P. R. Cerqueira, R. N. C. Pereira, Ó. L. S. Ramos, J. A. C. Teixeira, & A. A. Vicente (Eds.), *Edible packaging for fruits and vegetables in edible food packaging: Materials and processing technologies* (pp. 353–382). Boca Raton: CRC Press.

Mubarak, A., & Rao Engakanah, T. (2017). Aloe vera edible coating retains the bioactive compounds of wax apples (*Syzygium Samarangense*). *Malaysian Applied Biology, 46*(4), 141–148.

Muñoz-Labrador, A., Moreno, R., Villamiel, M., & Montilla, A. (2018). Preparation of citrus pectin gels by power ultrasound and its application as an edible coating in strawberries. *Journal of the Science of Food and Agriculture, 98*(13), 4866–4875. https://doi.org/10.1002/jsfa.9018.

Murmu, S. B., & Mishra, H. N. (2018). The effect of edible coating based on Arabic gum, sodium caseinate and essential oil of cinnamon and lemon grass on guava. *Food Chemistry, 245*, 820–828. https://doi.org/10.1016/j.foodchem.2017.11.104.

Nogueira, G. F., Soares, C. T., Cavasini, R., Fakhouri, F. M., & de Oliveira, R. A. (2019). Bioactive films of arrowroot starch and blackberry pulp: Physical, mechanical and barrier properties and stability to pH and sterilization. *Food Chemistry, 275*, 417–425. https://doi.org/10.1016/j.foodchem.2018.09.054.

Ortiz-Duarte, G., Pérez-Cabrera, L. E., Artés-Hernández, F., & Martínez-Hernández, G. B. (2019). Ag-chitosan nanocomposites in edible coatings affect the quality of fresh-cut melon. *Postharvest Biology and Technology, 147*, 174–184. https://doi.org/10.1016/j.postharvbio.2018.09.021.

Paul, S. K., Sarkar, S., Sethi, L. N., & Ghosh, S. K. (2018). Development of chitosan based optimized edible coating for tomato (*Solanum lycopersicum*) and its characterization. *Journal of Food Science and Technology, 55*(7), 2446–2456. https://doi.org/10.1007/s13197-018-3162-6.

Rangel-Marrón, M., Mani-López, E., Palou, E., & López-Malo, A. (2019). Effects of alginate-glycerol-citric acid concentrations on selected physical, mechanical, and barrier properties of papaya puree-based edible films and coatings, as evaluated by response surface methodology. *LWT–Food Science and Technology, 101*, 83–91. https://doi.org/10.1016/j.lwt.2018.11.005.

Robles-Flores, G., Abud-Arcgilla, M., Ventura-Canseco, L. M. C., Meza-Gordillo, R., Grajales-Lagunes, A., Ruiz-Cabrera, M. A., & Gutiérrez-Miceli, F. A. (2018). Development and evaluation of a film and edible coating obtained from the *Cajanus cajan* seed applied to fresh strawberry fruit. *Food and Bioprocess Technology, 11*(12), 2172–2181. https://doi.org/10.1007/s11947-018-2175-5.

Saberi, B., & Golding, J. B. (2018). Postharvest application of biopolymer-based edible coatings to improve the quality of fresh horticultural produce. In T. J. Gutiérrez (Ed.), *Polymers for food applications* (pp. 211–250). Cham: Springer. https://doi.org/10.1007/978-3-319-94625-2_9.

Salgado, P. R., Ortiz, C. M., Musso, Y. S., Di Giorgio, L., & Mauri, A. N. (2015). Edible films and coatings containing bioactives. *Current Opinion in Food Science, 5*, 86–92. https://doi.org/10.1016/j.cofs.2015.09.004.

Salinas-Roca, B., Soliva-Fortuny, R., Welti-Chanes, J., & Martín-Belloso, O. (2018). Effect of pulsed light, edible coating, and dipping on the phenolic profile and antioxidant potential of fresh-cut mango. *Journal of Food Processing and Preservation, 42*(5), e13591. https://doi.org/10.1111/jfpp.13591.

Shigematsu, E., Dorta, C., Rodrigues, F. J., Cedran, M. F., Giannoni, J. A., Oshiiwa, M., & Mauro, M. A. (2018). Edible coating with probiotic as a quality factor for minimally processed carrots. *Journal of Food Science and Technology, 55*(9), 3712–3720. https://doi.org/10.1007/s13197-018-3301-0.

Silva, A. F., Finkler, L., & Finkler, C. L. L. (2018). Effect of edible coatings based on alginate/pectin on quality preservation of minimally processed 'Espada' mangoes. *Journal of Food Science and Technology, 55*(12), 5055–5063. https://doi.org/10.1007/s13197-018-3444-z.

Sneha Nair, M., Saxena, A., & Kaur, C. (2018). Characterization and antifungal activity of pomegranate peel extract and its use in polysaccharide-based edible coatings to extend the shelf-life of capsicum (*Capsicum annuum* L.). *Food and Bioprocess Technology, 11*(7), 1317–1327. https://doi.org/10.1007/s11947-018-2101-x.

Soares, A. S., Ramos, A. M., Vieira, É. N. R., Vanzela, E. S. L., Oliveira, P. M., & Paula, D. A. (2018). Vacuum impregnation of chitosan-based edible coating in minimally processed pumpkin. *International Journal of Food Science and Technology, 53*(9), 2229–2238. https://doi.org/10.1111/ijfs.13811.

Tapia-Blácido, D. R., Maniglia, B. C., & Tosi, M. M. (2018). Transport phenomena in edible films. In T. J. Gutiérrez (Ed.), *Polymers for food applications* (pp. 149–192). Cham: Springer. https://doi.org/10.1007/978-3-319-94625-2_7.

Tesfay, S. Z., Magwaza, L. S., Mbili, N., & Mditshwa, A. (2017). Carboxyl methylcellulose (CMC) containing moringa plant extracts as new postharvest organic edible coating for avocado (*Persea americana* mill.) fruit. *Scientia Horticulturae, 226*(19), 201–207. https://doi.org/10.1016/j.scienta.2017.08.047.

Thakur, R., Pristijono, P., Golding, J. B., Stathopoulos, C. E., Scarlett, C. J., Bowyer, M., Singh, S. P., & Vuong, Q. V. (2018). Development and application of rice starch based edible coating to improve the postharvest storage potential and quality of plum fruit (*Prunus salicina*). *Scientia Horticulture, 237*, 59–66. https://doi.org/10.1016/j.scienta.2018.04.005.

Thakur, R., Pristijono, P., Bowyer, M., Singh, S. P., Scarlett, C. J., Stathopoulos, C. E., & Vuong, Q. V. (2019). A starch edible surface coating delays banana fruit ripening. *LWT – Food Science and Technology, 100*, 341–347. https://doi.org/10.1016/j.lwt.2018.10.055.

Todisco, K. M., Janzantti, N. S., Santos, A. B., Galli, F. S., & Mauro, M. A. (2018). Effects of temperature and pectin edible coatings with guava byproducts on the drying kinetics and quality of dried red guava. *Journal of Food Science and Technology, 55*(12), 4735–4746. https://doi.org/10.1007/s13197-018-3369-6.

Torres-León, C., Vicente, A. A., Flores-López, M. L., Rojas, R., Serna-Cock, L., Alvarez-Pérez, O. B., & Aguilar, C. N. (2018). Edible films and coatings based on mango (var. Ataulfo) byproducts to improve gas transfer rate of peach. *LWT–Food Science and Technology, 97*, 624–631. https://doi.org/10.1016/j.lwt.2018.07.057.

Valencia, G. A., & do Amaral Sobral, P. J. (2018). Recent trends on nano-biocomposite polymers for food packaging. In T. J. Gutiérrez (Ed.), *Polymers for food applications* (pp. 101–130). Cham: Springer. https://doi.org/10.1007/978-3-319-94625-2_5.

Vargas, M., Pastor, C., Chiralt, A., McClements, D. J., & González-Martínez, C. (2008). Recent advances in edible coatings for fresh and minimally processed fruits. *Critical Review in Food and Nutrition, 48*(6), 496–511. https://doi.org/10.1080/10408390701537344.

Ventura-Aguilar, R. I., Bautista-Baños, S., Flores-García, G., & Zavaleta-Avejar, L. (2018). Impact of chitosan based edible coatings functionalized with natural compounds on *Colletotrichum fragariae* development and the quality of strawberries. *Food Chemistry, 262*, 142–149. https://doi.org/10.1016/j.foodchem.2018.04.063.

Vivek, K., & Subbarao, K. V. (2018). Effect of edible chitosan coating on combined ultrasound and NaOCl treated kiwi fruits during refrigerated storage. *International Food Research Journal, 25*(1), 101–108.

Wang, X., Sun, X., Liu, H., Li, M., & Ma, Z. (2011). Barrier and mechanical properties of carrot puree films. *Food and Bioproduct Processing, 89*(2), 149–156. https://doi.org/10.1016/j.fbp.2010.03.012.

Won, J. S., Lee, S. J., Park, H. H., Song, K. B., & Min, S. C. (2018). Edible coating using a chitosan-based colloid incorporating grapefruit seed extract for cherry tomato safety and preservation. *Journal of Food Science, 83*(1), 138–146. https://doi.org/10.1111/1750-3841.14002.

Yan, J., Luo, Z., Ban, Z., Lu, H., Li, D., Yang, D., Aghdam, M. S., & Li, L. (2018). The effect of the layer-by-layer (LBL) edible coating on strawberry quality and metabolites during storage. *Postharvest Biology and Technology, 147*, 29–38. https://doi.org/10.1016/j.postharvbio.2018.09.002.

Yousuf, B., Qadri, O. S., & Srivastava, A. K. (2018). Recent developments in shelf-life extension of fresh-cut fruits and vegetables by application of different edible coatings: A review. *LWT – Food Science and Technology, 89*, 198–209. https://doi.org/10.1016/j.lwt.2017.10.051.

Zhang, L., Chen, F., Lai, S., Wang, H., & Yang, H. (2018). Impact of soybean protein isolate-chitosan edible coating on the softening of apricot fruit during storage. *LWT – Food Science and Technology, 96*, 604–611. https://doi.org/10.1016/j.lwt.2018.06.011.

Chapter 20
Smart and Active Edible Coatings Based on Biopolymers

Germán Ayala Valencia, Carla Giovana Luciano, and Alcilene R. Monteiro Fritz

Abstract In the last years, new concepts for edible coatings have been developed as a response to the demands of modern society, as well as the industrial production trends toward fresh, safe, wholesome, mildly preserved, convenient and quality food products with a longer shelf life. In this way, smart and active edible coatings have been developed as food packaging for several products. Smart edible coatings are used to monitor and inform food conditions to consumers in real time whereas active edible coatings are used to increase the shelf life in foods. Smart and active edible coatings are manufactured using biopolymers as solid matrix to immobilize additives with specific properties (e.g. pH sensitive, antimicrobial and antioxidant properties, among others). Recently, layer-by-layer (L-b-L) technique has been used to manufacture composite edible coatings. This technique is based on the alternate deposition of different biopolymers which enhance the physicochemical properties and activity of smart and active edible coatings. This chapter reviews the state of the art with regard to the use of L-b-L technique as well as the use of additives (including nanoparticles) to manufacture smart and active edible coatings. Especially, this chapter include information about the use of biopolymers for producing edible coatings by L-b-L; the use of additives (biological and synthetic) and nanoparticles to manufacture smart and active edible coatings, as well as prospects and limitations in smart and active edible coatings.

Keywords Additives · Food packaging · Food quality · Shelf life

G. A. Valencia (✉) · A. R. Monteiro Fritz
Department of Chemical and Food Engineering, Federal University of Santa Catarina, Florianópolis, SC, Brazil
e-mail: g.ayala.valencia@ufsc.br

C. G. Luciano
Department of Food Engineering, Faculty of Animal Science and Food Engineering, University of São Paulo, Pirassununga, SP, Brazil

T. J. Gutiérrez (ed.), *Polymers for Agri-Food Applications*,
https://doi.org/10.1007/978-3-030-19416-1_20

20.1 Introduction

The greatest challenge of the food industry is the deterioration of food products due to microbial growth and oxidation reactions such as degradation, enzymatic browning, and oxidative rancidity. One alternative to reduce food deterioration is to use films or edible coatings (Álvarez et al. 2017; 2018). Edible coatings constitute thin layers of material which act as a barrier against physical, chemical, and biological contaminants, as well as against water vapor, oxygen and moisture (Eça et al. 2014).

Biopolymers are good candidates to manufacture edible coatings for food industry (Valencia and Sobral 2018). In recent years, the development of novel biopolymer-based edible coatings containing smart and/or active additives has increased the shelf life of foods, improving food safety and quality as well as the convenience to consumers (Arnon-Rips and Poverenov 2018). Smart edible coatings are used to monitor and inform the food conditions to consumers in real time whereas active edible coatings are used to increase the shelf life in foods (Bracone et al. 2016; Ghaani et al. 2016).

Edible coatings based on single material often cannot answer numerous food packaging requirements. An alternative to improve edible coating properties is to develop composite coatings by the alternate deposition of different biopolymers to produce an edible coating, this technique is called layer-by-layer (L-b-L) (Arnon-Rips and Poverenov 2018).

This chapter aims to review the recent trends about the use of biopolymers to produce edible coatings by L-b-L; the use of additives (biological and synthetic) and nanoparticles to manufacture smart and active edible coatings, as well as prospects and limitations in smart and active edible coatings.

20.2 Application of L-b-L Technique to Manufacture Edible Coatings Based on Biopolymers

Edible coatings are mainly made from biodegradable polymers and are usually classified according to their structural material (Valencia and Sobral 2018). Currently, these coatings, based on several materials as proteins, lipids, polysaccharides or composites have been developed for their use as regular food packaging in an attempt to minimize damage to the environment caused by incorrect disposal of synthetic materials (Falguera et al. 2011; Gutiérrez et al. 2019; Herniou--Julien et al. 2019). The coatings with controlled release of active compounds is also of great interest. In this sense, nanostructured thin coatings have been well exploited, and they have showed a great potential. Far beyond the food industry, the application of these coatings ranges from developing of molecular devices and sensors to pharmaceutical and biomedical materials, and several techniques have been developed to reach an ultrathin coating (Ariga et al. 2007; Radeva et al. 2006). So, during the past decades, nano-engineered materials development has been widely studied.

According to Guzmán et al. (2017), functional materials built with bottom-up molecular or supramolecular assembly techniques gained attention, mainly because these techniques allow a great control over the composition and structure of the coatings, being much more accurate than bulk materials. So, this relatively new kind of manufacturing could enhance structural precision, which means an added-value for fields as design of functional biomaterials, biomedical engineering and biotechnology (McMahon et al. 2013; Richardson et al. 2016).

Among the techniques used to produce coatings, the traditional ones are Langmuir-Blodgett (LB) and the self-assembled monolayer (SAM) technique. However, these two proceedings have some restrictions, and they are not suitable for all the purposes. The LB technique is applicable on the production of multilayer coatings but needs some expensive devices and it is not appropriated to use on amphiphilic materials (Ariga et al. 2007; Fujii et al. 2010), while the SAM technique, although it can be applied to a wide range of substances, but is not proper for the manufacture of multilayer coatings (Love et al. 2005).

Another technique very studied in the past decades is the self-assembling, also known as L-b-L technique (Ban et al. 2018). This pervasive technique has received a growing interest, with an exponential growing number of publications, and it consists on the formation of a multilayer coating through the sequential adsorption of complementary molecules (Ariga et al. 2007; Guzmán et al. 2017). As compared to the traditional techniques, as LB and SAM, this step-by-step (L-b-L) approach is an easy, not expensive, and versatile process for multilayer coating formation, that allows different types of materials to be incorporated in the coating structure, considering most of the biomaterials have charged sites on the surface that are assembled by electrostatic interaction (Ariga et al. 2007; Srivastava and Kotov 2008). In addition, the equipment required to produce coatings by the L-b-L technique is simple and common to most laboratories, which makes this method even more attractive (Richardson et al. 2016).

In general, the L-b-L technique leads to the gradual growth of coatings with a very low thickness, involving the cyclical adsorption of opposite charge-materials onto specific substrates (Hu et al. 2017). Due to the specific features of the L-b-L assembly, a thin multilayer coating with a molecular level of thickness and composition control can be considered highly convenient to build (Kim et al. 2018a; Li et al. 2012; Richardson et al. 2016).

This process of multilayer formation was initially proposed by Iler (1966), describing the adsorption of opposite charge colloidal particles. In 1966, this author developed alternated layers of silica and alumina onto a smooth surface of a black plate glass, building layers of positively and negatively charged colloidal particles to form a coating. It was however only in 1992, that they arrived to develop a real multilayer coating by means of electrostatic opposite charge interaction. This author proved to be possible the deposition of polyelectrolytes on a substrate, ensuring coatings with well positioned individual bilayer, using electrostatic attraction as a driving force (Decher and Schmitt 1992). According to Decher and Schmitt (1992) also the use of polyelectrolytes is very advantageous in the manufacture of nanostructured coatings, guaranteeing good adherence and high quality of the coating.

The conventional L-b-L method for the production of coatings already has well-established protocols, and this technique generally relies on easy methodologies. This technique has also multiple advantages, and is suitable for the application on different fields, as sensors, drug delivery systems, catalysis and energy storage and conversion (Ban et al. 2018; Xiang et al. 2012).

As advantages of L-b-L, the possibility of using a wide range of materials can be mentioned. Almost every charged material can be used, including especially water-soluble proteins and polysaccharides, small organic molecules (natural or synthetic), nanotubes, dyes, enzymes, peptides and polypeptides (Srivastava and Kotov 2008; Xiao et al. 2016). Charged inorganic compounds (i.e. nanoparticles) are also suitable to be assembled by electrostatic adsorption (Murray and Kagan 2000). Besides, the possibility of precise thickness control in a nanoscale level, the immobilization of different biomolecules within the layered structure, the controlled release of active compounds, the increase on the coating adherence on the hydrophilic surface of the food, and a high defined control of the physicochemical properties, allow the L-b-L technique to be suitable to the formation of smart coatings that could be used for several functions and applications (Aloui and Khwaldia 2016; Lisdat 2017; Poverenov et al. 2014a; Sipahi et al. 2013; Srivastava and Kotov 2008). However, the conditions of this method must be well controlled, since some studies have proved that the result is highly influenced by the pH and salt concentration of the deposition solution, the concentration of the stratification material, the immersion time, the washing parameters, the charges density, the molecular weight of the materials, and other variables can cause differences in coating growth (Dubas and Schlenoff 1999; Nascimento et al. 2018). As it was placed, the fast development of this assembling technique has brought a lot of benefits in different research areas. Even so, some authors have affirmed that the unique advantages of the L-b-L assembling have not yet been fully exploited (Hu et al. 2017).

Although electrostatic interactions are the most exploited force studied in this technique, because it easily allows the construction of multilayer coatings with a high control (Hammond 2004). It is worth noting that there are other possible driving forces that can occur in the formation of L-b-L coatings. Assemblies based on ionic charge transfer (donor – acceptor interactions), coordinating bonding, hydrogen bonding, covalent attachment, biological recognition, and hydrophobic interactions have also been vigorously investigated (Borges and Mano 2014; Srivastava and Kotov 2008).

In a typical L-b-L formation process, first, a treatment on the substrate surface is necessary in order to create a positive or negative charge. After that, the charged substrates are successively immersed into solutions containing small molecules with an opposite charge (Fig. 20.1) (Shutava et al. 2019). Following the deposition of one layer it is often necessary to wash the material in order to remove the excess of free anions or cations and avoid molecular complexation phenomena, before the deposition of the next layer. Such steps, besides avoiding the contamination of the following adsorbing solutions, also allow the stabilization of the nanometric individual layers and the manufacture of coatings with fine and highly controlled compositions (Borges and Mano 2014). After repeated proceedings of dipping and wash, the coatings grow progressively on a linear or exponential manner on the substrate surface

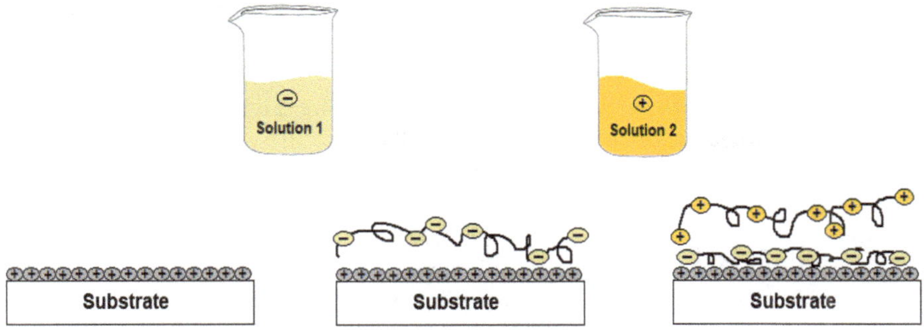

Fig. 20.1 Formation of L-b-L structure by electrostatic interaction

(Hu et al. 2017; Tan et al. 2003). According to Richardson et al. (2016), depending on the process, L-b-L buildup can be achieved by different deposition methods, among which the dip-coating, spin-coating, spraying and perfusion can be highlighted, and despite these coatings are frequently used on planar surfaces, it is also possible to be applied on porous substrates, spheres and cylindrical structures.

L-b-L technique has been the target of several researches, mainly for perishable foods as minimally processed vegetables or fresh-cut fruits (Arnon-Rips and Poverenov 2018), considering that thin coatings showed to be promising in several fields. In fact, the L-b-L electrodeposition could be perfectly used to coat highly hydrophilic food systems (Vargas et al. 2008). In this sense, various reviews can be found that highlight the L-b-L assemblies. Some of them are cited in the sequence.

Chitosan (Cs) and carboxymethyl cellulose were studied to recover fresh strawberries by using the L-b-L technique by Ban et al. (2018). These authors analyzed the changes on the fruit proteins regarding different treatments and found some promising results, with 34 proteins regulated by the L-b-L coatings.

Poverenov et al. (2014b) investigated coatings based on the L-b-L electrostatic deposition of gelatin and Cs, and compared this method with coatings based on blends of these two biopolymers. These authors studied the blended and L-b-L composite coatings on fresh-cut melon and concluded that the L-b-L coatings enhanced the water vapor permeability and had better elasticity and higher strength than the blended coatings. The L-b-L coatings presented also superior performance for the preservation of the melon texture.

A combination of alginate and Cs was also studied for the application in fresh-cut melons by using the L-b-L technique. In this case, the Cs was responsible for the reduction on the microbial contamination, and the alginate provided physical barrier in order to decrease the respiration rate (Poverenov et al. 2014a). The authors found that this technique provided a good adhesion of the coating on the melon matrix. The L-b-L coating improved also the gas exchange properties, suggesting that this simple method can improve quality and safety of food products (Poverenov et al. 2014a).

Moradi et al. (2015) produced L-b-L coatings based on pectin and Cs with the addition of lemon and peppermint essential oils in order to investigate the effectiveness of a multilayer edible coating to improve the quality and shelf-life of fish. The results obtained in this work demonstrated that this technique is suitable to be used in antimicrobial release systems, and could properly delay the growth of spoilage microorganisms, extending the fish shelf life.

Radeva et al. (2006) manufactured coatings by using the L-b-L technique with Cs and carboxymethyl cellulose (CMC). According to these authors, an irregular growth of the layers was observed and concluded that the adsorption of the polymers occurs only on the film surface, without diffusion inside and outside of the film bulk during deposition of each CMC/Cs bilayer.

Sipahi et al. (2013) studied the antimicrobial efficacy of L-b-L coatings in fresh-cut watermelon based on natural antimicrobial properties of the coating components requires further investigation, as well as their implications of using L-b-L for overall product quality.

20.3 Smart Edible Coatings

Smart edible coatings are self-adhesive labels attached to single packages which communicate the product's status to end users by means of an "ON/OFF" switching function on the package in response to changes in external/internal stimuli (Ghaani et al. 2016; Schaefer and Cheung 2018). Monitoring external/internal stimuli is used to inform the status of the product in term of its safety and quality by showing either food to be safe/unsafe or the freshness, ripeness or firmness (Pavelková 2013; Toro-Márquez et al. 2018). Smart edible coatings can thus contribute to improving HACCP (Hazard Analysis and Critical Control Points') and QACCP (Quality Analysis and Critical Control Points') systems (Biji et al. 2015).

Temperature and pH are the most important parameters that influence the shelf life of several foods (Biji et al. 2015; Wang et al. 2015b), therefore, most external/internal stimuli are monitored by means controlling temperature, pH or a derivative substance associated with a chemical or biological process influenced by temperature or pH (e.g. nutrient degradation and microbial growth) (Ghaani et al. 2016; Pavelková 2013; Schaefer and Cheung 2018; Wang et al. 2015b). Based on the principle of monitoring external/internal stimuli, smart edible coatings can be divided into indicators, data carriers and sensors (Fig. 20.2) (Biji et al. 2015; Ghaani et al. 2016; Wang et al. 2015b).

20.3.1 *Indicators*

Indicator systems are used to detect the presence of a substance in a qualitative or semi-quantitative form (Ghaani et al. 2016). Usually, indicator systems express a visible response such as color changes due to the redox oxidation of a biological compound or a physical deformation (Pavelková 2013; Wang et al. 2015b). These indicators are:

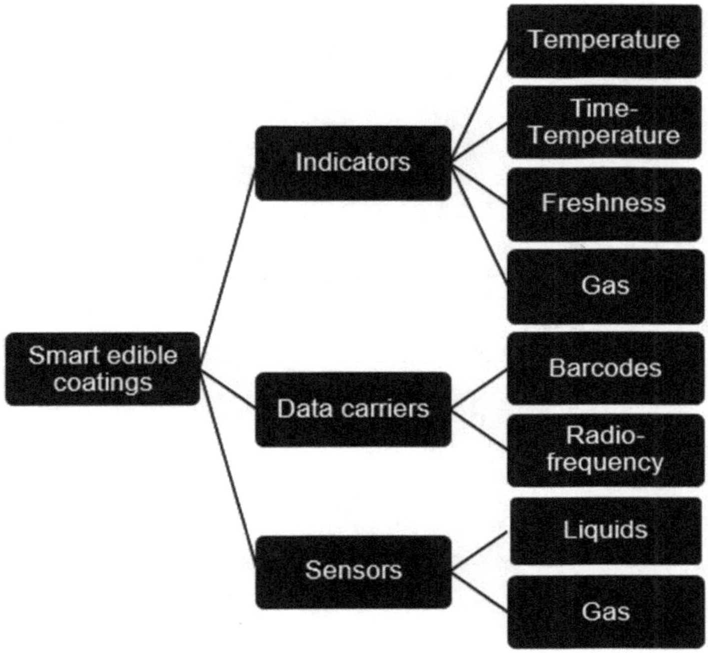

Fig. 20.2 Classification of smart edible coatings based on the principle of monitoring external/internal stimuli. Adapted from Ghaani et al. (2016)

20.3.1.1 Temperature Indicators (TIs)

TIs show whether foods have been cooled below or heated above a critical temperature, and has been associated to protein denaturation or microbial growth during freezing or defrosting processes (Vanderroost et al. 2014).

TIs use monomers during heat-sensitive polymerization such as acetylene groups ($R_1C \equiv C–C \equiv CR_2$) which are transformed into polydiacetylene ($R_1[C = C – C = C]_nR_2$), changing the color from red to blue when exposed to high temperature or radiation (Wang et al. 2015b). Azobenzene groups are also used with acetylene groups to manufacture TIs with reversibly or irreversibly color change in different temperature ranges (Wang et al. 2015b). Other compounds containing acetylene group such as 5,7-dodecadiyn-1,12-diolbis(octadecyl urethane), 5,7-dodecadiyn-1,12-diolbis(butoxycarbonyl urethane) and 2,4-hexadiyn-1,6-bis(alkylurea) are also used to manufacture smart edible coatings (Wang et al. 2015b).

Fresh-Check® and HEATmarker® from TempTime (New Jersey, USA) are some examples of smart edible coatings containing acetylene polymers. Fresh-Check® label consists of a small circle of a polymer surrounded by a printed reference ring. The circle color changes gradually from light to dark as the package is exposed to high temperatures (Fig. 20.3a). Similar to Fresh-Check®,

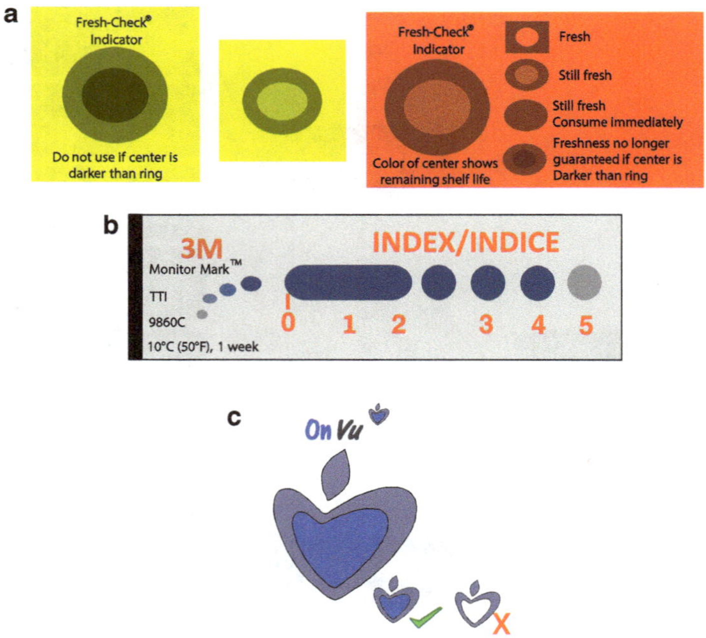

Fig. 20.3 (**a**) Schematic illustration for Fresh-Check® label. (**b**) example of Monitor Mark™ label. (**c**) example of OnVu™ label. Adapted from Ghaani et al. (2016)

HEATmarker® shows the color change of a light rectangular area with the heating (Pavelková 2013; Wang et al. 2015b). The main disadvantage of these materials is that polydiacetylene compounds pose a potential toxicity for food products, limiting future food packaging applications (Ghaani et al. 2016; Wang et al. 2015b).

Other TIs are based on glass transition (T_g) or melting (T_m) temperatures of the polymer in the label. Monitor Mark™ (St Paul, Minnesota, USA) is an example of temperature indicator based on T_g (Fig. 20.3b). In this label, a viscoelastic material (blue-dyed fatty acid ester) migrates into a porous film when exposed to temperatures higher than T_g. The colorimetric response is the distance of the diffusion and depends on temperature, the diffusing polymer's concentration and its T_g (Pavelková 2013; Wang et al. 2015b). An example of TIs based on T_m is the Bump Mark, this smart edible coating is based on a smooth gelatin coating deposited on a bumpy plastic sheet. When the food is exposed to high temperatures, gelatin coating becomes a liquid system and the bumpy surface is revealed, indicating that the food should not be consumed (Brunel University London 2014).

Some researchers have developed TIs at laboratory scale using anthocyanins and chlorophyll as sensitive compounds to temperature changes. These smart edible coatings were manufactured using Cs as matrix and showed irreversible visual color changes for a temperature range between 40 and 70 °C (Maciel et al. 2012a, b).

20.3.1.2 Time-Temperature Indicators (TTIs)

TTIs monitor temperature alterations over time, these labels use polymers or additives sensitive to light (photochromic), or even to oxygen which is monitored by means of redox reaction (Mohebi and Marquez 2015; Wang et al. 2015b).

TTIs based on photochromic compounds use spiropyran and spirooxazine compounds sensitive to light. Spiropyran and spirooxazine are colorless, absorbing the light between 200 and 400 nm. These compounds become colorful between 500 and 600 nm as a consequence of spiropyran and spirooxazine isomerization due to exposure to light at a controlled temperature (Kreyenschmidt et al. 2010; Wang et al. 2015b). The OnVu™ by Freshpoint (Switzerland) is the most popular smart edible coating based on photosensitive compounds (Fig. 20.3c), organic pigments, e.g. benzylpyridines, that change color, from blue to white with time at rates determined by temperature (Pavelková 2013).

Other TTIs used are the oxygen-sensitive compounds (redox reaction) such as anthraquinone and derivatives. The color of these compounds changes from uncolored to beige, being influenced by time and temperature. However, anthraquinones are expensive and classified as potential toxic substances, being applied only in a limited temperature range (Galagan and Su 2008; Wang et al. 2015b).

The use of nanoparticles is a recent alternative to manufacture TTIs. Wang et al. (2017) added hydrogen tetrachloroaurate trihydrated in alginate-based coatings and observed that the color of coatings changed from grey to red with time of exposure at high temperature (40 °C). The color change in alginate-based coatings was due to the formation of gold nanoparticles (AuNPs) from hydrogen tetrachloroaurate trihydrated, influenced by temperature and time. Several other examples of TTIs are still in the laboratory stage (Table 20.1), most researches have used enzymes as sensitive compounds to manufacture TTIs.

20.3.1.3 Freshness Indicators (FIs)

FIs are smart devices that enable the monitoring of the quality of food products throughout distribution, storage and marketing. Most FIs are used to provide direct information on the product's fresh quality such as seafood, fish, meat and fruits, this control is made by monitoring the microbial growth, chemical changes or gases involved in the ripening process (Ghaani et al. 2016; Mohebi and Marquez 2015).

ToxinGuard® (Toxin Alert, Canada) is an example of FIs for meat products, this label has antibodies printed on polyethylene edible coating which detects the presence of some microorganism such as *Escherichia coli, Listeria monocytogenes, Salmonella* spp., and *Campylobacter* spp. (Biji et al. 2015). Another example of FIs is SensorQ™ by Food Quality Sensor International (FQSI), this label has a green dye bromocresol immobilized on a polymeric edible coating. Bromocresol color change occurs when the amines interact with the label as a consequence of meat microbial deterioration (Fuertes et al. 2016).

Table 20.1 List of recent research focused on developing smart edible coatings

Classification	Polymeric matrix	Indicator	Sensitive	Reference
TTIs	Corn starch/ iodine	Amylase	Heat	Brizio and Prentice (2015)
	No matrix	Urease/ carbamide		Wu et al. (2013)
	No matrix	Lipase/glycerol/ tributyrate		Wu et al. (2015)
	Cellulose	Clorophenol red		Zabala et al. (2015)
	Trilaurin/ tripalmitin	Lipase		Tsironi et al. (2017)
FIs	Agar/potato starch	Anthocyanin	pH	Choi et al. (2017)
	Corn starch/ nanoclays			Gutiérrez (2018a); Gutiérrez and Alvarez (2018)
	Plantain starch and flour			Gutiérrez (2017); Gutiérrez, Suniaga, Monsalve, & García (2016b)
	Chitosan			Halász and Csóka (2018); Zhang et al. (2014); Yoshida et al. (2014)
	Starch/poly-vinyl alcohol			Liu et al. (2017); Zhai et al. (2017)
	Chitosan/corn starch			Silva-Pereira et al. (2015)
	Bacterial cellulose			Pourjavaher et al. (2017)
	Corn starch			Prietto et al. (2017)
	Chitosan/agarose			Wu et al. (2018)
	Cassava starch	Green tea/basil extracts		Medina-Jaramillo et al. (2017)
	Chitosan/κ-- carrageenan	Curcumin		Pereira and Andrade (2017); Liu et al. (2018)
GIs	Ethylcellulose	Anthocyanin/ poly-lysine	Carbon dioxide	Saliu and Della Pergola (2018)
	Alginate	Thionine	Oxygen	Vu and Won (2013)
	Polyaniline	Polyaniline	Volatile basic nitrogen	Wang et al. (2018)

TTIs: time-temperature indicators; FIs: freshness indicators; GIs: gas indicators

About the quality control of fresh fruit, RipeSense by RipSense™ and Ort Research is the first label that changes color to indicate the ripeness of the fruit. Initially, this label is red and then color changes to orange and finally yellow due to the label interaction with aroma components, ethylene and carbon dioxide produced during the ripening process (Fuertes et al. 2016; Ghaani et al. 2016).

Several other examples of FIs are still in the laboratory stage (Table 20.1), most researches have been focused on the anthocyanin applications to manufacturing FIs. Anthocyanins are sensitive to light, oxygen and temperature, and can reduce mechanical properties in coatings. Some papers have thus studied the anthocyanin stabilization (Gutiérrez and Alvarez 2018; Gutiérrez et al. 2017b; Pourjavaher et al. 2017) or the improving of physicochemical properties of biopolymers-based coatings (Ma et al. 2017), both using nanoparticles.

20.3.1.4 Gas Indicators (GIs)

GIs are used principally to monitoring toxic gases produced from decomposing process of food packaged into containers and that can endanger the health of consumers. Another application of GIs is to monitor the presence of oxygen and carbon dioxide levels, the first is important in vacuum packed foods, indicating that the package was sealed correctly or incorrectly, the second, for food packed under atmosphere modified (Fuertes et al. 2016; Ghaani et al. 2016).

GIs use redox dyes such as methylene blue, 2,6-dichloroindophenol, or *N,N,N′,N′*- tetramethyl-*p*-phenylenediamine combined with a reducing compound as reducing sugars and an alkaline compound as sodium hydroxide to guarantee the colorimetric response of label when exposed to the gas. Some commercial examples of GIs are Shelf Life Guard by UPM, Ageless Eye™ by Mitsubishi Gas Chemical Co., Vitalon® by Toagosei Chemical Inc., Freshilizer by Toppan Printing Co. and Tufflex GS by Sealed Air Ltd. (Ghaani et al. 2016). Some examples of GIs have been developed on the laboratory scale using ethylcellulose, alginate and polyaniline as a sensitive compound when exposed to gases (Table 20.1).

Most information about smart packaging using TIs, TTIs, FIs and GIs can be found in the specialized literature (Biji et al. 2015; Fang et al. 2017; Fuertes et al. 2016; Ghaani et al. 2016; Mohebi and Marquez 2015; Pavelková 2013; Purkayastha et al. 2017; Schaefer and Cheung 2018; Shukla et al. 2016; Wang et al. 2015b).

20.3.2 Data Carriers

Data carriers are edible coatings that have automatic identification devices, these devises improve the information flow within the food supply chain (Ghaani et al. 2016). Data carriers are rather intended for automatization, traceability, counterfeit protection or theft prevention, however, these systems do not communicate the product's status to end users (McFarlane and Sheffi 2003). Data carriers can be classified in barcode and radio frequency identification (Fig. 20.2) (Fuertes et al. 2016; Ghaani et al. 2016).

Barcode is a pattern of parallel spaces and bars arranged to represent 13 digits of data. Commercially, edible coatings are integrated with one-dimensional barcode, this barcode can contain some general information about the product such as version number, manufacturer, product type, among others (McFarlane and Sheffi 2003). Additional information can be added using a two-dimensional barcode (Ghaani et al. 2016; Schmidt et al. 2013).

Radio frequency identification (RFID) is a data carrier device used for support data information that can identify and locate a food product with a special tag that emits radio waves (Fuertes et al. 2016; Ghaani et al. 2016). Normally, RFID system are integrated by a tag having a microchip connected to a tiny antenna; a reader that emits and receives information in radio waves form; and a local network or web server that links the RFID hardware and enterprise applications (Bibi et al. 2017a; Ghaani et al. 2016).

RFID can be classified into two groups: passive and active. Passive RFID depends on the power supplied by the reader, whereas the active RFID is powered by an internal battery, which is used to run the microchip circuitry and to transmit a signal to the reader. A hybrid RFID is called as semi passive RFID, this data carrier uses a battery to maintain memory on the tag or to power the electronic system (Biji et al. 2015). As compared with barcode, RFID can be scanned automatically without human participation, in addition, RFID can be scanned in any orientation (McFarlane and Sheffi 2003).

Biopolymer-based coatings can be coupled in RFID systems, recently, Bibi et al. (2016; 2017b) manufactured coatings based on wheat gluten and observed that the electrical and dielectric properties of these materials were very relative humidity-sensitive. The same authors concluded that the dependence of both, dielectric permittivity and dielectric loss of wheat gluten coatings on relative humidity offered the possibility of using these materials for monitoring relative humidity in packaged food products.

Most information about smart packaging using barcode and RFID can be found in the specialized literature (Bibi et al. 2017a; Biji et al. 2015; Fuertes et al. 2016; Ghaani et al. 2016; McFarlane and Sheffi 2003; Mohebi and Marquez 2015; Schmidt et al. 2013).

20.3.3 Sensors

sensors are used in liquid or gas phase to quantify a specific compound called analyte by means of a chemical or physical interaction between the analyte and the sensor (Fig. 20.2) (Biji et al. 2015; Ghaani et al. 2016). Sensors are analytical devices based on direct spatial coupling between a receptor and a signal transducer equipped with an electronic amplifier (Malhotra et al. 2005; Thévenot et al. 1999).

The receptor is the sensing part of the sensor where the surface chemistry occurs as a consequence of interaction between the receptor with a specific analyte (Fig. 20.4) (Ghaani et al. 2016). The transducer is the device that converts a chemical signal from the receptor into a quantifiable electrical signal (Fig. 20.4). Transducers can be classified as electrical, electrochemical, optical, thermal, acoustic and piezoelectric. Finally, the amplifier is used to increase the electrical signal

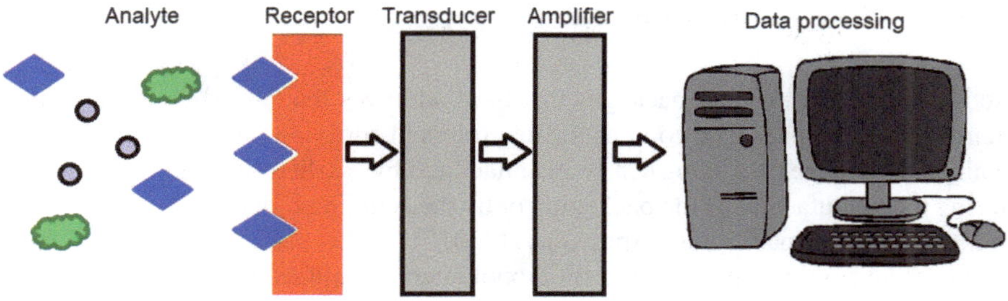

Fig. 20.4 Schematic of a sensor

that will be shown to end users by means of the software system (Fig. 20.4) (Malhotra et al. 2005; Thévenot et al. 1999).

Biosensors are sensors using a biological system as a receptor. These analytical devices have been applied immensely in food and defense processing industries, as well as in medical diagnostics, environmental monitoring and genetics (Malhotra et al. 2005). Generally, enzymes, antibodies, microorganisms, cell receptors, DNA, phage, tissues and biomimetic components have been used as receptors for the manufacture of biosensors (Adley 2014; Malhotra et al. 2005). Enzyme-based biosensors dominate the market. In food industry, these devices have been used to detect food borne pathogens, endotoxins, mycotoxins, pesticides, among others (Das et al. 2014; Kim et al. 2018b; Lv et al. 2018; Puligundla et al. 2012; Rubab et al. 2018; Selvaraj et al. 2015; Silva et al. 2018; Vidal et al. 2013; Wang et al. 2013; Yang et al. 2013; Ye et al. 2019).

Most enzyme-based biosensors are manufactured with biopolymers due to natural origin and their favorable interaction with living systems (Sawant 2017; Valencia and Sobral 2018). Biopolymers can be used for immobilization or attachment of enzymes to the biosensor interface by means of physical entrapment. Devi et al. (2013), e.g. immobilized xanthine oxidase on Cs films for the detection of xanthine in fish meat. In other studies, Valencia et al. (2014b) and Teepoo et al. (2017) immobilized horseradish peroxidase in Cs and Cs-gelatin coatings, respectively. The same authors concluded that these biosensors can be used to detect hydrogen peroxide in foods.

Nanoparticles can be used to manufacture biosensors, particularly, metal nanoparticles such as gold and iron nanoparticles are used for the immobilization of enzymes on the coating surface (Devi et al. 2013; Valencia et al. 2014b) or to increase the coating's conductivity (Valencia et al. 2014a).

In general, biosensors have a promising future for the detection of food components and contaminants in food matrices, however, some characteristics such as flexibility and miniaturization are necessaries to be incorporated in food packaging (Vanderroost et al. 2014). Most information about sensors and biosensors for food packaging can be found in the specialized literature (Adley 2014; Compagnone et al. 2017; Das et al. 2014; Sawant 2017; Silva et al. 2018; Vanderroost et al. 2014; Vidal et al. 2013; Ye et al. 2019).

20.4 Active Edible Coatings

Active edible coatings are packages that have additives that absorb or release substances from the packaged food or the environment surrounding the food (Yildirim et al. 2018). Active packaging can be manufactured by the direct addition of additives during the manufacture of the packaging or by the coupling of labels, sachets or pads in a conventional packaging (Ahmed et al. 2017). In this section will be discussed about active edible coatings based on biopolymers, manufactured with active additives during processing. Other active packaging using labels, sachets or pads as active material will not be discussed in this section, but they are available in the specialized literature (Ahmed et al. 2017; Otoni et al. 2016; Yildirim et al. 2018).

Based on the principle of active additive, biopolymer-based coatings with active properties can be classified in two groups: active edible coatings with antimicrobial/antioxidant properties and active edible coatings with scavenging properties.

20.4.1 Active Edible Coatings With Antimicrobial/Antioxidant Properties

Spoilage and pathogenic microorganisms, as well as food oxidation, are of major concern regarding food quality deterioration. Generally, microbial growth and oxidation of foods take place on the surface as a consequence of handling after processing and presence of oxygen in the package headspace, respectively (Ahmed et al. 2017; Gómez-Estaca et al. 2014).

An option to reduce microbial growth and increase oxidative stability in foods is to use edible coatings with antimicrobial/antioxidant properties. Edible coatings with antimicrobial properties are used to prolong the lag phase of microorganism, whereas edible coatings with antioxidant properties are used to control the oxygen level in foods in order to minimize lipid oxidation, microbial growth, discoloration, textural changes and production of off-flavors and toxic compounds (Ahmed et al. 2017).

Convectional additives with antimicrobial properties such as carbon dioxide, chlorine dioxide, sulfites, nitrites, organic acids and antibiotics, as well as conventional additives with antioxidant properties such as hydroxytoluene (BHT) or butylated hydroxyanisole (BHA) have been used to manufacture edible coatings with antimicrobial or antioxidant properties, respectively (Ahmed et al. 2017; Gómez-Estaca et al. 2014).

Toxicological problems from conventional additives cause great interest in the use of natural additives with antimicrobial/antioxidant properties in edible coatings (Boyaci et al. 2016). Hence, natural additives with antimicrobial/antioxidant properties obtained from plant source such as cinnamon (Abdollahzadeh et al. 2018; Santos et al. 2016), clove (Echeverría et al. 2018), curcuma (Bitencourt et al. 2014), boldo-do-chile, guaraná (Bonilla et al. 2018; Bonilla and Sobral 2016, 2017, 2018), garlic (Pérez-Córdoba et al. 2018), ginger (Noshirvani et al. 2017), limonene

(Chen et al. 2018), olive pomace (Crizel et al. 2018), oregano (Oliveira et al. 2017) and rosemary (Piñeros-Hernandez et al. 2017); or obtained by microbial metabolism such as nisin (Abdollahzadeh et al. 2018) and bacteriocins (Ahmad et al. 2017; Sidhu and Nehra 2017); or event enzymes (Boyaci et al. 2016) and oligomers (Castillo et al. 2017) have been extensively studied in food packaging technology. The principal disadvantage is that these compounds are heat-sensitive, which reduces their active properties during the manufacture of edible coatings (Ahmed et al. 2017).

Nanoparticles are another type of additive recently used by their antimicrobial/antioxidant properties in edible coatings. Particularly, silver (Abreu et al. 2015; Ortega et al. 2017, copper (Saldías et al. 2018), laponite (Li et al. 2015), titanium dioxide (Siripatrawan and Kaewklin 2018; Xie and Hung 2018; Zhang et al. 2017) and zinc oxide (Abdollahzadeh et al. 2018) nanoparticles have shown antimicrobial and even antioxidant properties. Nanoparticles are also thermally stable and they can be applied in coextrusion, extrusion or injection molding process to manufacture edible coatings (Ahmed et al. 2017).

20.4.2 Active Edible Coatings With Scavenging Properties

Controlling the gas concentrations in the package headspace is one of the keys to extending the shelf life of foods. Basically, oxygen, carbon dioxide and ethylene (climacteric foods) are the gases to be controlled in the package (Ahmed et al. 2017; Wang et al. 2015a). As described in Sect. 20.4.1., oxygen can induce food oxidation. Other gases such as carbon dioxide (too high concentration) and ethylene can result in anaerobic respiration and maturation of fresh foods, respectively (Wang et al. 2015b). Edible coatings can be used to absorb oxygen and excess of carbon dioxide, as well as to suppress the negative effects of ethylene in foods.

Most edible coatings are manufactured containing metal particles as oxygen absorbent, however, these materials cannot come into direct contact with food. Non-metallic oxygen scavengers such as unsaturated fatty acids, ascorbic acid, enzymatic and dyes are still in the laboratory stage (Ahmed et al. 2017; Yildirim et al. 2018).

In the last years, some researchers have manufactured edible coatings to absorb carbon dioxide and scavenging of ethylene. Wang et al. (2015b) manufactured agar-based edible coatings containing sodium carbonate (Na_2CO_3) and concluded that Na_2CO_3 did not alter carbon dioxide absorption in edible coating. Siripatrawan and Kaewklin (2018), Kaewklin et al. (2018) and Xie and Hung (2018) showed that titanium dioxide can be activated by UV light in edible coatings based on Cs, cellulose acetate, polycaprolactone and polylactic acid to scavenging ethylene in climacteric fruits. When illuminated with UV light, the electrons on the titanium dioxide surface are promoted and react with oxygen in the package, producing O_2^- and OH· which further oxidize ethylene to carbon dioxide and water as final products (Kaewklin et al. 2018).

For more information on this topic, the following literature is recommended: Ahmed et al. (2017), Atarés and Chiralt (2016) Gómez-Estaca et al. (2014), Ribeiro-Santos et al. (2017) and Yildirim et al. (2018).

20.5 Prospects and Limitations in Smart and Active Edible Coatings

With the aim of improving the edible coatings, some authors have used additives to improve physicochemical properties as well as to add active properties in the same coating. Li et al. (2015), e.g. developed gelatin-based coatings with laponite and observed that mechanical properties, water vapor permeability and water solubility of the coatings were significantly improved with the increase of the laponite concentration. In addition, laponite impeded lipid oxidation and decomposition of meat protein during storage due to the barrier capability of laponite. Other authors have added additives to manufacture biopolymer-based coatings with smart and active properties (Gutiérrez et al. 2016a). So, Medina-Jaramillo et al. (2017) developed starch-based coatings containing natural extracts such as green tea and basil. These authors concluded that green tea and basil extracts have antioxidant properties due to the high phenolic content. The same extracts showed also color changes when exposed to acid and basic pH due to the presence of chlorophyll and carotenoids. Finally, Liu et al. (2017) developed coatings based on starch/poly-vinyl alcohol containing anthocyanins and limonene and concluded that these coatings showed color changes between pH 1 and 14 as well as antimicrobial activity against *Bacillus subtilis*, *Aspergillus niger* and *Staphylococcus aureus*. The same authors concluded that the colorimetric and antimicrobial properties were due to the presence of anthocyanins and limonene in the coatings, respectively.

Regarding coating processing, the main disadvantage of edible coatings manufactured by L-b-L technique is the separation of layers. Recently, cold plasma has been used to increase the surface energy of polymers and biopolymers, enhancing adhesion and printing capacity in multilayer coatings (Pankaj et al. 2014). The cold plasma can also be used in biopolymer-based coating for the deposition of bioactive compounds, antimicrobials and barrier layers (Pankaj and Thomas 2016). The design for scaling of industrial processes should be studied using the L-b-L technique.

About additives with smart and active properties, it is necessary to check if these compounds proposed in the literature are in GRAS list (*Generally Recognized as Safe*) (Gutiérrez et al. 2017a). Additional studies should be conducted on the migration, accumulation in biosystems, toxicity, genotoxicity and carcinogenicity effects of additives with smart and active properties in coatings (Gutiérrez et al. 2019; Herniou--Julien et al. 2019). For that, it is necessary to quantify the specific migration limit permissible of each additive into foods (Valencia and Sobral 2018; Gutiérrez 2018b, c). Finally, it is necessary to study the encapsulation techniques to reduce the migration of additives from edible coatings to foods as well as to control release of additives without altering the sensorial properties of foods (Chen et al. 2018; Pérez-Córdoba et al. 2018).

20.6 Conclusions

Biopolymer-based edible coatings are probably the future materials that will be used in food packaging industry, considering that several biopolymers can be used in manufacture process of edible coatings. It however is necessary to understand as biopolymer-based edible coatings can become materials with important characteristics to food industry. L-b-L technique can be used to manufacture multilayer edible coatings containing additives from natural origin and nanoparticles with specific properties, opening a windows of news applications as smart and active packaging for food industry. Nonetheless, more studies must be addressed to understand: (i) layer separation in edible coatings manufactured by means of L-b-L technique, (ii) scalability of industrial processes using L-b-L technique, (iii) the migration of additives into drinks and foods as well as its risks associated with the potential ingestion of additives used to manufacture smart and active edible coatings.

Acknowledgements The authors would like to thank the Federal University of Santa Catarina (UFSC) and the University of São Paulo (USP) for their support.

Conflicts of Interest The author declares no conflict of interest.

References

Abdollahzadeh, E., Mahmoodzadeh Hosseini, H., & Imani Fooladi, A. A. (2018). Antibacterial activity of agar-based films containing nisin, cinnamon EO, and ZnO nanoparticles. *Journal of Food Safety, 38*(3), e12440. https://doi.org/10.1111/jfs.12440.

Abreu, A. S., Oliveira, M., De Sá, A., Rodrigues, R. M., Cerqueira, M. A., Vicente, A. A., & Machado, A. V. (2015). Antimicrobial nanostructured starch based films for packaging. *Carbohydrate Polymers, 129*, 127–134. https://doi.org/10.1016/j.carbpol.2015.04.021.

Adley, C. (2014). Past, present and future of sensors in food production. *Food, 3*(3), 491–510. https://doi.org/10.3390/foods3030491.

Ahmad, V., Khan, M. S., Jamal, Q. M. S., Alzohairy, M. A., Al Karaawi, M. A., & Siddiqui, M. U. (2017). Antimicrobial potential of bacteriocins: In therapy, agriculture and food preservation. *International Journal of Antimicrobial Agents, 49*(1), 1–11. https://doi.org/10.1016/j.ijantimicag.2016.08.016.

Ahmed, I., Lin, H., Zou, L., Brody, A. L., Li, Z., Qazi, I. M., Pavase, T. R., & Lv, L. (2017). A comprehensive review on the application of active packaging technologies to muscle foods. *Food Control, 82*, 163–178. https://doi.org/10.1016/j.foodcont.2017.06.009.

Aloui, H., & Khwaldia, K. (2016). Natural antimicrobial edible coatings for microbial safety and food quality enhancement. *Comprehensive Reviews in Food Science and Food Safety, 15*(6), 1080–1103. https://doi.org/10.1111/1541-4337.12226.

Álvarez, K., Famá, L., & Gutiérrez, T. J. (2017). Chapter 12. Physicochemical, antimicrobial and mechanical properties of thermoplastic materials based on biopolymers with application in the food industry. In M. Masuelli & D. Renard (Eds.), *Advances in physicochemical properties of biopolymers: Part 1* (pp. 358–400). Bentham Science Publishers. EE.UU. ISBN: 978-1-68108-454-1. eISBN: 978-1-68108-453-4. https://doi.org/10.2174/97816810845341170100I5.

Álvarez, K., Alvarez, V. A., & Gutiérrez, T. J. (2018). Chapter 3. Biopolymer composite materials with antimicrobial effects applied to the food industry. In V. K. Thakur & M. K. Thakur (Eds.), *Functional biopolymers* (pp. 57–96). Editorial Springer International Publishing. EE.UU. ISBN: 978-3-319-66416-3. eISBN: 978-3-319-66417-0. https://doi.org/10.1007/978-3-319-66417-0_3.

Ariga, K., Hill, J. P., & Ji, Q. (2007). Layer-by-layer assembly as a versatile bottom-up nanofabrication technique for exploratory research and realistic application. *Physical Chemistry Chemical Physics, 9*(19), 2319–2340. https://doi.org/10.1039/b700410a.

Arnon-Rips, H., & Poverenov, E. (2018). Improving food products' quality and storability by using layer by layer edible coatings. *Trends in Food Science and Technology, 75*, 81–92. https://doi.org/10.1016/j.tifs.2018.03.003.

Atarés, L., & Chiralt, A. (2016). Essential oils as additives in biodegradable films and coatings for active food packaging. *Trends in Food Science and Technology, 48*, 51–62. https://doi.org/10.1016/j.tifs.2015.12.001.

Ban, Z., Yan, J., Wang, Y., Zhang, J., Yuan, Q., & Li, L. (2018). Effects of postharvest application of chitosan-based layer-by-layer assemblies on regulation of ribosomal and defense proteins in strawberry fruit (*Fragaria × ananassa*). *Scientia Horticulturae, 240*, 293–302. https://doi.org/10.1016/j.scienta.2018.06.035.

Bibi, F., Guillaume, C., Vena, A., Gontard, N., & Sorli, B. (2016). Wheat gluten, a bio-polymer layer to monitor relative humidity in food packaging: Electric and dielectric characterization. *Sensors and Actuators, A: Physical, 247*, 355–367. https://doi.org/10.1016/j.sna.2016.06.017.

Bibi, F., Guillaume, C., Gontard, N., & Sorli, B. (2017a). A review: RFID technology having sensing aptitudes for food industry and their contribution to tracking and monitoring of food products. *Trends in Food Science and Technology, 62*, 91–103. https://doi.org/10.1016/j.tifs.2017.01.013.

Bibi, F., Guillaume, C., Gontard, N., & Sorli, B. (2017b). Wheat gluten, a bio-polymer to monitor carbon dioxide in food packaging: Electric and dielectric characterization. *Sensors and Actuators, B: Chemical, 250*, 76–84. https://doi.org/10.1016/j.snb.2017.03.164.

Biji, K. B., Ravishankar, C. N., Mohan, C. O., & Srinivasa Gopal, T. K. (2015). Smart packaging systems for food applications: A review. *Journal of Food Science and Technology, 52*(10), 6125–6135. https://doi.org/10.1007/s13197-015-1766-7.

Bitencourt, C. M., Fávaro-Trindade, C. S., Sobral, P. J. A., & Carvalho, R. A. (2014). Gelatin-based films additivated with curcuma ethanol extract: Antioxidant activity and physical properties of films. *Food Hydrocolloids, 40*, 145–152. https://doi.org/10.1016/j.foodhyd.2014.02.014.

Bonilla, J., & Sobral, P. J. A. (2016). Investigation of the physicochemical, antimicrobial and antioxidant properties of gelatin-chitosan edible film mixed with plant ethanolic extracts. *Food Bioscience, 16*, 17–25. https://doi.org/10.1016/j.fbio.2016.07.003.

Bonilla, J., & Sobral, P. J. A. (2017). Antioxidant and physicochemical properties of blended films based on gelatin-sodium caseinate activated with natural extracts. *Journal of Applied Polymer Science, 134*(7), 1–9. https://doi.org/10.1002/app.44467.

Bonilla, J., & Sobral, P. J. A. (2018). Gelatin-chitosan edible film activated with Boldo extract for improving microbiological and antioxidant stability of sliced Prato cheese. *International Journal of Food Science & Technology, 2017*, 1–8. https://doi.org/10.1111/ijfs.14032.

Bonilla, J., Poloni, T., & Sobral, P. J. A. (2018). Active edible coatings with Boldo extract added and their application on nut products: Reducing the oxidative rancidity rate. *International Journal of Food Science and Technology, 53*(3), 700–708. https://doi.org/10.1111/ijfs.13645.

Borges, J., & Mano, J. F. (2014). Molecular interactions driving the layer-by-layer assembly of multilayers. *Chemical Reviews, 114*(18), 8883–8942. https://doi.org/10.1021/cr400531v.

Boyaci, D., Korel, F., & Yemenicioğlu, A. (2016). Development of activate-at-home-type edible antimicrobial films: An example pH-triggering mechanism formed for smoked salmon slices using lysozyme in whey protein films. *Food Hydrocolloids, 60*, 170–178. https://doi.org/10.1016/j.foodhyd.2016.03.032.

Bracone, M., Merino, D., González, J., Alvarez, V. A., & Gutiérrez, T. J. (2016). Chapter 6. Nanopackaging from natural fillers and biopolymers for the development of active and intelligent films. In S. Ikram & S. Ahmed (Eds.), *Natural polymers: Derivatives, blends and composites* (pp. 119–155). New York. EE.UU. ISBN: 978-1-63485-831-1: Editorial Nova Science Publishers, Inc.

Brizio, A. P. D. R., & Prentice, C. (2015). Development of an intelligent enzyme indicator for dynamic monitoring of the shelf-life of food products. *Innovative Food Science and Emerging Technologies, 30*, 208–217. https://doi.org/10.1016/j.ifset.2015.04.001.

Brunel University London. (2014). 47 – A refreshing 'Best Before' Label. Retrieved from http://fifty.brunel.ac.uk/discovery-trail/47-a-refreshing-best-before-label/

Castillo, L. A., Farenzena, S., Pintos, E., Rodríguez, M. S., Villar, M. A., García, M. A., & López, O. V. (2017). Active films based on thermoplastic corn starch and chitosan oligomer for food packaging applications. *Food Packaging and Shelf Life, 14*.(Part B, 128–136. https://doi.org/10.1016/j.fpsl.2017.10.004.

Chen, Y., Shu, M., Yao, X., Wu, K., Zhang, K., He, Y., Nishinari, K., Phillips, G. O., Yao, Y., & Jiang, F. (2018). Effect of zein-based microencapsules on the release and oxidation of loaded limonene. *Food Hydrocolloids, 84*, 330–336. https://doi.org/10.1016/j.foodhyd.2018.05.049.

Choi, I., Lee, J. Y., Lacroix, M., & Han, J. (2017). Intelligent pH indicator film composed of agar/potato starch and anthocyanin extracts from purple sweet potato. *Food Chemistry, 218*, 122–128. https://doi.org/10.1016/j.foodchem.2016.09.050.

Compagnone, D., Di Francia, G., Di Natale, C., Neri, G., Seeber, R., & Tajani, A. (2017). Chemical sensors and biosensors in Italy: A review of the 2015 literature. *Sensors (Switzerland), 17*(4), 868. https://doi.org/10.3390/s17040868.

Crizel, T. d. M., Rios, A. d. O., Alves, V. D., Bandarra, N., Moldão-Martins, M., & Flôres, S. H. (2018). Active food packaging prepared with chitosan and olive pomace. *Food Hydrocolloids, 74*, 139–150. https://doi.org/10.1016/j.foodhyd.2017.08.007.

Das, A. P., Kumar, P. S., & Swain, S. (2014). Recent advances in biosensor based endotoxin detection. *Biosensors and Bioelectronics, 51*, 62–75. https://doi.org/10.1016/j.bios.2013.07.020.

Decher, G., & Schmitt, J. (1992). Buildup of ultrathin multilayer films by a self-assembly process: III. Consecutively alternating adsorption of anionic and cationic polyelectrolytes on charged surfaces. *Thin Solid Films, 210-211*(Part 2), 831–835. https://doi.org/10.1016/0040-6090(92)90417-a.

Devi, R., Yadav, S., Nehra, R., Yadav, S., & Pundir, C. S. (2013). Electrochemical biosensor based on gold coated iron nanoparticles/chitosan composite bound xanthine oxidase for detection of xanthine in fish meat. *Journal of Food Engineering, 115*(2), 207–214. https://doi.org/10.1016/j.jfoodeng.2012.10.014.

Dubas, S. T., & Schlenoff, J. B. (1999). Factors controlling the growth of polyelectrolyte multilayers. *Macromolecules, 32*(24), 8153–8160. https://doi.org/10.1021/ma981927a.

Eça, K. S., Sartori, T., & Menegalli, F. C. (2014). Films and edible coatings containing antioxidants–A review. *Brazilian Journal of Food Technology, 17*(2), 98–112. https://doi.org/10.1590/bjft.2014.017.

Echeverría, I., López-Caballero, M. E., Gómez-Guillén, M. C., Mauri, A. N., & Montero, M. P. (2018). Active nanocomposite films based on soy proteins-montmorillonite- clove essential oil for the preservation of refrigerated bluefin tuna (*Thunnus thynnus*) fillets. *International Journal of Food Microbiology, 266*, 142–149. https://doi.org/10.1016/j.ijfoodmicro.2017.10.003.

Falguera, V., Quintero, J. P., Jiménez, A., Muñoz, J. A., & Ibarz, A. (2011). Edible films and coatings: Structures, active functions and trends in their use. *Trends in Food Science and Technology, 22*(6), 292–303. https://doi.org/10.1016/j.tifs.2011.02.004.

Fang, Z., Zhao, Y., Warner, R. D., & Johnson, S. K. (2017). Active and intelligent packaging in meat industry. *Trends in Food Science and Technology, 61*(2), 60–71. https://doi.org/10.1016/j.tifs.2017.01.002.

Fuertes, G., Soto, I., Carrasco, R., Vargas, M., Sabattin, J., & Lagos, C. (2016). Intelligent packaging systems: Sensors and nanosensors to monitor food quality and safety. *Journal of Sensors*, ID 4046061, 1–8. https://doi.org/10.1155/2016/4046061.

Fujii, N., Fujimoto, K., Michinobu, T., Akada, M., Hill, J. P., Shiratori, S., Ariga, K., & Shigehara, K. (2010). The simplest layer-by-layer assembly structure: Best paired polymer electrolytes with one charge per main chain carbon atom for multilayered thin films. *Macromolecules, 43*(8), 3947–3955. https://doi.org/10.1021/ma100473j.

Galagan, Y., & Su, W. F. (2008). Fadable ink for time-temperature control of food freshness: Novel new time-temperature indicator. *Food Research International, 41*(6), 653–657. https://doi.org/10.1016/j.foodres.2008.04.012.

Ghaani, M., Cozzolino, C. A., Castelli, G., & Farris, S. (2016). An overview of the intelligent packaging technologies in the food sector. *Trends in Food Science and Technology, 51*, 1–11. https://doi.org/10.1016/j.tifs.2016.02.008.

Gómez-Estaca, J., López-de-Dicastillo, C., Hernández-Muñoz, P., Catalá, R., & Gavara, R. (2014). Advances in antioxidant active food packaging. *Trends in Food Science and Technology, 35*(1), 42–51. https://doi.org/10.1016/j.tifs.2013.10.008.

Gutiérrez, T. J. (2017). Surface and nutraceutical properties of edible films made from starchy sources with and without added blackberry pulp. *Carbohydrate Polymers, 165*, 169–179. https://doi.org/10.1016/j.carbpol.2017.02.016.

Gutiérrez, T. J. (2018a). Active and intelligent films made from starchy sources/blackberry pulp. *Journal of Polymers and the Environment, 26*(6), 2374–2391. https://doi.org/10.1007/s10924-017-1134-y.

Gutiérrez, T. J. (2018b). Are modified pumpkin flour/plum flour nanocomposite films biodegradable and compostable? *Food Hydrocolloids, 83*, 397–410. https://doi.org/10.1016/j.foodhyd.2018.05.035.

Gutiérrez, T. J. (2018c). Chapter 9. Biodegradability and compostability of food nanopackaging materials. In G. Cirillo, M. A. Kozlowski, & U. G. Spizzirri (Eds.), *Composite materials for food packaging* (pp. 269–296). WILEY-Scrivener Publisher. EE.UU. ISBN: 978–1–119-16020-5. https://doi.org/10.1002/9781119160243.ch9.

Gutiérrez, T. J., & Alvarez, V. A. (2018). Bionanocomposite films developed from corn starch and natural and modified nano-clays with or without added blueberry extract. *Food Hydrocolloids, 77*, 407–420. https://doi.org/10.1016/j.foodhyd.2017.10.017.

Gutiérrez, T. J., Guzmán, R., Medina Jaramillo, C., & Famá, L. (2016a). Effect of beet flour on films made from biological macromolecules: Native and modified plantain flour. *International Journal of Biological Macromolecules, 82*, 395–403. https://doi.org/10.1016/j.ijbiomac.2015.10.020.

Gutiérrez, T. J., Suniaga, J., Monsalve, A., & García, N. L. (2016b). Influence of beet flour on the relationship surface-properties of edible and intelligent films made from native and modified plantain flour. *Food Hydrocolloids, 54*. (Part B, 234–244. https://doi.org/10.1016/j.foodhyd.2015.10.012.

Gutiérrez, T. J., González Seligra, P., Medina Jaramillo, C., Famá, L., & Goyanes, S. (2017a). Chapter 14. Effect of filler properties on the antioxidant response of thermoplastic starch composites. In V. K. Thakur, M. K. Thakur, & M. R. Kessler (Eds.), *Handbook of composites from renewable materials* (pp. 337–370). WILEY-Scrivener Publisher. EE.UU. ISBN: 978-1-119-22362-7. https://doi.org/10.1002/9781119441632.ch14.

Gutiérrez, T. J., Ponce, A. G., & Alvarez, V. A. (2017b). Nano-clays from natural and modified montmorillonite with and without added blueberry extract for active and intelligent food nanopackaging materials. *Materials Chemistry and Physics, 194*, 283–292. https://doi.org/10.1016/j.matchemphys.2017.03.052.

Gutiérrez, T. J., Toro-Márquez, L. A., Merino, D., & Mendieta, J. R. (2019). Hydrogen-bonding interactions and compostability of bionanocomposite films prepared from corn starch and nano-fillers with and without added Jamaica flower extract. *Food Hydrocolloids, 89*, 283–293. https://doi.org/10.1016/j.foodhyd.2018.10.058.

Guzmán, E., Mateos-Maroto, A., Ruano, M., Ortega, F., & Rubio, R. G. (2017). Layer-by-Layer polyelectrolyte assemblies for encapsulation and release of active compounds. *Advances in Colloid and Interface Science, 249*, 290–307. https://doi.org/10.1016/j.cis.2017.04.009.

Halász, K., & Csóka, L. (2018). Black chokeberry (*Aronia melanocarpa*) pomace extract immobilized in chitosan for colorimetric pH indicator film application. *Food Packaging and Shelf Life, 16*, 185–193. https://doi.org/10.1016/j.fpsl.2018.03.002.

Hammond, P. T. (2004). Form and function in multilayer assembly: New applications at the nanoscale. *Advanced Materials, 16*(15), 1271–1293. https://doi.org/10.1002/adma.200400760.

Herniou--Julien, C., Mendieta, J. R., & Gutiérrez, T. J. (2019). Characterization of biodegradable/non-compostable films made from cellulose acetate/corn starch blends processed under reactive extrusion conditions. *Food Hydrocolloids, 89*, 67–79. https://doi.org/10.1016/j.foodhyd.2018.10.024.

Hu, Z., Huang, F., & Cao, Y. (2017). Layer-by-Layer assembly of multilayer thin films for organic optoelectronic devices. *Small Methods, 1*(12). https://doi.org/10.1002/smtd.201700264.

Iler, R. K. (1966). Multilayers of colloidal particles. *Journal of Colloid and Interface Science, 21*(6), 569–594. https://doi.org/10.1016/0095-8522(66)90018-3.

Kaewklin, P., Siripatrawan, U., Suwanagul, A., & Lee, Y. S. (2018). Active packaging from chitosan-titanium dioxide nanocomposite film for prolonging storage life of tomato fruit. *International Journal of Biological Macromolecules, 112*, 523–529. https://doi.org/10.1016/j.ijbiomac.2018.01.124.

Kim, J. H., Hong, W. s., & Oh, S. W. (2018a). Effect of layer-by-layer antimicrobial edible coating of alginate and chitosan with grapefruit seed extract for shelf-life extension of shrimp (*Litopenaeus vannamei*) stored at 4 °C. *International Journal of Biological Macromolecules, 120*.(Part B, 1468–1473. https://doi.org/10.1016/j.ijbiomac.2018.09.160.

Kim, Y. J., Kim, H. S., Chon, J. W., Kim, D. H., Hyeon, J. Y., & Seo, K. H. (2018b). New colorimetric aptasensor for rapid on-site detection of *Campylobacter jejuni* and *Campylobacter coli* in chicken carcass samples. *Analytica Chimica Acta, 1029*, 78–85. https://doi.org/10.1016/j.aca.2018.04.059.

Kreyenschmidt, J., Christiansen, H., Hübner, A., Raab, V., & Petersen, B. (2010). A novel photochromic time-temperature indicator to support cold chain management. *International Journal of Food Science and Technology, 45*(2), 208–215. https://doi.org/10.1111/j.1365-2621.2009.02123.x.

Li, G., Zhu, R., & Yang, Y. (2012). Polymer solar cells. *Nature Photonics, 6*(3), 153–161.

Li, X., Liu, A., Ye, R., Wang, Y., & Wang, W. (2015). Fabrication of gelatin-laponite composite films: Effect of the concentration of laponite on physical properties and the freshness of meat during storage. *Food Hydrocolloids, 44*, 390–398. https://doi.org/10.1016/j.foodhyd.2014.10.014.

Lisdat, F. (2017). Trends in the layer-by-layer assembly of redox proteins and enzymes in bioelectrochemistry. *Current Opinion in Electrochemistry, 5*(1), 165–172. https://doi.org/10.1016/j.coelec.2017.09.002.

Liu, B., Xu, H., Zhao, H., Liu, W., Zhao, L., & Li, Y. (2017). Preparation and characterization of intelligent starch/PVA films for simultaneous colorimetric indication and antimicrobial activity for food packaging applications. *Carbohydrate Polymers, 157*, 842–849. https://doi.org/10.1016/j.carbpol.2016.10.067.

Liu, J., Wang, H., Wang, P., Guo, M., Jiang, S., Li, X., & Jiang, S. (2018). Films based on κ-carrageenan incorporated with curcumin for freshness monitoring. *Food Hydrocolloids, 83*, 134–142. https://doi.org/10.1016/j.foodhyd.2018.05.012.

Love, J. C., Estroff, L. A., Kriebel, J. K., Nuzzo, R. G., & Whitesides, G. M. (2005). Self-assembled monolayers of thiolates on metals as a form of nanotechnology. *Chemical Reviews, 105*(4), 1103–1169. https://doi.org/10.1021/cr0300789.

Lv, M., Liu, Y., Geng, J., Kou, X., Xin, Z., & Yang, D. (2018). Engineering nanomaterials-based biosensors for food safety detection. *Biosensors and Bioelectronics, 106*, 122–128. https://doi.org/10.1016/j.bios.2018.01.049.

Ma, Q., Du, L., & Wang, L. (2017). Tara gum/polyvinyl alcohol-based colorimetric NH3indicator films incorporating curcumin for intelligent packaging. *Sensors and Actuators, B: Chemical, 244*, 759–766. https://doi.org/10.1016/j.snb.2017.01.035.

Maciel, V. B. V., Yoshida, C. M. P., & Franco, T. T. (2012a). Development of a prototype of a colourimetric temperature indicator for monitoring food quality. *Journal of Food Engineering, 111*(1), 21–27. https://doi.org/10.1016/j.jfoodeng.2012.01.037.

Maciel, V. B. V., Franco, T. T., & Yoshida, C. M. P. (2012b). Alternative intelligent material for packaging using chitosan films as colorimetric temperature indicators. *Polímeros, 22*(4), 318–324. https://doi.org/10.1590/S0104-14282012005000054.

Malhotra, B. D., Singhal, R., Chaubey, A., Sharma, S. K., & Kumar, A. (2005). Recent trends in biosensors. *Current Applied Physics, 5*(2), 92–97. https://doi.org/10.1016/j.cap.2004.06.021.

McFarlane, D., & Sheffi, Y. (2003). The impact of automatic identification on supply chain operations. *The International Journal of Logistics Management, 14*(1), 1–17. https://doi.org/10.1108/09574090310806503.

McMahon, R. E., Wang, L., Skoracki, R., & Mathur, A. B. (2013). Development of nanomaterials for bone repair and regeneration. *Journal of Biomedical Materials Research-Part B Applied Biomaterials, 101B*(2), 387–397. https://doi.org/10.1002/jbm.b.32823.

Medina-Jaramillo, C., Ochoa-Yepes, O., Bernal, C., & Famá, L. (2017). Active and smart biodegradable packaging based on starch and natural extracts. *Carbohydrate Polymers, 176*, 187–194. https://doi.org/10.1016/j.carbpol.2017.08.079.

Mohebi, E., & Marquez, L. (2015). Intelligent packaging in meat industry: An overview of existing solutions. *Journal of Food Science and Technology, 52*(7), 3947–3964. https://doi.org/10.1007/s13197-014-1588-z.

Moradi, L. T., Sharifan, A., & Larijani, K. (2015). Antimicrobial activity of lemon and peppermint essential oil in edible coating containing chitosan and pectin on rainbow trout (Oncorhynchus mykiss) fillets. *Journal of Medical Microbiology and Infectious Diseases, 3*(1–2), 38–43.

Murray, C., & Kagan, C. (2000). Synthesis and characterization of monodisperse nanocrystals and close-packed nanocrystal assembles. *Annual Review of Material Science, 30*, 545–610. https://doi.org/10.1146/annurev.matsci.30.1.545.

Nascimento, V., França, C., Hernández-Montelongo, J., Machado, D., Lancellotti, M., Cotta, M., Landers, R., & Beppu, M. (2018). Influence of pH and ionic strength on the antibacterial effect of hyaluronic acid/chitosan films assembled layer-by-layer. *European Polymer Journal, 109*, 198–205. https://doi.org/10.1016/j.eurpolymj.2018.09.038.

Noshirvani, N., Ghanbarzadeh, B., Gardrat, C., Rezaei, M. R., Hashemi, M., Le Coz, C., & Coma, V. (2017). Cinnamon and ginger essential oils to improve antifungal, physical and mechanical properties of chitosan-carboxymethyl cellulose films. *Food Hydrocolloids, 70*, 36–45. https://doi.org/10.1016/j.foodhyd.2017.03.015.

Oliveira, S. P., Bertan, L. C., De Rensis, C. M., Bilck, A. P., & Vianna, P. C. (2017). Whey protein-based films incorporated with oregano essential oil. *Polímeros, 27*(2), 158–164. https://doi.org/10.1590/0104-1428.02016.

Ortega, F., Giannuzzi, L., Arce, V. B., & García, M. A. (2017). Active composite starch films containing green synthetized silver nanoparticles. *Food Hydrocolloids, 70*, 152–162. https://doi.org/10.1016/j.foodhyd.2017.03.036.

Otoni, C. G., Espitia, P. J. P., Avena-Bustillos, R. J., & McHugh, T. H. (2016). Trends in antimicrobial food packaging systems: Emitting sachets and absorbent pads. *Food Research International, 83*, 60–73. https://doi.org/10.1016/j.foodres.2016.02.018.

Pankaj, S. K., & Thomas, S. (2016). Cold plasma applications in food packaging. In *Cold plasma in food and agriculture: Fundamentals and applications* (pp. 293–307). Elsevier Inc.

Pankaj, S. K., Bueno-Ferrer, C., Misra, N. N., Milosavljević, V., O'Donnell, C. P., Bourke, P., Keener, K. M., & Cullen, P. J. (2014). Applications of cold plasma technology in food packaging. *Trends in Food Science and Technology, 35*(1), 5–17. https://doi.org/10.1016/j.tifs.2013.10.009.

Pavelková, A. (2013). Time temperature indicators as devices intelligent packaging. *Acta Universitatis Agriculturae et Silviculturae Mendelianae Brunensis, 61*(1), 245–251. https://doi.org/10.11118/actaun201361010245.

Pereira, P. F., & Andrade, C. T. (2017). Optimized pH-responsive film based on a eutectic mixture-plasticized chitosan. *Carbohydrate Polymers, 165*, 238–246. https://doi.org/10.1016/j.carbpol.2017.02.047.

Pérez-Córdoba, L. J., Norton, I. T., Batchelor, H. K., Gkatzionis, K., Spyropoulos, F., & Sobral, P. J. A. (2018). Physico-chemical, antimicrobial and antioxidant properties of gelatin-chitosan based films loaded with nanoemulsions encapsulating active compounds. *Food Hydrocolloids, 79*, 544–559. https://doi.org/10.1016/j.foodhyd.2017.12.012.

Piñeros-Hernandez, D., Medina-Jaramillo, C., López-Córdoba, A., & Goyanes, S. (2017). Edible cassava starch films carrying rosemary antioxidant extracts for potential use as active food packaging. *Food Hydrocolloids, 63*, 488–495. https://doi.org/10.1016/j.foodhyd.2016.09.034.

Pourjavaher, S., Almasi, H., Meshkini, S., Pirsa, S., & Parandi, E. (2017). Development of a colorimetric pH indicator based on bacterial cellulose nanofibers and red cabbage (*Brassica oleraceae*) extract. *Carbohydrate Polymers, 156*, 193–201. https://doi.org/10.1016/j.carbpol.2016.09.027.

Poverenov, E., Danino, S., Horev, B., Granit, R., Vinokur, Y., & Rodov, V. (2014a). Layer-by-layer electrostatic deposition of edible coating on fresh cut melon model: Anticipated and unexpected effects of alginate-chitosan combination. *Food and Bioprocess Technology, 7*(5), 1424–1432. https://doi.org/10.1007/s11947-013-1134-4.

Poverenov, E., Rutenberg, R., Danino, S., Horev, B., & Rodov, V. (2014b). Gelatin-chitosan composite films and edible coatings to enhance the quality of food products: Layer-by-Layer vs. blended formulations. *Food and Bioprocess Technology, 7*(11), 3319–3327. https://doi.org/10.1007/s11947-014-1333-7.

Prietto, L., Mirapalhete, T. C., Pinto, V. Z., Hoffmann, J. F., Vanier, N. L., Lim, L. T., Guerra Dias, A. R., & da Rosa Zavareze, E. (2017). pH-sensitive films containing anthocyanins extracted from black bean seed coat and red cabbage. *LWT – Food Science and Technology, 80*, 492–500. https://doi.org/10.1016/j.lwt.2017.03.006.

Puligundla, P., Jung, J., & Ko, S. (2012). Carbon dioxide sensors for intelligent food packaging applications. *Food Control, 25*(1), 328–333. https://doi.org/10.1016/j.foodcont.2011.10.043.

Purkayastha, S., Biswal, A. K., & Saha, S. (2017). Responsive systems in food packaging. *Journal of Packaging Technology and Research, 1*(1), 53–64. https://doi.org/10.1007/s41783-017-0007-0.

Radeva, T., Kamburova, K., & Petkanchin, I. (2006). Formation of polyelectrolyte multilayers from polysaccharides at low ionic strength. *Journal of Colloid and Interface Science, 298*(1), 59–65. https://doi.org/10.1016/j.jcis.2005.12.010.

Ribeiro-Santos, R., Andrade, M., de Melo, N. R., & Sanches-Silva, A. (2017). Use of essential oils in active food packaging: Recent advances and future trends. *Trends in Food Science and Technology, 61*, 132–140. https://doi.org/10.1016/j.tifs.2016.11.021.

Richardson, J. J., Cui, J., Björnmalm, M., Braunger, J. A., Ejima, H., & Caruso, F. (2016). Innovation in layer-by-layer assembly. *Chemical Reviews, 116*(23), 14828–14867. https://doi.org/10.1021/acs.chemrev.6b00627.

Rubab, M., Shahbaz, H. M., Olaimat, A. N., & Oh, D. H. (2018). Biosensors for rapid and sensitive detection of Staphylococcus aureus in food. *Biosensors and Bioelectronics, 105*, 49–57. https://doi.org/10.1016/j.bios.2018.01.023.

Saldías, C., Díaz, D. D., Bonardd, S., Soto-Marfull, C., Cordoba, A., Saldías, S., Quezada, C., Radic, D., & Leiva, A. (2018). *In situ* preparation of film and hydrogel bio-nanocomposites of chitosan/fluorescein-copper with catalytic activity. *Carbohydrate Polymers, 180*. (October 2017), 200–208. https://doi.org/10.1016/j.carbpol.2017.10.018.

Saliu, F., & Della Pergola, R. (2018). Carbon dioxide colorimetric indicators for food packaging application: Applicability of anthocyanin and poly-lysine mixtures. *Sensors and Actuators, B: Chemical, 258*, 1117–1124. https://doi.org/10.1016/j.snb.2017.12.007.

Santos, D. C., Ribeiro-Santos, R., Ventura, L. A. F., Melo, N. R., Costa, B. S., Rojas, E. E. G., & Salgado, N. L. (2016). Antimicrobial activity studies and characterization of cellulose acetate films containing essential oils. *Italian Journal of Food Science, 28*(2), 248–257. https://doi.org/10.14674/1120-1770/ijfs.v28.

Sawant, S. N. (2017). Development of biosensors from biopolymer composites. In *Biopolymer composites in electronics* (pp. 353–383). Elsevier Inc. https://doi.org/10.1016/b978-0-12-809261-3.00013-9.

Schaefer, D., & Cheung, W. M. (2018). Smart packaging: Opportunities and challenges. *Procedia CIRP, 72*, 1022–1027. https://doi.org/10.1016/j.procir.2018.03.240.

Schmidt, M., Thoroe, L., & Schumann, M. (2013). RFID and barcode in manufacturing logistics: Interface concept for concurrent operation. *Information Systems Management, 30*(2), 100–115. https://doi.org/10.1080/10580530.2013.773801.

Selvaraj, J. N., Zhou, L., Wang, Y., Zhao, Y. J., Xing, F. G., Dai, X. F., & Liu, Y. (2015). Mycotoxin detection–Recent trends at global level. *Journal of Integrative Agriculture, 14*(11), 2265–2281. https://doi.org/10.1016/S2095-3119(15)61120-0.

Shukla, V., Kandeepan, G., Vishnuraj, M. R., & Soni, A. (2016). Anthocyanins based indicator sensor for intelligent packaging application. *Agricultural Research, 5*(2), 205–209. https://doi.org/10.1007/s40003-016-0211-0.

Shutava, T. G., Livanovich, K. S., & Sharamet, A. A. (2019). Layer-by-Layer films of polysaccharides modified with polyethylene glycol and dextran. *Colloids and Surfaces B: Biointerfaces, 173*, 412–420. https://doi.org/10.1016/j.colsurfb.2018.10.009.

Sidhu, P. K., & Nehra, K. (2017). Bacteriocin-nanoconjugates as emerging compounds for enhancing antimicrobial activity of bacteriocins. *Journal of King Saud University – Science,. In Press*. https://doi.org/10.1016/j.jksus.2017.12.007.

Silva, N. F. D., Magalhães, J. M. C. S., Freire, C., & Delerue-Matos, C. (2018). Electrochemical biosensors for *Salmonella*: State of the art and challenges in food safety assessment. *Biosensors and Bioelectronics, 99*, 667–682. https://doi.org/10.1016/j.bios.2017.08.019.

Silva-Pereira, M. C., Teixeira, J. A., Pereira-Júnior, V. A., & Stefani, R. (2015). Chitosan/corn starch blend films with extract from *Brassica oleraceae* (red cabbage) as a visual indicator of fish deterioration. *LWT – Food Science and Technology, 61*(1), 258–262. https://doi.org/10.1016/j.lwt.2014.11.041.

Sipahi, R. E., Castell-Perez, M. E., Moreira, R. G., Gomes, C., & Castillo, A. (2013). Improved multilayered antimicrobial alginate-based edible coating extends the shelf life of fresh-cut watermelon (*Citrullus lanatus*). *LWT – Food Science and Technology, 51*(1), 9–15. https://doi.org/10.1016/j.lwt.2012.11.013.

Siripatrawan, U., & Kaewklin, P. (2018). Fabrication and characterization of chitosan-titanium dioxide nanocomposite film as ethylene scavenging and antimicrobial active food packaging. *Food Hydrocolloids, 84*, 125–134. https://doi.org/10.1016/j.foodhyd.2018.04.049.

Srivastava, S., & Kotov, N. A. (2008). Composite Layer-by-layer (LBL) assembly withinorganic nanoparticles and nanowires. *Accounts of Chemical Research, 41*(12), 1831–1841. https://doi.org/10.1021/ar8001377.

Tan, H. L., McMurdo, M. J., Pan, G., & Van Patten, P. G. (2003). Temperature dependence of polyelectrolyte multilayer assembly. *Langmuir, 19*(22), 9311–9314. https://doi.org/10.1021/la035094f.

Teepoo, S., Dawan, P., & Barnthip, N. (2017). Electrospun chitosan-gelatin biopolymer composite nanofibers for horseradish peroxidase immobilization in a hydrogen peroxide biosensor. *Biosensors, 7*(4), 47. https://doi.org/10.3390/bios7040047.

Thévenot, D., Toth, K., Durst, R., & Wilson, G. (1999). Electrochemical biosensors: Recommended definitions and classification. *Pure and Applied Chemistry, 71*(12), 2333–2348. https://doi.org/10.1351/pac199971122333.

Toro-Márquez, L. A., Merino, D., & Gutiérrez, T. J. (2018). Bionanocomposite films prepared from corn starch with and without nanopackaged Jamaica (*Hibiscus sabdariffa*) flower extract. *Food and Bioprocess Technology, 11*(11), 1955–1973. https://doi.org/10.1007/s11947-018-2160-z.

Tsironi, T., Ronnow, P., Giannoglou, M., & Taoukis, P. (2017). Developing suitable smart TTI labels to match specific monitoring requirements: The case of *Vibrio* spp. growth during transportation of oysters. *Food Control, 73*.(Part A, 51–56. https://doi.org/10.1016/j.foodcont.2016.06.041.

Valencia, G. A., & Sobral, P. J. d. A. (2018). Recent trends on nano-biocomposite polymers for food packaging. In T. J. Gutiérrez (Ed.), *Polymers for food applications* (1st ed., p. 817). Springer. https://doi.org/10.1007/978-3-319-94625-2_5.

Valencia, G. A., Humberto, J., Llanos, R., Cristina, L., & Vercik, D. O. (2014a). Electronic transport in organic memories of chitosan with gold nanoparticles. *Journal of Polymer Engineering, 34*(2), 105–111. https://doi.org/10.1515/polyeng-2013-0142.

Valencia, G. A., Vercik, L. C. d. O., & Vercik, A. (2014b). A new conductometric biosensor based on horseradish peroxidase immobilized on chitosan and chitosan/gold nanoparticle films. *Journal of Polymer Engineering, 34*(7), 633–638. https://doi.org/10.1515/polyeng-2014-0072.

Vanderroost, M., Ragaert, P., Devlieghere, F., & De Meulenaer, B. (2014). Intelligent food packaging: The next generation. *Trends in Food Science and Technology, 39*(1), 47–62. https://doi.org/10.1016/j.tifs.2014.06.009.

Vargas, M., Pastor, C., Chiralt, A., McClements, D. J., & González-Martínez, C. (2008). Recent advances in edible coatings for fresh and minimally processed fruits. *Critical Reviews in Food Science and Nutrition, 48*(6), 496–511. https://doi.org/10.1080/10408390701537344.

Vidal, J. C., Bonel, L., Ezquerra, A., Hernández, S., Bertolín, J. R., Cubel, C., & Castillo, J. R. (2013). Electrochemical affinity biosensors for detection of mycotoxins: A review. *Biosensors and Bioelectronics, 49*, 146–158. https://doi.org/10.1016/j.bios.2013.05.008.

Vu, C. H. T., & Won, K. (2013). Novel water-resistant UV-activated oxygen indicator for intelligent food packaging. *Food Chemistry, 140*(1–2), 52–56. https://doi.org/10.1016/j.foodchem.2013.02.056.

Wang, F., Yang, Q., Kase, J. A., Meng, J., Clotilde, L. M., Lin, A., & Ge, B. (2013). Current trends in detecting non-O157 Shiga toxin–producing *Escherichia coli* in food. *Foodborne Pathogens and Disease, 10*(8), 665–677. https://doi.org/10.1089/fpd.2012.1448.

Wang, H. J., Jo, Y. H., An, D. S., Rhim, J. W., & Lee, D. S. (2015a). Properties of agar-based CO_2 absorption film containing Na_2CO_3 as active compound. *Food Packaging and Shelf Life, 4*, 36–42. https://doi.org/10.1016/j.fpsl.2015.03.004.

Wang, S., Liu, X., Yang, M., Zhang, Y., Xiang, K., & Tang, R. (2015b). Review of time temperature indicators as quality monitors in food packaging. *Packaging and Technology and Science, 28*(10), 839–867. https://doi.org/10.1002/pts.2148.

Wang, Y. C., Lu, L., & Gunasekaran, S. (2017). Biopolymer/gold nanoparticles composite plasmonic thermal history indicator to monitor quality and safety of perishable bioproducts. *Biosensors and Bioelectronics, 92*, 109–116. https://doi.org/10.1016/j.bios.2017.01.047.

Wang, W., Li, M., Li, H., Liu, X., Guo, T., Zhang, G., & Xiong, Y. (2018). A renewable intelligent colorimetric indicator based on polyaniline for detecting freshness of tilapia. *Packaging Technology and Science, 31*(3), 133–140. https://doi.org/10.1002/pts.2358.

Wu, D., Wang, Y., Chen, J., Ye, X., Wu, Q., Liu, D., & Ding, T. (2013). Preliminary study on time-temperature indicator (TTI) system based on urease. *Food Control, 34*(1), 230–234. https://doi.org/10.1016/j.foodcont.2013.04.041.

Wu, D., Hou, S., Chen, J., Sun, Y., Ye, X., Liu, D., Meng, R., & Wang, Y. (2015). Development and characterization of an enzymatic time-temperature indicator (TTI) based on Aspergillus niger lipase. *LWT-Food Science and Technology, 60*(2), 1100–1104. https://doi.org/10.1016/j.lwt.2014.10.011.

Wu, S., Wang, W., Yan, K., Ding, F., Shi, X., Deng, H., & Du, Y. (2018). Electrochemical writing on edible polysaccharide films for intelligent food packaging. *Carbohydrate Polymers, 186*, 236–242. https://doi.org/10.1016/j.carbpol.2018.01.058.

Xiang, Y., Lu, S., & Jiang, S. P. (2012). Layer-by-Layer self-assembly in the development of electrochemical energy conversion and storage devices from fuel cells to supercapacitors. *Chemical Society Reviews, 41*(21), 7291–7321. https://doi.org/10.1039/c2cs35048c.

Xiao, F. X., Pagliaro, M., Xu, Y. J., & Liu, B. (2016). Layer-by-layer assembly of versatile nanoarchitectures with diverse dimensionality: A new perspective for rational construction of multilayer assemblies. *Chemical Society Reviews, 45*(11), 3088–3121. https://doi.org/10.1039/c5cs00781j.

Xie, J., & Hung, Y. C. (2018). UV-A activated TiO_2 embedded biodegradable polymer film for antimicrobial food packaging application. *LWT-Food Science and Technology, 96*, 307–314. https://doi.org/10.1016/j.lwt.2018.05.050.

Yang, X., Kirsch, J., & Simonian, A. (2013). *Campylobacter* spp. detection in the 21st century: A review of the recent achievements in biosensor development. *Journal of Microbiological Methods, 95*(1), 48–56. https://doi.org/10.1016/j.mimet.2013.06.023.

Ye, Y., Guo, H., & Sun, X. (2019). Recent progress on cell-based biosensors for analysis of food safety and quality control. *Biosensors and Bioelectronics, 126*, 389–404. https://doi.org/10.1016/j.bios.2018.10.039.

Yildirim, S., Röcker, B., Pettersen, M. K., Nilsen-Nygaard, J., Ayhan, Z., Rutkaite, R., Radusin, T., Suminska, P., Marcos, B., & Coma, V. (2018). Active packaging applications for food. *Comprehensive Reviews in Food Science and Food Safety, 17*(1), 165–199. https://doi.org/10.1111/1541-4337.12322.

Yoshida, C. M. P., Maciel, V. B. V., Mendonça, M. E. D., & Franco, T. T. (2014). Chitosan biobased and intelligent films: Monitoring pH variations. *LWT-Food Science and Technology, 55*(1), 83–89. https://doi.org/10.1016/j.lwt.2013.09.015.

Zabala, S., Castán, J., & Martínez, C. (2015). Development of a time-temperature indicator (TTI) label by rotary printing technologies. *Food Control, 50*, 57–64. https://doi.org/10.1016/j.foodcont.2014.08.007.

Zhai, X., Shi, J., Zou, X., Wang, S., Jiang, C., Zhang, J., Huang, X., Zhang, W., Holmes, M., & Holmes, M. (2017). Novel colorimetric films based on starch/polyvinyl alcohol incorporated with roselle anthocyanins for fish freshness monitoring. *Food Hydrocolloids, 69*, 308–317. https://doi.org/10.1016/j.foodhyd.2017.02.014.

Zhang, X., Lu, S., & Chen, X. (2014). A visual pH sensing film using natural dyes from *Bauhinia blakeana Dunn. Sensors and Actuators, B: Chemical, 198*, 268–273. https://doi.org/10.1016/j.snb.2014.02.094.

Zhang, X., Xiao, G., Wang, Y., Zhao, Y., Su, H., & Tan, T. (2017). Preparation of chitosan-TiO2 composite film with efficient antimicrobial activities under visible light for food packaging applications. *Carbohydrate Polymers, 169*, 101–107. https://doi.org/10.1016/j.carbpol.2017.03.073.

Chapter 21
Active Edible Coatings for Fresh Fruits and Vegetables

Swarajya Laxmi Nayak, Shruti Sethi, R. R. Sharma, and Uma Prajapati

Abstract Fruits and vegetables are perishable products due to their high metabolic activities, water loss and susceptibility to pathogens. Thus, each year huge losses occur in horticultural products during handling and transportation. Various pre and postharvest treatments are available to attenuate these losses. Amongst these, use of edible coatings is a key means to enhance the storability of the product. The edible coatings are environment-friendly greener approach to mitigate the postharvest losses. Edible materials having suitable moisture and gas barrier properties are applied as a thin layer on the surface of the product to minimize the changes that occur due to biochemical processes. There is a paradigm shift in the development of coatings that are added with features such as improved functionality through incorporation of certain antimicrobials, texture enhancers and nutraceuticals. The use of active edible coatings is a novel approach to extend the shelf-life of the product by incorporating active ingredients into the polymer matrix, which can be consumed with the food, thus enhancing safety, and nutritional and sensory attributes. On the other hand, the smart edible packaging makes use of immobilized antimicrobial nanoparticles applied on the food surface to control the release of active ingredients that enhance the storage life of the product for desired duration. The materials used for the edible coating must be generally recognized as safe (GRAS) and must be approved by the FDA. The success of edible coatings for fresh products depends completely on the control of internal gas composition and ethanol fermentation, changes in color, firmness loss etc. This chapter will cover the recent advances in the development of coatings from different polymers incorporated with functional ingredients to improve quality and ease of use on fresh fruits and vegetables.

Keywords Active Edible Coating Additives · Antimicrobial · Antioxidant · Probiotic

---

S. L. Nayak · S. Sethi (✉) · R. R. Sharma · U. Prajapati
Division of Food Science and Postharvest Technology, Indian Council of Agricultural Research – Indian Agricultural Research Institute, New Delhi, India

T. J. Gutiérrez (ed.), *Polymers for Agri-Food Applications*,
https://doi.org/10.1007/978-3-030-19416-1_21

21.1 Introduction

Fruits and vegetables are a basket of nutrients required to keep ones's body healthy. However, they are very perishable and are victim of many physical, physiological, biochemical and pathological deteriorations. These factors affect the commercialization of the products resulting in huge economic losses (Gutiérrez et al. 2016). Since the purchase of any product depends on its aesthetic appeal, it must be attractive in terms of colour, firmness and the overall appearance. However, perishable horticultural products are prone to shrinkage, softening, colour change etc. during handling and storage. Of the several techniques to extend the shelf-life of fruits and vegetables, edible coatings have gained importance due to ease of use and fabrication according to the functional attribute desired. The application of these coatings on the product surface can be achieved by dipping, spraying or brushing. The coatings adhere to the food surface and create a modified atmosphere that provides a barrier to oxygen, moisture and the movement of solutes (Gutiérrez and Álvarez 2017; Tapia-Blácido et al. 2018). The eco-friendly polymer based edible coatings not only reduce the requirements of synthetic plastic packaging, but also uses the by-products of the food industry. The traditional use of edible coating was mainly focussed on minimizing water loss and delaying senescence through selective permeability to gases. However, the new generation of edible coatings is being specially designed to allow the incorporation and controlled release of nutraceuticals, antioxidants and natural antimicrobial agents through diverse fabrication techniques (Gutiérrez 2017, 2018a). These coatings are thin layers of edible material applied to the product surface and are used as an alternative to natural protective waxy coatings (Ansorena et al. 2018).

21.2 Historical Background of Edible Coatings

Edible coatings have been used for centuries to prevent the moisture migration, improve the food appearance and increase product shelf life. The use of wax coating on fruits by dipping is one of the oldest methods that was in vogue at the beginning of the twelfth century (Krochta and Mulder-Johnston 1997). The coating of citrus fruits with wax to retard desiccation was practiced in China in the twelfth and thirteenth centuries (Hardenburg 1967). In the 1930's, hot-melt paraffin waxes were commercially used as edible coatings for apples and pears (Park 1999). Since 1980's sucrose fatty acid esters (SFAEs), Semperfresh™, Tal Pro-long™, zein proteins are commercially used to coat fruits and vegetables, thus, the firmness, colour and weight during storage are retained (Dhall 2008). During fifteenth century, a protein-based edible film called Yuba was traditionally used in Asia to improve the appearance and preservation of some foods (Gennadios et al. 1993). In the nineteenth century, to prevent oxidation and rancidity of dried nuts during storage, sucrose was applied as an edible protective coating. The more important application of edible coatings so far concerns the use of an emulsion made of oil and waxes in water that is spread on fruits and vegetables to retard the moisture loss, improve their appearance, in terms of colour and sheen, and slow down the softening and start of mealiness (Debeaufort et al. 1998).

21.3 Functionality of Edible Coatings

As previously mentioned, edible coatings are thin layer of edible material that is applied to the product surface in addition to or as a replacement for natural protective waxy coatings, which provide a barrier against the movement of gas, moisture and solutes, and can be eaten along with the commodity (Dhall 2008). It is thus important that the composition of edible coatings must comply with food regulations (Guilbert et al. 1995). A good quality of edible coatings should meet some important criteria such as moisture barrier, solute or gas barrier, water/lipid solubility, good mechanical characteristics, nontoxicity, etc. The effective use of coatings depends greatly on temperature, thickness, alkalinity, type of coating and condition of fruits or vegetables (Park et al. 1994).

In addition to improving shelf life of fruits and vegetables, edible coatings also impart functionality to the product. The improvement in sensory properties of food can also be achieved by using an edible coating containing either colourants, flavours, anti-browning agents, antimicrobials and other functional substances, thereby making the products more attractive. Novel edible materials have been derived from many natural sources that are safe to use. Among all these applications, the use of edible coatings as active packaging stands out as a promising application for food packaging (Bracone et al. 2016). These are promising new approaches to extend the quality and shelf life of fruits and vegetables. Active packaging refers to the incorporation of additives into the package in order to maintain or extend the product quality and shelf life (Gutiérrez and Alvarez 2017a, b).

21.4 Properties of Edible Coating

The properties of edible coating depend mainly on molecular structure rather than chemical constitution and molecular size. Material of any kind cannot be used for coating purposes. Arvanitoyannis and Gorris (1999) detail certain specific requirements, such as:

1. It should not interfere with the quality of fruits and vegetables and should not impart an undesirable odour.
2. The coating material should be water resistant.
3. It should not reduce oxygen or build up excessive carbon dioxide to avoid anaerobic respiration.

4. It should have reduced water vapour permeability.
5. It should improve mechanical handling properties and appearance, maintain structural integrity, carry active agents such as antioxidants, vitamins, etc. and retain volatile flavour compounds.
6. It should be easily emulsifiable.
7. Materials must be dissolved in all types of solvent such as water, alcohol or a mixture of other solvents.
8. It must be economical.

21.5 Composition of Edible Coatings

It is known that edible coatings improve the food quality, since they are selective barriers to oxygen uptake, moisture transfer and loss of volatile aromas and flavours. The moisture barrier properties of edible coatings have been extensively studied by measuring their water vapour properties, because water plays a key role in spoilage reactions in foods. However, the barrier properties of edible coatings mainly depend on the material from which they are derived. Most of the edible coatings are derived from natural polymers having film-forming ability, such as polysaccharides, proteins and lipid (Álvarez et al. 2017). Edible coatings are produced from materials with film-forming ability. It includes polysaccharides such as cellulose, starch, chitin, chitosan (Cs), alginates, pectins etc., and proteins namely casein, whey protein, keratin, collagen, gelatin, soy protein, wheat gluten, corn-zein, peanut protein, cottonseed protein etc., lipids such as wax and oil-based coatings, monoglycerides, fatty acid resins and emulsions etc. (Gennadios et al. 1994; Gutiérrez and Alvarez 2017c; Gutiérrez 2018b; Merino et al. 2019). Their presence and abundance determine the barrier properties of material regarding oxygen, water vapour and carbon dioxide transfer in food systems.

21.6 Current Status and Recent Advances

Edible coatings are currently used in many food applications to improve handling, extend shelf life and modify appearance and flavour. It can be considered as a novel technique for carrying antimicrobials to the food surfaces of high moisture content, maintaining the flavour profile of foods with high aroma and retarding oxidation in intermediate and low moisture foods. Edible coatings also have the potential to maintain the quality of foods even after opening the package. There are multiple applications of edible coatings in the food industry. These include extending the shelf life of oxygen-sensitive foods, reducing packaging of whole and pre-cut fresh fruits and vegetables, extending product shelf life of frozen foods by reducing moisture loss, respiration and colour change, avoiding oxidation, as well as preventing moisture, aroma or colour migration (Marín et al. 2016). It is expected that the active packaging systems market has a promising future due to its integration in packaging materials or systems.

21.7 Applications of Active Edible Coatings on Fruits and Vegetables

The uses of edible coatings made from polysaccharides, proteins, lipids and composites containing several food additives, probiotics, antioxidants and oxidizing agents are discussed in this section. For some time, the main polysaccharides of interest for the coating fruits and vegetables have been cellulose, gums, starch, pectin and Cs (Saberi and Golding 2018). Polysaccharide coatings have a poor moisture barrier but have a good gas barrier and mechanical properties. Proteins from soy, milk, corn and wheat gluten are also used to prepare edible coatings (Garrido et al. 2018). Amongst them, zein is the only commercially produced protein used for such applications. Lipids such as bee wax, rice bran oil, paraffin wax and acetylated monoglycerides are also used in coating formulations. Lipids are hydrophobic and therefore, act as good moisture barriers. However, its mechanical properties are inferior to protein and polysaccharide-based films.

21.7.1 Edible Coatings as Carriers of Flavour, Colourant and Texturizer

As previously outlined, edible coatings can maintain and deliver desirable concentrations of functional ingredients to improve the product quality. Edible coatings can be incorporated with flavours, seasonings, vitamins, colourants and other beneficial plant-derived compounds. Table 21.1 briefly describes the use of edible coatings as carriers of flavours, colourants and texturizers.

21.7.2 Active Edible Coatings Carrying Nutraceuticals

Edible coatings also have the capacity to contain many active ingredients that could be used to enhance the nutritional and functional quality of food products (Gutiérrez and Álvarez 2016; Gutiérrez 2017). However, the integration of nutritional or nutraceutical ingredients into coatings of foods (Table 21.2) has been reported only by few workers, though there is growing interest in this area.

Table 21.1 Application of active edible coatings for imparting desirable flavour, colour and textural attributes

Matrix used	Applied on	Concentration	Effect	References
Alginate	Melon	Calcium lactate 2% w/v	Maintenance of firmness	Raybaudi-Massilia et al. (2008)
Alginate	Pineapple	Calcium chloride 2% w/v	Retention of internal liquids	Montero-Calderon et al. (2008)
Alginate and gellan	Fuji apple	1% *N*-acetyl cysteine and glutathione mixed with 2% w/v alginate and 0.5% w/v gellan	Maintenance of a* value	Rojas-Graü et al. (2007)
Cs	Raspberries	Calcium gluconate 5% w/v	Retention of textural quality	Han et al. (2004)
Whey protein concentrate	Apples	Calcium chloride 1% (w/v)	Retains firmness	Lee et al. (2003)

Table 21.2 Edible coatings as carriers of nutraceuticals

Produce	Coating used	Ingredient	Quantity of application	Effect	References
Papaya	Alginate	Ascorbic acid	Ascorbic acid @ 1.0 (w/v)	Retention of ascorbic acid in the fruit	Tapia et al. (2008)
Apple, papaya	Alginate and gellan	*Bifidobacterium lactis*	*Bifidobacterium lactis* @ 2.0 (w/v)	Maintenance of fruit quality up to 10 days during storage	Tapia et al. (2007)
Strawberries and raspberries	Cs-based coatings	Vitamin E	Cs containing 0.2% dl-α-tocopheryl acetate	Delayed change in colour, titratable acidity and pH of the fruits during cold storage	Han et al. (2004)

21.7.3 Edible Coatings as Carrier for Probiotic Organisms

Probiotics are live microorganisms, which at a given concentration can confer health benefits to the host (Sanders 2008). Probiotics help to support the growth of beneficial microorganisms, reduce potentially harmful bacteria and reinforce the body's natural defence mechanisms. Table 21.3 describes the beneficial use of edible coatings as carriers of probiotics in various fruits and vegetables. Such active packaging systems help to deliver novel health-enhanced products for the consumers.

Table 21.3 Application of probiotic-containing edible coatings on fruits and vegetables

Fruit/ Vegetable	Matrix used	Probiotic incorporated	Additives used	Effect	References
Carrot slices	Sodium alginate	*Lactobacillus acidophilus*	Calcium chloride	Control growth of *Bacillus cereus* and *Staphylococci.*	Shigematsu et al. (2018)
Grape	Hydroxypropylmethylcellulose, corn starch, sodium caseinate & pea protein	*Candida sake*	Oleic acid, span 80, tween 85, glycerol	Inhibit *Botrytis cinerea* (gray mould)	Marín et al. (2016)
Mandarin	Locust bean gum	*W. anomalua, M. pulcherrima, A. pullulans*	No additives used	Control mould caused by *P.digitatum* & *P. italicum*	Parafati et al. (2016)
Apple snacks	Methyl cellulose	*L. plantarum*	Fructooligosaccharides, ascorbic acid. Citric acid & sorbitol	Organoleptically good up to 90 days during storage	Tavera-Quiroz et al. (2015)
Orange	Sodium alginate & locust bean gum	*W. Anomalua*	Glycerol	Control green mould (*P. digitatum*)	Aloui et al. (2015)
Grape	Carboxymethylcellulose (CMC) & alginate	Brewer yeast	No additives were used	Showed good sensory quality up to 13 days	Yinzhe and Shaoying (2013)
Strawberry	Alginate	*C. laurentii*	Glycerol, palmitic acid, glycerol monostearate β-cyclodextrin.	Inhibit the growth of mould and maintain the quality up to 20 days	Fan et al. (2009)
Fresh-cut apples and Papayas	Alginate/Gellan	*B. lactis* Bb-12	Glycerol	Prevents decay up to 10 days	Tapia et al. (2007)
Tomato	Cs	*C. utilis*	Tween-80	Prevent the growth of *A. alternata* and *G. candidum*	Sharma et al. (2006)

21.7.4 Edible Coatings as Carrier of Antimicrobial Agents

Active compounds such as antimicrobials are incorporated into edible coatings to inhibit spoilage/pathogenic bacteria by maintaining effective concentrations of active compounds on food surfaces. There are several categories of antimicrobials that can potentially be incorporated into edible coatings: (1) organic acids, namely lactic, acetic, propionic, benzoic, sorbic, (2) fatty acid esters such as glyceryl monolaurate, (3) polypeptides such as peroxidase, lysozyme, lactoferrin, nisin, and (4) plant essential oils such as cinnamon, oregano, lemongrass and nitrites and sulphites, as listed in Table 21.4 (Álvarez et al. 2018). The incorporation of these antimicrobials can help in delivering safe food.

21.7.5 Active Edible Coatings Carrying Antioxidants

Antioxidants are free radical acceptors that delay, inhibit or interrupt the stage of propagation of autooxidation. These are used to protect against oxidative rancidity and enzymatic browning in fruits and vegetables. The incorporation of natural antioxidants into edible coatings can improve their functionality and applicability in foods. A wide range of natural antioxidants such as essential oils and plant extracts, as well as pure compounds, such as α-tocopherol and ascorbic acid have been incorporated into edible coatings to improve their bioactive properties. Table 21.5 shows the beneficial effect of antioxidant-containing edible coatings. In addition, antioxidants prevent the loss of vitamin C and the browning reactions, especially in the fresh-cut fruits and vegetables, thereby maintaining colour and improving the shelf-life and acceptability of the commodity.

21.7.6 Commercially Available Active Edible Coatings

Since the research on active edible coatings began, many products have been launched in the commercial market, which may or may not be specific to each product. Table 21.6 gives some of these formulations of active edible coatings that are available in the market.

21.8 Limitations for Use

Edible coatings are being developed with incorporation of active compounds such as antioxidants, antimicrobials, flavourings and nutraceuticals. Research is being done to maintain the mechanical properties, since they can be affected dramatically

Table 21.4 Examples of antimicrobials added into edible coatings applied on fruits and vegetables

Used on	Coating used	Antimicrobial compound added	Application	Effect	References
Capsicum	Cs and alginate	Pomegranate peel extract (PPE)	Cs (1% w/v) and alginate (2% w/v) combined with 1% w/v PPE	Control of *Colletotrichum gloeosporioides* during 20 days of low temperature storage	Nair et al. (2018a)
Strawberry	Cs	Caroteno proteins (CP)	Cs (1–3%) mixed with CP (0–3%)	Delayed yeast and mould growth during storage at 4 °C for 11 days	Hajji et al. (2018)
Blueberries	Sodium alginate (AL) and Cs	Apple fiber, orange fiber, inulin and oligofructose	Fiber used at concentrations of 7 g kg^{-1} with 20 g kg^{-1} Cs and 20 g kg^{-1} AL	Reduced decay rate by more than 50%, enhanced antioxidant properties, retention of fruit firmness during storage at 5 °C for 18 days	Alvarez et al. (2018)
Bell pepper	Cs	Lemongrass essential oil	Lemon grass oil @ 0.5% and 1.0% was incorporated into 0.5% and 1.0% of Cs Cs solution	Control of anthracnose both under *in vitro* and *in vivo* condition	Ali et al. (2015)
Satsuma mandarin	Cs	Clove oil	1% Cs combined with 0.5 mL/L clove oil	Inhibition of *Penicillium digitatum* both under *in vitro and in vivo* conditions	Shao et al. (2015)
Dates	Cs and locust bean gum	Citrus essential oils	Cs–2% (v/v) mixed with 2% citrus essential oil	Inhibition of *Aspergillus flavus* during storage of 12 days.	Aloui et al. (2014)
Broccoli	Cs	Essential oils and bioactive compounds	Cs solutions (2 g/100 mL) mixed with 1 mL essential oils and bioactive compounds	Control of *E. coli* and *L. monocytogenes*	Alvarez et al. (2013)
Pineapple	Sodium alginate and pectin	β-Cyclodextrin and trans-cinnamaldehyde)	Sodium alginate and carriers are added @ 2 g/100 g	Extension of shelf-life to 15 days at 4 °C by inhibiting microbial growth	Mantilla et al. (2013)
Papaya (fresh cut)	Cs and pectin	β-Cyclodextrin and trans-cinnamaldehyde	Microencapsulated β-cyclodextrin and trans-cinnamaldehyde complex (2 g/100 g)	Extending shelf life of fresh-cut papaya up to 15 days at 4 °C, higher retention of vitamin C and total carotenoids	Brasil et al. (2012)

Table 21.5 Coatings incorporating antioxidants

Used on	Matrix used	Additives	Effect	References
Guava	Cs and alginate	Pomegranate peel extract	Improved quality and shelf-life	Nair et al. (2018b)
Fuji apple slices	Cassava starch	Cinnamon bark essential oil & fennel essential oil	Higher total phenols and antioxidant activity	Oriani et al. (2014)
Mango (fresh-cut)	Alginate	Ascorbic acid & citric acid	Increased stability till 12 days at 4 °C	Robles-Sánchez et al. (2013)
Tomato	Rice starch	Coconut oil & green tea extract	Control of ripening and enhancement of total phenols and ascorbic acid content	Das et al. (2013)
Peach (fresh-cut)	Pectin	Cinnamon leaf oil	Increase in total phenolic and flavonoid content as well as antioxidant capacity	Ayala et al. (2013)
Table grapes	Pectin	Cinnamon leaf oil	Significant increases in the antioxidant capacity	Melgarejo-Flores et al. (2013)
Rose apple	Konjac glucomann	Pineapple fruit extracts (PE) from peel, pulp and core	Reduced PPO and POD activities with high retention of total phenols	Supapvanich et al. (2012)
Table grapes	Hydroxypropylmethylcellulose (HPMC) or Cs	Bergamot essential oil	Improved product appearance	Sanchez-Gonzalez et al. (2011)
Fresh-cut mango	Cassava starch	Citric acid	Delayed quality deterioration, retarded fruit respiration rate and better maintenance of colour	Chiumarelli et al. (2011)
Table grapes	HPMC	Ethanolic extract of propolis	Prevention of weight loss and browning	Pastor et al. (2011)
Sweet pepper	Cs	Cinnamon oil	Good sensory acceptability	Xing et al. (2011)
Pears	Cs	Rosemary extract	Reduced membrane permeability, vitamin C loss and weight loss	Xiao et al. (2010)
Lettuce and butternut squash	Sodium caseinate Cs and carboxymethyl cellulose	Oleoresins of rosemary, capsicum, cranberry and garlic	Maintained sensory acceptability	Ponce et al. (2008)
Pears (fresh-cut)	Alginate, gellan or pectin	*N*-acetylcysteine glutathione	Reduced vitamin C loss	Oms-Oliu et al. (2008)

Table 21.6 Commercially available active edible coatings

Coating	Composition
Semperfresh™	Sucrose esters with high proportion of short chain unsaturated fatty acid esters, sodium salts of CMC and mixed mono and diglycerides
Prolong	Mixture of sucrose fatty acid esters, sodium CMC and mono and diglycerides
Tal prolong	Mixture of sucrose fatty acid esters, sodium CMC and mono and diglycerides
Nature-seal™	Cellulose-based edible coating
Nu-coatFlo, Ban-seel	Sucrose esters of fatty acids and sodium salt of CMC
FreshSeal™	Polyvinyl alcohol, starch and surfactant
Citrashine	Sucrose ester and wax

with the incorporation of food grade additives. Therefore, the development of coatings with improved mechanical, functional and barrier properties is an important task. One of the main obstacles is the cost, which restricts its application in high value products. In addition to the cost, other limiting factors for the commercial use of edible coatings are the lack of materials with the desired functionalities, the investment cost for the installation of coating equipment, the strictness of the regulations and the difficulty of the production process. Despite these limitations, the food industry is looking for edible coatings that can be used in a broad spectrum of foods to add value to their products, increase product shelf life and/or reduce packaging. There is also a trend to develop new generation edible coatings which release active compounds using nanotechnological solutions such as nanoencapsulation and multilayer systems.

21.9 Regulatory Status for the Use of Edible Coatings

Since edible coatings are an integral part of the edible portion of food products, they should follow all required regulations pertinent to food ingredients. To maintain product safety and quality, all components, as well as functional additives should be food-grade and non-toxic materials. In addition, all process facilities should be acceptable for food processing and should strictly comply with current Good Manufacturing Practice (GMP). Acceptable ingredients for use in edible coatings should be Generally Recognized As Safe (GRAS), and used within any limitations specified by the Food and Drug Administration (FDA). According to the European Directive (1998) the ingredients that can be incorporated into the formulation of edible coatings are arabic and karaya gum, pectins, shellac, beeswax, candelilla wax, carnauba wax, lecithin, polysorbates, fatty acids and fatty acid salts. On the other hand, the FDA mentions other additives used as components of protective coatings applied to fresh fruits and vegetables such as morpholine, polydextrose, sorbitan monostearate, SFAEs, cocoa butter and castor oil (Vargas et al. 2008).

21.10 Conclusion

The trend of packaging research worldwide should be directed to produce active edible coatings that are affordable and can maintain high quality, wholesome, fresh-like products. These coatings are promising in postharvest management and will receive more attention in the near future. Natural ingredients have great potential for use as additives in edible coatings without the adverse effects associated with chemical preservatives. Future research should encourage additional studies to overcome the above limitations.

Acknowledgements The authors would like to thank colleagues and students who gave inputs for compiling this chapter.

Conflict of Interest The authors declare no conflict of interest.

References

Ali, A., Noh, N. M., & Mustafa, M. A. (2015). Antimicrobial activity of chitosan enriched with lemongrass oil against anthracnose of bell pepper. *Food Packaging Shelf Life, 3*, 56–61. https://doi.org/10.1016/j.fpsl.2014.10.003.

Aloui, H., Khwaldia, K., Licciardello, F., Mazzaglia, A., Muratore, G., Hamdi, M., & Restuccia, C. (2014). Efficacy of the combined application of chitosan and locust bean gum with different citrus essential oils to control postharvest spoilage caused by *Aspergillus flavus* in dates. *International Journal of Food Microbiology, 170*, 21–28. https://doi.org/10.1016/j.ijfoodmicro.2013.10.017.

Aloui, H., Licciardello, F., Khwaldia, K., Hamdi, M., & Restuccia, C. (2015). Physical properties and antifungal activity of bioactive films containing *Wicker hamomyces anomalus* killer yeast and their application for preservation of oranges and control of postharvest green mold caused by *Penicillium digitatum. International Journal of Food Microbiology, 200*, 22–30. https://doi.org/10.1016/j.ijfoodmicro.2015.01.015.

Alvarez, M. V., Ponce, A. G., & Moreira, M. (2013). Antimicrobial efficiency of chitosan coating enriched with bioactive compounds to improve the safety of fresh cut broccoli. *LWT–Food Science and Technology, 50*(1), 78–87. https://doi.org/10.1016/j.lwt.2012.06.021.

Álvarez, K., Famá, L., & Gutiérrez, T. J. (2017). Physicochemical, antimicrobial and mechanical properties of thermoplastic materials based on biopolymers with application in the food industry. In M. Masuelli & D. Renard (Eds.), *Advances in physicochemical properties of biopolymers: Part 1* (Vol. 2017, pp. 358–400). Bentham Science Publishers. EE.UU. ISBN: 978–1–68108-454-1. eISBN: 978–1–68108-453-4. https://doi.org/10.2174/9781681084534117010015.

Alvarez, M. V., Ponce, A. G., & Moreira, M. R. (2018). Influence of polysaccharide-based edible coatings as carriers of prebiotic fibers on quality attributes of ready-to-eat fresh blueberries. *Journal of the Science of Food and Agriculture, 98*(7), 2587–2597. https://doi.org/10.1002/jsfa.8751.

Álvarez, K., Alvarez, V. A., & Gutiérrez, T. J. (2018). Biopolymer composite materials with antimicrobial effects applied to the food industry. In V. K. Thakur & M. K. Thakur (Eds.), *Functional biopolymers* (pp. 57–96). Editorial Springer International Publishing.. EE.UU. ISBN: 978-3-319-66416-3. eISBN: 978-3-319-66417-0. pp. https://doi.org/10.1007/978-3-319-66417-0_3.

Ansorena, M. R., Pereda, M., & Marcovich, N. E. (2018). Edible films. In T. J. Gutiérrez (Ed.), *Polymers for food applications* (pp. 5–24). Cham: Springer. https://doi.org/10.1007/978-3-319-94625-2_2.

Arvanitoyannis, I., & Gorris, L. G. M. (1999). Edible and biodegradable polymeric materials for food packaging or coating. In J. C. Oliveira & F. A. R. Oliveira (Eds.), *Processing foods: Quality optimization and process assessment* (pp. 357–371). Boca Raton: CRC Press.

Ayala-Zavala, J. F., Silva-Espinoza, B. A., Cruz-Valenzuela, M. R., Leyva, J. M., Ortega-Ramírez, L. A., Carrazco-Lugo, D. K., Pérez-Carlón, J. J., Melgarejo-Flores, B. G., González-Aguilar, G. A., & Miranda, M. R. A. (2013). Pectin–cinnamon leaf oil coatings add antioxidant and antibacterial properties to fresh-cut peach. *Flavour and Fragrance Journal, 28*(1), 39–45. https://doi.org/10.1002/ffj.3125.

Bracone, M., Merino, D., González, J., Alvarez, V. A., & Gutiérrez, V. A. (2016). Nanopackaging from natural fillers and biopolymers for the development of active and intelligent films. In S. Ikram & S. Ahmed (Eds.), *Natural Polymers: Derivatives, blends and composites* (pp. 119–155). New YorkEE.UU. ISBN: 978-1-63485-831-1: Editorial Nova Science Publishers, Inc.

Brasil, I. M., Gomes, C., Puerta-Gomez, A., Castell-Perez, M. E., & Moreira, R. G. (2012). Polysaccharide-based multilayered antimicrobial edible coating enhances quality of fresh-cut papaya. *LWT–Food Science and Technology, 47*(1), 39–45. https://doi.org/10.1016/j.lwt.2012.01.005.

Chiumarelli, M., Ferrari, C. C., Sarantópoulos, C. I., & Hubinger, M. D. (2011). Fresh cut 'Tommy Atkins' mango pre-treated with citric acid and coated with cassava (*Manihot esculenta* Crantz) starch or sodium alginate. *Innovative Food Science and Emerging Technologies, 12*(3), 381–387. https://doi.org/10.1016/j.ifset.2011.02.006.

Das, D. K., Dutta, H., & Mahanta, C. L. (2013). Development of a rice starch-based coating with antioxidant and microbe-barrier properties and study of its effect on tomatoes stored at room temperature. *LWT–Food Science and Technology, 50*(1), 272–278. https://doi.org/10.1016/j.lwt.2012.05.018.

Debeaufort, F., Quezada-Gallo, J. A., & Voilley, A. (1998). Edible films and coatings: Tomorrow's packagings: A review. *Critical Reviews in Food Science and Nutrition, 38*(4), 299–313. https://doi.org/10.1080/10408699891274219.

Dhall, R. K. (2008). Advances in edible coatings for fresh fruits and vegetables: A review. *Critical Reviews in Food Science and Nutrition, 53*(5), 435–450. https://doi.org/10.1080/10408398.2010.541568.

European Directive-European Parliament and Council Directive N 98/72/EC. (1998). On food additive other than colors and sweeteners. Available from http.//ec.europa.eu/food/fs/sfp/addit flavor/ flav11en.pdf.

Fan, Y., Xu, Y., Wang, D., Zhang, L., Sun, J., Sun, L., & Zhang, B. (2009). Effect of alginate coating combined with yeast antagonist on strawberry (*Fragaria × ananassa)* preservation quality. *Postharvest Biology and Technology, 53*(1–2), 84–90. https://doi.org/10.1016/j.postharvbio.2009.03.002.

Garrido, T., Uranga, J., Guerrero, P., & de la Caba, K. (2018). The potential of vegetal and animal proteins to develop more sustainable food packaging. In: Gutiérrez, T. J. (Ed.). Polymers for food applications. pp 25–59. Springer, Cham. https://doi.org/10.1007/978-3-319-94625-2_3.

Gennadios, A., Weller, C. L., & Testin, R. F. (1993). Property modification of edible wheat, gluten-based films. *Transactions of the American Society of Agricultural Engineers, 36*, 465–470. https://doi.org/10.13031/2013.28360.

Gennadios, A., McHugh, T. H., Weller, C. L., et al. (1994). Edible coating and films based on protein. In J. M. Krochta, E. A. Balwin, & M. O. Niperos Carriedo (Eds.), *Edible coatings and films to improve. Food Quality* (pp. 201–277). Basel: Technomic Publishing Company.

Guilbert, S., Cuq, B., & Gontard, N. (1995). Technology and applications of edible protective films. *Packaging Technology and Science, 8*(6), 339–346. https://doi.org/10.1002/pts.2770080607.

Gutiérrez, T. J. (2017). Surface and nutraceutical properties of edible films made from starchy sources with and without added blackberry pulp. *Carbohydrate Polymers, 165*, 169–179. https://doi.org/10.1016/j.carbpol.2017.02.016.

Gutiérrez, T. J. (2018a). Active and intelligent films made from starchy sources/blackberry pulp. *Journal Polymers and the Environment, 26*(6), 2374–2391. https://doi.org/10.1007/s10924-017-1134-y.

Gutiérrez, T. J. (2018b). Biological macromolecule composite films made from Sagu starch and flour/poly(ε-caprolactone) blends processed by blending/thermo molding. *Journal Polymers and the Environment, 26*(9), 3902–3912. https://doi.org/10.1007/s10924-018-1268-6.

Gutiérrez, T. J., & Álvarez, K. (2016). Physico-chemical properties and *in vitro* digestibility of edible films made from plantain flour with added *Aloe vera* gel. *Journal of Functional Foods, 26*, 750–762. https://doi.org/10.1016/j.jff.2016.08.054.

Gutiérrez, T. J., & Álvarez, K. (2017). Transport phenomena in biodegradable and edible films. In M. A. Masuelli (Ed.), *Biopackaging* (pp. 58–88). Miami., EE.UU. ISBN: 978–1–4987-4968-8: Editorial CRC Press Taylor & Francis Group.

Gutiérrez, T. J., & Alvarez, V. A. (2017a). Properties of native and oxidized corn starch/polystyrene blends under conditions of reactive extrusion using zinc octanoate as a catalyst. *Reactive and Functional Polymers, 112*, 33–44. https://doi.org/10.1016/j.reactfunctpolym.2017.01.002.

Gutiérrez, T. J., & Alvarez, V. A. (2017b). Eco-friendly films prepared from plantain flour/PCL blends under reactive extrusion conditions using zirconium octanoate as a catalyst. *Carbohydrate Polymers, 178*, 260–269. https://doi.org/10.1016/j.carbpol.2017.09.026.

Gutiérrez, T. J., & Alvarez, V. A. (2017c). Cellulosic materials as natural fillers in starch-containing matrix-based films: A review. *Polymer Bulletin, 74*(6), 2401–2430. https://doi.org/10.1007/s00289-016-1814-0.

Gutiérrez, T. J., Guzmán, R., Medina Jaramillo, C., & Famá, L. (2016). Effect of beet flour on films made from biological macromolecules: Native and modified plantain flour. *International Journal of Biological Macromolecules, 82*, 395–403. https://doi.org/10.1016/j.ijbiomac.2015.10.020.

Hajji, S., Younes, I., Affes, S., Boufi, S., & Nasri, M. (2018). Optimization of the formulation of chitosan edible coatings supplemented with carotenoproteins and their use for extending strawberries postharvest life. *Food Hydrocolloids, 83*, 375–392. https://doi.org/10.1016/j.foodhyd.2018.05.013.

Han, C., Zhao, Y., Leonard, S. W., & Traber, M. G. (2004). Edible coatings to improve storability and enhance nutritional value of fresh and frozen strawberries (*Fragaria × ananassa*) and raspberries (*Rubus ideaus*). *Postharvest Biology and Technology, 33*(1), 67–78. https://doi.org/10.1016/j.postharvbio.2004.01.008.

Hardenburg, R. E. (1967). Wax and related coatings for horticultural products. A bibliography. *Agricultural Research Bulletin- USDA, 51*, 15.

Krochta, J. M. and Mulder-Johnston, C. D. (1997). Edible and biodegradable polymer films: Challenges and opportunities. *Food Technology, 51*, 61–74.

Lee, J. Y., Park, H. J., Lee, C. Y., & Choi, W. Y. (2003). Extending shelflife of minimally processed apples with edible coatings and antibrowning agents. *Lebensm Wiss U Technology, 36*(3), 323–329. https://doi.org/10.1016/s0023-6438(03)00014-8.

Mantilla, N., Castell-Perez, M. E., & Gomes, C. (2013). Multi-layered antimicrobial edible coating and its effect on quality and shelf-life of fresh-cut pineapple (*Ananas comosus*). *LWT–Food Science and Technology, 51*(1), 37–43. https://doi.org/10.1016/j.lwt.2012.10.010.

Marín, A., Cháfer, M., Atarés, L., Chiralt, A., Torres, R., Usall, J., & Teixidó, N. (2016). Effect of different coating-forming agents on the efficacy of the biocontrol agent *Candida sake* CPA-1 for control of *Botrytis cinerea* on grapes. *Biological Control, 96*, 108–119. https://doi.org/10.1016/j.biocontrol.2016.02.012.

Melgarejo-Flores, B. G., Ortega-Ramírez, L. A., Silva-Espinoza, B. A., González-Aguilar, G. A., Miranda, M. R. A., & Ayala-Zavala, J. F. (2013). Antifungal protection and antioxidant enhancement of table grapes treated with emulsions, vapors, and coatings of cinnamon leaf oil. *Postharvest Biology and Technology, 86*, 321–328. https://doi.org/10.1016/j.postharvbio.2013.07.027.

Merino, D., Gutiérrez, T. J., & Alvarez, V. A. (2019). Potential agricultural mulch films based on native and phosphorylated corn starch with and without surface functionalization with chitosan. *Journal Polymers and the Environment, 27*(1), 97–105. https://doi.org/10.1007/s10924-018-1325-1.

Montero-Calderon, M., Rojas-Grau, M. A., & Martin-Belloso, O. (2008). Effect of packaging conditions on quality and shelf-life of fresh-cut pineapple (*Ananas comosus*). *Postharvest Biology and Technology, 50*(2–3), 182–189. https://doi.org/10.1016/j.postharvbio.2008.03.014.

Nair, M. S., Saxena, A., & Kaur, C. (2018a). Characterization and antifungal activity of pomegranate peel extract and its use in polysaccharide-based edible coatings to extend the shelf-life of capsicum (*Capsicum annuum* L.). *Food and Bioprocess Technology, 11*(7), 1317–1327. https://doi.org/10.1007/s11947-018-2101-x.

Nair, M. S., Saxena, A., & Kaur, C. (2018b). Effect of chitosan and alginate based coatings enriched with pomegranate peel extract to extend the postharvest quality of guava (*Psidium guajava* L.). *Food Chemistry, 240*, 245–252. https://doi.org/10.1016/j.foodchem.2017.07.122.

Oms-Oliu, G., Soliva-Fortuny, R., & Martín-Belloso, O. (2008). Edible coatings with antibrowning agents to maintain sensory quality and antioxidant properties of fresh-cut pears. *Postharvest Biology and Technology, 50*(1), 87–94. https://doi.org/10.1016/j.postharvbio.2008.03.005.

Oriani, V. B., Molina, G., Chiumarelli, M., Pastore, G. M., & Hubinger, M. D. (2014). Properties of cassava starch-based edible coating containing essential oils. *Journal of Food Science, 79*(2), 189–194. https://doi.org/10.1111/1750-3841.12332.

Parafati, L., Vitale, A., Restuccia, C., & Cirvilleri, G. (2016). The effect of locust bean gum (LBG)-based edible coatings carrying biocontrol yeasts against *Penicillium digitatum* and *Penicillium italicum* causal agents of postharvest decay of mandarin fruit. *Food Microbiology, 58*, 87–94. https://doi.org/10.1016/j.fm.2016.03.014.

Park, H. J. (1999). Development of advanced edible coatings for fruits. *Trends in Food Science and Technology, 10*(8), 254–260. https://doi.org/10.1016/s0924-2244(00)00003-0.

Park, H. J., Chinnan, M. S., & Shewfelt, R. L. (1994). Edible corn-zein film coatings to extend storage life of tomatoes. *Journal of Food Processing and Preservation, 18*(4), 317–331. https://doi.org/10.1111/j.1745-4549.1994.tb00255.x.

Pastor, C., Sánchez-González, L., Marcilla, A., Chiralt, A., Cháfer, M., & González-Martínez, C. (2011). Quality and safety of table grapes coated with hydroxypropylmethylcellulose edible coatings containing propolis extract. *Postharvest Biology and Technology, 60*(1), 64–70. https://doi.org/10.1016/j.postharvbio.2010.11.003.

Ponce, A. G., Roura, S. I., del Valle, C. E., & Moreira, M. R. (2008). Antimicrobial and antioxidant activities of edible coatings enriched with natural plant extracts: *In vitro* and *in vivo* studies. *Postharvest Biology and Technology, 49*(2), 294–300. https://doi.org/10.1016/j.postharvbio.2008.02.013.

Raybaudi-Massilia, R. M., Martin-Belloso, O., & Mosqueda-Melgar, J. (2008). Edible alginate-based coating as carrier of antimicrobials to improve shelf-life and safety of fresh-cut melon. *International Journal of Food Science and Technology, 121*(3), 313–327. https://doi.org/10.1016/j.ijfoodmicro.2007.11.010.

Robles-Sánchez, R. M., Rojas-Graü, M. A., Odriozola-Serrano, I., González-Aguilar, G., & Martin-Belloso, O. (2013). Influence of alginate-based edible coating as carrier of anti-browning agents on bioactive compounds and antioxidant activity in fresh-cut kent mangoes. *LWT–Food Science and Technology, 50*(1), 240–246. https://doi.org/10.1016/j.lwt.2012.05.021.

Rojas-Graü, M. A., Tapia, M. S., Rodríguez, F. J., Carmona, A. J., & Martin-Belloso, O. (2007). Alginate and gellan-based edible coatings as carriers of antibrowning agents applied on fresh-cut fuji apples. *Food Hydrocolloids, 21*(1), 118–127. https://doi.org/10.1016/j.foodhyd.2006.03.001.

Saberi, B., & Golding, J. B. (2018). Postharvest application of biopolymer-based edible coatings to improve the quality of fresh horticultural produce. In T. J. Gutiérrez (Ed.), *Polymers for food applications* (pp. 211–250). Cham: Springer. https://doi.org/10.1007/978-3-319-94625-2_9.

Sánchez-González, L., Pastor, C., Vargas, M., Chiralt, A., González-Martínez, C., & Cháfer, M. (2011). Effect of hydroxypropylmethylcellulose and chitosan coatings with and without bergamot essential oil on quality and safety of cold-stored grapes. *Postharvest Biology and Technology, 60*(1), 57–63. https://doi.org/10.13031/2013.28360.

Sanders, M. E. (2008). Probiotics: Definition, sources, selection, and uses. *Clinical Infectious Diseases, 46*(2), 58–61. https://doi.org/10.1086/523341.

Shao, X., Cao, B., Xu, F., Xie, S., Yu, D., & Wang, H. (2015). Effect of postharvest application of chitosan combined with clove oil against citrus green mold. *Postharvest Biology and Technology, 99*, 37–43. https://doi.org/10.1016/j.postharvbio.2014.07.014.

Sharma, N., Verma, U., & Awasthi, P. (2006). A combination of the yeast *Candida utilis* and chitosan controls fruit rot in tomato caused by *Alternaria alternata* (Fr.). *The Journal of Horticultural Science and Biotechnology, 81*(6), 1043–1051. https://doi.org/10.1080/14620316.2006.11512170.

Shigematsu, E., Dorta, C., Rodrigues, F. J., Cedran, M. F., Giannoni, J. A., Oshiiwa, M., & Mauro, M. A. (2018). Edible coating with probiotic as a quality factor for minimally processed carrots. *Journal of Food Science and Technology, 55*(9), 3712–3720. https://doi.org/10.1007/s13197-018-3301-0.

Supapvanich, S., Prathaan, P., & Tepsorn, R. (2012). Browning inhibition in fresh-cut rose apple fruit cv. Taaptimjaan using konjac glucomannan coating incorporated with pineapple fruit extract. *Postharvest Biology and Technology, 73*, 46–49. https://doi.org/10.1016/j.postharvbio.2012.05.013.

Tapia, M. S., Rojas-Graü, M. A., Rodríguez, F. J., Ramírez, J., Carmona, A., & Martin-Belloso, O. (2007). Alginate- and gellan-based edible films for probiotic coatings on fresh-cut fruits. *Journal of Food Science, 72*(4), 190–196. https://doi.org/10.1111/j.1750-3841.2007.00318.x.

Tapia, M. S., Rojas-Grau, M. A., Carmona, A., Rodríguez, F. J., Soliva-Fortuny, R., & Martin-Belloso, O. (2008). Use of alginateand gellan-based coatings for improving barrier, texture and nutritional properties of fresh-cut papaya. *Food Hydrocolloids, 22*(8), 1493–1503. https://doi.org/10.1016/j.foodhyd.2007.10.004.

Tapia-Blácido, D. R., Maniglia, B. C., & Tosi, M. M. (2018). Transport phenomena in edible films. In T. J. Gutiérrez (Ed.), *Polymers for food applications* (pp. 149–192). Cham: Springer. https://doi.org/10.1007/978-3-319-94625-2_7.

Tavera-Quiroz, M. J., Romano, N., Mobili, P., Pinotti, A., Gómez-Zavaglia, A., & Bertola, N. (2015). Green apple baked snacks functionalized with edible coatings of methylcellulose containing Lactobacillus plantarum. *Journal of Functional Foods, 16*, 164–173. https://doi.org/10.1016/j.jff.2015.04.024.

Vargas, M., Pastor, C., Chiralt, A., McClements, D. J., & Gonzalez-Martinez, C. (2008). Recent advances in edible coatings for fresh and minimally processed fruits. *Critical Reviews in Food Science and Nutrition, 48*(6), 496–511. https://doi.org/10.1080/10408390701537344.

Xiao, C., Zhu, L., Luo, W., Song, X., & Deng, Y. (2010). Combined action of pure oxygen pretreatment and chitosan coating incorporated with rosemary extracts on the quality of fresh-cut pears. *Food Chemistry, 121*(4), 1003–1009. https://doi.org/10.1016/j.foodchem.2010.01.038.

Xing, Y., Li, X., Xu, Q., Yun, J., Lu, Y., & Tang, Y. (2011). Effects of chitosan coating enriched with cinnamon oil on qualitative properties of sweet pepper (*Capsicum annuum* L.). *Food Chemistry, 124*(4), 1443–1450. https://doi.org/10.1016/j.foodchem.2010.07.105.

Yinzhe, R., & Shaoying, Z. (2013). Effect of carboxymethyl cellulose and alginate coating combined with brewer yeast on postharvest grape preservation. ISRN Agronomy, 871396: 1–7. https://doi.org/10.1155/2013/871396.

Chapter 22
Active, Eco-Friendly and Edible Coatings in the Post-Harvest – A Critical Discussion

Alcilene R. Monteiro Fritz, Jéssica de Matos Fonseca, Thalles Canton Trevisol, Cristiane Fagundes, and Germán Ayala Valencia

Abstract Many researches about edible coating and applications on food have been developed during centuries. Beeswax was one of the first applications of edible coatings on apple and citrus fruits to prevent water loss and maintain the appearance of fruits in 12th and 13th centuries. Beyond the beewax, paraffin, carnauba and oils were also used as coating materials until the mid-twentieth century. In the 1930s, hot-melt paraffin waxes were marketed as edible coatings of fresh apples and pears. The studies showed only a comparison between the fruits with and without cover, but a prolongation in the conservation of the fruits with coating was observed. Thus, in the 1980s, several attempts were made to develop other materials that could be used to coat fresh fruits and vegetables to increase their shelf life. Research with natural biopolymers such as protein, starch, chitosan, whey protein, cellulose, lipids and others have made considerable progress in the development of fruit and vegetable coating. Thus, this chapter will discuss advances in edible coatings for whole and fresh cut fruits and vegetables. The positive and negative effect of the coating on physiology products such as repining, respiration rates, sugar, antioxidants, organic acid, texture and color enhancers. In addition, recent advances in the incorporation of food additives into coating solutions, composite coatings, coating application techniques and stages, as well as and ecotoxicity of coatings will be addressed in the present review.

Keywords Barrier properties · Biopolymers · Composites · Food additives · Food properties

A. R. Monteiro Fritz (✉) · J. de Matos Fonseca · T. C. Trevisol · G. A. Valencia
Department of Chemical and Food Engineering, Federal University of Santa Catarina, Florianópolis, SC, Brazil
e-mail: alcilene.fritz@ufsc.br

C. Fagundes
Department of Food Engineering, Catarinense Federal Institute, Concórdia, SC, Brazil

T. J. Gutiérrez (ed.), *Polymers for Agri-Food Applications*,
https://doi.org/10.1007/978-3-030-19416-1_22

22.1 Introduction

Fruits and vegetables are essential food in the human diet, mainly against obesity and prevent diseases such as diabetes and certain types of cancer. However, the consumption of fruits and vegetables is lower than that recommended by the World Health Organization (WHO) in many continents, such as Africa and North and South America. The low intake of fruits and vegetables is among the top 10 risk factors that contribute to early mortality, according to the evidence presented in the World Health Report 2003. On the other hand, around 1.3 billion tons of food are wasted or lost globally *per* year (Gustavsson et al. 2011), this includes 35% for fruits and vegetables in many countries, even in retail and consumption. The largest investments in research (around 95%) in the last 30 years have focused on increasing productivity and only 5% have been directed to reduce losses (Kader and Rolle 2004; Kader 2005; Kitinoja and AlHassan 2012).

Among the techniques proposed to reduce post-harvest loss are rapid cooling (remote control of field heat), product cleaning, sorting, sanitization, packaging, storage and distribution under refrigeration, designate Good Manufacture Practice (GMP), etc. Because fruits and vegetables are highly perishable, there are other alternatives available in the market and in the research phase such as fresh-cut processing, active and passive modified atmosphere and edible coating with and without bioactive compounds (Charles et al. 2003; Fagundes et al. 2015; Guerreiro et al. 2015a; Mohammadi et al. 2015; D'Aquino et al. 2016; Youssef et al. 2017; Gutiérrez 2017, 2018a).

The edible coating, active or not, are alternatives to reduce post-harvest loss and at the same time to prolong the shelf life of fresh and processed fruits and vegetables, beyond the other food products. In addition, active edible coatings using different active compounds such as essential oils (EOs) and emulsions have been tested such as added polymorphic curcumin within coatings (Sanphui et al. 2011), coatings from turmeric starch with naturally occurring curcumin have also been studied as active and friendly coatings on Foods (Tosati et al. 2017, 2018). In addition, nanotechnology has recently been applied to the fabrication of biocomposite coatings (Gutiérrez 2018b).

The positive and negative effects of coating on the physiology of products such as repining, respiration rates, sugar, antioxidants, organic acid, texture and color enhancers are important to select the most appropriate for each product.

Most of the recent advances in the development of edible coatings is the incorporation of active food additives in the coating solutions, composite coatings, coating-application techniques and stages, as well as in the coating ecotoxicity that will be treated in the present chapter.

22.2 Edible Coatings on Fruits and Vegetables: Effects and Interactions

Fresh fruits remain alive after harvest, so that respiration and transpiration occur during storage and, consequently, induce water loss and texture (Kader 2005). For this reason, some requirements to minimize the metabolic activities of the fruits and vegetables, such as barrier/permeability to gas and low temperature are required. However, due to the complexity of these products and the interactions between the coating and fruits and vegetables, more studies are required. The following sections, recent investigations on edible coating on whole and minimally processed fruits and vegetables are addressed.

22.3 Coating Application Techniques

Although edible coatings have been a target of study for decades, their functions such as the appearance enhancer (shiny skin), functional properties or preservatives can only be achieved when the coating application techniques are used properly. A suitable choice of these techniques will ensure that the active coating properties cause the desired benefit effect in all regions of the fruit and vegetables covered. This selection is based on the characteristics of the food, physicochemical and rheological properties of the coating (density, viscosity, surface tension), thickness coating required, and drying technology used (Andrade et al. 2012). The coating process consists in wetting the food surface with a coating solution followed by the adhesion stage, in which the coating solution penetrates the skin of fruits and vegetables. Thus, the high affinity between the coating solution and the food surface is essential to facilitate the spreadability of the coating on the surface, thus decreasing the time required for this procedure (Mittal 1977). Dipping, brushing, spraying, panning and fluidized bed are well-known coating application techniques (Askari et al. 2006; Bourtoom 2008; Galus and Kadzińska 2015; Youssef et al. 2017), while electrostatic spraying is a technique that has been used recently (Zhong et al. 2014). A schematic representation for each coating-application technique is shown in Fig. 22.1 and Table 22.1 presents studies using several techniques of coating application on whole or fresh-cut fruits and vegetables.

22.3.1 *Dipping*

Dipping is the most common laboratory-scale technique for coating foods due to its easy application, good coverage on food with an irregular surface, as well as being a cheap process compared to others techniques (Youssef et al. 2017). This method consists in immersing the fruit or vegetable into a tank that contains the coating solution (Fig. 22.1a). Some process parameters of the dipping technique are type of the solvent, temperature and viscosity of the coating solution, immersion and withdraw speed, dipping repetition and immersion time (Ahmadian-Yazdi et al. 2016; Yang et al. 2018). Despite the practicality of this method, some disadvantages can be highlighted: dilution of the coating solution, the residual accumulation of coating materials and the possible growth of microorganisms into dipping immersion tank (Youssef et al. 2017).

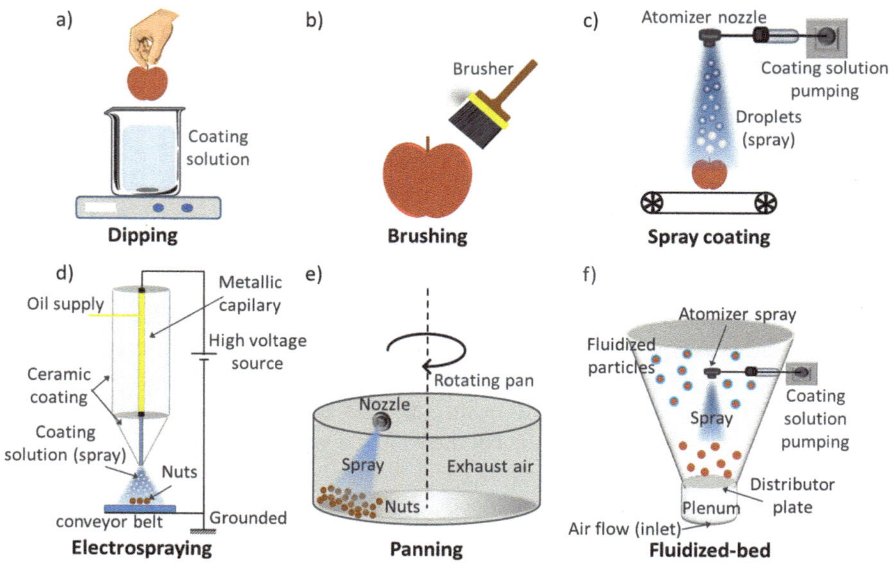

Fig. 22.1 Schematic representation of coating application techniques for fruits and vegetables: dipping (**a**), brushing (**b**), spray coating (**c**), electrospraying (**d**), panning (**e**) and fluidized bed (**f**). Figures (**d**) and (**e, f**) (based on Andrade et al. (2012); Khan et al. (2017) and Andrade et al. (2012))

To enhance the retention, uniformity and adherence of the coating solution, vacuum impregnation have been proposed to coat fruits and vegetables (Vargas et al. 2009). This method is performed by dipping the product into the coating solution, withdraw the food and drain the excess of solution, due to pressure changes, the fluid imprisoned in the food pores is exchanged by the external coating liquid. Various factors can influence the coating efficiency: duration of the vacuum period, vacuum pressure and atmospheric restoration time. In addition to its advantages, literature still lacks of studies to perform coating on food materials using this technique (Senturk Parreidt et al. 2018; Soares et al. 2018).

22.3.2 Brushing

Brushing is a simple technique that involves application of coatings on foods using a paint brusher (Poverenov et al. 2014) (Fig. 22.1b) or a brushing equipment (Nussinovitch et al. 2017), which can be used to polish the food surface (e.g. wax application).

Table 22.1 Effect of edible coating and active edible coating on whole and fresh-cut fruit and vegetables under the application of several techniques

Fruit/ vegetable	Material	Active additive[a]	Application Method	Main results	Reference
Apricot	Basil-seed Gum	Origanum essential oil	Dipping	Maintenance physicochemical quality and reduced total bacteria, mold and yeast count	Hashemi et al. (2017)
	Soybean protein isolate and chitosan (Cs)	No	Dipping	Decreased the weight loss, prevented the decrease in firmness and benefited the textural properties of the tissue and inhibit pectin degradation.	Zhang et al. (2018)
Asparagus	Polyvinylpyrrolidone	Silver nanoparticles	Dipping	Lower weight loss and bacteria, yeast, and mold growth and better color and texture	An et al. (2008)
Avocado	Carboxymethyl cellulose (CMC) and Cs	Moringa leaf extract	Dipping	Lower weight loss, respiration rate, and oxidation and higher firmness and phytochemical characteristics	Tesfay and Magwaza (2017)
	CMC	Moringa leaf and seed extract	Dipping	Lower weight loss, ethylene production, and respiration rate and reduction of fungal diseases and incidence	Tesfay et al. (2018)
Banana	Shellac and gelatin	No	Dipping	Reduction of color changes, acidity, total soluble sugar, weight loss and total mold/ yeast and increment on fruit firmness	Soradech et al. (2017)
	Rice starch and ι-carrageenan	No	Spraying	Lower respiration rate, ethylene production, weight loss and retained chlorophyll content and fruit firmness	Thakur et al. (2018a)
Bell pepper	Cs and gelatin	No	Brushing	Reduction of fruit deformation and weight loss, maintenance of biochemical parameters and inhibited microbial spoilage	Poverenov et al. (2014)
	Cs (nano)	*B. crassifolia* extract	Spraying	Reduction of weight loss, color changes, ethylene production, and fungi/yeast content and higher carotenoid and phenolic content	González-Saucedo et al. (2019)

(continued)

Table 22.1 (continued)

Fruit/ vegetable	Material	Active additive[a]	Application Method	Main results	Reference
Blackberry	Tapioca starch and beewax	No	Spraying	Increase of respiration rate, ethylene production, and weight loss and reduction of anthocyanin and phenol content	Pérez-Gallardo et al. (2015)
Broccoli	Cs and CMC	No	Dipping	Weight loss was reduced, fruit firmness was improved and color and chlorophyll content was maintained	Ansorena et al. (2011)
	Cs	No	Dipping	Inhibition of coliform and *E. coli* growth and yellowing and opening of the florets	Moreira et al. (2011)
Carrot	Cs	No	Dipping and vacuum impregnation	Increase of water vapour transmission and color maintenance and preservation of mechanical properties	Vargas et al. (2009)
	Alginate	Silver nanoparticles	Dipping	Maintained soluble protein content and reduced weight loss	Fayaz, A. et al. (2009)
	Alginate	Silver-montmorillonite nanoparticles	Dipping	Reduction of respiration rate and bacterial growth and increased the shelf life	Costa et al. (2012)
	Tapioca starch and hsian-tsao leaf gum	Cinnamon essential oil and grape seed extract	Dipping	Increase of respiration rate and reduced the whitening effect	Lai et al. (2013)
Cherry tomato	Hydroxypropyl methylcellulose (HPMC)	No	Dipping	Decreased the respiration rate	Tosati et al. (2015)
Cucumber	Cs (nano)	*Cinnamomum zeylanicum* essential oil	Dipping	Inhibit fungal development and increase the shelf life	Mohammadi et al. (2015)
	Cs (nano)	*Zataria multiflora* essential oil	Dipping	Reduction of respiration rate and weight loss, increase of firmness, color preservation and inhibition of bacteria, yeasts and mold growth	Mohammadi et al. (2016)
Eggplant	Soy protein and cysteine	No	Dipping	Reduction of weight loss and respiration rate and firmness and color preservation	Ghidelli et al. (2014)

Grape	k-carrageenan and HPMC	Cellulose nanofibers	Spraying	Mechanical properties stability, pH and maturity index reduction and water barrier improvement	Silva-Vera et al. (2018)
Green bean	Cs	Mandarin essential oil	Spraying	Antimicrobial activity against *L. innocua* and preservation of fruit's firmness and color	Severino et al. (2014)
Guavas	HPMC	No	Dipping	Reduction of firmness and weight loss and delay on ascorbic acid and sugar content and enzymatic activity	Vishwasrao and Ananthanarayan (2016)
Melon	Alginate	No	Dipping and vacuum impregnation	Improvement of fruit firmness	Parreidt et al. (2018)
	Cs	Ag nanoparticles	Dipping	Reduction of respiration rate, improvement of fruit's firmness and decrease of soluble solids content and antimicrobial effect against bacteria, yeast and mold	Ortiz-Duarte et al. (2019)
	Xantan gum	β-carotene	Dipping	Reduced respiration rate/better physicochemical properties	Zambrano-Zaragoza et al. (2017)
Orange	CMC	*Impatiens balsamina* extract	Dipping	Decay rate and weight loss were decreased and scavenger antioxidant and defense enzymes content increased	Zeng et al. (2013)
	CMC and Cs	No	Brushing	Reduction of weight and firmness loss	Arnon et al. (2014)
	Shellac, gelatin and persian gum	No	Dipping	Reduction of respiration rate, weight, and firmness loss	Khorram et al. (2017)
	Carnauba and beewax	Nanoclay (bentonite)	Dipping	Improvement of shelf-life and sensorial acceptance	Motamedi et al. (2018)
				Reduced respiration rate, ethylene production, weight and firmness loss, peel pitting, and fruit decay	Saberi et al. (2018)

(continued)

Table 22.1 (continued)

Fruit/ vegetable	Material	Active additive[a]	Application Method	Main results	Reference
Papaya	Whitemouth croaker protein isolate	Montmorillonite	Dipping	Reduction of weight loss, microbial growth and loss of firmness, lightness, and pH	Cortez-Vega et al. (2014)
Pear	HPMC and protein isolate	Potassium sorbate	Dipping	Reduction of weight and firmness loss, withhold biochemical factors, delay enzymatic activity and improvement of shelf-life	Dave et al. (2017)
Pepper and apple	Pullulan	*Satureja hortensis* extract	Dipping	Decrease of weight loss, maintenance of firmness and protection against wilting and wrinkling on surface	Kraśniewska et al. (2014)
Persimmon	Pectin	Potasium sorbate, nisin or sodium benzoate	Dipping	Browning inhibition and effective against mesophilic aerobic bacteria, psychrophilic aerobic bacteria moulds and yeasts	Sanchís et al. (2015)
Pineapple	Alginate and pectin	Trans-cinnamaldehyde	Dipping (layer-by-layer)	Psychrothrophics, yeast and, molds antimicrobial effect and preservation of juice leakage and texture	Mantilla et al. (2013)
	Alginate	Lemongrass essential oil	Dipping	Decrease of respiration rate, firmness and weight loss and total plate, yeast and mold counts and maintenance of color and sensory characteristics	Azarakhsh et al. (2014)
	Mucilage (linseed, nopal cactus and aloe), pullulan and Cs	No	Dipping (layer-by-layer)	Reduced weight and firmness loss, antimicrobial effect against *L. monocytogenes* and *S. typhi* and preservation of sensory properties	Treviño-Garza et al. (2017)
Plum	Alginate	No	Dipping	Decreased ethylene production, weight and firmness loss and delayed color and total acidity changes	Valero et al. (2013)
	Rice starch and ι-carrageenan	No	Dipping	Reduced weight loss and respiration rate and inhibited ethylene production	Thakur et al. (2018b)

Pomegranate	Cs, carnauba wax and resin wax	No	Dipping and brushing	Reduced respiration rate and weight loss and maintained quality and color parameters, total phenolics, flavonoids, and anthocyanins	Meighani et al. (2014)
Pumpkin	Tapioca starch and k-carrageenan	Potassium sorbate and $FeSO_4.7H_2O$	Dipping	Improvement of color parameters and ascorbic acid retention and increase of Fe accessibility	Genevois et al. (2016)
	Cs	No	Dipping and vacuum impregnation	Reduce of numbers of psychrotrophic and changes in water content, carotenoid content, microbial count, and, improvement shelf-life	Soares et al. (2018)
Rice	HPMC, methyl cellulose (MC) and zein (corn protein)	No	Spraying	Increase of hardness and toughness	Mridula and Pooja (2014)
Strawberry	CMC, HPMC and Cs	No	Dipping	Delayed weight loss, decay percentage, acidity, pH, total soluble solids and ascorbic acid and maintained phenols and anthocyanins	Gol et al. (2013)
	Cs	*Paeonia rockii* extract	Dipping	Slowed of native fungi growth, reduced weight loss and maintained ascorbic acid and phenol content	Pagliarulo et al. (2016)
	Alginate	Carvacrol and methyl cinnamate	Conventional and electrostatic spraying	Retention of firmness and color and reduction of weight loss and decay	Peretto et al. (2017)
	Quinoa protein and Cs	Thymol	Dipping	Reduce of respiration rate and mold and yeast growth	Robledo et al. (2018)
	CMC and Cs	No	Dipping (layer-by-layer)	Delayed volatiles decomposition and reduced firmness loss	Yan et al. (2019)
Sweet potato	Cassava starch	Potassium sorbate	Dipping	Browning prevention and increase of ascorbic acid and antioxidant capacity and reduction of weight loss	Ojeda et al. (2014)

(continued)

Table 22.1 (continued)

Fruit/ vegetable	Material	Active additive[a]	Application Method	Main results	Reference
Tangerine	Cs	Montmorillonite	Dipping	Reduction of respiration rate, weight loss	Xu et al. (2018)
Tomato	Pectin	Oregano essential oil	Brushing	Reduced fungal decay maintained phenol content and sensory aspects	Rodriguez-Garcia et al. (2016)
	Cs	*Caryocar brasiliense* Camb. Extract	Dipping	Reduced weight and firmness loss, color changes and mold and yeast growth	Breda et al. (2017)
	Mango kernel starch	No	Dipping	Reduction in weight loss andrestricted changes in soluble solids concentration, titratable acidity, ascorbic acid content, and firmness	Nawab et al. (2017)
	Cassava starch and Cs	Pomegranate peel extracts and *Lippia sidoides* Cham. essential oil	Dipping	Reduced weight loss and total soluble solids content	Araújo et al. (2018)
	Xanthan gum and candeuba wax	No	Dipping	Reduced firmness loss, mountainous of color and increase of lycopene content	Miranda-Linares et al. (2018)

[a]Excluding lipid, plasticizer and surfactant content

22.3.3 *Spray Coating*

Spray coating consists in a random and stochastic process, in which droplets are deposited on heated substrate surface using an atomizer or spray nozzle to atomize the liquids (Fig. 22.1c). The droplets can be dried, solidified individually or coalesce before solidification to form a thin solid layer (Huang et al. 2018). This technique offers advantages such as thickness control, homogenous coating and the possibility of multilayer applications that prevent contamination of the coating solution (Youssef et al. 2017). However, the drop generation (fluid behavior) is specific to each coating solution, which may require a different design of the atomizers, thus increasing costs. The main parameters of the process that must be controlled in an atomization are aperture size of atomizer nozzle, coating solution temperature (*in let*), gas flow and spray speed (Lee et al. 2011).

22.3.4 *Electrospraying*

Electrospraying is a simple and quick technique for coating and is adequate for the application of liquid coatings of food, since it is a hydrodynamic electrostatic atomization process with the ability to produce sub-micrometric droplets (Khan et al. 2017). In this process, coating solution is forced to pass through the nozzle connected to high-voltage source, and then this solution is electrostatically charged (Fig. 22.1d). This electrostatic force is sufficient to overcome the surface tension of the liquid, breaking the liquid jet into sub-micrometric droplets (atomization) (Peretto et al. 2017). A shear stress is generated during the break of coating solution, forming an aerosol cone, where the droplets repel each other avoiding their coalescence. These charged droplets are attracted by a grounded surface that leads to their discharge as well (Khan et al. 2017). The greatest advantage of the electrospraying is the improvement of the distribution of the coating on food surface and its low cost (Khan et al. 2013, 2014).

22.3.5 *Panning*

Pan coating is a technique used to apply a thin or thick successive layer, especially sugar syrup, on the candies. The coatings produced by panning can be of three types: soft panning, hard panning and chocolate panning. Soft panning consists in wetting the food particle into sugar syrup and dusting with sugar of other powder on surface food. This process is more used for candies as jujubes. On the other hand, in hard panning, food as gumballs, peanuts and chocolate candies are coated by the successive layers of the sugar syrup. The first layer is dried, so that other coatings are applied, resulting in a dense and compact cover. Lastly, chocolate panning consists of applying a chocolate layer around food like raisins, nuts, almonds and malt balls (Nieto 2009). The equipment for performing pan coating is composed by a rotating pan (Fig. 22.1e), where the coating solution is ladled or sprayed, so that the deposited food is uniformly coated. Forced air and high or environment temperature can be applied to dry the edible coating (Andrade et al. 2012).

22.3.6 *Fluidized Bed*

Fluidized bed is a technique originally developed by the pharmaceutical industry and nowadays its application was extended to food products such as peanuts (Lin and Krochta 2006). Although it is a costly technique, fluidized bed enables the use of lower surfactant concentrations than other coating application techniques such as dipping and panning (Andrade et al. 2012). This method typically consists in a spraying a coating solution or suspension onto food surface using a set of nozzles on the surface of the fluidized powders, thus forming a shell structure (Fig. 22.1f) (Lin and Krochta 2006). Fluidized beds can be classified into three different types: top spray, bottom spray and rotating fluidized bed, with the top-spray configuration being the more used by the food industries due to its higher performance. In this case, fluidization will occur when an upward fluid flow through a particle bed reaches enough current velocity capable to keep the particles without carrying them along the flow. The size of the particles generated by the fluidized bed is greater than 100 μm due to a higher aggregation or an insufficient stable fluidization state of smaller powder particles in the conventional fluidized bed (Andrade et al. 2012).

22.4 Influence of the Edible Coating on the Physicochemical Properties and Physiology of Whole and Fresh-Cut Fruits and Vegetables

The interest in the development of coatings based on biodegradable materials has increased mainly due to the demand for high quality food and environmental concerns about the disposal of non-renewable materials.

The amount of discarded packaging material has increased every year. Less than 5% of all plastics are recycling, which leads to a high accumulation of plastics in the environment. In addition, increasing consumer concerns about food safety led to the development of biodegradable, edible and renewable coatings suitable for food and nonfood packaging applications (Sutherland et al. 2010).

The prolongation of the shelf life of fresh-cut fruits and vegetables using edible coatings has attracted great interest, especially when using environmentally friendly and biodegradable materials (Álvarez et al. 2017; 2018). Several researchers have reported on the application of edible coatings to preserve fruits and vegetables (Zhang et al. 2018).

Fruits and vegetables are very perishable. During transportation and post-harvest storage these products undergo deterioration due to physiological reactions such as respiration, ethylene production and senescence. These reactions are directly related to storage conditions such as temperature, humidity and gas atmosphere. According to the physiological nature, the chemical composition and surface structure of fruits and vegetables, microbiological growth, as well as loss of water, texture, color and nutrients can occur during the post-harvest of the product (Yan et al. 2019; Thakur et al. 2018a).

Although the cooling temperature is the main method used for the storage and prolongation of fruits and vegetables, this technology has not been enough to extend the shelf life of the most perishable vegetables. Others alternatives have thus been studied to achieve this goal, including the use of edible coatings (Chen et al. 2014).

Studies have been shown to significantly reduce water loss at most temperatures and storage times in coated vegetables (see Table 22.1). Several edible coatings have been studied to coat whole and fresh-cut fruits and vegetables (Table 22.1). Fruit coatings began in the twelfth centuries with the application of Chinese wax. Nowadays, the basic ingredients of the coatings are lipids, polysaccharides and proteins. The extension of its shelf life depends on the interactions between coatings and fresh products. The influence of polysaccharide-based coatings, e.g. on weight loss is probably associated with the existence of hydroxyl groups that create hydrogen bonds within the coating matrix and with the cuticle in the shell. Alginate is another polysaccharide that is also used as an edible coating, since it has colloidal properties and can form insoluble polymers due to the cross-linking with calcium, thus preserving the cell wall of fruits and vegetables (Jiang 2013). Firmness is related with the non-degradation of the insoluble protopectins that maintain the cell wall. Thus, edible coating on fruits and vegetables probably delays the depolymerization of pectin due the reduction of O_2 and the high CO_2 concentration, which inhibits the activity the enzymes (Lin and Zhao 2007; Kumar and Sethi 2018).

The increase of the respiratory rate during the fruit ripening may be due to the enlargement of the stomata pores, through the ripening process leads the fruit to consume more oxygen (Thakur et al. 2018b). The use of edible coatings can create a modified atmosphere on the fruit surface with low concentrations of O_2 and CO_2, thus reducing the respiratory rate. The semi permeability of the membrane in the coated fruit reduces metabolic gas exchanges (O_2 and CO_2) through the coating, resulting in a deceleration in metabolic activity, including the effectiveness of oxidizing enzymes (Tesfay and Magwaza 2017). The reduction in the levels of ethylene production in fruits with edible coatings means that the coating provides an effective selective barrier of gases between the fruits and the surrounding atmosphere (Fig. 22.2). The semi-anaerobic conditions formed within the fruit can cause a reduced catalytic activity of 1-aminocyclopropane-1-carboxylic acid (ACC) oxidase, so that the ethylene production is effectively maintained by the coated fruit during storage (Deng et al. 2017). In addition, edible coating can suppress the metabolic activities that eventually led to the inhibition of anthocyanin synthesis and delayed the color development (Valero et al. 2013).

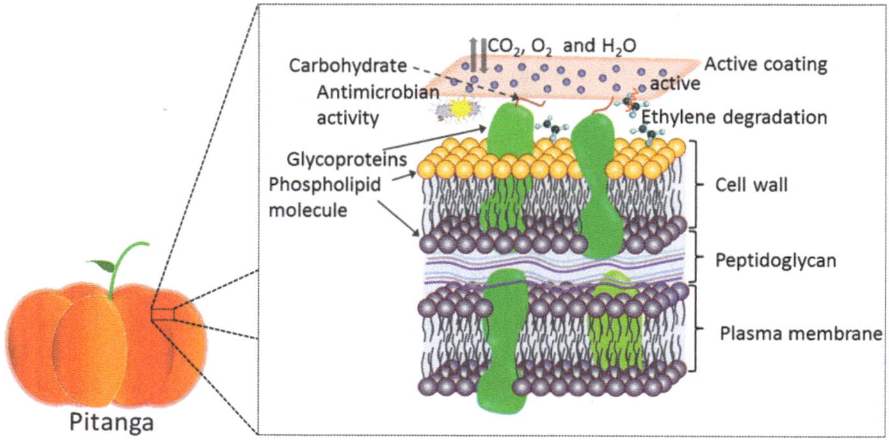

Fig. 22.2 Schematic representation of gas exchanges through coating for Pitanga (*Eugenia uniflora* L.) and the properties of the coating activity on the fruit surface

Tosati et al. (2015) carried out coatings of hydroxyl propyl methylcellulose (HPMC) to regulate the permeability to CO_2 and O_2 and respiration rate in coated cherry tomatoes (*Lycopersicon esculentum*). Cherry tomatoes coated with HPMC showed a lower respiration rate than uncoated tomatoes. A similar effect was also observed when decreasing temperature and O_2 consumption.

22.5 Active Edible Coating on Whole and Fresh-Cut Fruits and Vegetables: Physico-Chemical and Antimicrobial Properties and Respiration Rate in Response to the Interactions Between the Coating and the Products

The active edible coating is a type of coating that when it comes into contact with food changes the conditions of food and coating. Among the active compounds incorporated into coating by dipping are substances that avoid the development of fungal, microorganism or mold without modifying the sensory quality of the products, they also improve functional activities such as antioxidants, color and increase their shelf life (Eça et al. 2014; Kumari et al. 2017).

These advances in the active coating for fruits and vegetables in the literature (Table 22.1) are due to the growth of consumers of fresh and minimally processed products, during the last three decades in retail market. However, undesirable reactions such as browning, whiteness and water loss have limited their shelf life. To minimize the reactions of chemical, microbial and enzymatic deterioration, active edible coating with added bioactive compounds is an alternative to delay these reactions and extend the storage time these products (Robles-Sánchez et al. 2013; Azarakhsh et al. 2014; Kumari et al. 2017). Active edible coatings on the whole or fresh-cut products also reduce moisture loss and solute migration, respiration rate and, consequently physiological disorders (Galgano et al. 2015). However, active compounds may not be compatible with the sensory characteristic of whole or fresh-cut products.

22.6 Influence of the Active Edible Coating on Quality Parameters in Fruits and Vegetables

22.6.1 Physicochemical Properties

Ascorbic acid (AA) and citric acid (CA) are the natural antimicrobial antibowning agents of low cost that have been used in dipping coatings and have been applied on minimally processed fruits and vegetables (Robles-Sánchez et al. 2013; Raybaudi-Massilia et al. 2015; Uranga et al. 2019). In addition, both agents can provide nutrients associated with them such as calcium and vitamin D from coatings to fresh-cut products (Raybaudi-Massilia et al. 2015). AA and CA delay the browning because they inhibit the action of polyphenol oxidase, reducing o-quinones to diphenols. In this sense, the antimicrobial peptides (curcuminoids) from *Curcuman longa* and monoterpernes and sesquiterpenes from the EO have found many applications, including food preservation, since they can be incorporated into material packages or edibles coating to maintain the safety and quality of whole and fresh-cut fruits and vegetables (Antunes et al. 2012; Espitia et al. 2012, 2013). These bioactives can modify the structure of the biopolymers and improve their functionality and applicability (Eça et al. 2014). Table 22.1 shows the summary of the bioactive compounds added into edible coating on whole and fresh-cut fruits and vegetables in the last 5 years.

Eça et al. (2014) summarized coatings containing natural antioxidant from different matrices to improve the functionality and applicability on fresh-cut fruits. Zambrano-Zaragoza et al. (2017) incorporated nano sphere (Nsp) and nano capsules (Ncs) of β-carotene in xanthan gum coating made by dipping coating on fresh-cut melon (var. cantaloupe). The physicochemical properties from coated melon samples were better than uncoated samples. However, the water loss was lower for the coating samples containing Nsp and Ncs compared to the control and the coated samples without active compounds. Rojas-Graü et al. (2008) also observed that edible coating based on polysaccharide managed to reduce water loss from fresh-cut Fuji apple slices. Moreira et al. (2014) incorporated β-cyclodestrin and antimicrobial trans-cinnam aldehyde microencapsulated into chitosan (Cs) coating obtained by dipping to be applied to fresh-cut melon. Multilayer coating on fresh-cut melon and wrapped using polypropylene (Ziploc, Saran wrap and cheesecloth) helped to reduce loss of juice from melon samples and maintained its firmness for

15 days at 4 °C. The authors also reported that crosslinking of Cs and pectin with $CaCl_2$ prevented the water exudation from the coated melon samples. In fact, the maintenance of water content in the fresh products has been the most important requirement of the edible coatings. Mohammadi et al. (2015) coated fresh cucumbers with Cs nanoparticles loaded with *Cinnamomum zeylanicum* EO (CEO-CsN). It was reported in this study a lower weight loss in the CEO-CsN coated cucumbers compared to the uncoated samples or the samples without CEO. At least the shelf life of the cucumbers was prolonged for 12 days more than the control.

Alginate and pectin-based biopolymer coatings containing citral and eugenol were used by Guerreiro et al. (2017) to coat 'Bravo de Esmolfe' fresh-cut apples. The coated fresh fruits were treated in combination with three antibrowning agents: sodium chlorite, AA and CA. All antibrowning agents used were effective for all coated samples. However, samples submerged in AA and coated with alginate were more effective. The best results based on physicochemical and sensory quality were those combining alginate (1.0%)/eugenol (0.1%) plus AA. Probably, at this concentration, the contribution of the edible coating was more effective for the integrity of the tissue and the stiffness of the cell wall of the fresh-cut apples than the other formulations. Guerreiro et al. (2015) used the same formulations but without antibrowning agents to coat fresh-cut raspberries and strawberries. The best covers in terms of sensory attributes were obtained by using 2% alginate/0.5% citral. Rinaudo (2014) at least prepared twelve coating formulations based on alginate and pectin to coat whole apples, strawberries and raspberries. The best results were obtained using alginate coatings. This is possibly due to the ability to form crosslinked hydrogels in the presence of $CaCl_2$. However, due to the high sensitivity of the raspberry to the softening, the coating crosslinked from alginate and $CaCl_2$ could not avoid the moisturizing effect of the fruits.

More recently, the use of nanocomposite coatings has emerged as a new alternative to reduce the loss of postharvest. Barikloo and Ahmadi (2018) produced a nanocomposite packaging made from polypropylene and polyolefin elastomer with the addition of nanoclays for the conservation of strawberries with and without Cs coating. The weight loss was lower for the Cs-coated strawberries using the package with nanocomposites for 10 days, stored at 4 °C and 25 °C compared to the control samples. The results showed that the used obstacle technique: low temperature, nanocomposite packaging and Cs coating, allowed to reduce the loss of weight and maintain the color and firmness of the fruit from the first day of storage. The authors also concluded that the use of nanocomposite packaging in combination with Cs coatings resulted in the prolongation of the shelf life of strawberries in terms of physical, chemical and mechanical properties.

The use of edible coatings in conjunction with non-thermal technologies are being used as obstacle technologies to increase the shelf life of fruits and vegetables. Moreira et al. (2015) studied apple fiber-enriched gellan-gum prebiotic coatings on fresh-cut apples, together with the application of pulsed light. The authors concluded that the combined technologies contributed to extend the shelf life of fresh-cut apples. A positive synergistic effect on the extension of the useful life of whole mangoes was obtained by Salinas-Roca et al. (2016) using malic acid (MA)-

crosslinked alginate coatings obtained by dipping and then treated with pulsed light. The results in this study showed that the combination of both technologies allows to guarantee the quality and safety of fresh-cut mango during storage.

22.6.2 Effectivity of the Active Compounds Into Edible Coating for Food Safety

Mohammadi et al. (2016) evaluated CsN edible coatings containing *Zataria multiflora* EO on cucumbers in order to avoid the development of mold: *Phytophthora drechsleri*. The results in this study showed that the percentage of molded cucumbers without coating was 97.73% *versus* 3.25% and 10.95%, for the CEO-CsN and CsN-coated cucumbers, respectively, for 21 days at 10 °C.

Salvia-Trujillo et al. (2015) studied edible coatings based on sodium alginate loaded with conventional emulsion and nanoemulsion of lemon grass EO (LEO) to inactivate *Escherichia coli* on fresh-cut Fuji apples. Samples coated with sodium alginate without LEO reduced the microbial count from 10^6 CFU/g to 0.8 log unit. While samples coated with solutions containing 0.5 or 1.0% LEO inactivated *E. coli*. However, the nanoemulsion-loaded coating was more effective during storage for 2 weeks more than the conventional emulsion. Probably, the nano size of the oil drop penetrates faster than the micro size of the oil in fresh-cut apples and, consequently, into the bacterial cell.

Alginate and pectin based coatings, with added citral and eugenol were used on fresh-cut *Arbutus unedo* L. (Guerreiro et al. 2015c), raspberries (Guerreiro et al. 2015b) and strawberries (Guerreiro et al. 2015a), submerged separately in anti-browning agents: sodium chlorite, AA and CA. A greater effectiveness was obtained against aerobic mesophilic and psychrophilic bacteria, molds and yeasts using alginate-based coatings containing 0.2% eugenol 0.2% on *Arbutus unedo*. The citral or eugenol-loaded alginate/protein-based coatings applied on raspberry showed to be effective against mold and yeasts more than mesophilic aerobic microorganism. However, strawberry samples coated with alginate and pectin containing citral or eugenol EOs reduced mesophilic aerobic microorganism counts.

In recent years, functional edible biopolymer coatings with added silver nanoparticles (AgNPs) have been used on fruits and vegetables to improve the properties of these materials. Ortiz-Duarte et al. (2019) evaluated the effect of the Cs coating containing Ag nanocomposites on fresh-cut melon. The AgNP-loaded Cs coatings increased the antimicrobial activity against Gram-positive and Gram-negative bacteria compared to the samples only coated with Cs. It was also observed that the AgNP-loaded Cs coatings had greater antimicrobial activity than Cs coatings containing other metal ions.

Active edible coating has proven to be effective for the conversation of whole and minimally processed fruits and vegetables, since the interactions between the coatings and the products can act positively on the physicochemical attributes of the food and the inhibition of microbial growth. In addition, factors such as respiration rate and ethylene production are reduced during the postharvest of fruits and vegetables with the application of these edible coatings, thus increasing the shelf life of the food and reducing post-harvest losses. In summary, the main active coatings have been developed from the addition of metallic nanoparticles or EOs (Bracone et al. 2016; Gutiérrez et al. 2017).

22.6.3 *Effect of Active Edible Coatings on the Respiration Rate of Fruits and Vegetables*

Fruits and vegetables continue to live and breathe after the harvest. This process consumes O_2 and produces ethylene and CO_2, which accelerate the food ripening, thus reducing the shelf life of fruits and vegetables during storage. It is thus important and crucial to optimize the composition and properties of edible coatings in order to inhibit physiological changes (mainly loss of water and texture) associated with the exchange of gases during the respiration of stored fruits and vegetables (Fonseca et al. 2003; Baldwin 1994; Tosati et al. 2015; Gutiérrez and Álvarez 2017). Edible coatings in general lead to the obtaining of modified atmospheres within fruits and vegetables, thus positively affecting fruit and vegetable attributes such as flavor, firmness, color, texture, water loss, antioxidant compounds, sugars, organics acids, among others (Zambrano-Zaragoza et al. 2017; Baldwin 1994).

A higher respiration rate leads to a faster loss of some major components of fruits and vegetables such as acids, sugars and vitamins. The respiration process can also be minimized using low temperatures in order to reduce the metabolic process of the vegetable food and consequently the production of CO_2 and ethylene is reduced. It is very frequent, for this reason, that many research studies have been carried out applying edible coatings on fruits and vegetables at low storage temperature (around 5 °C). Fagundes et al. (2013) showed that the respiration rate of fresh-cut apples treated with 2.0% AA and 1.0% $CaCl_2$ decreased when stored at 2 °C, 5 °C and 7 °C and under modified passive atmosphere. The authors also observed a significant reduction ($p \leq 0.05$) in factors indicative of decomposition by respiration in fresh-cut apples by applying lower temperatures during storage. However, the reduction of storage temperature is not enough to decrease the respiration rate and ethylene production to prolong the shelf life of these products.

Zambrano-Zaragoza et al. (2017) reported a greater reduction in oxygen production from fresh-cut melon samples coated with xanthan gum with and without β-carotene Ncs. The β-carotene Ncs/xanthan gum-coated samples nonetheless showed the best results in terms of reducing of the oxygen rate by 49% compared to the controls (without coating). In general, the respiration rate of minimally processed products increased due to the stress caused by the cutting of whole fruits or vegetables.

Salvia-Trujillo et al. (2015) developed sodium alginate-based coatings containing emulsions and nanoemulsions of lemongrass EO (LEO) to coat Fuji apples. In this study, a reduction in respiration rate was observed by using emulsions and nanoemulsions containing 0.5% and 1% LEO. However, samples containing 0.1% emulsion or nanoemulsion without LEO increased the respiration rate in the fruit compared to the uncoated samples. According to the authors, probably the increase in the respiration rate in this case was due to a higher carbohydrate content available from the biopolymer coating, which can be metabolized during respiration.

The effectiveness of edible coatings on respiration, in terms of O_2 and CO_2 of fruits and vegetables, can be attributed to a lower gas permeability through the coating on the food surface. However, the properties of the active compounds within the biopolymer coatings can also influence gas diffusion and respiration. Figure 22.2 represents as an example the possible interactions that are given during transport phenomena in Pitanga (*Eugenia uniflora* L.) Brazilian fruits coated with edible biopolymers containing active compounds.

22.7 Migration of the Active Compounds from Coatings toward Food

Other important parameters of the coatings are the surface energy between the coating and the food surface, the interactions between the bioactive compounds and the polymer matrices, as well as the migration of the bioactive compounds from the coating to the food (Gutiérrez and González 2016, 2017; Gutiérrez et al. 2016; 2018; Herniou--Julien et al. 2019; Gutiérrez et al. 2019). According to Aloui and Khwaldia (2016), these coatings can release active compounds at controlled rates, which has allowed maintaining the quality and safety of fruits and vegetables during storage, and prolonging the shelf life of them. However, the released amount of the active compound must be controlled to achieve the desired effect. Figure 22.3 shows that the simulated release *in vivo* (Fig. 22.3a) and *in vitro* (Fig. 22.3b, simulant of foods) of the active compounds from the coating to the foods (e.g. Pitanga Brazilian fruit) depends on the type of coating material and the properties of the active compounds used. Nonetheless, parameters such as temperature, humidity, food composition (pH, lipid content and water content) and transport phenomena that occur within the coating are parameters that should also be considered in the studies. In this sense, the release rate of active compounds is a critical factor for food safety and quality.

A limitation of the use of active compounds into edible coatings is the alteration of the sensory characteristics of fruits and vegetables (or foods in general), bringing consequently the potential rejection of products from consumers. Murmu and Mishra (2018) evaluated the effect of edible coating based on gum Arabic and sodium caseinate containing different cinnamon and lemongrass EO concentrations on the physicochemical and antioxidant properties of stored whole guava. Good results were obtained at least during 35 days of storage by using the coatings containing 2% of both EOs in this study. However, these authors did not evaluate the diffusion kinetics of the EOs within the coatings to the food. In contrast, Mohammadi

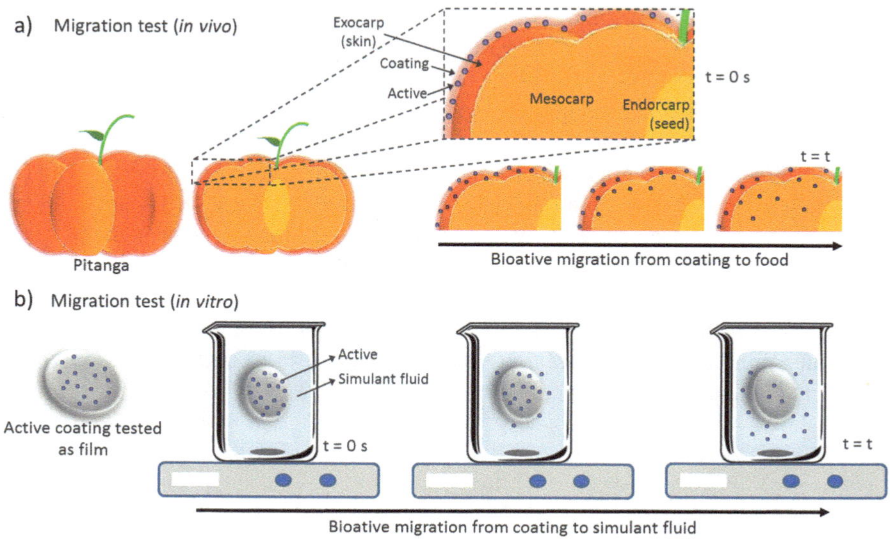

Fig. 22.3 Schematic representation of *in vivo* (**a**) and *in vitro* (**b**) tests of the migration of active compounds from coating to food or simulating fluid

et al. (2015) evaluated the CEO release kinetics from Cs edible coatings to a simulant medium such as ethanol. The authors initially observed a rapid diffusion of the CEO contained into the coatings to the similar medium, which was then sustained. However, these authors did not quantify the active compounds migrated from coating to food.

The release rate of carvacrol from zein/polylactic acid (PLA) fibers in an aqueous solution of ethanol (80 v/v) and in chloroform/*N*,*N*-Dimethyl formamide (9.1, v/v), respectively, were determined by Altan et al. (2018) using headspace gas chromatography-mass spectrometry (GC-MS). An initial burst release was observed followed by sustained release of carvacrol for both fibers. However, the zein fibers allowed a faster release of the carvacrol compared to the analogous fiber made of PLA. The migration mechanism was also investigated by Mohammadi et al. (2015) on CEO-loaded Cs nanoparticles, which were used to improve the shelf life of cucumber during cold storage.

The release rate of the active compounds must initially be rapid in order to inhibit microbial growth, since the microorganism count may increase before the antimicrobials penetrate the cell. However, the release rate of the active compound should not be so rapid as to increase the susceptibility of cross-contamination on the food surface. A burst release of the active compound should thus be initially given for the purpose of controlling the microbiological growth followed by a controlled and sustained release step of the active compound to regulate and maintain the minimum inhibitory concentration on target microorganism on the food surface (Uz and Altınkaya 2011; Altan et al. 2018). Other recent studies evaluating the effect of various edible coatings containing active compounds applied to different fruits and vegetables are highlighted in Table 22.1. However, the literature reported until 2018 in the web of science database, google academics, science direct and Research Gate, does not register studies carried out until now evaluating the *in vivo* migration rate of active compounds. For this reason, more studies should be conducted in this line.

22.8 Active Edible Coating Toxicity

Since the materials used for the coating of fruits or vegetables must have food quality and be edible (Dhall 2013). It has been assumed that the materials used for the development mainly of composite coatings also have the same quality. However, many studies have highlighted concerns related to the toxicity of compounds added to biopolymer coatings (Gutiérrez 2018c; 2019; Herniou--Julien et al. 2019).

As can be seen in Table 22.1, the natural extracts and the OEs have managed to prolong the shelf life of the food through its use in edible coatings. In addition, these active compounds have been classified as generally recognized as safe (GRAS). However, their application does not indicate the limit of toxicity when applied to food preservation. According to Acevedo-Fani et al. (2017) EOs and plant extracts can cause toxicological effects due to the "potential ability to interact with the human cellular membranes, in the same way that EOs acts on microbial membranes". Sánchez-González et al. (2011) reported some applications of these bioactive molecules in edible coatings, however, these compounds showed cytotoxicity, oral toxicity and skin allergies in *in vivo* studies.

Other *in vivo* studies have suggested mucosal irritation, vomiting, epigastric pain, when these substances are used in high concentrations (Riordan et al. 2002). Therefore, more studies are necessary to evaluate the maximum allowed quantities where EOs incorporating edible coatings do not cause damage or risk of toxicity or allergies to consumers.

The use of emerging nanotechnologies applied to food has taken place due to the greater reactivity of the compounds used, as well as the better bioavailability and mass transfer of them (Acevedo-Fani et al. 2017). Despite these advantages, the nanotoxicology of these materials is a worrying topic, since very little is known about the bioaccumulation of them, mainly during prolonged contact. Several factors can influence the toxicity of a nanomaterial (chemical composition, structure, reactivity, etc.), being necessary to take it into account when applying an edible coating (Acevedo-Fani et al. 2017). Divya et al. (2018) e.g. demonstrated that the Cs nanoparticles presented possible antifungal and antioxidant activity as coating agents. However, a cytotoxic effect was also observed in a dose-dependent effect against fibroblast cells.

Other nanoparticles frequently used in the coating materials are the AgNPs (An et al. 2008; Costa et al. 2012; Jiang et al. 2013; Ortiz-Duarte et al. 2019). However, Ag ions also have toxicity against the animal cells depending on the concentration, and few studies have investigated their effects on flora, fauna or the environment (Marchiore et al. 2017). Hadrup and Lam (2014) reviewed the oral toxicity of AgNPs on the human body. The authors detailed aspects of their digestion, organ influence and excretion. They have described that this metal seems to be distributed in all the organs investigated and presents a dose-dependence toxicity, causing weight loss and even death in animals. It is known that some nanoparticles (when eaten) can cross epithelial cells and interact or be bioaccumulated in various organs (Acevedo-Fani et al. 2017). In this sense, a consensus should be reached in the scientific community with the aim of making more effort in the study of the true potentials of the active compounds into the "edible coatings" in terms of the risks to consumers. Another important fact that must be considered is the environmental impact of the use of additives in the coating materials. Can an antimicrobial coating material influence the degradability of a fruit or vegetable? Given that concerns about nanotoxicology have been highlighted in recent years, what is the impact of the use of nanomaterials on edible coatings and on the environment? These questions have not yet been answered and should be studied in the future to evaluate which problems can improve the fauna and flora.

22.9 Remarks and Conclusions

The use of edible coatings to extend the shelf life of fruits and vegetables has been demonstrated for many years through scientific studies that indicate that the application of these materials allows to maintain the quality and safety of these products, thus reducing post-harvest losses. However, the new alternatives in coatings applied to food is directed to the use of nanocomposite materials, which can lead to the development of active, intelligent or functional packaging. Nonetheless, these composite materials could lead to problems of unintentional migration of substances on food, bringing as a potential consequence the migration of potentially toxic nanosubstances for consumer sand even placing the compostability of the material at risk. For this reason, more studies aimed at analyzing the toxicity and migration of nanocomposites should be developed in the future with the purpose of their application at the industrial level.

Acknowledgements The authors would like to thank the Federal University of Santa Catarina (UFSC) and Catarinense Federal Institute (IFC) for their support.

Conflicts of Interest The authors declare no conflict of interest.

References

Acevedo-Fani, A., Soliva-Fortuny, R., & Martín-Belloso, O. (2017). Nanostructured emulsions and nanolaminates for delivery of active ingredients: Improving food safety and functionality. *Trends in Food Science and Technology, 60*, 12–22. https://doi.org/10.1016/j.tifs.2016.10.027.

Ahmadian-Yazdi, M. R., Zabihi, F., Habibi, M., & Eslamian, M. (2016). Effects of process parameters on the characteristics of mixed-halide perovskite solar cells fabricated by one-step and two-step sequential coating. *Nanoscale Research Letters, 11*, 408–418. https://doi.org/10.1186/s11671-016-1601-8.

Aloui, H., & Khwaldia, K. (2016). Natural antimicrobial edible coatings for microbial safety and food quality enhancement. *Comprehensive Reviews in Food Science and Food Safety, 15*(6), 1080–1103. https://doi.org/10.1111/1541-4337.12226.

Altan, A., Aytac, Z., & Uyar, T. (2018). Carvacrol loaded electrospun fibrous films from zein and poly (lactic acid) for active food packaging. *Food Hydrocolloids, 81*, 48–59. https://doi.org/10.1016/j.foodhyd.2018.02.028.

Álvarez, K., Famá, L., & Gutiérrez, T. J. (2017). Physicochemical, antimicrobial and mechanical properties of thermoplastic materials based on biopolymers with application in the food industry. In M. Masuelli & D. Renard (Eds.), *Advances in physicochemical properties of biopolymers: Part 1* (pp. 358–400). Bentham science publishers. EE.UU. ISBN: 978-1-68108-454-1. eISBN: 978-1-68108-453-4. https://doi.org/10.2174/9781681084534117010015.

Álvarez, K., Alvarez, V. A., & Gutiérrez, T. J. (2018). Biopolymer composite materials with antimicrobial effects applied to the food industry. In V. K. Thakur & M. K. Thakur (Eds.), *Functional biopolymers* (pp. 57–96). Editorial Springer International Publishing. EE.UU. ISBN: 978-3-319-66416-3. eISBN: 978-3-319-66417-0. https://doi.org/10.1007/978-3-319-66417-0_3.

An, J., Zhang, M., Wang, S., & Tang, J. (2008). Physical, chemical and microbiological changes in stored green asparagus spears as affected by coating of silver nanoparticles-PVP. *LWT-Food Science and Technology, 41*(6), 1100–1107. https://doi.org/10.1016/j.lwt.2007.06.019.

Andrade, R. D., Skurtys, O., & Osorio, F. A. (2012). Atomizing spray systems for application of edible coatings. *Comprehensive Reviews in Food Science and Food Safety, 11*(3), 323–337. https://doi.org/10.1111/j.1541-4337.2012.00186.x.

Ansorena, M. R., Marcovich, N. E., & Roura, S. I. (2011). Impact of edible coatings and mild heat shocks on quality of minimally processed broccoli (*Brassica oleracea* L.) during refrigerated storage. *Postharvest Biology and Technology, 59*(1), 53–63. https://doi.org/10.1016/j.postharvbio.2010.08.011.

Antunes, M. D., Gago, C. M., Cavaco, A. M., & Miguel, M. G. (2012). Edible coatings enriched with essential oils and their compounds for fresh and fresh-cut fruit. *Recent Patents on Food, Nutrition and Agriculturee, 4*(2), 114–122. https://doi.org/10.2174/2212798411204020114.

Araújo, J. M. S., de Siqueira, A. C. P., Blank, A. F., Narain, N., & de Aquino Santana, L. C. L. (2018). A cassava starch–chitosan edible coating enriched with lippia sidoides Cham. Essential oil and pomegranate peel extract for preservation of italian tomatoes (*Lycopersicon esculentum* mill.) stored at room temperature. *Food and Bioprocess Technology, 11*(9), 1750–1760. https://doi.org/10.1007/s11947-018-2139-9.

Arnon, H., Zaitsev, Y., Porat, R., & Poverenov, E. (2014). Effects of carboxymethyl cellulose and chitosan bilayer edible coating on postharvest quality of citrus fruit. *Postharvest Biology and Technology, 87*, 21–26. https://doi.org/10.1016/j.postharvbio.2013.08.007.

Askari, G. R., Emam-Djomeh, Z., & Mousavi, S. M. (2006). Effects of combined coating and microwave assisted hot-air drying on the texture, microstructure and rehydration characteristics of apple slices. *Food Science and Technology International, 12*(1), 39–46. https://doi.org/10.1177/1082013206062480.

Azarakhsh, N., Osman, A., Ghazali, H. M., Tan, C. P., & Adzahan, N. M. (2014). Lemongrass essential oil incorporated into alginate-based edible coating for shelf-life extension and quality retention of fresh-cut pineapple. *Postharvest Biology and Technology, 88*, 1–7. https://doi.org/10.1016/j.postharvbio.2013.09.004.

Baldwin, E. A. (1994). Edible coatings for fresh fruits and vegetables: past, present, and future. In: Krochta JM, Baldwin EA, Nisperos-Carriedo MO, editors. Edible coatings and films to improve food quality. Lancaster, Pa.: Technomic Publishing Co., Inc. p 25–64.

Barikloo, H., & Ahmadi, E. (2018). Effect of nanocomposite-based packaging and chitosan coating on the physical, chemical, and mechanical traits of strawberry during storage. *Journal of Food Measurement and Characterization, 12*(3), 1795–1817. https://doi.org/10.1007/s11694-018-9795-3.

Bourtoom, T. (2008). Review article edible films and coatings: Characteristics and properties. *Structure, 15*, 237–248.

Breda, C. A., Morgado, D. L., de Assis, O. B. G., & Duarte, M. C. T. (2017). Effect of chitosan coating enriched with Pequi (*Caryocar brasiliense* Camb.) peel extract on quality and safety of tomatoes (*Lycopersicon esculentum* mill.) during storage. *Journal of Food Processing and Preservation, 41*(6), e13268. https://doi.org/10.1111/jfpp.13268.

Bracone, M., Merino, D., González, J., Alvarez, V. A., & Gutiérrez, T. J. (2016). Nanopackaging from natural fillers and biopolymers for the development of active and intelligent films. In S. Ikram & S. Ahmed (Eds.), *Natural polymers: Derivatives, blends and composites* (pp. 119–155). New York. EE.UU. ISBN: 978-1-63485-831-1: Editorial Nova Science Publishers, Inc.

Charles, F., Sanchez, J., & Gontard, N. (2003). Active modified atmosphere packaging of fresh fruits and vegetables: Modeling with tomatoes and oxygen absorber. *Journal of Food Science, 68*(5), 1736–1742. https://doi.org/10.1111/j.1365-2621.2003.tb12321.x.

Chen, Y.-L., Jiang, S. J., Feng, C. H., Wang, S. W., Lin, Y. H., & Liu, P. Y. (2014). Application of central composite design for the determination of exfoliating agents in cosmetics by capillary electrophoresis with electroosmotic flow modulation. *Analytical Letters, 47*(10), 1670–1682. https://doi.org/10.1080/00032719.2014.883523.

Cortez-Vega, W. R., Pizato, S., De Souza, J. T. A., & Prentice, C. (2014). Using edible coatings from Whitemouth croaker (*Micropogonias furnieri*) protein isolate and organo-clay nanocomposite for improve the conservation properties of fresh-cut 'Formosa' papaya. *Innovative Food Science and Emerging Technologies, 22*, 197–202. https://doi.org/10.1016/j.ifset.2013.12.007.

Costa, C., Conte, A., Buonocore, G. G., Lavorgna, M., & Del Nobile, M. A. (2012). Calcium-alginate coating loaded with silver-montmorillonite nanoparticles to prolong the shelf-life of fresh-cut carrots. *Food Research International, 48*(1), 164–169. https://doi.org/10.1016/j.foodres.2012.03.001.

D'Aquino, S., Mistriotis, A., Briassoulis, D., Di Lorenzo, M. L., Malinconico, M., & Palma, A. (2016). Influence of modified atmosphere packaging on postharvest quality of cherry tomatoes held at 20 °C. *Postharvest Biology and Technology, 115*, 103–112. https://doi.org/10.1016/j.postharvbio.2015.12.014.

Dave, R. K., Ramana Rao, T. V., & Nandane, A. S. (2017). Improvement of post-harvest quality of pear fruit with optimized composite edible coating formulations. *Journal of Food Science and Technology, 54*(12), 3917–3927. https://doi.org/10.1007/s13197-017-2850-y.

Deng, Z., Jung, J., Simonsen, J., & Zhao, Y. (2017). Cellulose nanomaterials emulsion coatings for controlling physiological activity, modifying surface morphology, and enhancing storability of postharvest bananas (*Musa acuminate*). *Food Chemistry, 232*, 359–368. https://doi.org/10.1016/j.foodchem.2017.04.028.

Dhall, R. K. (2013). Advances in edible coatings for fresh fruits and vegetables: A review. *Critical Reviews in Food Science and Nutrition, 53*(5), 435–450. https://doi.org/10.1080/10408398.2010.541568.

Divya, K., Smitha, V., & Jisha, M. S. (2018). Antifungal, antioxidant and cytotoxic activities of chitosan nanoparticles and its use as an edible coating on vegetables. *International Journal of Biological Macromolecules, 114*, 572–577. https://doi.org/10.1016/j.ijbiomac.2018.03.130.

Eça, K. S., Sartori, T., & Menegalli, F. C. (2014). Films and edible coatings containing antioxidants – A review. *Brazilian Journal of Food Technology, 17*(2), 98–112. https://doi.org/10.1590/bjft.2014.017.

Espitia, P. J. P., Pacheco, J. J. R., Melo, N. R. D., Soares, N. D. F. F., & Durango, A. M. (2013). Packaging properties and control of *Listeria monocytogenes* in bologna by cellulosic films incorporated with pediocin. *Brazilian Journal of Food Technology, 16*(3), 226–235. https://doi.org/10.1590/S1981-67232013005000028.

Espitia, P. J. P., Soares N de, F. F., Botti, L. C. M., Melo, N. R. D., Pereira, O. L., & Silva, W. A. D. (2012). Assessment of the efficiency of essential oils in the preservation of postharvest papaya in an antimicrobial packaging system. *Brazilian Journal of Food Technology, 15*(4), 333–342. https://doi.org/10.1590/S1981-67232012005000027.

Fagundes, C., Carciofi, B. A. M., & Monteiro, A. R. (2013). Estimate of respiration rate and physicochemical changes of fresh-cut apples stored under different temperatures. *Food Science and Technology, 33*(1), 60–67. https://doi.org/10.1590/S0101-20612013005000023.

Fagundes, C., Moraes, K., Pérez-Gago, M. B., Palou, L., Maraschin, M., & Monteiro, A. R. (2015). Effect of active modified atmosphere and cold storage on the postharvest quality of cherry tomatoes. *Postharvest Biology and Technology, 109*, 73–81. https://doi.org/10.1016/j.postharvbio.2015.05.017.

Fayaz, A. M., Balaji, K., Girilal, M., Kalaichelvan, P. T., & Venkatesan, R. (2009). Mycobased synthesis of silver nanoparticles and their incorporation into sodium alginate films for vegetable and fruit preservation. *Journal of Agricultural and Food Chemistry, 57*(14), 6246–6252. https://doi.org/10.1021/jf900337h.

Fonseca, S. C., Oliveira, F. A. R., Brecht, J. K., & Chau, K. V. (2003). PErforation-mediated modified atmosphere packaging: Influence of package geometry and perforation location on oxygen and carbon dioxide transfer. *Acta Horticulturae*, 333–336. https://doi.org/10.17660/actahortic.2003.600.47.

Galgano, F., Condelli, N., Favati, F., et al. (2015). Biodegradable packaging and edible coating for fresh-cut fruits and vegetables. *Italian Journal of Food Science, 27*, 1–20. https://doi.org/10.14674/1120-1770%2Fijfs.v27i1.70.

Galus, S., & Kadzińska, J. (2015). Food applications of emulsion-based edible films and coatings. *Trends in Food Science and Technology, 45*(2), 273–283. https://doi.org/10.1016/j.tifs.2015.07.011.

Genevois, C. E., De Escalada Pla, M. F., & Flores, S. K. (2016). Application of edible coatings to improve global quality of fortified pumpkin. *Innovative Food Science and Emerging Technologies, 33*, 506–514. https://doi.org/10.1016/j.ifset.2015.11.001.

Ghidelli, C., Mateos, M., Rojas-Argudo, C., & Pérez-Gago, M. B. (2014). Extending the shelf life of fresh-cut eggplant with a soy protein-cysteine based edible coating and modified atmosphere packaging. *Postharvest Biology and Technology, 95*, 81–87. https://doi.org/10.1016/j.postharvbio.2014.04.007.

Gol, N. B., Patel, P. R., & Rao, T. V. R. (2013). Improvement of quality and shelf-life of strawberries with edible coatings enriched with chitosan. *Postharvest Biology and Technology, 85*, 185–195. https://doi.org/10.1016/j.postharvbio.2013.06.008.

González-Saucedo, A., Barrera-Necha, L. L., Ventura-Aguilar, R. I., Correa-Pacheco, Z. N., Bautista-Baños, S., & Hernández-López, M. (2019). Extension of the postharvest quality of bell pepper by applying nanostructured coatings of chitosan with *Byrsonima crassifolia* extract (L.) Kunth. *Postharvest Biology and Technology, 149*, 74–82. https://doi.org/10.1016/j.postharvbio.2018.11.019.

Guerreiro, A. C., Gago, C. M. L., Faleiro, M. L., Miguel, M. G., & Antunes, M. D. (2015a). The use of polysaccharide-based edible coatings enriched with essential oils to improve shelf-life of strawberries. *Postharvest Biology and Technology, 110*, 51–60. https://doi.org/10.1016/j.postharvbio.2015.06.019.

Guerreiro, A. C., Gago, C. M. L., Faleiro, M. L., Miguel, M. G., & Antunes, M. D. (2017). The effect of edible coatings on the nutritional quality of 'bravo de Esmolfe' fresh-cut apple through shelf-life. *LWT, 75*, 210–219. https://doi.org/10.1016/j.lwt.2016.08.052.

Guerreiro, A. C., Gago, C. M. L., Faleiro, M. L., Miguel, M. G., & Antunes, M. D. (2015b). Raspberry fresh fruit quality as affected by pectin- and alginate-based edible coatings enriched with essential oils. *Scientia Horticulturae, 194*, 138–146. https://doi.org/10.1016/j.scienta.2015.08.004.

Guerreiro, A. C., Gago, C. M. L., Faleiro, M. L., Miguel, M. G., & Antunes, M. D. (2015c). The effect of alginate-based edible coatings enriched with essential oils constituents on *Arbutus unedo* L. fresh fruit storage. *Postharvest Biology and Technology, 100*, 226–233. https://doi.org/10.1016/j.postharvbio.2014.09.002.

Gustavsson, J., Cederberg, C., & Sonesson, U. (2011). *Global food losses and food waste: Extent, causes and prevention; study conducted for the international congress save food! at Interpack 2011, [16–17 May], Düsseldorf, Germany*. Rome: Food and Agriculture Organization of the United Nations.

Gutiérrez, T. J. (2017). Surface and nutraceutical properties of edible films made from starchy sources with and without added blackberry pulp. *Carbohydrate Polymers, 165*, 169–179. https://doi.org/10.1016/j.carbpol.2017.02.016.

Gutiérrez, T. J. (2018a). Active and intelligent films made from starchy sources/blackberry pulp. *Journal Polymers and the Environment, 26*(6), 2374–2391. https://doi.org/10.1007/s10924-017-1134-y.

Gutiérrez, T. J. (2018b). Biodegradability and compostability of food nanopackaging materials. In G. Cirillo, M. A. Kozlowski, & U. G. Spizzirri (Eds.), *Composites materials for food packaging* (pp. 269–296). Hoboken: Wiley. https://doi.org/10.1002/9781119160243.ch9.

Gutiérrez, T. J. (2018c). Are modified pumpkin flour/plum flour nanocomposite films biodegradable and compostable? *Food Hydrocolloids, 83*, 397–410. https://doi.org/10.1016/j.foodhyd.2018.05.035.

Gutiérrez, T. J., & Álvarez, K. (2017). Transport phenomena in biodegradable and edible films. In M. A. Masuelli (Ed.), *Biopackaging* (pp. 58–88). Miami., EE.UU. ISBN: 978–1–4987-4968-8: Editorial CRC Press Taylor & Francis Group.

Gutiérrez, T. J., & González, G. (2016). Effects of exposure to pulsed light on surface and structural properties of edible films made from cassava and taro starch. *Food and Bioprocess Technology, 9*(11), 1812–1824. https://doi.org/10.1007/s11947-016-1765-3.

Gutiérrez, T. J., & González, G. (2017). Effect of cross-linking with *Aloe vera* gel on surface and physicochemical properties of edible films made from plantain flour. *Food Biophysics, 12*(1), 11–22. https://doi.org/10.1007/s11483-016-9458-z.

Gutiérrez, T. J., González Seligra, P., Medina Jaramillo, C., Famá, L., & Goyanes, S. (2017). Effect of filler properties on the antioxidant response of thermoplastic starch composites. In V. K. Thakur, M. K. Thakur, & M. R. Kessler (Eds.), *Handbook of composites from renewable materials* (pp. 337–370). WILEY-Scrivener Publisher. EE.UU. ISBN: 978-1-119-22362-7. https://doi.org/10.1002/9781119441632.ch14.

Gutiérrez, T. J., Ollier, R., & Alvarez, V. A. (2018). Surface properties of thermoplastic starch materials reinforced with natural fillers. In V. K. Thakur & M. K. Thakur (Eds.), *Functional biopolymers* (pp. 131–158). Editorial Springer International Publishing. EE.UU. ISBN: 978-3-319-66416-3. eISBN: 978-3-319-66417-0. https://doi.org/10.1007/978-3-319-66417-0_5.

Gutiérrez, T. J., Suniaga, J., Monsalve, A., & García, N. L. (2016). Influence of beet flour on the relationship surface-properties of edible and intelligent films made from native and modified plantain flour. *Food Hydrocolloids, 54*, 234–244. https://doi.org/10.1016/j.foodhyd.2015.10.012.

Gutiérrez, T. J., Toro-Márquez, L. A., Merino, D., & Mendieta, J. R. (2019). Hydrogen-bonding interactions and compostability of bionanocomposite films prepared from corn starch and nano-fillers with and without added Jamaica flower extract. *Food Hydrocolloids, 89*, 283–293. https://doi.org/10.1016/j.foodhyd.2018.10.058.

Hadrup, N., & Lam, H. R. (2014). Oral toxicity of silver ions, silver nanoparticles and colloidal silver – A review. *Regulatory Toxicology and Pharmacology, 68*(1), 1–7. https://doi.org/10.1016/j.yrtph.2013.11.002.

Hashemi, S. M. B., Mousavi Khaneghah, A., Ghaderi Ghahfarrokhi, M., & Eş, I. (2017). Basil-seed gum containing Origanum vulgare subsp. viride essential oil as edible coating for fresh cut apricots. *Postharvest Biology and Technology, 125*, 26–34. https://doi.org/10.1016/j.postharvbio.2016.11.003.

Herniou--Julien, C., Mendieta, J. R., & Gutiérrez, T. J. (2019). Characterization of biodegradable/non-compostable films made from cellulose acetate/corn starch blends processed under reactive extrusion conditions. *Food Hydrocolloids, 89*, 67–79. https://doi.org/10.1016/j.foodhyd.2018.10.024.

Huang, J., Yuan, Z., Gao, S., Liao, J., & Eslamian, M. (2018). Understanding spray coating process: Visual observation of impingement of multiple droplets on a substrate. *Journal of Shanghai Jiaotong University (Science), 23*(1), 97–105. https://doi.org/10.1007/s12204-018-1914-0.

Jiang, T. (2013). Effect of alginate coating on physicochemical and sensory qualities of button mushrooms (*Agaricus bisporus*) under a high oxygen modified atmosphere. *Postharvest Biology and Technology, 76*, 91–97. https://doi.org/10.1016/j.postharvbio.2012.09.005.

Jiang, T., Feng, L., & Wang, Y. (2013). Effect of alginate/nano-ag coating on microbial and physicochemical characteristics of shiitake mushroom (*Lentinus edodes*) during cold storage. *Food Chemistry, 141*(2), 954–960. https://doi.org/10.1016/j.foodchem.2013.03.093.

Kader, A. A. (2005). Increasing food availability by reducing postharvest losses of fresh produce. *Acta Horticulturae*, 2169–2176. https://doi.org/10.17660/actahortic.2005.682.296.

Kader, A. A., & Rolle, R. S. (2004). *The role of post-harvest management in assuring the quality and safety of horticultural produce*. Rome: Food and Agriculture Organization of the United Nations.

Khan, M. K. I., Cakmak, H., Tavman, Ş., Schutyser, M., & Schroën, K. (2014). Anti-browning and barrier properties of edible coatings prepared with electrospraying. *Innovative Food Science & Emerging Technologies, 25*, 9–13. https://doi.org/10.1016/j.ifset.2013.10.006.

Khan, M. K. I., Mujawar, L. H., Schutyser, M. A. I., Schroën, K., & Boom, R. (2013). Deposition of thin lipid films prepared by electrospraying. *Food and Bioprocess Technology, 6*(11), 3047–3055. https://doi.org/10.1007/s11947-012-0974-7.

Khan, M. K. I., Nazir, A., & Maan, A. A. (2017). Electrospraying: A Novel technique for efficient coating of foods. *Food Engineering Reviews, 9*(2), 112–119. https://doi.org/10.1007/s12393-016-9150-6.

Khorram, F., Ramezanian, A., & Hosseini, S. M. H. (2017). Shellac, gelatin and Persian gum as alternative coating for orange fruit. *Scientia Horticulturae, 225*, 22–28. https://doi.org/10.1016/j.scienta.2017.06.045.

Kitinoja, L., & AlHassan, H. Y. (2012). Identification of appropriate postharvest technologies for small scale horticultural farmers and marketers in sub-saharan africa and South Asia – PART 1. Postharvest losses and quality assessments. *Acta Horticulturae*, 31–40. https://doi.org/10.17660/ActaHortic.2012.934.1.

Kraśniewska, K., Gniewosz, M., Synowiec, A., Przybył, J. L., Bączek, K., & Węglarz, Z. (2014). The use of pullulan coating enriched with plant extracts from *Satureja hortensis* L. to maintain pepper and apple quality and safety. *Postharvest Biology and Technology, 90*, 63–72. https://doi.org/10.1016/j.postharvbio.2013.12.010.

Kumar, P., & Sethi, S. (2018). Edible coating for fresh fruit: A review. *International Journal of Current Microbiology and Applied Sciences, 7*, 2619–2626. https://doi.org/10.20546/ijcmas.2018.705.303.

Kumari, M., Mahajan, H., Joshi, R., & Gupta, M. (2017). Development and structural characterization of edible films for improving fruit quality. *Food Packaging and Shelf Life, 12*, 42–50. https://doi.org/10.1016/j.fpsl.2017.02.003.

Lai, T. Y., Chen, C. H., & Lai, L. S. (2013). Effects of tapioca starch/decolorized Hsian-Tsao leaf gum-based active coatings on the quality of minimally processed carrots. *Food and Bioprocess Technology, 6*(1), 249–258. https://doi.org/10.1007/s11947-011-0707-3.

Lee, S. H., Heng, D., Ng, W. K., Chan, H. K., & Tan, R. B. (2011). Nano spray drying: A novel method for preparing protein nanoparticles for protein therapy. *International Journal of Pharmaceutics, 403*(1–2), 192–200. https://doi.org/10.1016/j.ijpharm.2010.10.012.

Lin, D., & Zhao, Y. (2007). Innovations in the development and application of edible coatings for fresh and minimally processed fruits and vegetables. *Comprehensive Reviews in Food Science and Food Safety, 6*(3), 60–75. https://doi.org/10.1111/j.1541-4337.2007.00018.x.

Lin, S. Y., & Krochta, J. M. (2006). Fluidized-bed system for whey protein film coating of peanuts. *Journal of Food Process Engineering, 29*(5), 532–546. https://doi.org/10.1111/j.1745-4530.2006.00081.x.

Mantilla, N., Castell-Perez, M. E., Gomes, C., & Moreira, R. G. (2013). Multilayered antimicrobial edible coating and its effect on quality and shelf-life of fresh-cut pineapple (*Ananas comosus*). *LWT – Food Science and Technology, 51*(1), 37–43. https://doi.org/10.1016/j.lwt.2012.10.010.

Marchiore, N. G., Manso, I. J., Kaufmann, K. C., Lemes, G. F., de Oliveira Pizolli, A. P., Droval, A. A., Brachta, L., Gonçalves, O. H., & Leimann, F. V. (2017). Migration evaluation of silver nanoparticles from antimicrobial edible coating to sausages. *LWT-Food Science and Technology 76(Part B), 76*, 203–208. https://doi.org/10.1016/j.lwt.2016.06.013.

Meighani, H., Ghasemnezhad, M., & Bakhshi, D. (2014). Effect of different coatings on post-harvest quality and bioactive compounds of pomegranate (*Punica granatum* L.) fruits. *Journal of Food Science and Technology, 52*(7), 4507–4514. https://doi.org/10.1007/s13197-014-1484-6.

Miranda-Linares, V., Escamilla-Rendón, P., Del Real-López, A., & Gonzalez-Reza, M. L. (2018). Solid lipid nanoparticles based edible coating for saladette tomato preservation. *Acta Horticulturae*, (1194), 305–312. https://doi.org/10.17660/actahortic.2018.1194.44.
Mittal, K. L. (1977). The role of the interface in adhesion phenomena. *Polymer Engineering and Science, 17*(7), 467–473. https://doi.org/10.1002/pen.760170709.
Mohammadi, A., Hashemi, M., & Hosseini, S. M. (2015). Chitosan nanoparticles loaded with *Cinnamomum zeylanicum* essential oil enhance the shelf life of cucumber during cold storage. *Postharvest Biology and Technology, 110*, 203–213. https://doi.org/10.1016/j.postharvbio.2015.08.019.
Mohammadi, A., Hashemi, M., & Hosseini, S. M. (2016). Postharvest treatment of nanochitosan-based coating loaded with *Zataria multiflora* essential oil improves antioxidant activity and extends shelf-life of cucumber. *Innovative Food Science & Emerging Technologies, 33*, 580–588. https://doi.org/10.1016/j.ifset.2015.10.015.
Moreira M del, R., Roura, S. I., & Ponce, A. (2011). Effectiveness of chitosan edible coatings to improve microbiological and sensory quality of fresh cut broccoli. *LWT–Food Science and Technology, 44*(10), 2335–2341. https://doi.org/10.1016/j.lwt.2011.04.009.
Moreira, M. R., Tomadoni, B., Martín-Belloso, O., & Soliva-Fortuny, R. (2015). Preservation of fresh-cut apple quality attributes by pulsed light in combination with gellan gum-based prebiotic edible coatings. *LWT–Food Science and Technology, 64*(2), 1130–1137. https://doi.org/10.1016/j.lwt.2015.07.002.
Moreira, S. P., de Carvalho, W. M., Alexandrino, A. C., de Paula, H. C. B., Rodrigues, M. D. C. P., de Figueiredo, R. W., Arraes Maia, G., Teixeira de Figueiredo, E. M. A., & Brasil, I. M. (2014). Freshness retention of minimally processed melon using different packages and multilayered edible coating containing microencapsulated essential oil. *International Journal of Food Science & Technology, 49*(10), 2192–2203. https://doi.org/10.1111/ijfs.12535.
Motamedi, E., Nasiri, J., Malidarreh, T. R., Kalantari, S., Naghavi, M. R., & Safari, M. (2018). Performance of carnauba wax-nanoclay emulsion coatings on postharvest quality of 'Valencia' orange fruit. *Scientia Horticulturae, 240*, 170–178. https://doi.org/10.1016/j.scienta.2018.06.002.
Mridula, D., & Pooja, J. (2014). Preparation of iron-fortified rice using edible coating materials. *International Journal of Food Science and Technology, 49*(1), 246–252. https://doi.org/10.1111/ijfs.12305.
Murmu, S. B., & Mishra, H. N. (2018). The effect of edible coating based on Arabic gum, sodium caseinate and essential oil of cinnamon and lemon grass on guava. *Food Chemistry, 245*, 820–828. https://doi.org/10.1016/j.foodchem.2017.11.104.
Nawab, A., Alam, F., & Hasnain, A. (2017). Mango kernel starch as a novel edible coating for enhancing shelf- life of tomato (*Solanum lycopersicum*) fruit. *International Journal of Biological Macromolecules, 103*, 581–586. https://doi.org/10.1016/j.ijbiomac.2017.05.057.
Nieto, M. B. (2009). Structure and function of polysaccharide gum-based edible films and coatings. In M. E. Embuscado & K. C. Huber (Eds.), *Edible films and coatings for food applications* (1st ed., pp. 57–112). New York: Springer.
Nussinovitch A, Tal Marmur I, Elkind Y (2017) Patent application publication. Pub. No.: US 2017/0139001 A1- Edible coating for plant matter. 1:32.
Ojeda, G. A., Sgroppo, S. C., & Zaritzky, N. E. (2014). Application of edible coatings in minimally processed sweet potatoes (*Ipomoea batatas* L.) to prevent enzymatic browning. *International Journal of Food Science and Technology, 49*(3), 876–883. https://doi.org/10.1111/ijfs.12381.
Ortiz-Duarte, G., Pérez-Cabrera, L. E., Artés-Hernández, F., & Martínez-Hernández, G. B. (2019). Ag-chitosan nanocomposites in edible coatings affect the quality of fresh-cut melon. *Postharvest Biology and Technology, 147*, 174–184. https://doi.org/10.1016/j.postharvbio.2018.09.021.
Pagliarulo, C., Sansone, F., Moccia, S., Russo, G. L., Aquino, R. P., Salvatore, P., Di Stasio, M., & Volpe, M. G. (2016). Preservation of strawberries with an antifungal edible coating using peony extracts in chitosan. *Food and Bioprocess Technology, 9*(11), 1951–1960. https://doi.org/10.1007/s11947-016-1779-x.

Parreidt, T. S., Schmid, M., & Müller, K. (2018). Effect of dipping and vacuum impregnation coating techniques with alginate based coating on physical quality parameters of cantaloupe melon. *Journal of Food Science, 83*(4), 929–936. https://doi.org/10.1111/1750-3841.14091.

Peretto, G., Du, W. X., Avena-Bustillos, R. J., Berrios, J. D. J., Sambo, P., & McHugh, T. H. (2017). Electrostatic and conventional spraying of alginate-based edible coating with natural antimicrobials for preserving fresh strawberry quality. *Food and Bioprocess Technology, 10*(1), 165–174. https://doi.org/10.1007/s11947-016-1808-9.

Pérez-Gallardo, A., García-Almendárez, B., Barbosa-Cánovas, G., Pimentel-González, D., Reyes-González, L. R., & Regalado, C. (2015). Effect of starch-beeswax coatings on quality parameters of blackberries (*Rubus* spp.). *Journal of Food Science and Technology, 52*(9), 5601–5610. https://doi.org/10.1007/s13197-014-1665-3.

Poverenov, E., Zaitsev, Y., Arnon, H., Granit, R., Alkalai-Tuvia, S., Perzelan, Y., Weinberg, T., & Fallik, E. (2014). Effects of a composite chitosan-gelatin edible coating on postharvest quality and storability of red bell peppers. *Postharvest Biology and Technology, 96*, 106–109. https://doi.org/10.1016/j.postharvbio.2014.05.015.

Raybaudi-Massilia, R., Vásquez, F., Reyes, A., Troncone, G., & Tapia, M. S. (2015). Novel edible coating of fresh-cut fruits: Application to prevent calcium and vitamin D deficiencies in children. *Journal of Scientific Research and Reports, 6*(2), 142–156. https://doi.org/10.9734/jsrr/2015/15964.

Rinaudo, M. (2014). Biomaterials based on a natural polysaccharide: Alginate. *TIP, 17*(1), 92–96. https://doi.org/10.1016/s1405-888x(14)70322-5.

Riordan, M., Rylance, G., & Berry, K. (2002). Poisoning in children 4: Household products, plants, and mushrooms. *Archives of Disease in Childhood, 87*(5), 403–406. https://doi.org/10.1136/adc.87.5.403.

Robledo, N., López, L., Bunger, A., Tapia, C., & Abugoch, L. (2018). Effects of antimicrobial edible coating of thymol nanoemulsion/quinoa protein/chitosan on the safety, sensorial properties, and quality of refrigerated strawberries (*Fragaria × ananassa*) under commercial storage environment. *Food and Bioprocess Technology, 11*(8), 1566–1574. https://doi.org/10.1007/s11947-018-2124-3.

Robles-Sánchez, R. M., Rojas-Graü, M. A., Odriozola-Serrano, I., González-Aguilar, G., & Martin-Belloso, O. (2013). Influence of alginate-based edible coating as carrier of antibrowning agents on bioactive compounds and antioxidant activity in fresh-cut Kent mangoes. *LWT-Food Science and Technology, 50*(1), 240–246. https://doi.org/10.1016/j.lwt.2012.05.021.

Rodriguez-Garcia, I., Cruz-Valenzuela, M. R., Silva-Espinoza, B. A., Gonzalez-Aguilar, G. A., Moctezuma, E., Gutierrez-Pacheco, M. M., Tapia-Rodriguez, M. R., Ortega-Ramirez, L. A., & Ayala-Zavala, J. F. (2016). Oregano (*Lippia graveolens*) essential oil added within pectin edible coatings prevents fungal decay and increases the antioxidant capacity of treated tomatoes. *Journal of the Science of Food and Agriculture, 96*(11), 3772–3778. https://doi.org/10.1002/jsfa.7568.

Rojas-Graü, M. A., Soliva-Fortuny, R., & Martín-Belloso, O. (2008). Effect of natural anti-browning agents on color and related enzymes in fresh-cut Fuji apples as an alternative to the use of ascorbic acid. *Journal of Food Science, 73*(6), S267–S272. https://doi.org/10.1111/j.1750-3841.2008.00794.x.

Saberi, B., Golding, J. B., Chockchaisawasdee, S., Scarlett, C. J., & Stathopoulos, C. E. (2018). Effect of biocomposite edible coatings based on pea starch and guar gum on nutritional quality of "Valencia" orange during storage. *Starch- Stärke, 70*(5–6), 1700299. https://doi.org/10.1002/star.201700299.

Salinas-Roca, B., Soliva-Fortuny, R., Welti-Chanes, J., & Martín-Belloso, O. (2016). Combined effect of pulsed light, edible coating and malic acid dipping to improve fresh-cut mango safety and quality. *Food Control, 66*, 190–197. https://doi.org/10.1016/j.foodcont.2016.02.005.

Salvia-Trujillo, L., Rojas-Graü, M. A., Soliva-Fortuny, R., & Martín-Belloso, O. (2015). Use of antimicrobial nanoemulsions as edible coatings: Impact on safety and quality attributes of fresh-cut *Fuji* apples. *Postharvest Biology and Technology, 105*, 8–16. https://doi.org/10.1016/j.postharvbio.2015.03.009.

Sánchez-González, L., Vargas, M., González-Martínez, C., Chiralt, A., & Cháfer, M. (2011). Use of essential oils in bioactive edible coatings: A review. *Food Engineering Reviews, 3*(1), 1–16. https://doi.org/10.1007/s12393-010-9031-3.

Sanchís, E., Ghidelli, C., Palou, L., Pérez-Gago, M. B., González, S., Sheth, C., & Mateos, M. (2015). Browning inhibition and microbial control in fresh-cut persimmon (*Diospyros Kaki* 'ROJO BRILLANTE') by apple pectin-based edible coatings. *Acta Horticulturae*, 335–341. https://doi.org/10.17660/actahortic.2015.1079.42.

Sanphui, P., Goud, N. R., Khandavilli, U. B. R., Bhanoth, S., & Nangia, A. (2011). New polymorphs of curcumin. *Chemical Communications, 47*(17), 5013–5015. https://doi.org/10.1039/c1cc10204d.

Senturk Parreidt, T., Schmid, M., & Müller, K. (2018). Effect of dipping and vacuum impregnation coating techniques with alginate based coating on physical quality parameters of cantaloupe melon. *Journal of Food Science, 83*(4), 929–936. https://doi.org/10.1111/1750-3841.14091.

Severino, R., Vu, K. D., Donsì, F., Salmieri, S., Ferrari, G., & Lacroix, M. (2014). Antibacterial and physical effects of modified chitosan based-coating containing nanoemulsion of mandarin essential oil and three non-thermal treatments against *Listeria innocua* in green beans. *International Journal of Food Microbiology, 191*, 82–88. https://doi.org/10.1016/j.ijfoodmicro.2014.09.007.

Silva-Vera, W., Zamorano-Riquelme, M., Rocco-Orellana, C., Vega-Viveros, R., Gimenez-Castillo, B., Silva-Weiss, A., & Osorio-Lira, F. (2018). Study of spray system spplications of edible coating suspensions based on hydrocolloids containing cellulose nanofibers on grape surface (*Vitis vinifera* L.). *Food and Bioprocess Technology, 11*(8), 1575–1585. https://doi.org/10.1007/s11947-018-2126-1.

Soares A de, S., Ramos, A. M., Vieira, É. N. R., Vanzela, E. S. L., de Oliveira, P. M., & Paula, D. D. A. (2018). Vacuum impregnation of chitosan-based edible coating in minimally processed pumpkin. *International Journal of Food Science & Technology, 53*(9), 2229–2238. https://doi.org/10.1111/ijfs.13811.

Soradech, S., Nunthanid, J., Limmatvapirat, S., & Luangtana-anan, M. (2017). Utilization of shellac and gelatin composite film for coating to extend the shelf life of banana. *Food Control, 73*(Pat B), 1310–1317. https://doi.org/10.1016/j.foodcont.2016.10.059.

Sutherland, W. J., Clout, M., Depledge, M. H., et al. (2010). A horizon scan of global conservation issues for 2010. *Trends in Ecology and Evolution, 25*(1), 1–7. https://doi.org/10.1016/j.tree.2009.10.003.

Tesfay, S. Z., & Magwaza, L. S. (2017). Evaluating the efficacy of moringa leaf extract, chitosan and carboxymethyl cellulose as edible coatings for enhancing quality and extending postharvest life of avocado (*Persea americana* mill.) fruit. *Food Packaging and Shelf Life, 11*, 40–48. https://doi.org/10.1016/j.fpsl.2016.12.001.

Tesfay, S. Z., Magwaza, L. S., Mditshwa, A., & Mbili, N. (2018). Carboxyl methylcellulose (CMC) incorporated with moringa leaf and seed extracts as new postharvest organic edible coating for avocado (*Persea americana* mill.) fruit. *Acta Horticulturae*, (1201), 161–168. https://doi.org/10.17660/actahortic.2018.1201.22.

Thakur, R., Pristijono, P., Bowyer, M., Singh, S. P., Scarlett, C. J., Stathopoulos, C. E., & Vuong, Q. V. (2018a). A starch edible surface coating delays banana fruit ripening. *LWT – Food Science and Technology, 100*, 341–347. https://doi.org/10.1016/j.lwt.2018.10.055.

Thakur, R., Pristijono, P., Golding, J. B., Stathopoulos, C. E., Scarlett, C. J., Bowyer, M., Singh, S. P., & Vuong, Q. V. (2018b). Development and application of rice starch based edible coating to improve the postharvest storage potential and quality of plum fruit (*Prunus salicina*). *Scientia Horticulturae, 237*, 59–66. https://doi.org/10.1016/j.scienta.2018.04.005.

Tosati, J. V., de Oliveira, E. F., Oliveira, J. V., Nitin, N., & Monteiro, A. R. (2018). Light-activated antimicrobial activity of turmeric residue edible coatings against cross-contamination of *Listeria innocua* on sausages. *Food Control, 84*, 177–185. https://doi.org/10.1016/j.foodcont.2017.07.026.

Tosati, J. V., Messias, V. C., Carvalho, P. I. N., Pollonio, M. A. R., Meireles, M. A. A., & Monteiro, A. R. (2017). Antimicrobial effect of edible coating blend based on turmeric starch residue and gelatin applied onto fresh frankfurter sausage. *Food and Bioprocess Technology, 10*(12), 2165–2175. https://doi.org/10.1007/s11947-017-1985-1.

Tosati, J. V., Oliveira, D., Lerin, L. A., & Monteiro, A. R. (2015). Respiration rate of cherry tomatoes and gas permeability of hydroxypropylmethyl cellulose-based coating. *International Journal of Emerging Technology and Advanced Engineering, 5*(3), 281–287.

Treviño-Garza, M. Z., García, S., Heredia, N., Alanís-Guzmán, M. G., & Arévalo-Niño, K. (2017). Layer-by-layer edible coatings based on mucilages, pullulan and chitosan and its effect on quality and preservation of fresh-cut pineapple (*Ananas comosus*). *Postharvest Biology and Technology, 128*, 63–75. https://doi.org/10.1016/j.postharvbio.2017.01.007.

Uranga, J., Puertas, A. I., Etxabide, A., Dueñas, M. T., Guerrero, P., & de la Caba, K. (2019). Citric acid-incorporated fish gelatin/chitosan composite films. *Food Hydrocolloids, 86*, 95–103. https://doi.org/10.1016/j.foodhyd.2018.02.018.

Uz, M., & Altınkaya, S. A. (2011). Development of mono and multilayer antimicrobial food packaging materials for controlled release of potassium sorbate. *LWT – Food Science and Technology, 44*(10), 2302–2309. https://doi.org/10.1016/j.lwt.2011.05.003.

Valero, D., Díaz-Mula, H. M., Zapata, P. J., Guillén, F., Martínez-Romero, D., Castillo, S., & Serrano, M. (2013). Effects of alginate edible coating on preserving fruit quality in four plum cultivars during postharvest storage. *Postharvest Biology and Technology, 77*, 1–6. https://doi.org/10.1016/j.postharvbio.2012.10.011.

Vargas, M., Chiralt, A., Albors, A., & González-Martínez, C. (2009). Effect of chitosan-based edible coatings applied by vacuum impregnation on quality preservation of fresh-cut carrot. *Postharvest Biology and Technology, 51*(2), 263–271. https://doi.org/10.1016/j.postharvbio.2008.07.019.

Vishwasrao, C., & Ananthanarayan, L. (2016). Postharvest shelf-life extension of pink guavas (*Psidium guajava* L.) using HPMC-based edible surface coatings. *Journal of Food Science and Technology, 53*(4), 1966–1974. https://doi.org/10.1007/s13197-015-2164-x.

Xu, D., Qin, H., & Ren, D. (2018). Prolonged preservation of tangerine fruits using chitosan/montmorillonite composite coating. *Postharvest Biology and Technology, 143*, 50–57. https://doi.org/10.1016/j.postharvbio.2018.04.013.

Yan, J., Luo, Z., Ban, Z., Lu, H., Li, D., Yang, D., Aghdam, M. S., & Li, L. (2019). The effect of the layer-by-layer (LBL) edible coating on strawberry quality and metabolites during storage. *Postharvest Biology and Technology, 147*, 29–38. https://doi.org/10.1016/j.postharvbio.2018.09.002.

Yang, J., Blawert, C., Lamaka, S. V., Snihirova, D., Lu, X., Di, S., & Zheludkevich, M. L. (2018). Corrosion protection properties of inhibitor containing hybrid PEO-epoxy coating on magnesium. *Corrosion Science, 140*, 99–110. https://doi.org/10.1016/j.corsci.2018.06.014.

Youssef, A. M., Assem, F. M., El-Sayed, S. M., Salama, H., & El-Salam, M. A. (2017). Utilization of edible films and coatings as packaging materials for preservation of cheeses. *Journal of Packaging Technology and Research, 1*(2), 87–99. https://doi.org/10.1007/s41783-017-0012-3.

Zambrano-Zaragoza, M. L., Quintanar-Guerrero, D., Del Real, A., Piñon-Segundo, E., & Zambrano-Zaragoza, J. F. (2017). The release kinetics of β-carotene nanocapsules/xanthan gum coating and quality changes in fresh-cut melon (cantaloupe). *Carbohydrate Polymers, 157*, 1874–1882. https://doi.org/10.1016/j.carbpol.2016.11.075.

Zeng, R., Zhang, A., Chen, J., & Fu, Y. (2013). Impact of carboxymethyl cellulose coating enriched with extract of *Impatiens balsamina* stems on preservation of 'Newhall' navel orange. *Scientia Horticulturae, 160*, 44–48. https://doi.org/10.1016/j.scienta.2013.05.015.

Zhang, L., Chen, F., Lai, S., Wang, H., & Yang, H. (2018). Impact of soybean protein isolate-chitosan edible coating on the softening of apricot fruit during storage. *LWT – Food Science and Technology, 96*, 604–611. https://doi.org/10.1016/j.lwt.2018.06.011.

Zhong, Y., Cavender, G., & Zhao, Y. (2014). Investigation of different coating application methods on the performance of edible coatings on mozzarella cheese. *LWT – Food Science and Technology, 56*(1), 1–8. https://doi.org/10.1016/j.lwt.2013.11.006.

Chapter 23
Chitosan Mono- and Bilayer Edible Coatings for Preserving Postharvest Quality of Fresh Fruit

Margherita Modesti, Luigi Zampella, and Milena Petriccione

Abstract Edible coatings have been widely used to improve the storability and quality of fruit during postharvest life. Chitosan (Cs)-based coatings are considered the best and safest edible and biological coatings for different types of fruits, due to their lack of toxicity, biodegradability, film-forming properties and antimicrobial action. Mono- and bilayer Cs coatings affect postharvest fruit quality and influence the level of oxidative stress in fresh fruits by controlling their physiological, biochemical and structural changes involved in degenerative processes correlated with fruit ripening and senescence. This chapter discusses the use of different mono- and multilayer Cs-based coatings to improve qualitative traits and control the postharvest oxidative stress on several fruits.

Keywords Antimicrobial · Oxidative stress

23.1 Introduction

In the supply chain of fresh fruits and vegetables, losses and wastes take place at different levels from farm to table (Porat et al. 2018). Many of them, which occur during postharvest handling, storage and processing, as a consequence many investigations in recent years have been developed with the objective of obtaining different strategies to reduce the post-harvest qualitative decay in fruits and vegetables (Brasil and Siddiqui 2018). Edible coatings represent an attractive and valid tool to prolong the postharvest life of fresh and ready-to-eat fruits, demonstrating good safety and optimum performance (Ali et al. 2018; Álvarez et al. 2018). A mono- or bi-layer edible coating, is that edible, thin and transparent material, formed directly

M. Modesti
Institute of Life Sciences, Scuola Superiore Sant'Anna, Pisa, Italy

L. Zampella · M. Petriccione (✉)
Council for Agricultural Research and Economics (CREA) – Research Centre for Olive, Citrus and Tree Fruit, Caserta, Italy
e-mail: milena.petriccione@crea.gov.it

T. J. Gutiérrez (ed.), *Polymers for Agri-Food Applications*,
https://doi.org/10.1007/978-3-030-19416-1_23

on fruit surface, thus providing a good and selective barrier to control the atmosphere composition within the fruit, delay metabolic processes and slow down microbial growth (Gutiérrez and Álvarez 2017; Hassan et al. 2018).

Biopolymers are widely used to make edible coatings due to their biocompatibility, biodegradability, low toxicity, easy availability and low cost (Garrison et al. 2016; Gutiérrez 2018; Ncama et al. 2018). Polysaccharides are primarily utilized as renewable biopolymers that have an important role in the biosphere, in fact, they are the structural elements of plants and animals (Thakur and Thakur 2016). The polysaccharides have an ordered and very compact hydrogen bond network structure with hydrophilic nature, the mains ones tested to realize edible coatings are Cs, starch, alginates, pullulan and xanthan (Pérez-Gago and Rhim 2014; Gutiérrez 2017; Álvarez et al. 2017).

23.2 Cs: Structure and Functions

Cs is a linear copolymer of β-1,4-linked D-glucosamine and *N*-acetyl-D- glucosamine, at commercial level with this term are indicated different heterogeneous polymers with several molecular weight (Mw), viscosity, degree of deacetylation and pKa (Merino et al. 2018b; Merino et al. 2019a, 2019b). Cs is a deacetylated derivate of chitin obtained from bioproduct of shellfish industry waste or the cell walls of some microorganisms and fungi used as functional materials applicable in different fields (Romanazzi et al. 2018).

Several studies have shown that the use of concentrations of aqueous acid solution (acetic, nitric, hydrochloric, perchloric and phosphoric acid) allows to obtain the complete Cs dispersion and influences their physical or antimicrobial properties (Lodhi et al. 2014; Fernandez-Saiz et al. 2009). Due to its chemical nature, a complete solubilization of Cs in acidic water (pKa smaller than 6.2), occurs when it reaches a maximum degree of protonation with a stoichiometry $[AcOH]/[Chit\text{-}NH_2] = 0.6$ (Rinaudo et al. 1999). Low pH values of Cs solution also improve its antimicrobial effectiveness, facilitating the electrostatic bonds between anionic carboxyl and phosphate groups of bacterial surface and positive amino ($-NH_2$) groups of Cs leading to the alteration of membrane properties and leakage of intracellular constituents of the microorganisms (Romanazzi et al. 2017; Merino et al. 2018a).

Mw of Cs influences the thickness of coating and the film-forming properties: coatings obtained with high Mw Cs are thicker compared to low Mw one (Jongsri et al. 2016). Antimicrobial property of Cs is also correlated to its Mw and degree of acetylation (DA), indeed, Cs polymer with lower Mw and DA has been shown to have a greater efficacy in controlling the rate of microbial growth (Sharma et al. 2016). Iriti and Faoro (2009) also suggested that a Cs concentration threshold allows to switch the induction of programmed cell death into cytotoxicity in cell host.

Cs has a low toxicity and, due to this property, is the first compound approved by European Union for plant protection (Reg. EU 2014/563) (Romanazzi et al. 2018).

23.3 Monolayer Cs-Based Coatings

Cs-based coatings are considered the best edible and biologically safe preservative coatings for different types of fruits, due to their lack of toxicity, biodegradability, film-forming properties and antimicrobial action (Ali et al. 2018; Romanazzi et al. 2018). This coating can form a semipermeable film on fruit surface, which act as a mechanical barrier that protects fruits from pathogen infection, as well as is an exogenous elicitor of host-defense responses (Bautista-Baños et al. 2006; Meng et al. 2008). Cs coating can also carry functional ingredients and active substances with antimicrobial, antioxidant and antibrowning properties, as well as dyes, nutrients that can increase the nutritional values and the stability of fruit during postharvest life (Saberi and Golding 2018).

23.3.1 Influence of Ripening Process on Cs Coated Fruit

Cs coatings delay ripening process by modifying the endogenous gas exchange such as CO_2, O_2 and C_2H_4, which affect the respiration rate in several fruits such as guava (Silva et al. 2018), mango (Silva et al. 2017), strawberry (Eshghi et al. 2014; Velickova et al. 2013; El Ghaouth et al. 1991), plum (Liu et al. 2014), peach (Li and Yu 2000), pear (Zheng et al. 1996), apple (Hu and Zou 1998; Zheng et al. 1996; Gardesh et al. 2016), table grape (Gao et al. 2013), thus improving the properties of these fruits during storage. The addition of several components in Cs-based coating such as peel pomegranate extract, nano-montmorillonite, lauroyl arginate ethyl, olive-waste extracts, lacquer wax, quinoa and soybean protein, sunflower oil can regulate gas barrier properties, stability, water-vapor sensitivity and restrict CO_2 production and utilization of O_2 in Cs-coated guava, apple, tangerine, blueberry, kiwifruit, apricot and table grape fruits (Abugoch et al. 2016; Hu et al. 2019; Khalifa et al. 2017; Nair et al. 2018; Sun et al. 2018; Xu et al. 2018; Zhang et al. 2018; Zhidong et al. 2018).

According to Silva et al. (2017), the slowdown of ripening process in Cs-coated fruit was accompanied by reprogramming of the carbon metabolism which controls the sugar levels, while a reduction in respiration rate could be due to a delay of senescence process (Ali et al. 2011). In Cs coated fruits, the reduction in respiration rate is due to some changes occurring in metabolic pathways such glycolysis and tricarboxylic acid cycle, as well as oxidative phosphorylation which lead to a low level in the ATP production and the beginning of anaerobic metabolism (Van Dongen et al. 2011; Silva et al. 2017).

The analysis carried out by scanning electron microscope has also shown that Cs coating smoothened the fruit pericarp by closing the lenticels, slowing down respiration and transpiration processes (Kumar et al. 2017; Dong et al. 2004). The partial blockage of the pores on pericarp in Cs-coated fruit has delayed and suppressed the onset of the respiratory peak (Eum et al. 2009; Kumar et al. 2017;

Velickova et al. 2013). Cs coating combined by low temperature contributes to reduce ethylene evolution rate during storage (Silva et al. 2017; 2018; Kumar et al. 2017; Li and Yu 2000).

Polyelectrolyte-macroion complex between Cs and cellulose nanocrystals (CH-CNCs) have been used to reduce hydrophobicity and increase the stability of the coating, while its properties are improved in terms of gas barrier, water vapor permeability and antibacterial and antioxidant properties.

Deng et al. (2018) demonstrated that CNC-Pickering oil-in-water emulsion system (CH-PCNCs) is a more stable system than CNC surfactant-stabilized emulsion (CH-CNC) and reinforces the Cs matrix, so the coating of fruits can be used to extend the postharvest storability during long term cold storage under high relative moisture of pears cv. Bartlett (Deng et al. 2017a, b; 2018). To guarantee a good protection of the fruit, a wetting of the coating on the entire surface of the fruit should be spread homogeneously. The surface wetting properties were investigated by Jung et al. (2016) on Cs coatings in contact with fruits, using contact angle determinations. The resulting coatings (CH-PCNC) showed a better wetting, i.e. low contact angle values were observed (~34.7°) compared to CH-CNC suspension (~38.1°) (Deng et al. 2018). In addition, CH-PCNC coating displayed lower surface tension values (~30.3 mN/m) compared to the critical tension of the fruit surface, suggesting a good wetting capacity and adhesion of the coatings onto the pear surfaces (Deng et al. 2018).

23.3.2 Physico-Chemical and Nutraceutical Features in Cs Coated Fruit

Cs-based coatings mainly exert their effects on several physiological processes that control the most important qualitative parameters such as total soluble solid content (TSS), titratable acidity (TA), firmness and nutraceutical compounds during fruit storage (Hassan et al. 2018). Taste is a fundamental trait in the consumer acceptability in stored fruit, influenced by the ratio between TSS and TA (Xin et al. 2017). Firmness is also another important attribute that influences the fruit quality during postharvest life, and it is judged and evaluated by consumers in the overall acceptance of fruits (Silva et al. 2018; Ali et al. 2011).

During fruit postharvest life, the TA losses could be due to the use of organic acids as substrates in the respiratory metabolism in the stored fruit, while TSS increase could be attributed to different factors, as well as breakdown of starch into sugar, decrease in respiration rate, hydrolysis of cell wall polysaccharides and increase of dry matter due to water loss (Petriccione et al. 2015a). Several studies have shown that Cs-based coatings slow down the changes related to ripening process, delaying the TSS increase and TA losses in stored fruit such as apricot (Zhang et al. 2018), tangerine (Xu et al. 2018), guava (Silva et al. 2018), kiwifruit (Hu et al. 2019), mango (Jongsri et al. 2017; Cissé et al. 2015), cherimoya (Liu et al. 2016), plum (Liu et al. 2014; Kumar et al. 2017), sweet cherry (Petriccione et al. 2015a), loquat (Petriccione et al. 2015b) and strawberry (Petriccione et al. 2015c; Treviño-Garza et al. 2015).

Cs coatings reduce the loss of firmness as reported by several studies in different fruit crops (Silva et al. 2017; 2018; Hong et al. 2012; Ali et al. 2011; Zhu et al. 2008; Abugoch et al. 2016; Drevinskas et al. 2017). Furthermore, Cs Mw influences TSS, TA and firmness as demonstrated by Jongsri et al. (2016) in mango fruit during cold storage at 25 °C for 15 days. High Mw Cs gives the best performance compared to medium or low Mw ones, improving these qualitative traits at the end of storage (Jongsri et al. 2016).

Fruit softening is due to the structural and non-structural modifications of carbohydrates in the cell walls during the ripening process, which causes the decline of the mechanical strength of the cell walls and cell-to-cell adhesion (Valero and Serrano 2010). Polygalacturonase (PG) and pectin methyl esterase (PME), cell wall-hydrolyzing enzymes, induce a modification of the pectin during fruit ripening (Ali et al. 2004). Loss of firmness could also be due to the cell turgor loss induced by transpiration (Mannozzi et al. 2018). Xin et al. (2017) showed that the firmest coated samples were those treated with a combination of Cs and nano-SiOx, which showed the least weight loss, suggesting that the reaffirmation of Chinese cherries occurred with low levels of weight loss, while softening was observed simultaneously with greater weight loss. These results agree with another study carried out by Paniagua et al. (2013) which displayed that moisture loss is the main cause of firmness changes during postharvest storage of blueberries.

Kumar et al. (2017) found a retention of firmness (~78%) in Cs-coated plum fruit, suggesting that coating induces changes in metabolic and enzymatic activities and, consequently, slows down the degradation process of cell wall components in the plums. Suppression of respiration rate in Cs-coated fruits could be a factor responsible to the reduction of PME enzyme activity, involved in cell wall degradation, during plum storage.

Valero et al. (2013) reported an enhancement in the activity of hydrolyzing enzymes followed by CO_2 production in climacteric fruit. Such findings have also been reported in Cs containing montmorillonite and lauroyl arginate ethyl (CML)-treated berries where the lower PG and PME enzyme activities were attributed to the inhibition of CO_2 production by CML (Sun et al. 2018). In agreement with these findings, the changes in pectin, PME and PG were accompanied by a dramatic decrease in protopectin content, and an increase in water-soluble pectin (WSP) in Cs-coated fruit, which is correlated with the increased PME and PG activities (Hu et al. 2019; Liu et al. 2016).

Jongsri et al. (2017) found that a coating obtained with Cs (1%) and combined with spermidine (0.1 ppm) significantly inhibited the increase in soluble pectin content and retained fruit firmness. This agrees to the results obtained by Liu et al. (2016), where the synergistic effects of a combined-coating, obtained by citric acid (20 mM) and Cs (1%), delayed the softening of cherimoya fruit stored at 15 °C for 10 days. This treatment also significantly inhibited the expression of genes related to the cell wall, including xyloglucan endotransglucosylase, PGs and expansins, with different patterns in the isoforms of these genes during storage.

Fruits are an important and rich source of nutraceutical compounds that play a fundamental role in disease prevention and health promotion (Kerch 2015; Septembre-Malaterre et al. 2018). In the last years, new processes and soft technologies have been developed to maintain high levels of bioactive compounds in stored fruit. Cs-coated plums maintained higher ascorbic acid and polyphenols levels throughout the storage period and displayed a high antioxidant activity compared to uncoated fruit (Kumar et al. 2017). It has been also reported the same trend with respect to total phenols in Cs-coated grapes (Sánchez-González et al. 2011) and to ascorbic acid content in ascorbic acid-Cs and citric acid-Cs coated plums, pomegranate and cherimoya fruits, respectively (Liu et al. 2014; 2016; Ozdemir and Gökmen 2017). It has also been observed in Cs coated plums that the suppression of respiratory activity allows to slow down the anthocyanin's synthesis, associated with postharvest ripening (Kumar et al. 2017; Liu et al. 2014).

A slower decrease in bioactive compounds content, such as ascorbic acid, polyphenols, anthocyanins and flavonoids was observed during storage in sweet cherry, loquat, kiwifruit, litchi, guava, pomegranate and strawberry fruits coated with Cs (Kaya et al. 2016; Khalifa et al. 2016; Petriccione et al. 2015a, b, c; Kumari et al. 2015; Munhuweyi et al. 2017; Jiang et al. 2018; Nair et al. 2018).

Higher content of ascorbic acid could be due to the low oxygen permeability in Cs-coated fruits, which led to the inhibition of the activity of enzymes involved in acid ascorbic oxidation (Kumar et al. 2017). Cs-based coating was found to be effective in restricting the losses of ascorbic acid on different fruits such as guava (Nair et al. 2018) and papaya (Ali et al. 2011), tangerine (Xu et al. 2018) or stimulating the resveratrol production in table grape (Freitas et al. 2015). A higher antioxidant activity was also revealed in Cs-coated fruit during storage (Nair et al. 2018; Munhuweyi et al. 2017; Khalifa et al. 2016).

23.3.3 Antimicrobial Features in Cs Coated Fruit

In last years, different strategies have been applied to contain postharvest decay caused by several species of bacteria and fungi. In the management of postharvest diseases is needed to contain the contaminations that take place on the fruit surfaces, the application of coatings with antimicrobial properties could be more effective because the added antimicrobials or the own matrix of Cs can be released gradually and have a high degree of selectivity (Grande-Tovara et al. 2018).

Cs can provoke the defenses of the host, which has been studied, monitoring the activities of the enzymes involved in the defense mechanisms, analyzing in turn the expression of the genes by means of a quantitative real-time polymerase chain reaction and evaluating more recently RNA-Seq in Cs-coated fruits (Romanazzi et al. 2018). Antimicrobial properties of Cs-based coatings have been widely analyzed by several authors through many *in-vitro* and *in-vivo* experiments against a wide range of plant pathogens. Cs coating is an alternative way of controlling microbial diseases, thus reducing the use of chemical compounds in the management of plant diseases and the levels of their residue onto fruits (Romanazzi et al. 2018).

Conflicting results have been obtained related to the correlations between the influence of Cs Mw and its antimicrobial activity (Romanazzi et al. 2017). Cs with low (LMWC, 3.4–51.3 kDa) and high Mw (HMWC, 136.8–342.0 kDa) showed different and particular levels of effectiveness against gray mold (*Botrytis cinerea*) of kiwifruit. LMWC significantly inhibited the development of gray mold, spore germination and mycelial growth compared to HMWC. This could be due to their greater ability to penetrate the cell walls of kiwifruit epidermal peel tissues and increase the expression levels of plant defense genes (Hua et al. 2019). Similarly, LMWC has greater antifungal activity than HMWC against *Penicillium italicum* due to its greater capacity to damage the calcium gradient required for cellular homeostasis and fungal surviva (Lee et al. 2016). Other studies confirmed a higher antifungal activity of LMWC against *Candida* spp. and *Fusarium oxysporium* and of HMWC against *Alternaria solani* and *Aspergillus niger* (Kulikov et al. 2014; Younes et al. 2014). Drevinskas et al. (2017) also displayed that Cs with Mws ranged from 28 to 1671 kDa had higher antibacterial activities than lower Mw Cs oligomers (Mws in range 1–22 kDa) and their effectiveness of inhibit the bacteria growth depended on the tested bacteria species (Drevinskas et al. 2017). Cs generally exerts a stronger antimicrobial action against the most sensitive yeasts and molds, followed by Gram-positive and Gram-negative bacteria (Grande-Tovara et al. 2018; Romanazzi et al. 2018).

Following Darolt et al. (2016) Cs-based coatings have only a preventive effect but not curative against the blue mold caused by *P. expansum* of apples cv. Fuji. Metabolic pathways involved in controlling blue mold decay in Cs-coated apples have been analyzed by two-dimensional electrophoresis followed mass spectrometry analysis. Changes of protein expression profiles revealed 20 differentially expressed proteins involved in defense responses, carbohydrate metabolism and energy production that could increase the resistance to pathogen infection in Cs-coated apples (Li et al. 2015). Cs coating inhibited *P. expansum* growth in jujube fruit as a function of concentration, which caused the initial collapse of the fungi plasma membrane, followed by its gradual disruption after treatment (Wang et al. 2014).

de Oliveira et al. (2014) extracted Cs from cell walls structural components of *Cunninghamella elegans* UCP 542 has been applied as a coating on the table grape (*Vitis labrusca* L.) cv. Isabella to control the *B. cinerea* and *P. expansum* growth in stored fruit at room or cold temperature.

Cs has been used as a carrier for the incorporation of microorganisms, such as bacteria, yeast and fungi, improving their survival rate in coated fruits. In the last years, yeasts have been regarded as biocontrol agents for the control of postharvest diseases of fruits and vegetables (Guimarães et al. 2018). *Pichia membranaefaciens* combined with Cs in coated citrus showed a synergic effect which inhibited the *C. gloesporioides* growth and induced morphological and structural changes of the pathogen (Zhou et al. 2016). In the same way, using an antagonist bacteria such as *Bacillus subtilis* ABS-S14 in Cs-based coating improved the protection against

P. digitatum Sacc. inhibiting fungi growth and stimulating the defense mechanisms of plant in citrus fruit (Waewthongrak et al. 2015).

Several studies have shown that adding elicitors to Cs-based coating can improve the ability to inhibit the development of fungal diseases in several fruits (Romanazzi et al. 2018). Jongsri et al. (2017) demonstrated a synergistic effect of Cs (1%) and spermidine (0.1 ppm) combination on mango 'Nam Dok Mai' fruit in the inhibition of the development of anthracnose disease caused by *Colletotrichum gloeosporioides*, respect to each individual component. Spermidine induced the enzyme activities such as chitinase, β-1,3-glucanase and peroxidase that participate in the defense mechanisms of the plant. Preharvest Cs-*g*-salicylic acid treatment in table grapes (*Vitis vinifera* L. 'Yongyou 1′) reduced the *B. cinerea* incidence in grape berries during storage at 0 °C for 42 days. A significant increase in phenylalanine ammonia lyase (PAL), chitinase, and β-1,3-glucanase activities and a higher TPC content were observed in CTS-*g*-SA-coated berries. These findings suggested that chitinase and β-1,3-glucanase contribute to lytic degradation of cell walls of fungi, while PAL participates in the synthesis of phenolic compounds with antifungal activity (Shen and Yang 2017).

Bacteriocins are active proteins with bactericidal activity against food spoiling bacteria and are considered as safe additives for food preservation (Sidhu and Nehra 2017). Aguayo et al. (2016) tested enterocin AS-48 to evaluate its anti-*Listeria* activity in Cs-coated fresh-cut apples inoculated with five *L. monocytogenes* strains. The addition of Bacteriocin of Cs did not significantly improve the antilisterial activity compared to Cs itself, probably due to the same load of two compounds competing for the binding sites on the bacteria. The addition of EDTA (20 mM) in combination with enterocin AS-48 (40 μg/ml) improved the inactivation of *L. monocytogenes* strains in Cs-coated apples, reducing viable counts below detection levels during storage early (Aguayo et al. 2016). Cs-based coating has also been used to stabilize the antimicrobial activity of lactoperoxidase system (LPOS) to extend the postharvest preservation of mangoes that inhibit the microbial growth of *Phomopsis*, *C. gloeosporioides* strains and *L. diplodia* (Cissé et al. 2015).

In last years, essential oils (EOs) have received great attention for their antimicrobial properties and have also been incorporated into Cs-based coatings due to their low potential to induce microorganisms' resistance (Yuan et al. 2016). The EO are several compounds produced during secondary metabolism of plant including terpenes and terpenoids, and rarely nitrogen- and sulfur-containing compounds, coumarins and homologues of phenylpropanoids (Nazzaro et al. 2017). Antimicrobial features of the EOs could be mainly attributed to the most abundant compounds together to synergy or antagonism with minor compounds (Grande-Tovara et al. 2018). The incorporation of EOs obtained from *Mentha piperita*, lavender, thyme, lemongrass, *Salvia officinalis*, enhanced the antimicrobial features of Cs-based coatings, inhibiting fungal decay in several fruits such as grape berries, pomegranate, table grapes, strawberries, guava and mango (Romanazzi et al. 2018).

23.3.4 Postharvest Oxidative Stress in Cs Coated Fruit

Reactive oxygen species (ROS), under steady state conditions are the byproducts of several metabolic pathways where ROS production and scavenging are in perfect equilibrium. ROS participate in the network of signaling pathways, which act as the main regulators of cellular and physiological responses. During fruit postharvest life, abiotic and biotic stress may increase the intracellular levels of ROS, such as H_2O_2, O_2^- and hydroxyl radicals, which can damage cell membranes and other biomolecules (protein, lipids, carbohydrates and DNA). ROS accumulation causes to the lipid membrane damage due to lipoxygenase (LOX) activity, leading to a rapid reduction of fruit quality and marketability. Malondialdehyde (MDA) and membrane permeability are used as two direct biomarkers to evaluate the lipid peroxidation and damage extent to cell membranes, respectively (Racchi 2013).

Cs-based coatings slow down fruit the senescence and quality deterioration, in part due to the delay in ROS generation. In several Cs-coated fruits during storage, a reduction in $O_2^{\bullet-}$ and H_2O_2 content has been demonstrated compared to untreated samples (Song et al. 2016; Liu et al. 2014; 2016; Hong et al. 2012). In loquat, a significant reduction around ~35% and ~69% in the $O_2^{\bullet-}$ and H_2O_2 content was observed in Cs/nano-silica treated fruit compared to untreated ones at the end of storage, respectively (Song et al. 2016). Similarly, Cs in combination with citric acid or ascorbic acid significantly reduced $O_2^{\bullet-}$production in coated cherimoya and plum fruit compared to untreated ones, respectively ($p \leq 0.05$) (Liu et al. 2014; 2016). In addition, the $O_2^{\bullet-}$ production rate of Cs-coated guava fruit was dependent on the different concentrations of Cs used (Hong et al. 2012). As a result, electrolyte leakage from control fruits was higher than Cs-coated loquat and jujube fruits during storage (Song et al. 2016; Zhang et al. 2014). Cs coatings reduce ROS accumulation and, consequently alleviate the lipid peroxidation of the membrane, as indicated by the lower MDA content in coated fruit at the end of storage (Song et al. 2016; Liu et al. 2014; 2016; Hong et al. 2012; Zhang et al. 2014; Silva et al. 2018; Adiletta et al. 2018; Pasquariello et al. 2015; Petriccione et al. 2015c; 2018).

ROS accumulation in fruit tissues is counteracted by antioxidant defense systems characterized by enzymatic and non-enzymatic components. Several studies have investigated the effects of Cs-based coatings of the antioxidant enzymes including superoxide dismutase (SOD, EC 1.15.1.1), ascorbate peroxidase (APX, EC 1.11.1.1), catalase (CAT, EC 1.11.1.6), etc. (Song et al. 2016; Liu et al. 2014; 2016; Hong et al. 2012; Zhang et al. 2014; Silva et al. 2018; Adiletta et al. 2018; Pasquariello et al. 2015; Petriccione et al. 2015c; 2018). These enzymes show different trends in fruit crops during storage.

SOD is the first ROS scavenging enzyme that catalyses the dismutation of radical superoxide to H_2O_2 and O_2. The detoxification of H_2O_2 in H_2O is mediated by CAT and APX, but the latter uses ascorbate as an electron donor. Cs-based coating increased the ability of scavenge excessive levels of ROS by improving the SOD, APX and CAT activity in loquat (Song et al. 2016), cherimoya (Liu et al. 2016), guava (Song et al. 2012), plum (Liu et al. 2014), sweet cherry (Pasquariello et al. 2015), strawberry (Petriccione et al. 2015c; Wang and Gao 2013), wine grape (Petriccione et al. 2018) and jujube (Zhang et al. 2014).

LOX activity causes the dioxygenation of polyunsaturated fatty acids which produce toxic hydroperoxy fatty acids and the consequent damages to cell membranes. Cs-coated fruit have shown significantly lower LOX activity than uncoated fruit during storage, suggesting that Cs coating contributed to the maintenance of membrane integrity and reduce skin and flesh color changes due to the preservation of cell compartmentalization (Pasquariello et al. 2015; Song et al. 2016; Adiletta et al. 2018; Petriccione et al. 2015c; 2018).

In addition, Cs-based coatings retard the fruit browning due to oxidation of phenolic substrates mediated by polyphenol oxidase (PPO). The combined treatment with citric acid or ascorbic acid and Cs coating allows the PPO activity to be delayed in cherimoya and plum fruit stored at 15 °C for 10 days and at 5 °C for 20 days, respectively (Liu et al. 2014; 2016). Likewise, Cs-coated strawberry, sweet cherry and loquat fruit showed a significantly lower PPO activity compared with uncoated ones and a higher polyphenol content due to a separation of PPO enzyme from its phenolic substrates (Adiletta et al. 2018; Petriccione et al. 2015a, b, c). Cs coatings have inhibited PPO activity and reduced browning of berries during wine grape partial dehydration (Petriccione et al. 2018). Cs-based coatings are thus effective in controlling the metabolism of ROS, thus improving the activities of antioxidant enzymes that reduce oxidative damage.

23.4 Multilayer Cs-Based Coatings

Multilayer edible coatings from Cs on fruit surfaces using polyelectrolytes have been made in recent years, using the layer-by-layer (LbL) deposition technique. Two solutions with oppositely charged polyelectrolytes are used in multiple dipping where each dipping step is separated by drying step to remove the excess of coating solution in whole or fresh-cut fruit (McClements et al. 2009; Vargas et al. 2008) (Fig. 23.1).

Several studies have shown that Cs-based multilayer coatings are valid tools to improve protection and prolong the shelf life in whole and ready-to-eat fruits. According to Sharma et al. (2016) bi-layer edible coatings combined the beneficial traits of each component in a new coating with superior characteristics. Table 23.1 lists several works about multilayer Cs-based edible coatings on different fruits.

23.4.1 Whole Fruit

Nanomultilayer coatings obtained by electrostatic LbL self-assembly, with five alternate layers of pectin and Cs, were used by Medeiros et al. (2012) to prolong the shelf-life of mango cv. 'Tommy Atkins'. This coating improved the gas flow barrier

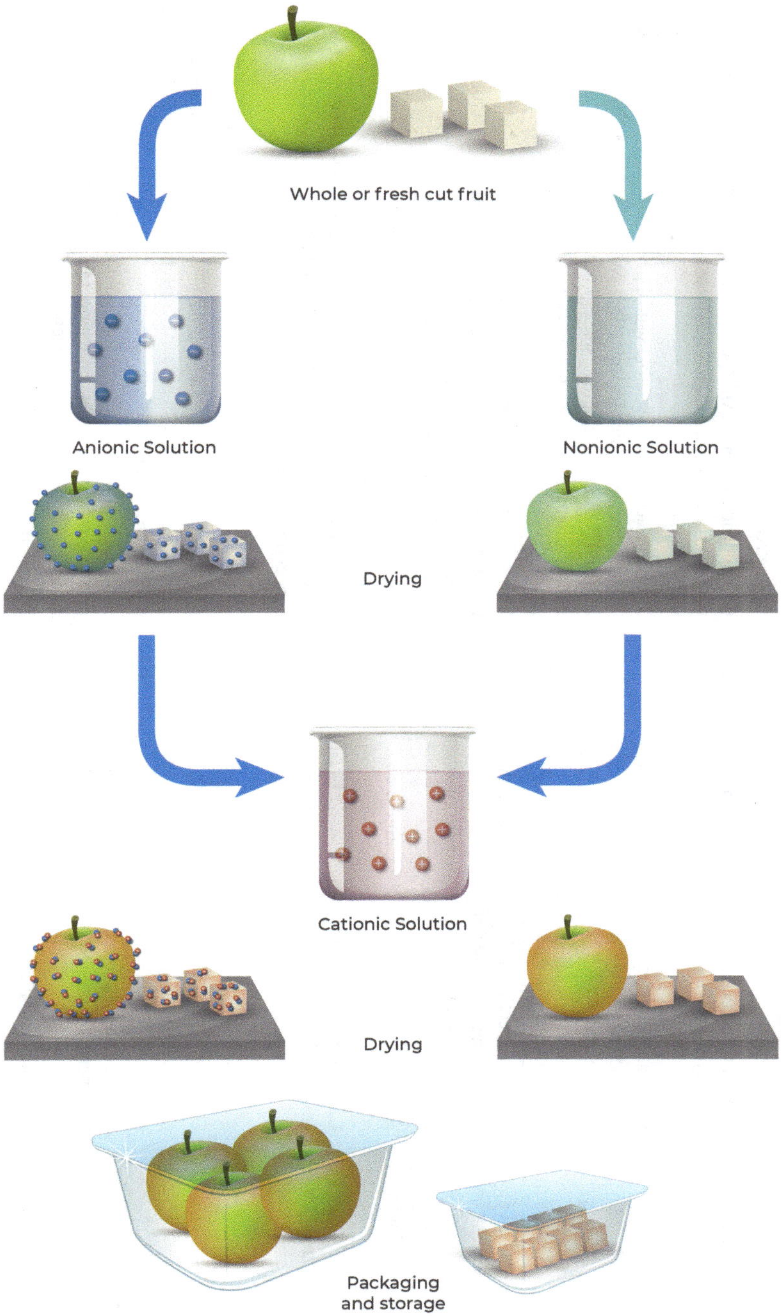

Fig. 23.1 Schematic representation of the main steps involved in the layer-by-layer edible coatings on whole and fresh-cut fruits

Table 23.1 Different studies carried out on Cs bilayer edible coatings in whole and ready-to-eat fruits

Fruit	Layer-by-Layer treatment	Storage conditions	Evaluated traits	Reference
Mango cv. 'Tommy Atkins' (whole fruit)	Chitosan and pectin (five nanolayers)	4 °C for 45 days and relative humidity of 93%	Oxygen and carbon dioxide permeabilities, water vapor permeability, mass loss, titratable acidity, total soluble solids.	Medeiros et a. (2012)
Papaya cv. 'Maradol' (fresh-cut fruit)	Chitosan, pectin, calcium chloride (cross-linking agent) and trans-cinnamaldehyde in beta-cyclodextrin	4 °C for 15 days	Titratable acidity, total soluble solids, pH, moisture content, firmness, juice leakage, flesh color, ascorbic acid and β-carotenes content, sensory testing, microbiological analysis, microscopic examination of coating	Brasil et al. (2012)
Cantaloupe (fresh-cut fruit)	Chitosan, pectin and trans-cinnamaldehyde in beta-cyclodextrin	4 °C for 15 days	Firmness, titratable acidity, total soluble solids, pH, moisture content, flesh color, ascorbic acid content, sensory analysis, microbial quality and growth, microscopic examination of coating.	Martiñon et al. (2014)
Cantaloupe cv. 'Yaniv' (fresh-cut fruit)	Alginate and chitosan	6 °C for 14 days	Gas-exchanges, texture, water vapor resistance, microbial growth.	Poverenov et al. (2014)
Mandarin (cvs. 'Or' and'Mor'), orange (cv. 'Navel') and grapefruit (cv. 'Star ruby') (whole fruit)	Carboxymethyl cellulose and chitosan	5 °C 4 weeks (mandarins and oranges) + 20 °C 5 days (shelf-life) 10 °C 4 weeks (grapefruit) + 20 °C 5 days (shelf-life)	Gas-exchanges, fruit gloss, firmness and weight loss, titratable acidity, total soluble solids, ethanol concentrations in the juice, sensory evaluations.	Arnon et al. (2014)
Mandarins cvs. 'Rishon' and 'Michal' (whole fruit)	Carboxymethylcellulose and chitosan	20 °C for 10 days and relative humidity of ~80–85%	Firmness, weight loss, ethanol concentration in juice, gas exchanges, disease incidence, fruit gloss, sensory evaluations, fruit ripening progression.	Arnon et al. (2015)
Mango cv. 'Tommy Atkins'	Chitosan and alginate (five nanolayers)	8 °C for 14 days and relative humidity of 93%	Weight loss, titratable acidity, total soluble solids, pH, ascorbic acid content, browning, microbiological analyses, malondialdehyde content.	Souza et al. (2015)
Pineapple (fresh-cut fruit)	Chitosan and pullulan, chitosan and mucilage (obtained by linseed, nopal cactus and aloe)	4 °C for 18 days	Weight loss, firmness, flesh color, titratable acidity, total soluble solids, pH, ascorbic acid content, microbiological and sensory parameters.	Treviño-Garza et al. (2017)

properties during storage (45 days), in addition, the coated fruit showed a higher value in mass loss and total soluble solids, whereas lower value in TA were obtained compared with uncoated ones. Uncoated samples displayed wrinkled appearance, microbial spoilage and flesh browning (Medeiros et al. 2012).

Different edible coatings based on layered polysaccharides were tested by Arnon et al. (2015) as a valid alternative to polyethylene-based commercial wax coating currently utilized to improve storage time of mandarin fruit. A bilayer obtained by a combination of carboxymethyl cellulose (CMC) (inner layer) and Cs (outer layer), gave the best performance without off-flavors insurgence. CMC/Cs bilayer provided a uniform and stable matrix due to CMC and imparted high gloss due to Cs. In addition, the coatings were water resistant and maintained their stability during exposure to cold storage and shelf life conditions. The fruits treated with bi-layered coating (CMC-Cs), using CMC 1.5% w/v and Cs 1.5% w/v, displayed a higher firmness and lower weight loss than single (CMC or Cs) layer-coated and uncoated ones. Physiological traits in by-layered coated mandarins were also improved, and an inhibition of the ripening progress was observed (Arnon et al. 2015).

Arnon et al. (2014) tested Cs-based bilayer coating, comprising CMC/Cs in order to preserve the postharvest quality of citrus fruits including mandarin (cvs. 'Or' and'Mor'), orange (cv. 'Navel') and grapefruit (cv. 'Star Ruby'). CMC/Cs bilayer edible coating was homogeneous, stable, inexpensive, safe and biodegradable. In addition, Cs in the bilayer coating prevented microbial growth and reduced the need to apply fungicides. This bilayer coating greatly enhanced the glossiness and appearance of the fruit, but was not effective with its water vapor-barrier properties to slow down weight loss. A cultivar-specific mandarin response to the application of bilayer Cs-based coating has been observed, in fact, 'Mor' cultivar accumulated more ethanol after coating and storage than 'Or', increasing the development of off-flavors (Arnon et al. 2014).

23.4.2 *Minimally Processed Fruit*

The first multilayered edible coating made of Cs and pectin, and using calcium chloride as cross-linking agent was realized by Brasil et al. (2012) to improve the shelf-life of fresh cut papaya at 4 °C up to 15 days. A trans-cinnamaldehyde–beta-cyclodextrins (TC-BCD) inclusion complex, microencapsulated by the freeze-drying method was incorporated in LbL coatings, resulting in a material with antimicrobial properties because TC is considered the guest antimicrobial molecule (Gouin 2004). TC-BCD complex improves some physical properties and hides unpleasant odors and off-flavors of TC (Del Valle 2004). LbL coatings made from Cs and pectin under acidic conditions with calcium ions formed a stable bilayer with uniform thickness (300 ± 1 μm) around the fruit surface, which reduced the juice leakage and improved the firmness of the coated fruit. The coating also maintained the physicochemical quality and nutraceutical compounds during storage, which delayed the respiration process and controlled the metabolic reactions. The multilayer edible coatings have reduced aerobic, psychrophilic, yeast and mold growth with significantly lower ($p \leq 0.05$) lower counts during storage (Brasil et al. 2012).

Martiñon et al. (2014) developed a multilayered edible coating based on Cs (2 g/100 g) and pectin (1 g/100 g) using as antimicrobial agent the encapsulated trans-cinnamaldehyde (2 g/100 g) to extend the shelf-life of fresh-cut cantaloupe during cold storage at 4 °C for 15 days. Coated fruit maintained the physico-chemical and sensorial properties up to 9 days of cold storage compared to 4 days for the uncoated ones. An increase of 5 days in the shelf-life extension is an important economic benefit of the ready-to-eat fruit industry (Mayen et al. 2007). The first LbL alginate–Cs coating was evaluated on fresh-cut melon cv. 'Yaniv' by Poverenov et al. (2014). Alginate substrate (inner layer) improved the fruit firmness, while Cs (outer layer) supplied antimicrobial protection against different microorganisms such as bacteria, yeast and molds. This coating also inhibited the formation of undesirable off-flavor volatiles and softening of the fruit during storage.

Cs-based nanomultilayer coating was realized by electrostatic LbL self-assembling on aminolyzed/charged PET using alginate and Cs to preserve the quality of fresh-cut mangoes stored at 8 °C for 14 days (Souza et al. 2015). Five nanolayers (Alg-Cs-Alg-Cs-Alg) were realized onto fresh-cut mangoes, exploiting the alternate deposition of alginate and Cs with opposite charge. Alginate-Cs nanolayers improved the water vapor barrier capacity, the weight loss from the fourth day of storage onwards. The coated fruits ready to eat showed a good physical-chemical quality, lower MDA content and browning compared to the uncoated ones. Based on microbiological analyses, the shelf life of fresh-cut mangoes could be extended up to 8 days at 8 °C compared with uncoated fresh-cut mangoes (<2 days) (Souza et al. 2015).

Other study carried out by Treviño-Garza et al. (2017) demonstrated as LbL coatings based on the combination of Cs with pullulan (nonionic polymer) and mucilages, obtained by linseed, nopal cactus and aloe, improved the postharvest life of packed fresh-cut pineapple stored at 4 °C for 18 days. Multilayer Cs-based coatings delay the weight loss, improving firmness and reducing browning of tissues during storage. A significant reduction ($p \leq 0.05$) in molds and yeasts was also recorded in the total count of aerobic and psychotropic microorganisms at the end of storage in fresh-cut coated pineapple. In particular, multilayer ECs manufactured with neutral polymers such as pullulan, allowed the Cs to conserve its antimicrobial characteristic due to the absence of electrostatic interactions. These films proved to be more effective in delaying microbial growth during storage (Treviño-Garza et al. 2017) (Fig. 23.1). Antimicrobial activity was also displayed shown against the growth of pathogenic microorganisms such as *L. monocytogenes* and *Salmonella typhi*. Significant differences ($p \leq 0.05$) in analyzed features, between treated pineapple samples compared to control ones, were registered from 6 days of cold storage (Treviño-Garza et al. 2017).

23.5 Conclusions

Mono- and bilayer Cs edible coatings are eco-friendly promising strategies that allow to improve the physicochemical and nutraceutical features during fruit postharvest life, prolonging their storage time. They can also reduce the microbial growth and improve the antioxidant enzymatic system involved in ROS detoxification. The positive attributes of Cs may also be enhanced by adding other compounds or combining them with other postharvest treatments.

Numerous studies have consistently shown that mono- and bilayer Cs edible coatings are valid tools to control the physiological, biochemical and structural changes involved in fruit ripening and senescence during storage.

Acknowledgements The authors are grateful to Dr. Giuseppina Adiletta for helpful suggestion and critical revision for this chapter. We thank to Dr. Francesco Caracciolo for his great help on figure graphic design and we also thank all collaborators who worked with us in various aspects on chitosan-based edible coating projects and most often they are coauthors of our publications.

Conflicts of Interest The authors declare no conflict of interest.

References

Abugoch, L., Tapia, C., Plasencia, D., Pastor, A., Castro-Mandujano, O., López, L., & Escalona, V. H. (2016). Shelf-life of fresh blueberries coated with quinoa protein/chitosan/sunflower oil edible film. *Journal of the Science of Food and Agriculture, 96*(2), 619–626. https://doi.org/10.1002/jsfa.7132.

Adiletta, G., Pasquariello, M. S., Zampella, L., Mastrobuoni, F., Scortichini, M., & Petriccione, M. (2018). Chitosan coating: A postharvest treatment to delay oxidative stress in loquat fruits during cold storage. *Agronomy, 8*, 54. https://doi.org/10.3390/agronomy8040054.

Aguayo, M. L., Burgos, M. J. G., Pulido, R. P., Gálvez, A., & López, R. L. (2016). Effect of different activated coatings containing enterocin AS-48 against *Listeria monocytogenes* on apple cubes. *Innovative Food Science and Emerging Technologies, 35*, 177–183. https://doi.org/10.1016/j.ifset.2016.05.006.

Ali, Z. M., Chin, L.-H., & Lazan, H. (2004). A comparative study on wall degrading enzymes, pectin modifications and softening during ripening of selected tropical fruits. *Plant Science, 167*(2), 317–327. https://doi.org/10.1016/j.plantsci.2004.03.030.

Ali, A., Muhammad, M. T. M., Sijam, K., & Siddiqui, Y. (2011). Effect of chitosan coatings on the physicochemical characteristics of Eksotika II papaya (*Carica papaya* L.) fruit during cold storage. *Food Chemistry, 124*(2), 620–626. https://doi.org/10.1016/j.foodchem.2010.06.085.

Ali, A., Yeoh, W. K., Forney, C., & Siddiqui, M. W. (2018). Advances in postharvest technologies to extend the storage life of minimally processed fruits andvegetables. *Critical Reviews in Food Science and Nutrition, 58*(15), 2632–2649. https://doi.org/10.1080/10408398.2017.1339180.

Álvarez, K., Lucía Famá, L., & Gutiérrez, T. J. (2017). Physicochemical, antimicrobial and mechanical properties of thermoplastic materials based on biopolymers with application in the food industry. In M. Masuelli & D. Renard (Eds.), *Advances in physicochemical properties of biopolymers: Part 1* (pp. 358–400). Bentham Science Publishers. EE.UU. ISBN: 978-1-68108-454-1. eISBN: 978-1-68108-453-4. https://doi.org/10.2174/9781681084534117010015.

Álvarez, K., Alvarez, V. A., & Gutiérrez, T. J. (2018). Biopolymer composite materials with antimicrobial effects applied to the food industry. In V. K. Thakur & M. K. Thakur (Eds.), *Functional biopolymers* (pp. 57–96). Editorial Springer International Publishing. EE.UU. ISBN: 978-3-319-66416-3. eISBN: 978-3-319-66417-0. https://doi.org/10.1007/978-3-319-66417-0_3.

Arnon, H., Zaitsev, Y., Porat, R., & Poverenov, E. (2014). Effects of carboxymethyl cellulose and chitosan bilayer edible coating on postharvest quality of citrus fruit. *Postharvest Biology and Technology, 87*, 21–26. https://doi.org/10.1016/j.postharvbio.2013.08.007.

Arnon, H., Granit, R., Porat, R., & Poverenov, E. (2015). Development of polysaccharides-based edible coatings for citrus fruits: A layer-by-layer approach. *Food Chemistry, 166*, 465–472. https://doi.org/10.1016/j.foodchem.2014.06.061.

Bautista-Baños, S., Hernández-Lauzardo, A. N., Velázquez-del Valle, M. G., Hernández-López, M., Ait Barka, E., Bosquez-Molina, E., & Wilson, C. L. (2006). Chitosan as a potential natural compound to control pre and postharvest diseases of horticultural commodities. *Crop Protection, 25*(2), 108–118. https://doi.org/10.1016/j.cropro.2005.03.010.

Brasil, I. M., & Siddiqui, M. (2018). Postharvest quality of fruitsand vegetables: An overview. In M. W. Siddiqui (Ed.), *Preharvest modulation of postharvest fruit and vegetable quality* (pp. 1–40). London: Elsevier, Inc.. https://doi.org/10.1016/B978-0-12-809807-3.00001-9.

Brasil, I. M., Gomes, C., Puerta-Gomez, A., Castell-Perez, M. E., & Moreira, R. G. (2012). Polysaccharide-based multilayered antimicrobial edible coating enhances quality of fresh-cut papaya. *LWT – Food Science and Technology, 47*(1), 39–45. https://doi.org/10.1016/j.lwt.2012.01.005.

Cissé, M., Polidori, J., Montet, D., Loiseau, G., & Ducamp-Collin, M. N. (2015). Preservation of mango quality by using functional chitosan lactoperoxidase systems coatings. *Postharvest Biology and Technology, 101*, 10–14. https://doi.org/10.1016/j.postharvbio.2014.11.003.

Darolt, J. C., Rocha Neto, A. C., & Di Piero, R. M. (2016). Effects of the protective, curative, and eradicative applications of chitosan against *Penicillium expansum* in apples. *Brazilian Journal of Microbiology, 47*(4), 1014–1019. https://doi.org/10.1016/j.bjm.2016.07.007.

de Oliveira, C. E. V., Magnani, M., de Sales, C. V., de Souza Pontes, A. L., Campos-Takaki, G. M., Stamford, T. C. M., & de Souza, E. L. (2014). Effects of chitosan from *Cunninghamella elegans* on virulence of post-harvest pathogenic fungi in table grapes (*Vitis labrusca* L.). *International Journal of Food Microbiology, 171*, 54–61. https://doi.org/10.1016/j.ijfoodmicro.2013.11.006.

Del Valle, E. M. M. (2004). Cyclodextrins and their uses: A review. *Process Biochemistry, 39*(9), 1033–1046. https://doi.org/10.1016/S0032-9592(03)00258-9.

Deng, Z., Jung, J., Simonsen, J., Wang, Y., & Zhao, Y. (2017a). Cellulose nanocrystal reinforced chitosan coatings for improving the storability of postharvest pears under both ambient and cold storages. *Journal of Food Science, 82*(2), 453–462. https://doi.org/10.1111/1750-3841.13601.

Deng, Z., Jung, J., & Zhao, Y. (2017b). Development, characterization, and validation of chitosan adsorbed cellulose nanofiber (CNF) films as water resistant and antibacterial food contact packaging. *LWT – Food Science and Technology, 83*, 132–140. https://doi.org/10.1016/j.lwt.2017.05.013.

Deng, Z., Jung, J., Simonsen, J., & Zhao, Y. (2018). Cellulose nanocrystals Pickering emulsion incorporated chitosan coatings for improving storability of postharvest bartlett pears (*Pyrus communis*) during long-term cold storage. *Food Hydrocolloids, 84*, 229–237. https://doi.org/10.1016/j.foodhyd.2018.06.012.

Dong, H., Cheng, L., Tan, J., Zheng, K., & Jiang, Y. (2004). Effect of chitosan coating on quality and shelf-life of peeled litchi fruit. *Journal of Food Engineering, 64*(3), 355–358. https://doi.org/10.1016/j.foodchem.2018.01.095.

Drevinskas, T., Naujokaitytė, G., Maruška, A., Kaya, M., Sargin, I., Daubaras, R., & Česoniene, L. (2017). Effect of molecular weight of chitosan on the shelf life and other quality parameters of three different cultivars of *Actinidia kolomikta* (kiwifruit). *Carbohydrate Polymers, 2017*(173), 269–275. https://doi.org/10.1016/j.carbpol.2017.06.002.

El Ghaouth, A., Arul, J., Ponnampalam, R., & Boulet, M. (1991). Chitosan coating effect on storability and quality of fresh strawberries. *Journal of Food Science, 56*(6), 1618–1631. https://doi.org/10.1111/j.1365-2621.1991.tb08655.x.

Eshghi, S., Hashemi, M., Mohammadi, A., Badii, F., Mohammadhoseini, Z., & Ahmadi, K. (2014). Effect of nanochitosan-based coating with and without copper loaded on physicochemical and bioactive components of fresh strawberry fruit (*Fragaria* ×*ananassa* Duchesne) during storage. *Food and Bioprocess Technology, 7*(8), 2397–2409. https://doi.org/10.1007/s11947-014-1281-2.

Eum, H. L., Hwang, D. K., Linke, M., & Lee, S. K. (2009). Influence of edible coating on quality of plum (*Prunus salicina* Lindl cv. 'Sapphire'). *European Food Research and Technology, 229*(3), 427–434. https://doi.org/10.1007/s00217-009-1054-8.

Fernandez-Saiz, P., Lagaron, J. M., & Ocio, M. J. (2009). Optimization of the biocide properties of chitosan for its application in the design of active films of interest in the food area. *Food Hydrocolloids, 23*(3), 913–921. https://doi.org/10.1016/j.foodhyd.2008.06.001.

Freitas, P. M., López-Gálvez, F., Tudela, J. A., Gil, M. I., & Allende, A. (2015). Postharvest treatment of table grapes with ultraviolet-C and chitosan coating preserves quality and increases stilbene content. *Postharvest Biology and Technology, 105*, 51–57. https://doi.org/10.1016/j.postharvbio.2015.03.011.

Gao, P., Zhu, Z., & Zhang, P. (2013). Effects of chitosan–glucose complex coating on postharvest quality and shelf life of table grapes. *Carbohydrate Polymers, 95*(1), 371–378. https://doi.org/10.1016/j.carbpol.2013.03.029.

Gardesh, A. S. K., Badii, F., Hashemi, M., Ardakani, A. Y., Maftoonazad, N., & Gorji, A. M. (2016). Effect of nanochitosan based coating on climacteric behavior and postharvest shelf-life extension of apple cv. *Golab Kohanz*. *LWT – Food Science and Technology, 70*, 33–40. https://doi.org/10.1016/j.lwt.2016.02.002.

Garrison, T. F., Murawski, A., & Quirino, R. (2016). Bio-based polymers with potential for biodegradability. *Polymers, 8*(7), 262. https://doi.org/10.3390/polym8070262.

Gouin, S. (2004). Microencapsulation: Industrial appraisal of existing technologies and trends. *Trends in Food Science and Technology, 15*(7–8), 330–347. https://doi.org/10.1016/j.tifs.2003.10.005.

Grande-Tovara, C. D., Chaves-Lopez, C., Serio, A., Rossi, C., & Paparella, A. (2018). Chitosan coatings enriched with essential oils: Effects on fungi involved in fruit decay and mechanisms of action. *Trends in Food Science & Technology, 78*, 61–71. https://doi.org/10.1016/j.tifs.2018.05.019.

Guimarães, A., Abrunhosa, L., Pastrana, L. M., & Cerqueira, M. A. (2018). Edible films and coatings as carriers of living microorganisms: A new strategy towards biopreservation and healthier foods. *Comprehensive Reviews in Food Science and Food Safety, 17*(3), 594–614. https://doi.org/10.1111/1541-4337.12345.

Gutiérrez, T. J. (2017). Chapter 8. Chitosan applications for the food industry. In S. Ahmed & S. Ikram (Eds.), *Chitosan: Derivatives, composites and applications* (pp. 183–232). WILEY-Scrivener Publisher.. EE.UU. ISBN: 978-1-119-36350-7. https://doi.org/10.1002/9781119364849.ch8.

Gutiérrez, T. J. (2018). Biodegradability and compostability of food nanopackaging materials. In G. Cirillo, M. A. Kozlowski, & U. G. Spizzirri (Eds.), *Composite materials for food packaging* (pp. 269–296). WILEY-Scrivener Publisher.. EE.UU. ISBN: 978-1-119-16020-5. https://doi.org/10.1002/9781119160243.ch9.

Gutiérrez, T. J., & Álvarez, K. (2017). Transport phenomena in biodegradable and edible films. In M. A. Masuelli (Ed.), *Biopackaging* (pp. 58–88). Miami,. EE.UU. ISBN: 978-1-4987-4968-8: Editorial CRC Press Taylor & Francis Group.

Hassan, B., Chatha, S. A. S., Hussain, A. I., Zia, K. M., & Akhtar, N. (2018). Recent advances on polysaccharides, lipids and protein based edible films and coatings: A review. *International Journal of Biological Macromolecules, 109*, 1095–1107. https://doi.org/10.1016/j.ijbiomac.2017.11.097.

Hong, K., Xie, J., Zhang, L., Sun, D., & Gong, D. (2012). Effects of chitosan coating on postharvest life and quality of guava (*Psidium guajava* L.) fruit during cold storage. *Scientia Horticulturae, 144*, 172–178. https://doi.org/10.1016/j.scienta.2012.07.002.

Hu, Y. W., & Zou, L. D. (1998). Effect of coating chitosan on storage of apple. *Commun Plant Physiol, 34*, 17–19.

Hu, H., Zhou, H., & Li, P. (2019). Lacquer wax coating improves the sensory and quality attributes of kiwifruit during ambient storage. *Scientia Horticulturae, 244*, 31–41. https://doi.org/10.1016/j.scienta.2018.09.026.

Hua, C., Li, Y., Wang, X., Su, M., Shi, W., Zhang, D., & Liu, Y. (2019). The effect of low and high molecular weight chitosan on the control of gray mold (*Botrytis cinerea*) on kiwifruit and host response. *Scientia Horticulturae, 246*, 700–709. https://doi.org/10.1016/j.scienta.2018.11.038.

Iriti, M., & Faoro, F. (2009). Chitosan as a MAMP, searching for a PRR. *Plant Signaling & Behavior, 4*(1), 66–68. https://doi.org/10.1016/j.plaphy.2008.08.002.

Jiang, X., Lin, H., Shi, J., Neethirajan, S., Lin, Y., Chen, Y., Wang, H., & Lin, Y. (2018). Effects of a novel chitosan formulation treatment on quality attributes and storage behavior of harvested litchi fruit. *Food Chemistry, 252*, 134–141. https://doi.org/10.1016/j.foodchem.2018.01.095.

Jongsri, P., Wangsomboondee, T., Rojsitthisak, P., & Seraypheap, K. (2016). Effect of molecular weights of chitosan coating on postharvest quality and physicochemical characteristics of mango fruit. *LWT – Food Science and Technology, 73*, 28–36. https://doi.org/10.1016/j.lwt.2016.05.038.

Jongsri, P., Rojsitthisak, P., Wangsomboondee, T., & Seraypheap, K. (2017). Influence of chitosan coating combined with spermidine on anthracnose disease and qualities of 'Nam Dok Mai' mango after harvest. *Scienta Horticulturae, 224*, 180–187. https://doi.org/10.1016/j.scienta.2017.06.011.

Jung, J., Deng, Z., Simonsen, J., Bastías, R. M., & Zhao, Y. (2016). Development and preliminary field validation of water-resistant cellulose nanofiber based coatings with high surface adhesion and elasticity for reducing cherry rain-cracking. *Scientia Horticulturae, 200*, 161–169. https://doi.org/10.1016/j.scienta.2016.01.016.

Kaya, M., Česoniene, L., Daubaras, R., Leskauskait, D., & Zabulion, D. (2016). Chitosan coating of red kiwifruit (*Actinidia melanandra*) for extending of the shelf life. *International Journal of Biological Macromolecules, 85*, 355–360. https://doi.org/10.1016/j.ijbiomac.2016.01.012.

Kerch, G. (2015). Chitosan films and coatings prevent losses of fresh fruit nutritional quality: A review. *Trends in Food Science and Technology, 46*, 159–166. https://doi.org/10.1016/j.tifs.2015.10.010.

Khalifa, I., Barakat, H., El-Mansy, H. A., & Soliman, S. A. (2016). Enhancing the keeping quality of fresh strawberry using chitosan- incorporated olive processing wastes. *Food Bioscience, 13*, 69–75. https://doi.org/10.1016/j.fbio.2015.12.008.

Khalifa, I., Barakat, H., El-Mansy, H. A., & Soliman, S. A. (2017). Preserving apple (*Malus domestica* var. Anna) fruit bioactive substances using olive wastes extract-chitosan film coating. *Information Processing in Agriculture, 4*(1), 90–99. https://doi.org/10.1016/j.inpa.2016.11.001.

Kulikov, S. N., Lisovskaya, S. A., Zelenikhin, P. V., Bezrodnykh, E. A., Shakirova, D. R., Blagodatskikh, I. V., & Tikhonov, V. E. (2014). Antifungal activity of oligochitosans (short chain chitosans) against some *Candida* species and clinical isolates of *Candida albicans*: Molecular weight–activity relationship. *European Journal of Medicinal Chemistry, 74*, 169–178. https://doi.org/10.1016/j.ejmech.2013.12.017.

Kumar, P., Sethia, S., Sharmah, R. R., Srivastav, M., & Varghese, E. (2017). Effect of chitosan coating on postharvest life and quality of plum during storage at low temperature. *Scientia Horticulturae, 226*, 104–109. https://doi.org/10.1016/j.scienta.2017.08.037.

Lee, C. G., Koo, J. C., & Park, J. K. (2016). Antifungal effect of chitosan as Ca^{2+} channel blocker. *European Journal of Plant Pathology, 32*(3), 242–250. https://doi.org/10.5423/ppj.oa.08.2015.0162.

Li, H., & Yu, T. (2000). Effect of chitosan on incidence of brown rot, quality and physiological attributes of postharvest peach fruit. *Journal of the Science of Food and Agriculture, 81*(2), 269–274. https://doi.org/10.1002/1097-0010(20010115)81:2<269::aid-jsfa806>3.0.co;2-f.

Li, H., Wang, Y., Liu, F., Yang, Y., Wu, Z., Cai, H., Zhang, Q., Wang, Y., & Li, P. (2015). Effects of chitosan on control of postharvest blue mold decay of apple fruit and the possible mechanisms involved. *Scientia Horticulturae, 186*, 77–83. https://doi.org/10.1016/j.scienta.2015.02.014.

Liu, K., Yuan, C., Chen, Y., Li, H., & Liu, J. (2014). Combined effects of ascorbic acid and chitosan on the quality maintenance and shelf life of plums. *Scientia Horticulturae, 176*, 45–53. https://doi.org/10.1016/j.scienta.2014.06.027.

Liu, K., Liu, J., Li, H., Yuan, C., Zhong, J., & Chen, Y. (2016). Influence of postharvest citric acid and chitosan coating treatment on ripening attributes and expression of cell wall related genes in cherimoya (*Annona cherimola* Mill.) fruit. *Scientia Horticulturae, 198*, 1–6. https://doi.org/10.1016/j.scienta.2015.11.008.

Lodhi, G., Kim, Y. S., Hwang, J. W., Kim, S. K., Jeon, Y. J., Je, J. Y., Ahn, C. B., Moon, S. H., Jeon, B. T., & Park, P. J. (2014). Chitooligosaccharide and its derivatives: Preparation and biological applications. *Bio Med Research International*, 654913. https://doi.org/10.1155/2014/654913.

Mannozzi, C., Tylewicz, U., Chinnici, F., Siroli, L., Rocculi, P., Dalla Rosa, M., & Romani, S. (2018). Effects of chitosan based coatings enriched with procyanidin by-product on quality of fresh blueberries during storage. *Food Chemistry, 251*, 18–24. https://doi.org/10.1016/j.foodchem.2018.01.015.

Martiñon, M. E., Moreira, R. G., Castell-Perez, M. E., & Gomes, C. (2014). Development of a multilayered antimicrobial edible coating for shelf-life extension of fresh-cut cantaloupe (*Cucumis melo* L.) stored at 4 °C. *LWT – Food Science and Technology, 56*(2), 341–350. https://doi.org/10.1016/j.lwt.2013.11.043.

Mayen, C. M., Marshall, M. I., & Lusk, J. (2007). Fresh cut melon e the money is in the juice. *Journal of Agricultural and Applied Economics, 39*(3), 597–609. https://doi.org/10.1017/S1074070800023294.

McClements, D. J., Decker, E. A., & Park, Y. (2009). Controlling lipid bioavailability through physicochemical and structural approaches. *Critical Reviews in Food Science and Nutrition, 49*(1), 48–67. https://doi.org/10.1080/10408390701764245.

Medeiros, B. G. D. S., Pinheiro, A. C., Carneiro-da-Cunha, M. G., & Vicente, A. A. (2012). Development and characterization of a nanomultilayer coating of pectin and chitosan—Evaluation of its gas barrier properties and application on 'Tommy Atkins' mangoes. *Journal of Food Engineering, 110*(3), 457–464. https://doi.org/10.1016/j.jfoodeng.2011.12.021.

Meng, X. H., Li, B. Q., Liu, J., & Tian, S. P. (2008). Physiological responses and quality attributes of table grape fruit to chitosan preharvest spray and postharvest coating during storage. *Food Chemistry, 106*(2), 501–508. https://doi.org/10.1016/j.foodchem.2007.06.012.

Merino, D., Gutiérrez, T. J., Mansilla, A. Y., Casalongué, C., & Alvarez, V. A. (2018a). Critical evaluation of starch-based antibacterial nanocomposites as agricultural mulch films: Study on their interactions with water and light. *ACS Sustainable Chemistry & Engineering, 6*(11), 15662–15672. https://doi.org/10.1021/acssuschemeng.8b04162.

Merino, D., Mansilla, A. Y., Gutiérrez, T. J., Casalongué, C. A., & Alvarez, V. A. (2018b). Chitosan coated-phosphorylated starch films: Water interaction, transparency and antibacterial properties. *Reactive and Functional Polymers, 131*, 445–453. https://doi.org/10.1016/j.reactfunctpolym.2018.08.012.

Merino, D., Gutiérrez, T. J., & Alvarez, V. A. (2019a). Potential agricultural mulch films based on native and phosphorylated corn starch with and without surface functionalization with chitosan. *Journal Polymers and the Environment, 27*(1), 97–105. https://doi.org/10.1007/s10924-018-1325-1.

Merino, D., Gutiérrez, T. J., & Alvarez, V. A. (2019b). Structural and thermal properties of agricultural mulch films based on native and oxidized corn starch nanocomposites. *Starch-Stärke, 71*(7–8), 1800341.https://doi.org/10.1002/star.201800341.

Munhuweyi, K., Oluwafemi, J. C., Lennox, C. L., van Reenen, A. J., & Opara, L. U. (2017). *In vitro* and *in vivo* antifungal activity of chitosan-essential oils against pomegranate fruit pathogens. *Postharvest Bioloy & Technology, 129*, 9–22. https://doi.org/10.1016/j.postharvbio.2017.03.002.

Nair, M. S., Saxena, A., & Kaur, C. (2018). Effect of chitosan and alginate based coatings enriched with pomegranate peel extract to extend the postharvest quality of guava (*Psidium guajava* L.). *Food Chemistry, 240*, 245–252. https://doi.org/10.1016/j.foodchem.2017.07.122.

Nazzaro, F. I. D., Fratianni, F., Coppola, R., & De Feo, V. (2017). Essential oils and antifungal activity. *Pharmaceuticals, 10*(4), 86. https://doi.org/10.3390/ph10040086.

Ncama, K., Magwaza, L. S., Mditshwa, A., & Tesfay, S. Z. (2018). Plant-based edible coatings for managing postharvest quality of fresh horticultural produce: A review. *Food Packaging and Shelf Life, 16*, 157–167. https://doi.org/10.1016/j.fpsl.2018.03.011.

Ozdemir, K. S., & Gökmen, V. (2017). Extending the shelf-life of pomegranate arils with chitosan-ascorbic acid coating. *LWT – Food Science and Technology, 76*. (Part A, 172–180. https://doi.org/10.1016/j.lwt.2016.10.057.

Paniagua, A. C., East, A. R., Hindmarsh, J. P., & Heyes, J. A. (2013). Moisture loss is the major cause of firmness change during postharvest storage of blueberry. *Postharvest Biology and Technology, 79*, 13–19. https://doi.org/10.1016/j.postharvbio.2012.12.016.

Pasquariello, M. S., Di Patre, D., Mastrobuoni, F., Zampella, L., Scortichini, M., & Petriccione, M. (2015). Influence of postharvest chitosan treatment on enzymatic browning and antioxidant enzyme activity in sweet cherry fruit. *Postharvest Biology and Technology, 109*, 45–56. https://doi.org/10.1016/j.postharvbio.2015.06.007.

Pérez-Gago, M. B., & Rhim, J. W. (2014). Edible coating and film materials: Lipid bilayers and lipid emulsions. In J. H. Han (Ed.), *Innovations in food packaging* (pp. 325–350). Amsterdam: Elsevier, Inc. https://doi.org/10.1016/B978-0-12-394601-0.00013-8.

Petriccione, M., De Sanctis, F., Pasquariello, M. S., Mastrobuoni, F., Rega, P., Scortichini, M., & Mencarelli, F. (2015a). The effect of chitosan coating on the quality and nutraceutical traits of sweet cherry during postharvest life. *Food Bioprocess Technology, 8*(2), 394–408. https://doi.org/10.1007/s11947-014-1411-x.

Petriccione, M., Pasquariello, M. S., Mastrobuoni, F., Zampella, L., Di Patre, D., & Scortichini, M. (2015b). Influence of a chitosan coating on the quality and nutraceutical traits of loquat fruit during postharvest life. *Scientia Horticulturae, 197*, 287–296. https://doi.org/10.1016/j.scienta.2015.09.051.

Petriccione, M., Mastrobuoni, F., Pasquariello, M. S., Zampella, L., Nobis, E., Capriolo, G., & Scortichini, M. (2015c). Effect of chitosan coating on the postharvest quality and antioxidant enzyme system response of strawberry fruit during cold storage. *Food, 4*(4), 501–523. https://doi.org/10.3390/foods4040501.

Petriccione, M., Pagano, L., Forniti, T., Zampella, L., Mastrobuoni, F., Scortichini, M., & Mencarelli, F. (2018). Postharvest treatment with chitosan affects the antioxidant metabolism and quality of wine grape during partial dehydration. *Postharvest Biology and Technology, 137*, 38–45. https://doi.org/10.1016/j.postharvbio.2017.11.010.

Porat, R., Lichter, A., Leon, A. T., Harker, R., & Buzby, J. (2018). Postharvest losses of fruit and vegetables during retail and in consumers' homes: Quantifications, causes, and means of prevention. *Postharvest Biology and Technology, 139*, 135–149. https://doi.org/10.1016/j.postharvbio.2017.11.019.

Poverenov, E., Danino, S., Horev, B., Granit, R., Vinokur, Y., & Rodov, V. (2014). Layer-by-layer electrostatic deposition of edible coating on fresh cut melon model: Anticipated and unexpected effects of alginate-chitosan combination. *Food and Bioprocess Technology, 7*(5), 1424–1432. https://doi.org/10.1007/s11947-013-1134-4.

Racchi, M. L. (2013). Antioxidant defenses in plants with attention to prunus and citrus spp. *Antioxidants, 2*(4), 340–369. https://doi.org/10.3390/antiox2040340.

Rinaudo, M., Pavlov, G., & Desbrieres, J. (1999). Influence of acetic acid concentration on the solubilization of chitosan. *Polymer, 40*(25), 7029–7032. https://doi.org/10.1016/S0032-3861(99)00056-7.

Romanazzi, G., Feliziani, E., Bautista-Baòos, S., & Sivakumar, D. (2017). Shelf life extension of fresh fruit and vegetables by chitosan treatment. *Critical Reviews in Food Science and Nutrition, 57*(3), 579–601. https://doi.org/10.1080/10408398.2014.900474.

Romanazzi, G., Feliziani, E., & Sivakumar, D. (2018). Chitosan, a biopolymer with triple action on postharvest decay of fruit and vegetables: Eliciting, antimicrobial and film-forming properties. *Frontiers in Microbiology, 9*, 2745. https://doi.org/10.3389/fmicb.2018.02745.

Saberi, B., & Golding, J. B. (2018). Postharvest application of biopolymer-based edible coatings to improve the quality of fresh horticultural produce. In T. J. Gutiérrez (Ed.), *Polymers for food applications* (pp. 211–250). Cham: Springer. https://doi.org/10.1007/978-3-319-94625-2_9.

Sánchez-González, L., Pastor, C., Vargas, M., Chiralt, A., & González-Martínez, C. (2011). Effect of hydroxypropyl methylcellulose and chitosan coatings with and without bergamot essential oil on quality and safety of cold-stored grapes. *Postharvest Biology and Technology, 60*(1), 57–63. https://doi.org/10.1016/j.postharvbio.2010.11.004.

Septembre-Malaterre, A., Remize, F., & Poucheret, P. (2018). Fruits and vegetables, as a source of nutritional compounds and phytochemicals: Changes in bioactive compounds during lactic fermentation. *Food Research International, 104*, 86–99. https://doi.org/10.1016/j.foodres.2017.09.031.

Sharma, S., Barman, K., & Siddiqui, M. W. (2016). Chitosan: Properties and roles in post harvest quality preservation of horticultural crops. In M. W. Siddiqui (Ed.), *Eco-Friendly technology for postharvest produce quality* (pp. 269–296). London: Elsevier, Inc. https://doi.org/10.1016/B978-0-12-804313-4.00009-8.

Shen, Y., & Yang, H. (2017). Effect of preharvest chitosan-g-salicylic acid treatment on postharvest table grape quality, shelf life, and resistance to Botrytis cinerea-induced spoilage. *Scientia Horticulturae, 224*, 367–373. https://doi.org/10.1016/j.scienta.2017.06.046.

Sidhu, P. K., & Nehra, K. (2017). Bacteriocin-nanoconjugates as emerging compounds for enhancing antimicrobial activity of bacteriocins. *Journal of King Saud University – Science (in press)*. https://doi.org/10.1016/j.jksus.2017.12.007.

Silva, G. M. C., Silva, W. B., Medeiros, D. B., Salvador, A. R., Cordeiro, M. H. M., da Silva, N. M., Santana, D. B., & Mizobutsi, G. P. (2017). The chitosan affects severely the carbon metabolism in mango (*Mangifera indica* L. cv. Palmer) fruit during storage. *Food Chemistry, 237*, 372–378. https://doi.org/10.1016/j.foodchem.2017.05.123.

Silva, W. B., Silva, G. M. C., Santana, D. B., Salvador, A. R., Medeiros, D. B., Belghith, I., da Silva, N. M., Cordeiro, M. H. M., & Mizobutsi, G. P. (2018). Chitosan delays ripening and ROS production in guava (*Psidium guajava* L.) fruit. *Food Chemistry, 242*, 232–238. https://doi.org/10.1016/j.foodchem.2017.09.052.

Song, H., Yuan, W., Jin, P., Wang, W., Wang, X., Yang, L., & Zhang, Y. (2016). Effects of chitosan/nano-silica on postharvest quality and antioxidant capacity of loquat fruit during cold storage. *Postharvest Biology and Technology, 119*, 41–48. https://doi.org/10.1016/j.postharvbio.2016.04.015.

Souza, S. J. F., Alves, A. I., Vieira, É. N. R., Vieira, J. A. G., Ramos, A. M., & Telis-Romero, J. (2015). Study of thermodynamic water properties and moisture sorption hysteresis of mango skin. *Food Science and Technology, 35*(1), 157–166. https://doi.org/10.1590/1678-457x.6557.

Sun, Z., Hao, J., Yang, H., & Chen, H. (2018). Effect of chitosan coatings enriched with lauroyl arginate ethyl and montmorillonite on microbial growth and quality maintenance of minimally processed table grapes (*Vitis vinifera* L. Kyoho) during cold storage. *Food and Bioprocess Technology, 11*(10), 1853–1862. https://doi.org/10.1007/s11947-018-2146-x.

Thakur, V. K., & Thakur, M. K. (2016). *Handbook of sustainable polymers: Processing and applications*. Singapore: Pan Stanford Publishing.

Treviño-Garza, M. Z., García, S., del Socorro Flores-González, M., & Arévalo-Niño, K. (2015). Edible active coatings based on pectin, pullulan, and chitosan increase quality and shelf life of strawberries (*Fragaria* × *ananassa*). *Journal of Food Science, 80*(8), M1823–M1830. https://doi.org/10.1111/1750-3841.12938.

Treviño-Garza, M. Z., García, S., Norma Heredia, N., Alanís-Guzmánc, M. A., & Arévalo-Niñoa, K. (2017). Layer-by-layer edible coatings based on mucilages, pullulan and chitosan and its effect on quality and preservation of fresh-cut pineapple (*Ananas comosus*). *Postharvest Biology and Technology, 128*, 63–75. https://doi.org/10.1016/j.postharvbio.2017.01.007.

Valero, D., & Serrano, M. (2010). *Postharvest biology and technology for preserving fruit quality*. Boca Raton: CRC-Taylor & Francis.

Valero, D., Mula-Diaz, H. M., & Zapata, P. J. (2013). Effect of alginate edible coating on preserving fruit quality in four plum cultivars during postharvest storage. *Postharvest Biology and Technology, 77*, 1–6. https://doi.org/10.1016/j.postharvbio.2012.10.011.

Van Dongen, J. T., Gupta, K. J., Ramírez-Aguilar, S. J., Araújo, W. L., Nunes-Nesi, A., & Fernie, A. R. (2011). Regulation of respiration in plants: A role for alternative metabolic pathways. *Journal of Plant Physiology, 168*(12), 1434–1443. https://doi.org/10.1016/j.jplph.2010.11.004.

Vargas, M., Pastor, C., Chiralt, A., McClements, D. J., & Gonzalez-Martinez, C. (2008). Recent advances in edible coatings for fresh and minimally processed fruits. *Critical Reviews in Food Science and Nutrition, 48*(6), 496–511. https://doi.org/10.1080/10408390701537344.

Velickova, E., Winkelhausen, E., Kuzmanova, S., & Alves, B. D. (2013). Impact of chitosan beeswax edible coatings on the quality of fresh strawberries (*Fragaria ananassa* cv Camarosa) under commercial storage conditions. *LWT – Food Science and Technology, 52*(2), 80–92. https://doi.org/10.1016/j.lwt.2013.02.004.

Waewthongrak, W., Pisuchpen, S., & Leelasuphakul, W. (2015). Effect of *Bacillus subtilis* and chitosan applications on green mould (*Penicilium digitatum* Sacc.) decay in citrus fruit. *Postharvest Biology and Technology, 99*, 44–49. https://doi.org/10.1016/j.postharvbio.2014.07.016.

Wang, S. Y., & Gao, H. (2013). Effect of chitosan-based edible coating on antioxidants, antioxidant enzyme system and postharvest fruit quality of strawberries (*Fragaria×ananassa* Duch.). *Food Science and Technology, 52*(2), 71–79. https://doi.org/10.1016/j.lwt.2012.05.003.

Wang, L., Wu, H., Qin, G., & Meng, X. (2014). Chitosan disrupts *Penicillium expansum* and controls postharvest blue mold of jujube fruit. *Food Control, 41*, 56–62. https://doi.org/10.1016/j.foodcont.2013.12.028.

Xin, Y., Chen, F. S., Lai, S. J., & Yang, H. S. (2017). Influence of chitosan-based coatings on the physicochemical properties and pectin nanostructure of Chinese cherry. *Postharvest Biology and Technology, 133*, 64–71. https://doi.org/10.1016/j.postharvbio.2017.06.010.

Xu, D., Qin, H., & Ren, D. (2018). Prolonged preservation of tangerine fruits using chitosan/montmorillonite composite coating. *Postharvest Biology and Technology, 143*, 50–57. https://doi.org/10.1016/j.postharvbio.2018.04.013.

Younes, I., Sellimi, S., Rinaudo, M., Jellouli, K., & Nasri, M. (2014). Influence of acetylation degree and molecular weight of homogeneous chitosans on antibacterial and antifungal activities. *International Journal of Food Microbiology, 185*, 57–63. https://doi.org/10.1016/j.ijfoodmicro.2014.04.029.

Yuan, G., Chen, X., & Li, D. (2016). Chitosan films and coatings containing essential oils: The antioxidant and antimicrobial activity, and application in food systems. *Food Research International, 89*.(Part 1, 117–128. https://doi.org/10.1016/j.foodres.2016.10.004.

Zhang, S., Yu, Y., Xiao, C., Wang, X., & Lei, Y. (2014). Effect of ultraviolet irradiation combined with chitosan coating on preservation of jujube under ambient temperature. *LWT – Food Science and Technology, 57*(2), 749–754. https://doi.org/10.1016/j.lwt.2014.02.046.

Zhang, L., Chen, F., Lai, S., Wang, H., & Yang, H. (2018). Impact of soybean protein isolate-chitosan edible coating on the softening of apricot fruit during storage. *LWT – Food Science and Technology, 96*, 604–611. https://doi.org/10.1016/j.lwt.2018.06.011.

Zheng, X. Q., Gong, M. B., & Wei, S. W. (1996). Effect of storage and preservation of chitosan on pear and apples. *Chinese Fruit, 2*, 16–19.

Zhidong, S., Jiasi, H., Huqing, Y., & Huiyun, C. (2018). Effect of chitosan coatings enriched with lauroylarginate ethyl and montmorillonite on microbial growth and quality maintenance of minimally processed table grapes (*Vitis vinifera* L. Kyoho) during cold storage. *Food and Bioprocess Technology, 11*(10), 1853–1862. https://doi.org/10.1007/s11947-018-2146-x.

Zhou, Y., Zhang, L., & Zeng, K. (2016). Efficacy of *Pichia membranaefaciens* combined with chitosan against *Colletotrichum gloeosporioides* in citrus fruits and possible modes of action. *Biological Control, 96*, 39–47. https://doi.org/10.1016/j.biocontrol.2016.02.001.

Zhu, X., Wang, Q. M., Cao, J. K., & Jiang, W. B. (2008). Effects of chitosan coating on postharvest quality of mango (*Mangifera indica* L. cv. Tainong) fruits. *Journal of Food Processing and Preservation, 32*(5), 770–784. https://doi.org/10.1111/j.1745-4549.2008.00213.x.

Chapter 24
Shellac-Based Coating Polymer for Agricultural Applications

Manee Luangtana-anan and Sontaya Limmatvapirat

Abstract Due to the interest in health and concern for the environment, this review work was concentrated on biodegradable and edible materials for the agricultural sector. Edible polymers are recognized as safe for human consumption. In agricultural industry, a film coating is required to prolong the shelf life of products, since moisture plays a critical role in their stabilities. Materials used in film coatings include proteins, polysaccharides and lipids. Shellac is a lipid polymer of interest due to an outstanding moisture barrier and an abundant natural resource of Thailand. However, the shellac film has some drawbacks due to the polymerization and strength of film. This book chapter will focus on the current work to solve the drawbacks of shellac. The application of shellac for film coating in agricultural products, as well as the use of surface free energy for the prediction of coating efficiency was also reviewed in this chapter.

Keywords Coating efficiency · Composite film · Polymerization · Surface free energy · Shellac

24.1 Introduction

A coating in the agricultural industry plays an important role in the supply chain and is an important part of the final process. Coatings are one of the emerging strategies for food-quality optimization. Fruits and vegetables are very perishable during the food supply chain such as agricultural production, postharvest management, processing, distribution and consumption, due to microbes, insects, respiration

M. Luangtana-anan (✉) · S. Limmatvapirat
Department of Pharmaceutical Technology, Faculty of Pharmacy, Silpakorn University, Nakhon Pathom, Thailand

Pharmaceutical Biopolymer Group (PBiG), Faculty of Pharmacy, Silpakorn University, Nakhon Pathom, Thailand
e-mail: luangtanaanan_m@su.ac.th

T. J. Gutiérrez (ed.), *Polymers for Agri-Food Applications*,
https://doi.org/10.1007/978-3-030-19416-1_24

and transpiration (Brasil and Siddiqui 2018; Gogo et al. 2017; Porat et al. 2018). The goals of coatings are thus to protect agricultural products from external factors that include oxygen and carbon dioxide (CO_2) contents, stress factor, ethylene ratios, temperature and microbiological activities and protect from internal factors such as the species, cultivar and its growth stage (Rahman et al. 2016; Saberi and Golding 2018). The quality factors of fresh products are important to ensure their commercialization. Therefore, the main purpose of a polymer when used as a coating is to increase the natural barrier of fruits and vegetables, and extend shelf life of fresh products. In addition to serving as a barrier, the coating polymer can be a carrier of food additives (e.g. antioxidants and antimicrobials), aroma compounds or nutritional substances (Fajardo et al. 2010; Martins et al. 2010; Perdones et al. 2012). Due to the interest in the health and the environment, biodegradable and edible polymers are selected as polymers for use as a coating. They are recognized as safe for human consumption and environmentally friendly. Biodegradable and edible polymers used in coatings include proteins, polysaccharides, and lipids (Gutiérrez 2018a). Shellac is a lipid polymer of interest due to an outstanding moisture barrier and an abundant natural resource of Thailand. However, shellac films have some drawbacks due to polymerization and strength of films (Limmatvapirat et al. 2004; Luangtana-anan et al. 2007). The aim of this book chapter was to describe the properties of shellac and highlight current approaches available to solve the weakness of shellac, and application of it as a coating polymer to extend the shelf life of agricultural products. The coating efficiency based on the knowledge of surface free energy was also described as a method to modify and improve the coating solution for the quality of agricultural products.

24.2 Shellac Properties

Shellac is a natural polymer prepared by uncomplicated processing. It is obtained from the secretion of the lac *Kerriar Lacca* insect, which grows on specific types of trees in China, India and Thailand. Shellac is composed of hard resin and soft resin of polyester and single ester with polyhydroxypolybasic acids such as aleuritic acid, jalaric acid, and laccijalaric acid. It has been widely used in several fields due to its high safety profile. It displays several properties as follows:

24.2.1 Water Vapor Protection

Shellac provides a good film-forming property and has a good water-protection capability which is highlighted by water vapor permeability coefficient (WVPC). Hagenmaier and Shaw (1991) investigated and reported that several shellac films had the lowest WVPC compared to cellulose, cellophane acetate and nylon-6 films. Shellac can be dissolved in various solvents such as alcohols and alkaline aqueous

Table 24.1 Water vapor permeability of shellac films fabricated by using different solvents and bases (Hagenmaier and Shaw 1991)

	Water vapor permeability coefficient (g/m^2.day.mmHg)			
%RH	Isopropanol	Ethanol	Mor*pholine*	NaOH
100	2.1±0.1	5.1±0.1		
84	1.8±0.4	3.8±0.2	11.0±2.1	
75	1.4±0.1		8.8±1.1	147±62
51	1.2±0.2		2.8±0.3	3.8±0.6
32	1.1±0.2		1.9±0.1	
11	0.86±0.04			

solution. The WVPC study of shellac films prepared by the use of isopropanol and ethanol as solvents exhibited the apparent higher water vapor protection than the films prepared from aqueous solutions, either morpholine or sodium hydroxide, depending on the relative humidity as shown in Table 24.1. The low water protection of films prepared from morpholine or sodium hydroxide is a result of high water solubility corresponding with the previous work of Limmatvapirat et al. (2007). The results from their studies indicated that shellac films made from ammonium hydroxide (AMN) and 2-amino-2-methyl-1-propanol (AMP) illustrated an increase in the water absorption and water solubility, particularly with the addition of AMP. They found that the water solubility of shellac films could be modified by mixing both salt forms at the proper ratio.

The water vapor permeability of shellac films is associated with wax content, the significant component of shellac, and other additive agents used to improve this property of films. The study of Luangtana-anan et al. (2007) reported that WVPC depended on types of plasticizers. The addition of polyethylene glycol (PEG), a water-soluble plasticizer, contributed to a remarkable rise in the WVPC, while the effects of other plasticizers including diethyl phthalate (DEP) and triacetin (TA), which were low in water solubility, there were no significant changes in WVPC values compared to plasticizer-free shellac films. In addition, the WVPC of all fabricated films tended to decrease after 3 months storage indicated the higher waterproofing of all shellacs as shown in Fig. 24.1 (Luangtana-anan et al. 2007).

24.2.2 *pH-Dependent Solubility*

The chemical structure of shellac polymer is composed of carboxylic groups. It can thus not be dissolved at low pH, but it would be initially dissolved at pH of around 7. As a result, shellac has been used in enteric coating to protect the disintegration of drug in the stomach (pH 1–5), and initiate the drug release in the small intestine (pH 5–8). The thickness of the coating film can be adjusted in order to allow no more than 10% of the drug to be released within 2 h in 0.1 N hydrochloric acid. Figure 24.2 illustrates the controlled release of shellac-coated paracetamol tablets both

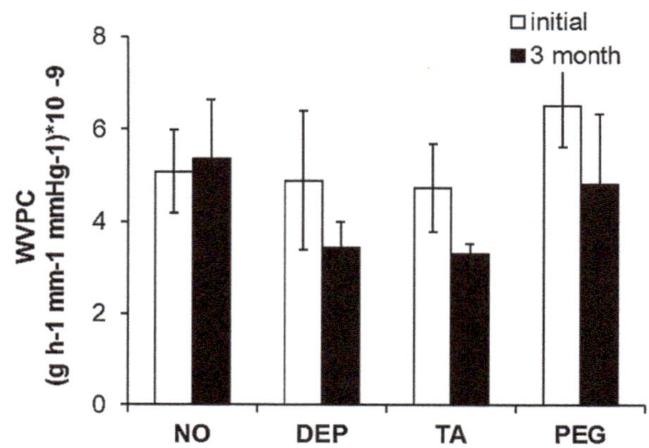

Fig. 24.1 Effect of storage time on WVPC of unplasticized (NO) and plasticized shellac films (DEP, TA, PEG) in acid form after storage at 40 °C, 75% relative humidity (RH) for 3 months (Luangtana-anan et al. 2007)

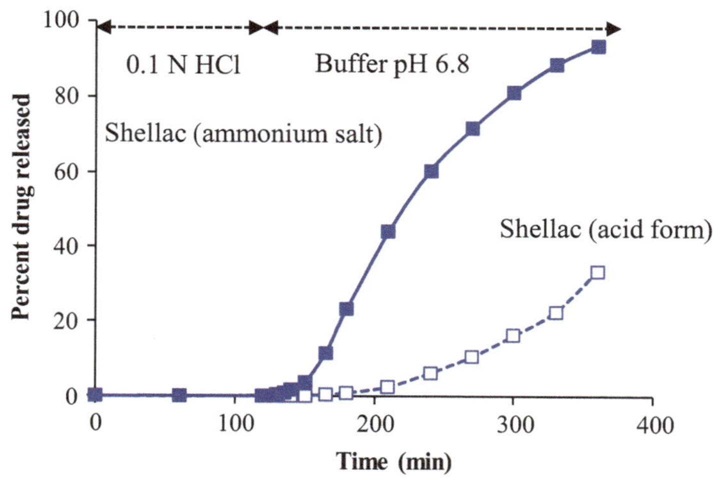

Fig. 24.2 Release of paracetamol from coated tablets using shellac in its acid and salt forms (Limmatvapirat et al. 2004)

in acid form (native form) and ammonium salt (AMN) form. All the shellac-coated paracetamol tablets displayed the delayed release profile in 0.1 N hydrochloric acid, simulating the stomach condition. Consequently, when the tablets were allocated in the pH 6.8 buffer solution, which simulated the small intestine environment, the release of paracetamol from AMN-form shellac-coated tablets tended to be superior than those from acid-form shellac-coated tablets. The results demonstrated that shellac in both forms can be used for enteric coating since the shellac can protect the drug in the acid environment for more than 2 h, fulfilling the requirement of USP 39. Nevertheless, the drug release was relatively low, due to the high pKa of shellac (6.9–7.5). Regarding the higher pH of the intestine, shellac has been used in the enteric coating designed for drug release in large intestine (Farag and Leopold 2011).

Shellac can also be applied in floating drug delivery system by remaining the matrix property in the acidic stomach; the drug would thus be released at the desired time (Chongcherdsak et al. 2013). Shellac has also been used as a medium protective coating, leading to the prolonged gastric retention time in the stomach. Regarding this pH-dependent property, shellac could also be applied in the agricultural field as a coating material. The seeds of the plant could be coated with shellac in order to provide the moisture barrier and extend the storage time. Before planting, the coated layer is easily removed by soaking it in alkali medium and the seeds are ready to be used.

24.2.3 Gloss

The gloss property, depending on the refractive index and the smooth surface condition, can affect the appearance of products. Based on its high refractive index (1.521–1.527) and smooth surface property, shellac film can be used for the coating of fruits, vegetables, medications, cosmetics and woods to enhance its appearance and attractiveness (Fig. 24.3). Khorram et al. (2017) found that oranges coated with various materials such as shellac, gelatin, Persian gum and commercial wax were shinier and glossier than the uncoated control group. The gloss of fruits was proportional to the amount of coating. The results also indicated that shellac-coated oranges had a polished look like those coated with commercial wax and Persian gum, while the gloss of gelatin-coated oranges seemed to be the lowest. However, the flavor of shellac-coated oranges probably changed as the amount of shellac coating increased. After a storage period, the cracking of shellac-coated orange peel did not clearly occur compared with gelatin and Persian gum coating. Shellac was thus

Fig. 24.3 Enhanced gloss of dragon fruits after coating with shellac

suggested to be used as a coating material due to its fast drying ability, not sticky and without odor (Khorram et al. 2017).

Generally, UV rays can affect the surface properties of coated products including water vapor permeability, mechanical characteristic and gloss property. Ghosh et al. (2015) evaluated the impact of UV rays on the gloss property of products coated with shellac and polyurethane. They found that the gloss tended to reduce slightly after the first 40 h of exposure to UV rays and remained constant for the period of exposure to UV rays for 100 h. This could imply that shellac and polyurethane can resist exposure to UV rays.

24.2.4 Thermal Conductivity and Surface Resistivity

Table 24.2 shows the thermal conductivity of several materials with differing thermal conductivity. Shellac's thermal conductivity is 0.25 W/(m × K), which is particularly like other polymers and plastics, and also provides a low thermal expansion coefficient. It can thus be applied as a thermal insulation. For example, aerogel containing shellac is classified as an efficient thermal protection. Surface resistivity of conductive materials is normally less than 10^5 ohms (or ohm/square), whereas surface resistivity of electrical insulators is commonly more than 10^{12} ohms. Shellac and Teflon have a surface resistivity greater than 10^{13} and 10^{16} ohms, while copper exhibits a very low surface resistivity (less than 1 ohm). Therefore, shellac can also be applied in electrical industries such as the production of wires and light bulbs due to its thermal and electrical insulation.

Table 24.2 Thermal conductivity of materials (The Engineering Toolbox; United States Department of Agriculture)

Material	Thermal conductivity W/(m×K)
Vacuum	0
Air	0.024
Fiber glass	0.04
Brick, insulating	0.15
Shellac	0.24
Nylon 6	0.25
Rosin	0.32
Epoxy	0.35
Polyethylene	0.33-0.51
Asbestos-cement board	0.744
Asphalt	0.75
Glass	1.05
Stainless steel	16
Copper	401
Diamond	1000

24.2.5 Adhesion

Shellac coating has a strong adhesion force with several surface materials, and is scratch resistant. The adhesion ability is dependent on the type of substrates. The adhesion force between shellac and copper surface is around 3300 pound/inch^2, which suggests its good copper adhesive. Shellac has thus been extensively used as a coating material based on the characteristics of strong adhesion, excellent water vapor barrier, gloss and insect protection. The work of Hoang-Dao et al. (2009) showed that shellac could be served as a component of dental coating to prevent tooth sensitivity due to its adhesive capacity. Shellac can also be applied efficiently on various types of fruits (both waxy skin and non-waxy skin) with the aims of gloss enhancement and fruit surface protection. The successful examples of fruits coated with shellac have been banana, apple and citrus fruits (Alleyne and Hagenmaier 2000; McGuire and Hagenmaier 2001; Soradech et al. 2017).

24.2.6 Thermal Behavior

The glass transition temperature (T_g) of shellac, is temperature at which shellac is converted from a hard and brittle state to a rubbery state or in the reverse transition, is ~38 °C. The onset temperature of melting and melting peak are around 53 °C and 57 °C, respectively, which do not differ notably from seed lac and raw lac, indicating that the impurities in shellac compound cannot affect its thermal property. In addition, the additional melting peak of shellac (and also seed lac and raw lac), arising from the melting of wax, is reported at ~73 °C. Nevertheless, the wax melting peak is apparently not observed in the de-waxed and bleached shellac due to the removal of wax during preparation process (Table 24.3). Besides, the thermal property of shellac is associated with the quality of raw materials, which vary according to the time and storage conditions and chemical structure. The modification of chemical structure by partial hydrolysis (Limmatvapirat et al. 2004), salt (Luangtana-anan et al. 2007) and ester formations (Limmatvapirat et al. 2008) could affect the melting point. Limmatvapirat et al. (2004) reported the modification of chemical structure by partial hydrolysis has no effect on melting property. The melting point of hydrolyzed shellac was ~55 °C, which was not different from that of the

Table 24.3 Thermal behaviors of lac products (Goswami and Saha 2000)

Lac products	Glass transition temperature (°C)	Onset temperature (°C)	Melting peak (°C)
Stick lac	39.3	54.0	56, 74.0
Seed lac	39.2	52.4	55.4, 73.4
Shellac	38.4	52.5	57.4, 73.4
Dewaxed lac	34.4	50.3	52.9
Bleached lac	41.9	49.2	55.3
Kiri	54.9	59.3	57.9, 72.0

unmodified shellac. Unmodified shellac after storage under accelerated conditions (40 °C and 75% RH) was found initially melted at the higher temperature of 80 °C, whereas the melting of hydrolyzed shellac could not be detected, even though the temperature increased up to 100 °C. The thermal performance altered during storage was attributed to the degradation of shellac, while the hydrolyzed shellac exhibited a severe thermal change as a result of partial hydrolysis. Shortening the shellac chains due to hydrolysis is not stable, leading to the significant change in thermal behaviors. The processing time and the concentration of base (e.g. sodium carbonate) used in the fabrication of seed lac should thus be appropriately adjusted to control the change in thermal property. Shellac obtained from natural resources without chemical modification is usually extracted in acid form. The endothermic peak of shellac (acid form) appeared at around 55 °C. The modification in salt forms using bases, including AMN, AMP, 2-amino-2-methyl-1, 3-propanediol or basic compound contributes to the change in thermal property proved by differential scanning calorimetry (DSC) as demonstrated in Fig. 24.4 (Limmatvapirat et al. 2007). The broad endothermic peaks of shellac in salt forms were evidently observed at higher temperature than that of the acid form, indicating the higher melting point of the salt form. The thermal behavior was also studied by hot stage microscope. The results indicated that the occurrence of the melting peak depended on the type and ratio of salt-forming agents. The shellac prepared in salt forms using the AMP and AMN mixture in 0:100 and 20:80 ratios showed partial melting on the surface, even though the temperature was increased to more than 100 °C. However, the salt forms of shellac (AMP:AMN) at the ratios of 40:60, 60:40, 80:20 and 100:0 were completely melted at 135, 115, 105 and 90 °C, respectively. It could be assumed

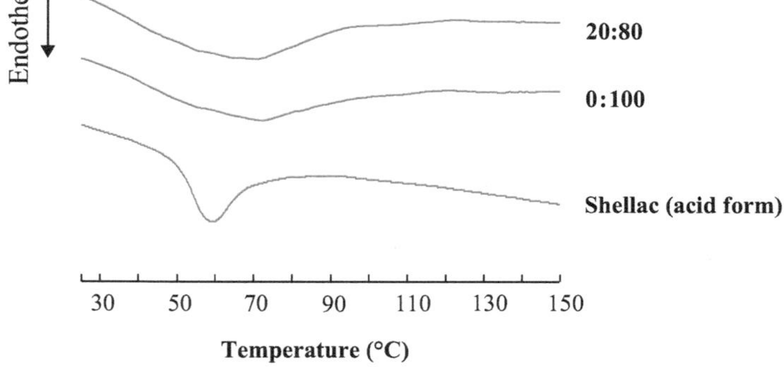

Fig. 24.4 DSC curves of acid form and various salt forms (AMP:AMN) of shellac (Limmatvapirat et al. 2007)

that the conversion of shellac from acid to salt form can contribute to an increase in the melting peak and a change in the thermoplastic and thermoset properties, leading to a difficulty in the shape molding. In contrast, the salt form of shellac provided and improved the solubility and stability. It can thus be efficiently applied in several products.

Polymers can be classified in terms of thermal properties into thermoplastic and thermoset polymers. Thermoplastic polymers melt under heating conditions and becomes rigid upon cooling. It is easy to reprocess without any changes in the chemical structure. Their linear chains show no crosslinking between polymer chains at the elevated temperature. Unlike thermoplastic, thermosetting polymer cannot be re-melted after molding, due to the formation of strong crosslinked bonds between the polymer chains, leading to thermal decomposition of chemical structure at high temperature. Shellac consists of hydroxyl and carboxyl groups which are subject to cross-linking, called polymerization. Shellac is a thermosetting polymer, which can be degraded, and can no longer be used after melting and cooling processes. However, under the appropriate temperature condition, shellac can be reprocessed without cross-linking formation, thus providing a thermoplastic characteristic. Due to the thermoplastic characteristic, a fluidity and strong adhesion of shellac has obtained, which allows a formation of several desired products such as phonograph records, stampers and grinding stones. Therefore, with proper control of melting temperature and time, shellac can demonstrate both thermoplastic and thermoset behaviors and can be widely used as a polymer of choice in various industries.

24.3 Drawbacks of Using Shellac as a Coating

Shellac polyester polymer consists of hydroxyl groups and carboxylic groups. Since these groups show a viability in interaction with their own side chains or with other side chains, these cause polymerization reactions and unstable shellacs, resulting in the changes in physicochemical properties of shellac (Limmatvapirat et al. 2005). The study of Luangtana-anan et al. 2007 reported that the physicochemical properties of films changed greatly during storage for 3 months as shown in Table 24.4.

Table 24.4 Properties of shellac in free acid form in storage for 3 months (Luangtana-anan et al. 2007)

Properties	Storage time (month)				
	0	0.5	1	2	3
AV (mg KOH/shellac)	79.5	82.2	80.7	75.7	66.7
Insoluble solid (g)	0.6	0.4	1.6	3.9	7.9
Stress (MPa)	2.48	1.21	–	–	–
Strain (%)	3.39	0.54	–	–	–
WVPC*10^{-9}(g h^{-1} mm^{-1} $mmHg^{-1}$)	5.09	4.91	6.41	7.36	5.36

The acid values decreased from 79.5 mg KOH/1 g shellac to 66.7 mg KOH/1 g shellac after 3 months of storage. The decrease in acid value was due to the esterification process resulting in an increase in insoluble solid (IS). The IS increased from 0.6 to 7.9 g after 3 months of storage. Even after being stored for a long time, shellac had protection against moisture. The WVPC remained low during 3-months of storage, WVPC was 5.09×10^{-9} g h^{-1} mm^{-1} $mmHg^{-1}$ and 5.36×10^{-9} g h^{-1} mm^{-1} $mmHg^{-1}$ at initial and 3 months, respectively. In another study by Soradech et al. (2013) the changes in physicochemical properties of shellac film upon storage for 6 months were shown to correspond with the study of Luangtana-anan et al. (2007). Several physicochemical properties have been changed such as acid value, IS, surface free energy, % polarity and mechanical properties. The acid value decreased from 76.0 to 58.3, while IS increased significantly from 0.7% to 51.7%. The surface free energy decreased from 37.2 to 34.3 mN/m and % polarity changed from 26.2% to 15% after 6 months of storage. The reduction in polarity was in agreement with the decrease in acid value. The puncture strength and % elongation decreased from 3.8 MPa to 0.5 MPa and 3.8% to 1.08% during 6 months of storage. The aging of shellac was also investigated by the work of Coelho et al. (2012). They reported the photodegradation of shellac under natural exposure to sunlight and artificial light which emits spectral rays above 295 nm, total flux of 2.6 mW/cm^2. They found that aleuritic acid was released from the backbone of shellac due to aging by the de-esterification process which is attributed to free acid available for cross-linking *via* hydrogen bonding with nearby molecules. The degradation process caused a change in the viscoelastic property of shellac. Aleuritic acid can thus be a marker for the aging of shellac. Apart from instability upon storage, the other main drawback of shellac as a coating is its weak mechanical properties. Luangtana-anan et al. (2007) found that the initial strength of shellac film was 2.48 MPa and the strain was 3.39%. The mechanical properties of shellac were greatly reduced during storage. The strength could not be detected after 15 days of storage, which contributed to the lack of suitability as a coating (Luangtana-anan et al. 2007). Significant reductions in mechanical properties on 3 months storage of shellac in AMN form was also reported in the study of Limmatvapirat et al. (2005).

24.4 Approaches to Solve the Drawbacks of Shellac as a Coating

Shellac has many advantages as a coating, since a coating with a glossy nature can be formed, as well as a superior barrier against moisture can be obtained compared with many natural polymers. It is also a biodegradable polymer and non-toxic. Although shellac has many advantages as a coating, the use of shellac in many industries has been declined due to the main problems of polymerization and weak mechanical properties. Several investigations have been carried out to solve the problems of instability. The carboxyl and hydroxyl groups of shellac are thus the target for the protection against polymerization, since both groups are sensitive to the esterification process causing polymerization and ISs. Our groups have studied many methods to improve the problem of polymerization, one of the approaches is to modify shellac in salt form to protect these free groups. The salts used were AMN and AMP (Limmatvapirat et al. 2007). The results have shown that stability is improved with the use of salt forms. Table 24.5 shows the changes in acid value and IS after storage for 3 months of coatings in free acid and both salt forms. The improvement was observed for shellac in salt forms indicated by the low changes in acid value and IS during storage. The ISs of shellac in AMN and AMP salts was around 1%, while acid form was 8%. This demonstrates that the carboxyl and hydroxyl groups can be linked with the AMN and AMP molecules, which results in no carboxyl and hydroxyl groups being available for cross-linking with other site chains. Polymerization consequently does not occur during storage and stabilized shellac is obtained due to the formation of salt forms (Luangtana-anan et al. 2007).

Another approach is through the formation of shellac phthalate by an annealing process at different temperatures and times. The annealing temperature and time affect the esterification process between hydroxyl groups of shellac and phthalate moiety. The proposed diagram for the formation of shellac phthalate is shown in Fig. 24.5. Figure 24.6 shows the higher acid value that was reported for a longer

Table 24.5 Changes in acid value (mg KOH/1g shellac) and insoluble solid (IS) of shellac in acid and salt forms during storage at 40 °C 75% RH. Adapted from Luangtana-anan et al. (2007)

Type	Properties	Storage time (month)				
Shellac		0	0.5	1	2	3
Acid	AV	79.5	82.2	80.7	75.7	66.7
AMN		76.2	76.8	80.7	81.6	80.6
AMP		65.1	75.7	75.4	74.1	68.9
Acid	IS (g)	0.6	0.4	1.6	3.9	7.9
AMN		0.1	0.3	0.3	0.6	1.3
AMP		0.3	0.8	0.4	1	0.5

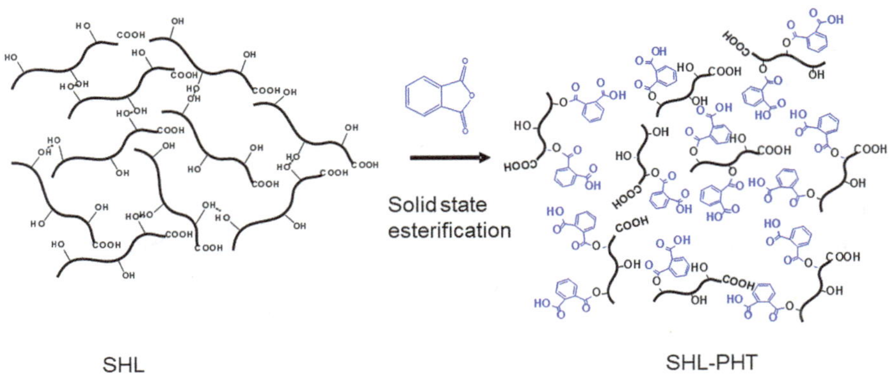

Fig. 24.5 Proposed diagram for the formation of shellac phthalate by annealing process (Panchapornpon et al. 2011

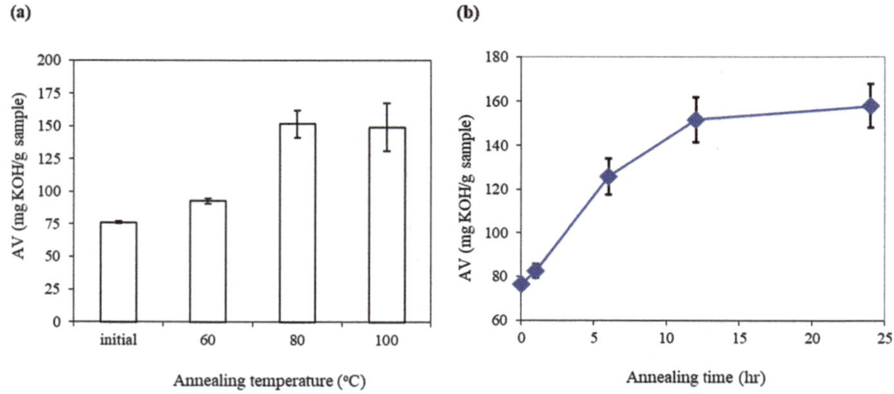

Fig. 24.6 Effect of annealing conditions on acid values (AV) of shellac samples: (**a**) annealing at various temperatures for 12 h, (**b**) annealing at 80 °C for 1, 6, 12 and 24 h (Panchapornpon et al. 2011)

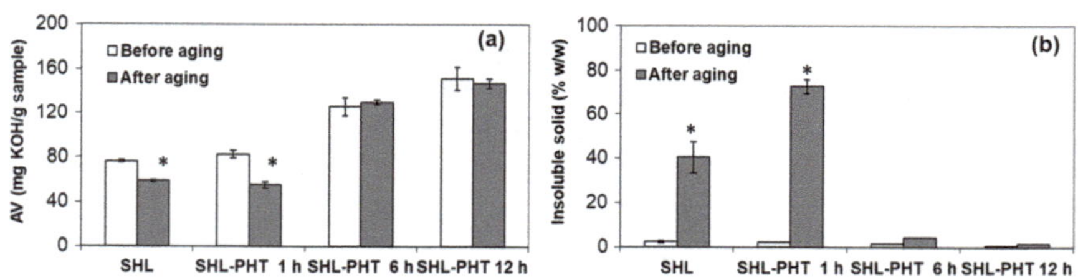

Fig. 24.7 Effect of aging by annealing at 80 °C for 24 h on acid value (AV, **a**) and insoluble solid (**b**) of native shellac (SHL) and shellac phthalate (SHL-PHT) prepared at 1, 6 and 12 h. (Panchapornpon et al. 2011)

time and temperature of the annealing process. The acid value increased from 75 to 150 for the shellac phthalate at 80 °C for 12 h. The increase in acid value suggested that phthalate moieties can react successfully with the hydroxyl groups of shellac molecules and shellac phthalate is formed. Different shellacs: native and derivative (SHL-PHT prepared at 1, 6 and 12 h), were also investigated, in terms of aging by annealing at 80 °C for 24 h. Figure 24.7 shows the aging effect of native and shellac derivative prepared at 1, 6 and 12 h. Shellac phthalate at 6 and 12 h exhibited low changes in acid value and IS upon heating at 24 h and 80 °C, while native and shellac-phthalate at 1 h showed significant changes. The low stability of SHL-PHT at 1 h could be a result of an insufficient esterification process during the preparation contributing to the high amount of unprotected hydroxyl groups. The IS of native and SHL-PHT at 1 h increased to 40% and 75%, respectively. The stabilized shellac was due to the protection of phthalate anhydride at hydroxyl groups of shellac, hence the difficulty in the polymerization.

The formation of shellac phthalate can thus prevent the polymerization of shellac in acid form. The formation of shellac phthalate can be prepared by the process of solid annealing with a suitable controlled temperature and time (Panchapornpon et al. 2011).

A combination of salt form between AMP and AMN was used to protect against polymerization of the sensitive groups. The proper ratio of salt composition could protect against the polymerization process for up to 6 months. At the ratio of 60:40, 80:20 and 100:0 of AMP:AMN could protect in the carboxylic groups, while at 20:80 and 0:100 a significant increase in insoluble acid was reported as shown in Fig. 24.8. This was a result of stronger binding capacity of AMP with the carboxylic group than AMN salt. The high AMP ratio, thus led to more stable composite polymer compared with the low AMP ratio (Limmatvapirat et al. 2007).

The application of plasticizers is another approach to protecting sensitive groups of polyester chains and has been investigated in many studies (Khairuddin et al. 2016; Luangtana-anan et al. 2007; Luangtana-anan et al. 2010). Plasticizers are commonly used for film forming, since they play an important role in improving flexibility of polymer chains, thus contributing to a higher strain. The incorporation of plasticizers can affect the continuity of a polymer chain, causing a modification in the functional properties of the polymer.

Several plasticizers have been used such as TA, PEG, glycerol, sorbitol, triethyl acetate (TEC), dibuthyl phathalate (DBP), diethyl phathalate (DEP), propylene glycol (PG), among others. The plasticizers can have a plasticizing or antiplasticizing effect. In general, the plasticizing effect improves the polymer properties in terms of mechanical property while the anti-plasticizing effect decreases the flexibility. Different types of film former are suitable for different plasticizers. Polyol plasticizers

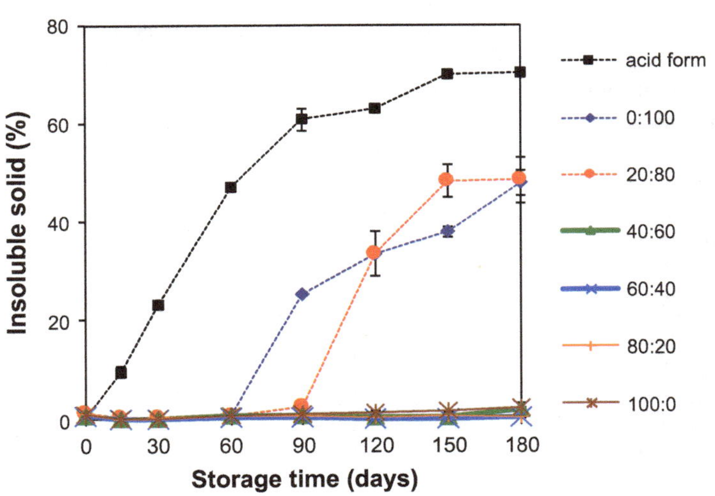

Fig. 24.8 Change of % insoluble solid of shellac in acid form and mixture of salt form between AMP:AMN after storage for 180 days (Limmatvapirat et al. 2007)

are generally used in gelatin film. It has been reported that glycerol is a good plasticizer for gelatin film because of its ability to reduce intermolecular hydrogen bonding, while increasing intermolecular spacing and reducing direct interaction. Because a small hydrophilic molecule of glycerol can be inserted between chains of proteins and, therefore, act as a plasticizer, while the other polyols, such as PG, ethylene glycol and diethylene glycol have a lower efficiency (Suderman et al. 2018). Plasticizers can affect the structure of polymers differently. The hydroxypropyl methylcellulose (HPMC) and hydroxyl propyl starch mixtures were investigated with different plasticizers: glycerol, PEG and PG. The blends plasticized with glycerol presented lamellar structure. All the plasticized blends showed crystallinity differently. The order of crystalline degree was glycerol > PEG > PG, whereas the compactness of the amorphous region of all samples decreased by plasticized with PEG and PG. Changes in the mechanical properties of these films after plasticization were in agreement with the changes in their multi-scale structures (Zhang et al. 2018). For the ethyl cellulose coating, DBP has been a suitable plasticizer compared with DEP due to its higher hydrophobicity, thus contributing to a homogeneous coating (Heng et al. 2003). PEG is generally used for shellac (Khairuddin et al. 2016; Luangtana-anan et al. 2007; 2010). The stabilization of shellac using 3 different plasticizers such as TA, DEP and PEG 400 after 3 months of storage is shown in Fig. 24.9. Different plasticizers have influenced differently the stability of the shellac. Plasticized shellac with DEP and TA affected greatly polymerization, whereas the addition of PEG 400 gave the greatest stability. The % IS of plasticized shellac with DEP, TA, and PEG 400 after 3 months of storage was 40%, 12%, and 0.5%, respectively. The highest instability of plasticized shellac with DEP was a result of inhomogeneous incorporation of DEP, which was confirmed by scanning electron microscope (Luangtana-anan et al. 2007). The larger molecule of PEG 400 can interrupt the polymer chain resulting in difficulty in the interaction between the active groups of shellac and, therefore, less polymerization. In addition, the high molecule of PEG400 showed less volatility unlike the low molecules, thus, the retained plasticizer was reported. The loss of plasticizers was also noted with Eudragit E film plasticized with TA and DEP during storage (Lin et al. 2000). PEG 400 was thus a suitable plasticizer for shellac film (Luangtana-anan et al. 2007).

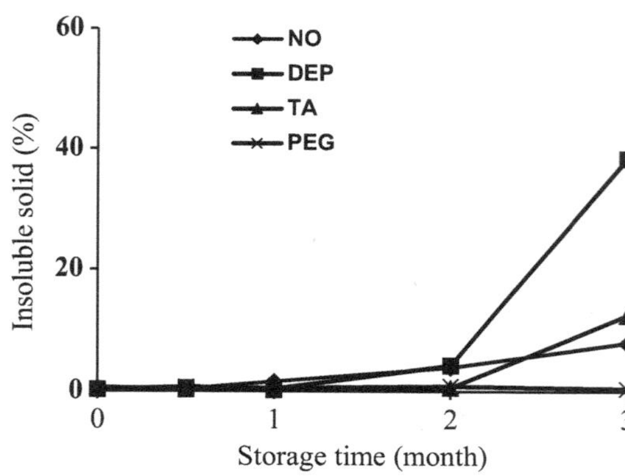

Fig. 24.9 Changes in acid values and insoluble solids after storage for 3 months of shellac in acid form plasticized with various plasticizers (No: unplasticized, DEP: diethyl phthalate, TA: triacetin, PEG: polyethylene glycol) (Luangtana-anan et al. 2007)

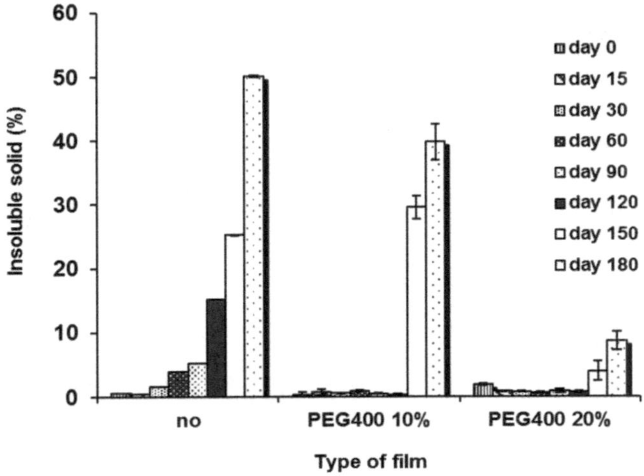

Fig. 24.10 Effect of concentration of PEG 400 on insoluble solid (%) of shellac film (Luangtana-anan et al. 2010)

The concentration of plasticizers influences the properties of polymer coatings (Oh and Luner 1999; Sirviö et al. 2018). PEG 400 at the 10% w/w concentration proved to be a suitable plasticizer. It was able to stabilize the shellac for 4 months, however, higher IS values appeared following the additional storage. Figure 24.10 shows the ISs of shellac plasticized with PEG 400 at 10% and 20% and without plasticizer during 6 months of storage. The higher concentration of PEG400 could improve the stability of shellac. The ISs of shellac plasticized with 10% and 20%w/w PEG400 was 40% and 8.0%, respectively, while unplasticized shellac was 50%. The results showed that the adequate concentration for stabilized shellac in acid form was 20% PEG400 (Luangtana-anan et al. 2010). The stabilization of shellac is not only dependent on type and concentration of plasticizer, but also molecular weight. The use of PEG 400 and 600 compared with unplasticized shellac on heating shellac at 125 °C for 10, 30, 90 and 180 min has been reported by Khairuddin et al. (2016). It was found that the stability of shellac decreased with heating time at 125 °C. At 30 min of heating time, PEG 400 had the most stable effect, whereas at 90 min and above PEG 600 was more stable as indicated by % ISs. Polymerization can also be indicated by a decrease in O-H group and an increase in C=O group (Limmatvapirat et al. 2007). The changes in O-H and C=O peak intensities was in accordance with the results of changes in % ISs. PEG 400 and 600 showed the lowest changes in O-H groups and the intensities of C=O groups at 30 min and 90 min, respectively, while unplasticized shellac exhibited significant changes. The influence of molecular weight of plasticizer on stabilization of shellac was also in agreement with our study (Luangtana-anan et al. 2010). The polymerization of shellac could be protected by the formation of hydrogen bonds between the carboxyl or hydroxyl groups of shellac and the hydroxyl groups of PEG. The diagram of protection is

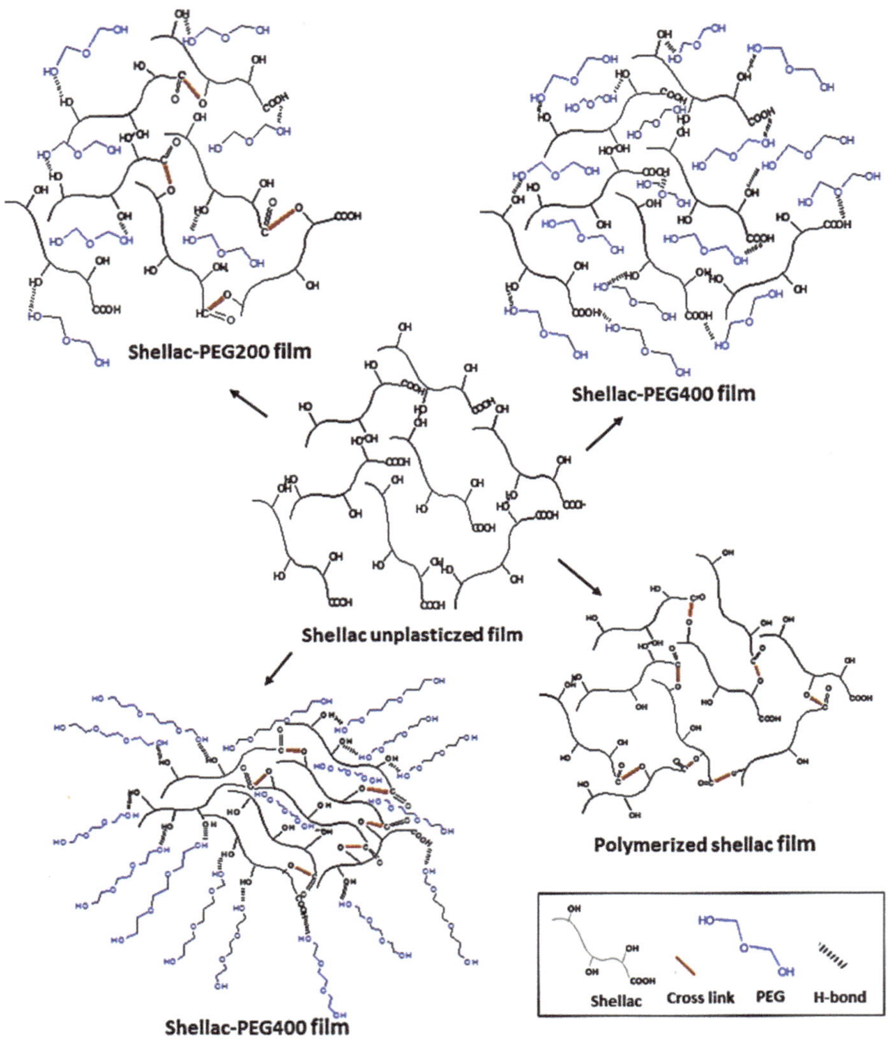

Fig. 24.11 Diagram of polymerization process of unplasticized and plasticized shellac film with PEG 200, 400 and 4000 (Luangtana-anan et al. 2010)

shown in Fig. 24.11. PEG 200 and 400 could prevent the polymer chains of shellac from the cross-linking formation of the esterification process, unlike the PEG 4000 which could not perform the protective capability efficiently due to the longer chain of PEG 4000, causing difficulties in insertion between the polymer networks. In addition, the high polarity of PEG 4000, indicated by a high surface free energy, tended to give a stronger hydrogen bonding, which resulted in the closest shellac

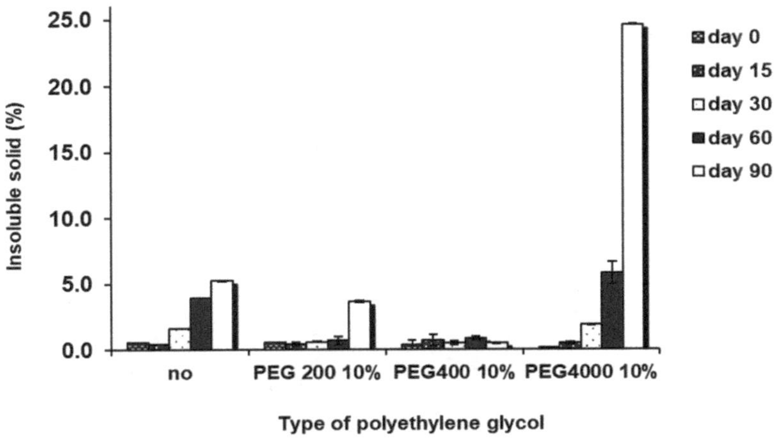

Fig. 24.12 Insoluble solid of several shellac films after storage for 3 months: unplasticized shellac (No) and plasticized with PEG 200, 400 and 4000 (Luangtana-anan et al. 2010)

chain network formed, thus leading to an increase in the polymerization. The surface free energy of PEG 200, 400, and 4000 was 28, 31, and 35 mJ/m^2, respectively (Luangtana-anan and Limmatvapirat 2010). The results of IS are shown in Fig. 24.12. The ISs in 3-month storage was PEG4000 > No > PEG 200 > PEG 400. In many studies, the method of using composite film to prevent the esterification process has been investigated (Asamo 2017; Soradech et al. 2013). Composite films are formed by a combination of two polymers. It is a very popular method as it is easy to prepare simply by mixing two polymers. The polymers can interact between polymer chains causing changes in their properties. Several polymers are used for the formation of composite films, among which are: gelatin, pectin, HPMC, etc. (Asamo 2017; Byun et al. 2012; Soradech et al. 2012). Shellac is a polyester, consisting of negatively charged groups from hydroxyl and carboxylic groups. Shellacs prepared in AMN form can interact through ionic bonds with positively charged polymers such as gelatin and pectin. The interaction of the COOH groups of shellac and the amine groups of gelatin was proved by FTIR as shown in Fig. 24.13.

The change at 1716 cm^{-1}, due to carbonyl stretching vibration of shellac and at 1663 cm^{-1} due to the C=O and C-N groups of gelatin, indicating that there is hydrogen bonding between shellac and gelatin, resulting in the formation of a composite film. In addition, the broad band spectrum of OH stretching increased with increasing gelatin concentration. These changes in the spectra supported the intermolecular interactions between shellac and gelatin and corresponded with the changes in acid values. The acid values of the shellac composite films decreased with the gelatin content as follows: 76.00, 70.65, 67.56, 64.43, 62.00 and 57.36 mg KOH/shellac at concentrations of 0,10, 20,30,40 and 50% gelatin, respectively (Soradech et al. 2012).

The formation of composite film based on shellac and gelatin can reduce polymerization indicating by the lower IS values as shown in Fig. 24.14a. The IS of shellac

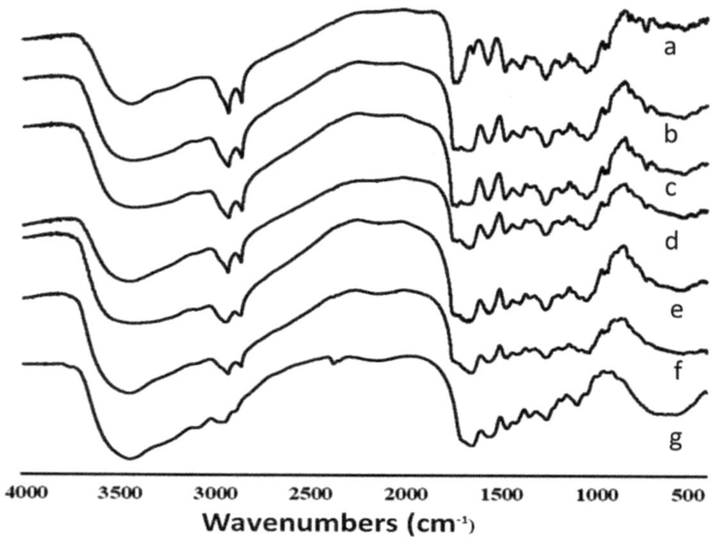

Fig. 24.13 FTIR spectroscopy of shellac film: (**a**) shellac and gelatin composite film with various concentrations of gelatin %w/w: 10 (**b**), 20 (**c**), 30(**d**), 40 (**e**), 50 (**f**) and gelatin film (**g**) (Soradech et al. 2012)

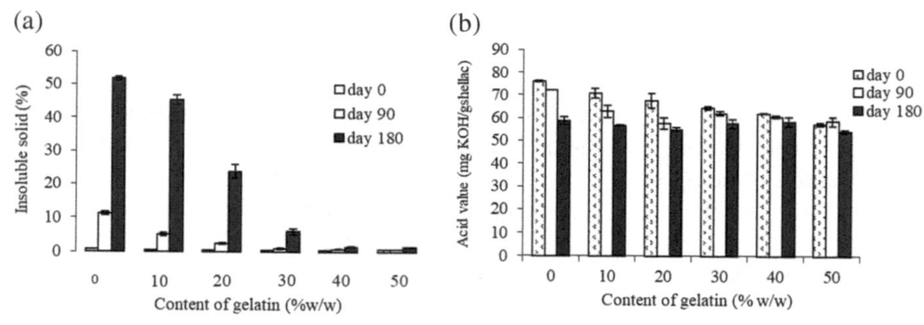

Fig. 24.14 The changes in insoluble solids (**a**) and acid values (**b**) of composite films in storage for 180 days (Soradech et al. 2013)

increased significantly after 3 months of storage. The ISs were 0.7, 11.2 and 51.7% for the shellac AMN at 0, 3 and 6 months of storage, respectively, while the composite film could prevent the esterification process, thus less ISs are formed. The protection capacity of gelatin is dependent on the concentration of gelatin. The lower concentrations of gelatin (10 and 20%) could not protect against the polymerization process resulting from the lower active groups (NH_3^+) of gelatin chains to protect all the carboxylic groups (-COOH) from the shellac molecules, whereas the high concentrations of gelatin could stabilize the shellac. For 6-month storage, high concentrations of gelatin showed a low IS value. The higher concentration of gelatin

Table 24.6 The change in free surface energy and % polarity of composite films after storage for 180 days (Soradech et al. 2013)

	Total free surface energy (mN/m)			Polarity (%)		
Gelatin %	Day 0	Day 90	Day 180	Day 0	Day 90	Day 180
0	37.28	34.96	34.32	26.23	25.66	14.98
10	39.10	37.47	37.33	29.39	25.33	24.11
20	40.00	38.56	38.07	31.88	26.82	24.95
30	40.83	39.49	38.53	35.66	29.17	26.42
40	41.46	39.61	38.84	39.85	29.59	28.17
50	43.30	40.26	40.04	42.82	30.35	27.52

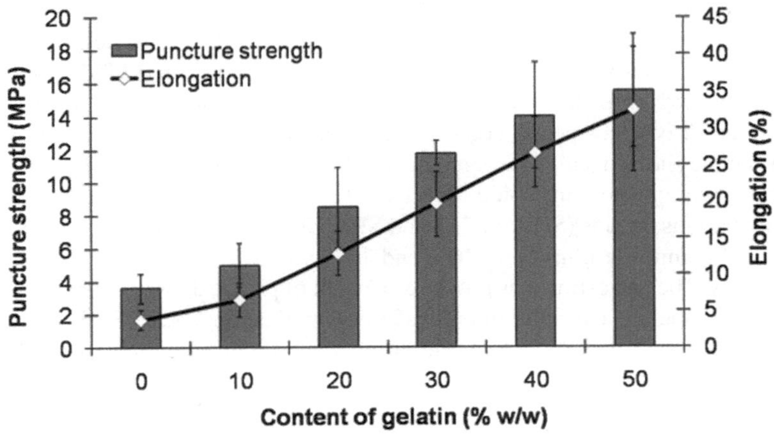

Fig. 24.15 Effect of gelatin content on mechanical properties of shellac and gelatin composite films (Soradech et al. 2012)

give higher NH_3^+ groups to protect the carboxylic groups of shellac, thus less polymerization (Soradech et al. 2013). The change in IS was in accordance with the change in acid value. High concentrations of gelatin exhibited a low decrease in acid value while low gelatin concentrations gave a high reduction in acid value after keeping for 6 months as shown in Fig. 24.14b. Table 24.6 shows the changes in surface free energy and % polarity of shellac and composite films during storage. A high decrease in surface free energy and % polarity was reported for the shellac. The improvement was reported with the composite film. The use of a second polymer to form a composite film can thus be applied to prevent the esterification process, resulting in less polymerization and retaining the desired properties of shellac film. The use of gelatin can not only prevent polymerization, but it can also enhance the mechanical properties of shellac. Since one of the drawbacks of shellac for coating is the strength of the film, the modification of shellac is required. The improvement was shown in Fig. 24.15. The incorporation of gelatin contributed to a significant increase in puncture strength and the values were in the range of 3.61–15.58 MPa as the gelatin content increased to 50% w/w. This was the result of the formation of hydrogen bonding between the carboxyl and hydroxyl groups of the shellac molecules and the amino and carboxyl groups of the gelatin molecules. An improvement in elongation was also noted with an increase in gelatin content due to a higher polarity of composite film, resulting in a greater water capture capacity, as indicated by the higher water content. The polarity increased from 26% to 42% as the gelatin content increase to 50%w/w and the water content increased from 8 to 15%. Gelatin has also been investigated to improve the strength of other polymers such as alginate, soy protein, keratin, etc. (Cao et al. 2007; Dong et al. 2006; Ramadoss et al. 2017). The gelatin can thus be applied in shellac film to enhance stability in terms of polymerization and mechanical properties.

Another second polymer for the formation of composite film is pectin. Pectin is a type of polysaccharide, consisting of D-galacturonic acid forming 1–4 α bonds with rhamnose sugar. There are several types of pectin depending on the degree of esterification. High methoxy pectin, (CU201) and low methoxy pectin (CU701) have been investigated and were showed to form a composite film with shellac at a 1:1 ratio. The use of all types of pectin could protect sensitive groups of shellac. The IS of single shellac showed a high IS percentage (40%) when stored for 90 days, whereas all types of composite films could reduce the IS values. At 90 days, shellac composites with various pectins (SHL:CU201 and SHL:CU701) exhibited a low % IS value. The ISs of composite films were 10% and 15% for SHL:CU201 and SHL:CU701, respectively. The protection was proved as a result of physical insertion into the polymer chain of shellac, causing difficulties in polymerization, a low IS value was thus obtained (Asamo 2017). Therefore, the formation of composite films using all types of pectin can protect against polymerization after storage for 90 days.

24.5 Application of Shellac as Coating of Agricultural Products

Several biodegradable polymers have been used to coat agricultural products such as polysaccharide, protein and lipid (Álvarez et al. 2017; Gutiérrez 2018b). Although there are several advantages in using these polymers, some drawbacks have been reported (Dehghani et al. 2018). The first two groups have the problem of low moisture protection (Gutiérrez et al. 2018; Herniou--Julien et al. 2019; Gutiérrez et al. 2019; Schmid and Müller 2019). The lipids have a high moisture barrier due to their low polarity, but they have a low cohesiveness, which contributes to a difficulty in forming independent polymers. The modification of these polymers (polysaccharides and proteins) is necessary to make them suitable for the coating, since they have hydrophilic properties, i.e. they are sensitive to moisture (Luangtana-anan et al. 2017; Ramos et al. 2016; Soradech 2010; Wihodo and Moraru 2013; Gutiérrez and Alvarez 2018; Gutiérrez 2018c, d). The pectin shows a hydrophilicity due to its high surface free energy (49.7 mN/m) and high polarity (49%), which leads to a high WVPC values: 7.9×10^{-7} g/mhPa (Luangtana-anan et al. 2017). Gelatin also shows a hydrophilicity (surface free energy - 47 mN/m and polarity - 44%), which leads to a high WVPC values: 1.21×10^{-7} g/mhPa and high-water content: 14%

(Soradech 2010). Therefore, the application of a second polymer with high protection against moisture is required to improve the weakness of these hydrophilic polymers (Ramos et al. 2016). Shellac is the polymer of choice due to its superior protection against moisture, and has been widely used as a coating (Phan The et al. 2008; Wei et al. 2015). Several examples on the use of shellac to solve the hydrophilic property of various polysaccharide and protein polymers have been reported (Byun et al. 2012; Luangtana-anan et al. 2017; Phan The et al. 2008; Saberi et al. 2017; Wei et al. 2015).

The use of agar and cassava starch as an edible film can be improved by the incorporation of shellac. The WVPC of agar and cassava starch films were significantly reduced with the incorporation of 50% shellac from 7.21×10^{-11} and 5.58×10^{-11} to 0.35×10^{-11} and 1.13×10^{-11} g/msPa, respectively. The coated surface exhibited hydrophobic characteristics, being observed by a high contact angle. The contact angle of agar coating increased from 82 ° to 92° and cassava starch coating increased from 51° to 82° with the addition of shellac (Phan The et al. 2008). The water contact angle (WCA) was directly correlated with the hydrophobic surface characteristics. The closest WCA to 90° represented a more hydrophobic surface. The moisture barrier of the pea starch/guar gum film was also improved by the addition of a shellac, stearic acid and tween 20. Tween 20 was required when shellac was incorporated with pea starch/guar gum to obtain a homogeneous coating (Saberi et al. 2017).

Konjac glucomanan is an environmentally friendly heteropolysaccharide that allows to obtain coatings. However, due to a large amount of hydroxyl groups in its structure contributes to high moisture sensitive properties, which is a main drawback for its use without modifications. The incorporation of shellac on konjac glucomanan film can improve its protection against moisture. WVPC was reduced from 13.55×10^{-11} g/msPa to 2.63×10^{-11} g/msPa, with the addition of shellac. The protection against moisture of konjac glucomanan shellac coating could be improved further with the addition of stearic acid (Wei et al. 2015). This improvement is due to a good distribution of stearic acid with the shellac solution, which contributes to an intimate interaction blends between the components of the konjac glucomannan and shellac coating.

HPMC is widely used in tablet coating as it provides protection against light and moisture, as well as masking the taste. It has the potential to replace synthetic plastics due to its good film-forming properties. However, pure HPMC has a limitation in food coating due to its high WVPC. A composite coating based on HPMC-shellac can improve WVPC only when used with a suitable emulsifier, since they cannot be mixed well without one. Saberi et al. (2017) reported lower WVPC values using lauric acid as an emulsifier, while stearic acid tended to show higher WVPC values. A composite coating with the optimum concentration of shellac is required to improve protection against moisture, since at higher shellac concentrations there is a lack of homogeneity of the coating and a decrease in moisture protection can be observed (Byun et al. 2012). Another study of shellac/gelatin composite films showed that the hydrophobic properties are obtained using higher concentrations of shellac, this was determined due to the increase of the WCA values and a reduction of the values of polarity percentage and surface free energy. WVPC and water content were reduced with a higher concentration of shellac (Soradech et al. 2012).

Table 24.7 Effects of storage at 40 °C and 75% RH for 90 days on wettability, surface free energy, % polarity, water content and WVPC of pectin/shellac composite films. (Luangtana-anan et al. 2007)

Content of shellac (%w/w)	Wettability (degree)		Surface free energy (mN/m)		Polarity (%)		Water content (%)		WVPC × 10^{-7} ($gh^{-1}m^{-1}Pa^{-1}$)	
	day 0	day 90	day 0	day 90	day 0	day 90	day 0	day 90	day 0	day 90
0	59.2	66.2	49.7	46.0	49.0	39.9	15.7	13.5	7.9	NA
10	62.5	70.5	46.3	44.3	45.8	32.8	12.5	12.2	7.0	NA
20	72.2	75.9	42.5	42.9	37.8	28.4	11.1	7.3	4.3	NA
30	74.7	81.1	40.8	40.1	32.7	24.8	9.1	6.6	3.6	NA
40	76.4	85.3	39.8	39.3	30.1	20.3	8.6	6.4	3.4	NA
50	78.3	88.6	39.4	37.4	27.9	20.1	7.1	5.0	2.0	1.4

Pectin has many advantages as a good film former, biodegradability and high mechanical property. Due to its very hydrophilic property, as a single polymer cannot be used for agricultural or pharmaceutical products. The study of Luangtana-anan et al. (2017) reported that pectin showed a high-water content (15%) and a high WVPC (7.89 ± 1.04 × 10^{-7} g/mhPa). The WVPC of pectin coating could not be detected after storage for 7 days. The addition of shellac can reduce the drawbacks of pectin. A reduction in hydrophilicity was found with the addition of shellac, which was indicated by a higher WCA and lower % polarity as shown in Table 24.7. In a 50% combination with shellac, the WVPC of the pectin could be improved, reducing the WVPC values from 7.89 × 10^{-7} g/mhPa to 2.0 × 10^{-7} g/mhPa. At low concentrations of shellac, the stability test at 40 °C and 75% RH for 90 days could not be achieved. The concentration of shellac required which could withstand moisture during 3-months storage was 50% (Luangtana-anan et al. 2017). Therefore, the weakness of all polysaccharide and protein coatings is the aspect of high-water affinity, which can be improved by incorporating the shellac.

In addition to its superior properties in protection against moisture, shellac is also excellent in protecting gas permeability as shown in Fig. 24.16. Shellac shows the best protection against gas, while polyethylene displays the highest permeability (Bai et al. 2003). In an initial study done by Alleyne and Hagenmaier (2000), shellac also showed less oxygen permeability than shellac/candelilla (Sh-cand) and shellac/carnauba wax (Sh-car) composite films. The protection against gas permeability provided the diffusion barrier that restricts flux of CO_2 and oxygen, which contributed to the low oxygen permeability in the fruit and high remaining of CO_2. The internal CO_2 and oxygen contents of shellac-coated apples were 10.1% and 8.9%, while the uncoated ones were 3.5% and 17.6%, respectively. Shellac showed the highest ethanol content as a result of its low permeability of ethanol than the composite coatings which leads to entrapment and accumulation of higher concentration of this compound in the shellac-coated fruit. The ethanol content was 32.4, 2.0, 5.1 and 3.5 μL/L for shellac, Sh-car, Sh-cand and uncoated apples, respectively. All coatings gave a

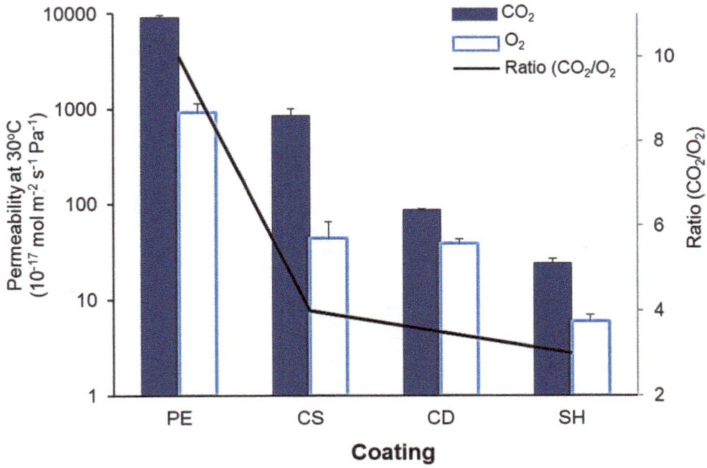

Fig. 24.16 Coating permeability to CO_2 and O_2. For coating treatments: PE = polyethylene, CD = Candelilla, CS = carnauba-shellac and SH = shellac (Bai et al. 2003)

lower weight loss, high firmness and low respiratory rate, which contributed to a lower oxygen permeability and higher remaining of CO_2 than uncoated apples. The gloss of all types of coating was higher than uncoated apples and the gloss was reduced with time. All the findings show that shellac-based coating has a high potential for protection against gas permeability.

Due to a high protection against gas permeability, a low internal oxygen content and high CO_2 accumulation is expected. Low oxygen flux can result in the anaerobic metabolism of the fruit, which causes the reduction of pyruvic acid to alcohol, and is attributed to the changes in odour and flavour (Alleyne and Hagenmaier 2000). The study of Mattheis et al. (1991) was about the effect of anaerobic metabolism of delicious apples stored at low oxygen content for 30 days. High levels of ethanol, acetaldehyde and several ethyl esters were reported in this study (Mattheis et al. 1991). A combination of shellac with some waxes is therefore applied to optimize gas permeability. This helps to control an adequate respiratory rate and contributes to a good quality fruit, prolonging the shelf life of products. The waxes used are carnauba wax (Jo et al. 2014) and candelilla wax (Alleyne and Hagenmaier 2000). Apart from waxes, an application of the other polymers to form a composite polymer is another means of controlling gas permeability. Polymers used for the formation of composite film are HPMC (Valencia-Chamorro et al. 2009) and pea starch/gua gum (Saberi et al. 2018).

Despite its qualities that offers a high level of protection against moisture and gas permeability, which plays a crucial role in the sensescene of products, shellacs' ability to protect against the microorganism growth is limited. The aid of another substance is needed for this purpose. Essential oils have been claimed to have an antifungal effect on fruits (Marei et al. 2012; Pérez-Alfonso et al. 2012). Essential oils are gaining popularity due to their volatile nature and friendliness with the

Table 24.8. % Control of microorganisms of blue mould (*P. italicum*) and green (*P. digitatum*) on orange using several coatings (Kouassi et al. 2012)

Substances	% blue mould control	% green mould control
Water	0	0
Ethanol	0	0
Shellac	0	0
Paraffin	0	0
Polyethylene	0	0
Shellac+carnauba wax	0	0
Cinnamon oil in ethanol	47	40
Shellac+cinnamon oil	100	100
Shellac+carnauba wax+cinnamon oil	83	75
Paraffin+cinnamon oil	70	62
Polyethylene+cinnamon oil	70	62

environment, which allows them to be used in small concentrations that are safe for consumption. They can be used to replace chemical fungicide. However, its volatile properties can be a disadvantage, resulting in low efficiency. The application of these essential oils with other coating polymers have been investigated to reduce the weakness of essential oils. Some essential oils used in shellac coatings are cinnamon oil, thymol and lemongrass oil (Jo et al. 2014; Kouassi et al. 2012; Pérez-Alfonso et al. 2012). Table 24.8 demonstrates the ability of several coatings to control fungi (*p italicum* and *p. digitatum*) on oranges. The findings showed that several polymers did not show a control of the microorganisms, while cinnamon oil in ethanol had a very low percentage of microorganism control as a result of the volatile properties of cinnamon oil. However, a combination of polymers and cinnamon oil could improve the % control of the microorganisms. Shellac provided the best combination with cinnamon oil. A control of 100% microorganisms for blue and green moulds was shown in the oranges, while the combination with paraffin or PEG showed some control but at a lower level.

The work on carnauba-shellac wax (CSW)-coated apples containing lemongrass oil (LO) nanoemulsion was prepared using high pressure homogenization. Coated apples reduced the emergence of bacteria, yeasts and moulds compared with uncoated apples. The improvement was reported with the addition of lemongrass oil as shown in Table 24.9. The addition of lemongrass oil decreased the bacterial growth to a higher extent than the CSW coating. Yeasts and moulds were observed after 1 month of storage with CSW coating, but was not detected in CSW/LO-coated apples after 5 months of storage. In addition, coatings could protect against weight loss and the loss of hardness of apples. The quality of apples could thus be improved during storage for more than 5 months with the addition of lemongrass oil (Jo et al. 2014).

In addition to essential oils for microorganism control, the application of modified coconut oil has been investigated for protection against bacterial growth, since it contains monolaurin which possesses antibacterial properties (Carpo et al. 2007;

Table 24.9 Change in populations (log CFU/g) of total aerobic bacteria, yeasts and moulds present on the coated 'Fuji' apples coated during storage (Jo et al. 2014)

Microorganism	Storage time (month)						
log CFU/g)		0	1	2	3	4	5
Total aerobic	Control	2.31	2.44	2.37	2.13	2.14	2.67
bacteria	CSW	1.77	1.41	1.67	1.23	1.10	1.86
	CSW-LO	1.68	1.34	1.33	1.22	<1.0	1.30
Yeast and moulds	Control	2.69	2.70	3.28	2.24	2.15	2.31
	CSW	ND	1.15	1.61	1.10	1.30	1.58
	CSW-LO	ND	ND	ND	ND	ND	ND

ND: Not detect.

Lieberman et al. 2006; Tangwatcharin and Khopaibool 2012). Sittipaisankul (2015) reported on water loss, firmness and the growth of anthracnose microorganisms on dragon fruit, which caused the low-quality fruit. The addition of modified coconut oil on shellac coating could prevent fungal growth. The findings showed that dragon fruits coated with shellac and uncoated one exhibited high microorganism percentages, while fruits coated with composite film (shellac and modified coconut) could prevent the fungal growth for 26 days (Sittipaisankul 2015). Jitareerat et al. (2018) studied the growth of anthracnose on mangoes, demonstrating that shellac coating containing modified coconut oil could control the microorganisms. In addition, the increase in improvement of the control of microorganisms was found with pretreatment of the fruit with hot water, followed by UV-C irradiation before coating with a shellac/modified coconut mixture (Jitareerat et al. 2018).

24.6 Composite Coatings Based on Shellac for Agricultural Products

Composite coatings are a combination of hydrocolloid components (either proteins or polysaccharides) and lipids. Composite coatings are formulated to combine the advantages of hydrocolloid and lipid components while masking or reducing the limitations of each. Since proteins or polysaccharides may not provide a barrier against moisture, shellac, a lipid component, possesses a high moisture barrier. Shellac can restrict the permeability of oxygen, causing anaerobic metabolism. Therefore, composite coatings are generally used. Some examples of where composite coatings based on shellac have been used to prolong the shelf life of banana and apple will be explained. The work of Soradech et al. (2017) investigated physicochemical changes on shellac/gelatin composite film-coated bananas. Changes in color, weight loss, firmness, titratable acidity and total soluble sugar (TSS) were significant observed in uncoated bananas during storage, while the composite film with or without PEG400 as a plasticizer could reduce the materialization of all these changes. Figure 24.17 shows the effect of different coatings on the color change of

Fig. 24.17 Physical aspect of bananas without and with film coating after storage at 25 °C for 30 days (1a = Uncoated, 1b = Coated with composite film, 1c = Coated with composite film + PEG 5%) (Soradech et al. 2017)

bananas after storage for 30 days. An extreme change in color was observed for uncoated bananas. After storage for 10 and 20 days, a change to yellow and then brown was observed, respectively. The degradation of chlorophyll, by the enzyme chlorophyllase, contributed to the color changes and an increase in the other pigments such as carotene and xanthophyll (Bico et al. 2009; Prabha and Bhagyalakshmi 1998). The changes of phenol to quinone by polyphenol oxidase (PPO) and a subsequent increase in macromolecules through polymerization, led to an increase in brown pigment (Bico et al. 2009; Toivonen and Brummell 2008). During storage, the uncoated banana exhibited high PPO activity and a low phenolic content due to the unprotected state of the sample against enzymatic reactions (Bico et al. 2009). The slight color changes in coated bananas was due to the greater protection of both formulations, resulting in a slower change in the chlorophyll and a decrease in the respiratory rate. In addition, the coated bananas showed smaller changes in weight loss and firmness (Newton, N) as shown in Fig. 24.18. The reduction in firmness was the result of water loss, disassembly and degradation of cell wall structure by the catalytic enzymes: polygalacturonase and pectinesterase (Toivonen and Brummell 2008; Zhou et al. 2011). The change in TSS is from the hydrolysis of starch and other carbohydrates producing total soluble sugars such as sucrose, glucose and fructose (Prabha and Bhagyalakshmi 1998). A lower TSS was reported for coated bananas. The tritable acid of all banana samples; uncoated, coated with composite film (CF) and composite film with PEG400 (CF + 5%PEG400), decreased significantly ($p < 0.01$) after storage for 30 days from ~110 mg malic acid/100 g sample to 45.63,

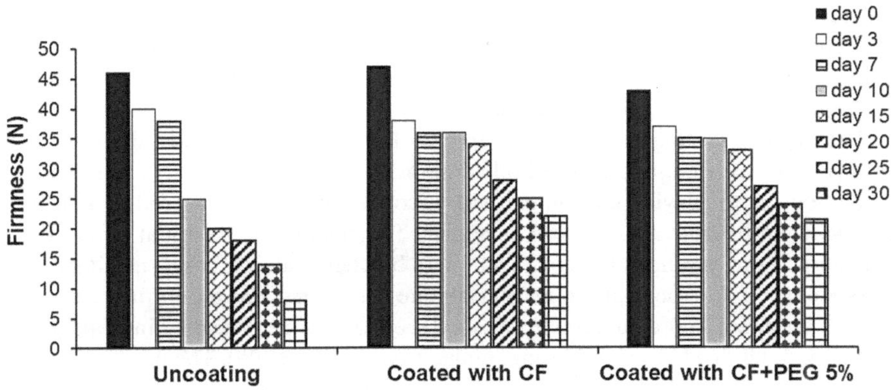

Fig. 24.18 Change in firmness of banana after storage at 25 °C for 30 days (Soradech et al. 2017)

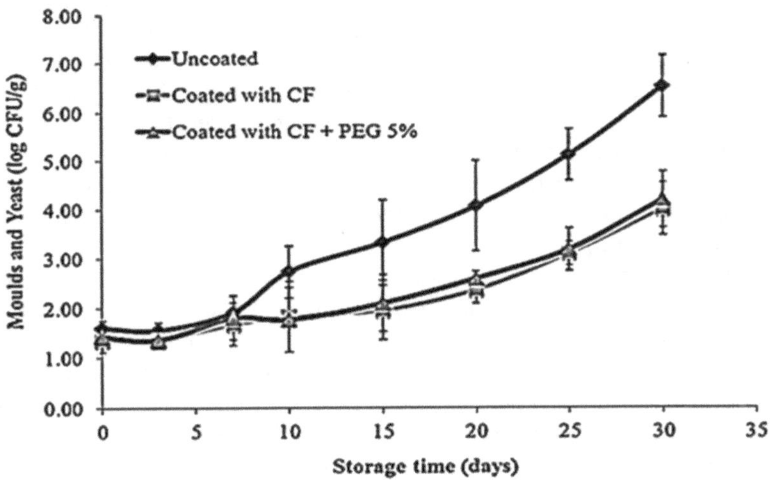

Fig. 24.19 Total mould and yeast counts on bananas with and without coatings after storage at 25 °C for 30 days (Soradech et al. 2017)

71.78 and 61.94 mg malic acid/100 g sample, respectively. Malic acid is the substrate for the enzymatic reactions of respiration. The reduction in respiratory rate due to the protection of coating solutions contributes to the minor change in tritable acid (Bico et al. 2009; Soradech et al. 2017). Figure 24.19 shows total molds and yeasts counts on bananas with and without film coatings after storage at 25 °C for 30 days. The protected bananas with each type of coating showed lower total molds/yeasts counts than uncoated samples after a 30-days storage. A smaller increase in total mould/yeast count was reported for both types of composite films compared with uncoated samples, and was in agreement with the lowest changes in all physical and chemical properties of the coated banana samples. The increase in total mould/yeast count of both coated bananas was well below the allowed limit of total mould/yeast counts (5 log CFU/g) for fruits and vegetables (Jacxsens et al. 2002). Therefore, the results show that bananas coated with unplasticized and plasticized composite films based on shellac and gelatin can extend the shelf life of bananas for 30 days compared to uncoated bananas.

Due to the growing consumer preferences for ready-to-use foods, the fresh-cut fruit industry is becoming increasingly significant. The main objective of fresh-cut technology is to provide consumers with a convenient and fresh product that has desirable nutritional and sensory qualities. The quality of fresh-cut produce is dependent on several factors such as biochemical defense mechanisms, stress-induced senescene programs and storage conditions. Fresh-cut fruits and vegetables are much more susceptible to spoilage than their corresponding full fruits and vegetables due to injuries during preparation (Brecht 1995). The principle problems with fresh-cut fruits involves changes in color and texture. Browning is caused by the action of PPO enzymes that interact with the phenols released during cutting to produce an undesirable brown color. The work in coating process thus must address the additional problem due to the cutting procedure. The softening has been attributed to acid hydrolysis and also to enzymatic degradation. Chauhan et al. (2011) coated the apple slices with various coatings such as bleached shellac, aloe gel (AG) and a composite solution of bleached shellac and AG. The coatings managed to reduce the respiratory rate, the ethylene synthesis rate, the oxidizing enzyme (PPO and peroxidase (POD)) and the leakage of electrolytes compared with uncoated apples. The reduction in respiratory rate was from the highest to the lowest in the order of shellac > shellac/AG composite > AG > uncoating. All the coatings prevent a significant loss of electrolytes from the surface, thus reducing the rate of senescence in the apple slices due to the decrease in respiratory rate. The application of sliced apple coating showed a significant reduction in the activities of these enzymes. The AG-coated samples showed a higher decrease in PPO (96.1 units/g/min) and POD activities (211.2 units/g/min) followed by composite and then shellac alone, whereas the uncoated apple slices showed the maximum activities as shown in Table 24.10. The firmness was significantly greater in the shellac-coated slices, followed by AG and composite-coated slices after 30 days of storage compared with untreated apple slices (Chauhan et al. 2011). The shellac/AG based composite coating can thus be applied to protect apple slices, contributing to prolong the shelf life.

Table 24.10 Effect of edible surface coatings on enzymatic activity (units/g/min) of apple slices (Chauhan et al. 2011)

Type of coating	Storage time (days)							
	0		10		20		30	
	PPO	POD	PPO	POD	PPO	POD	PPO	POD
Bleached shellac	105.7	230.7	127.6	245.8	136.8	268.9	175.6	297.8
AG	96.1	211.2	105.3	241.6	112.7	272.5	141.7	295.6
Bleached shellac + AG	98.1	221.5	136.9	256.8	150.8	286.4	187.9	307.2
Uncoated	122.8	288.5	198.4	320.6	–	–	–	–

24.7 Prediction of Coating Efficiency

Coating solutions require having certain important properties in order to form a coating that completely covers the surface of the substrates. The efficiency of a coating can be measured by the result of the coated substrate. In agricultural products, the measured parameters can be, e.g. the weight loss, color change, firmness, water and oxygen permeabilities and shelf life of the product. For pharmaceutical products, the measured parameters are the quality of dosage form such as release, protection against moisture, hardness and stability of dosage form etc. All these results are however indirect parameters. The direct measurement to detect the efficiency of a coating is through the determination of its surface free energy. Two parameters derived from surface free energy that can be used to predict the coating efficiency are the work of adhesion and spreading coefficients. Surface free energy is the energy required to expand surface area by 1 cm^2. Surface free energy is the characteristic of all materials that arise from the imbalance of force in the interface. The equipment currently available for measuring free energy is limited to measuring mobile substances, i.e. it can thus only be determined to liquid substances, not to solids. The equipment used for liquid substances is du Nouy tensiometer, the Wilhelmy plate technique, drop volume technique and pendant drop method, etc. The most commonly used technique to measure solid surface free energy is contact angle measurement (Luangtana-anan and Fell 1988; Gutiérrez and González 2017; Gutiérrez, Ollier, & Alvarez, 2018). At least two liquids with known components are used to measure the contact angle on the studied solid surface (Luangtana-anan and Fell 1987, 1988; Herniou--Julien et al. 2019). The surface free energy is calculated using several equations as shown below. Equation 24.1 is according to the method of Owens and Wendt (1969), Eq. 24.2 is according to method of Wu (1971). Total surface free energy is the summation between dispersion and polar forces (Eq. 24.3). The % polarity is then calculated according to Eq. 24.4.

$$\gamma_l\left(1+cos\theta\right) = 2\left(\gamma_s^d\gamma_l^d\right)^{1/2} + 2\left(\gamma_s^p\gamma_l^p\right)^{1/2} \tag{24.1}$$

$$\gamma_l\left(1+cos\theta\right) = 4\left(\frac{\gamma_s^d\gamma_l^d}{\gamma_s^d+\gamma_l^d} + \frac{\gamma_s^p\gamma_l^p}{\gamma_s^p+\gamma_l^p}\right) \tag{24.2}$$

$$\gamma_s = \gamma_s^d + \gamma_s^p \tag{24.3}$$

$$\%\text{polarity} = 100\left(\gamma_s^p,\gamma_s\right) \tag{24.4}$$

Where γ_s is the total surface free energy of a solid substrate, γ_s^p, γ_s^d are polar and dispersive forces of solid surface, respectively. γ_l^p, γ_l^d are polar and dispersive forces of standard liquid surface, respectively. θ is the contact angle of liquid formed on the solid surface.

Several studies have been used the contact angle measurement to determine the surface free energy of many types of fruits either using Eqs. 24.1 or 24.2 (Casariego et al. 2008; Ribeiro et al. 2007). The surface free energy, dispersion and polar forces of strawberries is 28.94, 22.99 and 5.95 mN/m, respectively (Ribeiro et al. 2007); for tomatoes is 28.71, 25.67 and 3.04 mN/m, respectively (Casariego et al. 2008), while for carrots is 26.48, 26.13 and 0.34 mN/m, respectively (Casariego et al. 2008). The fruit surfaces show a low surface free energy and polar force, thus indicating their hydrophobic nature. The hydrophobic nature of fruit skin is due to the presence of cell walls consisting of cutin and wax, which leads to a lower surface free energy and % polarity. Soradech (2010) also reported a low surface free energy and % polarity of mangos and bananas. The surface free energy, dispersion and polar forcers, as well as % polarity on bananas was 31.01, 28.92, 2.09 mN/m and 6.74% and for mangoes was 29.04, 27.57, 1.47 mN/m and 5.06%, respectively (Soradech 2010). The spreading coefficient (S) is calculated based on the difference between the work of adhesion (W_a) and the work of cohesion (W_c). Work of adhesion and cohesion is the work required to break adhesion force and cohesion force, respectively. The equations to calculate work of adhesion and work of cohesion are the following:

$$W_a = \gamma_S + \gamma_l - \gamma_{Sl} \quad (24.5)$$

$$W_c = 2\gamma_l \quad (24.6)$$

$$S = W_a - W_c \quad (24.7)$$

$$S = \gamma_S - (\gamma_l + \gamma_{Sl}) \quad (24.8)$$

Combining Eqs. 24.5 and 24.8 with Young's equation ($\gamma_s = \gamma_{sl} + \gamma_l cos\theta$), the work of adhesion and spreading coefficient are obtained as follows:

$$W_a = \gamma_l (cos\theta + 1) \quad (24.9)$$

$$S = \gamma_l (cos\theta - 1) \quad (24.10)$$

Soradech et al. (2013) reported that the increase in gelatin content showed a higher surface free energy from 37.28 to 43.3 mN/m with 50% gelatin content. The % polarity increased significantly from 26.23% to 42.82% with 50% gelatin content (Soradech et al. 2013). The increase in gelatin content contributed to a higher hydrophilicity, indicated by the high surface free energy and % polarity. The high hydrophilicity with the addition of gelatin provided less adhesion work and a lower spreading coefficient as shown in Table 24.11. Shellac is, thus, more suitable for coating bananas, due to its hydrophobic nature. A similar result was reported for mango coating (Soradech 2010).

The modified coating solution test is suitable for a given surface when using two types of surfaces: hydrophilic and hydrophobic. Glass represents a hydrophilic surface, while Teflon represents a hydrophobic surface. The surface free energy, dispersion force, polar force and % polarity of glass are 60.32, 11.41, 48.91 mN/m and

Table 24.11 Work of adhesion (*Wa*) and spreading coefficient (S) of shellac composite film with an increase in gelatin content on banana (*Soradech 2010; Soradech et al. 2013, 2017)

Content of gelatin	*Wa*	S	Surface free energy	% Polarity
(% w/w)	(mN/m)	(mN/m)	(mN/m)	
0	65.12 ± 0.53	–9.44 ± 0.53	37.28 ± 1.96	26.23
10	64.99 ± 0.70	–13.21 ± 0.70	39.10 ± 1.01	29.39
20	63.78 ± 1.22	–16.22 ± 1.22	40.00 ± 2.56	31.88
30	62.67 ± 1.35	–18.99 ± 1.35	40.83 ± 1.02	35.66
40	62.88 ± 0.71	–20.04 ± 0.71	41.46 ± 1.28	39.85
50	61.77 ± 0.95	–24.82 ± 0.95	43.30 ± 1.29	42.82
100	56.92 ± 0.94	–37.56 ± 0.94	47.24 ± 1.87*	46.16*

Table 24.12 Effect of gelatin content on the work of adhesion and spreading coefficient of shellac and gelatin composite film solution on glass and Teflon substrates (Soradech et al. 2012)

%w/w gelatin	Work of adhesion (mN/m)		Spreading coefficient (mN/m)	
	Glass	Teflon	Glass	Teflon
0	66.42	51.77	−8.14	−22.79
10	72.08	51.54	−6.12	−26.66
20	74.27	50.39	−5.73	−29.61
30	77.11	50.27	−4.55	−31.39
40	79.1	49.39	−3.44	−33.53
50	83.53	47.46	−3.07	−39.14

81.08%, while those surface of Teflon are 26.59, 26.23, 0.36 and 1.35%, respectively (Soradech et al. 2012).

Table 24.12 shows the work of adhesion and spreading coefficient of glass and Teflon with a composite film based on shellac and gelatin. An increase in gelatin composite film contributed to an increase in the work of adhesion and spreading coefficient for glass, whereas the opposite result was reported for the Teflon. The increase in the coating efficiency of glass corresponding with an increase in gelatin, was the result of a greater attraction between the high hydrophilic properties of the substrate and the high hydrophilic property of the coating solution when the gelatin was added. Teflon showed a low interaction, since the hydrophobic surface was not suitable for interaction with a composite coating with a high gelatin content (Soradech et al. 2012).

The modification of coating solutions used in the coating of several fruits to prolong shelf life have been studied widely. Vieira et al. (2016) reported that the wettability of a combination of chitosan (Cs) and *Aloe vera* on blueberry surface can be modified, as indicated by the spreading coefficient value. The spreading coefficient, a type of wetting, changed from - 89.8 to - 46.6 mN/m by the addition of tween 80 at the concentration of 0.1%w/v Cs. Since tween 80 is capable to reduce the surface tension of the coating solution, the work of cohesion for the coating solution is then reduced. This leads to an improved in compatibility between the solution and the blueberry skin's surface, and a greater wetting capacity of the coating solution. The value of spreading coefficient changed to the nearest from zero with the addition of tween 80. The increase in Cs and plasticizer concentration did not improve the wettability in this study and agreed with the work of Casariego et al. (2008). The increase in Cs and plasticizer concentrations caused a decrease in the wettability of Cs coating solution on the fruit skins, leading to a lower work of adhesion and spreading coefficient due to the low interaction between the coating and the fruit surface. This could be due to the hydrophobic property of fruit surfaces. The modification of a coating solution that makes it more hydrophilic does not contribute to a good interaction between the substrate and the coating solution. The work of Lima et al. (2010) reported a similar trend. These authors showed that an increase in galactomannan within a galactomannan/collagen coating solution reduced the work of adhesion and spreading coefficient of the hydrophobic substrate. The parameters of work of adhesion and spreading coefficient can also be applied in the optimization of coating solutions in pharmaceutical dosage form (Khan et al. 2001; Oh and Luner 1999). The coating solutions can thus be modified based on a knowledge of the surface free energy of both substrate surfaces. Through this process, the optimum coating solution can be achieved to prolong the shelf life of products.

24.8 Conclusions

The application of shellac on agricultural products is gaining interest due to its excellent protection against water and good film-forming properties, which produce glossy surfaces. Shellac coatings are environmentally friendly and can replace the use of synthetic coatings. Several methods (dipping, spraying, brushing and electro-spraying) have been simply applied a coating solution on food materials to protect them from microbial growths, to prolong their shelf life and improve other aspects of quality such as sensory attributes, appearance and freshness of the ingredients. Shellac can be used as a single coating or as a combination of various polymers or substances. The purpose of adding other substances is to control gas permeability because respiratory rate is a key factor in the senescence process of agricultural products. The respiratory rate varies depending on the type of agricultural products and coating solutions. The aim of respiratory rate control is to optimize the internal oxygen content and lower the CO_2 content. This has had the effect of decreasing the activities of the oxidizing enzymes that result in reductions in accumulated ethanol, ethylene synthesis, water loss and firmness. The other added substances are also to reduce microbial growth in order to prolong the quality of the fruit after post-harvest, since the shellac has a low antimicrobial activity. Furthermore, the addition of other compatible substances reduces the drawbacks of the shellac. Since shellac's main problems are instability due to polymerization and the weak mechanical properties, the modification of native shellac is thus required. Various approaches such as salt formations, composite formation and incorporation of some plasticizers are successfully applied in order to tackle the drawbacks of shellac. Therefore, shellac has a high potential to be a polymer of choice in coating of agricultural products. Research on the choice of the best coating can be explored in depth through knowledge of coating efficiency (adhesion work and spreading coefficient) based on surface free energy. The optimization of shellac coating in agricultural products can thus be achieved to extend fresh quality, the shelf life of products and ensure their marketability.

Acknowledgements The authors wish to thank the Faculty of Pharmacy, Silpakorn University for the facility supported and the proofreading service.

Conflicts of Interest The authors declare no conflict of interest.

References

Alleyne, V., & Hagenmaier, R. D. (2000). Candelilla-shellac: An alternative formulation for coating apples. *HortScience, 35*(4), 691–693. https://doi.org/10.21273/hortsci.35.4.691.

Álvarez, K., Famá, F., & Gutiérrez, T. J. (2017). Physicochemical, antimicrobial and mechanical properties of thermoplastic materials based on biopolymers with application in the food industry. In M. Masuelli & D. Renard (Eds.), *Advances in physicochemical properties of biopolymers: Part 1* (pp. 358–400). Bentham Science Publishers. EE.UU. ISBN: 978-1-68108-454-1. eISBN: 978-1-68108-453-4. https://doi.org/10.2174/97816810845341 17010015.

Asamo, S. (2017). Application of polymer blend between shellac and pectin as film coating material. Thesis of Master of Science, Silpakorn University.

Bai, J., Hagenmaier, R. D., & Baldwin, E. A. (2003). Coating selection for 'Delicious' and other apples. *Postharvest Biology and Technology, 28*(3), 381–390. https://doi.org/10.1016/S0925-5214(02)00201-6.

Bico, S. L. S., Raposo, M. F. J., Morais, R. M. S. C., & Morais, A. M. M. B. (2009). Combined effects of chemical dip and/or carrageenan coating and/or controlled atmosphere on quality of fresh-cut banana. *Food Control, 20*(5), 508–514. https://doi.org/10.1016/j.foodcont.2008.07.017.

Brasil, I. M., & Siddiqui, M. W. (2018). Postharvest quality of fruits and vegetables: An overview. In M. W. Siddiqui (Ed.), *Preharvest modulation of postharvest fruit and vegetable quality* (pp. 1–40). Academic. https://doi.org/10.1016/b978-0-12-809807-3.00001-9.

Brecht, J. (1995). Physiology of lightly processed fruits and vegetables. *HortScience, 30*(1), 18–22.

Byun, Y., Ward, A., & Whiteside, S. (2012). Formation and characterization of shellac-hydroxypropyl methylcellulose composite films. *Food Hydrocolloids, 27*(2), 364–370. https://doi.org/10.1016/j.foodhyd.2011.10.010.

Cao, N., Fu, Y., & He, J. (2007). Preparation and physical properties of soy protein isolate and gelatin composite films. *Food Hydrocolloids, 21*(7), 1153–1162. https://doi.org/10.1016/j.foodhyd.2006.09.001.

Carpo, B. G., Verallo-Rowell, V. M., & Kabara, J. (2007). Novel antibacterial activity of monolaurin compared with conventional antibiotics against organisms from skin infections: An in vitro study. *Journal of Drugs in Dermatology, 6*(10), 991–998.

Casariego, A., Souza, B. W. S., Vicente, A. A., Teixeira, J. A., Cruz, L., & Díaz, R. (2008). Chitosan coating surface properties as affected by plasticizer, surfactant and polymer concentrations in relation to the surface properties of tomato and carrot. *Food Hydrocolloids, 22*(8), 1452–1459. https://doi.org/10.1016/j.foodhyd.2007.09.010.

Chauhan, O. P., Raju, P. S., Singh, A., & Bawa, A. S. (2011). Shellac and aloe-gel-based surface coatings for maintaining keeping quality of apple slices. *Food Chemistry, 126*(3), 961–966. https://doi.org/10.1016/j.foodchem.2010.11.095.

Chongcherdsak, N., Aekthammarat, D., Limmatvapirat, C., & Limmatvapirat, S. (2013). Fabrication of shellac-based effervescent floating matrix tablet as a novel carrier for controlling of drug release. *Advanced Materials Research, 747*, 135–138. https://doi.org/10.4028/www.scientific.net/amr.747.135.

Coelho, C., Nanabala, R., Ménager, M., Commereuc, S., & Verney, V. (2012). Molecular changes during natural biopolymer ageing – The case of shellac. *Polymer Degradation and Stability, 97*(6), 936–940. https://doi.org/10.1016/j.polymdegradstab.2012.03.024.

Dehghani, S., Hosseini, S. V., & Regenstein, J. M. (2018). Edible films and coatings in seafood preservation: A review. *Food Chemistry, 240*, 505–513. https://doi.org/10.1016/j.foodchem.2017.07.034.

Dong, Z., Wang, Q., & Du, Y. (2006). Alginate/gelatin blend films and their properties for drug controlled release. *Journal of Membrane Science, 280*(1–2), 37–44. https://doi.org/10.1016/j.memsci.2006.01.002.

Fajardo, P., Martins, J. T., Fuciños, C., Pastrana, L., Teixeira, J. A., & Vicente, A. A. (2010). Evaluation of a chitosan-based edible film as carrier of natamycin to improve the storability of Saloio cheese. *Journal of Food Engineering, 101*(4), 349–356. https://doi.org/10.1016/j.jfoodeng.2010.06.029.

Farag, Y., & Leopold, C. S. (2011). Development of shellac-coated sustained release pellet formulations. *European Journal of Pharmaceutical Sciences, 42*(4), 400–405. https://doi.org/10.1016/j.ejps.2011.01.006.

Ghosh, M., Gupta, S., Kishan Kumar, V.S. (2015). Studies on the loss of gloss of shellac and polyurethane finishes exposed to UV Maderas: Ciencia y Tecnologia 17, 39–44. https://doi.org/10.4067/S0718-221X2015005000004.

Gogo, E. O., Opiyo, A. M., Hassenberg, K., Ulrichs, C., & Huyskens-Keil, S. (2017). Postharvest UV-C treatment for extending shelf life and improving nutritional quality of African indigenous leafy vegetables. *Postharvest Biology and Technology, 129*, 107–117. https://doi.org/10.1016/j.postharvbio.2017.03.019.

Goswami, D. N., & Saha, S. K. (2000). An investigation of the melting properties of different forms of lac by differential scanning calorimeter. *Surface Coatings International, 83*(7), 334–336. https://doi.org/10.1007/bf02692742.

Gutiérrez, T. J. (2018a). Polymers for food applications: News. In T. J. Gutiérrez (Ed.), Polymers for food applications (pp. 1–4). Cham: Springer. https://doi.org/10.1007/978-3-319-94625-2_1.

Gutiérrez, T. J. (2018b). Biological macromolecule composite films made from sagu starch and flour/poly(ε-caprolactone) blends processed by blending/thermo molding. *Journal Polymers and the Environment, 26*(9), 3902–3912. https://doi.org/10.1007/s10924-018-1268-6.

Gutiérrez, T. J. (2018c). Active and intelligent films made from starchy sources/blackberry pulp. *Journal Polymers and the Environment, 26*(6), 2374–2391. https://doi.org/10.1007/s10924-017-1134-y.

Gutiérrez, T. J. (2018d). Are modified pumpkin flour/plum flour nanocomposite films biodegradable and compostable? *Food Hydrocolloids, 83*, 397–410. https://doi.org/10.1016/j.foodhyd.2018.05.035.

Gutiérrez, T. J., & Álvarez, K. (2017). Biopolymers as microencapsulation materials in the food industry. In M. Masuelli & D. Renard (Eds.), *Advances in physicochemical properties of biopolymers: Part 2* (pp. 296–322). Bentham Science Publishers. EE.UU. ISBN: 978-1-68108-545-6. eISBN: 978-1-68108-544-9. https://doi.org/10.2174/9781681085449117010009.

Gutiérrez, T. J., & Alvarez, V. A. (2018). Bionanocomposite films developed from corn starch and natural and modified nano-clays with or without added blueberry extract. *Food Hydrocolloids, 77*, 407–420. https://doi.org/10.1016/j.foodhyd.2017.10.017.

Gutiérrez, T. J., & González, G. (2017). Effect of cross-linking with *Aloe vera* gel on surface and physicochemical properties of edible films made from plantain flour. *Food Biophysics, 12*(1), 11–22. https://doi.org/10.1007/s11483-016-9458-z.

Gutiérrez, T. J., Herniou--Julien, C., Álvarez, K., & Alvarez, V. A. (2018). Structural properties and in vitro digestibility of edible and pH-sensitive films made from Guinea arrowroot starch and wastes from wine manufacture. *Carbohydrate Polymers, 184*, 135–143. https://doi.org/10.1016/j.carbpol.2017.12.039.

Gutiérrez, T. J., Ollier, R., & Alvarez, V. A. (2018). Chapter 5. Surface properties of thermoplastic starch materials reinforced with natural fillers. In: *Functional Biopolymers*. Vijay Kumar Thakur, and Manju Kumari Thakur (Eds). Editorial Springer International Publishing.

EE.UU. ISBN: 978-3-319-66416-3. eISBN: 978-3-319-66417-0. pp. 131-158. https://doi.org/10.1007/978-3-319-66417-0_5

Gutiérrez, T. J., Toro-Márquez, L. A., Merino, D., & Mendieta, J. R. (2019). Hydrogen-bonding interactions and compostability of bionanocomposite films prepared from corn starch and nano-fillers with and without added Jamaica flower extract. *Food Hydrocolloids, 89*, 283–293. https://doi.org/10.1016/j.foodhyd.2018.10.058.

Hagenmaier, R. D., & Shaw, P. E. (1991). Permeability of shellac coatings to gases and water vapor. *Journal of Agricultural and Food Chemistry, 39*(5), 825–829. https://doi.org/10.1021/jf00005a001.

Heng, P. W., Sia, C. L. W., & Ong, K. (2003). Influence of storage conditions and type of plasticizers on ethylcellulose and acrylate films formed from aqueous dispersions. *Journal of Pharmacy and Pharmaceutical Sciences, 6*(3), 334–344.

Herniou--Julien, C., Mendieta, J. R., & Gutiérrez, T. J. (2019). Characterization of biodegradable/non-compostable films made from cellulose acetate/corn starch blends processed under reactive extrusion conditions. *Food Hydrocolloids, 89*, 67–79. https://doi.org/10.1016/j.foodhyd.2018.10.024.

Hoang-Dao, B. T., Hoang-Tu, H., Tran-Thi, N. N., Koubi, G., Camps, J., & About, I. (2009). Clinical efficiency of a natural resin fluoride varnish (shellac F) in reducing dentin hypersensitivity. *Journal of Oral Rehabilitation, 36*(2), 124–131. https://doi.org/10.1111/j.1365-2842.2008.01907.x.

Jacxsens, L., Devlieghere, F., & Debevere, J. (2002). Predictive modelling for packaging design: Equilibrium modified atmosphere packages of fresh-cut vegetables subjected to a simulated distribution chain. *International Journal of Food Microbiology, 73*(2–3), 331–341. https://doi.org/10.1016/S0168-1605(01)00669-9.

Jitareerat, P., Kalaya, S., Thoung, V.T., Uthairatanakij, A., & Limmatvapirat, S. (2018, 7–8 March). *Antifungal effects of ethanolic shellac-modified coconut oil (ES—MCO) combined with physical treatments against postharvest disease of mango and mangosteen.* Paper presented at the Proceedings of international symposium on innovative crop protection for sustainable protection, Gifu University.

Jo, W.-S., Song, H.-Y., Song, N.-B., Lee, J.-H., Min, S. C., & Song, K. B. (2014). Quality and microbial safety of 'Fuji' apples coated with carnauba-shellac wax containing lemongrass oil. *LWT-Food Science and Technology, 55*(2), 490–497. https://doi.org/10.1016/j.lwt.2013.10.034.

Khairuddin, Pramono, E., Utomo, S. B., Wulandari, V., Zahrotul, A. W., & Clegg, F. (2016). The effect of polyethylene glycol Mw 400 and 600 on stability of shellac waxfree. *Journal of Physics: Conference Series, 776*(1), 012054. https://doi.org/10.1088/1742-6596/776/1/012054.

Khan, H., Fell, J. T., & Macleod, G. S. (2001). The influence of additives on the spreading coefficient and adhesion of a film coating formulation to a model tablet surface. *International Journal of Pharmaceutics, 227*(1–2), 113–119. https://doi.org/10.1016/S0378-5173(01)00789-X.

Khorram, F., Ramezanian, A., & Hosseini, S. M. H. (2017). Shellac, gelatin and Persian gum as alternative coating for orange fruit. *Scientia Horticulturae, 225*, 22–28. https://doi.org/10.1016/j.scienta.2017.06.045.

Kouassi, K. H. S., Bajji, M., & Jijakli, H. (2012). The control of postharvest blue and green molds of citrus in relation with essential oil–wax formulations, adherence and viscosity. *Postharvest Biology and Technology, 73*, 122–128. https://doi.org/10.1016/j.postharvbio.2012.06.008.

Lieberman, S., Enig, M. G., & Preuss, H. G. (2006). A review of monolaurin and lauric acid: N*atural virucidal* and *bactericidal agents. Alternative and Complementary Therapies, 12*(6), 310–314. https://doi.org/10.1089/act.2006.12.310.

Lima, Á. M., Cerqueira, M. A., Souza, B. W. S., Santos, E. C. M., Teixeira, J. A., Moreira, R. A., & Vicente, A. A. (2010). *New edible coatings composed of galactomannans and collagen blends to improve the postharvest quality of fruits – Influence on fruits gas transfer rate. Journal of Food Engineering, 97*(1), 101–109. https://doi.org/10.1016/j.jfoodeng.2009.09.021.

Limmatvapirat, S., Limmatvapirat, C., Luangtana-anana, M., Nunthanid, J., Oguchi, T., Tozuka, Y., Yamamoto, K., & Puttipipatkhachorn, S. (2004). Modification of physicochemical and mechanical properties of shellac by partial hydrolysis. *International Journal of Pharmaceutics, 278*(1), 41–49. https://doi.org/10.1016/j.ijpharm.2004.02.030.

Limmatvapirat, S., Limmatvapirat, C., Puttipipatkhachorn, S., Nuntanid, J., & Luangtana-anan, M. (2007). Enhanced enteric properties and stability of shellac films through composite salts formation. *European Journal of Pharmaceutics and Biopharmaceutics, 67*(3), 690–698. https://doi.org/10.1016/j.ejpb.2007.04.008.

Limmatvapirat, S., Nunthanid, J., Puttipipatkhachorn, S., & Luangtana-anan, M. (2005). Effect of alkali treatment on properties of native shellac and stability of hydrolyzed shellac. *Pharmaceutical Development and Technology, 10*(1), 41–46. https://doi.org/10.1081/pdt-35897.

Limmatvapirat, S., Panchapornpon, D., Limmatvapirat, C., Nunthanid, J., Luangtana-anan, M., & Puttipipatkhachorn, S. (2008). Formation of shellac succinate having improved enteric film properties through dry media reaction. *European Journal of Pharmaceutics and Biopharmaceutics, 70*(1), 335–344. https://doi.org/10.1016/j.ejpb.2008.03.002.

Lin, S.-Y., Chen, K.-S., & Run-Chu, L. (2000). Organic esters of plasticizers affecting the water absorption, adhesive property, glass transition temperature and plasticizer permanence of eudragit acrylic films. *Journal of Controlled Release, 68*(3), 343–350. https://doi.org/10.1016/S0168-3659(00)00259-5.

Luangtana-anan, M., Limmatvapirat, S. (2010). *Effect of polyethylene glycol on physical properties and stability of shellac film*. Silpakorn University, Faculty of Pharmacy, Nakhon Pathom.

Luangtana-anan, M., Limmatvapirat, S., Nunthanid, J., Wanawongthai, C., Chalongsuk, R., & Puttipipatkhachorn, S. (2007). Effect of salts and plasticizers on stability of shellac film. *Journal of Agricultural and Food Chemistry, 55*(3), 687–692. https://doi.org/10.1021/jf061922+.

Luangtana-anan, M., Nunthanid, J., & Limmatvapirat, S. (2010). Effect of molecular weight and concentration of polyethylene glycol on physicochemical properties and stability of shellac film. *Journal of Agricultural and Food Chemistry, 58*(24), 12934–12940. https://doi.org/10.1021/jf1031026.

Luangtana-anan, M., Soradech, S., Saengsod, S., Nunthanid, J., & Limmatvapirat, S. (2017). Enhancement of moisture protective properties and stability of pectin through formation of a composite film: Effects of shellac and plasticizer. *Journal of Food Science, 82*(12), 2915–2925. https://doi.org/10.1111/1750-3841.13956.

Luangtana-anan, M., & Fell, J. T. (1988). Surface energetics of powders before and after compaction. *International Journal of Pharmaceutics, 41*(3), 237–240. https://doi.org/10.1016/0378-5173(88)90199-8.

Luangtana-anan, M., & Fell, J. T. (1987). Surface free energy determinations on powders. *Powder Technology, 52*(3), 215–218. https://doi.org/10.1016/0032-5910(87)80107-9.

Marei, G. I. K., Abdel Rasoul, M. A., & Abdelgaleil, S. A. M. (2012). Comparative antifungal activities and biochemical effects of monoterpenes on plant pathogenic fungi. *Pesticide Biochemistry and Physiology, 103*(1), 56–61. https://doi.org/10.1016/j.pestbp.2012.03.004.

Martins, J. T., Cerqueira, M. A., Souza, B. W. S., Md, C. A., & Vicente, A. A. (2010). Shelf life extension of ricotta cheese using coatings of galactomannans from nonconventional sources incorporating nisin against *Listeria monocytogenes*. *Journal of Agricultural and Food Chemistry, 58*(3), 1884–1891. https://doi.org/10.1021/jf902774z.

Mattheis, J. P., Buchanan, D. A., & Fellman, J. K. (1991). Change in apple fruit volatiles after storage in atmospheres inducing anaerobic metabolism. *Journal of Agricultural and Food Chemistry, 39*(9), 1602–1605. https://doi.org/10.1021/jf00009a012.

McGuire, R. G., & Hagenmaier, R. D. (2001). Shellac formulations to reduce epiphytic survival of coliform bacteria on citrus fruit postharvest. *Journal of Food Protection, 64*(11), 1756–1760. https://doi.org/10.4315/0362-028x-64.11.1756.

Oh, E., & Luner, P. E. (1999). Surface free energy of ethylcellulose films and the influence of plasticizers. *International Journal of Pharmaceutics, 188*(2), 203–219. https://doi.org/10.1016/S0378-5173(99)00224-0.

Owens, D. K., & Wendt, R. C. (1969). Estimation of the surface free energy of polymers. *Journal of Applied Polymer Science, 13*(8), 1741–1747. https://doi.org/10.1002/app.1969.070130815.

Panchapornpon, D., Limmatvapirat, C., Luangtana-anan, M., Nunthanid, J., Sriamornsak, P., & Limmatvapirat, S. (2011). Fabrication of thermally stabilized shellac through solid state reaction with phthalic anhydride. *Materials Letters, 65*(8), 1241–1244. https://doi.org/10.1016/j.matlet.2011.01.068.

Perdones, A., Sánchez-González, L., Chiralt, A., & Vargas, M. (2012). Effect of chitosan–lemon essential oil coatings on storage-keeping quality of strawberry. *Postharvest Biology and Technology, 70*, 32–41. https://doi.org/10.1016/j.postharvbio.2012.04.002.

Pérez-Alfonso, C. O., Martínez-Romero, D., Zapata, P. J., Serrano, M., Valero, D., & Castillo, S. (2012). The effects of essential oils carvacrol and thymol on growth of *Penicillium digitatum* and *P. italicum* involved in lemon decay. *International Journal of Food Microbiology, 158*, 101–106. https://doi.org/10.1016/j.ijfoodmicro.2012.07.002.

Phan The, D., Debeaufort, F., Luu, D., & Voilley, A. (2008). Moisture barrier, wetting and mechanical properties of shellac/agar or shellac/cassava starch bilayer bio-membrane for food applications. *Journal of Membrane Science, 325*(1), 277–283. https://doi.org/10.1016/j.memsci.2008.07.052.

Porat, R., Lichter, A., Terry, L. A., Harker, R., & Buzby, J. (2018). Postharvest losses of fruit and vegetables during retail and in consumers' homes: Quantifications, causes, and means of prevention. *Postharvest Biology and Technology, 139*, 135–149. https://doi.org/10.1016/j.postharvbio.2017.11.019.

Prabha, T. N., & Bhagyalakshmi, N. (1998). Carbohydrate metabolism in ripening banana fruit. *Phytochemistry, 48*(6), 915–919. https://doi.org/10.1016/S0031-9422(97)00931-X.

Rahman, M. M., Moniruzzaman, M., Ahmad, M. R., Sarker, B. C., & Khurshid Alam, M. (2016). Maturity stages affect the postharvest quality and shelf-life of fruits of strawberry genotypes growing in subtropical regions. *Journal of the Saudi Society of Agricultural Sciences, 15*(1), 28–37. https://doi.org/10.1016/j.jssas.2014.05.002.

Ramadoss, P., Thanigai Arul, K., Ramana Ramya, J., Rigana Begam, M., Sarath Chandra, V., & Manikandan, E. (2017). Enhanced mechanical strength and sustained drug release of gelatin/keratin scaffolds. *Materials Letters, 186*, 109–112. https://doi.org/10.1016/j.matlet.2016.09.095.

Ramos, M., Valdés, A., Beltrán, A., & Garrigós, C. M. (2016). Gelatin-based films and coatings for food packaging applications coatings. *Coatings, 6*(4), 41. https://doi.org/10.3390/coatings6040041.

Ribeiro, C., Vicente, A. A., Teixeira, J. A., & Miranda, C. (2007). Optimization of edible coating composition to retard strawberry fruit senescence. *Postharvest Biology and Technology, 44*(1), 63–70. https://doi.org/10.1016/j.postharvbio.2006.11.015.

Saberi, B., Chockchaisawasdee, S., Golding, J. B., Scarlett, C. J., & Stathopoulos, C. E. (2017). Development of biocomposite films incorporated with different amounts of shellac, emulsifier, and surfactant. *Food Hydrocolloids, 72*, 174–184. https://doi.org/10.1016/j.foodhyd.2017.05.042.

Saberi, B., Golding, J. B., Marques, J. R., Pristijono, P., Chockchaisawasdee, S., Scarlett, C. J., & Stathopoulos, C. E. (2018). Application of biocomposite edible coatings based on pea starch and guar gum on quality, storability and shelf life of 'Valencia' oranges. *Postharvest Biology and Technology, 137*, 9–20. https://doi.org/10.1016/j.postharvbio.2017.11.003.

Saberi, B., & Golding, J. B. (2018). Postharvest application of biopolymer-based edible coatings to improve the quality of fresh horticultural produce. In T. J. Gutiérrez (Ed.), *Polymers for food applications* (pp. 211–250). Cham: Springer. https://doi.org/10.1007/978-3-319-94625-2_9.

Schmid, M., Müller, K. (2019). Chapter 11 - Whey Protein-Based Packaging Films and Coatings. In: Deeth HC, Bansal N (eds) Whey Proteins. Academic Press, pp 407–437. https://doi.org/10.1016/B978-0-12-812124-5.00012-6.

Sirviö, J. A., Visanko, M., Ukkola, J., & Liimatainen, H. (2018). Effect of plasticizers on the mechanical and thermomechanical properties of cellulose-based biocomposite films. *Industrial Crops and Products, 122*, 513–521. https://doi.org/10.1016/j.indcrop.2018.06.039.

Sittipaisankul, P. (2015). Development of antimicrobial coating material from shellac-modified coconut oil for controlling of postharvest diseases. Master degree Thesis, Silpakorn University.

Soradech, S. (2010). Enhancement of mechanical properties and stability of shellac by formation of composite polymer with gelatin. Thesis of Master of Science, University of Silpakorn.

Soradech, S., Limatvapirat, S., & Luangtana-anan, M. (2013). Stability enhancement of shellac by formation of composite film: Effect of gelatin and plasticizers. *Journal of Food Engineering, 116*(2), 572–580. https://doi.org/10.1016/j.jfoodeng.2012.12.035.

Soradech, S., Nunthanid, J., Limmatvapirat, S., & Luangtana-anan, M. (2012). An approach for the enhancement of the mechanical properties and film coating efficiency of shellac by the formation of composite films based on shellac and gelatin. *Journal of Food Engineering, 108*(1), 94–102. https://doi.org/10.1016/j.jfoodeng.2011.07.019.

Soradech, S., Nunthanid, J., Limmatvapirat, S., & Luangtana-anan, M. (2017). Utilization of shellac and gelatin composite film for coating to extend the shelf life of banana. *Food Control, 73*. (Part B, 1310–1317. https://doi.org/10.1016/j.foodcont.2016.10.059.

Suderman, N., Isa, M. I. N., & Sarbon, N. M. (2018). The effect of plasticizers on the functional properties of biodegradable gelatin-based film: A review. *Food Bioscience, 24*, 111–119. https://doi.org/10.1016/j.fbio.2018.06.006.

Tangwatcharin, P., & Khopaibool, P. (2012). Activity of virgin coconut oil, lauric acid or monolaurin in combination with lactic acid against *Staphylococcus aureus*. *The Southeast Asian Journal of Tropical Medicine and Public Health, 43*, 969–985.

The Engineering Toolbox Thermal Conductivity of common Materials and Gases. (n.d.). https://goo.gl/rQc2EJ.

Toivonen, P. M. A., & Brummell, D. A. (2008). Biochemical bases of appearance and texture changes in fresh-cut fruit and vegetables. *Postharvest Biology and Technology, 48*(1), 1–14. https://doi.org/10.1016/j.postharvbio.2007.09.004.

United States Department of Agriculture Orange shellac. (n.d.). https://goo.gl/HVKofm. Accessed 8 Dec 2018.

Valencia-Chamorro, S. A., Pérez-Gago, M. B., del Río, M. Á., & Palou, L. (2009). Effect of antifungal hydroxypropyl methylcellulose (HPMC)–lipid edible composite coatings on postharvest decay development and quality attributes of cold-stored 'Valencia' oranges. *Postharvest Biology and Technology, 54*(2), 72–79. https://doi.org/10.1016/j.postharvbio.2009.06.001.

Vieira, J. M., Flores-López, M. L., de Rodríguez, D. J., Sousa, M. C., Vicente, A. A., & Martins, J. T. (2016). Effect of chitosan–Aloe vera coating on postharvest quality of blueberry (*Vaccinium corymbosum*) fruit. *Postharvest Biology and Technology, 116*, 88–97. https://doi.org/10.1016/j.postharvbio.2016.01.011.

Wei, X., Pang, J., Zhang, C., Yu, C., Chen, H., & Xie, B. (2015). Structure and properties of moisture-resistant konjac glucomannan films coated with shellac/stearic acid coating. *Carbohydrate Polymers, 118*, 119–125. https://doi.org/10.1016/j.carbpol.2014.11.009.

Wihodo, M., & Moraru, C. I. (2013). Physical and chemical methods used to enhance the structure and mechanical properties of protein films: A review. *Journal of Food Engineering, 114*(3), 292–302. https://doi.org/10.1016/j.jfoodeng.2012.08.021.

Wu, S. (1971). Calculation of interfacial tension in polymer systems. *Journal of Polymer Science Part C: Polymer Symposia banner, 34*(1), 19–30. https://doi.org/10.1002/polc.5070340105.

Zhang, L., Wang, X. F., Liu, H., Wang, Y., Yu, L., & Simon, G. P. (2018). Effect of plasticizers on microstructure, compatibility and mechanical property of hydroxypropyl methylcellulose/hydroxypropyl starch blends. *International Journal of Biological Macromolecules, 119*, 141–148. https://doi.org/10.1016/j.ijbiomac.2018.07.064.

Zhou, R., Li, Y., Yan, L., & Xie, J. (2011). Effect of edible coatings on enzymes, cell-membrane integrity, and cell-wall constituents in relation to brittleness and firmness of Huanghua pears (*Pyrus pyrifolia* Nakai, cv. Huanghua) during storage. *Food Chemistry, 124*(2), 569–575. https://doi.org/10.1016/j.foodchem.2010.06.075.

Chapter 25
Application of Bionanocomposites on Horticultural Products to Increase the Shelf Life

Adrija Choudhury, Peerzada Gh. Jeelani, Nishita Biswal, and Ramalingam Chidambaram

Abstract Post-harvest loss of horticultural products is mainly due to microbial contamination, over-ripening, losses during transit, press injuries and secondary infections due to moth/borer attacks. Traditional practices to reduce post-harvest losses include storing the products in paper and plastic boxes and cushioning them with paper. Key issues such as microbial contamination and over-ripening are not addressed in these measures. In recent years, bionanocomposites and polymers have increasingly been used as coating materials for post-harvest products to reduce microbial contamination, over-ripening and spoilage due to oxygen, carbon dioxide and water vapor exposure. Bionanocomposites are mixtures of biopolymers and nanoparticles (NPs), where the NPs increase the weak barrier and mechanical properties of the biopolymers. The advantages of using nanocoatings include better barrier performance, antimicrobial properties due to the presence of NPs such as silver, reduced material use and simpler film conversion. They are environment-friendly, unlike petroleum-based polymers. This chapter aims to analyse the recent trends in the use of bionanopolymers as coatings for horticultural products and their efficacy.

Keywords Biopolymers · Packaging · Postharvest loss

25.1 Introduction

Post-harvest loss of crops varies between 20–50% in developing countries. It can be caused by the inefficient transportation, storage and packing of the crops produced. These problems are usually solved with measures such as storing the product in cold rooms or packaging it in foams and plastics to prevent microbial

A. Choudhury · P. G. Jeelani · N. Biswal · R. Chidambaram (✉)
Department of Biotechnology, School of Bioscience & Technology,
Vellore Institute of Technology (VIT), Vellore, Tamilnadu, India
e-mail: cramalingam@vit.ac.in

T. J. Gutiérrez (ed.), *Polymers for Agri-Food Applications*,
https://doi.org/10.1007/978-3-030-19416-1_25

attacks and facilitate transportation. The materials used for traditional packaging of horticultural products include plastic and paper. Plastic is derived from petroleum, thus it is not a renewable resource (Gutiérrez and Alvarez 2017a, b, c, d; Gutiérrez 2018a). It is harmful to the environment since it contains toxic compounds such as nonylphenol (NP), polychlorinated biphenyls (PCBs), organic pesticides, such as polycyclic aromatic hydrocarbons (PAHs), dichlorodiphenyltrichloroethane (DDT), polybrominated diphenyl ethers (PBDEs) and bisphenol A (BPA) (Webb et al. 2013). Paper is derived from trees and huge demands for paper will lead to great deforestation. Therefore, a sustainable measure is generally sought to obtain a better package. One of those technologies is to use biopolymers (Álvarez et al. 2017). Commonly used biopolymers are starch and cellulose derivatives, poly-lactic acid (PLA), poly-caprolactone (PCL), poly-(butylene succinate) (PBS) and poly-hydroxybutyrate (PHB) (Rhim et al. 2013a). Incorporation of nanofillers such as clay, metal nanoparticles (NPs), nanocrystal cellulose has been experimentally proved to increase barrier properties and mechanical properties of the biopolymers (Pinto et al. 2014; Gutiérrez and Alvarez 2017e, 2018; Merino et al. 2018; Toro-Márquez et al. 2018; Gutiérrez et al. 2019; Herniou--Julien et al. 2019; Merino et al. 2019). The challenges remain to achieve complete dispersion of the NPs and increase the compatibility of two phases of the bionanocomposites (Rhim et al. 2013a).

25.2 Post-Harvest Loss: Causes and Its Control Measures

Agriculture is the backbone of Indian economy. It contributes around 14% to GDP. Increased agricultural production has enabled us to meet the ever-increasing food demands by growing population. After production, products must undergo various post-harvest operations. Each operation leads to some losses which decreases the availability of food. According to a study 3–6% of cereals, 4.3–6.1% of pulses are lost during harvest and post-harvest operations (Cardoen et al. 2015). Thus, there is a large gap between the amount of food produced and the amount of food that is available to people. The food security of a country can only be achieved when these losses can be prevented (Gutiérrez et al. 2016a). Post-harvest loss can be defined as the distortion in both quantity and quality of food from the time of harvest until it reaches consumers. The parameters on which food loss can be measured are reduced weight, low nutrient value, color change, texture etc. (Kumar et al. 2017). Although India is one of the tops most country in production of fruits and vegetables but not more than 2% of these products are processed. Approximately around 5–25% in developed countries and 25–50% in developing countries accounts for post-harvest losses in fruits and vegetables production.

25.2.1 Factors Influencing Post-Harvest Loss

All the factors responsible for post-harvest loss can be grouped into physical, physiological, mechanical and environmental factors. Fruits and vegetables have high moisture content and lack storage stability. Certain biological agents such as mites, insects also help in the loss of crops on a large scale. Environmental factors such as temperature, humidity and oxygen balance greatly influence post-harvest loss. Due to inadequate transportation on a large scale, the storage chain facilities destroy a substantial number of crops. The maximum post-harvest loss can be attributed to microbial spoilage caused by fungi, bacteria, molds, etc. (Kumar et al. 2017).

25.2.2 Control of Post-Harvest Losses

The control of postharvest produce can be divided into the following measures: chemical, gaseous and physical. Chemical measures usually target the retardation of browning reactions, inhibition of ethylene formation, reduction of water loss and respiration rates by chemicals such as organic acids, nitric oxide, sulphur dioxide, peroxyacetic acid, hydrogen peroxide (H_2O_2) etc. For gaseous treatments, ozone and 1-mehtylcyclopropene are used, packaging, heat treatments, cold storage and irradiation are used as physical treatments to combat post-harvest losses (Sandarani et al. 2018). Agricultural products must be packaged properly. The packaging helps reduce mechanical damage, unwanted physiological changes and deterioration during transport, storage and marketing (Álvarez et al. 2018). More infrastructural advances are needed, such as cold storage. Application of fertilizers and pesticides are detrimental for growth of crops and does little to control postharvest losses. The measures taken to combat post-harvest losses are described below:

25.2.2.1 Chemical and Gaseous Treatment

Chemical and gaseous treatments on the postharvest produce mainly target the enzymes required for ripening, oxidants in the produce and pathogenic organisms. Chemicals such as chlorine, ozone, H_2O_2 and sulphur dioxide are used for their anti microbial activities. Nitric oxide imparts antiethylene activity whereas the chemical 1-methylcyclopropene acts on the ethylene receptors to inhibit the binding of ethylene, thus as a result giving anti ethylene activities too. Calcium chloride on the other hand reduces injuries related to cold treatment by a suggested method of increasing firmness of the tissues (Sandarani et al. 2018).

25.2.2.2 Storage

The maintenance of the quality of the agricultural product along with the increase of its shelf life by preventing them from spoilage is referred as storage. Depending upon the use, various products after harvesting are stored under different conditions, e.g. the seeds that would be used for plantation in following year and the forage being processed for live stocks will need very different storage facilities. Vast ranges of storage structure are available for successful storage of horticultural products. Storage areas must be cool (refrigerated), shaded and ventilated (Kiaya 2014). Controlling the temperature and relative humidity (RH) during storage and transport of postharvest produce is one of the essential ways to combat losses (Kader 2003). Use of modified atmospheric packaging (MAP) along with the cold storage facilities is an increasing trend (Kader 2003). Storage is essential for following reasons:

- To increase the shelf life of perishable materials.
- For availability of food round the year.
- For large-scale processing.
- To maintain nutritional quality.
- To open the doors to the export business.

25.2.2.3 Packaging of Horticultural Produce

The food packaging industry is continuously changing along with advances in materials science and consumer demand. The packaging not only helps a uniform and effective distribution, but also facilitates the conservation of food and convenience for consumers. The packaging industry has become the third largest industry in the world (Robertson 1998). Packaging provides a favourable environment to the food by protecting them from mechanical shocks, vibrations, biological agents such as insects and mites, water vapor, dust etc. Traditional packaging materials are non-biodegradable and have come under scrutiny for the release of toxic chemicals into the packed food. Recently the focus has been shifted in developing functional packaging systems with advanced characteristics. Techniques such as MAP have been developed. This method modifies the atmosphere inside the packaging by controlling the respiration rates of the product and the gas transfer through the film used for packaging (Sandarani et al. 2018; Sousa-gallagher and Mahajan 2013). The main aim is to increase the levels of carbon dioxide (CO_2) inside the packaging which will have inhibitory effects on ethylene (Sandarani et al. 2018). It will also seek to eliminate oxygen (Sousa-gallagher and Mahajan 2013). Most of the commonly used plastic films are not suitable for MAP. If however they are used, micro-perforations are made to allow for exchange of gases (Sandarani et al. 2018). Therefore "breathable" polymers are used. These polymers can also be loaded with ethylene scavenging chemicals, NPs, antioxidants etc. In addition, better packaging helps reduce losses, maintains quality and increases shelf life. Fruits and vegetables require maximum oxygen transmission. Thus, they need to be packed in oxygen permeable material whereas processed food does not require such mass transfer. The existing packaging systems have some limitations such as a small margin for recycling, non-sustainability, in addition, low mechanical and barrier properties. The use of natural sources for packaging has raised a number of issues. They have poor mechanical and barrier properties and low sealability. Thus, over the years, industry and the academic world have come together to exploit the field of nanotechnology to solve these problems.

25.3 Potential Bionanocomposites for Food Packaging

Bionanocomposites are the materials having two or more phases, where one phase is a continuous phase which is mainly composed of biopolymers and the other is the discontinuous phase which is mainly composed of nanofillers (Othman 2014). Biopolymers have covalently bonded monomeric units. They are biodegradable which means that they can be broken down by innate microbes of the environment. The by-products generated by these microbes are ususally CO_2 and water which are usually not harmful to the environment (Othman 2014). Biopolymers can be divided into two major categories which are synthetic and natural. They can also be divided based on their sources. Biopolymers are environment friendly and ensure safety and quality of food, but the major setback is its weak mechanical and barrier properties. Therefore, to boost up the material properties of biopolymers, nanocomposite technology was initiated. Nanocomposites consist of one or mixture of polymers with either organic or inorganic nanofillers with dimension less than 100 nm. Studies claims that the incorporation of nanomaterials such as SiO_2, clay, TiO_2, nanocellulose (Mihindukulasuriya and Lim 2014) etc., have dramatically increased the mechanical and barrier properties of the biopolymers. It was found after extensive literature studies that properties such as thermal stability, toughness, stiffness, tensile strength and optical properties of polymers were improved after nanofiller incorporation (Fig. 25.1). The nanofillers used can be organic (clay, natural biopolymers) or inorganic (metal, metal-oxide) (Othman 2014).

25.3.1 Agar Based Nanocellulose

Agar is a very good reagent for preparation of biodegradable films for the purpose of using it as a packaging material (Shankar et al. 2014). Agar is isolated from red algae. It is a biocompatible and biodegradable and forms nanocomposites when proper nanofillers such as nanocrystalline cellulose (NCC) of are provided (Reddy and Rhim 2014). It can also form nanocomposites with clay, copper and silver NPs (AgNPs). Various studies have reported that the water vapor permeability (WVP) and mechanical stability improves by great extent with the use of such nanofillers (Rhim et al. 2013a; Bracone et al. 2016; Gutiérrez et al. 2017). They have high

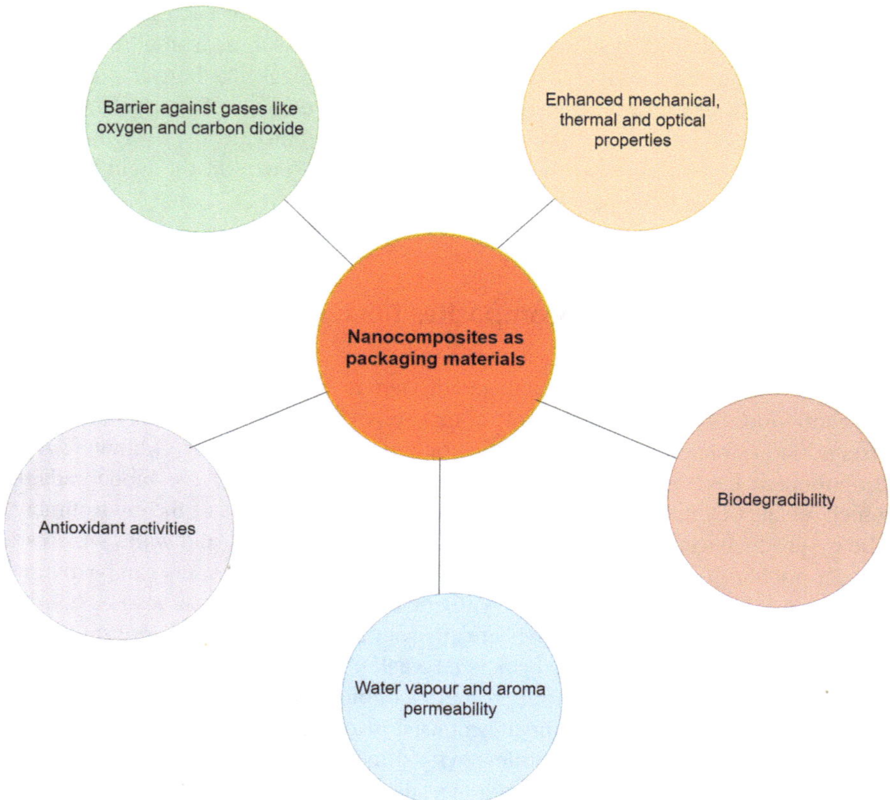

Fig. 25.1 Properties of bionanocomposites. It can be seen in Table 25.1 and Table 25.2 that there are many options for biopolymers and nanofillers that can be combined to form bionanocomposites. Here three types of bionanocomposites are discussed

Table 25.1 Classification of biopolymers according to their sources

	Natural	Synthetic			References
Source	Natural (From biomass)	Synthetic (By microbes)	Synthetic (From biomass)	Synthetic (From petroleum)	Othman (2014); Rhim et al. (2013a, b)
Biopolymer	Starch, whey protein, chitosan, agar, cellulose, gelatine, gluten, collagen	Poly-hydroxy alkanoates (PHA)	Poly-lactic acid (PLA)	Poly vinyl alcohol (PVA), PCL, Polyglycolic acid (PGA)	Othman (2014); Rhim et al. (2013a)

Table 25.2 Classification of nanofillers according to their chemical nature

Organic	Inorganic	References
Clay (Mt, Cloisite)	Metal (silver, copper, gold)	Conte et al. (2013); Lee et al. (2014); Othman (2014); Pinto et al. (2014); Rhim et al. (2011); Shin et al. (2014); Youssef et al. (2014)
Bionanopolymers (cellulose, chitosan)	Metal oxide (zinc oxide, magnesium oxide, titanium dioxide, silver oxide)	Kanmani and Rhim (2014); Martelli and Barros (2013); Othman (2014)
Antimicrobial agents (Nisin)		Imran et al. (2012); Othman (2014)

thermal stability. Cellulose is the most abundant biopolymer. This is because it is a key component of the plant cell wall. The agricultural wastes can thus be used to extract cellulose. Cellulose is a linear carbohydrate polymer. The D-glucopyranose units of cellusose are linked *via* β-1, 4 glycosidic-bond (Khan et al. 2012). NCC can be extracted by controlled acid hydrolysis of the cellulose source (Khan et al. 2012; Reddy and Rhim 2014).

NCC was prepared by Reddy and Rhim (2014) in the following way:

Cellulose fibers were hydrolyzed by 47% sulphuric acid for about 3 h with a 1:20 ratio of fiber to solution respectively by refluxing for 3 h at 60 °C under strong agitation. Excess distilled water was added to quench the reaction and then cooled to the room temperature. The suspensión of the treated fibres was centrifuged at 4000 rpm for 20 min using a bench-top centrifuge. Tthe supernatant was then discarded until it the suspensión exhibited turbidity. The fractions were washed, and the suspension sonicated by using an ultrasonic processor of high intensity for 5 min in a cold environment, preferabbly and ice bath. The suspension was then subjected to wáter dialysis. Dialysis was stopped once neutrality was attained. Finally, the suspension was freeze-dried to get NCC of the mulberry pulp fiber.

Nanocomposite film was prepared by Reddy and Rhim (2014) in the following way:

The isolated NCC was dispersed in 150 ml distilled water and stirred for about 1 h. This suspension was homogenized by further mixing at 12,000 rpm for 10 min. To this mix, 4 g of agar and 2 g of glycerol was mixed, and the solution stirred for about 30 min at 95 °C. The solution was then cast into Teflon film coated glass plate and conditioned in a constant temperature and humidity for at-least 2 days.

FTIR analysis proved that there was no change in the chemical nature of cellulose fragments after acid hydrolysis. Thermal degradation of the mulberry fibre and nanocellulose were observed in two steps. The initial degradation occured at 80 °C to 100 °C and the main degradation occured at the range of 200–370 °C which was attributed to degradation of cellulose. Thermal stability of the nanocomposite had three regions. The initial degradation was at 80–150 °C due to water evaporation. The second degradation was at 190–250 °C which was due to volatilization of plasticizers. The third degradation of the cellulose fragments occurred above 300 °C which was due to the thermal degradation of the biopolymer chain. The tensile strength of pristine agar film was found to be 33.3 MPa and there was an increase to 41.3 MPa seen in the composite film containing 3% (w/w) NCC. A decrease in tensile strength was found with more addition of NCC to agar. 3% NCC content in agar was also the optimum concentration to decrease WVP (Reddy and Rhim 2014).

25.3.2 Thermoplastic Starch-Clay Bionanocomposites

Starch is a semi crystalline polymer and its monomeric unit is glucopyranosyl units. Starch is present in two forms: amylose (linear) and amylopectin (branched) (Suárez and Gutiérrez 2017; Gutiérrez 2018b). It has α-1, 4 glycosidic linkage only in amylose and both α-1, 4 glycosidic linkage and α-1, 6 glycosidic linkage in amylopectin (Avella et al. 2005). Starch is one of the most easily available polymers in nature, it is completely biodegradable, and its resource material is very cheap (Gutiérrez et al. 2015a; Gutiérrez et al. 2015b; Gutiérrez et al. 2015c, b). However, the crude starch is present as granules. Thus, it poses difficulties to machine it into a thin film. To combat this problem, starch is gelatinized with 6–10% moisture under heat and pressure to form thermoplastic starch (TPS) (Park et al. 2002). TPS along cannot be used for food packaging. Therefore, environment friendly fillers such as clays are used to improve the property of TPS. Hydrophilic starch forms bionanocomposites with natural clay such as montmorillonite (Mt). The tensile strength of TPS increases after use of natural fillers. Clay is natural filler which is present in abundant and used as components in food, medical and cosmetics. The elastic modulus of TPS also increases by incorporation of clay and thus improves barrier properties and mechanical stability (Chen and Evans 2005).

Mt. clay is used as the nanofiller to prepare the bionanocomposite film. It is environment friendly and is present in large amount in nature (Wang et al. 2008). Park et al. (2002) prepared the TPS/clay hybrid vacuum dried the TPS/clay for 24 h at 80 °C, TPS and clay were mixed on a roller for 20 min at 110 °C. After preparation, the blends were placed in air tight polyethene bags to prevent moisture absorption.

Mt. was found to be non toxic to Caco-2 cells at a level of 5 μg/well in a 96 well plate (Wang et al. 2008). Four different clay types were used. Cloisite 6A, cloisite Na^+, cloisite 10A, cloisite 30B. It was found that TPS/cloisite Na^+ and TPS/cloisite 30B hybrid exhibited higher tensile strength than pure TPS. It was also seen that TPS/clay hybrids generally showed less WVP, with TPS/cloisite Na^+ showing the lowest WVP and TPS/cloisite 10A showing the highest among composite films (Park et al. 2002).

25.3.3 Metal/Metal Oxide-Carbohydrate Bionanocomposites

Metal NPs have gained considerable interest in their use as nanofillers in bionanocomposites. This is because the metals such as silver and zinc imparts antimicrobial properties to the bionanocomposite films (Rhim et al. 2013a). Metal oxides such as ZnO have attracted interest due to their biocompatibility and stability under difficult conditions (Mohammadi et al. 2013). This makes the metal and metal oxide NPs a very good candidate for the preparation of bionanocomposites. AgNPs can be readily incorporated in biopolymers such as starch and chitosan (Cs) and increase their mechanical, thermal properties and also imparting antimicrobial property. The antimicrobial properties of AgNPs vary according to factors such as the size, quantity and distribution of the particles, their interaction with the base material and also the degree of agglomeration (Rhim et al. 2013a). AgNPs exhibit a strong anti microbial property against broad range bacteria. It also improves the mechanical strength of bionanocomposites. There is a significant increase in surface hydrophobicity and a substantial increase in water vapor barrier properties was observed.

Synthesis of metal NPs are done conventionally with harsh reducing agents such as dimethyl formamide, triethanolamine, hydrazine, sodium borohydride, etc. (Rhim et al. 2013a). But a more environment friendly way is to synthesise it by using plant extracts.

AgNPs was prepared in 5 ml of *Aloe vera* extract where 2.5 mL of 30% ammonia solution was added to 5 mL of 10^{-2} M $AgNO_3$ (silver nitrate) solution. Water is added to make up the volume upto 50 ml. This changed the concentration of silver nitrate to 10^{-3} M. Formation of a faint yellow color after 24 h of the reaction indicates the formation of AgNPs. Characterisation by UV-vis absorbance and TEM measurements were done. Analysis of the extract revealed that the AgNPs formed are roughly spherical in shape (Chandran et al. 2006).

Agar/Ag bionanocomposite films were prepared by 4 g of agar was dissolved in a silver solution containing 1.2 g of glycerol (plasticiser). The solution was mixed thoroughly for 30 min at 95 °C with the help of a magnetic stirrer. After mixing period was over, it was cast into a Teflon film coated glass plate then dried for 24 h. The films were kept at a chamber of 50% RH and 25 °C for 48 h before analysis. The agar/Ag film exhibited initial thermal decomposition at a temperature of 90 °C and a second thermal decomposition at 324 °C (Rhim et al. 2013a). Due to the water retained in the film there is initial decomposition. The second decomposition is attributed to the decomposition of the agar base (Rhim et al. 2013a). The films having high amount of AgNP (1–2%) had a significant increase in the tensile strength compared to a control of only agar film (Rhim et al. 2013a). WVP also decreases upon incorporation of AgNP (Rhim et al. 2013a). Composites with an AgNP content of 0.5–2% showed significant antimicrobial properties against food borne pathogens such as *Listeria monocytogenes* (Gram-positive) and *Escherichia coli* O157:H7 (Gram-negative). The increasing concentration of Ag (Range- 0.5% to 2%) showed increasing antimicrobial properties, with 2% of Ag showing the highest bactericidal activity (Rhim et al. 2013a). Zinc oxide is also used to make bionanocomposite films. Out of the five zinc compounds listed as generally recognized as safe (GRAS) by the U.S. Food and Drug Administration (FDA, 21CFR182.8991), ZnO is among them, thus making it safe for use in packaging (Mohammadi et al. 2012).

ZnO/starch composite was prepared by ZnO nanorods were dispersed in water and mixed at 60 °C for 1 h. It was then subjected to sonication in an ultrasonic bath for 30 min. The solution was brought to room temperature and to this cooled solution 4% (w/w) starch was added. Sorbitol and glicerol 40% (w/w) of total solids were added in a 3:1 ratio respectively. The choice of plasticisers and their concentration was based on Abdorreza et al. (2011). The solution was then heated to 95 °C for 45 min to complete gelatinisation. Upon completion, the solution was cooled to room temperature and cast on Perspex plates. These were then put in a chamber having 25 °C and 50% RH.

With the same plasticizers, the control films were prepared but the NPs were not added. The dried films were removed and were stored at 50% RH and 23 °C until tested. Water absorption capacity of the nanocomposite film decreased as compared to the control. The Young modulus and tensile strength showed a significant increase in the bionanocomposite films. Due to increase water resistance of the bionanocomposite films, oxygen permeability also decreases. There was a decrease in the oxygen permeability of starch based films from 80 cm^3 μm/(m^2 day atm) to 50 cm^3 μm/(m^2 day atm) (Mohammadi et al. 2013). The increasing amount of ZnO nanorods shows increase in antimicrobial activity. A ZnO content of 5% (w/w) in starch shows highest level of inhibition (Mohammadi et al. 2013). This could be because of chemical interaction with the bacteria i.e. by penetration of bacterial cell by Zn ions and rendering them unviable or by generating H_2O_2 (Zhang et al. 2007). It could also be due to physical interaction where the nanorods could act as needles and pierce the bacterial cells (Abdorreza et al. 2011).

25.4 Aplications

Horticulture deals with the production of vegetables, fruits, ornamental plants and flowers. Horticultural products contain a lot of moisture in them which is why they are most prone to spoilage. In developing countries the percentage of horticultural products that goes to waste is around 20–30% (Yadollahi et al. 2010). One of the most common food pathogens is *Botrytis cinerea.* It is a voracious fungal pathogen which induces grey mold disease in many economically important plants both pre harvest and post harvest. It attacks the fruits and vegetables through direct penetration or through cuts caused while handling during cultivation (Yadollahi et al. 2010). The main requisites for a polymer to be used as a food packaging material are:

- Providing good barrier against gases such as CO_2 and oxygen.
- Providing better mechanical, optical and thermal properties.
- Providing a vapor and aroma barrier.

The traditional method of packaging horticultural products included wrapping them in plastic bags. These did not do much to prevent the spread of diseases such as gray mold. The plastic bags are usually made of high density polyethene (HDPE) which is non-biodegradable. With an increase in the production of horticultural products, there is an ever-increasing demand of packaging materials. It would do the environment much less harm to use materials which can be degraded easily without the release of toxic element to it. Since biodegradable polymers do not have good mechanical properties, they are reinforced with nanosized materials which increase their mechanical properties. The tensile strength of HDPE was found to be 15 MPa. The tensile strength (as indicated in Table 25.3) of starch-based clay and zinc oxide composites were reported to be 3.32 MPa and 6.62 MPa respectively whereas the tensile strength of agar based nanocellulose and silver NP composites were found to be 41.3 MPa and 53.44 MPa respectively. This shows that agar based bionanocomposites is much stronger than either of the starch-based composites as well as HDPE. Starch based bionanocomposites however present excellent biodegradability. The starch present in the composites gives a nutritive source for the microorganisms which will degrade the composite (Singh and Malshe 2017). Starch based clay composite gave very good WVP as indicated in Table 25.3. The nanoclay composites can provide substantial barrier against gases thus are used as packaging for fruits and vegetables (Singh and Malshe 2017). Antimicrobial properties can be imparted by bionanocomposites having AgNPs or zinc oxide NPs (Singh and Rattanpal 2014). The AgNPs, in addition to their antibacterial activity can also aid in the decomposition of ethylene (Singh and Rattanpal 2014). Ethylene is known to produce senescence, decay and softening in postharvest products (Pathak et al. 2017). The action of the AgNPs in the composite helps in slow ripening of the fruits and thus reduces spoilage. AgNPs also have known antimicrobial activities. The ZnO nanorods can interact chemically as well as physically with bacterial cells to impart bactericidal activity (Abdorreza et al. 2011; Zhang et al. 2007). Bionanocomposite coatings may have other active ingredients such as antioxidants or aroma-inducing compounds (Yadollahi et al. 2010).

Some applications of other types of bionanocomposites are:

(a) Cellulose based bionanocomposites

Cellulose is, as already mentioned a very abundant material in the world. Cellulose nanocrystals impart a very high tensile strength and mechanical properties to the bionanocomposites. On the other hand, they also have a low density. Cellulose obtained from different sources such as bacteria, cotton, other trees, underwent different processes to form bionanocomposites. Cellulose in association with plasticisier, clay and compatibiliser showed improved thermal stability and tensile strength modulus (Singh and Malshe 2017).

(b) Chitosan based bionanocomposites

Chitosan (Cs) is a polymer which has a random distribution of *N*-acetyl-glucosamine and β-(1–4)-D-glucosamine (Gutiérrez 2017a). Cs along with titanium dioxide (TiO_2) provides a good packaging for ethylene scavenging. TiO_2 can photodegrade ethylene and also act as an antimicrobial agent (Kaewklin et al. 2018). TiO_2 is

Table 25.3 Properties of the bionanocomposites

Nanocomposites	Thermal Stability	Tensile Strength	Oxygen permeability	Water absorption	Other properties	References
Agar/Nanocellulose (3%) composite	The first region at 80–150 °C the second degradation at 190–250 °C the third degradation above 300 °C	41.3 MPa	Decreases	WVP of 1.63×10^{-9} g m/(m^2 pa s)		Reddy and Rhim (2014)
TPS/clay (cloisite Na^+) composite	Increases	3.32 MPa	Decreases	TPS/clay (cloisite Na^+) composite had lowest WVP	Mt is non toxic	Park et al. (2002)
Agar/AgNP (2%) composite	First degradation- 90 °C and a second thermal decomposition at 324 °C.	53.44 MPa	Decreases	WVP of 1.47×10^{-9} g m/(m^2 pa s) decreases with increase in AgNP %.	Kills gram-positive (*L. monocytogenes*) and gram-negative (*E. coli* O157: H7).	Rhim et al. (2013a)
Starch/ZnO (5%) composite	Increases	6.62 MPa	48.20 cm^3 µm/(m^2 day atm)	1.32 g water/g dried film	Bactericidal to food borne pathogens	Mohammadi et al. (2013)

non-toxic, chemically stable, cheap and easily available. A study conducted by Kaewklin et al. (2018) analyzed the effects of the coating of Cs-based bionanocomposite films containing TiO_2 on Cherry Tomoto (*Solanum lycopersicum*), since ripening of tomato is controlled by factors such as climatic changes and ethylene (Kaewklin et al. 2018). Mechanism of action of Cs based TiO_2 bionanocomposite is as follows: The breakdown of ethylene takes place when the Cs-TiO_2 bionanocomposite is exposed to light in the UV range. The UV light excites the electrons of TiO_2 which then reacts with free water molecules or hydroxide ions to form hydroxyl radicals. The electrons then act on oxygen by reducing it to superoxide (O_2^-). These reactive species then acts on ethylene and oxidises it to form CO_2 and water molecules (Kaewklin et al. 2018). Cs, on the other hand has been shown to have antifungal properties. It was observed that Cs of various molecular weight, could inhibit the growth of the fungal pathogen *Botrytis cinerea*. Cs was also found to induce defence mechanisms in tomatoes (Yadollahi et al. 2010).

(c) Poly-lactic acid-based antioxidant films
Poly-lactic acid (PLA) is the polymer of lactic acid. It is a synthetically derived polymer from biomass. It has a high degree of transparency and has low crystallisation rate and thus it is easier to put it into different kinds of shapes. But PLA has a low gas barrier property, low glass transition property, less thermal stability and low toughness and ductility (Ramos et al. 2014). Therefore, when nano Mt. having high aspect ratio is blended along with PLA, the mechanical properties and the gas barrier becomes better. Ramos et al. (2014) blended Mt-reinforced PLA containing thymol to study the effects of this blend as an antioxidant film. Thymol is an antioxidant which is isolated from oregano and thyme essential oils. Thymol, having phenolic hydroxyl group can donate hydrogen atom to peroxyl radicals which then stabilises the phenoxyl radicals and thus terminate lipid peroxidation reactions (Ramos et al. 2014). The antioxidant activity was checked by the attacking activity on DPPH radicals. The highest level of inhibition of DPPH was found to be 83% (Ramos et al. 2014). Thus, this formulation containing 8% by weight of thymol and 2.5% by weight of D43B along with PLA shows good evidence of antioxidant activities.

25.5 Safety

There has been increased concern about the potential migration of NPs from the polymeric films in the recent years. Considering the commercial market where reinforced polymers for packaging have been successful for quite some time. However, relatively fewer studies have been conducted on the toxicity of these NPs for both humans and the ecosystem. NPs that are used to reinforce the polymers used as food contact materials are divided into four categories: improves the packaging property of the polymer (e.g. mechanical properties, barrier properties, etc.), gives additional active properties (e.g. antimicrobial properties), provides intelligent properties such as nanosensors, biodegradable NPs such as chitin/Cs NP (Gomes et al. 2016). Biodegradation can only occur when microbes and their enzymes act on the polymeric bonds (Gutiérrez 2018c). But the presence of NPs can either increase the rate of biodegradation by changing the crystallinity of the polymer or by making the polymer inaccessible to the microbes due to their anti microbial activity (Gomes et al. 2016). Due to the apparent benefits of reinforcing biopolymers with AgNPs, there is an increase in the use of AgNPs. But the studies conducted to determine migration of these NPs are very less (Metak et al. 2015). The less number of studies on NP migration is due to the fact that there are very less characterisation methods of the different types of the NPs and not enough methods for quantitative or qualitative studies (Gomes et al. 2016). A study by Amal M Metak et al. has found that AgNP migrated in insignificant amounts from reinforced containers than from reinforced polymeric film coatings where the AgNPs migrated in a quite a significant amount (<= 0.03 mg/L). Most of the studies conducted for the toxicity of engineered NP migration are for human models. There has been little study on the ecotoxicity of the ENPs from bionanocomposites (Gomes et al. 2016).

25.6 Future Trends

The use of conventional sources of starch, cellulose, clay, etc., such as potatoes, agricultural waste, etc. it will not be the only source through which bionocomposites are formed. Sources such as cashew to prepare starch/cashew nanocomposites (Pinto et al. 2014) or fish for the preparation of fish gelatin/Cs bionanocomposites (Fakhreddin et al. 2013) are some of the many emerging sources for bionanocomposite formation. In this manner, each food can have its own kind of food packaging owing to their unique properties. The bionanocompuestas films can have additional properties, sucha as edible or show spoilage of the food component, without the consumer having to search expiration dates. It can also have the property of showing nutritional factors or any type of bacterial spoilage (Gutiérrez and Álvarez 2016; Gutiérrez 2017b; Gutiérrez et al. 2018). These technologies are included in active and smart packages which, incorporating bionanocomposites technology, will revolutionize the packaging industry (Gutiérrez et al. 2016b; Gutiérrez 2018d, e).

25.7 Conclusion

The total post-harvest loss in India is estimated at around 92 Mton/year, which is quite alarming. To reduce this problem, proper food processing, packaging and storage is essential. The main objective of food packaging is to maintain wholesomeness, safety, quality of food and increase shelf life. There are many alternatives that open up for proper storage and packaging of food. Conventional packaging has its own disadvantages. Therefore, recent advances in bionanotechnology has opened use of bionanocomosites in food packaging to increase their mechanical strength, tensile strength, etc. Bionanocomposites and polymers have increasingly been used as coating materials for post-harvest products to reduce microbial contamination, over-ripening and spoilage due to oxygen, CO_2 and water vapor exposure. The advantages of using nanocoatings include better barrier performance and antimicrobial properties due to the presence of NPs such as silver, reduced material use and simpler film conversion. The agar-based nanopolymer, TPS-clay bionanocomposites, metal/carbohydrate bionanocomposites, etc., have proved to be very effective in reducing post-harvest losses.

Acknowledgments The authors wish to acknowledge Vellore Institute of Technology, Vellore, India for providing the opportunity to write this book chapter.

Conflicts of Interest The authors declare no conflict of interest.

References

Abdorreza, M. N., Cheng, L. H., & Karim, A. A. (2011). Food hydrocolloids effects of plasticizers on thermal properties and heat sealability of sago starch films. *Food Hydrocolloids, 25*(1), 56–60. https://doi.org/10.1016/j.foodhyd.2010.05.005.

Álvarez, K., Famá, L., & Gutiérrez, T. J. (2017). Physicochemical, antimicrobial and mechanical properties of thermoplastic materials based on biopolymers with application in the food industry. In M. Masuelli & D. Renard (Eds.), *Advances in physicochemical properties of biopolymers: Part 1* (pp. 358–400). Bentham Science Publishers. EE.UU. ISBN: 978-1-68108-454-1. eISBN: 978-1-68108-453-4. https://doi.org/10.2174/9781681084534117010015.

Álvarez, K., Alvarez, V. A., & Gutiérrez, T. J. (2018). Biopolymer composite materials with antimicrobial effects applied to the food industry. In V. K. Thakur & M. K. Thakur (Eds.), *Functional biopolymers* (pp. 57–96). Editorial Springer International Publishing.. EE.UU. ISBN: 978-3-319-66416-3. eISBN: 978-3-319-66417-0. https://doi.org/10.1007/978-3-319-66417-0_3.

Avella, M., De Vlieger, J. J., Emanuela, M., Fischer, S., Vacca, P., & Grazia, M. (2005). Biodegradable starch/clay nanocomposite films for food packaging applications. *Food Chemistry, 93*(3), 467–474. https://doi.org/10.1016/j.foodchem.2004.10.024.

Bracone, M., Merino, D., González, J., Alvarez, V. A., & Gutiérrez, T. J. (2016). Nanopackaging from natural fillers and biopolymers for the development of active and intelligent films. In S. Ikram & S. Ahmed (Eds.), *Natural polymers: Derivatives, blends and composites* (pp. 119–155). New York. EE.UU. ISBN: 978-1-63485-831-1: Editorial Nova Science Publishers, Inc.

Cardoen, D., Joshi, P., Diels, L., Sarma, P. M., & Pant, D. (2015). Resources, conservation and recycling agriculture biomass in India: Part 2. Post-harvest losses, cost and environmental impacts. *Resources, Conservation & Recycling, 101*, 143–153. https://doi.org/10.1016/j.resconrec.2015.06.002.

Chandran, S. P., Chaudhary, M., Pasricha, R., Ahmad, A., & Sastry, M. (2006). Synthesis of gold nanotriangles and silver nanoparticles using *Aloe vera* plant extract. *Biotechnology Progress, 22*(2), 577–583. https://doi.org/10.1021/bp0501423.

Chen, B., & Evans, J. R. G. (2005). Thermoplastic starch-clay nanocomposites and their characteristics. *Carbohydrate Polymers, 61*(4), 455–463. https://doi.org/10.1016/j.carbpol.2005.06.020.

Conte, A., Longano, D., Costa, C., Ditaranto, N., Ancona, A., Ciof, N., Scrocco, C., Sabbatini, L., Contò, F., & Del Nobile, M. A. (2013). A novel preservation technique applied to fiordilatte cheese. *Innovative Food Science & Emerging Technologies, 19*, 158–165. https://doi.org/10.1016/j.ifset.2013.04.010.

Countries, D. (2017). Reducing postharvest losses during storage of grain. *Foods, 6*(1), 8. https://doi.org/10.3390/foods6010008.
Fakhreddin, S., Rezaei, M., Zandi, M., & Farahmand, F. (2013). Preparation and functional properties of fish gelatin-chitosan blend edible films. *Food Chemistry, 136*(3–4), 1490–1495. https://doi.org/10.1016/j.foodchem.2012.09.081.
Gomes, V., Souza, L., & Fernando, A. L. (2016). Nanoparticles in food packaging: Biodegradability and potential migration to food—A review. *Food Packaging and Shelf Life, 8*, 63–70. https://doi.org/10.1016/j.fpsl.2016.04.001.
Gutiérrez, T. J. (2017a). Chitosan applications for the food industry. In S. Ahmed & S. Ikram (Eds.), *Chitosan: Derivatives, composites and applications* (pp. 183–232). WILEY-Scrivener Publisher. EE.UU. ISBN: 978-1-119-36350-7. https://doi.org/10.1002/9781119364849.ch8.
Gutiérrez, T. J. (2017b). Surface and nutraceutical properties of edible films made from starchy sources with and without added blackberry pulp. *Carbohydrate Polymers, 165*, 169–179. https://doi.org/10.1016/j.carbpol.2017.02.016.
Gutiérrez, T. J. (2018a). Biological macromolecule composite films made from Sagu starch and flour/poly(ε-caprolactone) blends processed by blending/thermo molding. *Journal Polymers and the Environment, 26*(9), 3902–3912. https://doi.org/10.1007/s10924-018-1268-6.
Gutiérrez, T. J. (2018b). Characterization and *in vitro* digestibility of non-conventional starches from guinea arrowroot and La Armuña lentils as potential food sources for special diet regimens. *Starch-Stärke, 70*(1–2). https://doi.org/10.1002/star.201700124.
Gutiérrez, T. J. (2018c). Biodegradability and compostability of food nanopackaging materials. In G. Cirillo, M. A. Kozlowski, & U. G. Spizzirri (Eds.), *Composite materials for food packaging* (pp. 269–296). WILEY-Scrivener Publisher. EE.UU. ISBN: 978-1-119-16020-5. https://doi.org/10.1002/9781119160243.ch9.
Gutiérrez, T. J. (2018d). Active and intelligent films made from starchy sources/blackberry pulp. *Journal Polymers and the Environment, 26*(6), 2374–2391. https://doi.org/10.1007/s10924-017-1134-y.
Gutiérrez, T. J. (2018e). Are modified pumpkin flour/plum flour nanocomposite films biodegradable and compostable? *Food Hydrocolloids, 83*, 397–410. https://doi.org/10.1016/j.foodhyd.2018.05.035.
Gutiérrez, T. J., & Álvarez, K. (2016). Physico-chemical properties and *in vitro* digestibility of edible films made from plantain flour with added *Aloe vera* gel. *Journal of Functional Foods, 26*, 750–762. https://doi.org/10.1016/j.jff.2016.08.054.
Gutiérrez, T. J., & Alvarez, V. A. (2017a). Films made by blending poly(ε-caprolactone) with starch and flour from sagu rhizome grown at the Venezuelan Amazons. *Journal Polymers and the Environment, 25*(3), 701–716. https://doi.org/10.1007/s10924-016-0861-9.
Gutiérrez, T. J., & Alvarez, V. A. (2017b). Properties of native and oxidized corn starch/polystyrene blends under conditions of reactive extrusion using zinc octanoate as a catalyst. *Reactive and Functional Polymers, 112*, 33–44. https://doi.org/10.1016/j.reactfunctpolym.2017.01.002.
Gutiérrez, T. J., & Alvarez, V. A. (2017c). Eco-friendly films prepared from plantain flour/PCL blends under reactive extrusion conditions using zirconium octanoate as a catalyst. *Carbohydrate Polymers, 178*, 260–269. https://doi.org/10.1016/j.carbpol.2017.09.026.
Gutiérrez, T. J., & Alvarez, V. A. (2017d). Data on physicochemical properties of active films derived from plantain flour/PCL blends developed under reactive extrusion conditions. *Data in Brief, 15*, 445–448. https://doi.org/10.1016/j.dib.2017.09.071.
Gutiérrez, T. J., & Alvarez, V. A. (2017e). Cellulosic materials as natural fillers in starch-containing matrix-based films: A review. *Polymer Bulletin, 74*(6), 2401–2430. https://doi.org/10.1007/s00289-016-1814-0.
Gutiérrez, T. J., & Alvarez, V. A. (2018). Bionanocomposite films developed from corn starch and natural and modified nano-clays with or without added blueberry extract. *Food Hydrocolloids, 77*, 407–420. https://doi.org/10.1016/j.foodhyd.2017.10.017.
Gutiérrez, T. J., Morales, N. J., Pérez, E., Tapia, M. S., & Famá, L. (2015a). Physico-chemical study of edible films based on native and phosphating cush-cush yam and cassava starches. *Food Packaging and Shelf Life, 3*, 1–8. https://doi.org/10.1016/j.fpsl.2014.09.002.

Gutiérrez, T. J., Morales, N. J., Tapia, M. S., Pérez, E., & Famá, L. (2015b). Corn starch 80:20 "waxy": Regular, "native" and phosphated, as bio-matrixes for edible films. *Procedia Materials Science, 8*, 304–310. https://doi.org/10.1016/j.mspro.2015.04.077.

Gutiérrez, T. J., Tapia, M. S., Pérez, E., & Famá, L. (2015c). Structural and mechanical properties of native and modified cush-cush yam and cassava starch edible films. *Food Hydrocolloids, 45*, 211–217. https://doi.org/10.1016/j.foodhyd.2014.11.017.

Gutiérrez, T. J., Tapia, M. S., Pérez, E., & Famá, L. (2015d). Edible films based on native and phosphated 80:20 waxy:Normal corn starch. *Starch-Stärke, 67*(1–2), 90–97. https://doi.org/10.1002/star.201400164.

Gutiérrez, T. J., Guzmán, R., Medina Jaramillo, C., & Famá, L. (2016a). Effect of beet flour on films made from biological macromolecules: Native and modified plantain flour. *International Journal of Biological Macromolecules, 82*, 395–403. https://doi.org/10.1016/j.ijbiomac.2015.10.020.

Gutiérrez, T. J., Suniaga, J., Monsalve, A., & García, N. L. (2016b). Influence of beet flour on the relationship surface-properties of edible and intelligent films made from native and modified plantain flour. *Food Hydrocolloids, 54*, 234–244. https://doi.org/10.1016/j.foodhyd.2015.10.012.

Gutiérrez, T. J., González Seligra, P., Medina Jaramillo, C., Famá, L., & Goyanes, S. (2017). Effect of filler properties on the antioxidant response of thermoplastic starch composites. In V. K. Thakur, M. K. Thakur, & M. R. Kessler (Eds.), *Handbook of composites from renewable materials* (pp. 337–370). WILEY-Scrivener Publisher. EE.UU. ISBN: 978-1-119-22362-7. https://doi.org/10.1002/9781119441632.ch14.

Gutiérrez, T. J., Herniou--Julien, C., Álvarez, K., & Alvarez, V. A. (2018). Structural properties and *in vitro* digestibility of edible and pH-sensitive films made from Guinea arrowroot starch and wastes from wine manufacture. *Carbohydrate Polymers, 184*, 135–143. https://doi.org/10.1016/j.carbpol.2017.12.039.

Gutiérrez, T. J., Toro-Márquez, L. A., Merino, D., & Mendieta, J. R. (2019). Hydrogen-bonding interactions and compostability of bionanocomposite films prepared from corn starch and nano-fillers with and without added Jamaica flower extract. *Food Hydrocolloids, 89*, 283–293. https://doi.org/10.1016/j.foodhyd.2018.10.058.

Herniou--Julien, C., Mendieta, J. R., & Gutiérrez, T. J. (2019). Characterization of biodegradable/non-compostable films made from cellulose acetate/corn starch blends processed under reactive extrusion conditions. *Food Hydrocolloids, 89*, 67–79. https://doi.org/10.1016/j.foodhyd.2018.10.024.

Imran, M., René, N., Jamshidian, M., Akhtar, M. J., Arab-tehrany, E., Jacquot, M., & Desobry, S. (2012). Food hydrocolloids microstructure and physico-chemical evaluation of nano-emulsion-based antimicrobial peptides embedded in bioactive packaging films. *Food Hydrocolloids, 29*(2), 407–419. https://doi.org/10.1016/j.foodhyd.2012.04.010.

Kader, A. A. (2003). A Perspective on Postharvest Horticulture (1978–2003). *HortScience, 38*(5), 1004–1008.

Kaewklin, P., Siripatrawan, U., Suwanagul, A., & Lee, Y. S. (2018). Active packaging from chitosan-titanium dioxide nanocomposite film for prolonging storage life of tomato fruit. *International Journal of Biological Macromolecules, 112*, 523–529. https://doi.org/10.1016/j.ijbiomac.2018.01.124.

Kanmani, P., & Rhim, J. (2014). Physicochemical properties of gelatin/silver nanoparticle antimicrobial composite films. *Food Chemistry, 148*, 162–169. https://doi.org/10.1016/j.foodchem.2013.10.047.

Khan, A., Khan, R. A., Salmieri, S., Le, C., Riedl, B., Bouchard, J., Chauve, G., Tan, V., Kamal, M. R., & Lacroix, M. (2012). Mechanical and barrier properties of nanocrystalline cellulose reinforced chitosan based nanocomposite films. *Carbohydrate Polymers, 90*(4), 1601–1608. https://doi.org/10.1016/j.carbpol.2012.07.037.

Kiaya, V. (2014). Post-harvest losses and strategies to reduce them, ACF International.

Kumar, D., & Kalita, P. (2017). Reducing postharvest losses during storage of grain crops to strengthen food security in developing countries. Foods, 6(1), 8.

Lee, J., Song, N., Jo, W., & Song, K. B. (2014). Effects of nano-clay type and content on the physical properties of sesame seed meal protein composite films. *International Journal of Food Science & Technology, 49*(8), 1869–1875. https://doi.org/10.1111/ijfs.12496.

Martelli, M. R., & Barros, T. (2013). Effect of chitosan nanoparticles and pectin content on mechanical properties and water vapor permeability of banana puree films. *Journal of Food Science, 78*(1), N98–N104. https://doi.org/10.1111/j.1750-3841.2012.03006.x.

Merino, D., Gutiérrez, T. J., Mansilla, A. Y., Casalongué, C., & Alvarez, V. A. (2018). Critical evaluation of starch-based antibacterial nanocomposites as agricultural mulch films: Study on their interactions with water and light. *ACS Sustainable Chemistry & Engineering, 6*(11), 15662–15672. https://doi.org/10.1021/acssuschemeng.8b04162.

Merino, D., Gutiérrez, T. J., & Alvarez, V. A. (2019). Structural and thermal properties of agricultural mulch films based on native and oxidized corn starch nanocomposites. *Starch-Stärke, 71*(7–8), 1800341. https://doi.org/10.1002/star.201800341.

Metak, A. M., Nabhani, F., & Connolly, S. N. (2015). Migration of engineered nanoparticles from packaging into food products. *LWT-Food Science and Technology, 64*(2), 781–787. https://doi.org/10.1016/j.lwt.2015.06.001.

Mihindukulasuriya, S. D. F., & Lim, L. (2014). Nanotechnology development in food packaging: A review. *Trends in Food Science & Technology, 40*(2), 149–167. https://doi.org/10.1016/j.tifs.2014.09.009.

Mohammadi, A., Karim, A., Mahmud, S., & Robal, M. (2012). Antimicrobial, rheological, and physicochemical properties of sago starch films filled with nanorod-rich zinc oxide. *Journal of Food Engineering, 113*(4), 511–519. https://doi.org/10.1016/j.jfoodeng.2012.07.017.

Mohammadi, A., Nassiri, R., Sheibani, S., Ariffin, F., & Karim, A. A. (2013). Preparation and characterization of bionanocomposite films filled with nanorod-rich zinc oxide. *Carbohydrate Polymers, 96*(1), 233–239. https://doi.org/10.1016/j.carbpol.2013.03.055.

Othman, S. H. (2014). Bio-nanocomposite materials for food packaging applications: Types of biopolymer and nano-sized filler. *Agriculture and Agricultural Science Procedia, 2*, 296–303. https://doi.org/10.1016/j.aaspro.2014.11.042.

Park, H., Li, X., Jin, C., Park, C., Cho, W., & Ha, C. (2002). Preparation and properties of biodegradable thermoplastic starch/clay hybrids. *Macromolecular Materials and Engineering, 287*(8), 553–558. https://doi.org/10.1002/1439-2054(20020801)287:8<553::aid-mame553>3.0.co;2-3.

Pathak, N., Caleb, O. J., Geyer, M., Herppich, W. B., Rauh, C., & Mahajan, P. V. (2017). Photocatalytic and photochemical oxidation of ethylene: Potential for storage of fresh produce—A review. *Food and Bioprocess Technology, 10*(6), 982–1001. https://doi.org/10.1007/s11947-017-1889-0.

Pinto, A. M. B., Santos, T. M., Alberto, C., Lima, J. R., Ito, E. N., & Azeredo, H. M. C. (2014). Starch-cashew tree gum nanocomposite films and their application for coating cashew nuts. *LWT-Food Science and Technology, 62*(1), 549–554. https://doi.org/10.1016/j.lwt.2014.07.028.

Ramos, M., Jiménez, A., Peltzer, M., & Garrigós, M. C. (2014). Development of novel nano-biocomposite antioxidant films based on poly (lactic acid) and thymol for active packaging. *Food Chemistry, 162*, 149–155. https://doi.org/10.1016/j.foodchem.2014.04.026.

Reddy, J. P., & Rhim, J. (2014). Characterization of bionanocomposite films prepared with agar and paper-mulberry pulp nanocellulose. *Carbohydrate Polymers, 110*, 480–488. https://doi.org/10.1016/j.carbpol.2014.04.056.

Rhim, J., Lee, S., & Hong, S. (2011). Preparation and characterization of agar/clay nanocomposite films: The effect of clay type. *Journal of Food Science, 76*(3), N40–N48. https://doi.org/10.1111/j.1750-3841.2011.02049.x.

Rhim, J. W., Park, H. M., & Ha, C. S. (2013a). Bio-nanocomposites for food packaging applications. *Progress in Polymer Science, 38*(10–11), 1629–1652. https://doi.org/10.1016/j.progpolymsci.2013.05.008.

Rhim, J. W., Wang, L. F., & Hong, S. I. (2013b). Food hydrocolloids preparation and characterization of agar/silver nanoparticles composite films with antimicrobial activity. *Food Hydrocolloids, 33*(2), 327–335. https://doi.org/10.1016/j.foodhyd.2013.04.002.

Robertson, G. L. (1998). *Food packaging, principles and practice* (2nd ed.). CRC Press.

Sandarani, M. D. J. C., Dasanayaka, D. C. M. C. K., & Jayasinghe, C. V. L. (2018). *Journal of Agricultural Science, 9*(1), 1–6.

Shankar, S., Teng, X., & Rhim, J. (2014). Properties and characterization of agar/CuNP bionanocomposite films prepared with different copper salts and reducing agents. *Carbohydrate Polymers, 114*, 484–492. https://doi.org/10.1016/j.carbpol.2014.08.036.

Shin, S., Kim, S., Lee, S., Park, K., & Han, J. (2014). Apple peel and carboxymethylcellulose-based nanocomposite films containing different nanoclays. *Journal of Food Science, 79*(3), E342–E353. https://doi.org/10.1111/1750-3841.12356.

Singh, N., Mahanwar, P. A., & Malshe, V. C. (2006). Biodegradable nanocomposites and their applications. PAINTINDIA-BOMBAY-, 56(7), 71.

Singh, N., Malshe V. (2017). *Biodegradable nanocomposites and their applications publication description.*

Singh, G., & Rattanpal, H. S. (2014). Use of nanotechnology in horticulture. *International Journal of Agricultural Sciences and Veterinary Medicine, 2*(1), 34–42.

Sousa-gallagher, M. J., & Mahajan, P. V. (2013). Integrative mathematical modelling for MAP design of fresh-produce: Theoretical analysis and experimental validation. *Food Control, 29*(2), 444–450. https://doi.org/10.1016/j.foodcont.2012.05.072.

Suárez, G., & Gutiérrez, T. J. (2017). Recent advances in the development of biodegadable films and foams from cassava starch. In C. Klein (Ed.), *Handbook on cassava: Production, potential uses and recent advances* (pp. 297–312). New York. EE.UU. ISBN: 978-1-53610-307-6: Editorial Nova Science Publishers, Inc.

Toro-Márquez, L. A., Merino, D., & Gutiérrez, T. J. (2018). Bionanocomposite films prepared from corn starch with and without nanopackaged Jamaica (*Hibiscus sabdariffa*) flower extract. *Food and Bioprocess Technology, 11*(11), 1955–1973. https://doi.org/10.1007/s11947-018-2160-z.

Wang, X., Du, Y., & Luo, J. (2008). Biopolymer/montmorillonite nanocomposite: Preparation, drug-controlled release property and cytotoxicity. *Nanotechnology, 19*(6), *065707*. https://doi.org/10.1088/0957-4484/19/6/065707.

Webb, H. K., Arnott, J., Crawford, R. J., & Ivanova, E. P. (2013). Plastic degradation and its environmental implications with special reference to poly(ethylene terephthalate). *Polymers, 5*(1), 1–18. https://doi.org/10.3390/polym5010001.

Yadollahi, A., Arzani, K., & Khoshghalb, H. (2010). The role of nanotechnology in horticultural crops postharvest management, in southeast asia symposium on quality and safety of fresh and fresh-cut produce. *ISHS Acta Horticulturae, 875*, 49–56. https://doi.org/10.17660/actahortic.2010.875.4.

Youssef, A. M., Abdel-Aziz, M. S., & El-sayed, S. M. (2014). International journal of biological macromolecules chitosan nanocomposite films based on ag-NP and au-NP biosynthesis by *Bacillus Subtilis* as packaging materials. *International Journal of Biological Macromolecules, 69*, 185–191. https://doi.org/10.1016/j.ijbiomac.2014.05.047.

Zhang, L., Jiang, Y., Ding, Y., Povey, M., & York, D. (2007). Investigation into the antibacterial behaviour of suspensions of ZnO nanoparticles (ZnO nanofluids). *Journal of Nanoparticle Research, 9*(3), 479–489. https://doi.org/10.1007/s11051-006-9150-1.

Chapter 26
Essential Oil Nanoemulsions as Antimicrobials and Antioxidants in Composite Food Packaging

Magda I. Pinzon, Leidy T. Sanchez, and Cristian C. Villa

Abstract The essential oils (OEs) are oily liquids extracted from plant materials that have been used as flavoring and antimicrobial agents in the food and agricultural industry. However, its application in food preservation has been limited due to its low water solubility. One way to incorporate OEs in food packaging is through the formation of nanoemulsions: drop size between 20 and 200 nm. Nanoemulsions are stable and transparent systems, which have been shown to have an increase in the biological activity of the encapsulated lipophilic compound. In recent years, the development of edible and biodegradable films incorporating EO nanoemulsions and their application in the preservation of food, especially fruits and vegetables has been of increasing interest. The objective of this chapter was to summarize and analyze recent publications on edible and biodegradable food packaging and its application in the conservation of fruits and vegetables.

Keywords Composite materials · Encapsulation

26.1 Introduction

One of the most important trends in recent years, both in food and material sciences, is the design and production of food packages that not only act as barriers or containers, but also have an active role in its conservation (Han et al. 2018; Werner et al. 2017; Yildirim et al. 2018). Active food packages are used in order to extend the shelf life of food products, while maintaining or improving the properties of the packaged foodstuffs. This is achieved by the inclusion of different types of substances that can be released into the food, decreasing or even stopping deterioration

M. I. Pinzon · L. T. Sanchez
Food Engineering Department, Faculty of Agroindustrial Sciences,
Quindío University, Armenia, Quindío, Colombia

C. C. Villa (✉)
Chemistry Department, Faculty of Basic Sciences and Technology,
Quindío University, Armenia, Quindío, Colombia
e-mail: ccvilla@uniquindio.edu.co

T. J. Gutiérrez (ed.), *Polymers for Agri-Food Applications*,
https://doi.org/10.1007/978-3-030-19416-1_26

process such as microbial growth and lipid oxidation, among others (Gutiérrez et al. 2016a, b; Atarés and Chiralt 2016; Bracone et al. 2016; Gutiérrez 2017, 2018a; Ribeiro-Santos et al. 2017; Toro-Márquez et al. 2018; Gutiérrez et al. 2018; Gutiérrez and Alvarez 2018). Despite there have been reports on the inclusion of synthetic antimicrobials and antioxidant molecules in active food packages, an effort has been made to reduce the use of these compounds, in favor of the natural molecules that can fulfill the same role in the food preservation. Among the natural products that are potentially used in food packaging, the essential oils (EOs) from aromatic plants have generated great interest, due to the well-known biological properties such as antioxidant, antimicrobial, insecticidal, antitumor and anti-inflammatory activities (Ribeiro-Santos et al. 2017). OEs are highly complex mixtures of hundreds of individual aroma compounds, although not fully oily, have a low water solubility. The EOs can be extracted from different parts of the plants such as flowers, herbs, leaves, fruits, seeds, rhizomes and roots (Asbahani et al. 2015; de Matos et al. 2019). The polymeric films incorporating EO are prepared most of the times by the addition of an appropriate amount of OE in the film-forming solution (FFS) followed by the removal of the solvent (usually water) by evaporation of the solvent, without the use of heat, resulting in the formation of a compact polymeric film (Ribeiro-Santos et al. 2017). Although, this process can be relatively easy to achieve, the EOs incorporation into polymeric films can be limited due to the low water solubility and high volatility. One way to incorporate EOs into polymer film matrices is by the formation of nano-emulsions, a type of emulsion with a drop size between 20 and 200 nm (Bilbao-Sainz et al. 2010; Komaiko and McClements 2016; Restrepo et al. 2018). Due to their small size, nanoemulsions tend to be optically stable and transparent systems, which allow to increase the activity of encapsulated OEs (Donsi and Ferrari 2016; Prakash et al. 2018). The nanoemulsion can be formed by two different approaches, the first is based on high energy homogenization using specialized mechanical devices, such as high shear mixers, high pressure homogenizers, sonicators or microfluidics. The second approach includes several low energy methods, based on a controlled mixture of all the nanoemulsion components through various experimental techniques including spontaneous emulsification, phase inversion temperature, phase inversion point and phase inversion composition (Komaiko and McClements 2016; Restrepo et al. 2018).

26.2 Nanoemulsion Incorporated in Edible Films

Edible films and coatings are commonly defined as continuous matrices made from biodegradable, removable and edible polymers, such as proteins, carbohydrates, gums and lipids (Gutiérrez et al. 2015a; b; Atarés and Chiralt 2016; Han et al. 2018; Gutiérrez 2018b; Otoni et al. 2017; Pinzon et al. 2018). It is worth noting that all the additives incorporated into the FFS including the components of the nanoemulsions must be 'generally recognized as safe' (GRAS) as food additive. The most

Table 26.1 Some of the biopolymers, stabilizing agents and OE most commonly used in the development of edible composite films

Biopolymers	Stabilizing agent	EOs
Starch and derivate	Tween 80, 60, 40,20	Cinnamon
Alginate and derivate	Span 60, 80	Lemongrass
Carboxymethyl cellulose (CMC)	Sodium caseinate	Citrus
Chitosan and derivate	Sodium dodecylsulfate	Mandarin
Pectin and derivate	Whey protein	Thymol
Whey protein		Carvacrol
Sodium caseinate		Clove
Gelatin		Oregano
Soy protein		

commonly used stabilizing agents for the formation of nanoemulsions are: small molecule surfactants (sugar ester surfactants, polyoxyethylene ether surfactants and ethoylated sorbitan esters), phospholipids, proteins and polysaccharides, although these latter two are not commonly suitable (McClements et al. 2009; McClements and Rao 2011). Table 26.1 shows the biopolymers, stabilizing agents and EO most commonly used in the formulation of composite edible films.

It has been reported that the inclusion of EO nanoemulsion in polymeric matrices can lead to several changes in its barrier, structural and mechanical properties, since the nanoemulsion components can interact with polymer molecules in different ways (Espitia et al. 2019). Barrier properties of edible films are mainly related to the permeability of water vapor, oxygen and other gases of importance in food packaging. Due the importance of water content in stored foods, water vapor permeability (WVP) is one of the main factors evaluated in composite edible films. Due to the hydrophobic nature of the EOs, the inclusion of them in the FFS allows repelling the water molecules when trying to cross through the edible film, thus the WVP values are reduced. Du et al. (2008) observed a decrease in WVP values for pectin composite films containing carvacrol nanoemulsions compared to pectin films without the addition of EO nanoemulsions. Similar behavior was reported for pectin films containing clove EO nanoemulsions (Sasaki et al. 2016), pectin/papaya puree composite films containing cynnamaldehyde EO nanoemulsions (Otoni et al. 2014) and sage EO nanoemulsion-loaded sodium alginate films (Acevedo-Fani et al. 2015). However, Otoni et al. (2016) reported that although EOs have a hydrophobic nature, they also have a plasticizing effect that could lead to a higher WVP values, since their hydrophobic nature and plasticizer are balanced in edible films. Restrepo et al. (2018), e.g. reported that the inclusion of lemongrass and rosemary EO nanoemulsions into banana starch films lead to a slight decrease in the WVP values of the film. The inclusion of EO nanoemulsions in edible films has also been shown to have different effects on the mechanical properties of edible films. Several authors have reported that due to the plasticizing effect of OEs in the polymer matrix, they interrupt the intermolecular structure and allow an increase in its elongation at break values (Acevedo-Fani et al. 2015; Du et al. 2009; Otoni et al. 2014; Restrepo et al. 2018;

Robledo et al. 2018b). In contrast, other authors have reported that EOs can have an anti-plasticizing effect in the polymeric films, thus increasing their stiffness and reducing their elongation at break values (Espitia et al. 2019).

As mentioned before, one of the main reasons for the development of edible films containing EOs nanoemulsions is the use of their antimicrobial and antioxidant activities for food preservation. Noori et al. (2018) studied the antimicrobial activity of ginger root EO nanoemulsion loaded-sodium caseinate films against *S. typhimurium* and *L. monocytogenes*. They reported that the composite films can be effective in the reducing the growth of both bacterial strains, with an even better performance than commercial films containing antimicrobials. The antioxidant activity of composite films was also higher than the sodium caseinate films alone. Acevedo-Fani et al. (2015) carried out an extensive studied of the antimicrobial activity of alginate films loaded with EO nanoemulsions from thyme, lemongrass and sage against *E. coli.* It was observed that only films with thyme EO nanoemulsions presented activity against the bacterial strain. The thymol EO nanoemulsion loaded-quinoa protein/chitosan composite films have also shown antifungal activity against *B. cinera,* a common fungus responsible of strawberries spoilage (Robledo et al. 2018b). Otoni et al. (2014) also reported the development of cinnamaldehyde EO nanoemulsion loaded-papaya/pectin composite films showing antimicrobial activity against Gram-positive bacteria: *L. monocytogenes* and *S. aureus.* Several authors have also reported on the use of edible films loaded EO nanoemulsions being applied to different food product, as shown in Table 26.2.

Recently, Robledo, López, Bunger, Tapia, and Abugoch (2018a) and Robledo et al. (2018b) reported that composite edible coatings made from quinoa protein – chitosan and thymol EO nanoemulsions can be used to improve the shelf-life of strawberries and cherry tomatoes, respectably. It was observed that the presence of thymol EO nanoemulsion reduced the fungal growth in both types of fruits with respect of uncoated and coated fruits without nanoemulsions during storage. The limited fungal growth in fruits treated with nanoemulsion coating was attributed to the controlled release of thymol EO from the polymer matrix to the food surface.

Table 26.2 Some recent applications of EO nanoemulsion loaded-edible films in stored food products

EO nanoemulsion	Polymeric matrix	Food product	References
Thymol	Quinoa protein/chitosan	Cherry tomatoes	Robledo et al. (2018b)
		Strawberries	Robledo et al. (2018a)
Lemongrass	Chitosan	Grape berries	Oh et al. (2017)
	Carnauba wax	Plums	Kim et al. (2013)
	Alginate	Fresh-cut fuji apples	Salvia-Trujillo et al. (2015)
Ginger	Sodium caseinate	Chicken breast fillets	Noori et al. (2018)
	Chitosan/montmorillonte	Poultry meat	Souza et al. (2018)
Basil	Alginate	Okra	Gundewadi et al. (2018)
Carvacrol	Whey protein/chitosan	Cucumber slices	Taştan et al. (2017)

One of the most used EOs in the development of composite edible films is lemongrass EO, due to its well-known antimicrobial activity against several types of bacteria. Salvia-Trujillo et al. (2013) used lemongrass EO nanoemulsion loaded-alginate composite films for the preservation of fresh-cut Fuji apples, observing that lemongrass EO nanoemulsion-loaded coating reduced growth of *E. coli* and the natural microbiota of the apple slices up to 2 weeks of storage. Kim et al. (2013) also reported that the incorporation of lemongrass EO nanoemulsions into carnauba wax coatings inhibited *E. coli* and *S. Typhimurium* in plums during storage. Similar results were reported by Oh et al. (2017) using a composite coating made from chitosan and lemongrass EO nanoemulsion for the conservation of grape berries.

Acknowledgements The authors would like to thank the Vicerrectoria de Investigaciones, Facultad de Ciencias Basicas y Tecnologias, Programa de Quimica and Programa de Ingenieria de Alimentos from Universidad del Quindio, for their support.

Conflicts of Interest The authors declare no conflict of interest.

References

Acevedo-Fani, A., Salvia-Trujillo, L., Rojas-Graü, M. A., & Martín-Belloso, O. (2015). Edible films from essential-oil-loaded nanoemulsions: Physicochemical characterization and antimicrobial properties. *Food Hydrocolloids, 47*, 168–177. https://doi.org/10.1016/j.foodhyd.2015.01.032.

Asbahani, A. E., Miladi, K., Badri, W., Sala, M., Addi, E. H. A., Casabianca, H., El Mousadik, A., Hartmann, D., Jilale, A., Renaud, F. N. R., & Elaissari, A. (2015). Essential oils: From extraction to encapsulation. *International Journal of Pharmaceutics, 483*(1), 220–243. https://doi.org/10.1016/j.ijpharm.2014.12.069.

Atarés, L., & Chiralt, A. (2016). Essential oils as additives in biodegradable films and coatings for active food packaging. *Trends in Food Science & Technology, 48*, 51–62. https://doi.org/10.1016/j.tifs.2015.12.001.

Bilbao-Sainz, C., Avena-Bustillos, R. J., Wood, D. F., Williams, T. G., & McHugh, T. H. (2010). Nanoemulsions prepared by a low-energy emulsification method applied to edible films. *Journal of Agricultural and Food Chemistry, 58*(22), 11932–11938. https://doi.org/10.1021/jf102341r.

Bracone, M., Merino, D., González, J., Alvarez, V. A., & Gutiérrez, T. J. (2016). Nanopackaging from natural fillers and biopolymers for the development of active and intelligent films. In S. Ikram & S. Ahmed (Eds.), *Natural polymers: Derivatives, blends and composites* (pp. 119–155). New York.. EE.UU. ISBN: 978-1-63485-831-1: Editorial Nova Science Publishers, Inc.

de Matos, S. P., Lucca, L. G., & Koester, L. S. (2019). Essential oils in nanostructured systems: Challenges in preparation and analytical methods. *Talanta, 195*, 204–214. https://doi.org/10.1016/j.talanta.2018.11.029.

Donsi, F., & Ferrari, G. (2016). Essential oil nanoemulsions as antimicrobial agents in food. *Journal of Biotechnology, 233*, 106–120. https://doi.org/10.1016/j.jbiotec.2016.07.005.

Du, W.-X., Olsen, C. W., Avena-Bustillos, R. J., McHugh, T. H., Levin, C. E., & Friedman, M. (2008). Storage stability and antibacterial activity against *Escherichia coli* O157:H7 of carvacrol in edible apple films made by two different casting methods. *Journal of Agricultural and Food Chemistry, 56*(9), 3082–3088. https://doi.org/10.1021/jf703629s.

Du, W. X., Olsen, C. W., Avena-Bustillos, R. J., McHugh, T. H., Levin, C. E., & Friedman, M. (2009). Effects of allspice, cinnamon, and clove bud essential oils in edible apple films on physical properties and antimicrobial activities. *Journal of Food Science, 74*(7), M372–M378. https://doi.org/10.1111/j.1750-3841.2009.01282.x.

Espitia, P. J. P., Fuenmayor, C. A., & Otoni, C. G. (2019). Nanoemulsions: Synthesis, characterization, and application in bio-based active food packaging. *Comprehensive Reviews in Food Science and Food Safety, 18*(1), 264–285. https://doi.org/10.1111/1541-4337.12405.

Gundewadi, G., Rudra, S. G., Sarkar, D. J., & Singh, D. (2018). Nanoemulsion based alginate organic coating for shelf life extension of okra. *Food Packaging and Shelf Life, 18*, 1–12. https://doi.org/10.1016/j.fpsl.2018.08.002.

Gutiérrez, T. J. (2017). Surface and nutraceutical properties of edible films made from starchy sources with and without added blackberry pulp. *Carbohydrate Polymers, 165*, 169–179. https://doi.org/10.1016/j.carbpol.2017.02.016.

Gutiérrez, T. J. (2018a). Active and intelligent films made from starchy sources/blackberry pulp. *Journal Polymers and the Environment, 26*(6), 2374–2391. https://doi.org/10.1007/s10924-017-1134-y.

Gutiérrez, T. J. (2018b). Polymers for food applications: News. In T. J. Gutiérrez (Ed.), *Polymers for food applications* (pp. 1–4). Cham. ISBN: 978-3-319-94624-5. eISBN: 978-3-319-94625-2: Editorial Springer International Publishing. https://doi.org/10.1007/978-3-319-94625-2_1.

Gutiérrez, T. J., & Alvarez, V. A. (2018). Bionanocomposite films developed from corn starch and natural and modified nano-clays with or without added blueberry extract. *Food Hydrocolloids, 77*, 407–420. https://doi.org/10.1016/j.foodhyd.2017.10.017.

Gutiérrez, T. J., Morales, N. J., Pérez, E., Tapia, M. S., & Famá, L. (2015a). Physico-chemical study of edible films based on native and phosphating cush-cush yam and cassava starches. *Food Packaging and Shelf Life, 3*, 1–8. https://doi.org/10.1016/j.fpsl.2014.09.002.

Gutiérrez, T. J., Tapia, M. S., Pérez, E., & Famá, L. (2015b). Structural and mechanical properties of native and modified cush-cush yam and cassava starch edible films. *Food Hydrocolloids, 45*, 211–217. https://doi.org/10.1016/j.foodhyd.2014.11.017.

Gutiérrez, T. J., Guzmán, R., Medina Jaramillo, C., & Famá, L. (2016a). Effect of beet flour on films made from biological macromolecules: Native and modified plantain flour. *International Journal of Biological Macromolecules, 82*, 395–403. https://doi.org/10.1016/j.ijbiomac.2015.10.020.

Gutiérrez, T. J., Suniaga, J., Monsalve, M., & García, N. L. (2016b). Influence of beet flour on the relationship surface-properties of edible and intelligent films made from native and modified plantain flour. *Food Hydrocolloids, 54*, 234–244. https://doi.org/10.1016/j.foodhyd.2015.10.012.

Gutiérrez, T. J., Herniou--Julien, C., Álvarez, K., & Alvarez, V. A. (2018). Structural properties and *in vitro* digestibility of edible and pH-sensitive films made from guinea arrowroot starch and wastes from wine manufacture. *Carbohydrate Polymers, 184*, 135–143. https://doi.org/10.1016/j.carbpol.2017.12.039.

Han, J.-W., Ruiz-Garcia, L., Qian, J.-P., & Yang, X.-T. (2018). Food packaging: A comprehensive review and future trends. *Comprehensive Reviews in Food Science and Food Safety, 17*(4), 860–877. https://doi.org/10.1111/1541-4337.12343.

Kim, I.-H., Lee, H., Kim, J. E., Song, K. B., Lee, Y. S., Chung, D. S., & Min, S. C. (2013). Plum coatings of lemongrass oil-incorporating carnauba wax-based nanoemulsion. *Journal of Food Science, 78*(10), E1551–E1559. https://doi.org/10.1111/1750-3841.12244.

Komaiko, J. S., & McClements, D. J. (2016). Formation of food-grade nanoemulsions using low-energy preparation methods: A review of available methods. *Comprehensive Reviews in Food Science and Food Safety, 15*(2), 331–352. https://doi.org/10.1111/1541-4337.12189.

McClements, D. J., & Rao, J. (2011). Food-grade nanoemulsions: Formulation, fabrication, properties, performance, biological fate, and potential toxicity. *Critical Reviews in Food Science and Nutrition, 51*(4), 285–330. https://doi.org/10.1080/10408398.2011.559558.

McClements, D. J., Decker, E. A., Park, Y., & Weiss, J. (2009). Structural design principles for delivery of bioactive components in nutraceuticals and functional foods. *Critical Reviews in Food Science and Nutrition, 49*(6), 577–606. https://doi.org/10.1080/10408390902841529.

Noori, S., Zeynali, F., & Almasi, H. (2018). Antimicrobial and antioxidant efficiency of nanoemulsion-based edible coating containing ginger (*Zingiber officinale*) essential oil and its effect on safety and quality attributes of chicken breast fillets. *Food Control, 84*, 312–320. https://doi.org/10.1016/j.foodcont.2017.08.015.

Oh, Y. A., Oh, Y. J., Song, A. Y., Won, J. S., Song, K. B., & Min, S. C. (2017). Comparison of effectiveness of edible coatings using emulsions containing lemongrass oil of different size droplets on grape berry safety and preservation. *LWT - Food Science and Technology, 75*, 742–750. https://doi.org/10.1016/j.lwt.2016.10.033.

Otoni, C. G., de Moura, M. R., Aouada, F. A., Camilloto, G. P., Cruz, R. S., Lorevice, M. V., Soares, N. F. F., & Mattoso, L. H. C. (2014). Antimicrobial and physical-mechanical properties of pectin/papaya puree/cinnamaldehyde nanoemulsion edible composite films. *Food Hydrocolloids, 41*, 188–194. https://doi.org/10.1016/j.foodhyd.2014.04.013.

Otoni, C. G., Avena-Bustillos, R. J., Olsen, C. W., Bilbao-Sáinz, C., & McHugh, T. H. (2016). Mechanical and water barrier properties of isolated soy protein composite edible films as affected by carvacrol and cinnamaldehyde micro and nanoemulsions. *Food Hydrocolloids, 57*, 72–79. https://doi.org/10.1016/j.foodhyd.2016.01.012.

Otoni, C. G., Avena-Bustillos, R. J., Azeredo, H. M. C., Lorevice, M. V., Moura, M. R., Mattoso, L. H. C., & McHugh, T. H. (2017). Recent advances on edible films based on fruits and vegetables—A review. *Comprehensive Reviews in Food Science and Food Safety, 16*(5), 1151–1169. https://doi.org/10.1111/1541-4337.12281.

Pinzon, M. I., Garcia, O. R., & Villa, C. C. (2018). The influence of *Aloe vera* gel incorporation on the physicochemical and mechanical properties of banana starch-chitosan edible films. *Journal of the Science of Food and Agriculture, 98*(11), 4042–4049. https://doi.org/10.1002/jsfa.8915.

Prakash, B., Kujur, A., Yadav, A., Kumar, A., Singh, P. P., & Dubey, N. K. (2018). Nanoencapsulation: An efficient technology to boost the antimicrobial potential of plant essential oils in food system. *Food Control, 89*, 1–11. https://doi.org/10.1016/j.foodcont.2018.01.018.

Restrepo, A. E., Rojas, J. D., Garcia, O. R., Sanchez, L. T., Pinzon, M. I., & Villa, C. C. (2018). Mechanical, barrier, and color properties of banana starch edible films incorporated with nanoemulsions of lemongrass (*Cymbopogon citratus*) and rosemary (*Rosmarinus officinalis*) essential oils. *Food Science and Technology International, 24*(8), 705–712. https://doi.org/10.1177/1082013218792133.

Ribeiro-Santos, R., Andrade, M., Melo, N. R. d., & Sanches-Silva, A. (2017). Use of essential oils in active food packaging: Recent advances and future trends. *Trends in Food Science & Technology, 61*, 132–140. https://doi.org/10.1016/j.tifs.2016.11.021.

Robledo, N., López, L., Bunger, A., Tapia, C., & Abugoch, L. (2018a). Effects of antimicrobial edible coating of thymol nanoemulsion/quinoa protein/chitosan on the safety, sensorial properties, and quality of refrigerated strawberries (*Fragaria* × *ananassa*) under commercial storage environment. *Food and Bioprocess Technology, 11*(8), 1566–1574. https://doi.org/10.1007/s11947-018-2124-3.

Robledo, N., Vera, P., López, L., Yazdani-Pedram, M., Tapia, C., & Abugoch, L. (2018b). Thymol nanoemulsions incorporated in quinoa protein/chitosan edible films; antifungal effect in cherry tomatoes. *Food Chemistry, 246*, 211–219. https://doi.org/10.1016/j.foodchem.2017.11.032.

Salvia-Trujillo, L., Rojas-Graü, M. A., Soliva-Fortuny, R., & Martín-Belloso, O. (2013). Effect of processing parameters on physicochemical characteristics of microfluidized lemongrass essential oil-alginate nanoemulsions. *Food Hydrocolloids, 30*(1), 401–407. https://doi.org/10.1016/j.foodhyd.2012.07.004.

Salvia-Trujillo, L., Rojas-Graü, M. A., Soliva-Fortuny, R., & Martín-Belloso, O. (2015). Use of antimicrobial nanoemulsions as edible coatings: Impact on safety and quality attributes of fresh-cut Fuji apples. *Postharvest Biology and Technology, 105*, 8–16. https://doi.org/10.1016/j.postharvbio.2015.03.009.

Sasaki, R. S., Mattoso, L. H. C., & de Moura, M. R. (2016). New edible bionanocomposite prepared by pectin and clove essential oil nanoemulsions. *Journal of Nanoscience and Nanotechnology, 16*(6), 6540–6544. https://doi.org/10.1166/jnn.2016.11702.

Souza, V. G. L., Pires, J. R. A., Vieira, É. T., Coelhoso, I. M., Duarte, M. P., & Fernando, A. L. (2018). Shelf life assessment of fresh poultry meat packaged in novel bionanocomposite of chitosan/montmorillonite incorporated with ginger essential oil. *Coatings, 8*(5), 177. https://doi.org/10.3390/coatings8050177.

Taştan, Ö., Pataro, G., Donsì, F., Ferrari, G., & Baysal, T. (2017). Decontamination of fresh-cut cucumber slices by a combination of a modified chitosan coating containing carvacrol nanoemulsions and pulsed light. *International Journal of Food Microbiology, 260*, 75–80. https://doi.org/10.1016/j.ijfoodmicro.2017.08.011.

Toro-Márquez, L. A., Merino, D., & Gutiérrez, T. J. (2018). Bionanocomposite films prepared from corn starch with and without nanopackaged Jamaica (*Hibiscus sabdariffa*) flower extract. *Food and Bioprocess Technology, 11*(11), 1955–1973. https://doi.org/10.1007/s11947-018-2160-z.

Werner, B. G., Koontz, J. L., & Goddard, J. M. (2017). Hurdles to commercial translation of next generation active food packaging technologies. *Current Opinion in Food Science, 16*, 40–48. https://doi.org/10.1016/j.cofs.2017.07.007.

Yildirim, S., Röcker, B., Pettersen, M. K., Nilsen-Nygaard, J., Ayhan, Z., Rutkaite, R., Radusin, T., Suminska, P., Marcos, B., & Coma, V. (2018). Active packaging applications for food. *Comprehensive Reviews in Food Science and Food Safety, 17*(1), 165–199. https://doi.org/10.1111/1541-4337.12322.

Chapter 27
Nanocellulose-Polymer Composites: Novel Materials for Food Packaging Applications

A. K. Bharimalla, P. G. Patil, S. Mukherjee, V. Yadav, and V. Prasad

Abstract Nanocellulose is a revolutionary bio-based nanomaterial that possesses remarkable properties and has potential application in different industries. As a biodegradable filler in the manufacture of composite materials, coating and self-standing thin films, it offers novel and promising properties. There are fewer revisions focused on the use of nanocellulose-impregnated composite materials for different food packaging applications. Researchers have reported that the use of nanocellulose as a reinforcement in biopolymers and synthetic polymers improves the mechanical and barrier properties of the composite material. In this chapter we provide an exhaustive review of recent advances in the synthesis of nanocellulose and its application as a filler to produce nanocomposites for food packaging.

Keywords Biodegradable composite · Green composite · Nanocellulose biopolymer

27.1 Introduction

Packaging is an essential means to preserve the food quality and safety from manufacturing to its final use by the consumer. Food packaging helps reduce food waste during handling, transportation and storage by protecting food from physical, chemical or biological damage. Packaging is important for both consumers and marketers, as it not only provides different information such as ingredients, product characteristics, nutritional content, storage, but also helps improve customer acceptability and increase the product sale. Therefore, packaging is now considered the most important marketing tool in the modern economy. Around \$839 billion was the total turnover of the packaging industry in terms of revenue generation in 2015 (Anonymous n.d.), which is projected to increase at a rate of 3.5% on an average annually between

A. K. Bharimalla · P. G. Patil · S. Mukherjee
ICAR-Central Institute for Research on Cotton Technology, Mumbai, India

V. Yadav · V. Prasad (✉)
Department of Chemistry, Institute of Science, Banaras Hindu University, Varanasi, India

T. J. Gutiérrez (ed.), *Polymers for Agri-Food Applications*,
https://doi.org/10.1007/978-3-030-19416-1_27

2015–2020, reaching a value of $998 billion in 2020 (Butschli 2016). The packaging sector is one of the fastest-growing sectors in India. India's share in global packaging is about 4%, which is equivalent to $32 billion per year in 2015 and is expected to reach $73 billion by 2020, with an annual growth rate of 18% (Anonymous n.d.). Factors such as the success in online marketing, the growth of organized retail, the increase in the consumption of ready-to-eat foods, the growth in flexible packaging market, the adoption of innovative packaging process and the development of machinery have contributed to the accelerated growth in the packaging industry. Of the total packaging sector, foods and beverages occupy the majority, accounting for 85%, followed by the plastic packaging market, which is expanding rapidly with a growth of 20–25% *per* year and is valued at 6.8 million of tons. The paper packaging industry reaches 7.6 million tons, and 10% is occupied by pharmaceutical products. Worldwide, among all packaging, food packaging has the highest share, i.e. $161 billion, followed by other packaging such as beverage $76 billion, pharmaceutical $21 billion, cosmetic $13.3 billion and another $153 billion (Alam 2013).

The food industry has used petroleum-based plastic materials such as polyethylene as the preferred packaging material since last few decades (Gutiérrez and Alvarez 2017a). These materials were perceived as flexible, cheap, safe and versatile (Tice 2003), but they possess disposal limitations with very little recycling. In addition, rising petroleum costs, environmental concerns for disposal, globalization of food supply combined with consumer preference for fresher, convenient and safer foods have led to the emergence of novel sustainable packaging materials. These materials are an eco-friendly alternative to synthetic ones (Gutiérrez and Alvarez 2017b, c). Continuous research is being conducted to meet the consumer demands for new and innovative food packaging materials and technologies to produce sustainable packaging materials (Merino et al. 2018a, b; 2019a, b). These materials have changed the consumer's perception about raw materials, their selection, processing and consumption for different end-use applications. This is due to direct and indirect links of raw materials with general sustainability, energy efficiency and especially, security of supply. There is substantial growth potential for sustainable packaging materials for food packaging such as bio-based packaging products, biopolymers, bioplastics.

Bio-based food packaging materials are derived from renewable agricultural or marine sources (Álvarez et al. 2017, 2018). In the current era, green composite with bio-based material has gained preference as compared to petroleum-based composite due to the advantage of being natural, ecological and renewable (Satyanarayan et al. 2009). It has two main benefits, as they allow for a healthier ecosystem as well as rural employment along with economic development of the agricultural community. The limitation of biocomposites in the packaging industry is due to their poor mechanical and barrier properties (Azeredo 2009). In addition to this, their thermal properties are not at par with petroleum-based product. Further, processability is also one of the biggest limitations with current machinery due to low thermal stability and poor mechanical strength of biocomposites. These limitations can be addressed using advanced materials in nano-dimension (Gutiérrez et al. 2017a; Bracone et al. 2016; Gutiérrez et al. 2019; Toro-Márquez et al. 2018; Gutiérrez 2018a; Gutiérrez and Alvarez 2018, 2017d).

Nano-dimensional materials have some advanced properties due to their defect less structure and high strength properties. The intervention of nanotechnology not only solves the problems related to the properties, but also the cost economics (Sorrentino et al. 2007) and the problems of disposal of packaging materials. Nanocellulose which is a biopolymer, renewable and abundant, can be a twenty-first century solution as a filler material for bio-based composite material to be used for packaging industry (Mittal 2011). A detailed analysis of the nanocellulose production and its compatibility with hydrophobic matrix, as well as the application to produce nanocomposite for food packaging, was carried out in this chapter.

The water vapor permeability (WVP) of food packaging material is a critical issue in biopolymers, because they absorb water, which degrades the food quality. Therefore, the recent thinking process is addressing issues such as high WVP, poor mechanical strength, low thermal stability, improved physico-chemical and recyclability properties by using nanocomposite technology (Sorrentino et al. 2007). Materials with improved structure and properties can be processed by controlling important chemical and physical interactions of composites governed by surface properties (Gutiérrez et al. 2018a). The nanocomposites show different properties than their individual constituents, since a change in the diameter of the particles, layer thickness or fibrous material diameter to nanometer range influences the surface area-to-volume ratio (Ochsner et al. 2009). Different nanofillers have the potential to improve the migration of gases and the flavor properties of plastic packaging and increase the shelf life of the product, e.g. carbon nanotubes, kaolinite, graphene, nanoclays (Arora and Padua 2010)(Gutiérrez et al. 2017b).

Consumer demand and awareness about environmental problems due to plastic packaging has led to the development of novel bio-based and edible packaging materials (Gutiérrez et al. 2015a, b, c, d). Such materials will not only reduce packaging waste, but will also improve the quality of the stored food and, therefore, increase the shelf life of the product. Many thermoplastic polymers as starch can be used to prepare biodegradable packaging (Tharanathan 2003). Others include cellulose, chitosan (Cs)/chitin, protein (animal, plant based) or lipids (animal, plant derived) etc (Gutiérrez 2017a).

'Active packaging' (AP), a novel concept that has shown a paradigm shift in food packaging since the last two decades by shifting the protection function of packaging from passive to active. AP is a system in which the product, the package and the environment interact in a positive way to prolong shelf life of the food. AP actively changes the condition of the package to prolong shelf life or improve food safety or sensory properties, while maintaining the quality of the food. Active packaging is designed to provide chemical and physical benefits. The active packaging elements can be divided into different categories as absorber/scavenger, releaser, remover, time-temperature indicator and antimicrobial system. For any food, fresh vegetables and fruits, the absorption systems can be used as active packaging components to remove undesired gases and substances (oxygen, carbon dioxide, moisture, ethylene, flavor or odor, etc.) in order to extend the shelf life and prevent spoilage. The releasers consist of emitters of ethanol and carbon dioxide, preservative agents etc., removers help to eliminate lactose, cholesterol etc. of food, while temperature control systems can be self-heating or isolating materials.

Intelligent packaging (IP) is another innovative packaging system capable of carrying out intelligent functions such as detection, sensing, registration, tracing, communication and application of scientific logic, to improve food quality, extend the shelf life of the product, increase safety, provide information and warn about possible problems (Yam et al. 2005). IP system can for e.g. carry out intelligent functions as a release of an antimicrobial to protect the food product from spoilage. IP helps improve food safety and achieve the desired food quality. They can be tags or labels attached to a primary or secondary packaging as barcodes, RFID tags, time/temperature/gas indicators or biosensors (Gutiérrez et al. 2018b; Gutiérrez 2018b, 2017b; Gutiérrez et al. 2016a, b). A series of IP enabled technologies are available incorporating antimicrobial substance in nano-form to process polymer composites to control microbial surface contamination of foods. The successful production of viable edible films and coatings from whey proteins and their multi-functions as antioxidants, carrier of antimicrobials or other nutraceuticals with a significant primary barrier and mechanical properties add value for commercial applications in food industries (Ramos et al. 2012). In addition, due to the increasing demand in consumer for minimally processed products without preservatives, the IP technology has gained much attention from the food industry, as it uses low level of preservatives in packaging materials and ensures a minimal contact with the food (Cha and Chinnan 2004).

Many researchers have incorporated antimicrobial substances, sensors in nanocellulose-based polymer composites for packaging applications to detect and prevent food spoilage. Fortunati et al. (2012a) developed multi-functional bio-nanocomposite films of poly(lactic acid) (PLA), cellulose nanocrystals (CNCs) and silver nanoparticles. These films showed a constant antibacterial effect against *Staphylococcus aureus and Escherichia coli*, with potential use in food packaging applications. El-Wakil et al. (2015) developed wheat gluten/nanocellulose/titanium dioxide composite papers for active food packaging. These coated papers showed good antimicrobial activity against Gram positive and negative bacteria, and yeast.

27.2 History of Packaging Industry

The need to store food arose with the progress in human civilization when humans began to travel long distances in search of food. The oldest packaging materials used for this purpose were pumpkin's, shells and leaves. Subsequently, humans devised different innovative nature-derived packages with an increase in the need for storage and packaging purposes, e.g. baskets woven from leaves and barks, hollowed out vegetables and wooden logs, different animal parts such as skin/hide, organs etc. During 1500 BC–500 AD, new packages such as ceramic vessels and amphorae arose in the Mediterranean region for the commercial transport of wine and other products. The different advances in food packaging over the years helped to protect food from external contact such as contamination by microbes, thus reducing its growth, in addition to its basic function as a means of consumption and storage. The additional packaging materials developed over the years prevent contact of food with change in air, pH and moisture, which can lead to microbial degradation of food (Raheem 2012).

Glassmaking started as a type of ceramics in 7000 BC and the glassmaking technique was industrialized around 1500 BC in Egypt. Around 7000 BC, colored water pots were used as containers to store food and water. By 1200 BC, the cups and bowls were made by pressing the glass into molds. Transparent glass was invented in the Christian era and its manufacturing process spread across the European continent for the next 1000 years. The technique was refined even more during eighteenth and nineteenth centuries. The first automatic machine to manufacture rotary bottles was invented and patented by Owens in the year 1889. During 1900s–1960, liquid products were stored mainly in glass containers.

The oldest form of flexible packaging was paper. The paper was invented in China during 200 BC–220 AD. The word paper is derived from the Greek word Papyrus. The paper-based packaging was first used by the Chinese to wrap foods with the use of treated mulberry bark during 1st and 2nd century BC. The papermaking technique was further refined over the next 1500 years and the knowledge travelled westwards from China to Central Asia, the Middle East, and Europe (Welt 2005). Papermaking technique was introduced in the United Kingdom in 1310, while it came Pennsylvania, USA in 1690 (Berger 2002).

Paperboard was first used to manufacture folding cartons in the 1800s, first paperboard carton/cardboard box was made in England in 1817 and its thinner type paperboard carton was later used for cereal storage. The use of paper in packaging began with the development of paper bags. The commercial production of paper bags began in Bristol, England in 1844. The corrugated boxes used widely today as shipping container to hold a series of smaller packages were developed in the 1850s (Risch 2009). In 1852, the machine to manufacture paper bags was invented and patented by Francis Wolle in United States of America (USA), which led to development of glued paper sacks and gusset design in 1870s, which are also used today. In 1870s, the world's first semi-flexible packaging was accidentally developed by Robert Gair, who invented the first automatic-made cardboard. These folding cartons used to date form the backbone of dry and processed food products. In 1905, machines were invented to produce in-line printed paper bags.

Initial metal-based packaging materials as boxes, cups and containers made from gold and silver were precious/too expensive for ordinary people to use in the ancient era. Other metals emerged progressively as common packaging materials with discovery of cheaper metals for the processing of strong alloys, coatings and thinner gauges and large-scale production (Hook and Heimlich 2011). In 1805, Nicholas Appert ("Father of Canning") proved that food could be sterilized and preserved for an extended time by boiling at high temperatures and then sealing in tin containers. Later, in the year 1810, inventor and merchant Peter Durand received a patent for developing a sealed cylindrical can made from tinplate for food storage. The first beverage can made of tin-plated steel was made by Kruger Beer in 1935, until then the drinks were available in glass bottles. In 1959, Coors introduced the aluminium can (which is currently used in carbonated beverages) and in the same year, Emral Fraze invented the pop tab. The first ring pull was introduced in 1963, which facilitated the opening of the can and provided ease of drinking directly from it. The stay tab which is a ring tab that stays attached to the can, was introduced in 1975.

Throughout the 20th century, paper and paperboard packaging increased in popularity and, subsequently, in the late 1970s and 1980s with the arrival of plastics as an important element in packaging, paper and paper-based products were replaced. The use of plastics in packaging began in the 1950s. The packaging industry is the largest user of plastics because more than 90% of flexible packaging is made of plastics and only 17% of rigid packaging. Generally, barrier resins are used for the preparation of plastic containers to improve product protection and make them more cost effective. Plastic materials made up of large organic molecules are useful products, since they are fluid, can be sealed with heat, are easy to print and can be integrated into production processes (Marsh and Bugusu 2007). The application of plastics in packaging has increased worldwide with an estimate of 280 metric tons (Paine and Paine 2012).

The molded deodorizing squeeze bottles were introduced in 1947 and in 1958 for packaging; while heat shrinkable films were developed from the mixture of styrene with synthetic rubber. Cellulose acetate first derived from wood pulp in 1900 and developed for photographic uses in 1909. New manufacturing protocols were also developed using various techniques such as forming, molding, casting and extrusion to prepare plastic products in large quantities (Plastic Make It Possible Report 2010).

In 1977, after the introduction of polyethylene terephthalate (PET) containers, it became easy to manufacture beverage containers, which are mainly used today. Other novel materials, such as transparent films and cellophanes, have been used as external wrappers to maintain the shape. By 1980, foods and other hot-fill products, such as jams, could also be packaged in PET. In 1986, aluminium trays were replaced by plastic microwavable trays. In 1996, metallocene-catalyzed polyolefin was introduced to reduce food waste. In 2000, the entry of PLA into packaging market signalled the return of biobased plastic (Vink et al. 2004).

Active packaging is a new area of research which involves the combination of food packaging materials with antimicrobial substances. With the advent of nanotechnology, it is much easier to functionalize polymer films based on nanoparticles with antimicrobial agents for food packaging. The nanocomposites are much more superior compared to conventional packaging materials in terms of barrier properties, mechanical strength and heat resistance (Herniou--Julien et al. 2019). The polymer nanocomposites for food packaging have been considered for the entire life cycle of the packaging material (Silvestre et al. 2011). The life cycle is a very important property in the packaging material from procurement, processing, transportation, delivery and disposal of raw materials (Chaffee and Yoros 2007).

27.3 Polymer Matrix

Fiber along with a suitable polymer matrix, provide good strength and adhesion to a composite material (Deo 2010). Generally, the matrix is deformed in the load application or load transfer. Therefore, when the matrix is reinforced with any fibrous filler, the applied load is transferred mostly to the fiber and provides an adequate load bearing capacity. The matrix redistributes the load to the surrounding fibers and restricts the buckling of the fibers due to compression (Outwater 2014). Polymer matrices are broadly divided into two categories: thermoset and thermoplastics.

27.3.1 Thermoplastic Polymer Matrix

Plastics that melt upon being heated and solidify upon cooling are known as thermoplastics which are generally solid at room temperature (Fig. 27.1). There is no chemical change in any number of melting and solidification of these plastics (Morena 1988). But a greater number of heating cycle degrades the polymer. Examples of thermoplastic polymers are unsaturated polyesters (UPs), polyetherimide, polyamide imide, polyphenylene sulfide, polyetherketone and liquid crystal polymers. Thermoplastics are mainly used in high volume industries, such as the automotive and aviation industries, etc. After adequate reinforcement with plastic, carbon or graphite fiber, they can be used instead of epoxy in the next generation aircrafts (Astrom 1997).

27.3.2 Thermoset Polymer Matrix

The polymeric materials after a chemical reaction are transformed from liquid to solid state, when the material is not cured, the materials are very small unlike molecules, i.e. monomers. After molecular cross-linking during the reaction, longer molecular chains are formed, leading to solidification (Fig. 27.2). These thermoset polymeric materials are permanent and irreversible. The different types of thermoset polymer matrices used in composites are bismaleimides, epoxy (epoxide), phenolic (PF), UP, polyimide, polyurethane (PU) and silicone (Astrom 1997).

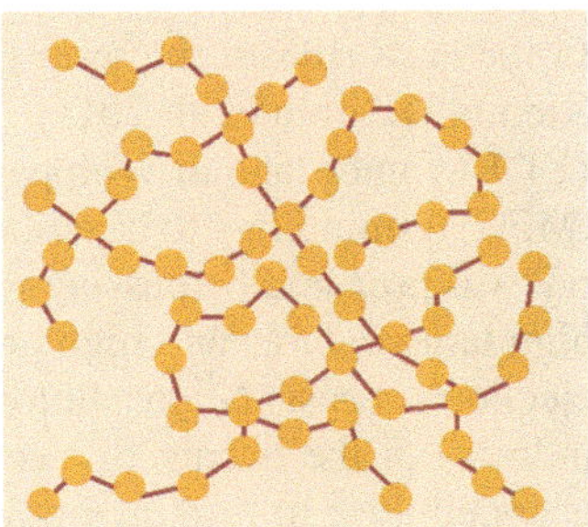

Fig. 27.1 Polymerized thermoplastic polymer

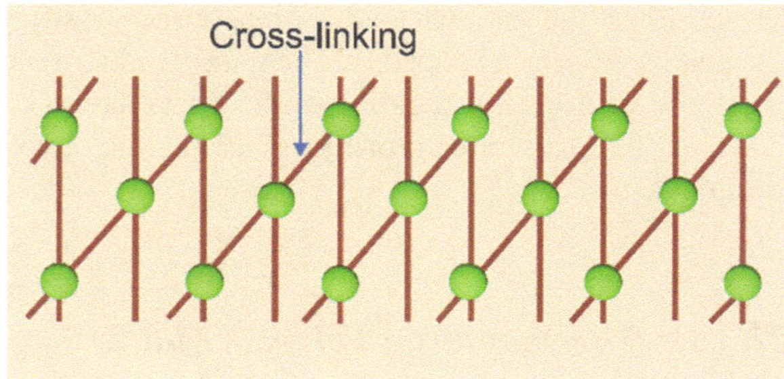

Fig. 27.2 Polymerized thermoset polymer

27.4 Cellulose

In 1839, the French chemist A. Payen used the term "cellulose" for the first time in his report of the French Academy (Payen and Hebd 1838). The French chemist Anselme Payen in 1838 described a resistant fibrous solid, which remained after treating various plant tissues with acids and ammonia (Klemm 2005). The molecular formula ($C_6H_{10}O_5$) was determined by elemental analysis and the term "cellulose" was used to refer it. It was used in the form of wood, cotton and other vegetable fibers as an energy source, for building materials and clothing for thousands years since its discovery (Doree 1947). Cellulose is a long-chain natural polymer made by linking of smaller molecules (Fig. 27.3). These links in the cellulose chain consist of sugar, β-D-glucose (Alemdar and Sain 2008). Cellulose is found in cotton and all ligoncellulosic biomasses. Ligoncellulosic agricultural by-products can be used to obtain cellulose, among them: the residues of wheat and cereals (Jahan et al. 2011), jute (Nelson et al. 2000), soybean husk (Bochek et al. 2003) flax fibers and flax straw (Bhattacharya et al. 2008), sugarcane bagasse (Istvan and Plackett 2010), corn, sorghum, barley, pineapple, sorghum, bananas, coconut crops, etc. (Hubbe et al. 2008). Despite its various use in the fiber, paper, films and polymer industries, the use of cellulose has generated interest in the processing of novel material applications due to its super functionalities, extremely large, active surface area and low cost (Turbak et al. 1983).

27.4.1 Nanocellulose

In the early 1980s, the term nanocellulose was publicly used for the first time by Klemm (Klemm et al. 2011). Turbak, Snyder and Sandberg in the late 1970s at the ITT Rayonier labs in Whippany, NJ, USA used the microfibrillated nanocellulose (MFC) terminology to describe a product prepared as a gel-like material. This material was prepared by passing wood pulp through a Gaulin-type milk homogenizer at high temperatures and high pressures impacting against a hard surface through the ejection process. Nanocellulose is a promising innovative material for the modern industry of the twenty first century. In future, synthetic fillers will be replaced by nanocellulose to reinforce polymer composites for use in the automotive industry, packaging and furniture production Jahan et al. (2011).

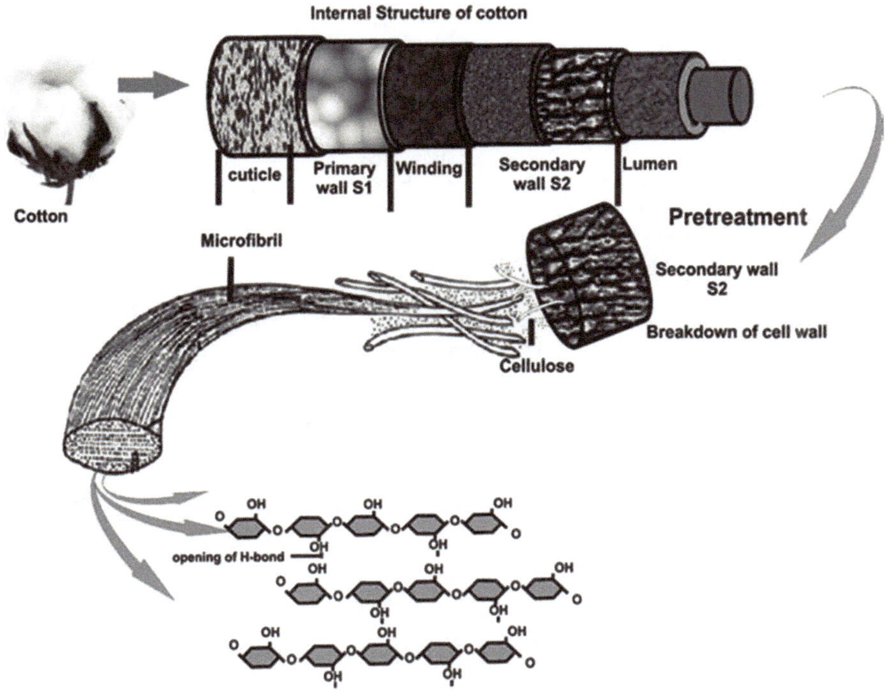

Fig. 27.3 Structural organization of the plant cell wall

Fig. 27.4 Image of nanocellulose prepared from wood pulp

The word 'nanocellulose' generally refers to cellulosic materials with a dimension in the nanometer range and look like gel if it is prepared from wood pulp as shown in Fig. 27.4. Depending upon the raw material, the production method and its size, nanocellulose is classified into three main categories (Iwamoto et al. 2009). The three main types of nanocellulose are cellulose nanocrystal (CNC), cellulose nanofibril (CNF) and bacterial nanocellulose (BNC). The nomenclature used by Klemm (2005) is given in Table 27.1.

Table 27.1 Classification of nanocellulose (Iwamoto et al. 2009; Gama et al. 2012; Cherian et al. 2010)

Type of nanocellulose	Synonyms	Typical sources	Average size	Images
Cellulose nanocrystal	Nanocrystalline cellulose, whiskers, rod like cellulose, microcrystals	Wood, cotton, hemp, flax, wheat straw, rice straw, mulberry bark, ramie, MCC, Avicel, tunicin, algae, bacteria, etc.	Diameter: 5–70 nm Length: 100–250 nm (from plant); 100 nm several micrometers (from cellulose of tunicates, algae, bacteria)	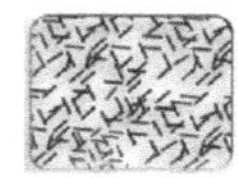
Nanofibrillated cellulose	Nanofibrils, microfibrils, nanofibrillated cellulose, microfibrillated cellulose	Wood, sugar beet, potato, tuber, hemp, flax etc.	Diameter: 5–60 nm Length: several micrometers	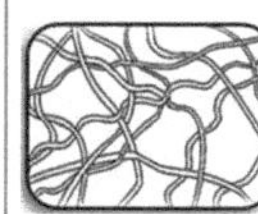
Bacterial nanocellulose	Microbial cellulose, biocellulose	Low-molecular-weight sugars and alcohols	Bacterial synthesis Diameter: 20–100 nm	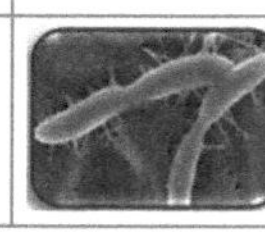

27.4.2 *Preparation and Properties of Nanocellulose*

The most general method for preparation of nanocellulose is top-down approach using physics, chemistry and chemo-mechanics. The detailed scheme of this method is shown in Fig. 27.5. BNC is being synthesized by a bottom-up method from the glucose by a family of bacteria (Iwamoto et al. 2009), yeast, algae, etc. In this chapter, we mainly focus our attention on nanocellulose produced by the top-down method from wood or agricultural/forest crops or residues, i.e. lignocellulosic biomass.

Several methods of preparing nanocellulose from cellulosic materials have been reported, such as steam explosion treatment (Mandal and Chakrabarty 2011), acid or alkaline hydrolysis (Moran et al. 2008; Henriksson et al. 2007), enzyme-assisted hydrolysis (Chen et al. 2011), as well as a combination of two or several of the aforementioned methods (Keeratiurai and Corredig 2009). High-pressure homogenization (HPH) due to its simplicity, high efficiency and no requirement of organic solvents is an efficient technology for biomass refining (Kaushik and Singh 2011). Typically MFC is produced as a suspension in water. The viscosity of the suspension changes during homogenization from a low-viscosity to a high. Normally, a 2% fiber suspended in water is used for the preparation of MFC. To minimize the size of the cellulose fibers and to avoid clogging during homogenization, pre-treatment of cellulose, such as steam explosion, microfluidization processor or other methods is essential (Lee et al. 2009a; Zimmermann et al. 2010; Brandt et al. 2010). Recently, ionic liquids at room temperature have emerged as new agents for pre-treatment, since they have an excellent dissolving capacity of cellulose (Li et al. 2012).

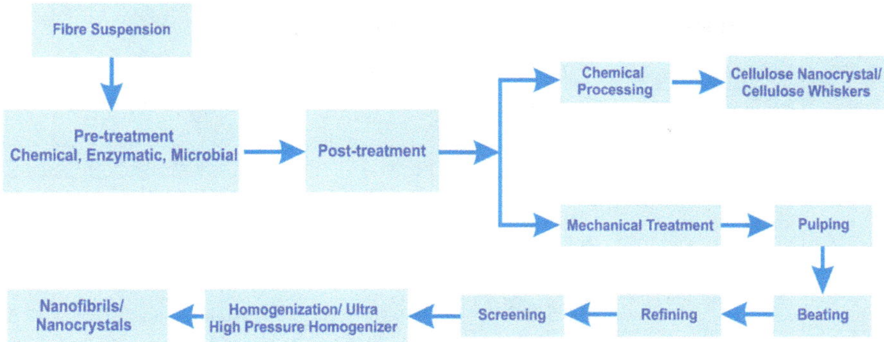

Fig. 27.5 Flow diagram for the synthesis of nanofibrils and nanocrystals by mechanical and chemical routes

At higher concentrations, the increase in viscosity during processing becomes too high to continue processing. Therefore, a schematic overview of approache to produce nanocellulosic materials in low concentration from fibers is provided in Fig. 27.5 (Siqueira et al. 2010; Lavoine et al. 2012).

Many intrinsic properties have been observed in MFC, such as low density, high chemical reactivity, high strength and modulus, and high transparency that make it attractive for applications (Nogi et al. 2009; Lee et al. 2009b; Paakko et al. 2007; Siro and Plackett 2010).

Recently, a great interest has been attracted to nanocellulose as a potential filler for use in nanocomposites. A wide range of polymer matrices can be incorporated with nanocellulose as waterborne polyurethane, poly(3-hydroxybutyrate), hydroxypropyl cellulose, poly(L-lactide), poly(3,4-ethylenedioxythiophene), polyvinyl acetate and poly(*o*-ethoxyaniline). Table 27.2 shows that the nanocellulose stiffness and tensile strength (TS) are comparable to those of aramid fiber (Kevlar) and better than glass fiber, which are used commercially to reinforce plastics.

Nanocellulose films exhibit high strength (> 200 MPa), high stiffness (~20 GPa), and high strain (12%). The strength/weight ratio of nanocellulose is 8 times more than stainless steel (Aulin et al. 2010). Nanocellulose composites can thus be used as coatings and films, paints, foams and packaging materials (Cai et al. 2011).

27.5 Nanocellulose-Polymer Composite Films

Nanocomposite is now an expanding research area that offers new materials with novel functional properties (Ruiz et al. 2000). Nanocomposites are made up of nano-size fillers with particular size, dimension and surface chemistry properties. Nanocellulose is used mainly with hydrocolloids such as starches, poly-saccharides, proteins, pectins and synthetic polymers. Different inorganic nanoparticles have

Table 27.2 Mechanical properties of various reinforcements

Sl. No.	Plant fibre/materials	Strength (GPa)	Modulus (GPa)	Reference
1	Cellulose micro/ nanofibril	10	150	Sakurada et al. (1962)
2	CNC	10	150	Revol et al. (1998)
3	Aramid fibers (Kevlar)	3–3.5	130	Denoyelle (2011)
4	Cotton	0.3–0.7	6–10	Cristaldi et al. (2010)
5	Kapok	0.093	4	Cristaldi et al. (2010)
6	Bamboo	0.57	27	Cristaldi et al. (2010)
7	Flax	0.5–0.9	50–70	Cristaldi et al. (2010)
8	Hemp	0.31–0.75	30–60	Cristaldi et al. (2010)
9	Jute	0.2–0.45	20–55	Cristaldi et al. (2010)
10	Aluminum wire	0.62	73	Bledzki and Gassan (1999)
11	Steel	0.54	200	Bledzki and Gassan (1999)
12	Kenaf	0.29–1.19	22–60	Bledzki and Gassan (1999)
13	Ramie	0.91	23	Cristaldi et al. (2010)
14	Abaca	0.012	41	Cristaldi et al. (2010)
15	Banana	0.53–0.92	27–32	Cristaldi et al. (2010)
16	Pineapple	0.413–1.627	60–82	Cristaldi et al. (2010)
17	Sisal	0.08–0.84	9–22	Cristaldi et al. (2010)
18	Coir	0.106–0.175	6	Cristaldi et al. (2010)
19	Ramie	0.4–0.94	61.4–128	Cristaldi et al. (2010)
20	E-glass	2–3.5	70	Westman et al. (2010)
21	Carbon	4	230–240	Westman et al. (2010)

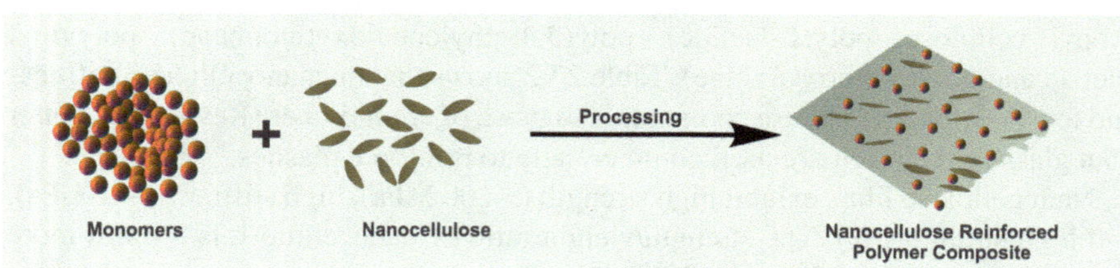

Fig. 27.6 Preparation of polymer–nanocellulose composite

been recognized as possible additives to improve the polymer performance (John and Thomas 2008). A polymer composite is a combination of a polymer matrix and a strong reinforcing phase, or filler (Fig. 27.6). Polymer nanocomposites are polymers (thermoplastics, thermosets or elastomers) that have been reinforced with small amounts (less than 5% by weight) of nanosized particles having high aspect ratios (L/D > 300) (Denault and Labrecque 2004).

27.5.1 Preparation of Nanocellulose-Based Composite Films

Polymers derived from petrochemical predominate in food packaging due to their easy processing, excellent barrier properties and low cost (García et al. 2004). The use of nanocellulose acting as an active substance carriers, such as antioxidants and antimicrobials helps to extend the food quality and safety, thus extending the shelf life of them. There is potential for the processing of nanocellulose-based polymer composites for preparing cheap and lightweight nanocomposites for food packaging applications. In the recent years, advanced research and development in the field of polymer science and the extensive use of advanced technology for the preparation of polymers have generated greater interest in the preparation and characterization of novel polymer materials and their composite films.

27.5.1.1 Solvent Casting Method

One of the simplest methods used for the preparation of polymer nanocomposites is the solvent casting method, since it requires simple processing equipment. The nanocellulose reinforcement in the solution casting is dispersed within a given medium (0.05–5 wt% solids), water or different organic media to prepare a homogenous nano-scale aqueous suspension by stirring at room temperature or using an autoclave reactor to mix at high temperatures and then polymer solutions are mixed with it. The composite films are subsequently produced with this mixture *via* casting on a suitable surface, followed by an evaporation/drying in oven under vacuum (Fig. 27.7) (Ghosh and Sain 2014).

This processing method strongly influences the mechanical properties of the resulting nanocomposite. It has been observed that solution casting of nanoscale reinforcements, especially with an aqueous latex dispersion matrix, produces superior results compared to melt-extruded composites due to better dispersion of nanoscale cellulose reinforcements and the possibility to form hydrogen bonds between reinforcements and matrix material (Favier et al. 1995). It is a low-temperature environmentally friendly process which requires a small amount of

Fig. 27.7 Solvent casting

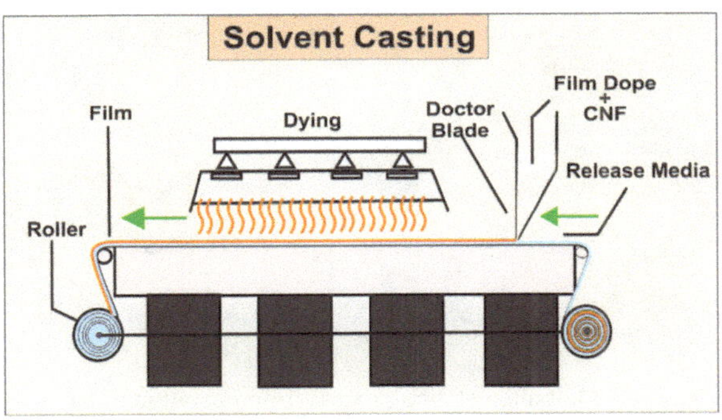

sample to produce films of uniform thickness. However, the method has several limitations, since it consumes time, is useful when a very small amount of reinforcement is required, is limited to the laboratory scale and can involve complex processes with a high energy consumption (Khoskhava 2014).

The solution casting method is used to process polymer composites with nanocellulose reinforcement in water soluble polymer matrices such as poly(oxyethylene) (PEO) (Samir et al. 2004), polyvinyl alcohol (PVOH) (Roohani et al. 2008), waterborne PU (Guangjun et al. 2008), starch (Teixeira et al. 2009) and water-insoluble polymer matrices as polypropylene (De Menezes et al. 2009), polyvinyl chloride (Chazeau et al. 1999), polyvinyl acetate (Garcia de Rodriguez et al. 2006), polycaprolactone (Siqueira et al. 2009) and PLA (Sanchez-Garcia and Lagaron 2010; Yu et al. 2008; Lin et al. 2009).

27.5.1.2 Melt Intercalation Process

This is the most promising and practical method for the preparation of polymer nanocomposites. Vaia, Ishii and Giannelis (1993) first used the melt intercalation process to process polymer composites in 1993. Since solvents are not required, this method is used in the polymer processing industry using compounding devices such as extruders or mixers. Melt intercalation is a top-down method of polymer processing. This simple economic and environmentally friendly processing method involves direct mixing of nanoreinforcement with molten polymer to optimize the polymer–nanomaterial interactions. The nanocomposite is formed when the polymer-filler mixture is hardened above the glass transition temperature (T_g) of the polymer (Fig. 27.8).

The polymer chains penetrate into the filler/reinforcement layers inducing intercalation. According to the compatibility between the surface of the layers and the polymer, either intercalated or exfoliated nanocomposites can be obtained (Kumar et al. 2009). This method results in normally lower levels of exfoliation

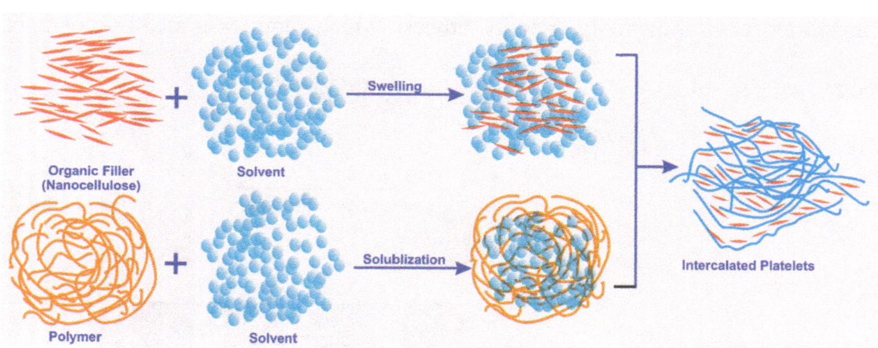

Fig. 27.8 Melt intercalation process for preparation of nanocomposite

compared to *in situ* polymerization. Polymers not suitable for solution intercalation, adsorption or *in situ* polymerization methods, can be used to prepare nanocomposites by this method (Hussain et al. 2006). The melting intercalation process of the polymer preparation reduces the interfacial tension and improves matrix-filler interactions. The method is more flexible and does not require chemical reaction or solvent (Liu et al. 1999; Karande 2013).

27.5.1.3 *In situ* Polymerization

The first method used to synthesize nanocomposites was *in situ* polymerization. The *in-situ* polymerization of monomers in the presence of cellulosic nanofillers is advantageous as compared to the traditional processing methods mentioned above, since a uniform dispersion of cellulosic nanofillers in the polymer matrix can be achieved by minimizing aggregation by their improved interaction with growing polymer molecules. This method also reduces the moisture absorption behavior of wrapped cellulosic nanofillers and improves the biodegradability of nanocomposites under compost conditions when required for specific applications such as packaging. However, this method is applicable only when the polymerization is carried out in liquid phase where liquid monomer molecules are polymerized in presence of nanocellulose filler. This process is conventionally used to synthesize thermoplastic nanocomposites, while for thermosets such as epoxies or UPs, a curing agent or peroxide is added to initiate the polymerization. Either the addition of curing agent or the increase in temperature can initiate polymerization for the thermosets (Fig. 27.9) (Messersmith and Giannelis 1995).

This processing method can be scaled up to an industrial level and can dramatically increase the industrial development of cellulosic nanocomposite products. Several researchers have used the *in situ* polymerization method to develop novel nanocellulose based composites with new potential applications, e.g. polymethylmethacrylate (PMMA) (Mabrouk et al. 2011; Maiti et al. 2013), polyacrylamide (PAAm) (Dufresne et al. 2003), PANI (Zhou et al. 2011), polyurethane (Lacerda et al. 2013), polypyrrhole (Nystrom et al. 2010).

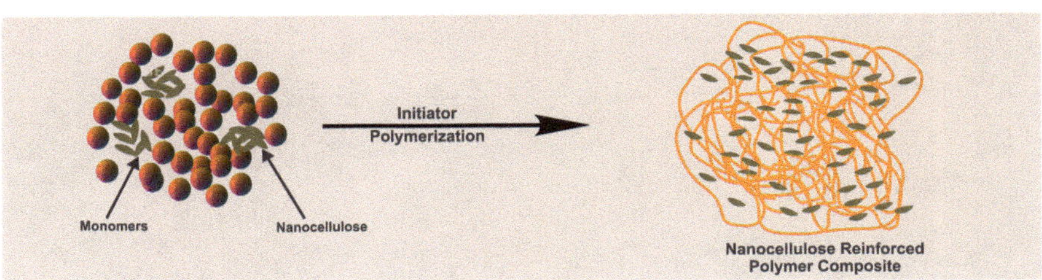

Fig. 27.9 Process for the preparation of nanocomposite by *in situ* polymerization

27.5.1.4 Ring-Opening Polymerization

The ring-opening polymerization (ROP) reaction is of particular interest because it finds wide application in thermostable polymers. By means of this technique, epoxy resin networks or cross-linked UPs can be synthesized, which are of great importance in the market of thermoset polymer-based composites.

ROP is a well-known technique to polymerize cyclic monomers such as lactones and lactides, where alcohol generally acts as an initiator. This method is used for the polymeric modification of cellulose and various cellulosic derivatives (Jerome and Lecomte 2008). Depending on monomers, catalysts, initiators used, the polymer is prepared through different mechanisms using ROP method (Fig. 27.10). Several scientists have used ROP for the preparation of nanocellulose-based polymer composites (Goffin et al. 2011; Peltzer et al. 2014; Habibi and Vignon 2008; Chen et al. 2009; Lonnberg et al. 2011; Braun et al. 2012). The nanocomposite sheets produced by ROP possess good mechanical strength.

The ROP approach was applied for the first time to the CNCs by Habibi et al. (2008) who grafted polycaprolactone onto the surface of CNCs using stannous octoate ($Sn(Oct)_2$) as a grafting and polymerization agent. Chen et al. (2009) and Lin et al. (2009) conducted similar ROP under microwave irradiation to improve graft efficiency and incorporated PCL-CNCs into the PCL matrix. Lonnberg et al. (2011) grafted PCL chains on nanofibrillated celluloses (NFCs) and the resultant PCL-grafted NFC films were prepared by thermoforming in the form of bilayer laminates. Oksman et al. (2006) reported a direct correlation between the length of grafted PCL chains and interfacial toughness of the resulting laminate.

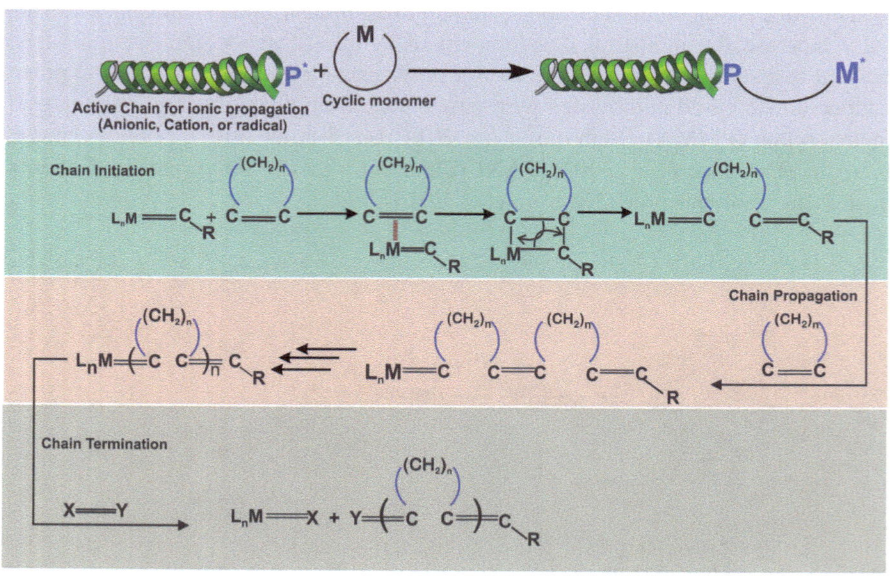

Fig. 27.10 Ring opening process for the preparation of nanocomposite

Extrusion methods are being used for the formation of nanofiber-reinforced polymer composites. Some reviews related to the preparation of composite films are highlighted below:

Oksman et al. (2006) reported on the preparation of PLA nanocomposites reinforced with cellulose whisker using melt extrusion method. In this method, the nanocellulose polymer composite was prepared by pumping nanocrystal suspension into the melt of the polymer during the extrusion process (Oksman et al. 2006). The authors observed that the hydrophilic nature of the cellulose results in the formation of additional hydrogen bonds between amorphous parts of the nanoparticles leading to an irreversible agglomeration during drying and aggregation in nonpolar matrices (Oksman et al. 2006). The film composites produced show a higher Tensile Strength (TS) as the proportion of modified cellulose increases and the elongation at break increases as the proportion of PLA increases. So modified cellulose plays a vital role in increasing the TS of composite films. These films can thus be used as a package to protect food from oxidation reaction and moisture (Sandeep et al. 2012).

Bruce et al. (2005) prepared composite materials using MFC from Swedish roots and different resins, including four types of acrylic and two types of epoxy resins. It was found that the stiffness and the strength of all the composites were significantly higher than the unmodified resins. Using this method, nanocomposites with good mechanical properties were prepared using vegetable pulp with a wide range of resins (Bruce et al. 2005). Low coefficient thermal expansion (CTE) is another beneficial feature of nanocellulose. It has been reported that the CTE values for nanocellulose are as low as 0.1 ppm/K, which is comparable to that of quartz glass. The low CTE for nanocellulose in combination with high strength and Young's Modulus (YM) makes it a potential reinforcement material for the manufacture of solar cells, actuators, flexible displays, electronic papers, panel sensors etc. (Nishino et al. 2004). Nogi and Yano (2008) prepared nanocomposites using transparent acrylic resin (with low YM) with 5% bacterial cellulose (BC). The ductile, foldable and transparent nanocomposite showed a low CTE and a high YM. A film stacking method was used to prepare PU-MFC composite materials, where the PU films and nonwoven cellulose fibril mats were stacked and compression molded (Seydibeyoglu and Oksman 2008). Wan et al. (2006) used PVOH reinforced with BC as a reinforcement material for medical device applications. The authors developed a PVOH-BC nanocomposite with mechanical properties over a wide range, thus making it appropriate to replace different tissues (Wan et al. 2006).

Khan et al. (2014) prepared methylcellulose (MC) films by using the casting methodology in a 1% aqueous solution containing 0.5% vegetable oil, 0.25% glycerol and 0.025% Tween-80. These authors studied the effect of gamma radiation on the nanocellulose-containing MC-based composites and observed that mechanical properties of the films increased slightly at low doses due to the reorientation of the nanocellulose fibers (NCFs), whereas barrier properties were improved even more. Dufresne et al. (2000) and Dufresne and Vignon (1998) prepared and reported that the biodegradability of modified potato thermoplastic starch (TPS) composite films containing MFC obtained by the solution casting methodology were preserved, and that MFC significantly reinforced the starch matrix, regardless of the plasticizer content, and increase in YM depended on filler content almost in linear manner. It was also observed that YM and TS were significantly improved in the nanocomposite films due to the uniform dispersion of the nanofibers in the polymer matrix. Azeredo et al. (2010) developed composite films by reinforcing the nanocellulose in Cs and glycerol as a plasticizer. Pereda et al. (2010) developed sodium caseinate films with nanocellulose by dispersing the fibrils into film-forming solutions, casting and drying. The composite films were found to be less transparent and more hydrophilic than pure sodium caseinate films. The caseinate films showed an initial rise and subsequent decrease in WVP with increase in filler content.

27.6 Characterization of Composite Films

Different characteristics such as surface morphology, mechanical, thermal and optical properties, as well as water and gas permeability using various instrumental techniques are conducted to evaluate the properties of nanocomposites. Microscopy helps to study and understand the distribution of nanocellulose in the polymer matrix and interfacial adhesion.

27.6.1 Scanning Electron Microscopy (SEM)

The morphological and macroscopic study of nanocomposites is carried out by using scanning electron microscopy (SEM). An electron beam scan focused across the surface of the sample results in signals that are converted into an image on a computer screen. The most widely used signals for imaging are secondary electrons, which are electrons that are excited from the sample molecules by the scanning electron beam. These secondary electrons give information about the texture of the surface and the dark regions in the obtained image mean that secondary electrons are prevented from reaching the detector. There are also other signals obtained when the beam strikes the sample, e.g. backscattered electrons, Auger electrons and X-rays (Zhou et al. 2006). Other methods used to study the morphology of nanocomposites are Atomic Force Microscopy (AFM), Fourier Transform Infrared (FTIR) spectroscopy, etc.

27.6.2 Permeability

Permeability is an important parameter for the evaluation of nanocellulose-reinforced composite films for packaging applications. The experiment with diffusion cells is used to determine the permeability of nanocellulose-reinforced film (van den Mooter et al. 1994). The cellulose nanoparticles confer improved barrier properties to the membranes used for packaging purposes. Due to the nanosize, gas molecules penetrate with difficulty into the crystalline domains of the cellulose nanoparticles. In addition, the cellulosic nanoparticles form a dense percolating network held together by strong inter-particle bonds, which reinforce their use in films. Meanwhile, CNFs have a strong gas barrier property compared to the CNCs. CNC's can form a dense hydrogen bonding network which can increase the gas barrier property of CNC-reinforced composite films. CNFs have tunable barrier properties which can be used to process high performance barrier films for packaging applications.

The mass transport rate of water/gases is expressed by the diffusion of specimens across a membrane according to Fick's first law, assuming that the linear concentration drop within the film (two-film theory). Using a partition coefficient K, the bulk concentrations in the diffusion cell chambers can be related to the concentrations on the surface of the film. Another assumption that is made is that the concentration of the diffused specimens present within the film is negligible compared to the total concentrations of the chamber. The complete derivation can be seen in the work of (van den Mooter et al. 1994). It is important to note that the permeability is scaled against the film thickness, resulting in the unit m^2/s.

27.6.3 Thermal Analysis

Thermal analysis is defined as the measurement of the physical and chemical properties of materials as a function of temperature. The two main thermal analysis techniques are thermogravimetric analysis (TGA) and differential thermal analysis (DTA). TGA is a method of thermal analysis in which changes in the physical and chemical properties of polymer materials are measured as a function of the increase in temperature (with constant heating rate) or as a function of time (with constant temperature and/or constant mass loss). The thermal decomposition property of nanocellulose-reinforced composite films is determined by TGA. The DTA measures the temperature difference between a sample and an inert reference material as a function of temperature. This method detects changes in heat content. Differential Scanning Calorimetry (DSC) is another closely related and modified version of DTA. The study of the thermal behavior of nanocomposites helps to evaluate the operational range of work of the nanocellulose-reinforced polymer composites as compared to the traditional composites. The nanocrystalline cellulose (NCC) reinforcement leads to an improvement in glass-rubber transition temperatures, melting point and thermal stability. Kaushik and Grewal (2011) studied the thermal behavior of the TPS/NCC composites for food packaging applications. The thermal degradation of the nanocomposites with 10% CNCs was studied using TGA under nitrogen atmosphere at a heating rate of 10 °C/min. The authors reported that the addition of CNCs had a significant effect on the activation energy for thermal degradation of the composite materials compared to the net matrix alone.

27.6.4 Dynamic Mechanical Analysis (DMA)

DMA is a viscoelastic technique that monitors changes in properties due to temperature and/or frequency or time change. The technique measures the total energy stored and dissipated in the material due to a dynamic stimuli. The viscoelastic properties are obtained from elastic and viscous responses. The elastic response is a measure of the energy stored in the material and yields the storage modulus (E0). On the other hand, the viscous response measures the energy dissipated in the material due to friction and internal movements, and yields the loss modulus (E00) (Barari et al. 2016). The effect of CNF reinforcement on the viscoelastic properties of epoxy composites was examined by Barari et al. (2016) using the DMA.

27.6.5 Differential Scanning Calorimetry (DSC)

DSC determines the energy changes within a material during the constant heating rate. These energy changes correspond to chemical reactions that occur in the sample or physical changes such as glass transition, crystallization, melting of crystals or sample decomposition (Sandler et al. 1998). The DSC can be considered as another thermal test method to measure the T_g of nanocellulose-based polymer composite samples. The T_g indicates the temperature at which polymers transform from hard state (glassy state) to soft (rubbery) state. The T_g increases with increasing chain stiffness and with increasing intermolecular attraction forces. Gray et al. (2018) studied DSC thermograms of low density polyethylene (LDPE)/TPS nanocomposites reinforced with CNCs. These authors reported that CNC increased the T_g and melting temperature (T_m) of LDPE/TPS blends, specially at low TPS content. This can be attributed to the strong interactions between hydroxyl groups of CNC with TPS and CNC increasing the crystallinity of TPS, shifting its melting point to higher temperatures. The authors also concluded that these LDPE/TPS/CNC composites can be a sustainable alternative to replace LDPE in food packaging application.

27.6.6 Thermal Stability

TGA is used to determine the thermal decomposition property of nanocellulose-reinforced composite films. The weight loss due to the formation of volatile products after degradation at high temperature is monitored as a function of temperature. When the sample is heated under an inert gas flow, a non-oxidative degradation occurs, while under the flow of air or oxygen an oxidative degradation occurs. In general, the use of clay as filler into the polymer matrix improves thermal stability by acting as a superior insulator and mass transport barrier to the volatile products generated during decomposition (Sinha and Okamoto 2003).

27.7 Properties of Polymer Nanocomposites

The composite materials exhibit outstanding mechanical properties with the mixture of nanocellulose and polymer matrix at low filler loading (Dufresne 2012). Mechanical solicitation is achieved within the polymer matrix under suitable condition by mechanically percolating a stiff network of nanoparticles. With the increase in the aspect ratio of CNCs, the stiffness of the percolating CNC increases (Bras et al. 2011), higher aspect ratio of the CNC is more important from the mechanical point of view, since it induces a decrease in the critical percolation threshold and stiffens the continuous network. In case of inhibition of formation of percolating nanoparticle network, only the high stiffness of crystalline cellulose in the nanoscale dimensions, the high aspect ratio and filler-matrix interactions are involved in the reinforcement phenomenon.

27.7.1 *Optical Properties*

The ultraviolet-visible (UV-Vis) spectrometer is used to calculate optical properties such as regular light transmittance (Tr) of nanocellulose-polymer composite films (Fig. 27.11). Fujisawa et al. (2012) reported that TEMPO oxidized cellulose nanofibrils (TOCNs) when reinforced in polystyrene (PS) provides superior reinforcement at low concentrations and improves optical transparency of PS composite film. Savadekar and Mhaske (2012) measured the Tr of the TPS/NCF films with a thickness of 0.60 mm using an UV-Vis spectroscope (UV-160A, Shimadzu, Japan) in a wavelength range of 200–800 nm. Soeta et al. (2015) prepared low-birefringent and highly tough poly(ethylene glycol) (PEG)-grafted CNF/cellulose triacetate (CTA) nanocomposite films. Due to the nanometric size effect of the TOCNs with a uniform width of ~3 nm, the PEG-TOCN/CTA nanocomposite films had a high transparency and a low birefringence with potential use in transparent optical films.

27.7.2 *Barrier Properties of Nanocomposite Films*

The food packaging materials used today are made from non-biodegradable polymeric materials which lead to serious environmental problems such as disposal, recycling etc. The preference for the use of these materials for food packaging polymers is their low cost, ease of processing and excellent barrier properties Lavoine et al. (2012). Biobased nanocellulose-based composites with improved barrier properties are desirable in our society to develop efficient, biodegradable and environmentally friendly packaging materials in the future. However, the low permeability of the polymers can be improved by reinforcing highly crystalline nature nanocellulose particles, since they can form a dense percolating network. In addition, there are strong particle-polymer–polymer molecular interactions, since these nano-sized particles have a greater ability to bond to the surrounding polymer

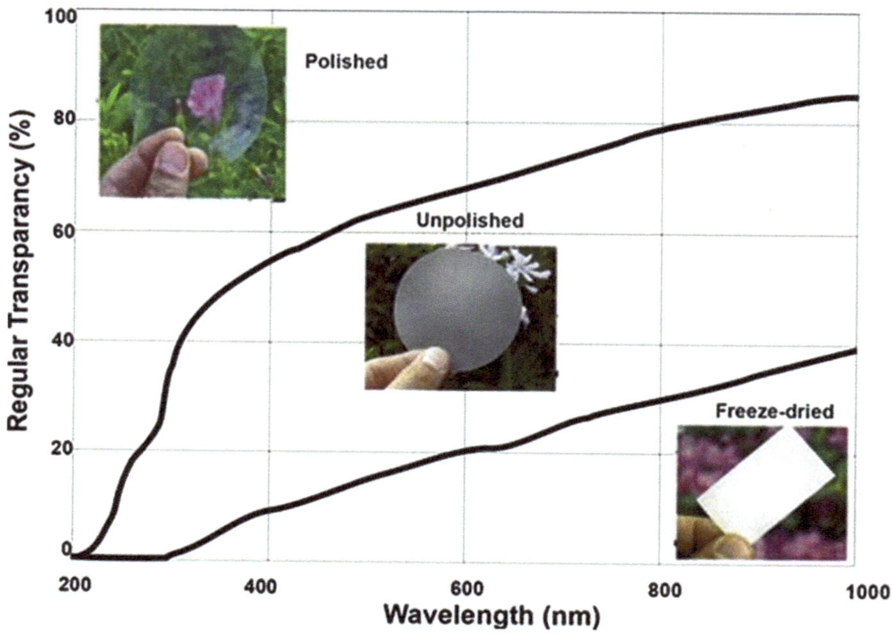

Fig. 27.11 Light transmittance of microfibrillated cellulose films (Lee et al. 2009a)

material, thereby reducing the chain segmental mobility and penetrant diffusivity (Dufresne 2013). Nanocomposite films prepared using nanocellulose as reinforcement prolong the shelf life of food and also improve the food quality, as they can serve as carriers of active substances such as antioxidants and antimicrobials (Andresen et al. 2007).

27.7.2.1 Oxygen Transmission Rate (OTR)

The nanocellulose can form a dense network with hydrogen bonds which can increase the gas barrier property of nanocellulose-reinforced composite films. Nanocellulose possesses tunable barrier properties which can be used to process high-performance oxygen barrier films for packaging applications (Nair et al. 2014).

Plackett et al. (2010) observed that the addition of 15 wt. % of CNFs substantially increased the oxygen barrier properties of amylopectin films. Saxena et al. (2010) produced a nanocomposite film with low oxygen permeability by casting an aqueous solution containing xylan, sorbitol and NCC. The oxygen permeability of the films prepared from xylan, sorbitol and 50% by weight of sulfonated CNC exhibited a significantly reduced oxygen permeability of 0.1799 $cm^3.\mu m/m^2.d.kPa$ compared with films prepared exclusively from xylan and sorbitol with an oxygen permeability of 189.16 $cm^3.\mu m/m^2.d.kPa$. Savadekar and Mhaske (2012) determined the OTR of the TPS/NCF films using an oxygen transmission rate (OTR) test machine (Labthink BTY-B1). After placing a film in a cell, the oxygen flow was introduced on one side of the films and the OTR was measured. The OTR in (cm^3/m^2 d Pa) was calculated from the mean OTR multiplied by the film thickness (μm) and divided by the oxygen gradient in the cell of the testing machine (1 kgf/cm^2).

Savadekar et al. (2012) determined the OTR of NFC/kappa-carrageenan (KCRG) composite film using an OTR test machine (Labthink BTY-B1). The authors reported that incorporation of 0.4 wt. % of NFC reinforcement led to reduced oxygen permeability of the KCRG/NFC composite compared to KCRG which has a high oxygen permeability. The incorporation of NFC led to the generation of tortuous path for the permeation of oxygen molecules into the KCRG biopolymer matrix, thus leading to a decrease in the OTR.

The PLA bionanocomposites containing 5 wt% of nanocrystals exhibited the highest oxygen barrier. The OTR for PLA nanocomposites with 5% w/w of unmodified CNCs was 17.4 ± 1.4 cm^3 mm m^{-2} day^{-1}, while that for CNCs modified with an ethoxylated nonylphenol acid phosphate ester in a 1/4 (wt/wt) the ratio was 15.8 ± 0.6 cm^3 mm m^{-2} day^{-1} (Fortunati et al. 2012b). The addition of 1 wt% of silver nanoparticles to the modified CNC-PLA composites further decreased the OTR to 12.6 ± 0.1 cm^3 mm m^{-2} day^{-1} (Fortunati et al. 2013). The OTR values of ternary systems consisting of PLA, PHB (poly hydroxybutyrate) and 5 wt% unmodified CNCs was 15.3 cm^3 mm m^{-2} day^{-1}, while for modified CNCs with an ethoxylated nonylphenol acid phosphate ester in a 1/1 (wt/wt) ratio was 13 cm^3 mm m^{-2} day^{-1} (Arrieta et al. 2014).

27.7.2.2 Water Vapor Permeability (WVP)/Moisture Vapor Transmission Rate (MVTR)

The water vapor transmission rate (WVTR) is an important property for packaging materials because it is essential to determine the shelf life of the products in the package and the low WVTR of the packaging material allows longer storage time and shelf life.

Svagan et al. (2009) used a dynamic vapor sorption apparatus from surface measurement systems to determine the water sorption kinetics of nanocellulose-reinforced plasticized starch nanocomposite film. The moisture diffusivity of nanocomposite films decreased rapidly with increasing nanofiber content and the diffusivity of the net cellulose network was comparatively very low. The reinforcement of natural biopolymers with CNCs reduced the WVTR of the resulting bionanocomposites. Saxena and Ragauskas (2009) prepared composite films using 10% sulfonated CNCs as reinforcement in xylan polymer. The authors reported a 74% reduction in the specific water transmission properties compared with the film without CNCs. Khan et al. (2014) reported that WVP decreased considerably in CNC-reinforced methyl cellulose composite films with increased in CNC content. The WVP of the control films (without CNCs) was 6.3 g.mm/m^2.day.kPa, while the nanocomposite films with 1 wt% CNC showed a WVP decreased of 4.7 g.mm/m^2. day.kPa. Static WVP of cellulose whisker-reinforced rubber nanocomposite films was determined by Bras et al. (2010) according to the standard. They estimated the moisture sorption test of cellulose whisker-rubber nanocomposite films at a relative humidity (RH) of about 75% at 25 °C using a saturated solution of sodium chloride.

Savadekar and Mhaske (2012) determined the WVTR values of TPS/NCF films gravimetrically according to the ASTM E96 method. Each test film was sealed at the top of the permeation cells containing distilled water. The permeation cells were placed in desiccators maintained at 0% RH. RH 0 was maintained using anhydrous calcium chloride ($CaCl_2$) in a cell. The composite films were cut into circles and sealed on the cell with melted paraffin. The water transferred through the film and absorbed by the desiccant was determined from the weight of the permeation cell. $CaCl_2$ was used as a desiccant. Each permeation cell was weighed in an interval of 24h. The WVTR was expressed in gm mm/cm^2 per day. They observed that WVTR decreased significantly with increasing content of NCF in NCF/TPS films. Seydibeyoglu and Oksman (2008) estimated the WVTR of the KCRG/NFC nanocomposite film and found that the high WVTR of KCRG decreases due to the dispersion of NFC in the KCRG matrix.

Song et al. (2014) conducted the WVTR tests on packaging paper coated with PLA/NCF composite material at 23 °C and 50% RH and at 37.8 °C and 90% RH according to the TAPPI standards. They reported that nanocomposite-coated paper samples showed a lower WVTR compared to packaging paper. Pereira et al. (2014) reinforced CNCs as fillers in PVOH matrix. The addition of 5 wt% CNCs decreased the WVP of pure PVOH films from 0.61 ± 0.04 g.mm/kPa.h.m^2 to 0.44 ± 0.01 g. mm/kPa.h.m^2. Reddy and Rhim (2014) reported a significant improvement in the water vapor barrier properties of agar composite films when reinforced with crystallized nanocellulose following the standard method of ASTM E96-95 with modification. The nanocellulose reinforcement (0.3 wt. %) lead to an improvement in the WVP of the composite film than pure agar film and, therefore, the resulting nanocomposite films could be used in completely biodegradable food packaging. The moisture vapor transmission rate (MVTR) tests were conducted gravimetrically on nanocellulose-reinforced semi-IPN composite films of PVA/PAAm using an ASTM method (ASTM Method 04.06:E96, 1983) by Mandal and Chakraborty (2015). These authors reported the highest barrier property for composite films with nanocellulose reinforcement of 5 wt. %.

Gray et al. (2018) reported that WVP coefficient and WVTR of the LDPE/TPS blends were greatly reduced by 1% CNC reinforcement, leading to better water vapor barrier properties. The results can be attributed to a good interfacial adhesion between the CNC and the starch which restricts the swelling and moisture diffusion of composites. In addition, the CNC loading at a certain level creates a dense and rigid network with hydrogen bonds which acts as a physical barrier for the transport of diffusing molecules.

27.7.3 Stiffness and Strength Properties

The stiffness and strength properties of nanocellulose polymer composites depend to a large extent on the shape of the nanocellulose. TS is the stress required to break the polymer material while being stretched. TS of the polymer increases with increasing saturation of the molecular weight at a given value. The nanocellulose upon the reinforcement improves the TS of the polymer composites as compared to traditional fillers. The TS of polymer materials was tested by using an universal testing machine by different researchers, namely Ching et al. (2015) reported that the YM of PVA-nanocellulose composite films increased with the reinforcement of small amount of nanocellulose due to the homogenous distribution of the nanofillers and the high interfacial surface area between the nanocellulose and the PVA matrix. YM is the ratio of stress to the strain in the linearly elastic region of polymer composite. It is a measure of the stiffness of the material and is measured using an Instron. Yakkan et al. (2015) characterized the YM of PP/CNF/Fusabond hybrid composite samples by tensile testing according to the ASTM D638 standard. They observed that there was 87% increase in the YM of PP/CNF composite (1.067 GPa) compared to that of pure polypropylene (0.570 GPa) indicating good dispersion of CNFs in the PP matrix with (0.1 wt %) loading of coupling agent Fusabond.

De Menezes et al. (2009) prepared a nanocomposite film by reinforcing cellulose nanowhiskers (CNWs) in highly hydrophobic semicrystalline commodity plastic LDPE. The nanocellulose whiskers were added at 5, 10 and 15 wt. % and the surface of the cellulose whiskers was chemically modified using organic acid chlorides such as stearoyl chloride. The homogeneity of the nanocomposite increased with the length of the grafted chains. The elongation at break increased when sufficiently long chains were grafted onto the surface of the nanoparticles. The TS of the film decreased by 23%, while the YM increased by 24% with respect to the LDPE when it was reinforced with 5 wt. % of nanocellulose whiskers.

Paralikar et al. (2008) observed an increase in TS and toughness in the membranes of the PVA matrix with NCC as a filler. The performance of nanocomposite further improved with the additional addition of poly(acrylic acid) (PAA). Ruiz et al. (2000) prepared NFC epoxy nanocomposites. These authors observed that the nanocomposites had the ability to be associated by means of hydrogen bonds and showed great aspect ratios. The benefits were associated with strong interactions between the epoxy network and the CNCs and the creation of a percolating network linked by hydrogen bonds between the NFC-epoxy. The best results were obtained with a 2% NFC reinforcement in epoxy resin with increased mechanical properties. Nakagaito and Yano (2008) also reported the effect of fiber content on the mechanical and thermal expansion properties of biocomposites based on NFC. A linear increase in YM with a fiber content of up to 40% was observed using a PF resin. The results also showed a correlation between the CTE relative to fiber content, which indicates the effective reinforcement reached by the NFC. The compression molding method was used to prepare composite with NFC sheets impregnated with phenol formaldehyde. The composites mechanical properties were YM (19 GPa) and bending strength up to 370 MPa.

27.7.4 Thermal Degradation

Very limited studies have been conducted to study the thermal degradation behavior of MFC-reinforced composite films. The thermal properties depend mainly on the MFC preparation process, as well as on the drying process Lavoine et al. (2012). It has been reported with respect to the first point that TEMPO-oxidized cellulose displays multiple degradation peaks Ruiz et al. (2000). Between 225–231 °C, maximum weight loss occurs (60–80%). Quievy et al. (2010) studied influence of the drying process on the thermal stability of MFC films obtained by homogenization. After freeze drying, the MFC gel formed a mat corresponding to the microfibril aggregates, in which some of the microfibrils remained distinct. The freezing step at 20 °C played an important role in the formation of agglomerates. In addition, this step even changes in the rheological properties of the MFC suspension.

27.7.5 Biodegradability

The degradability of the composites is important when a polymeric system is applied in daily life, since their degree of weight loss has a direct influence on the environment (Gutiérrez 2018c). The effect of different amounts of NCF on the weight loss rate of the film is done by the burial test in the soil. Different researchers have conducted studies of biodegradation of nanocellulose-reinforced polymer composites, since biodegradability is one of the key factors that influence the use of nanocellulose as a filler in the composites.

Bras et al. (2010) studied the effect of cellulose whiskers on the biodegradation of natural rubber (NR)/cellulose whisker nanocomposites. These authors observed that the cellulose whiskers significantly improved the biodegradation of the NR in the soil. Net NR lost about 19% of its weight after being buried in the soil for 4 weeks, while NR containing 7.5 and 12.5 wt% of cellulose whiskers lost about 62 and 71%, respectively, after the same period buried in the soil. Machado et al. (2012) developed biodegradable packaging using cassava starch plasticized with glycerol and nanocellulose extracted from coconut fiber. During storage, the packaging materials were oxidized, thus serving as a sacrificial material to minimize oxidation of the contents of the package. Abraham et al. (2012) studied the biodegradation of NR reinforced with nanocellulose by the vermi composting method with reference to the crosslinking of the NR matrix. They reported that the weight percentage of reinforced nanocellulose greatly influences the rate of biodegradation of the nanocomposite. The rate of biodegradation by vermi composting was comparatively higher in non-cross-linked composites than its cross-linked counterpart.

Maiti et al. (2013) studied the biodegradation behavior of PMMA/microcellulose crystalline (MCC) nanocomposites prepared by *in situ* polymerization (IPC) and *ex situ* dispersion method (EPC). Biodegradation studies of IPC and EPC films were performed in a simulated aerobic compost environment for 60 days. IPC showed a greater weight loss compared to the EPC and unreinforced PMMA. Arrieta et al. (2014) studied the degradability of PLA-PHB-CNC composites under composting conditions based on ISO 20200 standard. These authors observed that PLA-CNC and PLA-PHB-CNC lost more than 90% of the initial matter in 10 days; PLA and PLA-PHB-CNC in 14 days and PLA-PHB in 21 days. Luzi et al. (2015) studied the biodegradability of surfactant modified CNC (s-CNC)-reinforced PLA composites. They reported that the nanocomposite films disintegrated in less than 14 days and CNC modified with surfactant promoted the disintegration behavior. They summarized that films may have a potential application for short-term food packaging with low environmental impact. Gois et al. (2018) studied the biodegradation behavior of PLA nanocomposites with net (PLA/CNW) and surfactant modified CNWs (PLA/s-CNW) in garden soil. They reported that addition of surfactant accelerated the biodegradation of nanocomposites films and PLA/CNW/PEG1000 showed the fastest biodegradation rate.

27.8 Application of Polymer Nanocomposites in Food Packaging

The nanocellulose during the last few years has been used for the preparation of nanocomposites due to its reinforcing property (Table 27.3) Lavoine et al. (2014). The dimension of nanocellulose and its ability to build a strong entangled nanoporous network has fostered the emergence of high-value applications. Various materials have been cross-linked with nanocellulose by various approaches to obtain multifunctional properties such as improved coloration and dyeing, mechanical properties, barrier properties and WVP. Cs (a natural linear polysaccharide) is the second largest natural polysaccharide available which consists of 1,4-linked 2 amino-deoxy β–D glucan. It is a non-toxic, biodegradable, biofunctional and biocompatible material. Many researchers have reported that it has strong antifungal and antibacterial activities Darmadji and Izumimoto (1994). Films made from Cs have been used successfully as packaging material for the preservation of quality foods (Jo et al. 2001). It was found that 3–5% NCC-loaded Cs films gave the best TS values. The improvement in mechanical properties was mainly due to the formation of a percolating network and stronger filler-matrix interactions. NCC also causes improvement in the barrier properties by reducing the WVP and swelling properties. The study of the surface morphology of nanocomposite films indicated the homogeneous structure due to the adequate dispersion of NCC into the Cs matrix. Therefore, nanocomposite films produced by reinforcement of NCC would

Table 27.3 Application of nanocellulose based composites and their properties

Polymer	Type of NC	Products	Mechanical properties					Applications	Reference
			TS (MPa)	YM (MPa)	GBP (gm m^{-2} day kPa)	EB (%)	WVP (gm m^{-2} day)		
Chitosan/ Chitosan Sodium caseinate/ Nanocellulose/ Chitosan	NCC	NCC–chitosan nanocomposite films/ biodegradable films	55.3 — 245 (Increased)	4430 (Increased)	Decreased	47	3.31–2.23/0.23 ± 10.11/12.91	Transparent functional packaging/ edible food packaging nanoreinforcer materials in biodegradable packaging/ edible films for improving shelf life	Azeredo et al. (2010); Dehnad et al. (2014)
		Nanocomposite film	12.6 (Increased)	237.6 (Increased)	—	9.2	10.9×10^{13} Kg m/Pa s m^2	Edible food packaging	Pereda et al. (2010)
Methylcellulose (MC)	NC	NC–MC nanocomposite biodegradable films	—	—	50 kGy	—	6.34 (Decreased)	Biodegradable cellulosic based packaging	Khan et al. (2010)
Polycaprolactone MC matrix	NCC	NCC-reinforced nanocomposite antimicrobial diffusion films	18.7 + 3.7	124.1 + 14.6	—	—	—	Vegetable packages	Boumail et al. (2013)

Kappa-carrageenan	NFC	NFC–kappa-carrageenan based nanocomposite films	26	—	—	—	—	Food packaging	Savadekar et al. (2012)
Alginate biopolymer/ Alignate	Cellulose nanoparticles	Nanocomposite films	18.03–9	480	—	13	14	Food packaging	Abdollahi et al. (2013), (2010)
Kenaf/Corn starch	CNC	CNC–Starch nanocomposite films	3.5–8.2	326.1	—	38	8.68	Biobased packing/ Edible packaging	Piermaria et al. (2009)
Glucomannan, Pectin, Gelatin	NC	Nanocomposite films	68	14	—	7	3.3	Food packaging	Stuart et al. (2010)
Hydroxypropyl methylcellulose	MFC	MFC based nanocomposite films	—	—	—	—	6.39	Edible films, Transparent films	Zhu et al. (2011)
Caffeine based paper	MFC	MFC-coated papers	47 ± 4	7.3 ± 0.5	—	1.2 ± 0.1	—	Antimicrobial action for food-packaging application	Lavoine et al. (2014)
Vermiculite nanoplatelets + Nanocellulose fiber	Nanocellulose fibers	Biohybrid nanocomposite films	257	17.3	—	—	—	Flexible packaging	Aulin et al. (2012)
Thermoplastic starch/ chitosan	BC or NFC	BC/NFC based nanocomposite films	30	15	—	—	—	Functional packaging	Walker (2012)

(continued)

Table 27.3 (continued)

Polymer	Type of NC	Products	Mechanical properties					Applications	Reference
			TS (MPa)	YM (MPa)	GBP (gm m^{-2} day kPa)	EB (%)	WVP (gm m^{-2} day)		
Sugarcane bagasse nanofibers	Cellulose nanofibres	Cellulose nanocomposite film	140	12.8	—	12.8	5.2	Barrier and protective film in food packaging	Ghaderi et al. (2014)
Polylactic acid (PLA)	Cellulose nanofiber	Cellulose nanofiber–PLA nanocomposite films	3.6 GPa	71	—	—	—	Food packaging, water and milk bottles, degradable plastic bags/ green-based packaging materials/ short- term food packaging	Jonoobi et al. (2010)
LDPE	CNF	Glossy and transparent nanocomposite films	31	500	—	675	0.099	Film applications, like bread and frozen food bags, flexible lids, squeezable food bottles	Chambi and Grosso (2011)

HDPE	CNF	Transparent nanocomposite films	27.8	—	—	—	0.02	Bottles for milk, juice and water; margarine tubs, cereal box liners	Smith (1986)
PP	CNF	Nanocomposite film	43	1550	—	600	0.06	Thin film coatings/ Yogurt containers and margarine tubs, microwavable packaging	Azeredo et al. (2010)
PVC	CNF	Nanocomposite film	55	21	—	450	0.2	Films for improving shelf life	Azeredo et al. (2010)
Pea starch	CNF	Nanocomposite film	45	1584	—	94	66	Edible films for packaging	Azeredo et al. (2010)
Poly Propylene Carbocate PPC	CNW	Nanocomposite film	26.9	—	—	2960	—	Food packaging	Sainz et al. (2011)
Poly Vinyl Alcohol	CNF	Nanocomposite film	3.4	—	108.53	—	—	Packaging film	Tome et al. (2013)
Alginate	CNW	Nanocomposite film	22.4	290		8.25	1.64 g/m^2 s Pa	Food packaging	Abdollahi et al. (2013)

TS, tensile strength; *YM*, young's modulus; *GBP*, gas barrier properties; *EB*, elongation at break; *WVP*, water vapor permeability; *NCC*, nanocrystalline cellulose; *NC*, nanocellulose; *NFC*, nanofibrillated cellulose; *CNC*, cellulose nanocrystal; *CNW*, cellulose nanowhisker; *MFC*, microfibrillated cellulose; *HDPE*, high-density polyethylene; *PVC*, polyvinyl chloride; *CNF*, carbon nanofibre have a promising impact on food packaging due to their excellent barrier and mechanical properties. A higher concentration of cellulose significantly increased the YM and TS of the composite films, but decreased elongation. Antimicrobial diffusion films (ADFs) for food applications were developed by Boumil et al. (2013) which exhibited the highest TS on storage.

Savadekar et al. (2012) also successfully prepared NFC from short staple cotton fibers using a chemo-mechanical process. It was found in this study that it is important to provide an idea of the use and characteristics of NFC in KCRG for the preparation of biocomposite films.

Abdollahi et al. (2013) developed a biobased nanocomposite by incorporating nanocellulose obtained from acid hydrolysis using sulfuric acid into alginate biopolymer using solution casting method. With the increase in nanocellulose content from 0 to 5%, the TS of composite films increased, but with the further increase in the filler content TS decreased. Nevertheless, with the incorporation of nanocellulose, the transparency of the film decreased essentially at a high level (10%), which suggested the occurrence of a partial agglomeration of the fillers at 10% that coincided with mechanical and micro structural results.

A novel protocol to combine vermiculite nanoplatelets with NCF dispersion into functional biohybrid films was proposed by Aulin et al. (2012). The NCFs of 20 nm diameter and several micrometers long were mixed with high aspect ratio exfoliated vermiculite nanoplatelets through HPH. The resultant hybrid films obtained after solvent evaporation were stiff (YM of 17.3 GPa), strong (strength up to 257 MPa) and transparent.

The study of the oxygen barrier properties revealed an oxygen permeability of 0.07 cm^3 μm m^{-2} d^{-1} kPa^{-1} at 50% RH for the biohybrid films. In addition, the addition of nanoclay also significantly improved the water vapor barrier properties of biohybrid films. The excellent oxygen barrier and optical properties of films could be explored in a flexible packaging for oxygen sensitive devices such as thin film transistors or organic light emitting diode displays. It would also be a suitable packaging material for gas storage applications, barrier coatings/laminations etc. PLA is a versatile biodegradable polymer derived from natural sources. It is produced by ROP of lactide or by polycondensation of lactic acid. It is a suitable polymer matrix for the preparation of high-quality composite films by the incorporation of nanocellulose (Tsuji 2005).

Jonoobi et al. (2010) investigated the effect of CNFs in PLA matrix in terms of mechanical properties and dynamic mechanical properties for food packaging applications. The nanocomposite films with 5% CNF showed an increase in TS, YM and a better viscoelastic behavior.

Ghaderi et al. (2014) used cellulose nanofibers extracted from sugar cane bagasse to be employed in all cellulose nanocomposite (ACNC) films. The study showed that a very low value agricultural waste can be converted into a high-performance nanocomposite (TS 140 MPa). It was also found that the WVP of ACNC increased with more dissolution time. Therefore, due to the promising properties of ACNC, such as toughness, biobase, biodegradability and acceptable levels of WVP, they can be considered a multi-performance material with potential for application in the barrier and protective food packaging industry. The tensile properties of ACNC films are at least comparable or better than other biodegradable or nonbiodegradable films (Ghaderi et al. 2014).

The highest TS of 8.2 MPa was shown by a biocomposite having 6% CNC in thermoplastic cassava starch (TPCS). This suggests the stress transfer and interfacial interactions between the matrix phase and the filler, which is related to the high L/D and efficiency of the fiber treatment. The Kenaf fibers are also found to be compatible with agar and starch made from potato, and films was tested for their potential use in food packaging (Piermaria et al. 2009; Smith 1986).

PLA is used as a food packaging polymer for short-term products such as drinking cups, salad cups, containers, wrappers and laminated films (Sedlarik et al. 2012). But the gas and water vapor properties of the PLA make it insufficient for its use in some cases (Ljungberg and Wesslén 2002). Therefore, the preparation of PLA nanocomposites is a method to improve their properties. Song et al. (2014) prepared novel biodegradable composites by incorporating hydrophobic-modified NCF into a biodegradable PLA matrix. These authors suggested that addition of NCF improves the water vapor barrier property of the PLA matrix. The modified NCF/PLA composite at a concentration of 1 wt. % of NCF when coated on paper by a coating process reduced the WVTR of paper to 34 $g/m^2/d$ wrt from 1315$g/m^2/d$ of pure PLA. Paper is a promising green based packaging material. Similar effects on barrier of PLA/cellulose nanowhisker composites were also reported by Paralikar et al. (2008). Fortunati et al. (2013) prepared nanocomposite films by reinforcing NCC and silver nanoparticles in PLA matrix, which were found to be effective against Gram-positive (*S. aureus*) and Gram-negative (*E. coli*) bacteria and yeast.

Similarly, Reddy and Rhim (2014) prepared completely biodegradable food packaging material using CNCs and agar. The addition of CNCs lead to the improvement of the mechanical properties of the CNC/agar composite as it resulted in increased TS and YM and decreased in WVP.

The edible food packaging films were prepared by Pereda et al. (2010) by reinforcing the nanocellulose in the Cs matrix. Nanocellulose-based Cs composite films for antimicrobial meat packaging applications were prepared in a similar manner by Dehnad et al. (2014) and Khan et al. (2014). Ciesla et al. (2014) prepared PVA-starch-nanocellulose films for food packaging intended for radiation decontamination. They observed that the reinforcement of nanocellulose in starch-PVA films lead to an essential improvement in the mechanical properties of the films and a reduction in irradiation sensitivity as compared to starch-PVA films. They also noted that hydrophilicity of composite films can be reduced by irradiation.

27.9 Future Prospects and Challenges

Today, the current trend of packaging materials is that they are ecofriendly, biodegradable and not plastic. But, on the contrary, it is not so easy to avoid plastic due to its inherent quality, particularly in high population countries. However, the western world countries have initiated commercial production and the use of biobased material for packaging. But the problems are many and the availability of the raw material and supply chain management for a sustainable production is a great obstacle to satisfy the enormous needs of the market. The future packaging can be addressed through the proper management of materials that address environmental concerns, e.g. nano based composite materials, 100% biodegradable packaging, IP, innovative packaging using unutilized agromass residues to reduce carbon footprint etc. The petroleum-based packaging material can also be reinforced with biobased filler to facilitate biodegradation. Appropriate chemistry must be developed to be able to easily resolve the compatibility of the hydrophilic and hydrophobic properties of materials.

Another nearby field in food packaging is the use of stimuli-responsive polymer materials. These novel type of packaging materials can adapt to the surrounding environment and regulate the transport of molecules according to the reaction in response to external stimuli (Buonocore and Iannace 2013). To sustain life and maintain biological function, nature requires selectively tailored molecular assemblies and interfaces that provide a specific chemical function and structure, as well as a change in their environment (Stuart et al. 2010). Nanocellulose reinforcements offer potential advantages in specific properties related to their lower density and other advantages such as low CTE, transparency and barrier properties. NFCs have begun to receive additional attention as reinforcement material due to reductions in energy requirements to break down cellulose fibers to NFC.

Despite the continuous research in nanotechnology, there are many challenges that must be addressed in the future for economic production and, thus, ensure its efficient utilization as a filler/ reinforcement agent in different polymer composites. It is necessary to develop economical and profitable processes for the production of nanocellulose with optimal characteristics. The widespread application of these materials will require additional research to address problems related to their hydrophilic nature in many applications. Appropriate chemistry and techniques are needed to adequately disperse nanocellulose reinforcers or convert them into a useful form for incorporation into a variety of matrices and strongly bond them to it. A more efficient control of the structure at multiple scales is needed to tailor performance. The development of new analytical methods is necessary for the simulation of processing and also for the prediction of the mechanical properties of nanocellulose-based structures. The process modelling effort will be required to link the modelling of NFC distributions to optimize the properties. The appropriate applications requirement must be identified, investigated and demonstrated.

To date, the availability of significant amounts of nanocellulose has impeded faster research and development efforts on a large scale related to its use. If high-value applications can be found, it may be possible to integrate nanocellulose production into the materials flow of these bio-refineries, where it could potentially help improve economic announcements (Zhu et al. 2011) of commercial and government pilot plant-scale production facilities, which will probably improve the situation (Walker 2012).

27.10 Conclusion

Today's society is an expert in technology, but at the same time it is also concerned about its surrounding and its future impact. The modernization and sustainable growth that refer to any sector are complimentary to each other and every individual of mother earth wants to make this complimentary bonding stronger.

Due to more advances in composite materials, there is paradigm shift in material science. The packaging material is preferred more natural and composite based instead of monolithic material. Nanomaterials are also equally important to improve the functional character of the packaging material and reduce the material requirement in terms of material savings for sustainable use.

Nanocellulose has an enormous application in food packaging due to its competition with synthetic materials and infinite availability. Functional properties such as physical, chemical and barrier properties in case of nanocellulose composites strongly recommend that it would be the most suitable material for ecofriendly food packaging material in future. The problem remains with the hydrophilic nature of the nanocellulose material and the dispensing capacity with the matrix, which can be addressed by chemical and mechanical action or a combination of both. It is also understood that the nanocellulose composite often outweighs the cost, concern and with time the problems will be further refined and also solve the various problems facing the food packaging industry in present era.

Acknowledgements The authors are thankful to the Indian Council of Agricultural Research, New Delhi for supporting this work under World Bank funded National Agricultural Innovation Project (NAIP), through its sub-project entitled "Zonal Technology Management & Business Planning and Development Unit at Central Institute for Research on Cotton Technology, Mumbai".

Conflicts of Interest The authors declare no conflict of interest.

References

Abdollahi, M., Alboofetileh, M., Behrooz, R., Rezaei, M., & Miraki, R. (2013). Reducing water sensitivity of alginate bio-nanocomposite film using cellulose nanoparticles. *International Journal of Biological Macromolecules, 54*, 166–173. https://doi.org/10.1016/j.ijbiomac.2012.12.016.

Abraham, E., Elbi, P. A., Deepa, B., Jyotishkumar, P., Pothen, L. A., Narine, S. S., & Thomas, S. (2012). X-ray diffraction and biodegradation analysis of green composites of natural rubber/nanocellulose. *Polymer Degradation and Stability, 97*(11), 2378–2387. https://doi.org/10.1016/j.polymdegradstab.2012.07.028.

Alam, T. (2013). *Food packaging industry: Current, future prospects*. FnBnews.com. Available at: http://www. fnbnews.com/FB-Specials/Food-packaging-industry-Current-future-prospects.

Alemdar, A., & Sain, M. (2008). Biocomposites from wheat straw nanofibers: Morphology, thermal and mechanical properties. *Composites Science and Technology, 68*(2), 557–565. https://doi.org/10.1016/j.compscitech.2007.05.044.

Álvarez, K., Famá, L., & Gutiérrez, T. J. (2017). Chapter 12. Physicochemical, antimicrobial and mechanical properties of thermoplastic materials based on biopolymers with application in the food industry. In M. Masuelli & D. Renard (Eds.), *Advances in physicochemical properties of biopolymers: Part 1* (pp. 358–400). Bentham Science Publishers. EE.UU. ISBN: 978-1-68108-454-1. eISBN: 978-1-68108-453-4. https://doi.org/10.2174/9781681084534117010015.

Álvarez, K., Alvarez, V. A., & Gutiérrez, T. J. (2018). Chapter 3. Biopolymer composite materials with antimicrobial effects applied to the food industry. In V. K. Thakur & M. K. Thakur (Eds.), *Functional biopolymers* (pp. 57–96). Editorial Springer International Publishing. EE.UU. ISBN: 978-3-319-66416-3. eISBN: 978-3-319-66417-0. https://doi.org/10.1007/978-3-319-66417-0_3.

Andresen, M., Stenstad, P., Møretrø, T., Langsrud, S., Syverud, K., Johansson, L. S., & Stenius, P. (2007). Nonleaching antimicrobial films prepared from surface-modified microfibrillated cellulose. *Biomacromolecules, 8*(7), 2149–2155. https://doi.org/10.1021/bm070304e.

Anonymous. (n.d.). Emerging packaging markets potentials and trends in the packaging industries in India, Russia and China. Volume Available at: https://www.interpackalliance.com/cgi-bin/md_interpack/lib/all/lob/return_download.cgi/Factbook_GB.pdf?ticket=g_u_e_s_t&bid=4670&no_mime_type=0.

Arora, A., & Padua, G. W. (2010). Review: Nanocomposites in food packaging. *Journal of Food Science, 75*(1), 43–49. https://doi.org/10.1111/j.1750-3841.2009.01456.x.

Arrieta, M. P., Fortunati, E., Dominici, F., Rayon, E., Lopez, J., & Kenny, J. M. (2014). PLA-PHB/cellulose based films: mechanical, barrier and disintegration properties. *Polymer Degradation and Stability, 10*, 139–149. https://doi.org/10.1016/j.polymdegradstab.2014.05.010.

Astrom, B. T. (1997). Introduction to composites, composites manufacturing and other post-design issues. *Manufacturer Polymer Composites, 8*, 1–8.

Aulin, C., Gallstedt, M., & Lindstrom, T. (2010). Oxygen and oil barrier properties of microfibrillated cellulose films and coatings. *Cellulose, 17*(3), 559–574. https://doi.org/10.1007/s10570-009-9393-y.

Aulin, C., Salazar-Alvarez, G., & Lindstrom, T. (2012). High strength, flexible and transparent nanofibrillated cellulose–nanoclay biohybrid films with tunable oxygen and water vapor permeability. *Nanoscale, 4*, 6622–6628.

Azeredo, H. M., Mattoso, L. H., Avena-Bustillos, R. J. A., Filho, G. C., Munford, M. L., Wood, D., & Mchugh, T. H. (2010). Nanocellulose reinforced chitosan composite films as affected by nanofiller loading and plasticizer content. *Journal of Food Science, 75*(1), N1–N7. https://doi.org/10.1111/j.1750-3841.2009.01386.x.

Barari, B., Ellingham, T. K., Ghamhia, I. I., Pillai, K. M., El-Hajjar, R., Turng, L.-S., & Sabo, R. (2016). Mechanical characterization of scalable cellulose nano-fiber based composites made using liquid composite molding process. *Composites Part B, 84*, 277–284. https://doi.org/10.1016/j.compositesb.2015.08.040.

Berger, K. R. (2002). A brief history of packaging. ABE321, Agricultural and Biological Engineering Department, Florida Cooperative Extension Service, Institute of Food and Agricultural Sciences, University of Florida, pp. 1–5.

Bhattacharya, D., Germinario, L. T., & Winter, W. T. (2008). Isolation, preparation and characterization of cellulose microfibers obtained from bagasse. *Carbohydrate Polymers, 73*(3), 371–377. https://doi.org/10.1016/j.carbpol.2007.12.005.

Bledzki, A. K., & Gassan, J. (1999). Composites reinforced with cellulose based fibres. *Progress in Polymer Science, 24*(2), 221–274. https://doi.org/10.1016/S0079-6700(98)00018-5.

Bochek, A. M., Shevchuk, I. L., & Lavrentev, V. N. (2003). Fabrication of microcrystallinecellulose and powdered cellulose from short flax fiber and flax straw. *Russian Journal of Applied Chemistry, 76*(10), 1679–1682. https://doi.org/10.1023/b:rjac.0000015737.07117.12.

Boumail, A., Salmieri, S., Klimas, E., Tawema, P. O., Bouchard, J., & Lacroix, M. (2013). Characterization of trilayer antimicrobial diffusion films (ADFs) based on methylcellulose-polycaprolactone composites. *Journal of Agricultural and Food Chemistry, 61*(4), 811–821. https://doi.org/10.1021/jf304439s.

Bracone, M., Merino, D., González, J., Alvarez, V. A., & Gutiérrez, T. J. (2016). Chapter 6. Nanopackaging from natural fillers and biopolymers for the development of active and intelligent films. In S. Ikram & S. Ahmed (Eds.), *Natural polymers: derivatives, blends and composites* (pp. 119–155). New York. EE.UU. ISBN: 978-1-63485-831-1: Editorial Nova Science Publishers, Inc.

Brandt, A., Hallett, J. P., Leak, D. J., Murphy, R. J., & Welton, T. (2010). The effect of the ionic liquid anion in the pretreatment of pine wood chips. *Green Chemistry, 12*(4), 672–679. https://doi.org/10.1039/b918787a.

Bras, J., Hassan, M. L., Bruzesse, C., Hassan, E. A., El-Wakil, N. A., & Dufresne, A. (2010). Mechanical, barrier, and biodegradability properties of bagasse cellulose whiskers reinforced natural rubber nanocomposites. *Industrial Crops and Products, 32*(3), 627–633. https://doi.org/10.1016/j.indcrop.2010.07.018.

Bras, J., Viet, D., Bruzzese, C., & Dufresne, A. (2011). Correlation between stiffness of sheets prepared from cellulose whiskers and nanoparticles dimensions. *Carbohydrate Polymers, 84*(1), 211–215. https://doi.org/10.1016/j.carbpol.2010.11.022.

Braun, B., Dorgan, J. R., & Hollingsworth, L. O. (2012). Supra-molecular ecobionanocomposites based on polylactide and cellulosic nanowhiskers: Synthesis and properties. *Biomacromolecules, 13*(7), 2013–2019. https://doi.org/10.1021/bm300149w.

Bruce, D. M., Hobson, R. N., Farrent, J. W., & Hepworth, D. G. (2005). High-performance composites from low-cost plant primary cell walls. *Composites Part A: Applied Science and Manufacturing, 36*(11), 1486–1493. https://doi.org/10.1016/j.compositesa.2005.03.008.

Buonocore, G., Iannace, S. (2013, February/March). Molecular and supramolecu- lar design for active and edible packaging systems, food safety magazine, packaging.

Butschli, J. (2016). *Modest growth for global packaging demand through 2020*. Available at https://www.healthcarepackaging.com/article/trends-and-issues/global/modest-growth-global-packaging-demand-through-2020.

Cai, Z. J., Guang, Y., & Kim, J. (2011). Biocompatible nanocom- posites prepared by impregnating bacterial cellulose nanofibrils into poly(3-hydroxybutyrate). *Current Applied Physics, 11*(2), 247–249. https://doi.org/10.1016/j.cap.2010.07.016.

Cha, D. S., & Chinnan, M. J. (2004). Biopolymer based antimicrobial packaging: a review. *Critical Reviews in Food Science and Nutrition, 44*(4), 223–237. https://doi.org/10.1080/10408690490464276.

Chaffee, C., & Yoros, B.R. (2007). *Life cycle assessment for three types of grocery bags–Recyclable plastics, compo- stable, biodegradable plastic and recyclable paper*. Final Report, Boustead Consulting and Associates Limited: Ardmore.

Chambi, H. N. M., & Grosso, C. R. F. (2011). Mechanical and water vapor permeability properties of biodegradables films based on methycellulose, glucomannan, pectin and gelatin. *Ciencia a Tecnologia de Alimentos, 31*(3), 739–746. https://doi.org/10.1590/S0101-20612011000300029.

Chazeau, L., Cavaille, J. Y., Canova, G., Dendievel, R., & Boutherin, B. (1999). Viscoelastic properties of plasticized PVC reinforced with cellulose whiskers. *Journal of Applied Polymer Science, 71*(11), 1797–1808. https://doi.org/10.1002/(sici)1097-4628(19990314)71:11<1797::aid-app9>3.0.co;2-e.

Chen, G., Dufresne, A., Huang, J., & Chang, P. R. (2009). A Novel thermo-formable bionanocomposite based on cellulose-nanocrystal-*graft*-poly (ε-caprolactone). *Macromolecular Materials and Engineering, 294*(1), 59–67. https://doi.org/10.1002/mame.200800261.

Chen, W. S., Yu, H., Liu, Y. X., Chen, P., Zhang, M. X., & Hai, Y. F. (2011). Individualization of cellulose nanofibers from wood using high-intensity ultrasonication combined with chemical pretreatments. *Carbohydrate Polymers, 83*(4), 1804–1811. https://doi.org/10.1016/j.carbpol.2010.10.040.

Cherian, B. M., Leao, A. L., Souza, S. F., Thomas, S., Pothan, L. A., & Kottaisamy, M. (2010). Isolation of nanocellulose from pineapple leaf fibres by steam explosion. *Carbohydrate Polymers, 81*(3), 720–725. https://doi.org/10.1016/j.carbpol.2010.03.046.

Ching, Y. C., Rahman, A., Ching, K. Y., Sukiman, N. L., & Chuah, C. H. (2015). Preparation and characterization of polyvinyl alcohol composite reinforced with nanocellulose and nanosilica. *BioResources, 10*(2), 3364–3377.

Ciesla, K., Abramowska, A., Buczowski, M., Gluszewski Łuszewski, W., Nowicki, A., Boguski, J., Sartowska, B., & Strzelczak, G. (2014). Based on starch-PVA system and cellulose reinforced packaging materials prepared using radiation modification. Report of the 2nd RCM on Application of radiation technology in the development of advanced packaging materials for food products, 8–12 September, Bejaia, pp. 14–26.

Cristaldi, G., Latteri, A., Recca, G., & Cicala, G. (2010). Composites based on natural fibre fabrics. In P. D. Dubrovski (Ed.), *Woven fabric engineering* (pp. 317–342). https://doi.org/10.5772/10465.

Darmadji, P., & Izumimoto, M. (1994). Effect of chitosan in meat preservation. *Meat Science, 38*(2), 243–254. https://doi.org/10.1016/0309-1740(94)90114-7.

de Azeredo, H. M. C. (2009). Nanocomposites for food packaging applications. *Foodservice Research International, 42*(9), 1240–1253. https://doi.org/10.1016/j.foodres.2009.03.019.

De Menezes, A. J., Siqueira, G., Curvelo, A. A., & Dufresne, A. (2009). Extrusion and characterization of functionalized cellulose whiskers reinforced polyethylene nanocomposites. *Polymer, 50*(19), 4552–4563. https://doi.org/10.1016/j.polymer.2009.07.038.

Dehnad, D., Emam-Djomeh, Z., Mirzaei, H., Jafari, S. M., & Dadashi, S. (2014). Optimization of physical and mechanical properties for chitosan–nanocellulose biocomposites. *Carbohydrate Polymers, 105*, 222–228.

Denault, J., & Labrecque, B. (2004). Technology group on polymer nanocomposites–PNC-Tech. Industrial Materials Institute. National Research Council Canada, 75 de Mortagne Blvd. Boucherville, Québec, J4B 6Y4.

Denoyelle, T. (2011). Mechanical properties of materials made of nano-cellulose. In *Degree project in solid mechanics*. Stockholm: Royal Institute of Technology (KTH).

Deo, C. (2010). Preparation and characterization of polymer matrix composite using natural fiber lantana-camara. Doctoral dissertation. National Institute of Technology, Rourkela.

Doree, C. (1947). *The methods of cellulose chemistry including methods for the investigation of substances associated with cellulose in plant tissues* (p. 543). London: Chapman and Hall.

Dufresne, A. (2012). *Nanocellulose: From nature to high performance tailored materials*. Berlin: Walter de Gruyter GmbH & Co.

Dufresne, A. (2013). Nanocellulose: a new ageless bionanomaterial. *Materials Today, 16*(6), 220–227. https://doi.org/10.1016/j.mattod.2013.06.004.

Dufresne, A., & Vignon, M. R. (1998). Improvement of starch film performances using cellulose microfibrils. *Macromolecules, 31*(8), 2693–2696. https://doi.org/10.1021/ma971532b.

Dufresne, A., Dupeyre, D., & Vignon, M. R. (**2000**). Cellulose microfibrils from potato tuber cells: processing and characterization of starch-cellulose microfibril composites. *Journal of Applied Polymer Science, 76*(14), 2080–2092. https://doi.org/10.1002/(sici)1097-4628(20000628)76:14<2080::aid-app12>3.0.co;2-u.

Dufresne, A., Dupeyre, D., & Paillet, M. (2003). Lignocellulosic flour-reinforced poly (hydroxybutyrate-co-valerate) composites. *Journal of Applied Polymer Science, 87*(8), 1302–1315. https://doi.org/10.1002/app.11546.

El-Wakil, N. A., Hassan, E., Raga, A., Abou-Zeid, E., & Dufresne, A. (2015). Development of wheat gluten/nanocellulose/titanium dioxide nanocomposites for active food packaging. *Carbohydrate Polymers, 124*, 337–346. https://doi.org/10.1016/j.carbpol.2015.01.076.

Favier, V., Chanzy, H., & Cavaille, J. Y. (1995). Polymer nanocomposites reinforced by cellulose whiskers. *Macromolecules, 28*(18), 6365–6636. https://doi.org/10.1021/ma00122a053.

Fortunati, E., Armentano, I., Zhou, Q., Iannoni, A., Saino, E., Visai, L., Berglund, L. A., & Kenny, J. M. (2012a). Multifunctional bionanocomposite films of poly(lactic acid), cellulose nanocrystals and silver nanoparticles. *Carbohydrate Polymers, 87*(2), 1596–1605. https://doi.org/10.1016/j.carbpol.2011.09.066.

Fortunati, E., Peltzer, M., Armentano, I., Torre, L., Jimenez, A., & Kenny, J. M. (2012b). Effects of modified cellulose nanocrystals on the barrier and migration of PLA nano-composites. *Carbohydrate Polymers, 90*(2), 948–956. https://doi.org/10.1016/j.carbpol.2012.06.025.

Fortunati, E., Peltzer, M., Armentano, I., Jimenez, A., & Kenny, J. M. (2013). Combined effects of cellulose nanocrystals and silver nanoparticles on the barrier and migration properties of PLA nano-biocomposites. *Journal of Food Engineering, 118*(1), 117–124. https://doi.org/10.1016/j.jfoodeng.2013.03.025.

Fujisawa, S., Ikeuchi, T., Takeuchi, M., Saito, T., & Isogai, A. (2012). Superior reinforcement effect of TEMPO-oxidized cellulose nanofibrils in polystyrene matrix: optical, thermal, and mechanical studies. *Biomacromolecules, 13*(7), 2188–2194. https://doi.org/10.1021/bm300609c.

Gama, M., Gatenholm, P., & Klemm, D. (2012). *Bacterial nano cellulose: A sophisticated multi-functional material* (p. 304). Boca Raton: CRC Press. (ISBN 9781439869918).
Garcia de Rodriguez, N. L., Thielemans, W., & Dufresne, A. (2006). Sisal cellulose whiskers reinforced polyvinyl acetate nanocomposites. *Cellulose, 13*(3), 261–270. https://doi.org/10.1007/s10570-005-9039-7.
García, M. A., Pinotti, A., Martino, M. N., & Zaritzky, N. E. (2004). Characterization of composite hydrocolloid films. *Carbohydrate Polymers, 56*(3), 339–345. https://doi.org/10.1016/j.carbpol.2004.03.003.
Ghaderi, M., Mousavi, M., Yoursefi, H., & Labbafi, M. (2014). All-cellulose nanocomposite film made from bagasse cellulose nanofibers for food packaging application. *Carbohydrate Polymers, 104*, 59–65. https://doi.org/10.1016/j.carbpol.2014.01.013.
Ghosh, S. B., & Sain, M. (2014). The use of biobased nanofibres in composites. In O. Faruk & M. Sain (Eds.), *Biofiber reinforcements in composite materials* (pp. 594–595). Oxford: Elsevier Science & Technology. https://doi.org/10.1533/9781782421276.5.571.
Goffin, A. L., Raquez, J. M., Duquesne, E., Siqueira, G., Habibi, Y., Dufresne, A., & Dubois, P. (2011). Poly(ε-caprolactone) based nano-composites reinforced by surface-grafted cellulose nano-whiskers via extrusion processing: morphology, rheology and thermo-mechanical properties. *Polymer, 52*(7), 1532–1538. https://doi.org/10.1016/j.polymer.2011.02.004.
Gois, G. S., Andrade, M. F., Garcia, S. M. S., Vinhas, G. M., Santos, A. S. F., Medeiros, E. S., Oliveira, J. E., & de Almeida, Y. M. S. (2018). Soil biodegradation of PLA/CNW nanocomposites modified with ethylene oxide derivatives. *Materials Research*, 1–6. https://doi.org/10.1590/1980-5373-MR-2016-0960.
Gray, N., Hamzeh, Y., Kaboorani, A., & Abdulkhani, A. (2018). Influence of cellulose nanocrystal on strength and properties of low density polyethylene and thermoplastic starch composites. *Industrial Crops and Products, 115*, 298–305. https://doi.org/10.1016/j.indcrop.2018.02.017.
Guangjun, C., Wei, M., Chen, J., Huang, J., Dufresne, A., & Chang, P. R. (2008). Simultaneous reinforcing and toughening: new nanocomposites of waterborne polyurethane filled with low loading level of starch nanocrystals. *Polymer, 49*(7), 1860–1870. https://doi.org/10.1016/j.polymer.2008.02.020.
Gutiérrez, T. J. (2017a). Chapter 8. Chitosan applications for the food industry. In: *Chitosan: Derivatives, composites and applications*. S. Ahmed, and S. Ikram (Eds). WILEY-Scrivener Publisher. EE.UU. ISBN: 978-1-119-36350-7. pp. 183–232. https://doi.org/10.1002/9781119364849.ch8
Gutiérrez, T. J. (2017b). Surface and nutraceutical properties of edible films made from starchy sources with and without added blackberry pulp. *Carbohydrate Polymers, 165*, 169–179. https://doi.org/10.1016/j.carbpol.2017.02.016.
Gutiérrez, T. J. (2018a). Are modified pumpkin flour/plum flour nanocomposite films biodegradable and compostable? *Food Hydrocolloids, 83*, 397–410. https://doi.org/10.1016/j.foodhyd.2018.05.035.
Gutiérrez, T. J. (2018b). Active and intelligent films made from starchy sources/blackberry pulp. *Journal Polymers and the Environment, 26*(6), 2374–2391. https://doi.org/10.1007/s10924-017-1134-y.
Gutiérrez, T. J. (2018c). Chapter 9. Biodegradability and compostability of food nanopackaging materials. In G. Cirillo, M. A. Kozlowski, & U. G. Spizzirri (Eds.), *Composite materials for food packaging* (pp. 269–296). WILEY-Scrivener Publisher. EE.UU. ISBN: 978-1-119-16020-5. https://doi.org/10.1002/9781119160243.ch9.
Gutiérrez, T. J., & Alvarez, V. A. (2017a). Properties of native and oxidized corn starch/polystyrene blends under conditions of reactive extrusion using zinc octanoate as a catalyst. *Reactive and Functional Polymers, 112*, 33–44. https://doi.org/10.1016/j.reactfunctpolym.2017.01.002.
Gutiérrez, T. J., & Alvarez, V. A. (2017b). Eco-friendly films prepared from plantain flour/PCL blends under reactive extrusion conditions using zirconium octanoate as a catalyst. *Carbohydrate Polymers, 178*, 260–269. https://doi.org/10.1016/j.carbpol.2017.09.026.
Gutiérrez, T. J., & Alvarez, V. A. (2017c). Data on physicochemical properties of active films derived from plantain flour/PCL blends developed under reactive extrusion conditions. *Data in Brief, 15*, 445–448. https://doi.org/10.1016/j.dib.2017.09.071.

Gutiérrez, T. J., & Alvarez, V. A. (2017d). Cellulosic materials as natural fillers in starch-containing matrix-based films: A review. *Polymer Bulletin, 74*(6), 2401–2430. https://doi.org/10.1007/s00289-016-1814-0.

Gutiérrez, T. J., & Alvarez, V. A. (2018). Bionanocomposite films developed from corn starch and natural and modified nano-clays with or without added blueberry extract. *Food Hydrocolloids, 77*, 407–420. https://doi.org/10.1016/j.foodhyd.2017.10.017.

Gutiérrez, T. J., Morales, N. J., Pérez, E., Tapia, M. S., & Famá, L. (2015a). Physico-chemical study of edible films based on native and phosphating cush-cush yam and cassava starches. *Food Packaging and Shelf Life, 3*, 1–8. https://doi.org/10.1016/j.fpsl.2014.09.002.

Gutiérrez, T. J., Tapia, M. S., Pérez, E., & Famá, L. (2015b). Structural and mechanical properties of native and modified cush-cush yam and cassava starch edible films. *Food Hydrocolloids, 45*, 211–217. https://doi.org/10.1016/j.foodhyd.2014.11.017.

Gutiérrez, T. J., Tapia, M. S., Pérez, E., & Famá, L. (2015c). Edible films based on native and phosphated 80:20 waxy: Normal corn starch. *Starch-Stärke, 67*(1-2), 90–97. https://doi.org/10.1002/star.201400164.

Gutiérrez, T. J., Morales, N. J., Tapia, M. S., Pérez, E., & Famá, L. (2015d). Corn starch 80:20 "waxy":regular, "native" and phosphated, as bio-matrixes for edible films. *Procedia Materials Science, 8*, 304–310. https://doi.org/10.1016/j.mspro.2015.04.077.

Gutiérrez, T. J., Guzmán, R., Medina Jaramillo, C., & Famá, L. (2016a). Effect of beet flour on films made from biological macromolecules: native and modified plantain flour. *International Journal of Biological Macromolecules, 82*, 395–403. https://doi.org/10.1016/j.ijbiomac.2015.10.020.

Gutiérrez, T. J., Suniaga, J., Monsalve, A., & García, N. L. (2016b). Influence of beet flour on the relationship surface-properties of edible and intelligent films made from native and modified plantain flour. *Food Hydrocolloids, 54*, 234–244. https://doi.org/10.1016/j.foodhyd.2015.10.012.

Gutiérrez, T. J., González Seligra, P., Medina Jaramillo, C., Famá, L., & Goyanes, S. (2017a). Chapter 14. Effect of filler properties on the antioxidant response of thermoplastic starch composites. In V. K. Thakur, M. K. Thakur, & M. R. Kessler (Eds.), *Handbook of composites from renewable materials* (pp. 337–370). WILEY-Scrivener Publisher. EE.UU. ISBN: 978-1-119-22362-7. https://doi.org/10.1002/9781119441632.ch14.

Gutiérrez, T. J., Ponce, A. G., & Alvarez, V. A. (2017b). Nano-clays from natural and modified montmorillonite with and without added blueberry extract for active and intelligent food nanopackaging materials. *Materials Chemistry and Physics, 194*, 283–292. https://doi.org/10.1016/j.matchemphys.2017.03.052.

Gutiérrez, T. J., Ollier, R., & Alvarez, V. A. (2018a). Chapter 5. Surface properties of thermoplastic starch materials reinforced with natural fillers. In V. K. Thakur & M. K. Thakur (Eds.), *Functional biopolymers* (pp. 131–158). Editorial Springer International Publishing. EE.UU. ISBN: 978-3-319-66416-3. eISBN: 978-3-319-66417-0. https://doi.org/10.1007/978-3-319-66417-0_5.

Gutiérrez, T. J., Herniou--Julien, C., Álvarez, K., & Alvarez, V. A. (2018b). Structural properties and *in vitro* digestibility of edible and pH-sensitive films made from guinea arrowroot starch and wastes from wine manufacture. *Carbohydrate Polymers, 184*, 135–143. https://doi.org/10.1016/j.carbpol.2017.12.039.

Gutiérrez, T. J., Toro-Márquez, L. A., Merino, D., & Mendieta, J. R. (2019). Hydrogen-bonding interactions and compostability of bionanocomposite films prepared from corn starch and nano-fillers with and without added Jamaica flower extract. *Food Hydrocolloids, 89*, 283–293. https://doi.org/10.1016/j.foodhyd.2018.10.058.

Habibi, Y., & Vignon, M. R. (2008). Optimization of cellouronic acid synthesis by TEMPO-mediated oxidation of cellulose III from sugar beet pulp. *Cellulose, 15*(1), 177–185. https://doi.org/10.1007/s10570-007-9179-z.

Henriksson, M., Henriksson, G., Berglund, L. A., & Lindstrorn, T. (2007). An environmentally friendly method for enzyme-assisted preparation of microfibrillated cellulose (MFC) nanofibers. *European Polymer Journal, 43*(8), 3434–3441. https://doi.org/10.1016/j.eurpolymj.2007.05.038.

Herniou--Julien, C., Mendieta, J. R., & Gutiérrez, T. J. (2019). Characterization of biodegradable/non-compostable films made from cellulose acetate/corn starch blends processed under reactive extrusion conditions. *Food Hydrocolloids, 89*, 67–79. https://doi.org/10.1016/j.foodhyd.2018.10.024.
Hook, P., & Heimlich, J.E. (2011). A history of packaging CDFS-133, Ohio State University Fact Sheet, Community Development, Columbus. Available at: ohioline.osu.edu/cd-fact/0133.
Hubbe, M. A., Rojas, O. J., Lucia, L. A., & Sain, M. (2008). Cellulosic nanocomposites: A review. *BioResources, 3*(3), 929–980.
Hussain, F., Hojjati, M., Okamoto, M., & Gorga, R. E. (2006). Polymer-matrix nanocomposites, processing, manufacturing and application: An overview. *Journal of Composite Materials, 40*(17), 1511–1575. https://doi.org/10.1177/0021998306067321.
Istvan, S., & Plackett, D. (2010). Microfibrillated cellulose and new nanocomposite materials: A review. *Cellulose, 17*(3), 459–494. https://doi.org/10.1007/s10570-010-9405-y.
Iwamoto, S., Kai, W., Isogai, A., & Iwata, T. (2009). Elastic modulus of single cellulose microfibrils from tunicate measured by atomic force microscopy. *Biomacromolecules, 10*(9), 2571–2576. https://doi.org/10.1021/bm900520n.
Jahan, M. S., Saeed, A., Zhibin, H., & Ni, Y. (2011). Jute as raw material for the preparation of microcrystalline cellulose. *Cellulose, 18*(2), 451–459. https://doi.org/10.1007/s10570-010-9481-z.
Jerome, C., & Lecomte, P. (2008). Recent advances in the synthesis of aliphatic polyesters by ring-opening polymerization. *Advanced Drug Delivery Reviews, 60*(9), 1056–1076. https://doi.org/10.1016/j.addr.2008.02.008.
Jo, C., Lee, J. W., Lee, K. H., & Byun, M. W. (2001). Quality properties of pork sausage prepared with water- soluble chitosan oligomer. *Meat Science, 59*(4), 369–375. https://doi.org/10.1016/S0309-1740(01)00089-4.
John, M. J., & Thomas, S. (2008). Biofibres and biocomposites. *Carbohydrate Polymers, 71*(3), 343–364. https://doi.org/10.1016/j.carbpol.2007.05.040.
Jonoobi, M., Harun, J., Mathew, A. P., & Oksman, K. (2010). Mechanical properties of cellulose nanofiber (CNF) reinforced polylactic acid (PLA) prepared by twin screw extrusion. *Composites Science and Technology, 70*(12), 1742–1747. https://doi.org/10.1016/j.compscitech.2010.07.005.
Karande, V.S. (2013). *Polymer composites based on cellulosics nanomaterials*. Doctoral Thesis. 59. Available at: http://hdl.handle.net/10603/13586.
Kaushik, A., & Grewal, R. G. (2011). Thermal behaviour of nanocomposites based on glycerol plasticized thermoplastic starch and cellulose nanocrystallites. *AIP Conference Proceedings., 1393*(1), 353–354. https://doi.org/10.1063/1.3653755.
Kaushik, A., & Singh, M. (2011). Isolation and characterization of cellulose nanofibrils from wheat straw using steam explosion coupled with high shear homogenization. *Carbohydrate Polymers, 346*(1), 76–85. https://doi.org/10.1016/j.carres.2010.10.020.
Keeratiurai, M., & Corredig, M. (2009). Effect of dynamic high pressure homogenization on the aggregation state of soy protein. *Journal of Agricultural and Food Chemistry, 57*(9), 3556–3562. https://doi.org/10.1021/jf803562q.
Khan, R. A., Salmieri, S., Dussault, D., Calderon, J. U., Kamal, M. R., Safrany, A., & Lacroix, M. (2010). Production and properties of nanocellulose reinforced methylcellu- lose-based biodegradable films. *Journal of Agricultural and Food Chemistry, 58*(13), 7878–7885. https://doi.org/10.1021/jf1006853.
Khan, A., Salmieri, S., Fraschini, C., Bouchard, J., Riedl, B., & LAcroix, M. (2014). Genipin cross-linked nanocomposite films for the immobilization of antimicrobial agent. *ACS Applied Materials & Interfaces, 6*(17), 15232–15242. https://doi.org/10.1021/am503564m.
Khoskhava, V. (2014). *Polypropylene (PP) nanocomposites incorporating nanocrystalline cellulose (NCC)*.Ph. D. Thesis, McGill University, Montreal.
Klemm, D. (2005). Cellulose for fascinating biopolymer and sustainable raw material. *Polymer Science, 44*(22), 3358–3393. https://doi.org/10.1002/anie.200460587.
Klemm, D., Kramer, F., Moritz, S., Lindstrom, T., Ankerfors, M., Gray, D., & Dorris, A. (2011). Nanocelluloses: A new family of nature-based materials. *Angewandte Chemie, International Edition, 50*(24), 5438–5466. https://doi.org/10.1002/anie.201001273.

Kumar, A. P., Depan, D., Singh Tomer, N., & Singh, R. P. (2009). Nanoscale particles for polymer degradation and stabilization-Trends and future perspectives. *Progress in Polymer Science, 34*(6), 479–515. https://doi.org/10.1016/j.progpolymsci.2009.01.002.
Lacerda, P. S. S., Barros-Timmons, A. M. M. V., Freire, C. S. R., Silvestre, A. J. D., & Neto, C. P. (2013). Nanostructured composites obtained by ATRP sleeving of bacterial cellulose nanofibers with acrylate polymers. *Biomacromolecules, 14*(6), 2063–2073. https://doi.org/10.1021/bm400432b.
Lavoine, N., Desloges, I., Dufresne, A., & Bras, J. (2012). Microfibrillated cellulose – its barrier properties and applications in cellulosic materials: A review. *Carbohydrate Polymers, 90*(2), 735–764. https://doi.org/10.1016/j.carbpol.2012.05.026.
Lavoine, N., Desloges, I., & Bras, J. (2014). Microfibrillated cellulose coatings as new release systems for active packaging. *Carbohydrate Polymers, 103*, 528–537. https://doi.org/10.1016/j.carbpol.2013.12.035.
Lee, S. Y., Chun, S. J., Kang, I. A., & Park, J. Y. (2009a). Preparation of cellulose nanofibrils by high-pressure homogenizer and cellulose-based composite films. *Journal of Industrial and Engineering Chemistry, 15(1)*, 50–55. https://doi.org/10.1016/j.jiec.2008.07.008.
Lee, S. Y., Mohan, D. J., Kang, I. A., Doh, G. H., Lee, S., & Han, S. O. (2009b). Nanocellulose reinforced PVA composite films: Effects of acid treatment and filler loading. *Fibers and Polymers, 10*(1), 77–82. https://doi.org/10.1007/s12221-009-0077-x.
Li, J., Wei, X., Wang, Q., Chen, J., Chang, G., Kong, L., Su, J., & Liu, Y. (2012). Homogeneous isolation of nanocellulose from sugarcane bagasse by high pressure homogenization. *Carbohydrate Polymers, 90*(4), 1609–1613. https://doi.org/10.1016/j.carbpol.2012.07.038.
Lin, N., Chen, G., Huang, J., Dufresne, A., & Chang, P. R. (2009). Effects of polymer-grafted natural nanocrystals on the structure and mechanical properties of poly (lactic acid): A case of cellulose whisker-*graft*-polycaprolactone. *Journal of Applied Polymer Science, 113*(5), 3417–3425. https://doi.org/10.1002/app.30308.
Liu, L. M., Qi, Z. N., & Zhu, X. G. (1999). Studies on nylon 6/clay nanocomposites by melt-intercalation process. *Journal of Applied Polymer Science, 71*(7), 1133–1138. https://doi.org/10.1002/(sici)1097-4628(19990214)71:7<1133::aid-app11>3.0.co;2-n.
Ljungberg, N., & Wesslén, B. (2002). The effects of plasticizers on the dynamic mechanical and thermal properties of poly(lactic acid). *Journal of Applied Polymer Science, 86*, 1227.
Lonnberg, H., Fogelstrom, L., Zhou, Q., Hult, A., Berglund, L., & Malmstrom, E. (2011). Investigation of graft length impact on the interfacial toughness in a cellulose/poly (epsilon-caprolactone) bilayer laminate. *Composites Science and Technology, 71*(1), 9–12. https://doi.org/10.1016/j.compscitech.2010.09.007.
Luzi, F., Fortunati, E., Puglia, D., Petrucci, R., Kenny, J. M., & Torre, L. (2015). Study of disintegrability in compost and enzymatic degradation of PLA and PLA nanocomposites reinforced with cellulose nanocrystals extracted from Posidonia oceanica. *Polymer Degradation and Stability, 121*, 105–115. https://doi.org/10.1016/j.polymdegradstab.2015.08.016.
Mabrouk, A. B., Kaddami, H., Magnin, A., Belgacem, M. N., Dufresne, A., & Boufi, S. (2011). Preparation of nanocomposite dispersions based on cellulose whiskers and acrylic copolymer by miniemulsion polymerization: Effect of the silane content. *Polymer Engineering and Science, 51*(1), 62–70. https://doi.org/10.1002/pen.21778.
Machado, B. A. S., Nunes, I. L., Pereira, F. V., & Druzian, J. I. (2012). Development and evaluation of the effectiveness of biodegradable films of cassava starch with nanocellulose as reinforcement and yerba mate extract as an additive antioxidant. *Ciencia Rural, 42*(11), 2085–2091. https://doi.org/10.1590/S0103-84782012001100028.
Maiti, S., Sain, S., Ray, D., & Mitra, D. (2013). Biodegradation behaviour of PMMA/cellulose nanocomposites prepared by in-situ polymerization and ex-situ dispersion methods. *Polymer Degradation and Stability, 98*(2), 635–642. https://doi.org/10.1016/j.polymdegradstab.2012.11.011.
Mandal, A., & Chakrabarty, D. (2011). Isolation of nanocellulose from waste sugarcane bagasse (SCB) and its characterization. *Carbohydrate Polymers, 86*(3), 1291–1299. https://doi.org/10.1016/j.carbpol.2011.06.030.

Mandal, A., & Chakraborty, D. (2015). Characterization of nanocellulose reinforced semi-interpenetrating polymer network of poly(vinyl alcohol) and polyacrylamide composite films. *Carbohydrate Polymer, 134*, 240–250. https://doi.org/10.1016/j.carbpol.2015.07.093.

Marsh, K., & Bugusu, B. (2007). Food packaging – roles, materials, and environmental issues. *Journal of Food Science, 72*(3), R39–R55. https://doi.org/10.1111/j.1750-3841.2007.00301.x.

Merino, D., Mansilla, A. Y., Gutiérrez, T. J., Casalongué, C. A., & Alvarez, V. A. (2018a). Chitosan coated-phosphorylated starch films: Water interaction, transparency and antibacterial properties. *Reactive and Functional Polymers, 131*, 445–453. https://doi.org/10.1016/j.reactfunctpolym.2018.08.012.

Merino, D., Gutiérrez, T. J., Mansilla, A. Y., Casalongué, C., & Alvarez, V. A. (2018b). Critical evaluation of starch-based antibacterial nanocomposites as agricultural mulch films: Study on their interactions with water and light. *ACS Sustainable Chemistry & Engineering, 6*(11), 15662–15672. https://doi.org/10.1021/acssuschemeng.8b04162.

Merino, D., Gutiérrez, T. J., & Alvarez, V. A. (2019a). Potential agricultural mulch films based on native and phosphorylated corn starch with and without surface functionalization with chitosan. *Journal Polymers and the Environment, 27*(1), 97–105. https://doi.org/10.1007/s10924-018-1325-1.

Merino, D., Gutiérrez, T. J., & Alvarez, V. A. (2019b). Structural and thermal properties of agricultural mulch films based on native and oxidized corn starch nanocomposites. *Starch-Stärke, 71*(7–8), 1800341. https://doi.org/10.1002/star.201800341.

Messersmith, P. B., & Giannelis, E. P. (1995). Synthesis and barrier properties of poly(ε-caprolactone)-layered silicate nanocomposites. *Journal of Polymer Science Part A: Polymer Chemistry, 33*(7), 1047. https://doi.org/10.1002/pola.1995.080330707.

Mittal, V. (2011, September) Nanocomposites with biodegradable polymers synthesis, properties, and future perspectives. 1020. Print ISBN-13: 9780199581924, Published to Oxford Scholarship Online: https://doi.org/10.1093/acprof:oso/9780199581924.001.0001.

Moran, J. I., Alvarez, V. A., Cyras, V. P., & Vazquez, A. (2008). Extraction of cellulose and preparation of nanocellulose from sisal fibers. *Cellulose, 15*(1), 149–159. https://doi.org/10.1007/s10570-007-9145-9.

Morena, J. J. (1988). *Advanced composites mold making*. New York: Van Nostrand Reinhold. (ISBN 0-442-26414-3).

Nair, S. S., Zhu, J. Y., Deng, Y., & Ragauskas, A. J. (2014). High performance green barriers based on nanocellulose. *Sustainable Chemical Processes, 2*, 23. https://doi.org/10.1186/s40508-014-0023-0.

Nakagaito, A. N., & Yano, H. (2008). The effect of fiber content on the mechanical and thermal expansion properties of biocomposites based on microfibrillated cellulose. *Cellulose, 15*(4), 555–559. https://doi.org/10.1007/s10570-008-9212-x.

Nelson, Y. U., Edgardo, A. G., & Ana, A. W. (2000). Microcrystalline cellulose from soybean husk: Effects of solvent treatments on its properties as acetylsalicylic acid carrier. *International Journal of Pharmaceutics, 206*(1-2), 85–96. https://doi.org/10.1016/S0378-5173(00)00532-9.

Nishino, T., Matsuda, I., & Hirao, K. (2004). All-cellulose composite. *Macromolecules, 37*(20), 7683–7687. https://doi.org/10.1021/ma049300h.

Nogi, M., & Yano, H. (2008). Transparent nanocomposites based on cellulose produced by bacteria offer potential innovation in the electronics device industry. *Advanced Materials, 20*(10), 1849–1852. https://doi.org/10.1002/adma.200702559.

Nogi, M., Iwamoto, S., Nakagaito, A. N., & Yano, H. (2009). Optically transparent nanofiber paper. *Advanced Materials, 21*(16), 1595–1598. https://doi.org/10.1002/adma.200803174.

Nystrom, G., Mihranyan, A., Razaq, A., Lindstrom, T., Nyholm, L., & Stromme, M. (2010). A nanocellulose polypyrrole composite based on microfibrillated cellulose from wood. *The Journal of Physical Chemistry B, 114*(12), 4178–4182. https://doi.org/10.1021/jp911272m.

Ochsner, A., Ahmed, W., & Ali, N. (2009). *Nanocomposite coatings and nanocomposite materials* (pp. 326–396). Stafa-Zuerich: Trans Tech Publications Ltd..

Oksman, K., Mathew, A. P., Bondeson, D., & Kvien, I. (2006). Manufacturing process of cellulose whiskers/polylactic acid nanocomposites. *Composites Science and Technology, 66*(15), 776–2784. https://doi.org/10.1016/j.compscitech.2006.03.002.

Outwater, J. O. (2014). The mechanics of plastics reinforcement tension. *Modern Plastics*. http://www.automateddynamics.com/article/composite-basics/thermoset-vs-thermoplastic-composites.

Paakko, M., Ankerfors, M., Kosonen, H., Nykanen, A., Ahola, S., Osterberg, M., Ruokolainen, J., Laine, J., Larsson, P. T., Ikkala, O., & Lindstrom, T. (2007). Enzymatic hydrolysis combined with mechanical shearing and high-pressure homogenization for nanoscale cellulose fibrils and strong gels. *Biomacromolecules, 8*(6), 1934–1941. https://doi.org/10.1021/bm061215p.

Paine, F. A., & Paine, H. Y. (2012). *A handbook of food packaging* (p. 497). Springer.

Paralikar, S. A., Simonsen, J., & Lombardi, J. (2008). Poly (vinyl alcohol)/cellulose nanocrystal barrier membranes. *Journal of Membrane Science, 320*(1-2), 248–258. https://doi.org/10.1016/j.memsci.2008.04.009.

Payen, A., & Hebd, C. R. (1838). Mémoire sur la composition du tissu propre des plantes et du ligneux. *Comptes Rendus Hebdomadaires des Séances de l'Académie des Sciences, 7*, 1052–1056.

Peltzer, M., Pei, A., Zhou, Q., Berglund, L., & Jimenez, A. (2014). Surface modification of cellulose nanocrystals by grafting with poly(lactic acid). *Polymer International, 63*(6), 1056–1062. https://doi.org/10.1002/pi.4610.

Pereda, M., Amica, G., Racz, L., Norma, E., & Marcovich, N. E. (2010). Structure and properties of nanocomposite films based on sodium caseinate and nanocellulose fibers. *Journal of Food Engineering, 103*(1), 76–83. https://doi.org/10.1016/j.jfoodeng.2010.10.001.

Pereira, A. L. S., do Nascimento, D. M., Souza Filho, M. M., Morais, J. P. S., Vasconcelos, N. F., Feitosa, J. P. A., Brigida, A. I. S., & Rosa, M. F. (2014). Improvement of polyvinyl alcohol properties by adding nanocrystalline cellulose isolated from banana pseudostems. *Carbohydrate Polymers, 112*, 165–172. https://doi.org/10.1016/j.carbpol.2014.05.090.

Piermaria, J. A., Pinotti, A., Garcia, M. A., & Abraham, A. G. (2009). Films based on kefiran, an exopolysaccharide obtained from kefir grain: Development and characterization. *Food Hydrocolloid, 2*(3), 684–690. https://doi.org/10.1016/j.foodhyd.2008.05.003.

Plackett, D., Anturi, H., Hedenqvist, M., Ankerfors, M., Gallstedt, M., Lindstrom, T., & Siro, I. (2010). Physical properties and morphology of films prepared from microfibrillated cellulose and microfibrillated cellulose in combination with amylopectin. *Journal of Applied Polymer Science, 117*(6), 3601–3609. https://doi.org/10.1002/app.32254.

Plastic Make It Possible Report. (2010). *Plastic innovations in packaging through the decades.*, Available at: www.platicsmakeitpossible.com/2010/05. Accessed 2 Oct 2015.

Quievy, N., Jacquet, N., Sclavons, M., Deroanne, C., Paquot, M., & Devaux, J. (2010). Influence of homogenization and drying on the thermal stability of microfibrillated cellulose. *Polymer Degradation and Stability, 95*(3), 306–314. https://doi.org/10.1016/j.polymdegradstab.2009.11.020.

Raheem, D. (2012). Application of plastics and paper as food packaging materials – An overview. *Emirates Journal of Food and Agriculture, 25*(3), 177–188. https://doi.org/10.9755/ejfa.v25i3.11509.

Ramos, O. L., Fernandes, J. C., Silva, S. I., Pintado, M. E., & Malcata, F. X. (2012). Edible films and coatings from whey proteins: A review on formulation, and on mechanical and bioactive peptides. *Critical Reviews in Food Science and Nutrition, 52*(6), 533–552. https://doi.org/10.1080/10408398.2010.500528.

Reddy, J. P., & Rhim, J. W. (2014). Characterization of bionanocomposite films prepared with agar and paper-mulberry pulp nanocellulose. *Carbohydrate Polymers, 110*, 480–488. https://doi.org/10.1016/j.carbpol.2014.04.056.

Revol, J. F., Godbout, L., & Gray, D. G. (1998). Solid self-assembled films of cellulose with chiral nematic order and optically variable properties. *Journal of Pulp and Paper Science, 24*(5), 146–149.

Risch, S. J. (2009). Food packaging history and innovations. *Journal of Agricultural and Food Chemistry, 57*(18), 8089–8092. https://doi.org/10.1021/jf900040r.

Roohani, M., Habibi, Y., Belgacem, N. M., Ebrahim, G., Karimi, A. N., & Dufresne, A. (2008). Cellulose whiskers reinforced polyvinyl alcohol copolymers nanocomposites. *European Polymer Journal, 44*(8), 2489–2498. https://doi.org/10.1016/j.eurpolymj.2008.05.024.

Ruiz, M. M., Cavaille, J. Y., Dufresne, A., Gerard, J. F., & Graillat, C. (2000). Processing and characterization of new thermoset nanocomposites based on cellulose whiskers. *Composite Interfaces, 7*(2), 117–131. https://doi.org/10.1163/156855400300184271.

Sainz, C. B., Bras, J., & Williams, T. (2011). HPMC reinforced with different cellulose nano-particles. *Carbohydrate Polymers, 86*(4), 1549–1557. https://doi.org/10.1016/j.carbpol.2011.06.060.

Sakurada, I., Nukushina, Y., & Ito, T. (1962). Experimental determination of elastic modulus of crystalline regions in oriented polymers. *Journal of Polymer Science, 57*(165), 651–660. https://doi.org/10.1002/pol.1962.1205716551.

Samir, M. A. S. A., Alloin, F., Sanchez, J.-Y., & Dufresne, A. (2004). Cellulose nanocrystals reinforced poly(oxyethylene). *Polymer, 45*(12), 4149–4157. https://doi.org/10.1016/j.polymer.2004.03.094.

Sanchez-Garcia, M., & Lagaron, J. (2010). On the use of plant cellulose nanowhiskers to enhance the barrier properties of polylactic acid. *Cellulose, 17*(5), 987–1004. https://doi.org/10.1007/s10570-010-9430-x.

Sandeep, S. L., Viveka, S., Madhu Kumar, D. J., & Nagaraja, G. K. (2012). Preparation and properties of composite films from modified cellulose fibre-reinforced with PLA. *Der Pharma Chemica, 4*(1), 159–168.

Sandler, S. R., Karo, W., Bonesteel, J.-A., & Pearce, E. M. (1998). Differential scanning calorimetry. In *Polymer synthesis and characterization – A laboratory manual* (Vol. 27, pp. 120–128). San Diego: Academic.

Satyanarayan, K. G., Arizaga, G. G. C., & Wypych, F. (2009). Biodegradable composites based on lignocellulosic fibers an overview. *Progress in Polymer Science, 34*, 982–1021.

Savadekar, N. R., & Mhaske, S. T. (2012). Synthesis of nano cellulose fibers and effect on thermoplastics starch based films. *Carbohydrate Polymers, 89*(1), 146–151. https://doi.org/10.1016/j.carbpol.2012.02.063.

Savadekar, N. R., Karande, V. S., Vigneshwaran, N., Bharimalla, A. K., & Mhaske, S. T. (2012). Preparation of nano cellulose fibers and its application in kappa-carrageenan based film. *International Journal of Biological Macromolecules, 51*(5), 1008–1013. https://doi.org/10.1016/j.ijbiomac.2012.08.014.

Saxena, A., & Ragauskas, A. J. (2009). Water transmission barrier properties of biodegradable films based on cellulosic whiskers and xylan. *Carbohydrate Polymers, 78*, 357–360. https://doi.org/10.1016/j.carbpol.2009.03.039.

Saxena, A., Elder, T. J., Kenvin, J., & Ragauskas, A. J. (2010). High oxygen nanocomposite barrier films based on xylan and nanocrystalline cellulose. *Nano-Micro Letters, 2*(4), 235–241. https://doi.org/10.1007/bf03353849.

Sedlarik, V., Otgonzul, O., Kitano, T., Gregorova, A., Hrabalova, M., Junkar, I., Cvelbar, U., Mozetic, M., & Saha, P. (2012). Effect of phase arrangement on solid state mechanical and thermalproperties of polyamide 6/polylactide based co-polyester blends. *Journal of Macromolecular Science Part B, 51*(5), 982–1001. https://doi.org/10.1080/00222348.2011.610265.

Seydibeyoglu, M. O., & Oksman, K. (2008). Novel nanocomposites based on polyurethane and microfibrillated cellulose. *Composites Science and Technology, 68*(3-4), 908–914. https://doi.org/10.1016/j.compscitech.2007.08.008.

Silvestre, C., Duraccio, D., & Cimmino, S. (2011). Food packaging based on polymer nanomaterials. *Progress in Polymer Science, 36*(12), 1766–1782. https://doi.org/10.1016/j.progpolymsci.2011.02.003.

Sinha, S. R., & Okamoto, M. (2003). Polymer/layered silicate nanocomposites: A review from preparation to processing. *Progress in Polymer Science, 28*(11), 1539–1641. https://doi.org/10.1016/j.progpolymsci.2003.08.002.

Siqueira, G., Bras, J., & Dufresne, A. (2009). Cellulose whiskers versus microfibrils: Influence of the nature of the nanoparticle and its surface functionalization on the thermal and mechanical properties of nanocomposites. *Biomacromolecules, 10*(2), 425–432. https://doi.org/10.1021/bm801193d.

Siqueira, G., Bras, J., & Dufresne, A. (2010). Cellulosic bionanocomposites: A review of preparation, properties and applications. *Polymers, 2*(4), 728–765. https://doi.org/10.3390/polym2040728.

Siro, I., & Plackett, D. (2010). Microfibrillated cellulose and new nanocomposite materials: a review. *Cellulose, 17*(3), 459–494. https://doi.org/10.1007/s10570-010-9405-y.

Smith, S. A. (1986). Polyethylene, low density. In M. Bakker (Ed.), *The wiley encyclopedia of packaging technology* (pp. 514–523). New York: Wiley.

Soeta, H., Fujisawa, S., Saito, T., Berglund, L., & Isogai, A. (2015). Low-birefringent and highly tough nanocellulose-reinforced cellulose triacetate. *ACS Applied Materials & Interfaces, 7*(20), 11041–11046. https://doi.org/10.1021/acsami.5b02863.

Song, Z., Xiao, H., & Zhao, Y. (2014). Hydrophobic-modified nano-cellulose fiber/PLA biodegradable composites for lowering water vapor transmission rate (WVTR) of paper. *Carbohydrate Polymers, 111*, 442–448. https://doi.org/10.1016/j.carbpol.2014.04.049.

Sorrentino, A., Gorrasi, G., & Vittoria, V. (2007). Potential perspectives of bio-nanocomposites for food packaging applications. *Trends in Food Science and Technology, 18*(2), 84–95. https://doi.org/10.1016/j.tifs.2006.09.004.

Stuart, M. A. C., Hyck, W. T. S., Genzer, J., Muller, M., Ober, C., Stamm, M., Sukhorukov, G. B., Szleifer, I., Tsukruk, V. V., Urban, M., Winnik, F., Zauscher, S., Luzinov, I., & Minko, S. (2010). Emerging applications of stimuli-responsive polymer materials. *Nature Materials, 9*, 101–113. https://doi.org/10.1038/nmat2614.

Svagan, A. J., Hedenqvist, M. S., & Berglund, L. (2009). Reduced water vapour sorption in cellulose nanocomposites with starch matrix. *Composites Science and Technology, 69*(3–4), 500–506. https://doi.org/10.1016/j.compscitech.2008.11.016.

Teixeira, E. D. M., Pasquini, D., Curvelo, A. A., Corradini, E., Belgacem, M. N., & Dufresne, A. (2009). Cassava bagasse cellulose nanofibrils reinforced thermoplastic cassava starch. *Carbohydrate Polymers, 78*(3), 422–431. https://doi.org/10.1016/j.carbpol.2009.04.034.

Tharanathan, R. N. (2003). Biodegradable films and composite coatings: Past, present and future. *Trends in Food Science and Technology, 14*(3), 71–78. https://doi.org/10.1016/S0924-2244(02)00280-7.

Tice, P. (2003). *Packaging materials 4. Polyethylene for food packaging applications*. International Life Sciences Institute Report. Website: http://orig.ilsi.org/file/PM4_Polyethylene.pd.

Tome, L. C., Fernandes, S. C. M., Perez, D. S., Sadocco, P., Silvestre, A. J. D., Neto, C. P., Marrucho, I. M., & Carmen, S. R. F. (2013). The role of nanocellulose fibers, starch and chitosan on multipolysaccharide based films. *Cellulose, 20*(4), 1807–1818. https://doi.org/10.1007/s10570-013-9959-6.

Toro-Márquez, L. A., Merino, D., & Gutiérrez, T. J. (2018). Bionanocomposite films prepared from corn starch with and without nanopackaged Jamaica (*Hibiscus sabdariffa*) flower extract. *Food and Bioprocess Technology, 11*(11), 1955–1973. https://doi.org/10.1007/s11947-018-2160-z.

Tsuji, H. (2005). Poly(lactide) stereocomplexes: formation, structure, properties, degradation, and applications. *Macromolecular Bioscience, 5*(7), 569–597. https://doi.org/10.1002/mabi.200500062.

Turbak, A. F., Snyder, F. W., & Sandberg, K. R. (1983). Microfibrillated cellulose, a new cellulose product: Properties, uses and commercial potential. *Journal of Applied Polymer Science: Applied Polymer Symposium, 37*, 815–827.

Vaia, R. A., Ishii, H., & Giannelis, E. P. (1993). Synthesis and properties of two-dimensional nanostructure by direct intercalation of polymer melts in layered silicates. *Chemistry of Materials, 5*(12), 1694–1696. https://doi.org/10.1021/cm00036a004.

van den Mooter, G., Samyn, C., & Kinget, R. (1994). Characterization of colon-specific azo polymers: A study of the swelling properties and the permeability of isolated polymer films. *International Journal of Pharmaceutics, 111*(2), 127–136. https://doi.org/10.1016/0378-5173(94)00102-2.

Vink, E. T. H., Rabago, K. R., Glassner, D. A., Springs, B., O'Connor, R. P., Kolstad, J., & Gruber, P. R. (2004). The sustainability of NatureWorksTM polylactide polymers and IngeoTM polylactide fibers: An update of the future. *Macromolecular Bioscience, 4*(6), 551–564. https://doi.org/10.1002/mabi.200400023.

Walker, C. (2012). Thinking small is leading to big changes, in Paper360°TAPPI, 8–13.

Wan, W. K., Hutter, J. L., Millon, L. E., & Guhados, G. (2006). Bacterial cellulose and its nanocomposites for biomedical applications. *ACS Symposium Series, 938*, 221–241.

Welt, B. (2005). *A brief history of packaging*. Institute of Food and Agricultural Sciences, Document ABE321, University of Florida.

Westman, M.P., Laddha, S.G., Fifield, L.S., Kafentzis, T. A., & Simmons, K.L. (2010). *Natural fiber composites: A review*. Prepared for the U.S. Department of Energy under Contract DE-AC05–76RL01830, PNNl-99352.

Yakkan, E., Uysalman, T., Atagur, M., Kara, H., Sever, K., Yildirim, A., Girginer, B., & Seydibeyoglu, M. O. (2015). Nanocellulose-polypropylene nanocomposites enhanced with fusabond coupling agent. *Journal of Bartin Faculty of Forestry, 20*(3), 491–502.

Yam, K. L., Takhistov, P. T., & Miltz, J. (2005). Intelligent packaging: Concepts and applications. *Journal of Food Science, 70*(1), R1–R10. https://doi.org/10.1111/j.1365-2621.2005.tb09052.x.

Yu, J. F. A., Dufresne, A., Gao, S., Huang, J., & Chang, P. R. (2008). Structure and mechanical properties of poly(lactic acid) filled with (starch nanocrystal) -*graft*-poly(ε-caprolactone). *Macromolecular Materials and Engineering, 293*(9), 763–770. https://doi.org/10.1002/mame.200800134.

Zhou, W., Apkarian, R.P., Wang, Z.L., Joy, D. (2006). Fundamentals of scanning electron microscopy. In: Zhou, W.W., Zhong, L. ed. *Scanning microscopy for nanotechnology – Techniques and applications*, Springer Verlag, pp. 1–40. https://doi.org/10.1007/978-0-387-39620-0_1

Zhou, C., Wua, Q., Yue, Y., & Zhang, Q. (2011). Application of rod-shaped cellulose nanocrystals in polyacrylamide hydrogels. *Journal of Colloid and Interface Science, 353*(1), 116–123. https://doi.org/10.1016/j.jcis.2010.09.035.

Zhu, J. Y., Sabo, R., & Luo, X. (2011). Integrated production of nano-fibrillated cellulose and biofuel (Ethanol) by enzymatic fractionation of wood fibers. *Green Chemistry, 13*(5), 1339–1344. https://doi.org/10.1039/c1gc15103g.

Zimmermann, T., Bordeanu, N., & Strub, E. (2010). Properties of nanofibrillated cellulose from different raw materials and its reinforcement potential. *Carbohydrate Polymers, 79*(4), 1086–1093. https://doi.org/10.1016/j.carbpol.2009.10.045.

Index

T. J. Gutiérrez (ed.), *Polymers for Agri-Food Applications*,
https://doi.org/10.1007/978-3-030-19416-1

C

Q

R

S

FSC
www.fsc.org
MIX
Papier | Fördert
gute Waldnutzung
FSC® C083411

Kolibri 360 GmbH
Ferdinand-Jühlke-Straße 7
99095 Erfurt, Deutschland
produktsicherheit@kolibri360.de